Geophysical Monograph Series

Geophysical Monograph Series

197 **Auroral Phenomenology and Magnetospheric Processes: Earth and Other Planets** *Andreas Keiling, Eric Donovan, Fran Bagenal, and Tomas Karlsson (Eds.)*
198 **Climates, Landscapes, and Civilizations** *Liviu Giosan, Dorian Q. Fuller, Kathleen Nicoll, Rowan K. Flad, and Peter D. Clift (Eds.)*
199 **Dynamics of the Earth's Radiation Belts and Inner Magnetosphere** *Danny Summers, Ian R. Mann, Daniel N. Baker, and Michael Schulz (Eds.)*
200 **Lagrangian Modeling of the Atmosphere** *John Lin (Ed.)*
201 **Modeling the Ionosphere-Thermosphere** *Jospeh D. Huba, Robert W. Schunk, and George V. Khazanov (Eds.)*
202 **The Mediterranean Sea: Temporal Variability and Spatial Patterns** *Gian Luca Eusebi Borzelli, Miroslav Gacic, Piero Lionello, and Paola Malanotte-Rizzoli (Eds.)*
203 **Future Earth – Advancing Civic Understanding of the Anthropocene** *Diana Dalbotten, Gillian Roehrig, and Patrick Hamilton (Eds.)*
204 **The Galápagos: A Natural Laboratory for the Earth Sciences** *Karen S. Harpp, Eric Mittelstaedt, Noemi d'Ozouville, and David W. Graham (Eds.)*
205 **Modeling Atmospheric and Oceanic Flows: Insightsfrom Laboratory Experiments and Numerical Simulations** *Thomas von Larcher and Paul D. Williams (Eds.)*
206 **Remote Sensing of the Terrestrial Water Cycle** *Venkat Lakshmi (Ed.)*
207 **Magnetotails in the Solar System** *Andreas Keiling, Caitriona Jackman, and Peter Delamere (Eds.)*
208 **Hawaiian Volcanoes: From Source to Surface Rebecca** *Carey, Valerie Cayol, Michael Poland, and Dominique Weis (Eds.)*
209 **Sea Ice: Physics, Mechanics, and Remote Sensing** *Mohammed Shokr and Nirmal Sinha (Eds.)*
210 **Fluid Dynamics in Complex Fractured-Porous Systems** *Boris Faybishenko, Sally M. Benson, and John E. Gale (Eds.)*
211 **Subduction Dynamics: From Mantle Flow to Mega Disasters** *Gabriele Morra, David A. Yuen, Scott King, Sang Mook Lee, and Seth Stein (Eds.)*
212 **The Early Earth: Accretion and Differentiation** *James Badro and Michael Walter (Eds.)*
213 **Global Vegetation Dynamics: Concepts and Applications in the MC1 Model** *Dominique Bachelet and David Turner (Eds.)*
214 **Extreme Events: Observations, Modeling and Economics** *Mario Chavez, Michael Ghil, and Jaime Urrutia-Fucugauchi (Eds.)*
215 **Auroral Dynamics and Space Weather** *Yongliang Zhang and Larry Paxton (Eds.)*
216 **Low-Frequency Waves in Space Plasmas** *Andreas Keiling, Dong-Hun Lee, and Valery Nakariakov (Eds.)*
217 **Deep Earth: Physics and Chemistry of the Lower Mantle and Core** *Hidenori Terasaki and Rebecca A. Fischer (Eds.)*
218 **Integrated Imaging of the Earth: Theory and Applications** *Max Moorkamp, Peter G. Lelievre, Niklas Linde, and Amir Khan (Eds.)*
219 **Plate Boundaries and Natural Hazards** *Joao Duarte and Wouter Schellart (Eds.)*
220 **Ionospheric Space Weather: Longitude and Hemispheric Dependences and Lower Atmosphere Forcing Timothy Fuller-Rowell,** *Endawoke Yizengaw, Patricia H. Doherty, and Sunanda Basu (Eds.)*
221 **Terrestrial Water Cycle and Climate Change Natural and Human-Induced Impacts** *Qiuhong Tang and Taikan Oki (Eds.)*
222 **Magnetosphere-Ionosphere Coupling in the Solar System** *Charles R. Chappell, Robert W. Schunk, Peter M. Banks, James L. Burch, and Richard M. Thorne (Eds.)*
223 **Natural Hazard Uncertainty Assessment: Modeling and Decision Support** *Karin Riley, Peter Webley, and Matthew Thompson (Eds.)*
224 **Hydrodynamics of Time-Periodic Groundwater Flow: Diffusion Waves in Porous Media** *Joe S. Depner and Todd C. Rasmussen (Auth.)*
225 **Active Global Seismology** *Ibrahim Cemen and Yucel Yilmaz (Eds.)*
226 **Climate Extremes** *Simon Wang (Ed.)*
227 **Fault Zone Dynamic Processes** *Marion Thomas (Ed.)*
228 **Flood Damage Survey and Assessment: New Insights from Research and Practice** *Daniela Molinari, Scira Menoni, and Francesco Ballio (Eds.)*
229 **Water-Energy-Food Nexus – Principles and Practices** *P. Abdul Salam, Sangam Shrestha, Vishnu Prasad Pandey, and Anil K Anal (Eds.)*
230 **Dawn–Dusk Asymmetries in Planetary Plasma Environments** *Stein Haaland, Andrei Rounov, and Colin Forsyth (Eds.)*
231 **Bioenergy and Land Use Change** *Zhangcai Qin, Umakant Mishra, and Astley Hastings (Eds.)*
232 **Microstructural Geochronology: Planetary Records Down to Atom Scale** *Desmond Moser, Fernando Corfu, James Darling, Steven Reddy, and Kimberly Tait (Eds.)*
233 **Global Flood Hazard: Applications in Modeling, Mapping and Forecasting** *Guy Schumann, Paul D. Bates, Giuseppe T. Aronica, and Heiko Apel (Eds.)*
234 **Pre-Earthquake Processes: A Multidisciplinary Approach to Earthquake Prediction Studies** *Dimitar Ouzounov, Sergey Pulinets, Katsumi Hattori, and Patrick Taylor (Eds.)*
235 **Electric Currents in Geospace and Beyond** *Andreas Keiling, Octav Marghitu, and Michael Wheatland (Eds.)*
236 **Quantifying Uncertainty in Subsurface Systems** *Celine Scheidt, Lewis Li, and Jef Caers (Eds.)*
237 **Petroleum Engineering** *Moshood Sanni (Ed.)*
238 **Geological Carbon Storage: Subsurface Seals and Caprock Integrity** *Stephanie Vialle, Jonathan Ajo-Franklin, and J. William Carey (Eds.)*
239 **Lithospheric Discontinuities** *Huaiyu Yuan and Barbara Romanowicz (Eds.)*
240 **Chemostratigraphy Across Major Chronological Eras** *Alcides N.Sial, Claudio Gaucher, Muthuvairavasamy Ramkumar, and Valderez Pinto Ferreira (Eds.)*
241 **Mathematical Geoenergy:Discovery, Depletion, and Renewal** *Paul Pukite, Dennis Coyne, and Daniel Challou (Eds.)*
242 **Ore Deposits: Origin, Exploration, and Exploitation** *Sophie Decree and Laurence Robb (Eds.)*
243 **Kuroshio Current: Physical, Biogeochemical and Ecosystem Dynamics** *Takeyoshi Nagai, Hiroaki Saito, Koji Suzuki, and Motomitsu Takahashi (Eds.)*
244 **Geomagnetically Induced Currents from the Sun to the Power Grid** *Jennifer L. Gannon, Andrei Swidinsky, and Zhonghua Xu (Eds.)*
245 **Shale: Subsurface Science and Engineering** *Thomas Dewers, Jason Heath, and Marcelo Sánchez (Eds.)*
246 **Submarine Landslides: Subaqueous Mass Transport Deposits From Outcrops to Seismic Profiles** *Kei Ogata, Andrea Festa, and Gian Andrea Pini (Eds.)*
247 **Iceland: Tectonics, Volcanics, and Glacial Features** *Tamie J. Jovanelly*
248 **Dayside Magnetosphere Interactions** *Quigang Zong, Philippe Escoubet, David Sibeck, Guan Le, and Hui Zhang (Eds.)*

Geophysical Monograph 249

Carbon in Earth's Interior

Craig E. Manning
Jung-Fu Lin
Wendy L. Mao
Editors

This Work is a co-publication of the American Geophysical Union and John Wiley and Sons, Inc.

AGU 100
ADVANCING EARTH AND SPACE SCIENCE

WILEY

This Work is a co-publication between the American Geophysical Union and John Wiley & Sons, Inc.

This edition first published 2020 by John Wiley & Sons, Inc., 111 River Street, Hoboken, NJ 07030, USA and the American Geophysical Union, 2000 Florida Avenue, N.W., Washington, D.C. 20009

Published under the aegis of the AGU Publications Committee

Brooks Hanson, Executive Vice President, Science
Lisa Tauxe, Chair, Publications Committee
For details about the American Geophysical Union visit us at www.agu.org.

Wiley Global Headquarters
111 River Street, Hoboken, NJ 07030, USA
For details of our global editorial offices, customer services, and more information about Wiley products visit us at www.wiley.com.

Library of Congress Cataloging-in-Publication data is available.

Cover Images: © Valerio Cerantola
Cover design by Wiley

Set in 10/12pt Times New Roman by SPi Global, Pondicherry, India

10 9 8 7 6 5 4 3 2 1

CONTENTS

CONTRIBUTORS

Evan Abramson
Department of Earth and Space Sciences, University of Washington, Seattle, Washington, USA

Anton Arefiev
Sobolev Institute of Geology and Mineralogy, Novosibirsk, Russia; *and* Vereshchagin Institute for High Pressure Physics, Moscow, Russia; *and* Novosibirsk State University, Novosibirsk, Russia

Yuanfei Bi
Department of Civil and Environmental Engineering, George Washington University, Washington, DC, USA

Roberto Bini
LENS, European Laboratory for Non-linear Spectroscopy, Firenze, Italy; *and* CNR-ICCOM, Istituto di Chimica dei Composti OrganoMetallici, Firenze, Italy; *and* Dipartimento di Chimica "Ugo Schiff," Università di Firenze, Firenze, Italy

Eglantine Boulard
Sorbonne Université, UMR CNRS 7590, Muséum National d'Histoire Naturelle, IRD, Institut de Minéralogie, Physique des Matériaux et Cosmochimie-IMPMC, Paris, France

Peter A. Canovas, III
Group Exploring Organic Processes in Geochemistry (GEOPIG), Arizona State University, Tempe, Arizona, USA; *and* School of Earth and Space Exploration, Arizona State University, Tempe, Arizona, USA

Boxiao Cao
Department of Civil and Environmental Engineering, George Washington University, Washington, DC, USA

Razvan Caracas
CNRS, Ecole Normale Supérieure de Lyon, Université de Lyon, Laboratoire de Géologie de Lyon, Lyon, France; *and* CEED, The Center for Earth Evolution and Dynamics, University of Oslo, Oslo, Norway

Bin Chen
Hawaii Institute of Geophysics and Planetology, University of Hawaii at Manoa, Honolulu, Hawaii, USA

Daniele Cherniak
Department of Earth and Environmental Sciences, Rensselaer Polytechnic Institute, Troy, New York, USA

Ronald Cohen
Extreme Materials Initiative, Carnegie Institution for Science, Washington, DC, USA; *and* Department für Geo- und Umweltwissenschaften, Ludwig Maximilians Universität München, München, Germany

Isabelle Daniel
Université Lyon, Université Lyon 1, ENS de Lyon, CNRS, UMR 5276, Laboratoire de Géologie de Lyon, Villeurbanne, France

Rajdeep Dasgupta
Department of Earth, Environmental and Planetary Sciences, Rice University, Houston, Texas, USA

Larissa Dobrzhinetskaya
Department of Earth Sciences, University of California–Riverside, Riverside, California, USA

Junjie Dong
Department of Earth and Planetary Sciences, Harvard University, Cambridge, Massachusetts, USA

Megan S. Duncan
Department of Geosciences, Virginia Tech, Blacksburg, Virginia, USA

Kamil Dziubek
LENS, European Laboratory for Non-linear Spectroscopy, Firenze, Italy; *and* CNR-ICCOM, Istituto di Chimica dei Composti OrganoMetallici, Firenze, Italy

James Eguchi
Department of Earth, Environmental, and Planetary Sciences, Rice University, Houston, Texas, USA; *and* Department of Earth, Planetary, and Space Sciences, University of California–Los Angeles, Los Angeles, California, USA

Guillaume Fiquet
Sorbonne Université, UMR CNRS 7590, Muséum National d'Histoire Naturelle, IRD, Institut de Minéralogie, Physique des Matériaux et Cosmochimie-IMPMC, Paris, France

Suyu Fu
Department of Geological Sciences, Jackson School of Geosciences, University of Texas at Austin, Austin, Texas, USA

Nir Goldman
Physical and Life Sciences Directorate, Lawrence Livermore National Laboratory, Livermore, California, USA; *and* Department of Chemical Engineering, University of California–Davis, California, USA

Alexander F. Goncharov
Geophysical Laboratory, Carnegie Institution for Science, Washington, DC, USA

Federico A. Gorelli
CNR-INO, Istituto Nazionale di Ottica, Firenze, Italy; *and* LENS, European Laboratory for Non-linear Spectroscopy, Firenze, Italy

Meghan Guild
School of Earth and Space Exploration, Arizona State University, Tempe, Arizona, USA

François Guyot
Sorbonne Université, UMR CNRS 7590, Muséum National d'Histoire Naturelle, IRD, Institut de Minéralogie, Physique des Matériaux et Cosmochimie-IMPMC, Paris, France

Naohisa Hirao
Japan Synchrotron Radiation Research Institute, Sayo, Japan

Yoshinori Ito
Department of Earth and Planetary Material Sciences, Graduate School of Science, Tohoku University, Sendai, Japan

Kirill Ivanov
The Zavaritsky Institute of Geology and Geochemistry, Ekaterinburg, Russia

Seiji Kamada
Department of Earth and Planetary Material Sciences, Graduate School of Science, Tohoku University, Sendai, Japan

Yoshio Kono
Geophysical Laboratory, Carnegie Institution of Washington, Argonne, Illinois, USA; *and* Geodynamics Research Center, Ehime University, Ehime, Japan

Matthew Kroonblawd
Physical and Life Sciences Directorate, Lawrence Livermore National Laboratory, Livermore, California, USA

Vladimir Kutcherov
KTH Royal Institute of Technology, Stockholm, Sweden; *and* Gubkin University, Moscow, Russia

Jie Li
Department of Earth and Environmental Sciences, University of Michigan, Ann Arbor, Michigan, USA

Tianshu Li
Department of Civil and Environmental Engineering, George Washington University, Washington, DC, USA

Yuan Li
State Key Laboratory of Isotope Geochemistry, Guangzhou Institute of Geochemistry, Chinese Academy of Sciences, Guangzhou, China

Jung-Fu Lin
Department of Geological Sciences, Jackson School of Geosciences, University of Texas at Austin, Austin, Texas, USA

Yu Lin
Stanford Institute for Materials and Energy Sciences, SLAC National Accelerator Laboratory, Menlo Park, California, USA

Konstantin Litasov
Sobolev Institute of Geology and Mineralogy, Novosibirsk, Russia; *and* Novosibirsk State University, Novosibirsk, Russia; *and* Vereshchagin Institute for High Pressure Physics, Moscow, Russia

Jiachao Liu
Department of Geological Sciences, Jackson School of Geosciences, University of Texas at Austin, Austin, Texas, USA

Sergey S. Lobanov
GFZ German Research Center for Geosciences, Potsdam, Germany

Craig E. Manning
Department of Earth, Planetary, and Space Sciences, University of California–Los Angeles, Los Angeles, California, USA

Wendy L. Mao
Department of Geological Sciences, Stanford University, Stanford, California, USA; *and* Photon Science, SLAC National Accelerator Laboratory, Menlo Park, California, USA

Matteo Masotta
Dipartimento di Scienze della Terra, Università di Pisa, Pisa, Italy

Juliette Maurice
Dipartimento di Scienze della Terra, Università degli Studi di Milano, Milano, Italy

Marco Merlini
Dipartimento di Scienze della Terra, Università degli Studi di Milano, Milano, Italy

Sula Milani
Dipartimento di Scienze della Terra, Università degli Studi di Milano, Milano, Italy

Masaaki Miyahara
Department of Earth and Planetary Material Sciences, Graduate School of Science, Tohoku University, Sendai, Japan; *and* Department of Earth and Planetary Science, Graduate School of Science, Hiroshima University, Higashi-Hiroshima, Japan

Elena Mukhina
Skolkovo Institute of Science and Technology, Moscow, Russia

Michelle Muth
Department of Earth, Environmental and Planetary Sciences, Rice University, Houston, Texas, USA; *and* Department of Earth Sciences, University of Oregon, Eugene, Oregon, USA

Alexandra Navrotsky
Peter A. Rock Thermochemistry Laboratory and NEAT ORU, University of California–Davis, California, USA

Yasuo Ohishi
Japan Synchrotron Radiation Research Institute, Sayo, Japan

Eiji Ohtani
Department of Earth and Planetary Material Sciences, Graduate School of Science, Tohoku University, Sendai, Japan

Shin Ozawa
Department of Earth and Planetary Material Sciences, Graduate School of Science, Tohoku University, Sendai, Japan

Sulgiye Park
Department of Geological Sciences, Stanford University, Stanford, California, USA

John Percival
Geological Survey of Canada, Ottawa, Ontario, Canada

Ivan Podborodnikov
Sobolev Institute of Geology and Mineralogy, Novosibirsk, Russia; *and* Vereshchagin Institute for High Pressure Physics, Moscow, Russia; *and* Novosibirsk State University, Novosibirsk, Russia

Stefano Poli
Dipartimento di Scienze della Terra "Ardito Desio," Università degli Studi di Milano, Milano, Italy

Sergey Rashchenko
Sobolev Institute of Geology and Mineralogy, Novosibirsk, Russia; *and* Novosibirsk State University, Novosibirsk, Russia

Takeshi Sakai
Geodynamics Research Center, Ehime University, Matsuyama, Japan

Tatsuya Sakamaki
Department of Earth and Planetary Material Sciences, Graduate School of Science, Tohoku University, Sendai, Japan

Mario Santoro
CNR-INO, Istituto Nazionale di Ottica, Firenze, Italy; *and* LENS, European Laboratory for Non-linear Spectroscopy, Firenze, Italy

Piergiorgio Scarlato
Istituto Nazionale di Geofisica e Vulcanologia, Roma, Italy

Demetrio Scelta
LENS, European Laboratory for Non-linear Spectroscopy, Firenze, Italy; *and* CNR-ICCOM, Istituto di Chimica dei Composti OrganoMetallici, Firenze, Italy

Morgan Schaller
Department of Earth and Environmental Sciences, Rensselaer Polytechnic Institute, Troy, New York, USA

Erwin Schettino
Dipartimento di Scienze della Terra "Ardito Desio", Università degli Studi di Milano, Milano, Italy; *and* Instituto Andaluz de Ciencias de la Tierra, CSIC-Universidad de Granada, Granada, Spain

Aleksandr Serovaiskii
Gubkin University, Moscow, Russia

Anton Shatskiy
Sobolev Institute of Geology and Mineralogy, Novosibirsk, Russia; *and* Novosibirsk State University, Novosibirsk, Russia; *and* Vereshchagin Institute for High Pressure Physics, Moscow, Russia

Everett L. Shock
Group Exploring Organic Processes in Geochemistry (GEOPIG), Arizona State University, Tempe, Arizona, USA; *and* School of Earth and Space Exploration, Arizona State University, Tempe, Arizona, USA; *and* School of Molecular Sciences, Arizona State University, Tempe, Arizona, USA; *and* Center for Fundamental and Applied Microbiomics, Arizona State University, Tempe, Arizona, USA

Natalia Solomatova
CNRS, Ecole Normale Supérieure de Lyon, Université de Lyon, Laboratoire de Géologie de Lyon, Lyon, France

Vincenzo Stagno
Dipartimento di Scienze della Terra, Sapienza Universita' di Roma, Italy; *and* Istituto Nazionale di Geofisica e Vulcanologia, Roma, Italy

Veronica Stopponi
Dipartimento di Scienze della Terra, Sapienza Universita' di Roma, Italy

Dimitri Sverjensky
Department of Earth and Planetary Sciences, Johns Hopkins University, Baltimore, Maryland, USA

Suguru Takahashi
Department of Earth and Planetary Material Sciences, Graduate School of Science, Tohoku University, Sendai, Japan

Oliver Tschauner
Department of Geoscience, University of Nevada–Las Vegas, Las Vegas, Nevada, USA

Alberto Vitale Brovarone
Dipartimento di Scienze della Terra, Università degli studi di Torino, Torino, Italy; *and* Sorbonne Université, Museum National d'Histoire Naturelle, UMR CNRS 7590, IRD, Institut de Minéralogie, des Physique de Matériaux et de Cosmochimie, Paris, France

Jianwei Wang
Department of Geology and Geophysics, Louisiana State University, Baton Rouge, Louisiana, USA

Bruce Watson
Department of Earth and Environmental Sciences, Rensselaer Polytechnic Institute, Troy, New York, USA

Choong-Shik Yoo
Department of Chemistry and Institute of Shock Physics, Washington State University, Pullman, Washington, USA

Feng Zhu
Hawaii Institute of Geophysics and Planetology, University of Hawai'i at Mānoa, Honolulu, Hawaii, USA

PREFACE

Carbon in Earth's fluid envelopes—the atmosphere and hydrosphere—plays a fundamental role in our planet's climate system. It is also essential for the origin and evolution of life, for a large fraction of the energy we use, and for the multitude of carbon-based materials so essential to the modern world. Yet the source and original quantity of carbon in our planet is uncertain (Marty et al., 2013), as are the identities and relative importance of early chemical processes associated with planetary differentiation (e.g., the moon-forming impact, core formation, the onset of plate tectonics). Numerous lines of evidence point to the early and continuing exchange of substantial carbon between Earth's surface and its interior (Dasgupta, 2013), such as information carried by subducted carbon trapped in diamonds, mantle-derived magmas rich in carbon, carbonate-bearing rocks found in fossil subduction zones, and springs carrying deeply sourced carbon-bearing gases (Burton et al., 2013; Jones et al., 2013; Ni & Keppler, 2013; Shirey et al., 2013). Although quantifying the input and output fluxes is challenging, there is little doubt that a substantial amount of carbon resides in our planet's interior (Dasgupta and Hirschmann, 2010, Kelemen & Manning, 2015).

These uncertainties arise in part from continuing difficulties in establishing the forms, transformations, and movements of carbon in Earth's interior. The present volume provides a snapshot of recent work aimed at improving this picture. It presents research aimed at understanding the physical and chemical behavior of carbon-bearing materials at conditions relevant to Earth's interior – behavior that ultimately dictates the availability of this element so important to processes near our planet's surface.

The papers in this volume are a mix of reviews and reports of current research on the structure, stability, reactivity, and dynamics of carbon-based materials relevant to natural systems, as well as to allied substances that carry carbon, and the complex interactions between moving fluids, magmas, and rocks in Earth's interior. Carbon materials of Earth and planetary interest are found in a wide range of structural states (Hazen et al., 2013; Oganov et al., 2013). Of the many transformations between these states, one of the most profound is that induced by change from sp^2 to sp^3 bonding of carbon in a structure. This transformation occurs in native carbon (graphite to diamond), in CO_2 ices, carbonate minerals, and hydrocarbons. In Chapter 1, Lobanov and Goncharov review this transformation in a subset of these materials. A key point is that the sp^2-sp^3 change leads to higher coordination number and is promoted by high pressure, and is therefore encountered at the extreme pressures of planetary interiors. However, as shown by Tschauner (Chapter 2), diamond remains the only naturally sampled material that preserves carbon in sp^3-bonded sites. Tschauner reviews carbonaceous inclusions found in terrestrial diamonds delivered to the surface from the mantle, in some cases at high residual pressures. The crystalline forms run the gamut of carbon oxidation states: native carbon and carbides; oxidized carbon in CO_2 ices and carbonate minerals; and, not discussed by Tschauner, rare hydrocarbon inclusions as well (e.g., Sobolev et al., 2019), though the origins of such materials have in the past been ascribed to later, shallower processes.

Carbon's cosmochemical abundance and chemical behavior favor carbon as a potential light element in the core. If present, carbon would likely be strongly partitioned into the inner core, as Fe-carbide. While early work favored Fe_3C (cementite) as the likely inner-core carbide, recent studies advanced the idea that this phase is not stable at inner core conditions, and Fe_7C_3 is instead the more likely inner-core carbide. Takahashi et al. (Chapter 3) performed new experiments that show that Fe_3C is stable to inner core conditions. Both carbide phases may be present in the inner core and could be consistent with seismological observations.

Chen and Wang (Chapter 4) review the structure and physical properties of carbon-bearing Fe-Ni liquids at conditions relevant to planetary cores. Where present, carbon may play an important role in controlling structural transformation in Fe-Ni-C liquids.

Comparatively little carbon can be incorporated into silicates, and the mechanism(s) for accommodating even small amounts is poorly known. Navrotsky et al. (Chapter 5) discuss silicate-rich ceramics that incorporate carbon via substitution of C for O in the silica tetrahedron, the fundamental building block of the rock-forming silicate minerals and the structural backbone of silicate melts. Geologic pathways for production of such materials may include large impact events, and these materials may be precursors for some puzzling natural occurrences of silicon carbide and carbonado.

Oxidized carbon, as CO_2, is important to a wide range of geologic processes from the surface to the interior of Earth, and potentially other solar system objects and exoplanets. It is therefore essential to understand the behavior of CO_2 itself at elevated pressure and temperature. The properties and transformations of CO_2 gas, liquid, and supercritical fluid are relatively well understood compared to CO_2 ices. As with H_2O, compression of CO_2 at very low to very high temperature produces a wide range of ice structures, which display a remarkable variety of bonding environments that suggest surprising possibilities for the forms and transformations of CO_2 in planetary interiors. Chapters 6 and 7, by Santoro et al. and Yoo, present overviews of the current state of knowledge of high-pressure CO_2 phases and their structures and properties. Despite years of aggressive investigation, the equilibrium phase diagram remains elusive. Metastable states and surprising forms such as high-pressure amorphous phases persist, likely owing to a complex energy landscape with multiple local minima and challenging kinetics (e.g., Machon et al., 2014), as has recently been illustrated by Tulk et al. (2019) for H_2O ices. Nevertheless, it is clear that CO_2 phase space contains a rich variety of molecular ices that give way at high pressure to a polymerized, extended covalent structure, CO_2-V, in which *sp*^{3} carbon is tetrahedrally coordinated by oxygen in a silica-like structure. This structure raises the possibility of solid solution with SiO_2, but this has yet to be conclusively verified. The contrasting interpretations of some of the features and phases of the CO_2 system in the two chapters attests to the challenges of working on this important but kinetically sluggish and energetically complex chemical system.

Li et al. (Chapter 8) explore the role of carbon surfaces on H_2O ice and methane clathrate crystallization. Using classical molecular dynamics, they find that ice nucleation and growth depends strongly on the chemistry, crystallinity, and topography of the nucleating surface. Gas hydrates initially nucleate as amorphous clusters, but crystallinity increases with the size of the hydrate. The picture is highly complex on the molecular scale, and there appear to be numerous pathways for hydrate growth.

The primary solid storage site for oxidized carbon is in carbonate minerals. The carbonate minerals exhibit a wide range in structures and bonding environments for carbon, as seen in CO_2. Merlini et al. (Chapter 9) review research over the last 10–15 years that reveals the complex pressure and temperature dependence of the crystal chemistry of carbonate minerals. From Earth's surface to the mid mantle, an essential building block of carbonate minerals is the trigonal CO_3^{-2} ion. In addition to pressure-induced transformations such as calcite to aragonite, and aragonite to post-aragonite, arrays of carbonate ions exhibit many subtle changes in geometry that give rise to a host of subtly different stable and metastable mineral polymorphs. At pressures of the mid-mantle and greater, trigonal coordination of C by O gives way to tetrahedral coordination, with attendant transformation to crystal structures featuring CO_4^{-4} rings and chains.

At Earth's surface and in the crust, the most abundant carbonate mineral is $CaCO_3$ calcite. As with other minerals, calcite can be a rich repository of information about its environment of formation, but it is relatively underexploited in this regard. Building on their previous work on how volatile elements can be retained in calcite to provide information on ancient gas and fluid chemistry, Cherniak et al. (Chapter 10) present new results on nitrogen diffusivity in calcite. The data demonstrate that N is readily retained in calcite that does not suffer metamorphism at >500°C, or deformation, or alteration. This raises the prospects that ancient calcites could be mined for information about atmospheric evolution and the geologic nitrogen cycle.

Fe-Mg carbonates may be the most prevalent carbonate materials in the mid to lower mantle. Boulard et al. (Chapter 11) review the *sp*^{2}-*sp*^{3} structural transformation in $(Mg,Fe)CO_3$. They highlight the potential importance of Fe^{3+} carbonates: Fe disproportionation may be important to stabilizing carbonate minerals at these great depths. In addition to the change in coordination due to the *sp*^{2}-*sp*^{3} transition in carbon, Fe-Mg carbonates also exhibit an important transformation due to the spin transition of iron. Liu et al. (Chapter 12) review various experimental and theoretical methodologies in the investigation of this phenomenon and show that this transition in carbonates likely occurs between 50 and 80 GPa along the representative mantle geotherm. A substantial decrease in volume of up to 10%, shear wave splitting anisotropy, and deformation textures raise the possibility of seismic detectability. Na-Ca carbonates may also be important in certain subducted lithologies. Chapter 13 by Rashchenko et al. reviews the wide variety of crystal structures of high-pressure Na-Ca carbonates.

The daunting variety of carbonate crystal structures leads to an immensely challenging problem in working out the stable phase relations among carbonate minerals, and between carbonates and other oxides. Litasov et al. (Chapter 14) make a valiant effort to systematically evaluate the phase relations in unary, binary, and ternary carbonate systems relevant to conditions of Earth's mantle. However, phase relations in carbonate systems alone are insufficient to assess carbon phase equilibria in the mantle. Even for oxidizing conditions, the presence of additional minerals in mantle lithologies controls the distribution and nature of carbon hosts. Li et al (Chapter 15) show that at conditions of the mantle transition zone

(15 GPa and 1200°C), aragonite will react with wadsleyite in model slab lithologies to produce magnesite, Ca perovskite, and periclase. Rates of reaction are enhanced by the presence of H_2O. Because the solidus temperature of magnesite-bearing lithologies is higher, transfer of carbon from aragonite to calcite by this reaction mechanism has the effect of promoting transport of carbon deeper into the mantle.

The solubility of carbon in terrestrial magmas is a complex function of pressure, temperature, bulk composition, and oxygen fugacity. Moreover, carbon in magmas occurs in various forms. Solomatova et al. (Chapter 16) review bulk carbon solubility and the speciation of magmatic carbon based on recent insights from molecular dynamics calculations. Computational studies are especially important given the extreme challenges faced by experimentalists in inferring carbon speciation in quenched glasses, especially from very high pressure. Solomatova et al. show that molecular dynamics studies return trends in solubility and speciation that are similar to those derived experimentally, while revealing evidence for novel polymerization of carbon at very high pressures.

The solubility and speciation of carbon in high-pressure liquids is especially important for the deep carbon cycle, as melts produced from the slab afford one of the most effective ways of returning subducted carbon to the exosphere. Two chapters present new experimental results that drive home this point. Muth et al. (Chapter 17) investigated the solubility and speciation of carbon in hydrous rhyolitic melts that can be expected from sediment and slab melting along some slab-top geotherms. They find an important variation with Na number, defined as Na/(Na+K). All else equal, carbon solubility and the fraction of CO_3^{-2} relative to molecular CO_2 increase with Na number. An empirical model suggests that such melts could readily deliver the carbon found in subduction zone volcanic systems at plausible fractional contributions of slab melts to mantle wedge–derived basalts.

The low melting temperature of Ca-rich carbonated systems is highlighted by Schettino and Poli (Chapter 18). They find that model lithologies approximating pelagic limestones yield evidence for the presence of a hydrous carbonated liquid at temperatures as low as 850°C at 4.2 and 6 GPa. Such liquids would represent exceptionally efficient transport agents in subduction zone settings.

The viscosities of nominally anhydrous carbonate-rich melts at upper mantle pressures are very low, consistent with rapid ascent rates of even very small melt fractions. However, such melts are also extremely reactive and will therefore change composition upon ascent, in part by becoming more silica rich. Stagno et al. (Chapter 19) determined the viscosity of carbonate-silicate liquids at high pressure. Viscosities are about an order of magnitude higher than those of pure carbonate liquids at similar conditions, which will lead to comparatively lower ascent rates and, by virtue of increasing melt fraction, shorter residence times.

Mixtures of water and carbon dioxide are arguably the primary solvent components for fluids in the Earth's crust and upper mantle. Abramson (Chapter 20) reviews models of H_2O–CO_2 mixing behavior, informed by new data at high pressures. Though such fluids have historically been modeled as strictly molecular mixtures, phase relations and spectroscopic observations require that the topology of the miscibility gap is locally significantly impacted by reaction of CO_2 and H_2O to form bicarbonate in the fluid phase. Taking this into account poses major challenges for equations of state for mixed fluids.

Some of the carbon in crustal and mantle fluids derives from dissolution of carbonate minerals during high-pressure metamorphism, and this dissolution will be impacted by other important solutes such as alkali halides. Eguchi et al. (Chapter 21) experimentally determined calcite solubility in H_2O with varying concentrations of a range of alkali halides (NaCl, KCl, LiCl, CsCl). Rising salt concentration enhances calcite solubilty no matter the identity of the salt, but the extent of enhancement increases with decreasing ionic radius of the alkali cation.

In the experiments of Eguchi et al., the fO_2 was sufficiently high that calcite dissolution likely produced only oxidize carbonate species. However, it is increasingly being recognized that organic solutes may be important in many deep-fluid settings. Sverjensky et al. (Chapter 22) show that the chemistry of aqueous organic solutes changes profoundly with depth in the Earth. In shallow geologic fluids such as oil field brines and geothermal systems, the chemistry of aqueous organic solutes is dominated by kinetic inhibition of formation and interaction with methane. However, in deeper crustal and mantle settings, a closer approach to equilibrium predominates, which leads to aqueous species with a range of oxidation states intermediate between CH_4 and CO_2. Given appropriate conditions, phase separation to form a coexisting hydrocarbon fluid may occur.

Of the shallower environments, oceanic hydrothermal systems are especially important to aqueous organic chemistry, as they afford favorable environments for abiotic synthesis of life-essential amino acids (Ménez et al., 2018). In such settings, polypeptide synthesis is key to the formation of more complex biomolecules. Kroonblawd and Goldman (Chapter 23) performed molecular dynamics simulations to explore the pathways for aqueous glycine oligomerization at hydrothermal vent conditions. They find that relatively low temperatures of ~100°C provide optimal conditions for oligoglycine formation.

Moving deeper, one environment in which aqueous organic solutes may be much more important than previously thought is in subduction zones. Guild and Shock (Chapter 24) use thermodynamic modeling to evaluate the abundance and distribution of aqueous organic solutes in subduction zone fluids relevant to equilibration with mantle mineral assemblages. They find that organic species are important even at fO_2 of quartz-fayalite-magnetite, and become more so as fO_2 decreases. Both C1 and C2 species are stable, and their abundances increase when potential kinetic limitations on methane formation are taken into account. Canovas and Shock (Chapter 25) further explore aqueous organic chemistry during subduction, in this case with a view to evaluating the energetics of the citric acid cycle. They show that energetics may be favorable for supporting a biosphere deeper in subduction zones than previously thought. Kutcherov et al. (Chapter 26) report on experiments interpreted to have produced hydrocarbons at mantle conditions. They hypothesize a deep hydrocarbon cycle that tracks the fate of these hydrocarbons in the mantle.

Bringing things full circle, Park et al. (Chapter 27) examine the compression behavior of diamondoids, nanoclusters of sp^3 bonded carbon terminated by hydrogen. These hydrocarbon molecules, housed in a diamond-like structure, are found in natural petroleum, have potentially important material properties, and could represent an unexpected pathway to diamond growth at high pressure from subducted kerogen (e.g., Plank & Manning, 2019).

The papers in this volume represent an outgrowth of a decade of research partly stimulated by the Deep Carbon Observatory. While the past decade has seen major advances in our understanding of carbon in planetary interiors, it is clear that much remains to be done to understand the forms, transformations, and movements of carbon at extreme conditions.

Craig E. Manning
Jung-Fu Lin
Wendy L. Mao

REFERENCES

Burton, M.R., Sawyer, G.M., Granieri, D. (2013). Deep carbon emissions from volcanoes. *Rev. Mineral Geochem. 75*, 323–354.

Dasgupta, R. (2013). Ingassing, storage, and outgassing of terrestrial carbon through geologic time. *Reviews in Mineralogy and Geochemistry*, *75*, 183–229.

Dasgupta, R., Hirschmann, M.M. (2010). The deep carbon cycle and melting in Earth's interior. *Earth and Planetary Science Letters*, *298*, 1–13.

Hazen, RM, Downs RT, Jones AP, Kah L. (2013). Carbon mineralogy and crystal chemistry. *Reviews in Mineralogy and Geochemistry*, *75*, 7–46.

Jones, AP, Genge M, Carmody L. (2013). Carbonate melts and carbonatites. *Reviews in Mineralogy and Geochemistry*, *75*, 289–322.

Kelemen, PB, Manning CE. (2015). Reevaluating carbon fluxes in subduction zones, what goes down, mostly comes up. *Proceedings of the National Academy of Sciences*, *112*, E3997–E4006.

Machon, D., Meersman, F., Wilding, M., Wilson, M., McMillan, P. (2014). Pressure-induced amorphization and polyamorphism: Inorganic and biochemical systems. *Progress in Materials Science*, *61*, 216–282.

Marty, B., Alexander, C.O., Raymond, S.N. (2013). Primordial origins of Earth's carbon. *Reviews in Mineralogy and Geochemistry*, *75*, 149–181.

Ménez, B., Pisapia, C., Andreani, M., Jamme, F., Vanbellingen, Q.P., Brunelle, A., Richard, L., Dumas, P., Réfrégiers, M. (2018). Abiotic synthesis of amino acids in the recesses of the oceanic lithosphere. *Nature*, *564*, 59.

Ni, H., Keppler, H. (2013). Carbon in silicate melts. *Reviews in Mineralogy and Geochemistry*, *75*, 251–287.

Oganov, A.R., Hemley, R.J., Hazen, R.M., Jones, A.P. (2013). Structure, bonding, and mineralogy of carbon at extreme conditions. *Reviews in Mineralogy and Geochemistry*, *75*, 47–77.

Plank, T.A, Manning, C.E. (2019). Subducting carbon. *Nature*, *574*, 343–352.

Shirey, S.B, Cartigny, P., Frost, D.J, Keshav, S., Nestola, F., Nimis, P., Pearson, D.G., Sobolev, N.V., Walter, M.J. (2013). Diamonds and the geology of mantle carbon. *Reviews in Mineralogy and Geochemistry*, *75*, 355–421.

Sobolev, N.V., Tomilenko, A.A., Bul'bak, T.A., Logvinova, A.M. (2019). Composition of hydrocarbons in diamonds, garnet, and olivine from diamondiferous peridotites from the Udachnaya Pipe in Yakutia, Russia. *Engineering*, *5*, 471–478.

Tulk, C.A., Molaison, J.J., Makhluf, A.R., Manning, C.E., Klug, D.D. (2019). Absence of amorphous forms when ice is compressed at low temperature. *Nature*, *569*, 542–545.

1

Pressure-Induced sp^2-sp^3 Transitions in Carbon-Bearing Phases

Sergey S. Lobanov[1] and Alexander F. Goncharov[2]

ABSTRACT

Carbon-bearing phases show a rich variety of structural transitions as an adaptation to pressure. Of particular interest is the crossover from sp^2 carbon to sp^3 carbon, as physical and chemical properties of carbon in these distinct electronic configurations are very different. In this chapter we review pressure-induced sp^2-sp^3 transitions in elemental carbon, carbonates, and hydrocarbons.

1.1. INTRODUCTION

Understanding of pressure-induced phase transitions in carbon-bearing phases is fundamental for physics and chemistry of dense condensed matter and has important implications for Earth and planetary sciences. Carbon is unique in that it can form compounds with three distinct electronic configurations (we will focus on sp^2 and sp^3 configurations here) depending on the way *s* and *p* atomic orbitals of carbon interact with each other. The different geometrical arrangement of chemical bonds typical of compounds with sp^2-bonded and sp^3-bonded carbon (Figure 1.1, inset) results in threefold (lower steric repulsion) and fourfold (higher steric repulsion) coordination, respectively. High pressure is an efficient driver for sp^2-sp^3 transitions as it promotes denser crystal structures and allows negating the steric repulsion of neighboring atoms in sp^3 phases. More often than not, material properties over such transitions change in a radical manner. It is most obvious in the case of pure carbon where graphite and diamond represent archetypal examples of sp^2-bonded and sp^3-bonded crystals, which are extremely dissimilar in the crystal and electronic structure, and hence in the properties such as, for example, electrical conductivity, elasticity, and vibrational anisotropy (Oganov et al., 2013).

Carbon prefers the sp^2-configuration at low pressure, although there are exceptions (e.g. CH_4), because of the small carbon atomic radius (70 pm) that is not sufficient to reduce steric repulsion among the atoms in its coordination. The increase of coordination number upon the sp^2-sp^3 transition always entails longer carbon-to-neighbor bonds in the sp^3-phase to compensate for the increase in steric repulsion (Prewitt & Downs, 1998). As a result, significant atomic rearrangements are necessary for sp^2-sp^3 phase transitions with large energy barriers typical of such crossovers (Powles et al., 2013). High-temperature (*T*) conditions are often required to reach the thermodynamic ground state, such as in the case of graphite and diamond. Consequently, many carbon-bearing phases can be preserved outside their thermodynamic stability range. This intrinsic metastability of sp^3 carbons finds many useful scientific and technological applications. Natural diamonds, despite being metastable at ambient conditions, often secure unique samples of the Earth's mantle due to their chemical inertness (Kopylova et al., 2010; Wirth et al., 2014; Pearson et al., 2014; Smith et al., 2016; Nestola et al., 2018; Tschauner et al., 2018). The sluggish kinetics of the graphite-to-diamond transition also helps to decipher the complex *P-T* history of high-pressure metamorphic complexes and mantle xenoliths (De Corte et al., 2000; Hwang et al., 2001; Massonne, 2003; Korsakov et al., 2010; Mikhailenko et al., 2016; Shchepetova et al., 2017). The rich polymorphism of carbonates driven by the diverse carbon bonding patterns presents another example with relevance to geosciences as it creates

[1] *GFZ German Research Center for Geosciences, Potsdam, Germany*

[2] *Geophysical Laboratory, Carnegie Institution for Science, Washington, DC, USA*

Carbon in Earth's Interior, Geophysical Monograph 249, First Edition. Edited by Craig E. Manning, Jung-Fu Lin, and Wendy L. Mao.

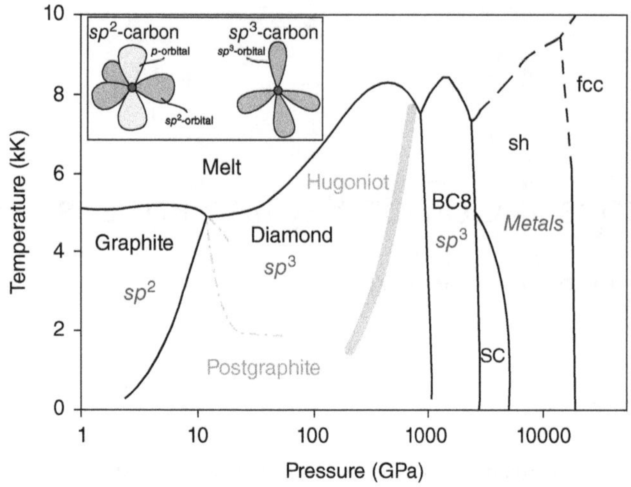

Figure 1.1 Phase diagram of carbon at extreme *P-T* conditions. BC8, sc (simple cubic), sh (simple hexagonal), and fcc (face-centered cubic) are theoretically predicted carbon structures at $P > 1$ TPa that have never been observed in experiment. Inset: The electronic structure of sp^2 and sp^3-hybridized carbon. See electronic version for color representation of the figures in this book.

pathways for transporting carbon into the deep mantle. Aside from the geological importance, the sp^2-sp^3 polymorphism of carbon-bearing phases is of great technological interest as it allows accessing a rich range of diverse physical and chemical properties in an isochemical setting (Field, 1992). This chapter reviews sp^2-sp^3 transitions in carbon-bearing phases that include pure carbon, carbonates, and hydrocarbons.

1.2. ELEMENTAL CARBON

Graphite and diamond are the only two stable allotropes of carbon at $P < \sim 1000$ GPa (Figure 1.1). However, a variety of metastable phases are known with a broad range of kinetic stability, which is a unique feature of carbon, making its phase diagram very complex, especially at near-ambient conditions (e.g. Bundy et al., 1996). In addition, the melting curves of the carbon polymorphs remain uncertain in the limit of high pressure and high temperature, where obtaining direct experimental data is challenging. Here we update the previous reviews (Oganov et al., 2013; Bundy et al., 1996; Goncharov, 1987; Hazen et al., 2013), mainly concentrating on the phase relations at extreme conditions and focusing on materials of interest to geosciences.

1.2.1. Metastable Phases

Lonsdaleite is a naturally occurring diamond polytype that was first described in Canyon Diablo meteorite (Frondel & Marvin, 1967; Hanneman et al., 1967) and has been used since to pinpoint the *P-T* conditions of the impact. Lonsdaleite has a hexagonal crystal structure that has a slightly higher energy than cubic diamond. As such, the formation of this phase (e.g. from graphite) must follow a certain kinetic route (Erskine & Nellis, 1991; Xie et al., 2017; Bundy & Kasper, 1967), making its occurrence difficult to uniquely identify in both natural and synthetic samples. This resulted in conflicting reports concerning the microstructure of samples synthesized at static pressure conditions, questioning the uniqueness of lonsdaleite as a structurally discrete material (Németh et al., 2014; Shiell et al., 2016). Further, the existence of a nanometer (nm) scale structural complexity in lonsdaleite makes its characterization complex even at ambient conditions. As a result of its structural complexities, the mechanism of lonsdaleite formation in early shock wave experiments (e.g. Erskine & Nellis, 1991) was difficult to resolve. In recent years, however, the development of in situ X-ray diffraction (XRD) characterization of dynamic processes on a very fast time scale (ps to ns) allowed for better understanding of the lonsdaleite formation and the carbon phase diagram (Rygg et al., 2012; Gupta et al., 2012; Gauthier et al., 2014). Nevertheless, ambiguities still exist. For example, a direct transition from graphite to lonsdaleite at 50 GPa was inferred by Turneaure et al. (2017), while Kraus et al. (2016) reported on a different sequence that includes a transition to cubic diamond at 50 GPa and subsequently to lonsdaleite at 170 GPa. These works stimulate the development of theoretical models, which also remain contradictory concerning the mechanisms of graphite-to-diamond transformations (Xie et al., 2017; Pineau, 2013; Mundy et al., 2008; Khaliullin et al., 2011).

As with lonsdaleite, other metastable carbon phases may serve as indicators of shock metamorphism *P-T* conditions. Several kinetically accessible carbon materials are formed by compression of graphite (and other metastable sp^2-bonded carbons) at room temperature. Without sufficient thermal energy to activate the transition to diamond-like phases, the system relaxes in structures that represent a compromise between the thermodynamic stimulus and kinetic hindrance. Cold compression (at 300 K) to above 15 GPa, well into the stability field of diamond, results in an sp^3-bonded form of carbon (termed postgraphite here) as suggested by a decrease in the electrical conductivity, appearance of optical transparency, and changes in inelastic X-ray scattering spectra (Goncharov et al., 1989; Utsumi & Yagi, 1991; Mao et al., 2003). XRD measurements on this sp^3-bearing carbon are not conclusive in the structural determination (Mao et al., 2003; Wang et al., 2012) because only powder-like data was examined and the peaks of the high-pressure phase were weak and broad. Raman spectroscopy, which is generally sensitive to the type of chemical bonding, is also not decisive as sp^3 carbon has a two orders of magnitude smaller scattering cross-section compared to sp^2 carbon when the spectra are excited in the visible spectral range (Ferrari, 2002). The Raman spectra of cold-compressed graphite (Wang et al., 2012; Goncharov et al., 1990; Xu et al., 2002) show a large broadening of the C-C stretching mode of the sp^2-bonded carbon in the high-pressure phase. However, positive spectroscopic identification of sp^3-bonded carbon has not been achieved because the first-order diamond peak of the diamond anvils and the disorder-induced D-band of sp^2-bonded carbon both obscure the spectral range of interest. Overall, the XRD and Raman data suggest the presence of disorder in postgraphite phase(s). The persistence of the sp^2 fingerprints in its Raman spectra upon cold compression indicates that postgraphite still contains sp^2-bonded carbon, which is inconsistent with the theoretically predicted M-carbon phase (Li et al., 2009) usually considered the best match to the experiment. Interestingly, the transition to postgraphite is reversible at 300 K (Utsumi & Yagi, 1991; Mao et al., 2003; Wang et al., 2012), but this phase can be quenched at low temperatures <100 K (Miller et al., 1997). A very similar phase can be obtained from carbon nanotubes (albeit at higher pressures than from graphite), which appears quenchable to ambient conditions (Wang et al., 2004).

Annealing of a cold-compressed graphite at T >1000 K and P ~20 GPa results in a pressure-quenchable phase that is structurally different from postgraphite and shows similarities to lonsdaleite (Utsumi & Yagi, 1991). Glassy carbon cold-compressed to 50 GPa and heated to ~1800 K produces an amorphous and pressure-quenchable sp^3-bonded phase (Zeng et al., 2017; Yin et al., 2011). These results highlight the importance of the initial carbon microstructure on the dense high-pressure synthesis products, which has the potential for creating new ultralight and ultrastrong materials (Hu et al., 2017). Theoretical description of all these metastable carbon phases (see Shi et al., 2018, and references therein) remains largely unsatisfactory.

1.2.2. Phase Diagram at High Temperature and in the TPa Pressure Range

In the limit of high temperature, the graphite-to-diamond transition is well established and sharp (Bundy et al., 1996). In contrast, the theoretically predicted sp^2-sp^3 transition in molten carbon is sluggish and the results appear sensitive to the level of theory used in the computation (e.g. first-principles vs. classical) (Ghiringhelli et al., 2004). With regard to solid–solid transformations and melting at very high pressures, carbon is unique in the group IV of the periodic table in that it has no p-electrons in the core, allowing the valence p-electrons to be closer to the nucleus, making stronger directional chemical bonds (Yin & Cohen, 1983). As a result, the diamond structure has a much larger P-T stability range compared to its Si and Ge sp^3 analogs. Unlike Si and Ge, the melting curve of diamond is positive (above the graphite-diamond-melt triple point) (Bundy et al., 1996), and the sp^3-bonded BC8 phase stable at P > ~1 TPa is not metallic (cf. β-tin Si) but semiconducting (Correa et al., 2006; Martinez-Canales et al., 2012). According to theoretical calculations (Wang et al., 2005; Benedict et al., 2014), the slope of this diamond melting curve changes from positive to negative above 450 GPa due to the softening of the transverse acoustic branches near the L and X symmetry points (Correa et al. 2006). Dynamic compression experiments detected the melting of diamond close to the expected range of P-T conditions, inferred a crossover in the slope of its melting line, and provided evidence of the second diamond-BC8-liquid triple point (Brygoo et al., 2007; Knudson et al., 2008; Eggert et al., 2009). At higher pressures, theory predicts that the BC8 structure undergoes metallization (Correa et al., 2006) or transforms to a metallic simple cubic (sc) phase above 2.7 TPa, and then further to metallic simple hexagonal (sh) and cubic fcc structures (Figure 1.1) (Martinez-Canales et al., 2012; Benedict et al., 2014). Recent ramp compression experiments that relied on single-surface velocity measurements did not record discontinuities that could be attributed to the predicted phase transition (Smith et al., 2014). In the absence of direct structural characterization, however, any robust conclusion seems premature.

1.3. CARBONATES

1.3.1. (Mg,Fe) Carbonates

To the best of our knowledge, Skorodumova et al. (2005) provided the first computational evidence of thermodynamic stability of sp^3-$MgCO_3$ with a pyroxene structure ($C2/c$) at P > 113 GPa with CO_4^{2-} tetrahedral groups arranged in chains. This study, however, explored only the relative stabilities of several possible sp^3-carbonate structures as suggested by crystallographic and chemical similarities with Mg-bearing silicates. The first unconstrained search for the stable structures in the $MgCO_3$ system by evolutionary crystal structure prediction algorithms revealed an sp^2-sp^3 transition at 82 GPa (Oganov et al., 2008) with a $C2/m$ structure being ~0.2 eV more favorable than the pyroxene structure initially proposed by Skorodumova et al. (2005). This sp^3-$MgCO_3$ phase was termed magnesite-II (phase II) with a wide stability field up to 138 GPa where another sp^3 modification of $MgCO_3$ ($P2_1$, phase III) was predicted to take over (Oganov et al., 2008). More structures, however, were proposed in the 90–120 GPa range as energetically competitive by Panero and Kabbes (2008) based on a first principles evaluation of several potential structures. Pickard and Needs (2015) revised the phase diagram of $MgCO_3$ using ab initio random structure searching technique to find new stable *P*-1 (85-101 GPa) and $P2_12_12_1$ (P >144 GPa) $MgCO_3$ structures with sp^3 carbon in addition to the *C2*/*m* structure proposed earlier (Oganov et al., 2008). Most recently, another polymorph of sp^3-$MgCO_3$ ($P2_1$) was predicted at P > 143 GPa (Yao et al., 2018). Figure 1.2 summarizes theoretical predictions for the $MgCO_3$ (and other) system(s).

These first-principles predictions stimulated the experimental search for pressure-induced sp^2-sp^3 transitions in carbonates. Boulard et al. (2011) attempted to synthesize sp^3-$MgCO_3$ in a diamond anvil cell (DAC) using synchrotron XRD as a probe to detect the new phases. At 80 GPa and ~2400 K, the authors observed new XRD peaks that could not be attributed to the starting materials ($MgCO_3$ or MgO + CO_2). This new phase was not temperature-quenchable and only the low-pressure magnesite structure could be observed after cooling (Boulard et al., 2011). Nonetheless, on the basis of Le Bail–type crystallographic refinement of high-temperature XRD patterns, the authors concluded that they successfully synthesized magnesite II with a crystal structure based on three CO_4-tetrahedra linked into $(C_3O_9)^{6-}$ rings

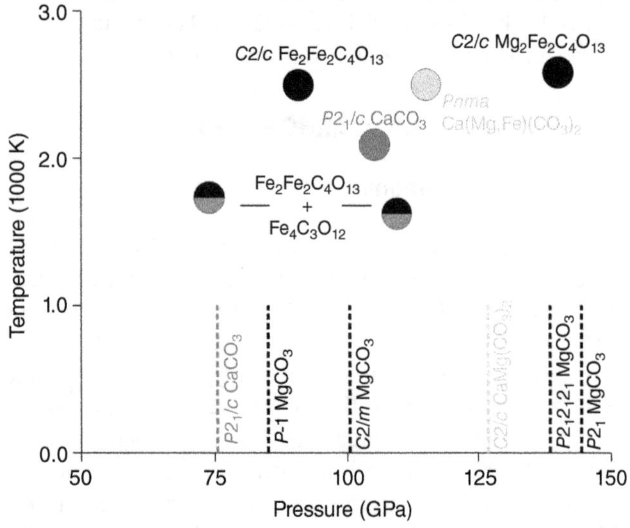

Figure 1.2 Pressure-temperature conditions of sp^2-sp^3 phase transitions in carbonates. Theoretical predictions (calculations at 0 K) are shown by dashed lines. Only the most recent reports are shown: $MgCO_3$ (Pickard & Needs, 2015; Yao et al., 2018), $CaMg(CO_3)_2$ (Solomatova & Asimow, 2017), and $CaCO_3$ (Pickard & Needs, 2015). Circles depict the synthesis conditions of sp^3-phases in (Mg,Fe)- (Merlini et al., 2015), (Ca,Mg,Fe)- (Merlini et al., 2017), and Ca-carbonate (Lobanov et al., 2017) systems. See electronic version for color representation of the figures in this book.

(Boulard et al., 2011). A crystal structure with $P2_1/c$ symmetry allowed for improved Le Bail refinement over the predicted *C2*/*m* $MgCO_3$ structure.

Similar experimental evidence was put forward in support of the sp^3-(Fe, Mg)CO_3 synthesis from $(Fe_{0.75}Mg_{0.25})$ CO_3 or $(Mg_{0.6}Fe_{0.4})$ + CO_2 after laser heating to ~2000 K at 80 GPa and to ~2850 K at 105 GPa, respectively (Boulard et al., 2011). Regardless of the starting components, the measured XRD patterns showed similarities to that observed in pure $MgCO_3$, and the authors concluded that the same crystal structure formed in these two systems. In the iron-containing system, however, the synthesized phase was not only temperature-quenchable but also pressure-quenchable, and the recovered samples were analyzed by electron energy loss spectroscopy, which showed a peak at 290.7 eV (assigned to $C_3O_9^{6-}$) as opposed to 290.3 eV in reference sp^2-siderite (Boulard et al., 2011). However, it is unclear whether the authors' interpretation is unequivocal as the reported energy resolution (width of the zero-loss peak at its half height) was 0.6 eV.

In a subsequent work, Boulard et al. (2015) reported on the room-temperature infrared spectrum of $(Fe_{0.75}Mg_{0.25})$ CO_3 that was presynthesized at ~2100 K at 103 GPa. The synthesis of *C2*/*m* (Fe,Mg)CO_3, as well as of a high-pressure polymorph of $CaMn_2O_4$-type Fe_3O_4, was reported based on synchrotron XRD. One should bear in mind, however, that reliable identification of these low symmetry structures at P ~100 GPa based solely on a Le Bail–type refinement is challenging. For example, the identification of the Fe_3O_4 structure could not have been possible because iron oxides show a complex high *P*-*T* behavior (Bykova et al., 2016), the details of which were not known at the time of the Boulard et al. (2015) work. The stable structure of Fe_3O_4 is different from the one used for Le Bail refinement in Boulard et al. (2015), which undermines the robustness and the overall conclusions of that work.

Merlini et al. (2015) provided a more reliable insight into the crystal structures of sp^3-(Fe,Mg) carbonates by means of single-crystal structure solution methods finding that sp^2-$(Fe_{0.75}Mg_{0.25})$ CO_3 transforms to a *C2*/*c* $Mg_2Fe_2(C_4O_{13})$ together with a *C2*/*m* $Fe_{13}O_{19}$ at P ~135 GPa and T ~2650 K (Figure 1.2). The new carbonate phase contained tetrahedrally coordinated sp^3-hybridized carbon linked into corner-shared truncated C_4O_{13} chains and apparently remained stable on decompression down to 40 GPa. Importantly, this study provided the first evidence that the composition of the (Mg,Fe) carbonate may change over the sp^2-sp^3 transition, which had not been considered by first-principles computations. A similar single-crystal XRD technique was subsequently applied to the $FeCO_3$ endmember uncovering two distinct sp^3 carbonates at P > ~70 GPa (Cerantola et al., 2017). According to this study, $Fe_4(C_3O_{12})$ and $Fe_2Fe_2(C_4O_{13})$ coexist in the temperature limit of ~1500–2200 K, while only $Fe_2Fe_2(C_4O_{13})$ is present at $T > \sim 2200$ K. The crystal structure of $Fe_4(C_3O_{12})$ involves isolated CO_4 tetrahedra, while $Fe_2Fe_2(C_4O_{13})$ contains truncated chains of CO_4 tetrahedra (Figure 1.3) and is isostructural with $Mg_2Fe_2(C_4O_{13})$ found earlier by Merlini et al. (2015).

Overall, the case of (Mg,Fe) carbonates shows that they exhibit a remarkably complex crystallographic behavior over the sp^2-sp^3 transition. Initial theoretical predictions of the stable sp^3-$MgCO_3$ did find some experimental support in both Mg carbonates and (Mg,Fe) carbonates. However, subsequent advanced crystallographic studies of (Mg,Fe) carbonates and Fe carbonates unraveled a complex behavior that always involves a change in the carbonate chemical composition, which was not anticipated from the earlier ab initio computations. For these reasons, earlier experimental reports on sp^3-(Mg,Fe)CO_3 (Boulard et al., 2011, 2015, 2012) must be deemed unreliable.

1.3.2. Ca Carbonates and Ca(Mg,Fe) Carbonates

The theoretical search for an sp^2-sp^3 transition in $CaCO_3$ using evolutionary crystal structure prediction algorithms yielded an orthorhombic pyroxene-type ($C222_1$) structure at $P > 137$ GPa (Oganov et al., 2006). Ono et al. (2007) showed that laser-heating of $CaCO_3$ at $P > 130$ GPa results in a new phase that is structurally consistent with the predicted $C222_1$ $CaCO_3$. To the best of our knowledge, the synthesis of sp^3-$CaCO_3$ by Ono et al. (2007) is the first experimental report of an sp^3 carbonate. The crystal structure of this phase, however, was challenged by more recent computations that found that $P2_1/c$ $CaCO_3$ is ~0.2 eV/f.u. more energetically favorable than the $C222_1$ model and is stable at $P > 76$ GPa (Figure 1.2) and up to at least 160 GPa (Pickard & Needs, 2015; Yao et al., 2018). Both $C222_1$ and $P2_1/c$ $CaCO_3$ are based on CO_4-tetrahedra linked into 1D pyroxene-like chains with only slight differences in the arrangement of CO_4-groups. All vertex-sharing helices in the $C222_1$ structure are right-handed, while half helices in the $P2_1/c$ $CaCO_3$ are left-handed.

To test these theoretical predictions, Lobanov et al. (2017) used synchrotron XRD and Raman spectroscopy as probes for the sp^2-sp^3 transition. Crystallographic data collected after the heating of $CaCO_3$ to ~2000 K at 105 GPa provided evidence of the sp^2-sp^3 transition but were not sufficient to discriminate between the proposed $C222_1$ and $P2_1/c$ $CaCO_3$ models. Raman spectra measured before and after the heating, however, were clearly distinct. A characteristic C-O stretching vibration appears after the heating with a frequency that is ~20% lower than that in sp^2 $CaCO_3$ prior to the heating (Figure 1.4a), consistent with the longer C-O bond length in CO_4 tetrahedra. This spectral feature was reproduced independently via ab initio computations of the vibrational spectrum of sp^3 $CaCO_3$ and provided strong evidence in favor of the $P2_1/c$ model of $CaCO_3$ (Figure 1.4b). Altogether, the combined crystallographic and spectroscopic approach proved successful in detecting the sp^2-sp^3 transition in $CaCO_3$ and provided a reliable reference for the frequency of C-O stretching in CO_4 tetrahedra, which can be used to identify sp^2-sp^3 transitions in other carbonates (Lobanov et al., 2017).

Because of the chemical reactions with mantle minerals, $CaMg(CO_3)_2$ is a more realistic composition for the mantle carbonate. Merlini et al. (2017) reported on the synthesis of sp^3 $Ca(Mg_{0.6},Fe_{0.4})(CO_3)_2$ (dolomite-IV) at 115 GPa and 2500 K. The crystal structure of dolomite-IV was solved from single-crystal XRD data and contains CO_4 tetrahedra linked into threefold C_3O_9 rings

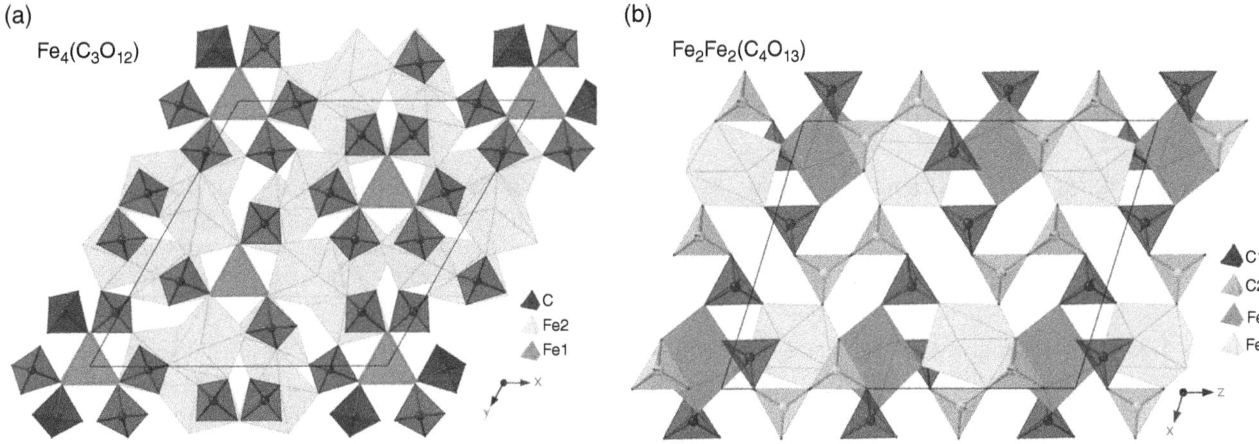

Figure 1.3 Crystal structures of $Fe_4(C_3O_{12})$ and $Fe_2Fe_2(C_4O_{13})$ after Cerantola, et al. (2017). Oxygen atoms are not shown for clarity. Unit cells are shown by thin black lines. See electronic version for color representation of the figures in this book.

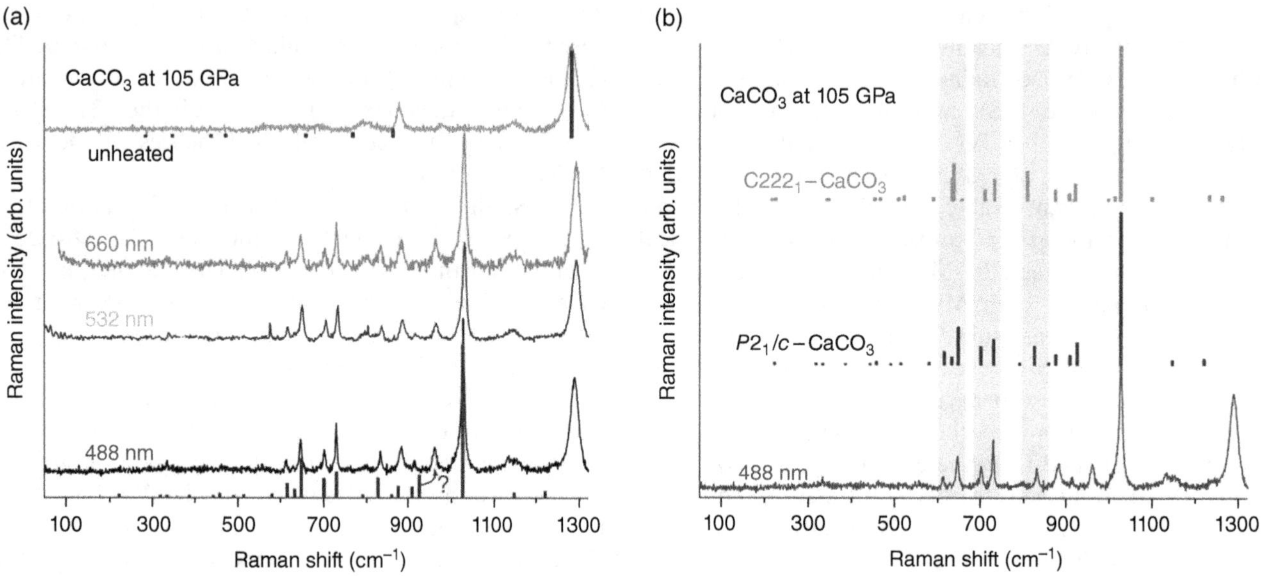

Figure 1.4 (a) Raman spectra of $CaCO_3$ at 105 GPa collected with 488, 532, and 660 nm excitations. Vertical bars are computed Raman modes of *P*2$_1$/*c*-$CaCO_3$ (bottom) and post-aragonite $CaCO_3$ (top). (b) Experimental spectrum of $CaCO_3$ laser-heated at 105 GPa in comparison with the theoretical spectra of *P*2$_1$/*c* and *C*222$_1$-$CaCO_3$ at 105 GPa as computed by LDA-DFT. Shaded areas are guides to compare the computed spectra with experiment. Adapted from Lobanov et al., 2017. See electronic version for color representation of the figures in this book.

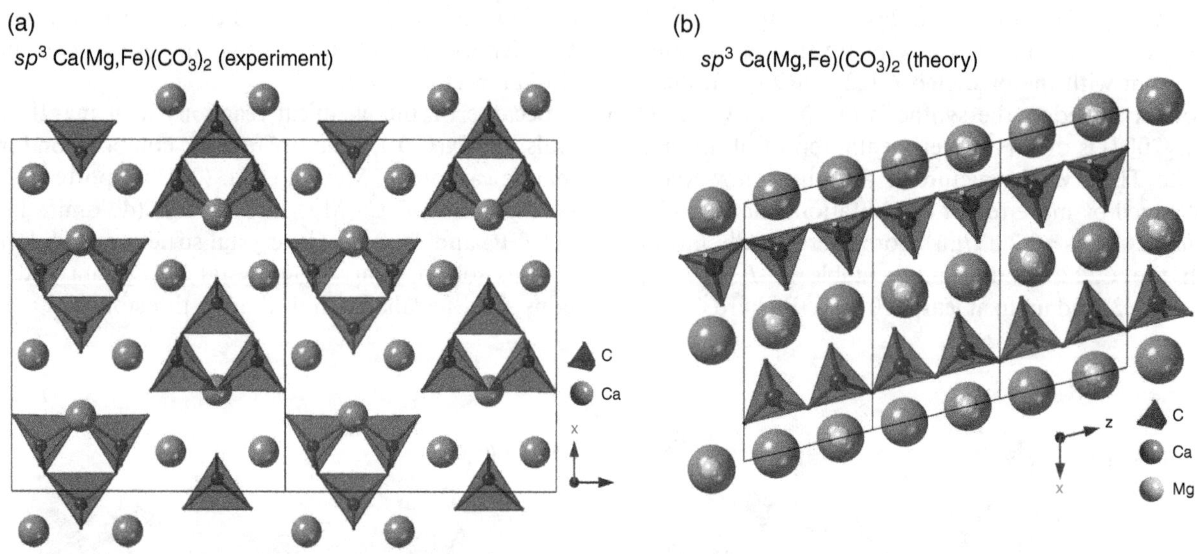

Figure 1.5 (a) Crystal structure of *sp*^{3}-Ca(Mg$_{0.6}$,Fe$_{0.4}$)$(CO_3)_2$ (dolomite-IV) after Merlini, et al. (2017). Mg and Fe atoms are not shown as they occupy identical crystallographic positions to Ca. (b) Theoretically predicted crystal structure of *sp*^{3}-$CaMg(CO_3)_2$ (Solomatova & Asimow, 2017). In a and b, oxygen atoms are not shown for clarity. Unit cells are shown by thin lines. See electronic version for color representation of the figures in this book.

(Figure 1.5a). Interestingly, Ca, Mg, and Fe cations occupy identical positions in the lattice, suggesting that moderate addition of iron does not bear important effects on the *sp*^{3}-$CaMg(CO_3)_2$ crystal structure. The presence of C_3O_9 rings in the dolomite-IV structure (Merlini et al., 2017) is an important similarity to the predicted magnesite-II phase (Oganov et al., 2008). However, an evolutionary search for *sp*^{3}-$CaMg(CO_3)_2$ yielded a different motif with CO_4-tetrahedra linked into pyroxene-like chains (Solomatova & Asimow, 2017), similarly to what has been predicted and experimentally confirmed for *sp*^{3}-$CaCO_3$. Interestingly, Raman spectra of dolomite-III at

$P > 63$ GPa, another polymorph of $CaMg(CO_3)_2$ stable at 36–115 GPa (Merlini et al., 2017), show a new band appearing as a low-frequency shoulder of the C-O symmetric stretching in CO_3 groups (Vennari & Williams, 2018). This shoulder was interpreted to be due to an increase in carbon coordination (3+1); thus, it may indicate a sluggish sp^2-sp^3 transition in $CaMg(CO_3)_2$ at room temperature.

In summary, $CaCO_3$ is the only sp^3 carbonate for which theoretical and experimental data on the structure of the sp^3-bonded phase are in agreement. The crystal structures of more complex sp^3 carbonates, such as in the (Mg,Fe) and (Ca,Mg) systems, have been solved directly from single crystal XRD data, but the agreement between these structures and those from theoretically predicted models is yet to be seen. Since the experimentally observed structures are chemically complex and contain a large number of atoms per cell, it is possible that earlier theoretical simulations on chemically simple carbonate systems did not sample unit cells with a sufficient number of atoms. On the other hand, it is possible that the experimentally observed phases are not representative of the ground state sp^3 structures because of severe chemical and thermal gradients that are typical of laser-heated DACs, especially when iron is present in the system. Subsequent studies employing multiple characterization techniques will be of significant value to the topic of sp^2-sp^3 transitions in carbonates.

1.4. HYDROCARBONS

High-pressure behavior of the carbon-hydrogen systems is relevant to planetary sciences as hydrocarbons are an important mantle component of icy giant planets (Hubbard, 1981). Application of high external pressure to hydrocarbons, which are typically soft molecular crystals, forces the molecules to shorten intermolecular distances, making intermolecular and intramolecular forces comparable (Hemley & Dera, 2000). At short enough intermolecular distances, molecules of unsaturated hydrocarbons (i.e. with π-bonds in the chemical structure) are forced to react with each other (polymerize) with the formation of conjugated molecular structures. Polymerization reactions in unsaturated hydrocarbons involve breaking of unsaturated π-bonds with the formation of σ-bonds between the adjacent molecules. As such, pressure-induced polymerization of unsaturated hydrocarbons can be viewed as an sp^2-sp^3 phase transition (Drickamer, 1967). This process is typically irreversible and hydrocarbon polymers are available for ex situ characterization after decompression.

Pressure-induced polymerization of unsaturated hydrocarbons has been extensively characterized by Raman and IR spectroscopy, which detected signatures of sp^3-bonded chemical structures at high pressure. Many examples of pressure-induced polymerization in unsaturated hydrocarbons have been reported in the literature (see Schettino & Bini, 2007, for a more thorough review). Below we provide only a handful of these. High-density polyethylene with a large fraction of sp^3 carbon can be obtained by compressing ethylene (C_2H_4) in a DAC at room temperature to ~3.6 GPa (Chelazzi et al., 2004). Solid acetylene polymerizes at 3.5 GPa with the *trans*-opening of the triple bond (Aoki et al., 1988; Santoro et al., 2003). Benzene, a prototype of aromatic compounds, polymerizes at $P > 23$ GPa (Prizan et al., 1990) with the formation of a one-dimensional sp^3 carbon nanomaterial (Fitzgibbons et al., 2015). Although the mechanism of benzene polymerization remains controversial (Wen et al., 2011), it is generally agreed that the process is governed by the pressure-activated π-π electron density overlap of the neighboring molecules, as initially suggested by Drickamer (1967). Ciabini et al. (2007) showed that benzene polymerization is triggered at a critical C-C distance of ~2.6 Å between adjacent C_6H_6 molecules. These experimentally observed polymerization reactions have also been reproduced by ab initio computations and confirm the importance of intermolecular separation and molecular orientation (Ciabini et al., 2007; Bernasconi et al., 1997).

The conjugated nature of the recovered products may give rise to peculiar physical properties. For example, a very high electrical conductivity (~100 Ω^{-1} cm^{-1}) was observed for polyacetylene (Clark & Hester, 1991). Hydrocarbons are expected to form polymeric states not only at relatively low pressures and room temperature but also at P-T conditions representative of planetary interiors (Lee & Scandolo, 2011; Chau et al., 2011; Sherman et al., 2012; Lobanov et al., 2013) where polymerized hydrocarbons may promote the high electrical conductivity expected of the icy mantles of Uranus and Neptune (Stanley & Bloxham, 2004). As such, knowledge of the conditions of hydrocarbon polymerization at extreme pressure and temperature is critical for our understanding of the icy giant planet interiors.

Many theoretical and experimental studies have explored the high P-T stability of hydrocarbons. Thermodynamic and molecular dynamics modeling of hydrocarbon systems showed that conjugated hydrocarbons (molecular and polymeric) are more stable at high pressure than at 1 atm (Sherman et al., 2012; Kenney et al., 2002; Zhang & Duan, 2009; Spanu et al., 2011; Gao et al., 2010). This is consistent with experimental reports on hydrocarbon synthesis from methane in laser-heated DACs (Lobanov et al., 2013; Hirai et al., 2009; Kolesnikov et al., 2009). Interestingly, unsaturated hydrocarbons can be formed from methane at $P > 24$ GPa and $T > 1500$ K (Lobanov et al., 2013), as observed by Raman spectroscopy at 300 K after laser heating. This, however, is inconsistent with the apparent instability of unsaturated carbon-carbon double and triple bonds at high pressure (as discussed above). It is likely that Raman spectroscopy at 300 K provides a distorted snapshot of the hydrocarbon fluid at high temperature and that unsaturated carbon-carbon bonds form upon quenching of a highly dissociated fluid. Recent laser-shock experiments on hydrocarbons provided an in situ characterization of hydrocarbon fluids at high *P-T* and confirmed that such fluids are nonmolecular at $P < 100$–150 GPa and $T < \sim 3000$ K (Kraus et al., 2017). Further, this nonmolecular hydrocarbon fluid undergoes dissociation into diamond and hydrogen at $P > 150$ GPa and $T > 3000$ K. As such, methane dissociation to diamond and hydrogen at high *P-T* can be viewed as an sp^3-sp^3 transition.

ACKNOWLEDGMENTS

This work was supported by the Deep Carbon Observatory. SSL acknowledges the support of the Helmholtz Young Investigators Group CLEAR (VH-NG-1325) and Deutsche Forschungsgemeinschaft within the research group FOR2125 CarboPaT.

REFERENCES

Aoki, K., Usuba, S., Yoshida, M., Kakudate, Y., Tanaka, K., & Fujiwara, S. (1988). *J. Chem. Phys.*, *89*, 529.

Benedict, L. X., Driver, K. P., Hamel, S., Militzer, B., Qi, T. T., Correa, A. A., Saul, A., & Schwegler, E. (2014). *Phys. Rev. B*, *89*, 224109.

Bernasconi, M., Chiarotti, G. L., Focher, P., Parrinello, M., & Tosatti, E. (1997). *Phys. Rev. Lett.*, *78*, 2008.

Boulard, E., et al. (2012). *Journal of Geophysical Research*, *117*, B02208.

Boulard, E., Gloter, A., Corgne, A., Antonangeli, D., Auzende, A. L., Perrillat, J. P., Guyot, F., & Fiquet, G. (2011). *Proc. Natl. Acad. Sci. U.S.A.*, *108*, 5184.

Boulard, E., Pan, D., Galli, G., Liu, Z. X., & Mao, W. L. (2015). *Nature Comm.*, *6*, 6311.

Brygoo, S., Henry, E., Loubeyre, P., Eggert, J., Koenig, M., Loupias, B., Benuzzi-Mounaix, A., & Rabec Le Bundy, F. P., Bassett, W. A., Weathers, M. S., Hemley, R. J., Mao, H. U., & Goncharov, A. F. (1996). *Carbon*, *34*, 141.

Bundy, F. P., & Kasper, J. S. (1967). *J. Chem. Phys.* *46*, 3437.

Bykova, E., et al. (2016). *Nat. Commun.*, *7*, 10661.

Cerantola, V., et al. (2017). *Nat. Commun.*, *8*, 15960.

Chau, R., Hamel, S., & Nellis, W. J. (2011). *Nat. Commun.*, *2*, 1198.

Chelazzi, D., Ceppatelli, M., Santoro, M., Bini, R., & Schettino, V. (2004). *Nat. Mater.*, *3*, 470.

Ciabini, L., Santoro, M., Gorelli, F. A., Bini, R., Schettino, V., & Raugei, S. (2007). *Nat. Mater.*, *6*, 39.

Clark, R. J. H., & Hester, R. E. *Spectroscopy of advanced materials*. Chichester: Wiley, Advances in Spectroscopy, 19.

Correa, A. A., Bonev, S. A., & Galli, G. (2006). *Proc. Natl. Acad. Sci. U.S.A.*, *103*, 1204.

De Corte, K., Korsakov, A., Taylor, W. R., Cartigny, P., Ader, M., & De Paepe, P. (2000). *Isl. Arc*, *9*, 428.

Drickamer, H. G. (1967). *Science*, *156*, 1183.

Eggert, J. H., Hicks, D. G., Celliers, P. M., Bradley, D. K., McWilliams, R. S., Jeanloz, R., Miller, J. E., Boehly, T. R., & Collins, G. W. (2009). *Nat. Phys.*, *6*, 40.

Erskine, D. J., & Nellis, W. J. (1991). *Nature*, *349*, 317.

Ferrari, A. C. (2002). *Diamond Relat. Mater.*, *11*, 1053.

Field, J. E. (1992). *The properties of natural and synthetic diamond.* San Diego: Academic Press.

Fitzgibbons, T. C., Guthrie, M., Xu, E. S., Crespi, V. H., Davidowski, S. K., Cody, G. D., Alem, N., & Badding, J. V. (2015). *Nat. Mater.*, *14*, 43.

Frondel, C., & Marvin, U. B. (1967). *Nature*, *214*, 587.

Gao, G. Y., Oganov, A. R., Ma, Y. M., Wang, H., Li, P. F., Li, Y. W., Iitaka, T., & Zou, G. T. (2010). *J. Chem. Phys.*, *133*, 144508.

Ghiringhelli, L. M., Los, J. H., Meijer, E. J., Fasolino, A., & Frenkel, D. (2004). *Phys. Rev. B*, *69*, 100101.

Gloahec, M. (2007). *Nat. Mater.*, *6*, 274.

Goncharov, A. F. (1987). *Sov. Phys. Uspekhi*, *30*, 525.

Goncharov, A. F., Makarenko, I. N., & Stishov, S. M. (1989). *Zh. Eksper. Teor. Fiziki*, *96*, 670.

Gauthier, M. et al. (2014). *Rev. Sci. Instrum.*, *85*, 11E616.

Goncharov, A. F., Makarenko, I. N., & Stishov, S. M. (1990). *High Pressure Res.*, *4*, 345.

Gupta, Y. M., Turneaure, S. J., Perkins, K., Zimmerman, K., Arganbright, N., Shen, G., and Chow, P. (2012). *Rev. Sci. Instrum.*, *83*, 123905.

Hanneman, R. E., Strong, H. M., & Bundy, F. P. (1967). *Science*, *155*, 995.

Hazen, R. M., Downs, R. T., Jones, A. P., & Kah, L. (2013). *Rev. Mineral. Geochem.*, *75*, 7.

Hemley, R. J., & Dera, P. (2000). *Rev. Mineral. Geochem.*, *41*, 335.

Hirai, H., Konagai, K., Kawamura, T., Yamamoto, Y., & Yagi, T. (2009). *Phys. Earth Planet. Inter.*, *174*, 242.

Hu, M. et al. (2017). *Sci. Adv.*, *3*, e1603213.

Hubbard, W. B. (1981). *Science*, *214*, 145.

Hwang, S. L., Shen, P., Chu, H. T., Yui, T. F., & Lin, C. C. (2001). *Earth Planet. Sci. Lett.*, *188*, 9.

Kenney, J. F., Kutcherov, V. A., Bendeliani, N. A., & Alekseev, V. A. (2002). *Proc. Natl. Acad. Sci. U.S.A.*, *99*, 10976.

Khaliullin, R. Z., Eshet, H., Kühne, T. D., Behler, J., & Parrinello, M. (2011). *Nat. Mater.*, *10*, 693.

Knudson, M. D., Desjarlais, M. P., & Dolan, D. H. (2008). *Science*, *322*, 1822.

Kolesnikov, A., Kutcherov, V. G., & Goncharov, A. F. (2009). *Nature Geosci.*, *2*, 566.

Kopylova, M., Navon, O., Dubrovinsky, L., & Khachatryan, (2010). *G. Earth Planet. Sci. Lett.*, *291*, 126.

Korsakov, A. V., Perraki, M., Zedgenizov, D. A., Bindi, L., Vandenabeele, P., Suzuki, A., & Kagi, H. (2010). *J. Petrol.*, *51*, 763.

Kraus, D., et al. (2016). *Nat. Commun.*, *7*, 10970.

Kraus, D., et al. (2017). *Nat. Astron.*, *1*, 606.

Lee, M. S., & Scandolo, S. (2011). *Nat. Commun.*, *2*, 1184.

Li, Q., Ma, Y., Oganov, A. R., Wang, H., Wang, H., Xu, Y., Cui, T., Mao, H.-K., & Zou, G. T. (2009). *Phys. Rev. Lett.*, *102*, 175506.

Lin, Y., Zhang, L., Mao, H.-K., Chow, P., Xiao, Y., Baldini, M., Shu, J., & Mao, W. L. (2011). *Phys. Rev. Lett.*, *107*, 175504.

Lobanov, S. S., et al. (2017). *Phys. Rev. B*, *96*, 104101.

Lobanov, S. S., Chen, P. N., Chen, X. J., Zha, C. S., Litasov, K. D., Mao, H. K., & Goncharov A. F., (2013). *Nature Comm.*, *4*, 2446.

Mao W. L., et al. (2003). *Science*, *302*, 425.

Martinez-Canales, M., Pickard, C. J., & Needs, R. J. (2012). *Phys. Rev. Lett.*, *108*, 045704.

Massonne, H. J. (2003). *Earth Planet. Sci. Lett.*, *216*, 347.

Merlini, M., Cerantola, V., Gatta, G. D., Gemmi, M., Hanfland, M., Kupenko, I., Lotti, P., Muller, H., & Zhang, L. (2017). *Am. Mineral.*, *102*, 1763.

Merlini, M., Hanfland, M., Salamat, A., Petitgirard, S., & Muller, H. (2015). *Am. Mineral.*, *100*, 2001.

Mikhailenko, D. S., Korsakov, A. V., Zelenovskiy, P. S., & Golovin, A. V. (2016). *Am. Mineral.*, *101*, 2155.

Miller, E. D., Nesting, D. C., & Badding, J. V. (1997). *Chem. Mater.*, *9*, 18.

Mundy, C. J., Curioni, A., Goldman, N., Will Kuo, I. F., Reed, E. J., Fried, L. E., & Ianuzzi, M. (2008). *J. Chem. Phys.*, *128*, 184701.

Németh, P., Garvie, L. A. J., Aoki, T., Dubrovinskaia, N., Dubrovinsky, L., & Buseck, P. R. (2014). *Nat. Commun.*, *5*, 5447.

Nestola, F., et al. (2018). *Nature*, *555*, 237.

Oganov, A. R., Glass, C. W., & Ono, S. (2006). *Earth Planet. Sci. Lett.*, *241*, 95.

Oganov, A. R., Hemley, R. J., Hazen, R. M., & Jones, A. P. (2013). *Rev. Mineral. Geochem.*, *75*, 47.

Oganov, A. R., Ono, S., Ma, Y. M., Glass, C. W., & Garcia, A. (2008). *Earth Planet. Sci. Lett.*, *273*, 38.

Ono, S., Kikegawa, T., & Ohishi, Y. (2007). *Am. Mineral.*, *92*, 1246.

Panero, W. R., & Kabbes, J. E. (2008). *Geophys. Res. Lett.*, *35*, L14307.

Pearson, D. G., et al. (2014). *Nature*, *507*, 221.

Pickard, C. J., & Needs, R. J. (2015). *Phys. Rev. B*, *91*, 104101.

Pineau, N. (2013). *J. Chem. Phys. C*, *117*, 12778.

Powles, R. C., Marks, N. A., Lau, D. W. M., McCulloch, D. G., & McKenzie, D. R. (2013). *Carbon*, *63*, 416.

Prewitt, C. T., & Downs, R. T. (1998). *Rev. Mineral. Geochem.*, 37, *283* (1998).

Pruzan, P., Chervin, J. C., Thiery, M. M., Itie, J. P., Besson, J. M., Forgerit, J. P., & Revault, M. (1990). *J. Chem. Phys.*, *92*, 6910.

Rygg, J. R., et al. (2012). *Rev. Sci. Instrum.*, *83*, 113904.

Santoro, M., Ciabini, L., Bini, R., & Schettino, V. (2003). *J. Raman Spectrosc.*, *34*, 557.

Schettino, V., & Bini, R. (2007). *Chem. Soc. Rev.*, *36*, 869.

Shchepetova, O. V., Korsakov, A., Mikhailenko, D., Zelenovskiy, P., Shur, V., & Ohfuji, H. (2017). *J. Raman Spectrosc.*, *48*, 1606.

Sherman, B. L., Wilson, H. F., Weeraratne, D., & Militzer, B. (2012). *Phys. Rev. B*, *86*, 224113.

Shi, X., He, C. R., Pickard, C. J., Tang, C. C., & Zhong, J. (2018). *Phys. Rev. B*, *97*, 014104.

Shiell, T. B., McCulloch, D. G., Bradby, J. E., Haberl, B., Boehler, R., & McKenzie, D. R. (2016). *Sci. Rep.*, *6*, 37232.

Skorodumova, N. V., Belonoshko, A. B., Huang, L., Ahuja, R., & Johansson, B. (2005). *Am. Mineral.*, *90*, 1008.

Smith, E. M., Shirey, S. B., Nestola, F., Bullock, E. S., Wang, J. H., Richardson, S. H., & Wang, W. Y. (2016). *Science*, *354*, 1403.

Smith, R. F., et al. (2014). *Nature*, *511*, 330.

Solomatova, N. V., & Asimow, P. D. (2017). *Am. Mineral.*, *102*, 210.

Spanu, L., Donadio, D., Hohl, D., Schwegler, E., & Galli, G. (2011). *Proc. Natl. Acad. Sci. U.S.A.*, *108*, 6843.

Stanley, S., & Bloxham, J., (2004). *Nature*, *428*, 151.

Tschauner, O., et al. (2018). *Science*, *359*, 1136.

Turneaure, S. J., Sharma, S. M., Volz, T. J., Winey, J. M., & Gupta, Y. M. (2017). *Sci. Adv.*, *3*, eaao3561.

Utsumi, W., & Yagi, T. (1991). *Science*, *252*, 1542.

Vennari, C. E., & Williams, Q. (2018). *Am. Mineral.*, *103*, 171.

Wang, X., Scandolo, S., & Car, R. (2005). *Phys. Rev. Lett.*, *95*, 185701.

Wang, Y., Panzik, J. E., Kiefer, B., & Lee, K. K. M. (2012). *Sci. Rep.*, *2*, 520.

Wang, Z., et al. (2004). *Proc. Natl. Acad. Sci. U.S.A.*, *101*, 13699.

Wen, X. D., Hoffmann, R., & Ashcroft, N. W. (2011). *J. Am. Chem. Soc.*, *133*, 9023.

Wirth, R., Dobrzhinetskaya, L., Harte, B., Schreiber, A., & Green, H. W. (2014). *Earth Planet. Sci. Lett.*, *404*, 365.

Xie, Y.-P., Zhang, X.-J., & Liu, Z.-P. (2017). *J. Am. Chem. Soc.*, *139*, 2545.

Xu, J., Mao, H.-K., & Hemley, R. J. (2002). *J. Phys.: Cond. Mater.*, *14*, 11549.

Yao, X., Xie, C., Dong, X., Oganov, A. R., & Zeng, Q. F. (2018). *Phys. Rev. B*, *98*, 014108.

Yin, M. T., & Cohen, M. L. (1983). *Phys. Rev. Lett.*, *50*, 2006.

Zeng, Z., et al. (2017). *Nat. Commun.*, *8*, 322.

Zhang, C., & Duan, Z. H. (2009). *Geochim. Cosmochim. Acta*, *73*, 2089.

2

High-Pressure Carbonaceous Phases as Minerals

Oliver Tschauner

ABSTRACT

High-pressure carbonaceous minerals establish a link between one of the most abundant chemical elements and processes that occur at high pressures, either in the interior of the Earth or during shock-metamorphosis in space or on Earth. Diamonds from the Earth's mantle carry inclusions of carbides, carbonates, methane, and carbon-dioxide. These inclusions elucidate diamond formation and the carbon recycling in the mantle. This chapter focuses on these inclusion minerals themselves, the means of detection, and the reconstruction of pressure-temperature conditions of their formation in the mantle.

2.1. INTRODUCTION

Understanding the evolution of the interior of Earth requires correlation of geochemical and geophysical data. However, geochemistry and geophysics probe different time spans and, probably, also different spatial scales of reservoirs (Helffrich & Wood, 2001; van Keken et al., 2002): The differentiation of chemical elements in Earth often occurs on timescales that are beyond the survival of distinguishable seismic features of the geophysical heterogeneities that have caused them, such as a particular subducted slab or mantle plume. In addition, noticeable geochemical reservoirs in the mantle may not even correspond to a particular zone or region in Earth which can be detected through seismic waves but are finely dispersed in the mantle or simply reflect a temporal evolution of isotopic signatures. Correlation becomes the more difficult the deeper parts of the mantle shift into scope: In absence of large bodies of rocks obducted from great depth, individual grains of minerals that were entrapped in diamonds are the only direct witnesses of processes in the deeper Earth, and the mineralogical and chemical information obtained from them has to provide the constraints on geochemical deep-mantle signatures and their correlation with seismic features and with geodynamic models. Minerals that originate from sublithospheric mantle may have transformed back to phases more stable at shallower depth but may still carry the chemical signatures of their origin because diamond is a chemically inert host material of exceptionally high confining pressures and temperatures (Brenker et al., 2002; Harte et al., 1999; Shirey et al., 2013; D. Smith et al., 2018; Sobolev, 1997; Stachel et al., 2000; Walter et al., 2011). In rare cases, diamonds conserve high-pressure minerals in their original structure by supporting high residual pressures of these inclusions that prevent transformation to lower-pressure polymorphs (Howell et al., 2012; Navon, 1991; Nestola et al., 2018; Pearson et al., 2014; Tschauner et al., 2018). The residual pressure of these inclusions corresponds to the end point of a thermodynamic path that connects current residual pressure at 300 K with the pressure and temperature of entrapment of this inclusion. Because of the high plastic yield limit, low compressibility, and low thermal expansivity of diamond, these paths are nearly isochoric (Navon, 1991, 2017; Tschauner et al., 2018), although thermoelastic corrections for diamond have to be made or the approach of a corrected isochore is substituted by the use of isomekes (Angel et al., 2014; Nasdala et al., 2003). This overall simple, monotonic path makes the reconstruction of the conditions of entrapment of these inclusions comparatively much easier than in case of inclusions in other minerals. For the same reason, inclusions in diamond differ from inclusions in garnets or other minerals, which have much lower plastic deformation limits than diamond by supporting inclusions with

Department of Geoscience, University of Nevada–Las Vegas, Las Vegas, Nevada, USA

Carbon in Earth's Interior, Geophysical Monograph 249, First Edition. Edited by Craig E. Manning, Jung-Fu Lin, and Wendy L. Mao.

residual pressures in the range of several GPa. Exhumation speed may be estimated from isotope-data of inclusions whose pressure has been determined in prior geochemical analysis or from the nitrogen-aggregation state of the host diamond (Taylor et al., 1990). Both methods come with challenges: Inclusions with high residual pressure appear to be generally small (micrometer-scale) because larger inclusions cause accordingly higher local stresses in the surrounding diamond and result in cracking and annealing at shallow depth. However, isotope analysis of very small inclusions is very difficult. The analysis of nitrogen aggregation states is in principle straightforward. However, many diamonds are not homogeneous containing zones of different nitrogen concentration and aggregation. Overall, the assessment of exhumation speed is less straightforward than the assessment of pressure and temperature of the source region.

As it will be shown, high-pressure carbonaceous phases play an important role in assessing the global carbon cycle; they are conserved within and connected to the formation of diamonds, and they place diamond formation into the context of the global carbon cycle.

2.1.1. High-Pressure Minerals in Meteorites and Impactites

High-pressure minerals also occur (and are actually found more frequently than in diamonds) in a very different environment: Meteorites and terrestrial crustal rocks that have been subjected to severe shock-metamorphism during asteroid impacts (Sharp & DeCarli, 2006; Stöffler et al., 1991) contain veins and pockets of minerals that otherwise only occur in the deep mantle of rocky planets like Earth. A comprehensive list of meteorite minerals, including shock-metamorphic minerals, can be found in Rubin and Ma (2017). A list of the more recently discovered high-pressure minerals is given by Ma (2018). Carbon and carbonaceous species are abundant in the solar system, and the occurrence of diamond in carbonaceous chondrites is established (Anders & Zinner, 1993). Moissanite (SiC-2H) (Anders & Zinner, 1993) and other carbides have been found as well in iron-meteorites, ureilites, enstatite-chondrites, achondrites, and in CAIs (calcium-aluminum rich inclusions that formed under highly reducing conditions early in the solar system). Most of these minerals have not formed at high pressure, and the reader is referred to the recent comprehensive account of meteorite minerals by Rubin and Ma (2017). Besides diamond, lonsdaleite (the 2H polytype of diamond) has been reported but was recently discredited (Nemeth et al., 2014). Furthermore, chaoite and other structurally uncharacterized dense forms of carbon were reported from terrestrial impact craters (El Goresy & Donnay, 1968; El Goresy et al., 2003). At the high pressures and temperatures in the isobaric core of large impacts carbonates dissociate (Schulte et al., 2009), whereas the decaying shockwave in the more remote regions of the bedrock is not known to cause irreversible transformations of carbonates to high-pressure forms. Thus, with diamond as the only well-characterized impact-generated high-pressure phase of a carbonaceous species (Anders & Zinner, 1993; Hough et al., 1995), we do not further discuss shock-related dense carbonaceous phases here. Novel micrometer and submicrometer structural and spectroscopic probes may reveal more impact-related carbonaceous phases in the future. Occurrence and formation of diamonds from impacts, meteorites, and interplanetary and interstellar dust are interesting topics by themselves but more fit for a monograph about diamond rather than carbon in general.

Within the context of this monograph about carbon in natural systems we dedicate this chapter to two particular types of terrestrial occurrences of carbonaceous minerals: (a) naturally occurring dense polymorphs of C-rich phases that are stable or that form at high pressure and (b) occurrences of carbonaceous minerals that retain high residual pressures as inclusions in diamond.

We use the term *carbonaceous* in reference to phases with carbon as constituting element. We discuss a few cases where carbon as minor component in a solid high-pressure phase induces a characteristic change in structure. This chapter is about high-pressure carbonaceous minerals. Minerals are crystalline phases by definition (Nickel & Grice, 1998); hence, state and properties of dense liquid or fluid carbonaceous phases are beyond our scope.

We follow the conventional classification scheme of minerals by chemical classes starting with elemental minerals, alloys, and carbides, and progressing to sulfides, oxides, and so on (Dana, 1869). While this scheme has become problematic because it largely ignores structural relations between minerals of different composition and fundamental differences in chemical bonds of phases of same composition, it causes no significant issues in discussing a topic that is already restricted by its focus on carbon as constituent atomic species and may serve here as a known and accepted framework. Consequently, the number of subchapters is limited to elements, carbides, oxides, and carbonates. Unless stated otherwise, we henceforth reserve the term *diamond* here to natural terrestrial diamonds that are not impact related.

2.2. CARBONACEOUS PHASES AS INCLUSION IN DIAMOND: ELEMENTS AND CARBIDES

2.2.1. Diamond in Diamond: Indication for Methane Breakdown in the Lithospheric Mantle

The first carbonaceous phase that occurs as inclusion in diamonds is, seemingly paradoxically, diamond. Terrestrial diamonds are composed of single crystal domains on the scale of tens of micrometers to millimimeters or more. Mosaicity of diamonds varies extensively (see, for instance, Shirey et al., 2013), and many diamonds exhibit zones or kernels of different age, orientation, and domain sizes (Boyd et al., 1994; Shirey et al., 2013). In addition, domains of very fine-grained diamond are found within single crystal diamonds. Polycrystalline diamond is found in the surrounding of inclusions of other minerals, at the surface of single crystal diamonds, and other locations where mechanical damage disrupted the original large-scale crystalline order. A distinct occurrence of polycrystalline diamond is very fine-grained, isolated, polycrystalline diamond inclusions within diamond single crystals. These inclusions are fully encapsulated in well-crystallized single crystal domains of low strain. The grain size ranges from 10 nm to ¼ μ m based on the Scherrer equation that correlates noninstrumental powder diffraction peak width with average grain size. The spatial extension of these inclusions ranges from a few to a few tens of micrometers. To our knowledge, the pressure of these fine polycrystalline diamond inclusions is always ambient to nearly ambient and not different from that of the surrounding single crystal diamond host lattice beyond uncertainty. An example of polycrystalline diamond in single crystal diamond is given in Figure 1.1. It is indicative that these pockets of fine-grained polycrystalline diamond are a result of breakdown of C-bearing fluid after entrapment. In particular, breakdown of methane and subsequent diffusion of H out of the inclusions into point defects of the surrounding diamond host lattice seems a plausible explanation of this occurrence (see section 2.2.2.1). This process also accounts for efficient pressure relaxation between inclusion and host crystal. Other than for heavier atomic or molecular species for H_2, the diamond host lattice does not necessarily define a closed system. If we accept methane breakdown as origin of these inclusions, the pressure-temperature regime of their formation is constrained by the graphite-diamond phase boundary and the methane breakdown reaction. In subduction-zone environments or the mantle, decomposition of methane as component of C-H-O fluids depends on the activity of the fluid components (Stachel & Luth, 2015). However, the breakdown of encapsulated methane is constrained by pressure and temperature only, albeit that proton diffusion into the surrounding diamond lattice may affect mass balance. The relation between methane breakdown and plausible P-T regimes for formation and ascent of the host diamond is shown in Figure 2.1b: Plausible P-T points are defined through the intersection of the breakdown reaction (Lobanov et al., 2013) and the pressure-temperature evolution in a cold slab at 14 GPa, 1255 K as upper pressure bound, with the conditions of a hot slab at 5.7 GPa, 1350 K as lower bound or with the graphite-diamond boundary at 4.6 GPa, 1360 K as ultimate lower bound. Both lower bound estimates require interpolation of the decomposition line to conditions where experimentally mixtures of diamond, hydrogen, and higher alkanes had been observed (dashed blue line in Figure 2.1b; Lobanov et al., 2013). These constraints have to be placed into context with information from other inclusions in the same diamond to turn into a useful assessment of P-T paths. Currently, there is no good account how these occurrences of fine-grained diamond inclusions relate to type and origin of their host diamonds and about their abundance, because it requires synchrotron microdiffraction to detect these polycrystalline diamond inclusions within diamonds.

Graphite is a common inclusion in natural diamonds. In many cases chemically distinct inclusions in diamond are surrounded by graphitic damage zones, thus resulting in an optically dark appearance of these inclusions independent of their intrinsic optical absorption of visible light (Harris & Vance, 1972). Graphite and its occurrence in diamond as retrograde transformation product (Harris & Vance, 1972) are not within the scope of this chapter. We note that carbon phases of structure and bonding intermediate between graphite and diamond (Mao et al., 2003; Oganov et al., 2013) are suspected to occur also in such damage zones in diamonds but have not yet been firmly proved to exist in nature.

2.2.2. Carbides: Possible Indicators of Reducing Conditions in the Transition Zone and Lower Mantle

2.2.2.1. Methane

Methane is a component of C-H-O rich fluids in the mantle that are related to diamond formation (Stachel & Luth, 2015). Possible mechanisms are discussed in other chapters of this book. Here we focus on indications for methane or methane-bearing phases as crystalline inclusions in diamond. Methane crystallizes at 89.7 K in an fcc lattice (Hemley & Dera, 2000; Press, 1972), which at 8 GPa, 300 K, undergoes major structural rearrangement into a larger cubic structure with carbon atoms equivalent to those of α-Mn (Maynard-Casely et al., 2014). This high-pressure phase is called "methane B." This phase of methane appears to be stable to at least 25 GPa (Maynard-Casely et al., 2014). However, on Earth methane as solid phase is commonly found as clathrate with H_2O: Methane hydrates I and II are typical clathrate structures with van der Waals forces confining the methane guest molecule within the H-bonded ice network (Hemley & Dera, 2000; Loveday & Nelmes, 2008) and are abundant in permafrost soils and at the slope of continental shelfs (Kvenvolden, 1993). Above 2.4 GPa a different crystalline phase in the methane-water system replaces these clathrates (Loveday et al., 2001a, 2001b; Machida et al., 2007). Methane-hydrate III is a so-called "stuffed ice" (Loveday et al., 2001b): Other than in the clathrates, there is H

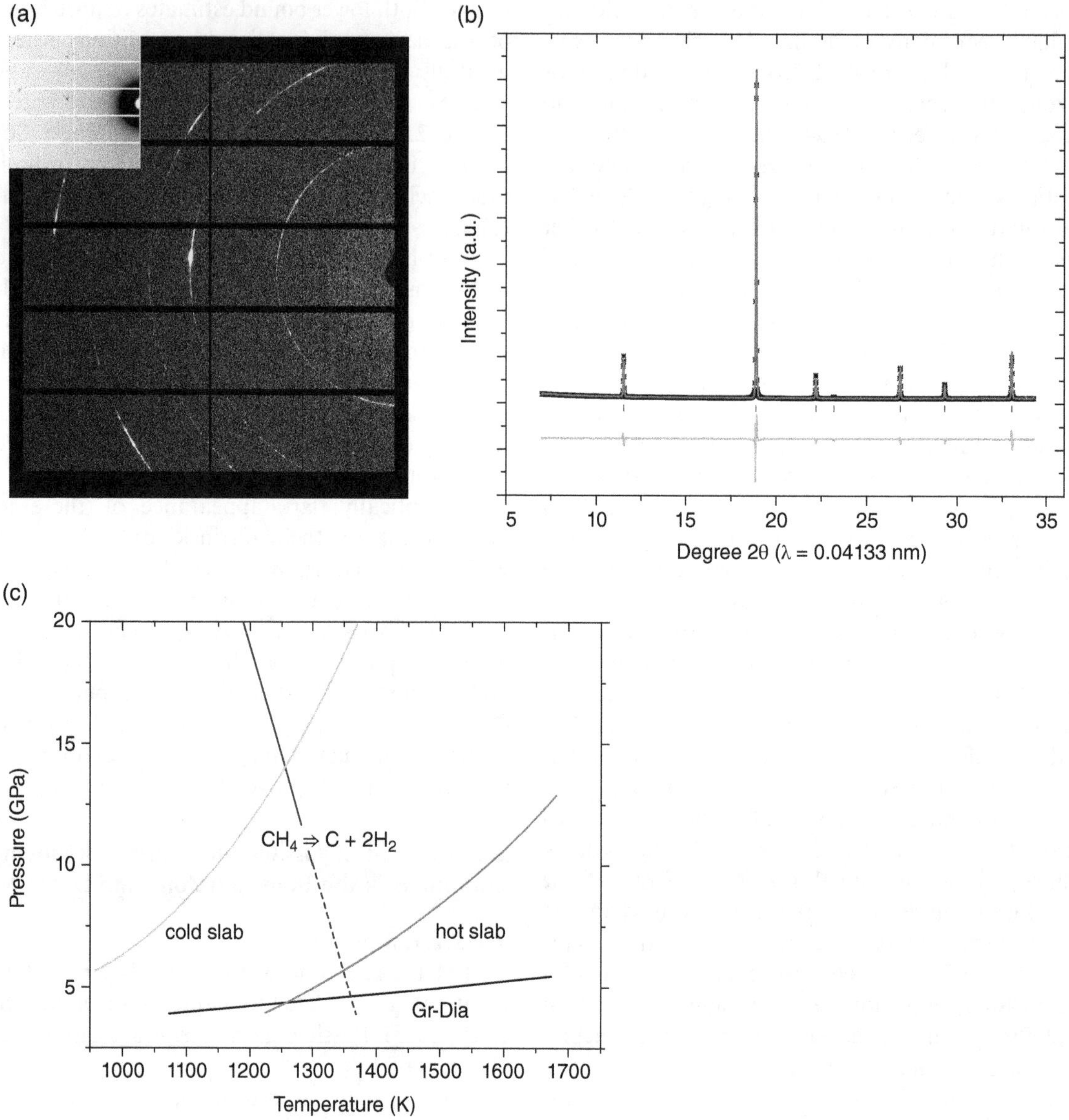

Figure 2.1 (a) Diffraction image of polycrystalline diamond within a single crystal diamond from Orapa, Botswana. The diffraction image was obtained during an overview diffraction map with $2 \times 4\,\mu m^2$ focused x-ray beam of 0.4133 Å wavelength at beamline 13-IDD, GSECARS, APS, Argonne National Laboratory. A Pilatus CdTe detector was used for data acquisition. In order to discriminate the diffraction signal from the intense elastic and inelastic diffuse scattering of the host diamond, an inclusion-free frame from within 20 μm distance from the inclusion was subtracted from this diffraction frame. Thus, the Bragg reflections from the host diamond are not visible. The strong texture of the polycrystalline diamond inclusion is clearly visible as intensity variation along the Debye fringes. Inset: Diffraction image before background frame subtraction. The single crystal diamond reflections are clearly visible. (b) Integrated diffraction pattern (black crosses) along with a whole profile refinement (red line) and residual of fit (green line). Bragg angles of diamond reflections are indicated as tickmarks. No excess pressure beyond a 0.1 GPa uncertainty is measurable. (c) Relation of the methane breakdown reaction (Lobanov et al., 2013) with the graphite-diamond phase boundary and pressure-temperature conditions in cold and hot subducted slabs (after Thomson et al., 2016). Intersection points mark possible conditions for diamond growth and ascent but additional geobarometric and thermometric constraints are necessary to determine which regime has actually occurred. See electronic version for color representation of the figures in this book.

bonding between methane and H_2O. The resulting structure is much denser than methane-hydrate I or II and can be well described as a distorted ice-Ih structure with methane molecules residing in the channels of this structure (Loveday et al., 2001b). Structures and properties of methane-hydrates are described in detail by Loveday and Nelmes (2008).

As of yet, direct evidence for methane in diamond comes from gas release into mass spectrometers upon crushing diamond specimens and from optical spectroscopy. E. Smith et al. (2015, 2016, 2018) reported Raman shifts of C-H bands at 2914 cm^{-1} and assigned them to methane. Infrared spectroscopy of diamonds shows frequent occurrence of C-H absorption bands that are related to alkanes (Navon, 1989) within the frequency range of 2915–3000 cm^{-1} (Figure 2.2). A frequently observed, additional C-H absorption band at 3107 cm^{-1} is assigned to protons that were trapped on nitrogen-related point defects in the diamond lattice and formed a chemical bond with N (Goss et al., 2014). This band is very sharp, consistent with an isolated chemical bond that allows for slowly decaying phonon states separated from the energy range of diamond phonons (defect band).

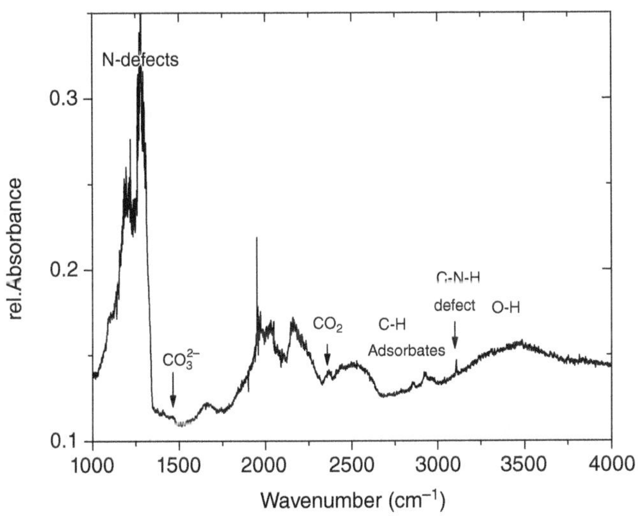

Figure 2.2 Infrared transmission spectrum of a doubly polished platelet of a fibrous diamond from Orapa, Botswana. The spectrum was taken from the fibrous rim of the specimen with 60×80 μ m^2 slits with 1.5 cm^{-1} spectral resolution at beamline 1.4, ALS, LBNL, Berkeley and has not been reported previously. The most prominent absorption feature are N-defect bands at 1000–1300 cm^{-1} and bands from two phonon processes of the regular diamond lattice between about 1800 and 2600 cm^{-1}. C-O, C-H, and O-H related bands are indicated. The energy of the "nu"$_3$ (C-O asymmetric stretching band) of CO_2 at 2365(1) cm^{-1} corresponds to 2.7(2) GPa, based on the pressured dependence of this mode in solid CO_2 (Hanson & Jones, 1981). Similarly, the carbonate asymmetric stretching mode at 1465 cm^{-1} corresponds to an elevated pressure, but the interpretation is less straightforward than for solid CO_2. For instance, for dolomite the energy of this band corresponds to a pressure around 2–3 GPa (Efthimiopoulos et al., 2017), but for calcite it is ~1 GPa (Koch-Mueller et al., 2016). See electronic version for color representation of the figures in this book.

IR-spectra of diamonds often show multiple bands around 2850–2950 cm^{-1}. These bands are related to surface contamination rather than to hydrocarbon inclusions within diamond (Navon, 1989).

Bands in the 2950 to 3050 cm^{-1} range can be assigned to methane at ambient pressure to several GPa or to other hydrocarbons at ambient to elevated pressures. Further studies are required to understand the state and the residual pressure of CH_4 in diamond. While methane is a likely component of diamond-forming fluids (see review by Shirey et al., 2013), it is important to know if C-H bands in diamond adhere to free alkanes and at which residual pressures they occur, or are bound to water as clathrates. We mentioned above that occurrences of fine-grained diamond inclusions within single crystal diamond can be interpreted as result of breakdown of entrapped methane (Figure 2.1).

2.2.2.2. Transition Metal Carbides and Carbonaceous Iron

Carbides of transition metals, notably of Fe, have been reported as rare occurrences in diamonds (Bulanova et al., 2010; Jacob et al., 2004; Kaminsky & Wirth, 2011; Mikhail et al., 2014). Mikhail et al. (2014) and E. Smith et al. (2016) found indications that these occurrences are not exotic anomalies. Inclusion of iron and iron carbides are plausibly connected with the formation process of sublithospheric diamonds and in particular with the formation of CLIPPIR, a specific type of diamonds (E. Smith et al., 2016). Tschauner et al. (2018) observed inclusions of carbonaceous iron in a number of diamonds of different origin and observed that some of these inclusions maintain high residual pressures, similar to inclusions of ice-VII, ilmenite, and calcite, reported in the same study.

The particular interest in iron and iron-carbide inclusions comes from their potential role in deep carbon recycling. In an experimental study Rohrbach and Schmidt (2011) found indication that diamond formation, carbon subduction, and the depth dependence of the mantle redox state are intrinsically connected. It is generally assumed that the mantle becomes more reducing with increasing depth. In the shallow upper mantle the dominant oxidation state of iron is 2+. Pressure in the range of tens of GPa favors disproportionation of ferrous iron into ferric and metallic iron (Frost et al., 2004; Frost & McCammon, 2008). Garnet inclusions in sublithospheric diamonds exhibit overall higher amounts of ferric iron (as skiagite) (Kiseeva et al., 2018) in consistency with the experimental studies. Rohrbach and Schmidt (2011) proposed that native iron in the transition zone and lower mantle reacts with carbonate from fluids that are released from subducted slabs. The reaction results in formation of diamond and iron carbides. Consequently, inclusions of iron and iron carbides in diamond are not reflecting unusual processes but may be remnants of this "redox freezing" process.

The solubility of carbon in iron varies largely with temperature and structure. Pressure on the scale of several 10 GPa was found to suppress solubility (Lord et al., 2009). Pressure also affects the stability field of iron carbides and favors different stoichiometries than does ambient pressure (Lord et al., 2009). The carbon-rich side of the iron-carbon phase diagram at high pressure has been studied extensively in relation with the search for the light element in the Earth's out core (Lord et al., 2009; Dasgupta & Hirschmann, 2010, Chen et al., 2014). The iron-rich side of the Fe-C phase diagram is well known at ambient pressure, but little work has been dedicated to it at elevated pressure (Narygina et al., 2011). In nature, iron carbides have been found in meteorites (see review by Rubin & Ma, 2017), in terrestrial volcanites whose conduit intersected coal deposits (Irmer, 1920), in fulgurites (Essene & Fisher, 1986), and as inclusions in diamond (Jacob et al., 2004; Kaminsky & Wirth, 2011; Mikhail et al., 2014; E. Smith et al., 2016). As the other occurrences do not involve pressures on a GPa level, we only discuss the occurrences in diamond.

Cohenite, Fe_3C, has been found in diamonds (Jacob et al., 2004; Mikhail et al., 2014; E. Smith et al., 2016). Kaminsky and Wirth (2011) also reported haxonite, $Fe_{23}C_6$, and di-iron-carbide as inclusions in diamonds from Juina, Brazil. Occurrence of cohenite places a possible constraint on P-T conditions of entrapment: Pressures above 8 GPa favor formation of Fe_7C_3 at high carbon activity (Lord et al., 2009). Thus, cohenite has either been entrapped at lower pressures or formed upon retrograde transformation from Fe_7C_3 where excess carbon was resorbed by the surrounding diamond. E. Smith et al. (2016) reported occurrences of inclusions with taenite and cohenite intergrowth in CLIPPIR diamonds. In this case, Fe is in excess, and cohenite is exsolution product of a more carbonaceous alloy or melt, whereas the surrounding diamond was not significantly involved in formation of these phases other than through methane formation. Smith et al. argue that the cohenite-iron inclusion point to formation of the host diamonds in the deep transition zone because they observe in other CLIPPIR diamonds majoritic garnets and calcium silicate phases that are possible retrograde transformation products of $CaSiO_3$-perovskite (E. Smith et al., 2016).

This study as well as earlier observations of taenite and cohenite (Jacob et al., 2004; Kaminsky & Wirth 2011; Mikhail et al., 2014) are based on analyses of inclusions exposed by cutting or FIB milling.

A different approach was chosen recently by Tschauner et al. (2018) using in-situ synchrotron micro-X-ray diffraction and X-ray fluorescence spectroscopy for detection and identification of submicroscopic inclusions. While this work was primarily aiming to examine former fluid-inclusions in diamonds such as ice-VII, the same method also allowed for identifying micrometer-scale inclusions of other minerals, including iron, iron-nickel, and Fe-carbides (Tschauner et al., 2018). As in the case of ice-VII, in-situ micro-XRD allows for assessing the residual pressure of these inclusions and reconstructing the pressure and temperature of their entrapment based on nearly isochoric release paths (Navon 1991, 2017; Tschauner et al., 2018) or isomekes (Angel et al., 2014; Nasdala et al., 2003). In most cases the iron and carbide inclusions detected by synchrotron micro-X-ray diffraction and fluorescence resided at pressures below 1 GPa, but a smaller set of inclusions was found at residual pressures of 1 to 7 GPa. These rather high residual pressures are also consistent with the redox-freezing model, although it has to be noted that presently other pathways of reduction of ferrous or ferric species as side-product of reduction of carbonate to carbon cannot be excluded. For instance, ice-VII inclusions in diamonds are remnants of diamond-forming fluids, but probably they do not represent the bulk composition of those fluids (Tschauner at al., 2018).

We note that both taenite (fcc-iron, γ-iron) and iron (bcc-iron, α-iron) are observed, although so far only in different host diamonds. For clarification we recall here that mineral names are defined by structure and composition of the dominant endmember (Nickel & Grice, 1998). For taenite, this endmember is iron in fcc structure, although taenite usually contains few to several at% Ni (Rubin & Ma, 2017). For iron, the endmember is also Fe but in bcc structure, although in many natural occurrences it contains noticeable amounts of Ni ("kamacite"). In synchrotron micro-X-ray fluorescence measurements, I found that taenite inclusions in diamond contain Ni at levels of few to 30 at%, but iron contains ~1 at% Ni or less. However, this observation is based on about a dozen occurrences and it is not clear if this holds as a general pattern for iron inclusions in diamond.

In this context we note that "hcp-Fe" (ε-Fe) has actually been found in serpentinized peridotites and CAIs (calcium-aluminum-rich inclusions in chondrites) and is an approved mineral with the name hexaferrum (Mochalov et al., 1998; Rubin & Ma, 2017). In both types of occurrences hexaferrum is stabilized at ambient conditions through high amounts of Ir and Ru. So far, it has not been reported as inclusion in diamonds, where it would define release from lower mantle conditions.

The carbon-content of μ m-scale inclusions of taenite and iron is hard to assess through microchemical analysis.

However, the same structural distortion as in synthetic carbonaceous iron and iron-nickel is observed for many of these iron-metal inclusions, and Rietveld refinement of relative electron density at nonequivalent lattice sites is consistent with the presence of a light element. The saturation limit of C in fcc-iron and Fe-Ni has been examined by Narygina et al. (2011) and was found to be <0.9 at% at 9 GPa beyond which cohenite exsolution occurs upon cooling. The C content of iron has been found to be <4at% at 9 GPa, above which cohenite exsolution occurs, whereas at 20 GPa exsolution occurs at lower concentrations (Narygina et al., 2011). SEM-EDS analyses of exposed iron inclusions do not indicate the presence of Si. An example of a diffraction pattern of iron as diamond inclusion is given in Figure 2.3. We note that the iron inclusions in diamonds that we identified were always fine grained and textured. The d-spacings of a few single crystal reflections in the same pattern can be matched to cohenite and similar patterns with smooth powder-like iron patterns, and a few spots of d-spacings that match cohenite have been observed several times. Although the occurrence of diffraction features in the same diffraction image does not strictly imply coexistence, we note that the intergrowth of taenite with cohenite reported by E. Smith et al. (2016) generates an equivalent kind of diffraction patterns. Isolated inclusions of cohenite were generally found to be more coarse grained (few crystallites on micrometer scale within one inclusion) and consistently at pressures below 1 GPa. This low to ambient residual pressure may be owed to inelastic deformation and cracking of the host diamond around the carbide inclusions. Cohenite has a much higher bulk modulus than fcc- or bcc-iron (Chen et al., 2014), and a nearly isochoric release path from the conditions of the transition zone would imply much higher residual pressures than for metallic iron. We recall that some occurrences of cohenite in diamond may be products of the backreaction of Fe_7C_3, in which case the condition of a nearly conserved inclusion volume did not avail and the present state is the result of pressure release upon backreaction.

Finally, we note that other carbides have been found in concentrates from ultra-high-pressure terrains such as the Luobusha complex in Tibet. In particular, qushongite (WC), yarlongite, and tongbaite (Cr_3C_2) have been reported (Fang et al., 2009; Shi et al., 2009). The actual mechanism of their formation is not clear but their occurrence in rocks that have been obducted from deep lithospheric or even sublithospheric mantle (Dobrzhinetzkaya et al., 2009) justifies their mentioning here. It is also noteworthy that tantalcarbide (Ta,Nb)C was found in placers in the Ural mountains already at the beginning of the 20th century (originally interpreted as native Ta; von John, 1910), where possible synthetic origin can be precluded. Although tantalcarbide and niobocarbide (Jedwab, 2009) are derived from orogenic mantle peridotites, they are not necessarily mantle minerals. Formation under reducing conditions at low pressure during serpentinization cannot

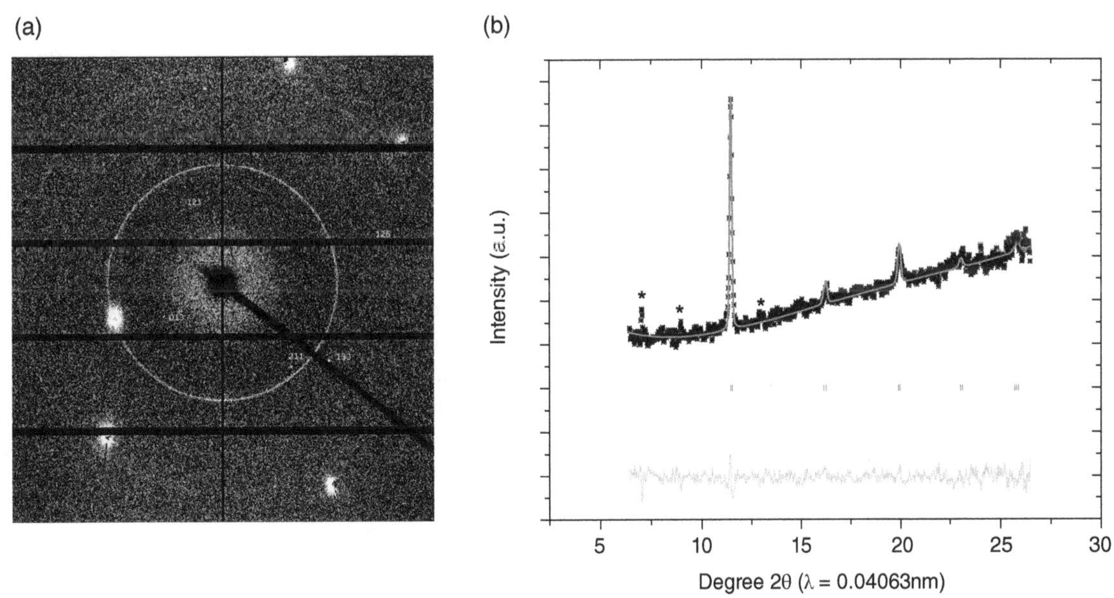

Figure 2.3 (a) Diffraction image of suessite as inclusion in a diamond from Orapa, Botswana. The image was obtained and treated in the same way as the one shown in Figure 2.1. Few isolated indexed single crystal reflections match cohenite. (b) Integrated diffraction pattern (black) with Rietveld refined modeled pattern of tetragonally distorted iron (red) and residual of fit (green). Bragg angles of allowed reflections are indicated by tickmarks. See electronic version for color representation of the figures in this book.

be excluded in absence of additional information about their content of trace elements and carbon isotopes.

2.3. OXIDES AND CARBONATES

As in the case of carbonaceous iron and iron carbides, the occurrence of CO_2 and carbonates as inclusions in diamonds bears witness of the formation process of diamond and the process of carbon recycling in the mantle. Extensive studies on carbonaceous fluid inclusions in diamonds (Izreali et al., 2001; Jablon & Navon, 2016) and experimental work on carbonate and carbonate-rich silicate melts have been conducted (Litasov et al., 2011; Podnorodnikov et al., 2001) to the purpose of elucidating these processes. Both molecular carbon dioxide and carbonate were detected in diamond by means of their characteristic absorption bands in transmission infrared spectroscopy (Navon, 1991). The energy of the band related to asymmetric stretching of the carbonate C-O bond ranges from 1430 to 1445 cm^{-1}. Frequently, a doublet or multiplet is observed (Navon, 1989; see also Figure 2.3), which indicates either carbonate inclusions at different residual pressures or of different composition and structure. In addition, a carbonate C-O bending mode is observed at 876 ± 1.5 cm^{-1} shifted to 878 to 879 cm^{-1} in diamonds where the carbonate C-O stretching band is shifted to 1445 cm^{-1} (Navon, 1989), which is consistent with elevated residual pressure in the GPa-range. We discuss the residual pressure of carbonate inclusions in diamonds further below.

2.3.1. Carbon Dioxide

Carbon dioxide exhibits a large variety of phases at different pressures and temperatures (Iota et al., 1999, 2007; Tschauner et al., 2001; Datchi et al., 2009, 2012, 2014; Litasov et al., 2011; Santoro et al., 2012). CO_2 crystallizes as dry ice (CO_2-I) at 194.5 K and ambient pressure and becomes stable at ambient temperature at 0.5 GPa (Bridgman, 1914). The structure is an fcc-like arrangement of CO_2 molecules (Simon & Peters, 1980) whose librational modes stabilize the cubic lattice. At pressures in the GPa range, CO_2 undergoes a transition to a metastable low-symmetric arrangement of molecules (CO_2-II) which transforms to orthorhombic CO_2-III at ~ 10 GPa (Aoki et al., 1994). This spontaneous structural distortion is a result of the suppression of librational modes upon compression. At above 23 GPa and temperatures above 2000 K, CO_2 assumes a crystalline network structure, CO_2-V. This drastic change in structure and chemical bonding was first observed by Iota et al. (1999) by means of Raman spectroscopy and subsequently characterized as formation of a cristobalite-like structure (Datchi et al., 2012; Santoro, 2012). In the intermediate pressure-temperature regime between the stability field of CO_2-I and CO_2-V, a variety of different phases has been observed with little consistency in structure and stability fields (Datchi et al., 2012, 2014; Iota et al., 1999, 2007; Santoro et al., 2012). This includes different reconfigurations of molecules to denser packing (Datchi et al., 2012, 2014) as well as, possibly, unbent CO_2 molecules in a rutile-like arrangement (Iota et al., 2007). Various studies (Tschauner et al., 2001; Takafuji et al., 2006; Litasov et al., 2011) found that above 25–30 GPa, CO_2 decomposes along a negative Clapeyron slope that intersects the present average mantle geotherm between 900 and 1100 km depth. Given the range of residual pressures found in diamond inclusions (Navon, 1991; Schrauder & Navon, 1993; Tschauner et al., 2018), CO_2-I,-II, -III, and -IV may occur as inclusions in diamond.

Both bending and asymmetric stretching related bands of molecular CO_2 were observed, often in diamonds that also contain carbonate. Interestingly, the IR absorption band of the vibron (that is: the C-O asymmetric stretching vibration of molecular CO_2) of CO_2 in diamonds has an energy of around 2350 cm^{-1} or higher (2365 cm^{-1} in the example given in Figure 2.3) and is usually split. This energy is noticeably above the vibron energy of dry ice (CO_2-I) at ambient pressure (interpolated to 300 K; Hanson & Jones, 1981). The splitting of the vibron indicates its interaction with other phonon states and is also characteristic for CO_2 phases at elevated pressure (Hanson & Jones 1981). It was noticed that at least in some cases, the CO_2 bending mode energies do not match the the energies of the vibron for any pressure and any known phase of CO_2 (Hainschwang et al., 2008). A possible explanation is that CO_2 is not occurring as free phase but captured in a clathrate or "filled ice." Currently, there is no diffraction-based structural information about this dense CO_2 or CO_2-bearing phase.

In general, the vibron energy is a good scale of pressure as it is rather independent on the structural configuration of CO_2 molecules, due to its nature as intramolecular vibration (Hanson & Jones, 1981). Consequently, Navon (1991) and Schrauder and Navon (1993) used this information to estimate current residual pressure of CO_2-bearing inclusions. In a second step, they estimated the conditions of entrapment of CO_2 based on isochoric paths that connect the current residual pressure and temperature with the pressure and temperature of entrapment. They obtained pressures of between 4 and 7 GPa where these isochores of CO_2 intersect plausible geotherms of the lithospheric mantle underneath the localities of the CO_2-hosting diamonds. Hereby they assumed that CO_2 forms a phase by itself. These pressures and temperatures are at the upper limit of CO_2 stability in eclogite (Thomson et al., 2016). CO_2 as inclusion in diamond does not necessarily represent high CO_2 concentrations in diamond-forming environments but may represent a transient state or by-product of the breakdown of carbonate and subsequent diamond formation: Forming diamond can trap reaction educts and transient products as inclusions. For instance, methane and molecular hydrogen have been observed as inclusions in synthetic diamonds (E. Smith & Wang, 2016) and ice-VII inclusions in natural sublithospheric diamonds are likely such by-products of diamond formation during fluid-rock interaction rather than implying presence of an H_2O-dominated fluid (Tschauner et al., 2018).

2.3.2. Carbonates

Calcite, magnesite, siderite, eitelite $Na_2Mg(CO_3)_2$, nahkolite $NaHCO_3$, shortite $Na_2Ca_2(CO_3)_3$, and nyerereite $Na_2Ca(CO_3)_2$ have been observed as inclusions in diamonds. While these minerals are not genuinely high-pressure phases, it has been argued that they have formed in the mantle along with their host diamonds. Infrared absorption bands of some carbonate inclusions in diamond are consistent with elevated sustained pressures (Figure 2.3). Recently, a magnesian calcite inclusion at a residual pressures of 8.5±0.5 GPa was reported (Tschauner et al., 2018). This direct as well the indirect petrographic evidence for carbonate formation at pressures of several GPa makes it important to discuss these occurrences here. We start with nonquenchable high-pressure polymorphs of these carbonates, which have not yet been observed but may occur as inclusions in diamonds; then we briefly summarize observed occurrences of carbonate inclusions and conclude with the reconstruction of the entrapment conditions of a high-pressure calcite inclusion in a diamond.

Pure calcite undergoes a sequence of transformations to high-pressure structures: At 2 GPa, calcite undergoes a distortive transition to calcite-II (Merill & Bassett, 1975), which at 4 and 9 GPa transforms to calcite-III and calcite-IV, respectively, both of which assume very different configurations of Ca and carbonate groups (Merlini et al., 2012, 2014), though they can still be considered as highly distorted NaCl-type lattices, like calcite itself. At elevated temperature and pressure, calcite transforms to aragonite, which itself undergoes structural transitions upon compression (Oganov et al., 2006, 2008; D. Smith et al., 2018). As pure phases, magnesite and dolomite appear to be stable up to pressures in the 70–80 GPa range (Fiquet et al., 1994; Oganov et al., 2006, 2008, 2013; Pickard & Needs, 2015).

In principle, calcite-II, -III, -IV, aragonite, and postaragonite may be expected as inclusions in diamonds over the pressure range that has been found for CO_2 (Schrauder & Navon, 1993), ice-VII (Tschauner et al., 2018), magnesian calcite, and ilmenite (Tschauner et al., 2018), but none of these phases of $CaCO_3$ has been observed so far; the principal reason is the Mg content of calcite entrapped in diamond. The magnesite component stabilizes the calcite structure due to its high negative excess volume (Redfern & Angel, 1999; Buob et al., 2006). With increasing pressure of the calcite inclusions, an increase of Mg content is observed to levels where calcite-dolomite segregation occurs at low pressure. Experimental studies indicate that dolomite, magnesite, and Na carbonates replace calcite as stable carbonate phase in eclogite over pressures of 3 to above 15 GPa (Thomson et al., 2016), but inclusions in diamond may be more enriched in less compatible elements such as Ca over Mg or K over Na.

Siderite has been observed as inclusion of diamond but at nearly ambient pressure. Hence, it may be a precipitate at a crack. This also holds for numerous occurrences of Mg-poor calcite whose residual pressure is ambient or nearly ambient.

Studies based on Raman spectroscopy and electron microscopy (Bulanova et al., 2010; Kaminsky et al., 2009; Kaminsky & Wirth, 2011) have found a number of alkaline carbonates as inclusions in diamond, which highlights the role of carbonaceous melt in formation and growth of diamonds. These minerals are eitelite $Na_2Mg(CO_3)_2$, nahkolite $NaHCO_3$, shortite $Na_2Ca_2(CO_3)_3$, and nyerereite $Na_2Ca(CO_3)_2$, which have also been found as fluid inclusions in olivine xenocrysts from kimberlites (Golovin & Sharygin, 2007) and appear to have stability fields in carbonaceous melts in the range of a few GPa (Podnorodnikov et al., 2018). These carbonates are also found in carbonatitic volcanites. Sodic carbonates can also form at much higher pressure in eclogitic paragenesis (Thomson et al., 2017).

So far, residual pressures of calcite inclusions have been found to range between 1 and 9 GPa. Figure 2.4a shows the diffraction pattern of one of these inclusions (Tschauner et al., 2018). The composition is $(Ca_{0.75(2)}Mg_{0.25(2)})CO_3$. As described in Tschauner et al. (2018), the residual pressure can be estimated based on the compositional dependence of the bulk moduli and volumes of carbonates in the calcite and dolomite series (Redfern & Angel, 1999). Thermoelastic parameters were calculated by a combination of ab-initio and finite-temperature methods. The residual pressure was corrected for elastic relaxation of the surrounding diamond (Angel et al., 2014). This corrected pressure was used as endpoint of a quasi-isochoric release path, where volume changes of the diamond were taken into account (Tschauner et al., 2018). Figure 2.4 shows the upper and lower bound of the release path. The path intersects the solidus of carbonated MORB around its maximum between 11 and 13 GPa (Thomson et al., 2017). This pressure-temperature regime is a plausible upper bound for entrapment of the carbonate and diamond formation. With this tentative fix point, the limits of the entrapment path of magnesian

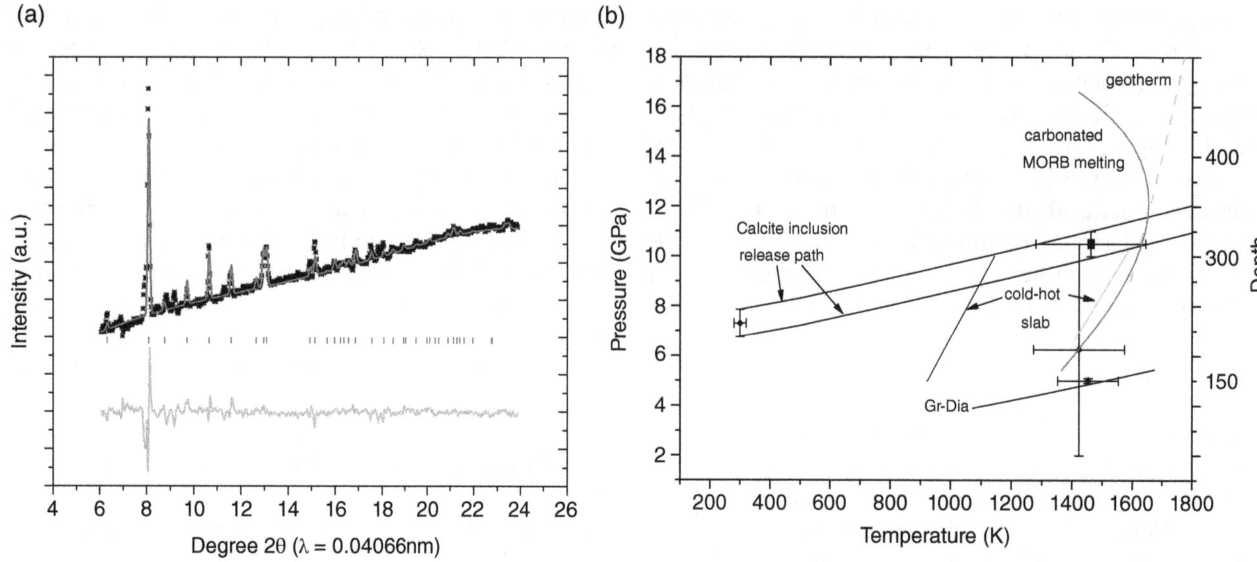

Figure 2.4 (a) Integrated diffraction pattern (black crosses), Rietveld refined calcite model pattern (red line), and residual of fit (green line) of a magnesian calcite inclusion in a cloudy diamond from Shandong, China (after Tschauner et al., 2018, Supplement). Tick marks indicate the Bragg angles of allowed reflections of calcite. The Rietveld refinement was used to determine unit cell parameters and Ca/Mg ratio. The phase contrast between Ca and Mg is sufficiently large to constrain this ratio to within 8% of the Mg content. (b) Reconstruction of entrapment conditions of this magnesian calcite. Black lines: Upper and lower bound of the release path for the inclusion. The possible breakdown of magnesian calcite into dolomite and magnesite (Buob et al., 2006) is not considered. Black diamonds: current residual pressure corrected for elastic relaxation of diamond (Tschauner et al., 2018), geobarometric and geothermometric estimate of growth of clinopyroxene with ilmenite exsolution lamellae (Gao et al., 2008), estimate of entrapment of an olivine inclusion in a diamond from the same area (Yin et al., 2017). Red line: solidus of carbonated MORB (Thomson et al., 2016); blue and orange lines: bounds for cold and hot subduction (after Thomson et al., 2016). Dashed green line: average mantle reference geotherm (after Brown & Shankland, 1981). See electronic version for color representation of the figures in this book.

calcite define a range of 1300 to 1650 K and 11 to 13 GPa, just above the boundary between upper mantle and transition zone. We compare this result with an estimate of pressure-temperature regime based on ilmenite exsolution in clinopyroxenes (Gao et al., 2008) and an independent estimate of entrapment conditions of an olivine crystal in diamond (Yin et al., 2017). The geobarometric estimate from ilmenite exsolution in clinopyroxene varies significantly between a minimal pressure of 2 GPa and 8 GPa and temperatures of 1300 to 1600 K. The estimated entrapment of an olivine inclusion does not provide an uncertainty in pressure. The three data points taken together constrain pressures and temperatures of subduction and ascent of material such as subduction to 11–13 GPa or deeper, release of carbonate-rich melt and subsequent melting of shallower peridotitic mantle, in agreement with scenarios that have been proposed for the North China Craton from geochemical data (Gao et al., 2008). However, reconstruction of entrapment conditions of diamond inclusions (Figure 2.4b; Yin et al., 2017) along with geobarometric estimates promise better constraints on the actual P-T conditions of these different steps.

CONCLUSION

Carbonaceous mineral inclusions at high residual pressures and high-pressure carbonaceous minerals provide important constraints on formation of diamond and the deep carbon cycle of Earth. Both reduced carbonaceous species, such as carbides, and oxidized species, such as carbonates, allow for correlating experimental petrological constraints on formation and stability with actual occurrences of carbonates and carbides in deep, diamantiferous mantle. As of yet, diamond, methane, transition metal carbides, CO_2, and various carbonates have been found either directly as minerals through microchemical and microstructural analysis, or have been detected through optical spectroscopy of inclusions in diamonds. The actual structural state of CH_4 and CO_2 either as pure phases as inclusions or as clathrates or icy phases is not yet known. Iron carbides, carbonaceous iron, and carbonate minerals have been identified. Some of these inclusions sustain still-high residual pressures, which can be used to reconstruct the pressures and temperatures of their entrapment. These results can be combined with geochemical and geophysical modeling of processes in the mantle.

ACKNOWLEDGMENTS

The author would like to thank the two reviewers and the editor for their helpful comments. The author also acknowledges support through NSF-EAR 1838330. Data presented were obtained at beamlines 13-IDD, GeoSoilEnviroCARS (The University of Chicago), 16-IDC, HPCAT, Advanced Photon Source (APS), Argonne National Laboratory, and 1.4 at the Advanced Light Source, which is a DOE Office of Science User Facility under contract no. DE-AC02-05CH11231. HPCAT operations are supported by DOE-NNSA's Office of Experimental Sciences. GeoSoilEnviroCARS is supported by the National Science Foundation – Earth Sciences (EAR-1634415) and Department of Energy – GeoSciences (DE-FG02-94ER14466). The Advanced Photon Source is a U.S. Department of Energy (DOE) Office of Science User Facility operated for the DOE Office of Science by Argonne National Laboratory under Contract No. DE-AC02-06CH11357. The publication of this chapter and this monograph is supported through the Deep Carbon Observatory.

REFERENCES

Anders. E., & Zinner, E. (1993). Interstallar grains in primitive meteorites: Diamond, silicon-carbide, and graphite. *Meteoritics*, *28*(4), 490–514.

Angel, R. J., et al. (2014). Geobarometry from host-inclusion systems: The role of elastic relaxation. *American Mineralogist*, *99*(10), 2146–2149.

Aoki, K., et al. (1994).Crystal-structure of the high-pressure phase of solid CO2. *Science*, *263*(5145), 356–358.

Boyd, S. R., et al. (1994). Modelling the growth of natural diamonds. *Chemical Geology*, *116*, 29–42.

Brenker, F. E., et al. (2002). Exhumation of lower mantle inclusions in diamond: ATEM investigation of retrograde phase transitions, reactions and exsolution. *Earth and Planetary Science Letters*, *198*(1-2), 1–9.

Bridgman, P. W. (1914). Change of phase under pressure: I. The phase diagram of eleven substances with especial reference to the melting curve. *Physical Review*, *3*(3), 153–203.

Brown, J. M., & Shankland, T. J. (1981). Thermodynamic parameters in the Earth as determined from seismic profiles. *Geophysical Journal of the Royal Astronomical Society*, *66*(3), 579–596.

Bulanova, G., et al. (2010). Mineral inclusions in sublithospheric diamonds from Collier 4 kimberlite pipe, Juina, Brazil: Subducted protoliths, carbonated melts and primary kimberlite magmatism. *Contrib. Mineral. Petrol.*, *160*, 489–510.

Buob, A., et al. (2006) Experiments on CaCO3-MgCO3 solid solutions at high pressure and temperature. *American Mineralogist*, *91*, 435-440.

Chen, B., et al. (2014). Hidden carbon in Earth's inner core revealed by shear softening in dense Fe7C3. *Proceedings of the National Academy of Sciences of the United States of America*, *111*(50), 17755–17758.

Dana, J. D. (1869). *System of mineralogy*. 5th ed. New York: Wiley & Sons.

Dasgupta, R., & Hirschmann, M. M. (2010). The deep carbon cycle and melting in Earth's interior. *Earth and Planetary Science Letters*, *298*(1-2), 1–13.

Datchi, F. et al. (2009). Structure of carbon dioxide phase IV: Breakdown of the intermediate bonding state scenario. *Physical Review Letters*, *103*(18), Article Number: 185701.

Datchi, F. et al. (2012). Structure of polymeric carbon dioxide CO2-V. *Physical Review Letters*, *1089*(12), 125701.

Datchi, F. et al. (2014). Structure and compressibility of the high-pressure molecular phase II of carbon dioxide. *Physical Review B*, *89*(14), Article Number: 144101.

Dobrzhinetskaya, L. F., et al. (2009). High-pressure highly reduced nitrides and oxides from chromitite of a Tibetan ophiolite. *Proceedings of the National Academy of Sciences of the United States of America*, *106*(46), 19233–19238.

Efthimiopoulos, I., et al. (2017). Combined high-pressure and high-temperature vibrational studies of dolomite: Phase diagram and evidence of a new distorted modification. *Physics and Chemistry of Minerals*, *44*(7), 465–476.

El Goersy, A., & Donnay, G. (1968). A new allotropic form of carbon from Ries crater. *Science*, *161*(3839), 363.

El Goresy, A., et al. (2003). A new natural, super-hard, transparent polymorph of carbon from the Popigai impact crater, Russia. *Comptes Rendus Geoscience*, *335*(12), 889–898.

Essene, E. J., & Fisher, D. C. (1986) Lightning strike fusion: Extreme reduction and metal-silicate liquid immiscibility. *Science*, *234*(4773), 189–193.

Fang, Q., et al. (2009). Qusongite (WC): A new mineral, locality: Luobusa ophiolite, Qusong County, Tibet, China. *American Mineralogist*, *94*, 387–290.

Fiquet, G., et al. (1994). High-pressure X-ray diffraction study of carbonates: MgCO3, CaMg(CO3)2, and CaCO3. *American Mineralogist*, *79*, 15–23.

Frost, D. J., et al. (2004). Experimental evidence for the existence of iron-rich metal in the Earth's lower mantle. *Nature*, *428*(6981), 409–412.

Frost, D. J., & McCammon, C. A. (2008). The redox state of Earth's mantle. *Annual Review of Earth and Planetary Sciences*, *36*, 389–420.

Gao, S., et al. (2008). Recycling deep cratonic lithosphere and generation of intraplate magmatism in the North China Craton. *Earth and Planetary Science Letters*, *270*, 41–53.

Golovin, A. V., & Sharygin, V. V. (2007). Petrogenetic analysis of fluid and melt inclusions in minerals from mantle xenoliths from the Bele pipe basanites (North Minusa depression). *Russian Geology and Geophysics*, *48*(10), 811–824.

Goss, J. P., et al. (2014). Identification of the structure of the 3107 cm(-1) H-related defect in diamond. *Journal of Physics: Condensed Matter*, *26*(14), Article Number: 145801. doi: 10.1088/0953-8984/26/14/145801.

Hainschwang, T., et al. (2008). HPHT treatment of CO2 containing and CO2-related brown diamonds. *Diamond and Related Materials*, *17*(3), 340–351.

Hanson, R. C., & Jones, L. H. (1981). Infrared and Raman studies of pressure effects on the vibrational-modes of solid CO2. *Journal of Chemical Physics*, *75*(3), 1101–1112.

Harris, J. W., & Vance, E. R. (1972). Induced graphitization around crystalline inclusions in diamond. *Contributions to Mineralogy and Petrology*, *35*(3), 227–234.

Harte, B., et al. (1999). Lower mantle mineral associations in diamond crystals from Sao Luiz, Brazil. In Yingwei Fei, C. M. Bertka, & B. O. Mysen (Eds.), *Mantle petrology: Field observations and high pressure experimentation: A tribute to Francis R. (Joe) Boyd* (Spec. Publ. 6, pp. 125–153). The Geochemical Society.

Helffrich, G. R., & Wood, B. J. (2001). The Earth's mantle. *Nature*, *412*(6846), 501–507.

Hemley, R. J., & Dera, P. (2000). Molecular crystals. *Reviews in Mineralogy & Geochemistry*, *41*, 335–419.

Hough R. M., et al. (1995). Diamond and silicon-carbide in impact melt rock from the Ries impact crater. *Nature*, *378*(6552), 41–44.

Howell, D., et al. (2012). Inclusions under remnant pressure in diamond: A multi-technique approach. *European Journal of Mineralogy*, *24*(4), 563–573.

Iota, V., et al. (1999). Quartzlike carbon dioxide: An optically nonlinear extended solid at high pressures and temperatures. *Science*, *283*(5407), 1510–1513.

Iota, V., et al. (2007). Six-fold coordinated carbon dioxide VI. *Nature Materials*, *6*(1), 34–38.

Irmer, W. (1920). Der Basalt des Bühls bei Kassel und seine Einschlüsse von Magnetit, Magnetkies und gediegen Eisen. *Abhandlungen der Senckenbergischen Naturforschenden Gesellschaft*, *13*, 91–108.

Izraeli, E. S., et al. (2001). Brine inclusions in diamonds: A new upper mantle fluid. *Earth and Planetary Science Letters*, *187*, 323–332.

Jablon, B. M., & Navon, O. (2016). Most diamonds were created equal. *Earth and Planetary Science Letters*, *443*, 41–47.

Jacob, D. E., et al. (2004). Cohenite, native iron and troilite inclusions in garnets from polycrystalline diamond aggregates. *Contributions to Mineralogy and Petrology*, *146*(5), 566–576.

Jedwab, J. (1990). Native ruthenium in tantalum carbide concentrates from the Ural Mountains, USSR. *Mineralogy and Petrology*, *43*(2), 137–146.

Kaminsky, F. V., & Wirth, R. (2011). Iron-carbide inclusions on lower-mantle diamond from Juina, Brazil. *Canadian Mineralogist*, *49*(2), 555–572.

Kaminsky, F. V., & Wirth, R. (2017). Nitrides and carbonitrides from the lowermost mantle and their importance in the search for Earth's "lost" nitrogen. *American Mineralogist*, *102*(8), 1667–1676.

Kaminsky, F., Wirth, R., Matsyuk, S., Schreiber, A., & Thomas, R. (2009). Nyerereite and nahcolite inclusions in diamond: Evidence for lower-mantle carbonatitic magmas. *Mineralogical Magazine*, *73*(5), 797–816. doi: 10.1180/minmag.2009.073.5.797

Kiseeva, E. S., et al. (2018). Oxidized iron in garnets from the mantle transition zone. *Nature Geoscience*, *11*(2), 144–147.

Koch-Mueller, M., et al. (2016). Phase transitions in the system CaCO3 at high P and T determined by in situ vibrational spectroscopy in diamond anvil cells and first-principles simulations. *Physics and Chemistry of Minerals*, *43*(8), 545–561.

Kvenvolden, K. A. (1993). Gas hydrates: Geological perspective and global change. *Reviews of Geophysics*, *31*(2), 173–187.

Litasov, K. D., et al. (2011). Crossover from melting to dissociation of CO2 under pressure: Implications for the lower mantle. *Earth and Planetary Science Letters*, *309*(3–4), 318–323.

Lobanov, S. S., et al. (2013). Carbon precipitation from heavy hydrocarbon fluid in deep planetary interiors. *Nature Communications*, *4*, Article Number: 2446.

Lord, O. T., et al. (2009). Melting in the Fe-C system to 70 GPa. *Earth and Planetary Science Letters*, *284*(1–2), 157–167.

Loveday, J. S., et al. (2001a). Stable methane hydrate above 2 GPa and the source of Titan's atmospheric methane. *Nature*, *410*, 661–663.

Loveday, J. S., et al. (2001b). Transition from cage clathrate to filled ice: The structure of methane hydrate III. *Physical Review Letters*, *87*, 215501-1.

Loveday, J. S., & Nelmes, R. J. (2008). High-pressure gas hydrates. *Physical Chemistry Chemical Physics*, *10*(7) 937–950.

Ma, C. (2018). A closer look at shocked meteorites: Discovery of new high-pressure minerals. *American Mineralogist*, *103*(10), 1521–1522.

Machida, S.-I., et al. (2007). Raman-spectra of methane hydrate up to 86 GPa. *Physics and Chemistry of Minerals*, *34*, 31–35. doi 10.1007/s00269-006-0126-6.

Mao, W. L., et al. (2003) Bonding changes in compressed superhard graphite. *Science*, *302*(5644), 425–427.

Maynard-Casely H. E., et al. (2014). The crystal structure of methane B at 8 GPa: An α-Mn arrangement of molecules. *Journal of Chemical Physics*, *141*, 234313. doi: 10.1063/1.4903813.

Merrill, L., & Bassett, W. A. (1975). Crystal-structure of CaCO3(II), a high-pressure metastable phase of calcium-carbonate. *Acta Crystallographica Section B – Structural Science*, *31*, 345–350.

Merlini, M., et al. (2012). CaCO3-III and CaCO3-VI, high-pressure polymorphs of calcite: Possible host structures for carbon in the Earth's mantle. *Earth and Planetary Science Letters*, *333*, 265–271.

Merlini, M., et al. (2014). Evidence of interspersed co-existing CaCO3-III and CaCO3-IIIb structures in polycrystalline CaCO3 at high pressure. *Mineralogical Magazine*, *78*(2), 225–233.

Mikhail, S., et al. (2014). *Geochemistry Geophysics Geosystems*, *15*, 855–866.

Mochalov, A. G., et al. (1998). Hexaferrum (Fe, Ru) (Fe, Os) (Fe, Ir)-A new mineral. *Zap Vseross Mineral Obshch*, *127*, 41–51.

Nasdala, L., et al. (2003). Spectroscopic 2D-tomography: Residual pressure and strain around mineral inclusions in diamonds. *European Journal of Mineralogy*, *15*, 931–935. doi: 10.1127/0935-1221/2003/0015-0931.

Navon, O. (1989). Chemical and Mineralogical Characterization of Microinclusions in diamonds. PhD Thesis, California Institute of Technology 1989.

Navon, O. (1991). High internal pressures in diamond fluid inclusions determined by infrared-absorption. *Nature*, *353*(6346), 746–748.

Navon O., et al. (2017). Solid molecular nitrogen (delta-N-2) inclusions in Juina diamonds: Exsolution at the base of the transition zone. *Earth and Planetary Science Letters, 464*, 237–247.

Narygina, O., et al. (2011). Phase relations in Fe–Ni–C system at high pressures and temperatures. *Physics and Chemistry of Minerals, 38*, 203–214. doi 10.1007/s00269-010-0396-x.

Nemeth, P., et al. (2014). Lonsdaleite is faulted and twinned cubic diamond and does not exist as a discrete material. *Nature Communications, 5*, Article Number 5447.

Nestola, F., et al. (2018). CaSiO3 perovskite in diamond indicates the recycling of oceanic crust into the lower mantle. *Nature, 560*. doi:10.1038/nature25972.

Nickel, E. H., & Grice, J.D. (1998). The IMA commission on new minerals and mineral names: Procedures and guidelines on mineral nomenclature. *Canad. Min., 36*, 913–926.

Oganov, A. R., et al. (2006). High-pressure phases of CaCO3: Crystal structure prediction and experiment. *Earth and Planetary Science Letters, 241*(1–2), 95–103.

Oganov, A. R., et al. (2008). Novel high-pressure structures of MgCO3, CaCO3 and CO2 and their role in Earth's lower mantle. *Earth and Planetary Science Letters, 273*(1–2), 38–47.

Oganov, A. R., et al. (2013). Structure, bonding, and mineralogy of carbon at extreme conditions. Carbon in Earth,book series: *Reviews in Mineralogy, & Geochemistry, 75*, 47–77.

Pearson, D. G., et al. (2014). Hydrous mantle transition zone indicated by ringwoodite included within diamond. *Nature, 507*(7491), 221–223.

Pickard, C. J., & Needs, R. J. (2015). Structures and stability of calcium and magnesium carbonates at mantle pressures. *Physical Review B, 91*(10), Article Number: 104101.

Press W. (1972). Structure and phase transitions of solid heavy methane (CD4). *Journal of Chemical Physics, 56*, 2597.

Podborodnikov, I. V., Shatskiy, A., Arefiev, A. V., Bekhtenova, A., Litasov, K. D. (2019). New data on the system Na2CO3-CaCO3-MgCO3 at 6 GPa with implications to the composition and stability of carbonatite melts at the base of continental lithosphere. *Chemical Geology, 515*, 50–60. doi:10.1016/j.chemgeo.2019.03.027

Redfern, S.A.T., & Angel, R. J. (1999). High-pressure behaviour and equation of state of calcite, CaCO3. *Contributions to Mineralogy and Petrology, 134*(1), 102–106.

Rohrbach, A., & Schmidt M. W. (2011). Redox freezing and melting in the Earth's deep mantle resulting from carbon-iron redox coupling. *Nature, 449*, 456–458.

Rubin A. E., & Ma. C. (2017). Meteoritic minerals and their origins. *Chemie der Erde – Geochemistry, 77*(3), 325–385.

Santoro, M., et al. (2012). Partially collapsed cristobalite structure in the non molecular phase V in CO2. *Proceedings of the National Academy of Sciences of the United States of America, 109*(14), 5176–5179.

Schrauder, M., & Navon, O. (1993). Solid carbon-dioxide in a natural diamond. *Nature, 365*(6441), 42–44.

Schulte, P., et al. (2009). A dual-layer Chicxulub ejecta sequence with shocked car-bonates from the Cretaceous-Paleogene (K-Pg) boundary, Demerara Rise, western Atlantic. *Geochimica et Cosmochimica Acta, 73*(4), 1180–1204.

Sharp, T. G., & DeCarli, P. S. (2006). Shock effects in meteorites. In *Meteorites and the Early Solar System II* (pp. 653–677). Tucson: University of Arizona Press.

Shi, N., et al. (2009). Yarlongite: A new metallic carbide mineral. *Acta Geologica Sinica – English Edition, 83*(1), 52–56.

Shirey, S. B., et al. (2013). Diamonds and the geology of mantle carbon. Carbon in Earth, book series: *Reviews in Mineralogy, & Geochemistry, 75*, 355–421.

Simon, A., & Peters, K. (1980). Single-crystal refinement of the structure of carbon-dioxide. *Acta Crystallographica Section B – Structural Science, 36*, 2750–2751.

Smith, D., et al. (2018a). Postaragonite phases of CaCO3 at lower mantle pressures. *Physical Review Materials, 2*(1), Article Number: 013605.

Smith, E.M., et al. (2016). Large gem diamonds from metallic liquid in Earth's deep mantle. *Science, 354* (6318), 1403–1405.

Smith, E. M., et al. (2018b). Blue boron-bearing diamonds from Earth's lower mantle. *Nature, 560*(7716), 84–85.

Smith, E. M., & Wang, W. (2016). Fluid CH4 and H2 trapped around metallic inclusions in HPHT synthetic diamond. *Diamond and Related Materials, 68*, 10–12.

Sobolev, N. V., et al. (1997). Mineral inclusions in diamonds from the Sputnik kimberlite pipe, Yakutia. *Lithos, 39*(3–4), 135–157.

Stachel, T., et al. (2000). Kankan diamonds (Guinea) II: Lower mantle inclusion parageneses. *Contribution to Mineralogy and Petrology, 140*(1), 16–27.

Stachel, T., & Luth R. W. (2015). Diamond formation: Where, when and how? *Lithos, 220–223*, 200–220.

Stoffler, D., et al. (1991). Shock metamorphism of ordinary chondrites. *Geochimica et Cosmochimica Acta, 55*(12), 3845–3867.

Takafuji, N., et al. (2006). Decarbonation reaction of magnesite in subducting slabs at the lower mantle. *Physics and Chemistry of Minerals, 33*(10), 651–654.

Taylor, W. R., Jaques, A. L., & Ridd, M. (1990). Nitrogen-defect aggregation characteristics of some Australasian diamonds: Time-temperature constraints on the source regions of pipe and alluvial diamonds. *American Mineralogist, 75*, 1290–1310.

Thomson, A. R., et al. (2016). Slab melting as a barrier to deep carbon subduction. *Nature, 529*(7584), 76–78.

Tschauner, O., et al. (2001). New transformations of CO2 at high pressures and temperatures. *Physical Review Letters, 87*(7), Article Number: 075701.

Tschauner, O., et al. (2018). Ice-VII inclusions in diamonds: Evidence for aqueous fluid in Earth's deep mantle. *Science, 359*(6380), 1136–1138.

van Keken, P. E., et al. (2002). Mantle mixing: The generation, preservation, and destruction of chemical heterogeneity. *Annual Review of Earth and Planetary Sciences, 30*, 493–525.

von John, W. (1910). Native tantalum. *Nature, 83*, 398.

Walter, M. J., et al. (2011). Deep mantle cycling of oceanic crust: Evidence from diamonds and their mineral inclusions. *Science, 334*(6052), 54–57.

Yin, Z., et al. (2017). Inclusions of alpha-quartz, albite and olivine in a mantle diamond. *Gondwana Research, 44*, 228–235.

3

Phase and Melting Relations of Fe_3C to 300 GPa and Carbon in the Core

Suguru Takahashi[1], Eiji Ohtani[1], Takeshi Sakai[2], Seiji Kamada[1], Shin Ozawa[1], Tatsuya Sakamaki[1], Masaaki Miyahara[1,3], Yoshinori Ito[1], Naohisa Hirao[4], and Yasuo Ohishi[4]

ABSTRACT

Carbon is a plausible candidate for a light element in the Earth's core. Here, we show that Fe_3C melts incongruently to form Fe_7C_3 and liquid at least up to 200 GPa based on in-situ X-ray diffraction measurements and textural observations of the recovered samples, and Fe_3C is stable at least up to 300 GPa at high temperatures. The C content of the liquid coexisting with Fe_7C_3 decreases and the Fe-Fe_3C eutectic composition shifts toward the Fe-rich direction with increasing pressure. The present result revealed that both Fe_3C and Fe_7C_3 are plausible constituents in the inner core.

3.1. INTRODUCTION

Seismological studies (Dziewonski & Anderson, 1981) show that the inner core is less dense than pure iron under the core conditions (Mao et al., 1990); therefore, it has been accepted that the inner core comprises iron along with lighter elements (Birch, 1964; Poirier, 1994). The density deficit of the core is estimated to be 3%–10% for the outer core (Anderson & Isaak, 2002; Shanker et al., 2004) and 2.5%–9% for the inner core (Dubrovinsky et al., 2000; Dewaele et al., 2006). Carbon is one of the major candidates for the light elements contained in the Earth's core (Hirayama et al., 1993; Wood, 1993; Dasgupta & Walker, 2008; Chi et al., 2014) because carbon has a high abundance in the solar system (Anders & Grevesse, 1989) and it can dissolve significantly in molten iron, even at ambient and upper mantle pressures (Wood, 1993; Dasgupta & Walker, 2008). Although several studies on the physical and chemical properties of iron carbides have been carried out at high pressure, there are still very limited numbers of experimental works on the phase and melting relationships in the Fe-C system to the core conditions.

Wood (1993) estimated the phase diagrams of the Fe-C system up to the pressure of the core mantle boundary (CMB) and suggested that about 2–4 wt.% of carbon could be stored in the core during the core formation stage. He suggested that the Earth's inner core is composed solely of Fe_3C. Nakajima et al. (2009) reported the phase relationships in the Fe-C system up to 30 GPa based on textural observations, chemical analysis of the recovered samples, and in-situ X-ray diffraction experiments using a Kawai-type multianvil apparatus. According to Nakajima et al. (2009), Fe_3C melts incongruently to form Fe_7C_3 and a liquid phase at 30 GPa. They proposed that Fe_7C_3 is present in the inner core. Lord et al. (2009) determined the melting temperatures of Fe_3C and Fe_7C_3 and the Fe-Fe_3C eutectic temperature up to 70 GPa from the temperature plateau obtained while increasing the laser power in a laser-heated diamond anvil cell (DAC). However, there were obvious discrepancies between the melting curves of Fe_3C and Fe_7C_3

[1] *Department of Earth and Planetary Material Sciences, Graduate School of Science, Tohoku University, Sendai, Japan*

[2] *Geodynamics Research Center, Ehime University, Matsuyama, Japan*

[3] *Department of Earth and Planetary Science, Graduate School of Science, Hiroshima University, Higashi-Hiroshima, Japan*

[4] *Japan Synchrotron Radiation Research Institute, Sayo, Japan*

Carbon in Earth's Interior, Geophysical Monograph 249, First Edition. Edited by Craig E. Manning, Jung-Fu Lin, and Wendy L. Mao.

reported by Nakajima et al. (2009) and Lord et al. (2009). In addition, Lord et al. predicted that Fe_3C decomposes to Fe and Fe_7C_3 and suggested that the inner core is composed of Fe + Fe_7C_3. On the other hand, Sata et al. (2010) reported that Fe_3C is stable up to at least 187 GPa at 300 K. In this study, high-pressure and high-temperature experiments using a double-sided laser-heated DAC were performed up to the pressure conditions of the inner core to investigate the phase and melting relationships of Fe_3C. The melting temperature of Fe_3C was determined from in-situ X-ray diffraction experiments and textural observations on recovered samples. This study aimed to reveal the phase and melting relationships of the Fe-C system under core conditions and to discuss carbon storage in the Earth's inner core.

3.2. EXPERIMENTAL PROCEDURE

We have conducted two types of experiments, in-situ X-ray diffraction experiments for determination of the phase and melting relations of Fe_3C, and a quenching experiment at 59 GPa and 2850K for textural observations of the recovered sample.

3.2.1. Sample Preparations

The starting material for the in-situ X-ray diffraction experiments was a powdered sample of cementite Fe_3C, which was synthesized from a stoichiometric mixture of iron and graphite at 3 GPa and 1273 K for 12 h using a 3000-ton Kawai-type multianvil apparatus installed at Tohoku University. The synthesized Fe_3C was confirmed to be a single phase of cementite by X-ray diffraction and SEM observations, that is, it was a single phase of cementite Fe_3C without chemical zoning and interstitial phases.

The starting material used for the quench experiment was a mixture of Fe and Fe_3C synthesized by the above procedure. The powdered mixture was laser heated using a Nd:YAG laser (λ = 1.064 μm) in an Ar atmosphere and quenched rapidly, forming a fine dendritic texture with a typical size <1 μm. The starting materials thus synthesized were analyzed based on the procedure by Dasgupta & Walker (2008) and Walker et al. (2013). After polishing the samples, the carbon content was analyzed using a wavelength-dispersive spectrometer with an EPMA (JXA-8800; JEOL, Tokyo, Japan). The calibration standard, which is cementite without zoning or interstitial phases synthesized at 3 GPa and 1273 K, was coated with aluminium. An acceleration voltage of 15 kV and a beam current of 90 nA were employed to analyze the carbon content. An LDE1H crystal in the EPMA was used to measure the carbon content in the sample. We used a defocused beam (50 μm in diameter) to perform the quantitative analysis of the bulk starting material for the quench experiment. The starting mixture for the quench experiment contained 5.2 wt.% carbon (see Table 3.1), which is slightly enriched in Fe compared to Fe_3C.

Table 3.1 The compositions of the starting material for the quenched experiment, and the quenched liquid and coexisting Fe_7C_3 recovered from 59.4 GPa and 2850 K. Numbers in parentheses are uncertainties on the last digits. The uncertainty of the composition of the quenched liquid is the compositional variation in the multiple analysis of the quenched melt region (see Figure 3.3).

		Fe (wt.%)	C (wt.%)	Total
Starting material		94.6 (3)	5.2 (4)	99.8
Recovered sample	Fe_7C_3	91.2 (1)	8.1 (1)	99.3
	Quenched liquid	93.9 (7)	4.7 (4)	98.6

3.2.2. In-Situ X-Ray Diffraction Experiments

The in-situ X-ray diffraction experiments were conducted at the BL10XU beamline at the SPring-8 facility (Hyogo, Japan) (Ohishi et al., 2008). DAC experiments were performed using a symmetric-type diamond anvil cell. Diamond anvils with various culet sizes from 40 to 350 μm were used, depending on the required experimental pressure conditions. Rhenium or tungsten gasket was pre-indented to a thickness of 10–70 μm, and a 20–120 μm diameter hole was drilled in the gasket to form a sample chamber. A foil from the powdered Fe_3C sample was prepared using a cold-compression technique. The sample foil was sandwiched between either NaCl or SiO_2 glass layers and placed in the sample chamber. The NaCl or SiO_2 layers served as a pressure-transmitting medium and as a thermal insulator.

The sample was heated using a double-sided laser heating technique employing an SPI fiber laser (λ = 1.070 μm). The shape of the fibre laser beam was adjusted to a flat-top beam using the beam-shaping system at the BL10XU beamline. This technique enabled us to decrease the temperature gradient across the sample and achieve a homogenous heating of the area with a diameter of approximately 30 μm. The temperature was determined by fitting the emission spectra from the surface of the heated sample to Planck's radiation law as a grey body formula using the typical wavelength range between 600 and 800 nm (Shen et al., 1996). The wavelength of the monochromatic X-ray beam and the distance between the sample and the X-ray detector were calibrated using X-ray diffraction patterns from CeO_2 based on the double cassette method. The typical wavelength of the X-rays was in the range of 0.4140–0.4224 Å, which was determined with a precision of 0.04%. The X-ray beam was collimated to a diameter of 15 μm. An IP detector (RAXIS-IV; Rigaku, Tokyo, Japan) and a CCD detector (SMART APEX; Bruker AXS, Madison, WI) were used to collect the angle-dispersive X-ray diffraction data. Exposure time for the IP and CCD was 3 min and 1–10 s, respectively. The examples of the diffraction patterns are given in Figures 3.1 and 3.2. The onset of melting of the sample was determined from the disappearance of the X-ray diffraction peaks and reappearance of the peaks after quenching at high pressure, as described in previous works (Kamada et al., 2010; Campbell et al., 2007; Morard et al., 2008).

Each integrated X-ray diffraction pattern, along with the 2θ angle (i.e., one dimensional X-ray diffraction profile), was analyzed using an IP analyzer and PD Indexer software package programmed by Y. Seto.

The experimental pressures were determined based on the lattice parameters of NaCl at ambient and high temperatures using the equation of state of NaCl with the B2 structure (Fei et al., 2007). An average of the lattice parameters at high temperature of the sample and those at ambient temperature at the diamond anvil surface was employed to determine the generated pressure using the procedure described by Campbell et al. (2009).

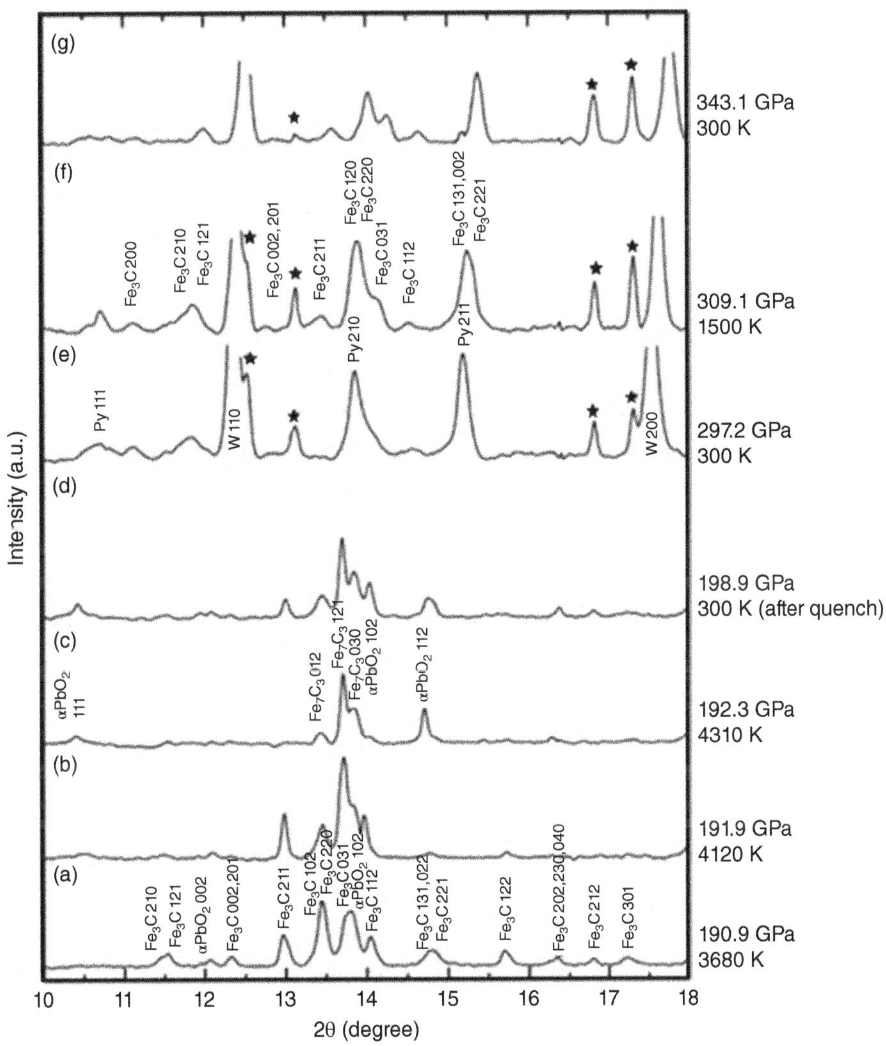

Figure 3.1 Typical examples of X-ray diffraction patterns of Fe_3C at (a)–(d) 191–199 GPa and (e)–(g) 297–343 GPa, respectively. (a) Fe_3C was a stable subsolidus phase at 190 GPa and T > 3000 K. (b) After further heating, Fe_3C peaks were observed up to 4120 K. (c) These peaks disappeared and Fe_7C_3 peaks were observed at 4310 K. (d) After quenching, the Fe_3C peaks reappeared. This result indicates that Fe_3C incongruently melted to form Fe_7C_3 and a liquid phase. (e) The Fe_3C peaks were broad before heating at 297 GPa. (f) During heating, the Fe_3C peaks became sharper at 1500 K. (g) Fe_3C was observed as a stable subsolidus phase up to 343 GPa. Key: * denotes peaks derived from outside the sample chamber since the peak positions are fixed and do not shift with pressure and temperature. See electronic version for color representation of the figures in this book.

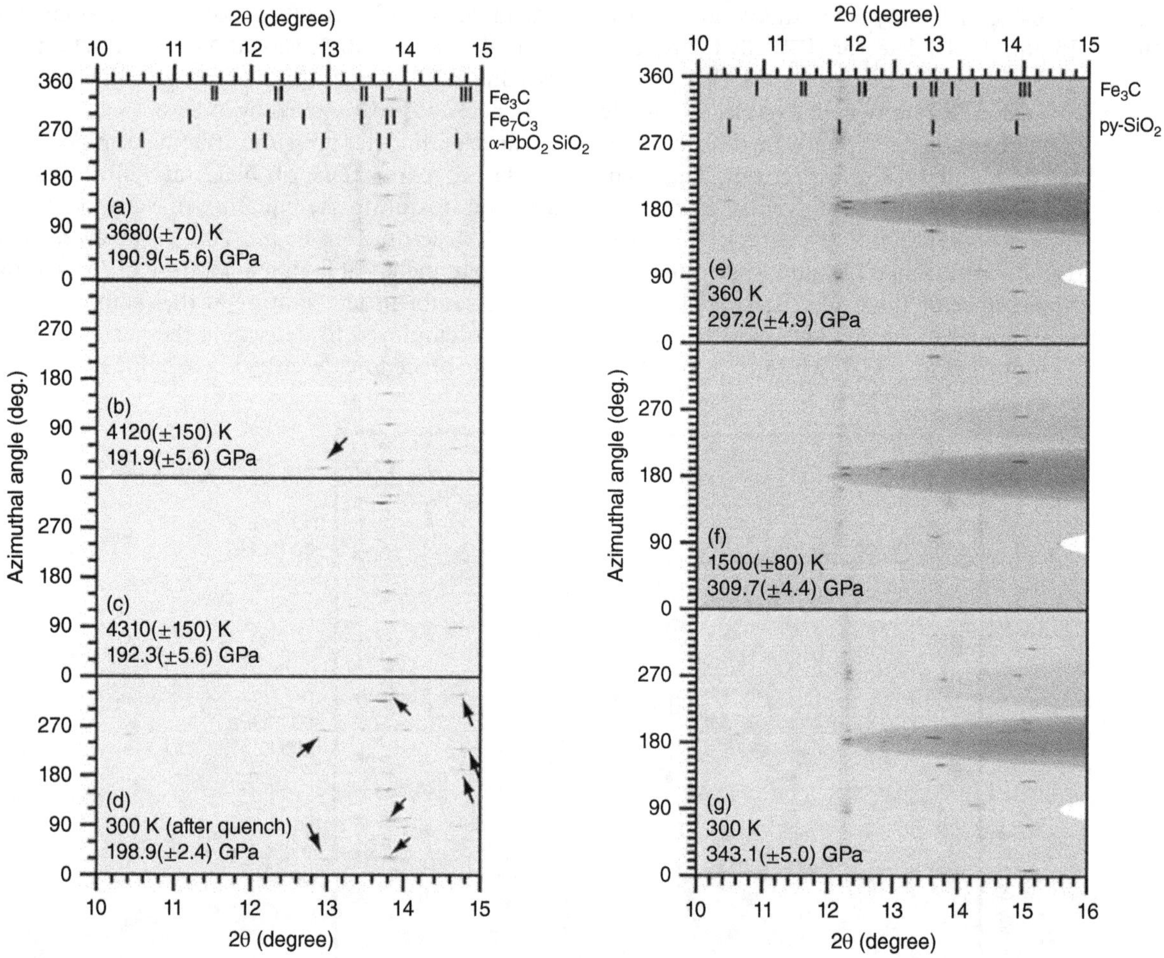

Figure 3.2 IP (a, b, c, d) and CCD (e, f, g) data obtained at high pressure and temperature. Arrows in the figures are the diffraction peaks from SiO_2 pressure medium (α-PbO_2 type SiO_2 and py, pyrite type SiO_2). These profiles correspond to one dimensional pattern shown in Figure 3.1. The peak positions of Fe_3C, Fe_7C_3, and pyrite SiO_2 are based on the equation of state of these phases by Sata et al. (2010), Chen et al. (2012), and Kuwayama et al. (2011), respectively. See electronic version for color representation of the figures in this book.

In addition, in the experiments with SiO_2 glass as the pressure-transmitting medium, pressures were estimated using the equation of state of Fe_3C (Sata et al., 2010) with thermal expansion calculated by Vočadlo et al. (2002). The results of the experiments using two different pressure media were consistent with each other as shown in Table 3.2.

3.2.3. Quench Experiment and Sample Analysis

The sample sandwiched between SiO_2 glass layers was compressed in a DAC and heated using a double-sided laser heating technique employing an SPI fiber laser (λ = 1.070 μm) with a flat top beam. The pressure was determined at room temperature using the ruby fluorescence method (Dewaele et al., 2004) and/or the pressure shift of the edge of the T_{2g} Raman band of the culet of the diamond anvil (Akahama & Kawamura, 2004). The pressure values determined by the two methods were consistent with each other. After decompression, the heated region of the recovered sample was cut by an FIB system (JEM-9320FIB; Jeol) (Miyahara et al., 2008). The texture of the prepared sample was observed using an FE-SEM (JSM-7001F; JEOL) at Tohoku University. We observed the coexistence of Fe_7C_3 and a quenched melt phase in the recovered samples. Subsequently, the chemical compositions of the coexisting phases were analyzed using the wavelength-dispersive mode of an FE-EPMA (JXA-8530F; JEOL) at Tohoku University. An accelerating voltage of 15 kV and a beam current of 30 nA were employed to analyze the carbon content, and an LDE2H crystal in the EPMA was used to measure the carbon content in the sample. We determined the calibration curve from the relationship between the carbon content

Table 3.2 Experimental conditions and observed phases.

RUN #	T (K)	P (GPa)	Observed Phases	VNaCl (Å^3)	VSiO$_2$ (Å^3)
In-situ X-ray diffraction experiments					
Fe3C350_061	300	68.0(9)[a]	Fe_3C, NaCl(B2)	22.78(8)	
Fe3C350_063	1360(120)	72.7(24)	Fe_3C, NaCl(B2)	22.67(2)	
Fe3C350_064	1710(200)	74.0(32)	Fe_3C, NaCl(B2)	22.66(5)	
Fe3C350_065	1770(240)	73.2(33)	Fe_3C, NaCl(B2)	22.75(12)	
Fe3C350_066	2090(70)	74.1(40)	Fe_3C, NaCl(B2)	22.76(10)	
Fe3C350_067	2270(60)	74.5(44)	Fe_3C, NaCl(B2)	22.78(10)	
Fe3C350_068	2450(70)	74.5(49)	Fe_3C, NaCl(B2)	22.83(9)	
Fe3C350_069	2630(90)	74.7(53)	Fe_3C, NaCl(B2)	22.87(9)	
Fe3C350_070	2810(80)	74.3(57)	Fe_3C, NaCl(B2)	22.96(4)	
Fe3C350_071	3020(130)	72.5(61)	Fe_7C_3 (+L), NaCl(B2)	23.18(6)	
Fe3C350_072	3310(110)	73.4(68)	Fe_7C_3 (+L), NaCl(B2)	23.18(15)	
Fe3C350_073	3420(140)	75.3(71)	NaCl(B2)	23.05(1)	
Fe3C350_074	300	68.5(5)	Fe_3C, Fe_7C_3, NaCl(B2)	22.73(4)	
Fe3C3502_030	300	47.9(16)	Fe_3C, NaCl(B2)	24.84(20)	
Fe3C3502_031	1500(50)	50.1(26)	Fe_3C, NaCl(B2)	25.04(13)	
Fe3C3502_032	1820(50)	50.7(33)	Fe_3C, NaCl(B2)	25.10(16)	
Fe3C3502_033	2010(50)	51.4(37)	Fe_3C, NaCl(B2)	25.08(12)	
Fe3C3502_034	2270(90)	51.8(43)	Fe_3C, NaCl(B2)	25.14(12)	
Fe3C3502_035	2430(110)	52.0(46)	Fe_3C, NaCl(B2)	25.17(5)	
Fe3C3502_036	2550(100)	53.1(49)	Fe_3C, NaCl(B2)	25.09(3)	
Fe3C3502_038	2790(100)	51.4(54)	Fe_7C_3 (+L), NaCl(B2)	25.41(20)	
Fe3C3502_039	3000(170)	50.4(58)	Fe_7C_3 (+L), NaCl(B2)	25.64(20)	
Fe3C3502_040	3340(150)	53.6(66)	NaCl(B2)	26.35(8)	
Fe3C3502_041	300	45.3(9)	Fe_3C, Fe_7C_3, NaCl(B2)	25.17(12)	
Fe3C130_038	300	122.0(9)	Fe_3C, NaCl(B2)	19.48(4)	
Fe3C130_039	1900(80)	138.5(39)	Fe_3C, NaCl(B2)	19.02(4)	
Fe3C130_040	2220(220)	142.2(47)	Fe_3C, NaCl(B2)	18.92(2)	
Fe3C130_041	2260(200)	142.6(48)	Fe_3C, NaCl(B2)	18.91(5)	
Fe3C130_042	2690(130)	143.5(59)	Fe_3C, NaCl(B2)	18.94(19)	
Fe3C130_043	300	134.2(32)	Fe_3C, NaCl(B2)	18.97(13)	
Fe3C130_044	2860(90)	151.3(64)	Fe_3C, NaCl(B2)	18.67(19)	
Fe3C130_045	3600(200)	153.6(83)	Fe_3C, NaCl(B2)	18.69(16)	
Fe3C130_046	3600(260)	154.5(83)	Fe_3C, NaCl(B2)	18.65(17)	
Fe3C130_047	3660(150)	152.4(84)	Fe_3C, NaCl(B2)	18.74(11)	
Fe3C130_048	3260(300)	154.7(74)	Fe_3C, NaCl(B2)	18.60(6)	
Fe3C130_049	300	144.1(27)	Fe_3C, NaCl(B2)	18.60(10)	
Fe3C130_050	300	146.3(34)	Fe_3C, NaCl(B2)	18.52(12)	
Fe3C130_051	2980(200)	149.2(67)	Fe_3C, NaCl(B2)	18.76(9)	
Fe3C130_052	3620(180)	154.7(83)	Fe_3C, NaCl(B2)	18.65(2)	
Fe3C130_055	4270(460)	157.0(100)	Fe_7C_3 (+L), NaCl(B2)	18.66(5)	
Fe3C130_056	3910(260)	153.5(90)	Fe_7C_3 (+L), NaCl(B2)	18.73(14)	
Fe3C130_057	300	142.4(22)	Fe_7C_3 (+L), NaCl(B2)	18.66(8)	
Fe3C050_002	300	183.4(22)	Fe_3C, NaCl(B2)	17.35(6)	
Fe3C050_003	1920(90)	209.1(41)	Fe_3C, NaCl(B2)	16.83(1)	
Fe3C050_004	3070(130)	217.7(73)	Fe_3C, NaCl(B2)	16.74(4)	
Fe3C050_005	2740(150)	226.1(64)	Fe_3C, NaCl(B2)	16.51(4)	
Fe3C050_008	2020(50)	214.8(44)	Fe_3C, NaCl(B2)	16.71(4)	
Fe3C050_009	2710(70)	219.6(63)	Fe_3C, NaCl(B2)	16.66(2)	
Fe3C050_012	4100(150)	229.0(101)	Fe_3C, NaCl(B2)	16.57(3)	
Fe3C050_014	3000(80)	231.5(71)	Fe_3C, NaCl(B2)	16.42(9)	
Fe3C050_015	2810(40)	228.3(66)	Fe_3C, NaCl(B2)	16.47(6)	
Fe3C050_016	300	217.4(13)	Fe_3C, NaCl(B2)	16.50(3)	
Fe3C100_010	1980(40)	187.3(56)	Fe_3C, SiO_2(α-PbO_2 type)		

(Continued)

Table 3.2 (Continued)

RUN #	T (K)	P (GPa)	Observed Phases	VNaCl (Å³)	VSiO$_2$ (Å³)
Fe3C100_011	2860(70)	189.2(56)	Fe_3C, SiO_2(α-PbO_2 type)		
Fe3C100_012	3680(70)	190.9(56)	Fe_3C, SiO_2(α-PbO_2 type)		
Fe3C100_013	4120(150)	191.9(56)	Fe_3C, SiO_2(α-PbO_2 type)		
Fe3C100_015	4310(150)	192.3(56)	Fe_7C_3 (+L), SiO_2(α-PbO_2 type)		
Fe3C100_016	300	198.9(24)	Fe_3C, Fe_7C_3, SiO_2(α-PbO_2 type)		
Fe3C040_018	300	297.2(49)	Fe_3C, SiO_2(pyrite type)		59.49(4)
Fe3C040_020	300	309.7(30)	Fe_3C, SiO_2(pyrite type)		58.72(19)
Fe3C040_022	1500(80)	309.7(44)	Fe_3C, SiO_2(pyrite type)		59.17(44)
Fe3C040_023	300	306.8(57)	Fe_3C, SiO_2(pyrite type)		58.81(18)
Fe3C040_025	300	313.1(31)	Fe_3C, SiO_2(pyrite type)		58.44(14)
Fe3C040_027	300	319.8(34)	Fe_3C, SiO_2(pyrite type)		58.50(13)
Fe3C040_029	1580(100)	325.8(23)	Fe_3C, SiO_2(pyrite type)		58.37(19)
Fe3C040_030	300	327.4(61)	Fe_3C, SiO_2(pyrite type)		58.18(15)
Fe3C040_032	300	321.6(83)	Fe_3C, SiO_2(pyrite type)		58.14(6)
Fe3C040_033	300	328.1(23)	Fe_3C, SiO_2(pyrite type)		57.94(10)
Fe3C040_034	300	324.8(31)	Fe_3C, SiO_2(pyrite type)		58.27(2)
Fe3C040_035	300	322.4(28)	Fe_3C, SiO_2(pyrite type)		58.16(13)
Fe3C040_038	300	322.8(27)	Fe_3C, SiO_2(pyrite type)		58.03(24)
Fe3C040_040	300	324.2(101)	Fe_3C, SiO_2(pyrite type)		58.18(44)
Fe3C040_042	300	333.6(69)	Fe_3C, SiO_2(pyrite type)		57.67(10)
Fe3C040_043	300	343.1(50)	Fe_3C, SiO_2(pyrite type)		57.30(37)
Quenched experiment					
Fe3C350lab_001	2850(200)	59.4(60)[b]	Fe_3C, Fe_7C_3, Quenched L, SiO_2		

Numbers in parentheses are uncertainties on the last digits.
[a] The pressures were determined by the equations of state B2 (Fei et al., 2007) for the runs using the NaCl pressure medium, and the equation of state of Fe_3C (Sata et al., 2010) with the thermal expansion (Vočadlo et al., 2002) for the runs with the SiO_2 glass pressure medium.
[b] The pressure was determined by the pressure dependency of the edge of T_{2g} Raman band of the diamond anvil (Akahama & Kawamura, 2004) and ruby fluorescence (Dewaele et al., 2004).

and the C Kα count rates in the standards employed (Fe and Fe_3C). The standard of Fe_3C was synthesized at 3 GPa and 1273 K. Fe_3C synthesized at this condition contains no compositional zoning nor impurities and is thus close to stoichiometry, according to Walker et al. (2013). An image of the recovered sample is shown in Figure 3.3 and the results of chemical analyses are summarized in Table 3.1.

3.3. RESULTS

3.3.1. In-Situ X-Ray Diffraction Experiments

The experiments were carried out in the pressure range 50–340 GPa and the temperature range 300–4300 K. The diffraction patterns of Fe_3C at pressures around 200 GPa and >300 GPa are shown in Figures 3.1 and 3.2. We observed Fe_3C up to 191.9 GPa and 4120 K. After heating to higher temperatures, the X-ray diffraction peaks of Fe_3C disappeared, and diffraction peaks assignable to Fe_7C_3 were observed at 192.3 GPa and 4310 K. After quenching, the diffraction peaks of Fe_3C reappeared. This indicates clearly incongruent melting of Fe_3C to Fe_7C_3 and liquid, as was reported in previous studies at low pressure around 7~14 GPa (Nakajima et al., 2009; Litasov et al., 2013; Walker et al., 2013). At higher pressure conditions, we observed X-ray diffraction peaks ascribable to Fe_3C up to 325 GPa at 1500 K and up to 343.1 GPa at 300 K (Figure 3.2). These results are consistent with preliminary data on the Fe_3C stability at 360 GPa and temperatures up to 5500 K (Tateno et al., 2010), in which a carbide was formed by reaction of Fe with diamond anvils. Although Liu et al. (2016) reported that Fe_3C decomposes to Fe and Fe_7C_3 at pressures above 160 GPa, we clearly observed the existence of Fe_3C up to 325 GPa and high temperature without decomposition, as shown in Figures 3.1 and 3.2. Chen et al. (2018) reported the sound velocities of Fe_3C up to core pressures. Our results together with previous experiments (Gao et al., 2011; Chen et al., 2014; Chen et al., 2018) indicate that Fe_3C with the cementite structure is likely to be stable in the inner core pressure and temperature conditions, although we need further detailed studies to confirm it. The experimental conditions and results are given in Table 3.2.

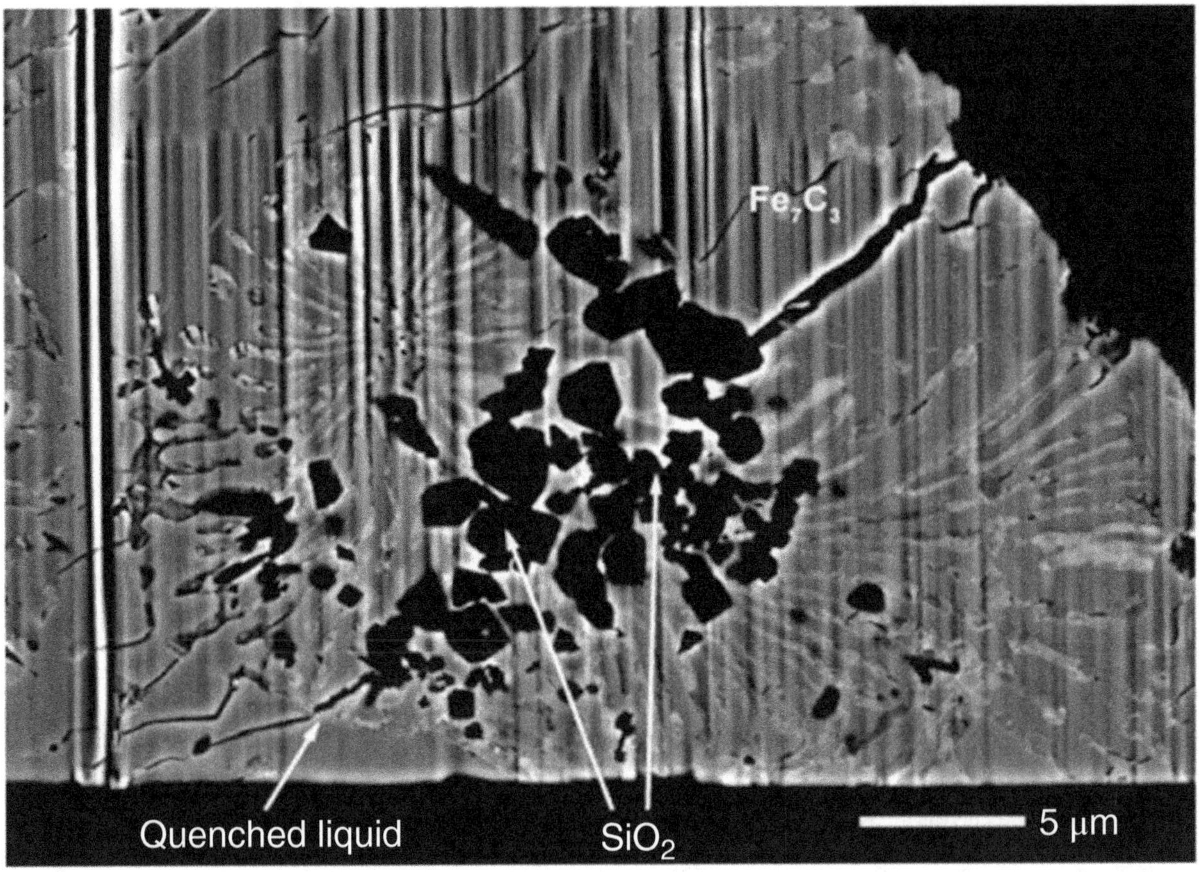

Figure 3.3 A backscattered electron microscope image of a sample recovered from 59.4 GPa and 2850 K. The area enclosed by the red dashed line shows a quenched liquid phase. SiO_2 was used as the pressure-transmitting medium. The quenched liquid is surrounded by Fe_7C_3 grains. The region outside Fe_7C_3 grains is composed of Fe_7C_3 (dark) and Fe_3C (bright) grains with granular textures. The compositions of the quenched liquid and Fe_7C_3 are given in Table 3.2. This image indicates incongruent melting of Fe_3C. See electronic version for color representation of the figures in this book.

3.3.2. Textual Observations and Chemical Analysis of the Recovered Sample

We performed textural observations of a recovered sample quenched from 59.4 GPa and 2850 K for a starting material with a bulk carbon content of 5.2 wt.%, which is slightly more enriched in Fe than the stoichiometric Fe_3C, as shown in Table 3.1. The quench texture of this run is shown in Figure 3.3. The carbon content in the quenched liquid phase was measured to be 4.7 wt.%, lower than the carbon content of the starting material. In addition, we observed Fe_7C_3 as a liquidus phase. This result confirms incongruent melting of Fe_3C to Fe_7C_3 and liquid at 59.4 GPa.

3.4. DISCUSSION

The experimental results are summarized in Figure 3.4. Our results of incongruent melting temperature of Fe_3C were fitted using Simon's equation as follows (Simon & Glatzel, 1929):

$$T_m = T_m^R \times \left(\frac{P - P^R}{A} + 1 \right)^{\frac{1}{C}}$$

where T_m and T_m^R are the melting temperature at an experimental pressure (P) and a reference pressure (P^R), respectively, and A and C are fitting parameters. This equation is employed for convenience of melting curve fitting following previous authors (Lord et al., 2009; Terasaki et al., 2011). The melting curve was fitted with the reference, P^R= 0 GPa and T_m^R = 1500 K GPa, which is the metastable incongruent melting temperature of Fe_3C at 1 bar (Benz & Elliott, 1961). The fitting parameters obtained here are $A = 18.9 \pm 3.3$ and $C = 2.4 \pm 0.2$. The incongruent melting temperature of Fe_3C was estimated to be 5130 ± 180 K at 330 GPa by extrapolation to the ICB pressure as shown in Figure 3.4. The fitted curve is consistent with the core isentropes and temperature at ICB estimated based on melting of mantle materials and

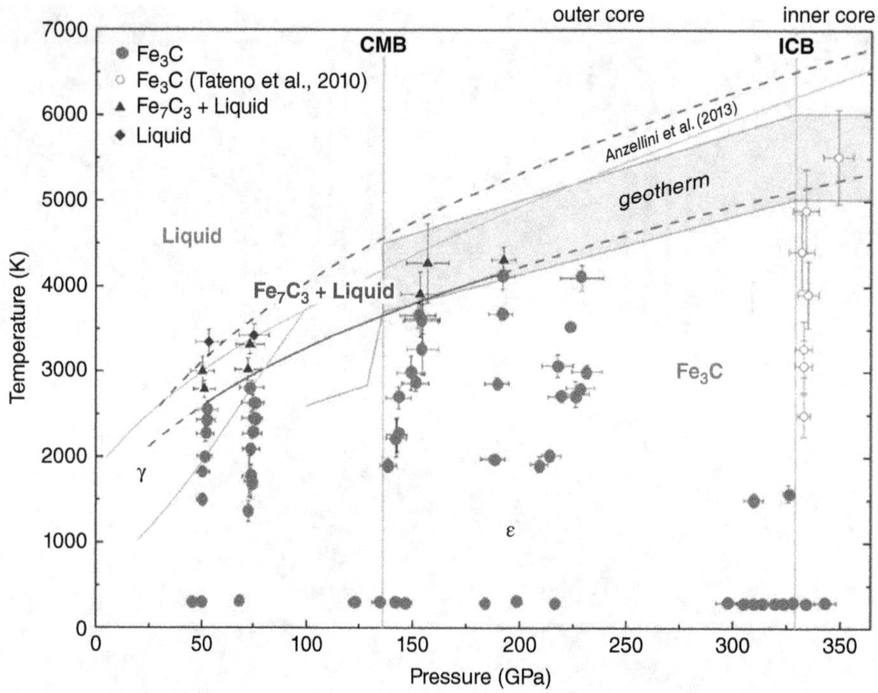

Figure 3.4 Phase diagram of Fe_3C. The solid circles, triangles, and diamonds represent Fe_3C, Fe_7C_3 + liquid, and liquid, respectively. The open circles show Fe_3C as products of the reaction between iron and the diamond anvil, as observed by Tateno et al. (2010). The grey lines show the melting and phase boundary of Fe (Anzellini et al., 2013). The solid curve is the solidus temperature fitted using Simon's equation (Simon & Glatzel, 1929). The curve represents the approximate boundaries between Fe_3C and Fe_7C_3 + liquid. The curve was fitted with the reference P^R = 0 GPa and T_m^R = 1500 K GPa (Benz & Elliott, 1961), and parameters A = 18.9 ± 3.3 and C = 2.4 ± 0.2. The liquidus curve is shown as a dashed curve, which is expressed by the parameters A = 18.9 and C= 2.4, which are assumed to be the same as the parameters of the solidus curve, and T_m^R = 1798 K and P^R= 7.7 GPa (Terasaki et al., 2014). The solidus temperature at the ICB is estimated to be 5130 K. The shaded area represents the temperature distribution of the core (Nimmo, 2015). See electronic version for color representation of the figures in this book.

energetics of the core, that is, heat flux from the core (Nimmo, 2015).

Figure 3.5 shows the phase and melting relationships of the Fe-C system from 14 GPa (Nakajima et al., 2009) to 200 GPa. The composition of the liquid formed by incongruent melting of Fe_3C at 60 GPa determined in this work (Table 3.1) is Fe-rich in composition compared to that at 14 GPa. The compositions of the peritectic liquids coexisting with Fe_7C_3 shift toward the Fe-rich side at pressures from 14 GPa to 60 GPa. This result indicates that the Fe-Fe_3C eutectic composition will also change to Fe-rich in composition, as has been shown in previous studies (Lord et al., 2009; Fei & Brosh, 2014) expanding compositional fields of liquidus Fe_3C and Fe_7C_3.

Lord et al. (2009) proposed that both hcp-Fe and Fe_7C_3 are stable as subsolidus phases and are candidates for the inner core. However, our results clearly reveal that Fe_3C is a stable carbon-bearing compound, although it melts incongruently into Fe_7C_3 and liquid at temperatures above 5130 ± 180 at 330 GPa (Figure 3.4). Therefore, both Fe_3C and Fe_7C_3 are potential stable carbon-bearing phases in the inner core.

Several authors have estimated the carbon content in the core. Wood (1993) estimated the concentration of carbon in the core to be 2–4 wt.% based on high solubility of carbon in liquid Fe and high abundance of carbon in C1 chondritic meteorites. McDonough (2003) estimated a carbon content of 0.2 wt.% from his estimation of bulk Earth composition and mass balance calculations between the mantle and core. Fiquet et al. (2009) estimated the carbon content in the core to be about 1 wt.% based on the sound velocity of Fe_3C and the preliminary reference Earth model (PREM) inner core. Mookherjee et al. (2011) estimated the maximum carbon content in the inner core to be 1.5 wt.% based on sound velocity measurements and ab initio calculations on Fe_7C_3. According to recent experiments at 1–3 GPa and 1773–2073 K, the partition coefficient of carbon between metal and silicate melts varies between 500 and 5400, which is different from previous estimations and

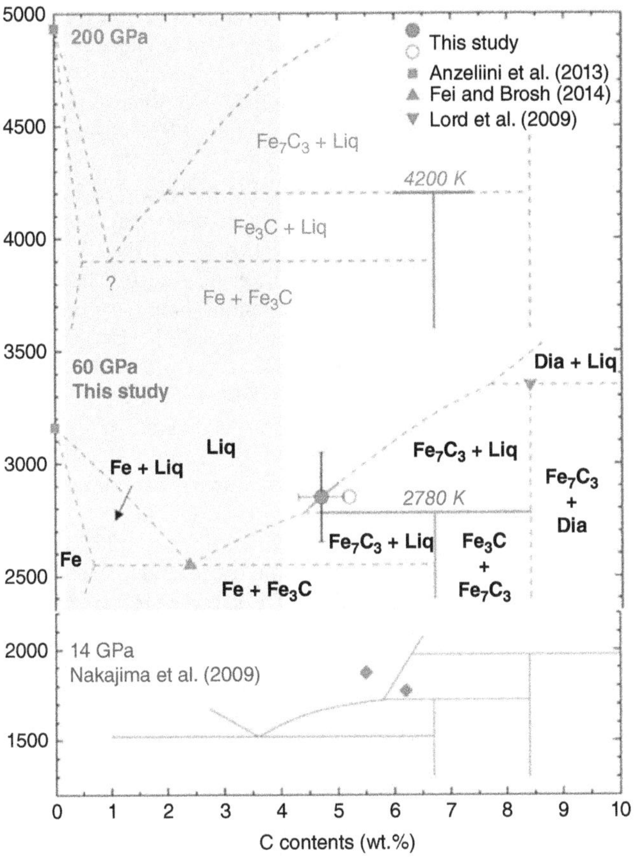

Figure 3.5 The Fe-rich portion of the Fe-C phase diagram up to 200 GPa inferred from our experiments in comparison with experiments and thermodynamic calculations at 14 GPa (Nakajima et al., 2009). The solid and open circles represent the carbon content of the quenched melt coexisting with Fe_7C_3 and the composition of the starting material in the quenching experiment at 60 GPa, respectively. The grey diamond symbols show the melt compositions at 14 GPa (Nakajima et al., 2009). The melting temperatures of Fe, solid squares, were based on the data from Anzellini et al. (2013). The eutectic composition at 60 GPa, a solid triangle, was estimated to have about 2.4 wt.% C based on the pressure shift of the Fe-Fe_3C eutectic composition by Fei and Brosh (2014). The peritectic temperature was estimated by the Simon's equation fitting of the experimental data (Figure 3.2). The shaded area represents the possible bulk carbon content in the Earth's core (Wood, 1993; McDonough, 2003; Fiquet et al., 2009). Compared to the Fe-C phase diagram at 14–60 GPa, the composition of the quenched melt coexisting with Fe_7C_3 is likely to be shifted toward the Fe-rich side with increasing pressure (Nakajima et al., 2009; Fei & Brosh, 2014). See electronic version for color representation of the figures in this book.

indicates strong partitioning to the metal, comparable with partitioning of highly siderophile elements (Chi et al., 2014). If the Earth was accreted from C1 chondrite parental bodies with high carbon content, up to 4 wt.% of carbon can be easily concentrated in the Earth's core. Therefore, the carbon content in the Earth's core is likely to be in the range 0.2–4 wt.%. The equation of state of Fe_7C_3 showed that the density of the inner core can well be explained by a single phase of Fe_7C_3 (Nakajima et al., 2011) or a mixture of 90-60 vol.% of Fe_7C_3 and 10-40% of hcp-Fe (Chen et al., 2012). The equation of state of Fe_3C determined previously (Sata et al., 2010; Litasov et al., 2013) indicates similar compression behaviors of Fe_3C to Fe_7C_3. Thus, the inner core can be explained by Fe_7C_3 or Fe_3C within the experimental uncertainty of the equation of state, assuming the ICB temperature ~5000 K. Finally, sound velocity measurements of Fe_7C_3 (Chen et al., 2014; Prescher et al., 2015) indicate that it can explain low V_S of the inner core without application of specific crystallization conditions (Mattesini et al., 2013; Yoshida et al., 1996) or premelting effects (Martorell et al., 2013). Indeed, Fe_3C is the next plausible compound to explain low V_S in the inner core (Gao et al., 2011; Gao et al., 2008).

The previous and present melting relationships of the Fe-Fe_3C system shown in Figure 3.5 indicate that the peritectic liquid coexisting with Fe_7C_3 shifts toward the Fe-rich side with increasing pressure. The eutectic liquid composition of the system is likely to shift toward a similar direction with increasing pressure. If Fe carbide is a liquidus phase during crystallization of the inner core, the initial crystallizing phase from the liquid core may be Fe_7C_3, followed by Fe_3C, and then by a mixture of hcp-Fe and Fe_3C at the latest stage. Crystallization of hcp-Fe and Fe_3C from the eutectic liquid provides no compositional convection in the outer core, because no light element is expelled by the crystallization. The latent heat due to crystallization will only be the energy source to create geodynamo in the outer core.

Thus, the distribution of these phases in the inner core could account for its internal structure, as revealed by seismology of the innermost inner core (Ishii & Dziewonski, 2002; Cao & Romanowicz, 2007). Our results suggest the innermost inner core of Fe_7C_3, which is consistent with a recent compression study (Nakajima et al., 2011; Chen et al., 2012). According to elasticity of Fe_7C_3 and Fe_3C determined by ab initio calculations, both carbides are highly anisotropic (Mookherjee et al., 2011; Mookherjee, 2011) and could explain anisotropy in the inner core even better than hcp-Fe. Caracas (2017), on the other hand, argued that the anisotropy of hcp-iron increases with increasing carbon content. Lai et al. (2018) showed experimentally a strong anisotropy of Fe_7C_3 up to 80 GPa and 800 K, and argued that high temperature may significantly influence the elastic anisotropy. A recent seismological study (Wang et al., 2015) suggested that both the inner core and the innermost inner core have different types of seismic anisotropy. If Fe_3C and Fe_7C_3 crystallized in the core, they could cause the anisotropic structure within the inner core. We should notice that detailed quantitative arguments on carbide layering in the inner core with its anisotropy is not possible at present due to insufficient experimental data on the phase relations of the Fe-C system under the inner core conditions and should be evaluated in future studies.

Crystallization of carbon-bearing phases would concentrate carbon in the inner core, resulting in a depletion of carbon in the outer core. Other light elements, such as H, O, Si, and S need to be present to satisfy the observed density of the outer core, since the outer core has a smaller density compared to the inner core. Although the multicomponent systems are not yet studied at high pressure to reveal liquidus phases during the inner core crystallization, carbides are very promising candidates if we compare melting temperatures of different compounds under the core conditions (FeSi by Lord et al., 2010; $Fe_{72.5}Si_{27.5}$ by Fischer et al., 2011; $Fe_{90.1}Si_{9.9}$ by Asanuma et al., 2010; $Fe_{75}O_5S_{20}$ by Terasaki et al., 2011). In the Fe-C-S system, iron carbide would be the liquidus phase, which could coexist with a sulfur-rich liquid, although the experiments are limited to 7 GPa (Dasgupta et al., 2009) and need to be confirmed at higher pressures.

ACKNOWLEDGMENTS

We thank I. Ohira, Y. Shibazaki, N. Nishitani, T. Sakairi, A. Suzuki, and M. Murakami for their technical assistance and useful discussions. We also appreciate K. D. Litasov for his advice to improve the manuscript. This work was supported by a Grant-in-Aid for Scientific Research from the Ministry of Education, Culture, Science, Sport, and Technology of Japan (Nos. 18104009, 22000002, and 15H05748) to EO. ST was supported by the Research Fellowship for Young Scientists of Japan Society for the Promotion of Science. The in-situ X-ray diffraction experiments were performed with the approval of the Japan Synchrotron Radiation Research Institute (JASRI) (Proposal Nos. 2012B1560; 2013A1496; 2013B0104; 2014A0104).

AUTHOR CONTRIBUTIONS

ST, EO, TS, IM, SK, and TS performed high-pressure experiments. ST, SO, MM, YI, KS, and NA carried out chemical analyses. NH and YO supported the in-situ XRD experiments at SPring-8. ST, EO, and TS wrote the paper. All authors discussed the results and commented on the manuscript.

REFERENCES

Akahama, Y., & Kawamura, H. (2004). High pressure Raman spectroscopy of diamond anvils to 250 GPa: Method for pressure determination in the multimegabar pressure range. *Journal of Applied Physics*, *96*, 3748–3751.

Anders, E., & Grevesse, N. (1989). Abundances of the elements: Meteoritic and solar. *Geochimica et Cosmochimica Acta*, *53*, 197–214.

Anderson, O. L., & Isaak, D. G. (2002). Another look at the core density deficit of Earth's outer core. *Physics of the Earth and Planetary Interior*, *131*, 19–27.

Anzellini, S., Dewaele, A., Mezouar, M., Loubeyre, P., & Morard, G. (2013). Melting of iron at Earth's inner core boundary based on fast X-ray diffraction. *Science*, *340*, 464–466.

Asanuma, H., Ohtani E., Sakai, T., et al. (2010). Melting of iron-silicon alloy up to the core-mantle boundary pressure: Implications to the thermal structure of the Earth's core. *Physics and Chemistry of Minerals*, *37*, 353–359.

Benz, M. G., & Elliott, J. F. (1961). The austenite solidus and revised iron-carbon diagram. *Transaction of AIME*, *221*, 323–331.

Birch, F. (1964). Density and composition of mantle and core. *Journal of Geophysical Research*, *69*, 4377–4388.

Campbell, A. J., Danielson, L., Righter, K., et al. (2009). High pressure effects on the iron–iron oxide and nickel–nickel oxide oxygen fugacity buffers. *Earth and Planetary Science Letters*, *286*, 556–564.

Campbell, A. J., Seagle, C. T., Heinz, D. L., et al. (2007). Partial melting in the iron-sulfur system at high pressure: A synchrotron X-ray diffraction study. *Physics of the Earth and Planetary Interior*, *162*, 119–128.

Cao, A., & Romanowicz, B. (2007). Test of the innermost inner core models using broadband PKIKP travel time residuals. *Geophysical Research Letters*, *34*, L08303.

Caracas, R. (2017). The influence of carbon on the seismic properties of solid iron, *Geophysical Research Letters*, *44*, 128–134. doi:10.1002/2016GL071109

Chen, B. Gao, L., Lavina, B., et al. (2012). Magneto-elastic coupling in compressed Fe_7C_3 supports carbon in Earth's inner core. *Geophysical Research Letters*, *39*, L18301.

Chen, B., Lai, X., Li, J. et al. (2018). Experimental constraints on the sound velocities of cementite Fe_3C to core pressures. *Earth and Planetary Science Letters*, *494*, 164–171.

Chen, B., Li., Z., Zhang, et al. (2014). Hidden carbon in Earth's inner core revealed by shear softening in dense Fe_7C_3. *Proceedings of the National Academy of Sciences of the United States of America*, *111*, 17755–17758.

Chi, H., Dasgupta, R., Duncan, M., & Shimizu, N. (2014). Partitioning of carbon between Fe-rich alloy melt and silicate melt in a magma ocean: Implications for the abundance and origin of volatiles in Earth, Mars, and the Moon. *Geochimica et Cosmochimica Acta*, *139*, 447–471.

Dasgupta, R., Buono, A., Whelan, G., & Walker, D. (2009). High-pressure melting relations in Fe-C-S systems: Implications for formation, evolution, and structure of metallic cores in planetary bodies. *Geochimica et Cosmochimica Acta*, *73*, 6678–6691.

Dasgupta, R., & Walker, D. (2008). Carbon solubility in core melts in a shallow magma ocean environment and distribution of carbon between the Earth's core and the mantle. *Geochimica et Cosmochimica Acta*, *72*, 4627–4641.

Dewaele A., Loubeyre, P., & Mezouar, M. (2004). Equations of state of six metals above 94 GPa. *Physical Review B*, *70*, 094112.

Dewaele, A., Loubeyre, P., Occelli, M., et al. (2006). Quasihydrostatic equation of state of iron above 2 Mbar. *Physical Review Letters*, *97*, 215504–215507.

Dubrovinsky, L. S., Saxena, S. K., Tutti, F., et al. (2000). In situ X-ray study of thermal expansion and phase transition of iron at multimegabar pressure. *Physical Review Letters*, *84*, 1720–1723.

Dziewonski, A. M., & Anderson, D. L. (1981). Preliminary reference Earth model. *Physics of the Earth and Planetary Interior*, *25*, 297–356.

Fei, Y., & Brosh, E. (2014). Experimental study and thermodynamic calculations of phase relations in the Fe-C system at high pressure. *Earth and Planetary Science Letters*, *408*, 155–162.

Fei, Y., Ricolleau, A., Frank, M., et al. (2007). Toward an internally consistent pressure scale. *Proceedings of the National Academy of Sciences of the United States of America*, *104*, 9182–9186.

Fiquet, G., Badro, J., Gregoryanz, E., Fei, Y., & Occelli, F. (2009). Sound velocity in iron carbide (Fe_3C) at high pressure: Implications for the carbon content of the Earth's inner core. *Physics of the Earth and Planetary Interiors*, *172*, 125–129.

Fischer, R. A., Campbella, A. J., Reamana, D. M., et al. (2011). Phase relations in the Fe-FeSi system at high pressures and temperatures. *Earth and Planetary Science Letters*, *373*, 54–64.

Gao, L., Chen, B., Wang, J. et al. (2008). Pressure-induced magnetic transition and sound velocities of Fe_3C: Implications for carbon in the Earth's inner core. *Geophysical Research Letters*, *35*, L17306, doi:17310.11029/12008gl034817

Gao, L., Chen, B., Zhao, J. et al. (2011). Effect of temperature on sound velocities of compressed Fe_3C, a candidate component of the Earth's inner core. *Earth and Planetary Science Letters*, *309*, 213–220.

Hirayama, Y., Fujii, T., & Kurita, K. (1993). The melting relation of the system, iron and carbon at high pressure and its bearing on the early stage of the Earth. *Geophysical Research Letters*, *20*, 2095–2098.

Ishii, M., & Dziewonski, A. M. (2002). The innermost inner core of the Earth: Evidence for a change in anisotropic behavior at the radius of about 300 km. *Proceedings of the National Academy of Sciences of the United States of America*, *99*, 14026–14030.

Kamada, S., Terasaki, H., Ohtani, E., et al. (2010). Phase relationships of the Fe-FeS system in conditions up to the Earth's outer core. *Earth and Planetary Science Letters*, *294*, 94–100.

Kuwayama, Y., Hirose, K., Sata, N., & Ohishi, Y. (2011). Pressure-induced structural evolution of pyrite-type SiO_2. *Physics and Chemistry of Minerals*, *38*, 591–597.

Lai, X., Zhur, F., Liu, J. et al. (2018). The high-pressure anisotropic thermoelastic properties of a potential inner core carbon-bearing phase, Fe_7C_3, by single-crystal X-ray diffraction. *American Mineralogist*, *103*, 1568–1574. https://doi.org/10.2138/ am-2018-6527

Litasov, K. D., Sharygin, I. S., Dorogokupets, P. I., et al. (2013) Thermal equation of state and thermodynamic properties of iron carbide Fe_3C to 31 GPa and 1473 K. *Journal of Geophysical Research: Solid Earth*, *118*, 5274–5284.

Liu, J., Lin, J.-F., Prakapenka, V. B., et al. (2016). Phase relations of Fe_3C and Fe_7C_3 up to 185GPa and 5200K: Implication for the stability of iron carbide in the Earth's core. *Geophysical Research Letters*, *43*. https://doi:10.1002/2016GL071353

Lord, O. T., Walter, M. J., Dasgupta, R., et al. (2009). Melting in the Fe-C system to 70 GPa. *Earth and Planetary Science Letters*, *284*, 157–167.

Lord, O. T., Walter, M. J., Dobson, D. P., et al. (2010). The FeSi phase diagram to 150 GPa. *Journal of Geophysical Research*, *15*, B06208. https://doi:10.1029/2009JB006528.

Mao, H. K., Wu, Y., Chen, L. C., & Shu, J. F. (1990). Static compression of iron to 300 GPa and $Fe_{0.8}Ni_{0.2}$ alloy to 260 GPa: Implications for composition of the core. *Journal of Geophysical Research*, *95*, 21737–21742.

Martorell, B., Vočadlo, L., Brodholt, J., & Wood, I.G. (2013). Strong premelting effect in the elastic properties of hcp-Fe under inner-core conditions. *Science*, *342*, 466–468.

Mattesini, M., Belonoshko, A., Tkalčić, H., et al. (2013). Candy wrapper for the Earth's inner core. *Scientific Reports*, *3*, 2096. https://doi: 2010.1038/srep02096

McDonough, W. F. (2003). Compositional model for the Earth's core. In R. W. Carlson, (Ed.), *Treatise on Geochemistry 2* (pp. 547–568). Amsterdam: Elsevier.

Miyahara, M., Sakai, T., Ohtani, E., et al. (2008). Application of FIB system to ultra-high-pressure Earth science. *Journal of Mineralogical and Petrological Sciences*, *103*, 88–93.

Mookherjee, M. (2011). Elasticity and anisotropy of Fe_3C at high pressures. *American Mineralogist*, *96*, 1530–1536.

Mookherjee, M., Nakajima, Y., Steinle-Neumann, G., et al. (2011). High pressure behaviour of iron carbide (Fe_7C_3) at inner core conditions. *Journal of Geophysical Research*, *116*, B04201.

Morard, G., Andrault, D., Guignot, N., et al. (2008). In situ determination of Fe-Fe_3S phase diagram and liquid structural properties up to 65 GPa. *Earth and Planetary Science Letters*, *272*, 620–626.

Nakajima, Y., Takahashi, E., Sata, N., et al. (2011). Thermoelastic property and high-pressure stability of Fe_7C_3: Implication for iron-carbide in the Earth's core. *American Mineralogist*, *96*, 1158–1165.

Nakajima, Y., Takahashi, E., Suzuki, T., & Funakoshi, K. (2009). "Carbon in the core" revisited. *Physics of the Earth and Planetary Interior*, *174*, 202–211.Nimmo, F. (2015). Energetics of the core. In G. Schubert (Ed.), *Treatise on Geophysics* (2nd ed.). Amsterdam: Elsevier.

Ohishi, Y., Hirao, N., Sata, N., Hirose, K., & Takata, M. (2008). Highly intense monochromatic X-ray diffraction facility for high-pressure research at SPring-8. *High Pressure Research*, *28*, 163–173.

Poirier, J. P. (1994). Light elements in the earth's outer core: A critical review. *Physics of the Earth and Planetary Interior*, *85*, 319–337.

Prescher, C., Dubrovinsky, L., Bykova, E., et al. (2015). High Poisson's ratio of Earth's inner core explained by carbon alloying. *Nature Geoscience*, *8*, 220–223.

Sata, N., Hiose, K., Hirose, K., et al. (2010). Compression of FeSi, Fe_3C, $Fe_{0.95}O$, and FeS under the core pressures and implication for light element in the Earth's core. *Journal of Geophysical Research*, *115*, B09204.

Shanker, J., Singh, B. P., & Srivastava, S. K. (2004). Volume-temperature relationship for iron at 330 GPa and Earth's core density deficit. *Physics of the Earth and Planetary Interior*, *147*, 333–341.

Shen, G., Mao, H. K., & Hemley, R. J. (1996). Laser-heated diamond anvil cell technique: Double-sided heating with multimode Nd:YAG laser. In *Advanced Materials '96* (pp. 149–152) (Proc. 3rd NIRIM International Symposium on Advanced Materials).

Simon, F., & Glatzel, G. (1929). Remarks on fusion pressure curve. *Zeitschrift für anorganische und allgemeine Chemie*, *178*, 517–522.

Tateno, S. Hirose, K., Ohishi, Y., & Tatsumi, Y. (2010). The structure of iron in Earth's inner core. *Science*, *330*, 359–361.

Terasaki, H., Kamada, S., Sakai, T., et al. (2011). Liquidus and solidus temperatures of a Fe-O-S alloy up to the pressures of the outer core: Implication for the thermal structure of the Earth's core. *Earth and Planetary Science Letters*, *394*, 559–564.

Terasaki, H. Shibazaki, Y., Nishida, K. et al. (2014). Repulsive nature for hydrogen incorporation to Fe_3C up to 14 GPa. *ISIJ International*, *54*, 2637–2642.

Vočadlo, L., Brodholt, J., Dobson, D. P., et al. (2002). The effect of ferromagnetism on the equation of state of Fe_3C studied by first-principles calculations. *Earth and Planetary Science Letters*, *203*, 567–575.

Walker, D., Dasgupta, R., Li, J. & Buono, A. (2013). Nonstoichiometry and growth of some Fe carbides. *Contribution to Mineralogy and Petrology*, *166*, 935–957.

Wang, T., Song, X., & Xia, H. H. (2015). Equatorial anisotropy in the inner part of Earth's inner core from autocorrelation of earthquake coda. *Nature Geoscience*, *8*, 224–227.

Wood, B. J. (1993). Carbon in the core. *Earth and Planetary Science Letters*, *117*, 593–607.

Yoshida, S., Sumita, I.,& Kumazawa, M. (1996). Growth model of the inner core coupled with the outer core dynamics and the resulting elastic anisotropy. *Journal of Geophysical Research: Solid Earth*, *101*, 28085–28103.

4

Structure and Properties of Liquid Fe-C Alloys at High Pressures by Experiments and First-Principles Calculations

Bin Chen[1] and Jianwei Wang[2]

ABSTRACT

It is believed that the core formation processes sequestered a large majority of Earth's carbon into its metallic core. Incorporation of carbon to liquid iron may significantly influence its properties under physicochemical conditions pertinent to the deep magma ocean and thus the chemical evolution of terrestrial planets and moons. Compared to available experimental data on the physical properties of crystalline iron alloys under pressure, there is a remarkable lack of data on the properties of liquid iron-rich alloys, due to experimental challenges. Here we review experimental and computational results on the structure and properties of iron or iron-nickel liquids alloyed with carbon upon compression. These laboratory data provide an important foundation on which the interpretation of ultrahigh pressure laboratory data and the verification of theoretical data will have to be based. The low-pressure data can be used to validate results from theoretical calculations at the same conditions, and high-pressure calculations can be used to estimate and predict liquid properties under core conditions. Availability of the liquid properties of Fe-C liquids will provide essential data for stringent tests of carbon-rich core composition models for the outer core.

4.1. INTRODUCTION

As the remotest and most dynamic region of the planet and accounting for 18% of the total planetary volume, the liquid core of the Earth is estimated to be mostly composed of iron (Fe) and 5% to 10% nickel (Ni), with a minor fraction of light impurities that lower its density and melting point with respect to pure Fe (Birch, 1964; Li & Fei, 2014; McDonough & Sun, 1995; Poirier, 1994). The estimates of the core density deficit relative to solid Fe vary between 6% and 10% for the outer core (O. L. Anderson & Isaak, 2002; Anderson & Ahrens, 1994; Fei et al., 2016; Jephcoat & Olson, 1987; Shanker et al., 2004; Stevenson, 1981; Stixrude, Cohen, et al., 1997). Knowledge on the nature and abundance of the light alloying elements (S, O, C, Si, H) in the core has significant implications for planetary accretion and differentiation history, bulk composition and chemical evolution of the planet, and dynamic processes in the outer core for the generation of the Earth's magnetic field (Li & Fei, 2014, and references therein; Stevenson, 2010). Furthermore, the composition of the core holds the key for answering the fundamental questions in geophysics and planetary sciences: "How has the core evolved to its current state?" and "What is the role of the core in the planet's internal processes and dynamics?" As in the Earth, the presence of liquid cores has been identified in other planets such as Mercury (Margot et al., 2007) and Mars (Fei & Bertka, 2005), as well as the moons of planets in the solar system such as the Moon

[1] *Hawaii Institute of Geophysics and Planetology, University of Hawaii at Manoa, Honolulu, Hawaii, USA*

[2] *Department of Geology and Geophysics, Louisiana State University, Baton Rouge, Louisiana, USA*

Carbon in Earth's Interior, Geophysical Monograph 249, First Edition. Edited by Craig E. Manning, Jung-Fu Lin, and Wendy L. Mao.

(Weber et al., 2011) and Ganymede (Bland et al., 2008; Hauck et al., 2006). The advent of space exploration of terrestrial-like planets and moons also gives rise to tremendous interest in the internal structures of these bodies, and particularly the possible existence of cores and their composition and internal dynamics.

Carbon is the fourth most abundant element in the solar system but is depleted by three orders of magnitude in the silicate Earth. Most of Earth's carbon resides in the deep interior, but the fact that carbon is readily outgassed makes these estimates difficult and controversial. A significant fractionation of Earth's carbon may have partitioned into the core during differentiation (Chen & Li, 2016; Dasgupta & Walker, 2008; Nakajima et al., 2009). Carbon has been suggested to be among the principal light elements in the core, partly due to its cosmochemical abundance, occurrence of Fe carbide phases in meteorites, and high affinity and solubility in Fe-Ni liquids during core-mantle differentiation (Chen & Li, 2016; Wood, 1993; Wood et al., 2013). Recent advances in high-pressure and synchrotron techniques have permitted both static and dynamic high-pressure experiments on crystalline iron alloys under pressure-temperature (P-T) conditions found at the center of the Earth (Li & Fei, 2014). For the solid inner core, the phase stability, thermal equation-of-state, and sound velocities of relevant solid Fe-carbide phases, under or approaching core conditions, have been extensively investigated (Chen et al., 2012, 2018; Fiquet et al., 2009; Gao et al., 2008; Lai et al., 2018; Liu et al., 2016; Mookherjee, 2011; Mookherjee et al., 2011; Nakajima et al., 2009; Prescher et al., 2012, 2015; Sata et al., 2010). In contrast, for much of the liquid outer core, such experimental investigations and discussion are still significantly lacking, largely due to the huge gap in the P-T space accessible to experiment at actual core conditions. Furthermore, there are very limited experimental data on thermoelastic properties of Fe-C liquids even at low pressures (Kuwabara et al., 2015; Lai et al., 2017; Shibazaki et al., 2015; Shibazaki & Kono, 2018; Terasaki et al., 2006, 2010). These properties of iron-light-element liquids are usually investigated under 10 GPa in the laboratory, far below the expected conditions of the outer core.

The elastic and viscoelastic properties of liquid Fe alloys, such as the density, sound velocities, attenuation, and viscosity, are a prerequisite for the interpretation of the behavior of the outer core and for constructing core composition models. To account for the density deficit for the outer core, and to understand buoyancy effects accompanying inner core crystallization and hence core dynamics, it is critical to measure the density of Fe-rich liquids under high P-T conditions in the laboratory. Recent advances in experimental techniques allow measurement of the liquid properties at relatively low pressures. For instance, using sink/float or X-ray absorption methods, the densities of several liquid Fe-rich alloys have been measured under static high-pressure conditions (mostly <10 GPa), i.e. Fe-S liquids (Nishida et al., 2008; Sanloup, Guyot, Gillet, Fiquet, Mezouar, et al., 2000; Tateyama et al., 2011), Fe-C liquids (Sanloup et al., 2011; Terasaki et al., 2010), and Fe-Si liquids (Sanloup et al., 2004; Tateyama et al., 2011; Yu & Secco, 2008) (see Table 4.1).

Likewise, recent advances in high-performance computation permit calculations of structure, equations of state, viscosity, and sound velocity of melts at high temperatures and pressures, using first-principles molecular dynamics (MD) simulations (Lai et al., 2017; Stixrude, Cohen, et al., 1997). Simulations of systems of in the order of 1000 atoms and up to tens of pico-seconds become practical and are expected to be adequate to calculate the properties of the melts with acceptable accuracies. First-principles simulations of Fe-Ni-C liquids are, however, limited in the literature (Alfé, 2010; Alfé et al., 2000; Belashchenko, 2014; Côté et al., 2012; Desjarlais, 2013; Lai et al., 2017; Sobolev & Mirzoev, 2013; Vočadlo et al., 2000; Zhang & Yin, 2012). In concert with current high P-T experimental capabilities, it is desirable to systematically investigate the density, sound velocity, viscosity, and structure of Fe-Ni-C liquids under moderate pressures by experiments and approaching core conditions by theoretical calculations, an approach to avoid long extrapolations of low-pressure properties in order to have a better understanding of the carbon inventory in planetary cores. The ambient and low-pressure experimental data can serve as a baseline for higher pressure materials properties and can be directly applied to the modeling for planetary cores of smaller bodies. The low-pressure experimental data can also be used to benchmark the systematic biases that often occur

Table 4.1 Existing studies on density of liquid iron and its alloys with carbon.

Source	Composition	P-T Range	Method
Jimbo & Cramb (1993)	Fe-C liquid (0–4 wt% C)	0 GPa, 1523–1823 K	Sessile drop profile
Terasaki et al. (2010)	Fe_3C	3.6–9.5 GPa, 1973 K	X-ray absorption
Sanloup et al. (2011)	Fe-5.7 wt% C	2–8 GPa, 2273 K (corrected)	X-ray absorption
Shimoyama et al. (2013)	Fe-3.5 wt% C	≤6.8 GPa, 2200 K	X-ray absorption
Morard et al. (2017)	Fe-C liquid (2.5–3.0 wt% C)	6–58 GPa, 3200 K	Diffuse scattering

in the first-principles calculations, such as the tradeoffs between the accuracy and duration of the simulations. After being benchmarked against lower pressure experimental results, first-principles calculations can then be effectively and reliably carried out to explore more relevant P-T ranges for the core.

In this chapter, we first discuss recent results on the structure of the Fe-Ni-C liquids at high pressures. We then discuss the structural effects on the elastic and viscoelastic properties of Fe-Ni-C liquids, from the perspectives of laboratory experiments and computational techniques, in order to establish a systematic dataset on the density, sound velocity, viscosity, and structure of the liquids in a previously unexplored pressure-temperature-composition sector.

4.2. STRUCTURE OF FE-C AND FE-NI-C LIQUIDS

One of the most remarkable observations in physical sciences is that most substances (liquid or solid state) with a well-defined chemical composition can transform from one structure to another simply by varying thermodynamic conditions such as pressure and temperature. Physical properties of the substances on the mascroscopic scale are intimately controlled by their structure at the atomistic scale. In planetary sciences, knowledge of the structural evolution of iron liquids as core-forming materials upon compression is essential for understanding the evolution and dynamics of terrestrial planets and moons, as the liquid properties are governed by the liquid structure. However, the pressure-induced structural transitions of iron liquids have not been thoroughly investigated, largely due to the subtle changes of the liquid structure and the ambiguities of the reported liquid structural data in most previous studies (Sanloup, Guyot, Gillet, Fiquet, Hemley, et al., 2000; Shen et al., 2004; Shibazaki et al., 2015).

4.2.1. Experimental Results

Based on X-ray diffraction measurements and the appearance of two distinct subpeaks of the second peak in the experimental pair distribution function (PDF), Sanloup, Guyot, Gillet, Fiquet, Hemley, et al. (2000) reported a structural change in liquid Fe from *bcc*-like to *bcc*-*fcc*-like local order in the vicinity of the δ-γ-liquid triple point at ~5 GPa. This splitting, however, was not observed in the other two measurements of liquid Fe up to 6.4 GPa (Kono et al., 2015) and 58 GPa (Shen et al., 2004). The incorporation of carbon in liquid Fe may promote the structural transition. A liquid structural transition in Fe-3.5 wt%C liquid was inferred from peak distance changes in their experimentally determined PDFs at ~5 GPa (Kono et al., 2015). Structural transitions in liquids have also been inferred from changes in physical and chemical behaviors of liquid Fe alloys at high pressures. Both Fe-5.7 wt%C (Sanloup et al., 2011) and Fe-3.5 wt%C (Shimoyama et al., 2013) liquids showed marked changes in their densities at ~5 GPa. The structural transition in the metallic liquids was also thought to be intimately related to the observed nonlinear systematics in the metal-silicate partitioning behaviors for trace elements (i.e., Ni, Co, and W) (Sanloup et al., 2011). Compared to the studies of solid Fe-carbides (Chen et al., 2014; Fang et al., 2009; Fiquet et al., 2009; Gao et al., 2008, 2011; Li et al., 2002; Lin et al., 2004; Litasov et al., 2014; Litasov et al., 2013; Mookherjee, 2011; Mookherjee et al., 2011; Prescher et al., 2015; Sata et al., 2010; Scott et al., 2001), relatively limited experimental work has been conducted on the Fe-C or Fe-Ni-C liquids. Occurrence of structural changes in liquids Fe-alloys in the P-T-X space such as a polyamorphic transition at ~5 GPa inevitably makes the extrapolation of their physical properties to core conditions irrelevant. Therefore, it is crucial to probe the high P-T structure of Fe-rich alloys considered for the outer core, in order to profoundly understand their elastic and viscoelastic properties and geochemical consequences, e.g., the partitioning coefficients of elements between liquid silicates and liquid metallic phase (Sanloup et al., 2011).

Recently, the structure change of Fe or Fe-Ni liquid alloyed with various amounts of carbon has been experimentally investigated at high pressures (Lai et al., 2017; Shibazaki et al., 2015). Based on multiangle energy-dispersive X-ray diffraction measurements, a liquid structure transition at ~5 GPa was inferred from the observations of peak distance changes in experimentally determined PDFs in Fe-3.5 wt% C liquids (Shibazaki et al., 2015) and $Fe_{90}Ni_{10}$-5 wt%C and $Fe_{90}Ni_{10}$-3 wt% C (Lai et al., 2017). Notably, the derived PDFs for $Fe_{90}Ni_{10}$-5 wt% C and $Fe_{90}Ni_{10}$-3 wt% C liquids up to 7.3 GPa and 1773 K show changes to the rich features of the second nearest neighbors peaks at ~5 GPa, which were used as primary indication of a polyamorphic liquid structural transition for both liquids (Figure 4.1) (Lai et al., 2017). Unlike Si, which is found to be substitutionally incorporated into Fe, carbon, as well as a small amount of S, is alloyed with liquid Fe through interstitial incorporation (Shibazaki & Kono, 2018). Detailed structural information of the liquids, however, could not be directly derived solely from the experimental data. A model analysis based on atomic packing motifs is consistent with the result from first-principles MD calculations, which suggest that the transition is related to the three-atom polyhedral connection of Fe-Ni atomic motifs in the high-pressure liquids at >5 GPa (Figure 4.1 and Figure 4.2) (Lai et al., 2017). Such a change in intermediate range structure has a significant effect on the properties

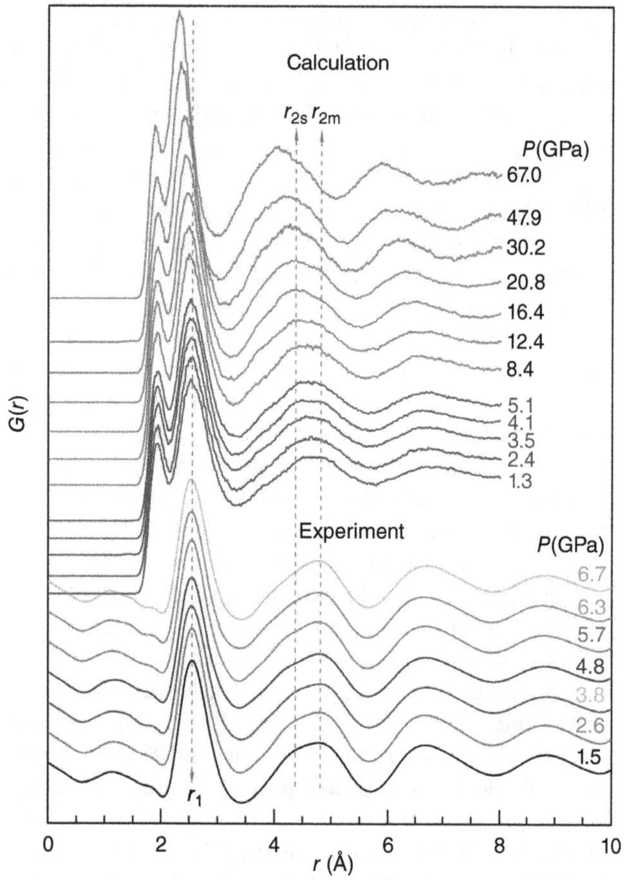

Figure 4.1 Reduced pair distribution function $G(r)$ of $Fe_{90}Ni_{10}$-5 wt% C liquid at high pressures and temperatures by experiments up to 7 GPa (Lai et al., 2017) and molecular dynamics calculations up to 67 GPa. The dashed lines with arrows serve as guides for the r_1 and r_2 (r_{2s}, r_{2m}) positions of $G(r)$ of the $Fe_{90}Ni_{10}$-5 wt% C liquid at low pressures. See electronic version for color representation of the figures in this book.

of the liquid because a three-atom connection (a triangle) between polyhedra could geometrically induce a structural and dynamical stability (Figure 4.2).

4.2.2. Computational Results

First-principles MD simulation shows that short-range order of carbon atoms in Fe-C liquids changes significantly with carbon concentration (0 to 5 wt% carbon), varying from simple carbon solution in iron to carbon forming local sites like the distorted prismatic sites in Fe_3C (Sobolev & Mirzoev, 2013). MD simulations indicate that liquid structure and transition are not affected by replacing Fe with 10 at% Ni in the $Fe_{90}Ni_{10}$-5 wt% C liquid (Lai et al., 2017). For the $Fe_{90}Ni_{10}$-5 wt% C liquid, the short-range structure in the pressure range from 0–67 GPa (Figure 4.2) shows the coordination number in the Fe/Ni-Fe/Ni subsystem around 12, indicative of a nearly close-packed structure. The coordination number in the C-Fe/Ni subsystem increases from 6.5 to 8.5, indicating an approximately octahedral to cubic transition. The Fe/Ni-Fe/Ni bond length is significantly compressed as compared to the C-Fe/Ni bond. The mixing of Fe, Ni, and C are not ideal, which is based on their pair distribution functions. The preferential clustering between the atoms is pressure dependent. Such subtle miscibility changes with pressure between the elements are possibly related to changes in the local chemical and structural environment but may particularly influence other physical properties. Based on the analysis of the partial configurationally decomposed distribution function, the intermediate-range structures of both Fe/Ni-Fe/Ni and C-Fe/Ni subsystems change substantially, which can be characterized by a significant increase of three-atom shared polyhedra. The computational results suggest a close correlation between the structural change and an increased bulk modulus, decreased diffusion coefficient, and increased shear viscosity from the polyamorphic transition. Both miscibility and clustering of the liquid component elements are structural characteristics that appear to have a significant impact on the physical properties of the liquid.

The synergy between experiments and calculations provides critical structural information about the liquids for the interpretation and understanding of the intrinsic nature of the pressure-induced structural rearrangements of the liquids and the resultant effects on the liquid properties. The structural changes of liquid iron alloys upon compression, as exemplified by the Fe-Ni-C liquids in this study and others, can have a significant effect on the liquid properties, such as viscosity, sound velocity, and density. The liquid structure change may also influence the partitioning behaviors of major or trace elements and isotopic fractionation between metal and silicates during planetary core formations. Knowledge of the equation of the state and polyamorphic transitions in core-forming liquid iron alloys is thus crucial for our understanding of the chemical evolution of terrestrial planets and moons.

4.3. RELATIONSHIP BETWEEN PROPERTY AND STRUCTURE AND GEOPHYSICAL IMPLICATIONS

Incorporation of light elements such as carbon into liquid Fe is known to have significant effect on the macroscopic properties of the liquids, which are dictated by their atomic-scale structures at high pressures (Shibazaki & Kono, 2018). Light elements are known to reduce the densities of their alloys with liquid Fe at high pressures (Morard et al., 2014; Nishida et al., 2008; Nishida et al., 2016; Sanloup et al., 2004; Shimoyama et al., 2013; Tateyama et al., 2011; Terasaki et al., 2010). For Fe-C liquids, the densities of Fe-5.7 wt% C (Sanloup et al., 2011) and Fe-3.5 wt% C (Shimoyama et al., 2013) both showed marked changes at ~5 GPa (Figure 4.3), which is the pressure at which both experiments and computations

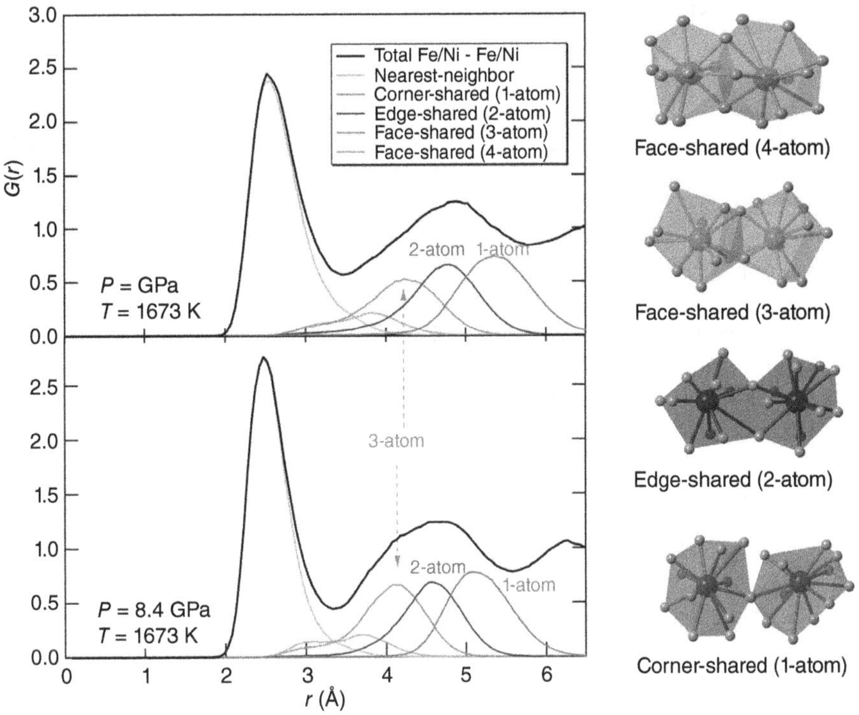

Figure 4.2 The total pair distribution functions (PDFs) of the Fe/Ni-Fe/Ni pair in the $Fe_{91}Ni_9$-5 wt% C liquid and the decomposed PDFs for the second nearest neighbor shells at (a) 1.3 GPa, 1673 K and (b) 8.4 GPa, 1673 K from MD simulations, modified after Lai et al. (2017). The color-coded polyhedra are 1-atom corner-shared (red), 2-atom edge-shared (blue), 3-atom face-shared (magenta), and 4-atom distorted face-shared (green) connections of Fe/Ni atom with their second nearest neighbors. See electronic version for color representation of the figures in this book.

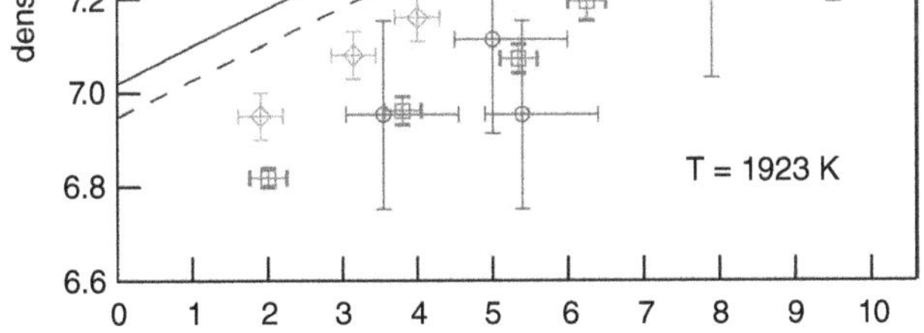

Figure 4.3 The effect of carbon alloying with Fe on the density of Fe-C liquids, in comparison with density of liquid Fe (W. W. Anderson & Ahrens, 1994). See electronic version for color representation of the figures in this book.

inferred a liquid structural transition (Lai et al., 2017; Shibazaki et al., 2015). Effect of the liquid atomic structure on the properties is also reflected on the equation of the state in a subtle manner. Two separate piece-wise fittings of the data of pressure as a function of volume of $Fe_{90}Ni_{10}$-5 wt% C liquid at pressures from 0 to 5 GPa and 5 to 67 GPa result in a better overall fitting statistic than fitting of the data in whole pressure range from 1 to 67 GPa, consistent with the liquid-liquid polyamorphic transition at a volume of ~11 Å^3/atom around ~5 GPa. Furthermore, the residual fluctuation from the fitting of a single Birch-Murnaghan equation of state are closely correlated with the coordination number of three-atom shared polyhedra truncates in the pressure range, which suggests that there is a direct correlation between the intermediate-range structure of the liquid and fluctuations in its equation of state. This correlation is further highlighted by the fact that a greater number of three-atom shared polyhedra at higher pressures is correlated to a higher bulk modulus at smaller volumes based on the result of the piece-wise fitting of Birch-Murnaghan equation of state.

On the other hand, incorporation of light elements such as C in the liquid Fe may increase or decrease the V_P (Shibazaki & Kono, 2018). Substitutional incorporation of Si in liquid Fe as characterized by the shortening r_1 and r_2 distances in the PDFs may increase the V_P. In contrast, the expansion of r_1 and r_2 distances by the interstitial incorporation of C and a small amount of S may decrease the V_P of the liquids (Shibazaki & Kono, 2018). Due to the challenges of experiments on liquids, most experiments on liquid properties and structure were conducted at low pressures (<10 GPa) and the above structure V_P relationship may not be valid at higher pressure conditions. Indeed, V_P of liquid Fe-16 at% C was reported to be higher than that of pure liquid Fe (Nakajima et al., 2015), contradictory to the empirical relationship determined at lower pressures (Shibazaki & Kono, 2018).

The distinct effect of C on the r_1 and r_2 distances of liquid Fe and the structural evolution with pressure may also control other physical and chemical properties of liquid Fe. Viscosity of liquid Fe alloys at high pressure is another important physical property that is closely related to the internal dynamics of the outer core. Estimates of the viscosity of Earth's outer core varies in the orders of magnitude, particularly between observation and theory (Secco, 1995). Laboratory measurements indicated that the interstitial incorporation C into liquid Fe has little effect on the viscosity at <5 GPa (Terasaki et al., 2006). The structural evolution of the Fe-Ni-C liquids as explained by the increased fraction of three-atom cluster connections at ~5 GPa may significantly affect the viscosity of the liquids, which still awaits further experimental verification. For small terrestrial planets or moons, i.e. Mercury, Ganymede, and Moons, the structural change of the Fe-C liquids may occur at depths in the liquid portion of their cores. The resultant changes in liquid properties such as density and viscosity may lead to distinct layered stratifications and convection in the liquid cores.

The structural change of the Fe-Ni-C liquids may also influence chemical properties such as partitioning behaviors of major or trace elements and isotopic fractionation between metal and silicates during the formation of planetary cores. The nonlinear systematics in the metal-silicate partitioning behavior for trace elements (i.e., Ni, Co, and W) were also assumed to result from the polyamorphic transition, which was inferred from a liquid compressibility change in the Fe-C liquid (Sanloup et al., 2011) and is consistent with the experimental and computational study on the structural change of Fe-C liquid at ~5 GPa (Lai et al., 2017; Shibazaki et al., 2015). The structural change may also render the distinct geochemical imprints (i.e., iron and carbon isotopes) of the bulk silicate Earth left behind by the core formation: bulk silicate Earth carbon is isotopically distinct (–5 ‰) from that in other terrestrial planets or planetary bodies (i.e., Mars and Vesta) and primitive undifferentiated meteorites (chondrites) (–20 ± 5 ‰) (Wood et al., 2013). The knowledge of the structural changes and potential polyamorphic transitions in core-forming liquid iron alloys is thus crucial for our understanding of the chemical evolution of terrestrial planets and moons (Lai et al., 2017).

4.4. CONCLUSIONS

An outer core compositional model requires reliable material properties of iron-nickel liquids alloyed with lighter elements under core conditions. Probing the high P-T behavior of the iron-rich liquids through synergies between laboratory experiments and theoretical calculations could help improve our understanding of the physics and chemistry of the core, particularly of the liquid outer core. This approach provides a tractable methodology and offers a new and practical interplay between experiments and first-principles-based modeling for studying the liquid properties under outer core conditions that are difficult to reach with the current static high-P capabilities. A key component of this methodology is that all details of the structural and electronic information are fully accessible from the computations while experiments are used to validate the computed properties under experimentally accessible conditions. The new data from this methodology would add additional constraints on core composition and shed new light on the deep interiors of the Earth, as well as terrestrial planets and moons. Such a method can be particularly of value in determining if there are unanticipated changes in the large pressure range encountered in the Earth's outer core similar to the polyamorphic transition around ~5 GPa or changes of intermediate structure and clustering involving light elements such as C. Recent studies based on such methodology show promising results.

ACKNOWLEDGMENTS

We thank the two anonymous reviewers for providing constructive reviews. This work was supported by the NSF grant (EAR-1565708) to BC and the NSF grant (EAR-1565678) to JW.

REFERENCES

Alfé, D. (2010). Iron at Earth's core conditions from first principles calculations. *Reviews in Mineralogy and Geochemistry*, *71*(1), 337–354. https://doi.org/10.2138/rmg.2010.71.16

Alfé, D., Kresse, G., & Gillan, M. J. (2000). Structure and dynamics of liquid iron under Earth's core conditions. *Physical Review B*, *61*, 132–142.

Anderson, O. L., & Isaak, D. G. (2002). Another look at the core density deficit of Earth's outer core. *Physics of the Earth and Planetary Interiors*, *131*, 19–27.

Anderson, W. W., & Ahrens, T. J. (1994). An equation of state for liquid iron and implications for the Earth's core. *Journal of Geophysical Research: Solid Earth*, *99*, 4273–4284.

Belashchenko, D. K. (2014). Computer simulation of liquid metals. *Physics-Uspekhi*, *56*(12), 1176–1216. https://doi.org/10.3367/UFNe.0183.201312b.1281

Birch, F. (1964). Density and composition of mantle and core. *Journal of Geophysical Research-Solid Earth*, *69*(20), 4377–4388. https://doi.org/10.1029/JZ069i020p04377

Bland, M. T., Showman, A. P., & Tobie, G. (2008). The production of Ganymede's magnetic field. *Icarus*, *198*(2), 384–399. https://doi.org/10.1016/j.icarus.2008.07.011

Chen, B., Gao, L., Lavina, B., Dera, P., Alp, E. E., Zhao, J., & Li, J. (2012). Magneto-elastic coupling in compressed Fe_7C_3 supports carbon in Earth's inner core. *Geophysical Research Letters*, *39*, L18301. https://doi.org/10.1029/2012GL052875

Chen, B., Lai, X., Li, J., Liu, J., Zhao, J., Bi, W., et al. (2018). Experimental constraints on the sound velocities of cementite Fe_3C to core pressures. *Earth and Planetary Science Letters*, *494*, 164–171. https://doi.org/10.1016/j.epsl.2018.05.002

Chen, B., & Li, J. (2016). Carbon in the core. In *Deep Earth* (Vol. 207, pp. 277–288). Hoboken, NJ: John Wiley & Sons. https://doi.org/10.1002/9781118992487.ch22

Chen, B., Li, Z., Zhang, D., Liu, J., Hu, M. Y., Zhao, J., et al. (2014). Hidden carbon in Earth's inner core revealed by shear softening in dense Fe_7C_3. *Proceedings of the National Academy of Sciences*, *111*(50), 17755–17758. https://doi.org/10.1073/pnas.1411154111

Côté, A. S., Vočadlo, L., & Brodholt, J. P. (2012). Ab initio simulations of iron-nickel alloys at Earth's core conditions, *345–348*(C), 126–130. https://doi.org/10.1016/j.epsl.2012.06.025

Dasgupta, R., & Walker, D. (2008). Carbon solubility in core melts in a shallow magma ocean environment and distribution of carbon between the Earth's core and the mantle. *Geochimica et Cosmochimica Acta*, *72*, 4627–4641. https://doi.org/10.1016/j.gca.2008.06.023

Desjarlais, M. P. (2013). First-principles calculation of entropy for liquid metals. *Phys. Rev. E*, *88*(6). https://doi.org/10.1103/PhysRevE.88.062145

Fang, C. M., van Huis, M. A., & Zandbergen, H. W. (2009). Structural, electronic, and magnetic properties of iron carbide Fe_7C_3 phases from first-principles theory. *Physical Review B*, *80*, 224108+. https://doi.org/10.1103/PhysRevB.80.224108

Fei, Y., & Bertka, C. (2005). The interior of Mars, *308*, 1120–1121.

Fei, Y., Murphy, C., & Shibazaki, Y. (2016). Thermal equation of state of hcp-iron: Constraint on the density deficit of Earth's solid inner core. Retrieved from http://onlinelibrary.wiley.com/doi/10.1002/2016GL069456/full

Fiquet, G., Badro, J., Gregoryanz, E., Fei, Y., & Occelli, F. (2009). Sound velocity in iron carbide (Fe_3C) at high pressure: Implications for the carbon content of the Earth's inner core. *Physics of the Earth and Planetary Interiors*, *172*(1–2), 125–129. https://doi.org/10.1016/j.pepi.2008.05.016

Gao, L., Chen, B., Wang, J., Alp, E. E., Zhao, J., Lerche, M., et al. (2008). Pressure-induced magnetic transition and sound velocities of Fe_3C: Implications for carbon in the Earth's inner core. *Geophysical Research Letters*, *35*, L17306. https://doi.org/10.1029/2008GL034817

Gao, L., Chen, B., Zhao, J., Alp, E. E., Sturhahn, W., & Li, J. (2011). Effect of temperature on sound velocities of compressed Fe_3C, a candidate component of the Earth's inner core. *Earth and Planetary Science Letters*, *309*(3–4), 213–220. https://doi.org/10.1016/j.epsl.2011.06.037

Hauck II, S. A., Aurnou, J. M., & Dombard, A. J. (2006). Sulfur's impact on core evolution and magnetic field generation on Ganymede. *Journal of Geophysical Research: Planets*, *111*(59), E09008.

Jephcoat, A., & Olson, P. (1987). Is the inner core of the Earth pure iron? *Nature*, *325*(6102), 332.

Jimbo, I., & Cramb, A. W. (1993). The density of liquid iron-carbon alloys. *Metallurgical Transactions B*, *24*(1), 5–10. https://doi.org/10.1007/BF02657866

Kono, Y., Kenney-Benson, C., Shibazaki, Y., Park, C., Shen, G., & Wang, Y. (2015). High-pressure viscosity of liquid Fe and FeS revisited by falling sphere viscometry using ultrafast X-ray imaging. *Physics of the Earth and Planetary Interiors*, *241*, 57–64. https://doi.org/10.1016/j.pepi.2015.02.006

Kuwabara, S., Terasaki, H., Nishida, K., Shimoyama, Y., Takubo, Y., Higo, Y., et al. (2015). Sound velocity and elastic properties of Fe-Ni and Fe-Ni-C liquids at high pressure. *Physics and Chemistry of Minerals*, *43*(3), 1–8. https://doi.org/10.1007/s00269-015-0789-y

Lai, X., Chen, B., Wang, J., Yoshio, K., Kono, Y., & Zhu, F. (2017). Polyamorphic transformations in Fe-Ni-C liquids: Implications for chemical evolution of terrestrial planets. *Journal of Geophysical Research: Solid Earth*, *122*(12), 9745–9754. https://doi.org/10.1002/2017JB014835

Lai, X., Zhu, F., Liu, J., Zhang, D., Hu, Y., Finkelstein, G. J., et al. (2018). The high-pressure anisotropic thermoelastic properties of a potential inner core carbon-bearing phase, Fe_7C_3, by single-crystal X-ray diffraction. *American Mineralogist*, *103*(10), 1568–1574. https://doi.org/10.2138/am-2018-6527

Li, J., & Fei, Y. (2014). Experimental constraints on core composition. In H. D. Holland & K. K. Turekian (Eds.), *Treatise on geochemistry* (pp. 527–557). Oxford: Elsevier. https://doi.org/10.1016/B978-0-08-095975-7.00214-X

Li, J., Mao, H. K., Fei, Y., Gregoryanz, E., Eremets, M., & Zha, C.-S. (2002). Compression of Fe_3C to 30 GPa at room temperature. *Physics and Chemistry of Minerals*, *19*(3), 166–169. https://doi.org/10.1007/s00269-001-0224-4

Lin, J.-F., Struzhkin, V. V., Struzhkin, V. V., Mao, H., Hemley, R. J., Chow, P., et al. (2004). Magnetic transition in compressed Fe_3C from x-ray emission spectroscopy. *Physical Review B*, *70*(2), 212405. https://doi.org/10.1103/PhysRevB.70.212405

Litasov, K. D., Sharygin, I. S., Dorogokupets, P. I., Shatskiy, A., Gavryushkin, P. N., Sokolova, T. S., et al. (2013). Thermal equation of state and thermodynamic properties of iron carbide Fe_3C to 31 GPa and 1473 K. *Journal of Geophysical Research: Solid Earth*, *118*(1), 5274–5284. https://doi.org/10.1002/2013JB010270

Litasov, K. D., Sharygin, I. S., Shatskii, A. F., Gavryushkin, P. N., Dorogokupets, P. I., Sokolova, T. S., et al. (2014). P-V-T equations of state for iron carbides Fe_3C and Fe_7C_3 and their relationships under the conditions of the Earth's mantle and core, *453*(2), 1269–1273. https://doi.org/10.1134/S1028334X13120192

Liu, J., Li, J., & Ikuta, D. (2016). Elastic softening in Fe_7C_3 with implications for Earth's deep carbon reservoirs. *Journal of Geophysical Research: Solid Earth*, *121*(3), 1514–1524. https://doi.org/10.1002/2015JB012701

Margot, J. L., Peale, S. J., Jurgens, R. F., Slade, M. A., & Holin, I. V. (2007). Large longitude libration of Mercury reveals a molten core, *316*(5825), 710–714. https://doi.org/10.1126/science.1140514

McDonough, W. F., & Sun, S.-S. (1995). The composition of the Earth, *120*, 223–253. https://doi.org/10.1016/0009-2541(94)00140-4

Mookherjee, M. (2011). Elasticity and anisotropy of Fe_3C at high pressures. *American Mineralogist*, *96*(10), 1530–1536. https://doi.org/10.2138/am.2011.3917

Mookherjee, M., Nakajima, Y., Steinle-Neumann, G., Glazyrin, K., Wu, X., Dubrovinsky, L. S., et al. (2011). High-pressure behavior of iron carbide (Fe_7C_3) at inner core conditions. *Journal of Geophysical Research: Solid Earth*, *116*(B4), 91. https://doi.org/10.1029/2010JB007819

Morard, G., Andrault, D., Antonangeli, D., & Bouchet, J. (2014). Properties of iron alloys under the Earth's core conditions. *Comptes Rendus Geoscience*, *346*(5), 130–139. https://doi.org/10.1016/j.crte.2014.04.007

Morard, G., Nakajima, Y., Andrault, D., Antonangeli, D., Auzende, A. L., Boulard, E., et al. (2017). Structure and density of Fe-C liquid alloys under high pressure. *Journal of Geophysical Research: Solid Earth*, *122*(10), 7813–7823. https://doi.org/10.1002/2017JB014779

Nakajima, Y., Imada, S., Hirose, K., Komabayashi, T., Ozawa, H., Tateno, S., et al. (2015). Carbon-depleted outer core revealed by sound velocity measurements of liquid iron-carbon alloy. *Nature Communications*, *6*, 8942. https://doi.org/10.1038/ncomms9942

Nakajima, Y., Takahashi, E., Suzuki, T., & Funakoshi, K.-I. (2009). "Carbon in the core" revisited. *Physics of the Earth and Planetary Interiors*, *174*(1–4), 202–211. https://doi.org/10.1016/j.pepi.2008.05.014

Nishida, K., Suzuki, A., Terasaki, H., Shibazaki, Y., Higo, Y., Kuwabara, S., et al. (2016). Towards a consensus on the pressure and composition dependence of sound velocity in the liquid Fe-S system. *Physics of the Earth and Planetary Interiors*, *257*(C), 230–239. https://doi.org/10.1016/j.pepi.2016.06.009

Nishida, K., Terasaki, H., Ohtani, E., & Suzuki, A. (2008). The effect of sulfur content on density of the liquid Fe–S at high pressure. *Physics and Chemistry of Minerals*, *35*(7), 417–423. https://doi.org/10.1007/s00269-008-0236-4

Poirier, J.-P. (1994). Light elements in the Earth's outer core: A critical review. *Physics of the Earth and Planetary Interiors*, *85*(3–4), 319–337. https://doi.org/10.1016/0031-9201(94)90120-1

Prescher, C., Dubrovinsky, L. S., Bykova, E., Kantor, A., Nakajima, Y., Miyajima, N., et al. (2015). High Poisson's ratio of Earth's inner core explained by carbon alloying. *Nature Geoscience*, *8*(3), 220–223. https://doi.org/10.1038/ngeo2370

Prescher, C., Dubrovinsky, L. S., McCammon, C., Glazyrin, K., Nakajima, Y., Kantor, A., et al. (2012). Structurally hidden magnetic transitions in Fe_3C at high pressures. *Physical Review B*, *85*(14), 140402. https://doi.org/10.1103/PhysRevB.85.140402

Sanloup, C., Fiquet, G., Gregoryanz, E., Morard, G., & Mezouar, M. (2004). Effect of Si on liquid Fe compressibility: Implications for sound velocity in core materials. *Geophysical Research Letters*, *31*(7), n/a-n/a. https://doi.org/10.1029/2004GL019526

Sanloup, C., Guyot, F., Gillet, P., Fiquet, G., Hemley, R. J., Mezouar, M., & Martinez, I. (2000). Structural changes in liquid Fe at high pressures and high temperatures from synchrotron X-ray diffraction. *Europhysics Letters (EPL)*, *52*(2), 151–157. https://doi.org/10.1209/epl/i2000-00417-3

Sanloup, C., Guyot, F., Gillet, P., Fiquet, G., Mezouar, M., & Martinez, I. (2000). Density measurements of liquid Fe-S alloys at high-pressure. *Geophysical Research Letters*, *27*(6), 811–814. https://doi.org/10.1029/1999GL008431

Sanloup, C., van Westrenen, W., Dasgupta, R., Maynard-Casely, H., Maynard-Casely, H., & Perrillat, J. P. (2011). Compressibility change in iron-rich melt and implications for core formation models. *Earth and Planetary Science Letters*, *306*(1–2), 118–122. https://doi.org/10.1016/j.epsl.2011.03.039

Sata, N., Hirose, K., Shen, G., Nakajima, Y., Ohishi, Y., & Hirao, N. (2010). Compression of FeSi, Fe_3C, $Fe_{0.95}O$, and FeS under the core pressures and implication for light element in the Earth's core. *Journal of Geophysical Research*, *115*(B9), 91. https://doi.org/10.1029/2009JB006975

Scott, H. P., Williams, Q., Williams, Q., & Knittle, E. (2001). Stability and equation of state of Fe_3C to 73 GPa: Implications for carbon in the Earth's core. *Geophysical Research Letters*, *28*(9), 1875–1878.

Secco, R. A. (1995). Viscosity of the outer core. *Mineral Physics and Crystallography, A Handbook of Physical Constants*, *2*, 218–226.

Shanker, J., Singh, B. P., & Srivastava, S. K. (2004). Volume-temperature relationship for iron at 330 GPa and the Earth's core density deficit. *Physics of the Earth and Planetary Interiors*, *147*(4), 333–341.

Shen, G., Prakapenka, V. B., Rivers, M. L., & Sutton, S. R. (2004). Structure of liquid iron at pressures up to 58 GPa. *Physical Review Letters*, *92*(18), 185701.

Shibazaki, Y., & Kono, Y. (2018). Effect of silicon, carbon, and sulfur on structure of liquid iron and implications for structure-property relations in liquid iron-light element alloys. *Journal of Geophysical Research: Solid Earth*, *123*, 4697–4706. https://doi.org/10.1029/2018JB015456

Shibazaki, Y., Kono, Y., & Fei, Y. (2015). Microscopic structural change in a liquid Fe-C alloy of ~5 GPa. *Geophysical Research Letters*, *42*(13), 5236–5242. https://doi.org/10.1002/2015GL064271

Shimoyama, Y., Terasaki, H., Ohtani, E., Urakawa, S., Takubo, Y., Nishida, K., et al. (2013). Density of Fe-3.5 wt% C liquid at high pressure and temperature and the effect of carbon on the density of the molten iron. *Physics of the Earth and Planetary Interiors*, *224*, 77–82. https://doi.org/10.1016/j.pepi.2013.08.003

Sobolev, A., & Mirzoev, A. (2013). Ab initio studies of the short-range atomic structure of liquid iron-carbon alloys. *Journal of Molecular Liquids*, *179*, 12–17. https://doi.org/10.1016/j.molliq.2012.11.019

Stevenson, D. J. (1981). Models of the Earth's core. *Science*, *214*(4521), 611–619.

Stevenson, David J. (2010). Planetary magnetic fields: Achievements and prospects. *Space Science Reviews*, *152*(1), 651–664. https://doi.org/10.1007/s11214-009-9572-z

Stixrude, L., Cohen, R. E., & Singh, D. J. (1997). Iron at high pressure: Linearized-augmented-plane-wave computations in the generalized-gradient approximation. *Physical Review B*, *50*(9), 6442–6445.

Stixrude, L., Wasserman, E., & Cohen, R. E. (1997). Composition and temperature of Earth's inner core. *Journal of Geophysical Research-Solid Earth*, *102*(B), 24729–24739. https://doi.org/10.1029/97JB02125

Tateyama, R., Ohtani, E., Terasaki, H., Nishida, K., Shibazaki, Y ., Suzuki, A., & Kikegawa, T. (2011). Density measurements of liquid Fe-Si alloys at high pressure using the sink-float method. *Physics and Chemistry of Minerals*, *38*(1), 801–807. https://doi.org/10.1007/s00269-011-0452-1

Terasaki, H., Nishida, K., Shibazaki, Y., Sakamaki, T., Suzuki, A., Ohtani, E., & Kikegawa, T. (2010). Density measurement of Fe_3C liquid using X-ray absorption image up to 10 GPa and effect of light elements on compressibility of liquid iron. *Journal of Geophysical Research: Solid Earth*, *115*(B6), B06207. https://doi.org/10.1029/2009JB006905

Terasaki, H., Suzuki, A., Ohtani, E., Nishida, K., Sakamaki, T., & Funakoshi, K. (2006). Effect of pressure on the viscosity of Fe-S and Fe-C liquids up to 16 GPa. *Geophysical Research Letters*, *33*(22), 2124. https://doi.org/10.1029/2006GL027147

Vočadlo, L., Alfé, D., Price, G. D., & Gillan, M. J. (2000). First principles calculations on the diffusivity and viscosity of liquid Fe-S at experimentally accessible conditions. *Physics of the Earth and Planetary Interiors*, *120*, 145–152.

Weber, R. C., Lin, P. Y., Garnero, E. J., Williams, Q., & Lognonne, P. (2011). Seismic detection of the lunar core. *Science*, *331*(6015), 309–312. https://doi.org/10.1126/science.1199375

Wood, B. J. (1993). Carbon in the core. *Earth and Planetary Science Letters*, *117*, 593–607.

Wood, B. J., Li, J., & Shahar, A. (2013). Carbon in the core: Its influence on the properties of core and mantle. *Review in Mineralogy and Geochemistry*, *75*(1), 231–250. https://doi.org/10.2138/rmg.2013.75.8

Yu, X., & Secco, R. A. (2008). Equation of state of liquid Fe–17 wt% Si to 12 GPa. *High Pressure Research*, *28*(1), 19–28.

Zhang, Y., & Yin, Q.-Z. (2012). Carbon and other light element contents in the Earth's core based on first-principles molecular dynamics. *Proceedings of the National Academy of Sciences*, *109*(48), 19579–19583.

5

A Geologic Si-O-C Pathway to Incorporate Carbon in Silicates

Alexandra Navrotsky[1]**, John Percival**[2]**, and Larissa Dobrzhinetskaya**[3]

ABSTRACT

Geologic and planetary processes are punctuated by sudden cataclysmic events, and planetary evolution is irrevocably changed by impacts and intense seismic and magmatic/volcanic activity. Such events often are associated with or generate high temperature, high pressure, and low oxygen fugacity. Their traces in the accessible geologic record are not pristine but altered by subsequent petrologic reactions. Evidence from the thermochemistry of synthetic materials, largely studied in a materials science context, in Si-O-C and M-Si-O-C-H systems under reducing conditions can be used to propose some possible rare but significant reactions, together called a geologic Si-O-C pathway, involving carbon-containing silicate melts, glasses, and amorphous materials. The substitution of carbon for oxygen in the first coordination shell of silicon provides a reducing local environment for the formation of metals, carbides, and silicides. Grains of these refractory compounds may persist long after the main carbon-containing silicate phase has transformed and disappeared. Such relict refractory materials may be markers of impact events and unusual volcanism. Anomalies in minor phases, trace elements, and textures in settings ranging from ultra-high pressure metamorphic rocks to impact craters to carbonado diamonds may be linked to the transient presence of carbon-rich silicate phases generated under reducing conditions from initially carbon-rich target rocks and/or impactors.

5.1. INTRODUCTION AND BACKGROUND

The chemical state of carbon coexisting with or dissolved in silicates at various conditions of temperature, pressure, and oxygen fugacity is critical to modeling the global carbon cycle and to understanding geologic and planetary processes. Under relatively oxidizing conditions, carbon can enter silicate melts, glasses, and crystals as carbonate (CO_3^{2-}) groups, and such equilibria have been studied extensively (Duncan et al., 2017; Kim et al., 2018; Kohn et al., 1991; Le Voyer et al., 2017; Mysen 2017). Molecular CO_2 can be confined in the pores and channels of open framework structures such as zeolites. Such porous phases and their surfaces, when containing guest molecules, can undergo high-pressure phase transitions that are different from those in guest-free zeolites (Haines et al., 2010; Yong et al., 2018).

The substitution of carbon for silicon in tetrahedral sites of crystalline, glassy, and molten silicates has been a subject of speculation for many years. Computational and experimental studies have confirmed polymeric and tetrahedral framework (cristobalite-like) forms of carbon dioxide at pressures above 40 GPa (Oganov et al., 2008; Park et al., 2003; Seto et al., 2010). Carbonates containing tetrahedral CO_4^{2-} groups have also been proposed to exist above 80 GPa (Oganov et al., 2008). Although there is increasing evidence for such phases, their pressures of formation lie beyond the range of most petrologic processes. Nor can one directly address to what extent carbon can substitute for tetrahedral silicon in multicomponent silicates, given the large difference in Si-O and C-O tetrahedral bond lengths. One must then ask whether other more energetically affordable ways of incorporating carbon into silicates at lower pressure exist, especially

[1] *Peter A. Rock Thermochemistry Laboratory and NEAT ORU, University of California–Davis, Davis, California, USA*

[2] *Geological Survey of Canada, Ottawa, Ontario, Canada*

[3] *Department of Earth Sciences, University of California–Riverside, Riverside, California, USA*

Carbon in Earth's Interior, Geophysical Monograph 249, First Edition. Edited by Craig E. Manning, Jung-Fu Lin, and Wendy L. Mao.

under reducing conditions where CO_2 is destabilized and carbides can form.

The materials science and petrology communities have sought to put carbon into silica glass and silicates (Dasgupta et al., 2013; Fine & Stolper, 1986; Pantano et al., 1999). The direct incorporation of carbon into silica by high temperature reaction (including melting) is limited both by the low reactivity of silica and graphite in the solid state and by the carbothermic reduction of silica by graphite (Hutchinson et al., 1988) to form silicon carbide and carbon monoxide gas at temperatures above 1500 °C by the reaction

$$SiO_2 + 3C = SiC + 2CO(\text{gas}). \quad (5.1)$$

These difficulties have been circumvented by synthetic routes involving organosilicon compounds, whose controlled polymerization and pyrolysis at 500–1400 °C have led to a series of compositionally controlled "polymer-derived ceramics" or PDCs (Colombo et al., 2010; Mera et al. 2013) containing silicon, oxygen, and carbon (see Figure 5.1), as well as more complex systems containing those elements plus nitrogen, boron, hydrogen, and smaller amounts of other metals (Mera et al., 2013; Ionescu et al., 2013). Though amorphous by X-ray diffraction, PDCs show complex interpenetrating nanostructures of silica-rich and carbon-rich domains. The domains are thin, fractal, and intergrown. Within the silica-rich domains, carbon substitutes not for silicon but for oxygen, leading to "mixed tetrahedra" (SiC_nO_{4-n}) with silicon at the center and zero to four carbon atoms in the first coordination shell, as documented by extensive spectroscopic studies, especially by nuclear magnetic resonance techniques (Mera et al., 2013; Widgeon et al., 2010) (see Figure 5.1). The carbon in the silica-rich domains is not uniformly distributed but tends to be enriched near their interfaces with the carbon-rich domains.

PDCs persist to about 1500 °C, until the carbothermic reduction reaction (Equation 5.1) sets in (Hutchison et al., 1988). It was initially assumed that the PDCs were thermodynamically metastable and their lack of decomposition at high temperature below that of carbothermic reduction was a result of kinetic hindrance. However, calorimetric studies of their enthalpies of formation showed that many PDC compositions are stable in enthalpy relative to a mixture of crystalline SiO_2, SiC, and C (Mera et al., 2013; Varga et al., 2007). Since the entropies of the disordered and nanostructured PDCs are almost certainly higher than those of a mixture of crystalline components, the negative enthalpies of formation coupled with positive entropies of formation make them thermodynamically stable, having lower Gibbs free energy of formation than an isochemical

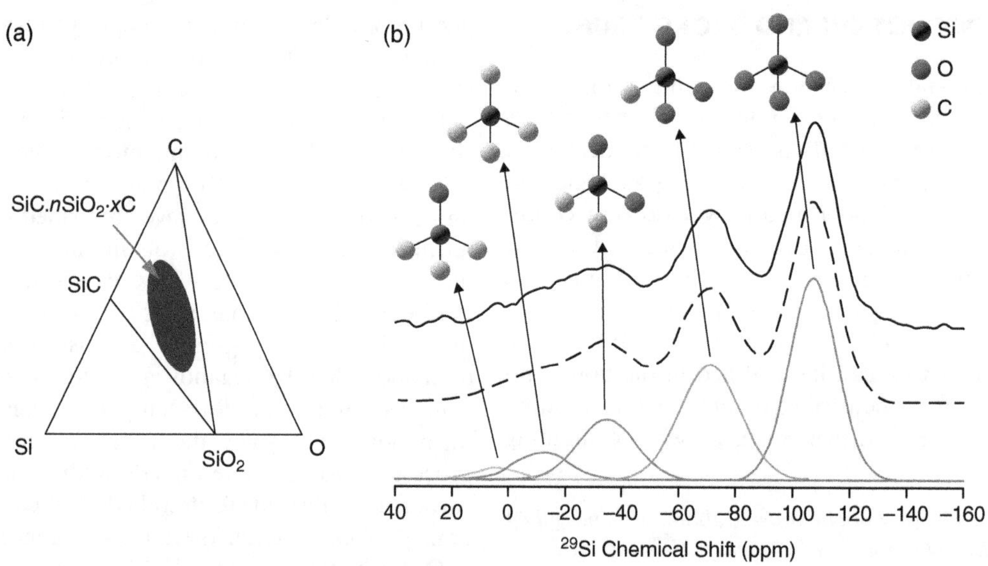

Figure 5.1 (a) Approximate composition range (mole fraction basis) over which polymer-derived ceramic (PDC) phases form in the Si-O-C system. (b) ^{29}Si MAS NMR spectra of a polysiloxane-derived SiOC PDC of composition $SiO_{1.5}C_{0.68}$ pyrolyzed at 1100 °C in argon. The experimental spectrum is shown by the solid bold curve on top, followed by the simulated spectrum (dashed curve), and the individual simulated components (bottom, solid curves). Peaks from right to left correspond to SiO_4, SiO_3C, SiO_2C_2, SiC_4, and $SiOC_3$ groups, as modeled by deconvolutions, with the $SiOC_3$ and SiC_4 groups not always clearly separated in the experimental data (Mera et al., 2013). This figure illustrates the power of NMR to distinguish mixed-bond tetrahedra. See electronic version for color representation of the figures in this book.

mixture of crystalline components. This observation led us to speculate that analogous carbon-containing glassy, liquid, or crystalline phases, with carbon substituting for oxygen rather than for silicon, might form under reducing conditions in the crust and mantle. Reducing conditions are obviously required to minimize the formation of CO_2 or carbonates (Kadik et al., 2004; Sen et al., 2013). Very high pressure is not required, but moderate pressure would disfavor the carbothermic reduction reaction, which produces a gas phase and would thus extend the stability field of the silicon oxycarbide phases to higher temperatures. Later diamond cell and multianvil experiments in the Navrotsky group on mixtures of silica glass, silicon carbide, and graphite were inconclusive because of very slow reactions, even at temperatures of 1500–2000 °C and pressures of 5–10 GPa and therefore were not published. Thus, it is not clear at present how pressure affects the stability and decomposition of PDCs or whether, if gas evolution and crystallization are suppressed, they can melt to produce carbon-containing silicate liquids with mixed-bond tetrahedra. The discussion below suggests that PDC-like solids can be produced in several ways: by condensation from a vapor, by cooling of a multicomponent supercritical fluid from very high temperature and pressure, or from a silicate melt at low oxygen fugacity and moderate temperature and pressure.

Motivated by the observed thermodynamic stability of PDCs, we conclude that the polymer route was a kinetic and mechanistic convenience, allowing a synthetic pathway to a stable product. Thus, the product should be attainable by other simpler means not involving organometallic polymer precursors. Laser ablation of a silica glass target in an acetylene atmosphere (Tavakoli et al., 2015) produced a powder that had many of the attributes of the PDC materials, though it was fine grained and contained more hydrogen. It contained mixed-bond SiC_nO_{4-n} tetrahedra and had comparable energetic stability to PDCs of similar composition. Its method of synthesis, rapid formation of a plume of vapor and particles followed by rapid collection on a cold substrate, has some similarity to a geologic impact process, and the observations made us think of the role of Si-O-C melts, glasses, and amorphous solids in impact phenomena.

It is possible that C can substitute for O in amorphous PDC-like materials other than pure silica, with implications for reduced silicate melts in the Earth's mantle (Sen et al., 2013). For example, carbon substitutes for oxygen in lithium silicate systems of less than full polymerization, extending the possible range of compositions for PDC-like materials to be more geologically relevant in terms of Q-speciation (Qn (n = 0 to 4) being the average number of bridging oxygens in a silicate tetrahedron) (Sen et al., 2013). Later studies synthesized PDCs with Mg and Mg+Ca (Ionescu et al., 2018). Although phase separation and crystallization were complications,

Table 5.1 Crystalline phases formed upon annealing of metal-containing Si-O-C polymer-derived ceramics (Ionescu et al., 2013).

Metal Added	Phases Observed
Zr	$ZrSiO_4$
Hf	$HfSiO_4$
Mn	$MnSiO_3$
Lu	Lu_2O_3, $Lu_2Si_2O_7$
Sn	Sn
Fe	Fe_3Si, Fe_3Si_3 + SiC
V	V_8C_7

substitution of C for O was indeed seen, including in the more depolymerized glass phase of lower silica content that coexists with almost pure silica in the carbon-free MgO-CaO-SiO_2 system. The energetic stability of these alkaline earth PDCs, determined by calorimetry, is generally comparable to that of other Si-O-C materials studied earlier. In addition, some of these compositions appear bioactive toward hydroxyapatite formation, while others are not. This may have relevance to medical implants and perhaps even to the origin of life and its evolution.

We are currently working on laser ablation studies on a sintered crystalline forsterite (Mg_2SiO_4) pellet. The product of ablation in acetylene is black, amorphous to XRD, and appears to be close to forsterite in composition. Preliminary results suggest some carbon incorporation.

In another set of studies, a number of different metals were incorporated into the polymer, which was then pyrolyzed; initial metal incorporation in the PDC was followed at higher temperature by crystallization of secondary phases (Ionescu et al., 2013). The results are given in Table 5.1. Metals, oxides, or silicides are crystallized, controlled by thermodynamic stability at the C–CO equilibrium. Thus, the metals whose oxides are stable at the C–CO equilibrium (Zr, Hf, Mn, Lu) form oxides that react further with the silica component to form silicates, Sn remains as elemental tin, Fe forms silicides, and V forms carbides. The formation of ternary oxides and the potential competition between carbide and silicide formation appear to be controlled by the phase(s) of lowest free energy, suggesting equilibrium is attained relatively rapidly.

5.2. HYPOTHESIS: THE GEOLOGIC SI-O-C PATHWAY

Impacts and some deep-seated magmatic processes produce local very high temperatures and locally reducing conditions that can generate transient silica-rich vapors, supercritical fluids, or melts containing all the constituents of the host rocks, their associated fluids, and the impactor. These short-lived products can contain significant carbon substituting for oxygen and may be precursors to geologic Si-O-C materials whose carbon is available for further chemical reaction, without the need to equilibrate with an external carbon-bearing fluid. Such geologic Si-O-C materials may be similar to materials made in the laboratory by PDC and laser ablation routes that contain silica-rich and carbon-rich nanodomains. The geologic Si-O-C materials can precipitate small amounts of uncommon oxides, metals, carbides, carbon, silicon, and silicides. These widely disseminated crystalline grains reflect thermodynamic equilibrium under the pressure, temperature, and low oxygen fugacity conditions of the initial event. High pressure is not required for their formation, and the oxygen fugacity is, to a first approximation, that of the C-CO equilibrium. The silica-rich nanodomains can undergo relatively rapid alteration, and their metamorphosed products may not be readily identifiable, but the refractory carbon, metal, carbide, and silicide grains may be able to survive cycles of alteration through geologic time (analogous to refractory zircon grains), sometimes preserving morphological evidence of having been molten or crystallizing from a vapor, melt, or glass. These refractory grains of highly reduced phases provide strong evidence for the initial cataclysmic impact or magmatic event. For brevity, the sequence of events described above is referred to as the geologic Si-O-C pathway.

5.2.1. Geologic Evidence for the Si-O-C Pathway

The geologic Si-O-C pathway provides a mechanism for formation and evolution of carbon-containing silicates under internally controlled low oxygen fugacity and having an internal source of carbon, without the need for (though not precluding) ultra-high-pressure conditions. Ahead, we briefly discuss some petrologic observations and geologic scenarios that may be consistent with evolution along the geologic Si-O-C pathway.

Impact events are obvious candidates. Giant impacts, relatively common in the late stages of planetary accretion, can produce significant melting and/or vaporization of large fractions of a planet (Cuk et al., 2016). The condensation of silica-rich (supercritical) vapor can produce a melt rich not just in rock-forming mineral components but also in metals and carbon coming from the impactor (e.g., a metal-rich bolide or a planetary impactor with a metallic core) and from the initially vaporized carbonates, fluids, and organic matter in the host rocks. The initial ultra-high temperatures (thousands of degrees) and pressures (tens to hundreds of GPa) dissipate rapidly, but the extensive melt formed, termed a *magma ocean* if it covers much or all of the planet's surface, initially will contain a relatively homogeneous distribution of all the above components. As it cools, such a melt may transform into a glass that contains carbon-rich and silicate-rich domains, precipitating metals, carbides, and silicides, maintaining low oxygen fugacity at least locally, and evolving along the geologic Si-O-C pathway suggested above. Alternatively, the supercritical fluid or vapor may condense directly to an amorphous material (glass), which then undergoes similar evolution.

Though we have no direct petrologic evidence of the likely largest impact on Earth, namely the Moon-forming event, there is evidence to support the geologic Si-O-C pathway from a number of smaller (though still large and cataclysmic) impact events. The Sudbury structure in Ontario, Canada, is the third-largest known impact crater on Earth. Both the commercial interest in nickel and the accessibility of the crater have allowed extensive field, petrologic, and geochemical studies (Dietz, 1964; Grieve et al., 1991, 2010; Lightfoot, 2017; Ubide et al., 2017). Of particular interest is the likelihood that the initial impact initiated volcanic activity involving melting of a large amount of crust over a relatively long time, as well as interaction of the volcanic activity with seawater; see Ubide et al. (2017) for a review. These authors also argue that impact-triggered volcanism may be a more general planetary phenomenon (and better preserved) on Mercury, Venus, and Mars than on Earth.

From the present point of view, we seek evidence of carbon incorporation in the initial impact melt and in subsequent volcanic melts. The impact basin fill at Sudbury, referred to as the Onaping formation, is a complex 1.4–1.6 km thick brecciated mélange consisting of several distinct members, ranging in color from black (carbon-rich) to green (the color suggesting high ferrous ion concentration consistent with reducing conditions) (Grieve et al., 2010). Each member contains varying amounts of vitric clasts (suggested to be glass or to have come from glass), lithic clasts and shards (showing crystallinity and complex textures), and cross-cutting dikes. The average silica content is about 63 wt%. Grieve et al. (2010) argue for water being essential to the postimpact history to produce the observed lithologies and textures, but that issue is tangential to the major hypothesis in the present work. The Onaping formation contains variable carbon content, ranging from 0.5 to 6 wt% in different members, but some carbon appears everywhere. The carbon is isotopically light, and its concentration correlates with that of iridium. Various biogenic, sedimentary, hydrothermal, and other sources for the carbon have been proposed, but none is fully convincing.

The geologic Si-O-C pathway may explain some of the above observations. The initial carbon could have come from the bolide, the host rocks, although carbon-rich sediments are absent from the target sequence, or from some combination thereof. Regardless of the carbon source, the high temperature and reducing conditions produced by the impact, coupled with the silica-rich host rocks, could have produced geologic Si-O-C materials, perhaps initially in the molten state. Their subsequent evolution through the nanostructured ceramic stage could have produced the observed vitric silica-rich fragments and graphite. The ubiquitous but varying distribution of this carbon, not well correlated with other geochemical, petrologic, or textural indicators, suggests its presence in the initial melt. Its variation may reflect locally different subsequent histories, perhaps related to volcanism and reactions with water. The reported presence of lonsdaleite, impact diamonds, and fullerenes (Lightfoot, 2017) also needs to be considered. These could form from the carbon initially in the geologic Si-O-C melt or glass.

The synthesis of diamond-like carbon outside the diamond stability field has been well documented in the materials field, e.g., by chemical vapor deposition or by the action of chlorine gas on silicon carbide at atmospheric pressure and temperatures as low as 1000 °C (Gogotsi et al., 2001). Thus, the presence of diamond-like phases is not necessarily an indicator of high pressure, and their absence is not a reliable indicator of low pressure, since diamond may transform or decompose after initial formation. An unanswered question is whether, and under what conditions, diamond, rather than graphite, can crystallize from the Si-O-C melt, vapor, or nanoceramic. If nanodiamond forms, it may be further stabilized with respect to graphite by surface energy effects. This topic needs further experimental investigation, but the tentative conclusion is that one cannot exclude the formation of metastable diamond at low pressure by the geologic Si-O-C pathway.

Libyan Desert glass is widely distributed over an area of approximately 4000 km^2. Its origin is generally attributed to a 29 Ma impact event, but no source crater has been definitely identified. Pratesi et al. (2002), using transmission electron microscopy, identified spherules arising from silicate-silicate glass immiscibility. They also found 5–50 nm thick graphite ribbons polygonalized into closed structures up to 200 nm in diameter. Such textures are quite similar to those in annealed PDCs in the Mg-Ca-Si-O-C system (Ionescu et al., 2018). Thus, these observations are consistent with a process involving impact into silica-rich sediments, formation of a geologic Si-O-C melt, cooling to a geologic Si-O-C glass or nanoceramic, and phase separation into two silicate glass phases plus a carbon-rich phase. If the Libyan Desert glass is sufficiently unaltered, one might find the presence of mixed SiC_nO_{4-n} tetrahedra by NMR or other techniques.

Widely distributed, altered "impact spherules" have been reported to be potentially associated with the 2.02 Ga Vredefort impact structure in South Africa (Kamo et al., 1996). They are also found in a dolomitic layer in the Ketilidian orogeny in Greenland (Chadwick et al., 2001), in which spherules contain concentrations of carbonaceous material, and in the Zaonega Formation in Karelia, which contains spherules with bituminous material (Huber et al., 2014). A nanodiamond-rich layer spanning three continents is suggested to be associated with an impact contemporaneous with the start of the Younger Dryas climate fluctuation and extinction event about 12,800 years ago (Kinzie et al., 2014). This layer also contains carbon spherules that, however, could be fire ash. Their very recent age and possibly relatively unaltered condition make the silicate spherules, carbon spherules, and nanodiamonds a very promising assemblage for further study and possible detection of mixed-bond tetrahedra suggested by the geologic Si-O-C pathway.

Carbonado is a rare and enigmatic form of diamond-like carbon. As summarized by Haggerty (2014), it is 3.8 Ga old, polycrystalline, and porous but strongly bonded and super hard. It contains silicate and other mineral inclusions consistent with crustal rocks and their alteration products, and inclusions of metals, SiC, FeC, TiN, and new ternary Cu, Ti nitrides. The latter group of inclusions is indicative of strongly reducing conditions and possibly very high temperature. Textural evidence suggests that some of the carbonado diamond surfaces may have been molten, again suggesting very high temperatures and pressures. However, the porosity argues against a high-pressure origin. Considering the contradictions in all conventional models of terrestrial carbonado formation, Haggerty turned to possible extraterrestrial origin. He suggested that carbonado was delivered to the early Earth from sources such as carbon-rich exoplanets or the interiors of large gaseous planets.

Does the geologic Si-O-C pathway remove some of the inherent contradictions in proposed terrestrial origins for carbonado? Is an impact origin that generates carbonado on Earth rather than bringing it previously formed in an extraterrestrial impactor possible? Although each carbonado is massive compared to other diamonds, the total amount of carbon tied up in them is small because they are rare. Thus, their rarity may reflect the need of the bolide to strike a local carbon-rich lithology. The reduced carbide and silicide phases are consistent with phase equilibria controlled by the C-CO equilibrium at 1000–1500 °C, and indeed mirror those seen in the PDCs. The porosity of 10% or more may represent the removal by alteration of the silica-rich phase in the nanoceramic, leaving some behind as silicate mineral inclusions. In PDCs, the interfaces between silica and carbon domains concentrate the mixed bonds, and their removal may indeed lead to surface smoothing and reconstruction of the remaining carbon, without actual melting. PDCs also readily incorporate nitrogen and hydrogen (Colombo et al., 2010), providing a precursor for their incorporation into carbonado. Studies of the annealing and weathering of the PDC analogues may produce unique textures and size distributions; whether these suggest carbonado is not known. Thus, the geologic Si-O-C pathway needs further experimental study in the context of carbonado formation, as it may provide a tenable mechanism of formation of such carbon nanocomposites at relatively low temperature and pressure. Furthermore, the geologic Si-O-C pathway may provide a mechanism for carbonado formation in carbon-rich planets as well.

A final example is the occurrence of silicon carbide (moissanite) containing inclusions of silicon and complex silicide phases related to $FeSi_2Ti$, Si_2TiV_5, and $FeSi_2$ (Dobrzhinetskaya et al., 2018). The SiC is hosted in a pyroclastic tuff derived from alkali basalt formations in northern Israel. The latter may be related to hot spot volcanism (Eppelbaum et al., 2004). In preparation of hand specimens for study by Raman spectroscopy and transmission electron microscopy by Dobrzhinetskaya et al. (2018), extreme care was taken to avoid contamination, and the moissanite and its inclusions are considered to be native to the tuff. It is striking that the iron and vanadium silicides, though more complex in composition than those in simple binary systems, are very similar to those found to crystallize from iron- and vanadium-doped polymer-derived ceramics (see Table 5.1). The silicide inclusions often show a spherical morphology consistent with crystallization from a melt. The authors consider melting in silicon carbide–metal silicide systems and conclude that the molten alloy droplets could have formed at temperatures similar to those of basalt melting. It is generally assumed that the SiC needed to be transported as xenocrysts from the upper mantle or transition zone where they formed under extremely reducing high-pressure conditions to the shallower depth of active volcanism. However, Dobrzhinetskaya et al. (2018) suggest that "the studied SiC formed as a result of an interaction of a small volume of reducing fluids and crustal materials (SiO_2) possibly available from the walls of the alkaline basalt reservoir." The proposed geologic Si-O-C pathway can take such a reducing fluid, or a graphitic or organic carbon source, and by dissolving carbon in the silicate melt (basalt) retain reducing conditions through the evolution of carbon-containing melt to glass to nanophase ceramic and the subsequent precipitation of silicon carbide and its associated carbide and silicide inclusions. Neither melting of the carbide or silicide nor very high pressure is required. The refractory carbides and silicides do not equilibrate with the surrounding magma as it crystallizes under normally high oxygen fugacity, but remain as indicators of their formation. Such local pockets of low oxygen fugacity at moderate temperatures and pressures, leading to the geologic Si-C-O pathway, may explain the occurrence of small grains of highly reduced refractory metal and carbide phases in many geologic settings without requiring their transport from extreme depth. In some cases, these refractory inclusions may have survived from earlier cycles of petrologic reactions.

Presolar dust particles have also been suggested to contain nanodiamond, graphite, titanium carbide, and other phases (Nguyen & Messenger, 2011) that may suggest condensation and evolution along a similar Si-O-C pathway. It is also tempting to speculate that in carbon-rich exoplanets containing some silicates, the Si-O-C pathway could be more prevalent.

5.3. UNANSWERED QUESTIONS AND FUTURE RESEARCH DIRECTIONS

Laboratory experiments and computational approaches need to address the following key questions. (1) Can silicate melts or glasses containing significant carbon concentrations resulting from the substitution of carbon for oxygen in mixed-bond tetrahedra be synthesized either directly or by the heating of PDC-like nanoceramics? Low oxygen fugacity (several orders of magnitude below the Fe/FeO buffer), moderately high temperature (1500–2000 °C), and pressure just high enough to suppress the gas formation that is a driving force for carbothermic reduction of silica (1–5 GPa) would be good starting points for such a study. (2) Can diamond (as micron-sized or larger crystals, nanodiamond, or with carbonado-like textures) be formed by the evolution of the Si-O-C nanoceramics? Can such diamond form and persist metastably outside the normal diamond stability field? (3) How do the Si-O-C nanoceramics transform, metamorphose, or weather under geologic conditions in contact with minerals and fluids? Do the silica-rich nanodomains disappear while the carbon-rich domains remain (as moissanite, graphite, or diamond)? Do the carbides and silicides formed by metal addition persist? Such experiments can be performed starting with synthetic and well-characterized PDC starting materials.

The occurrences described above have several features in common that make them candidates for a geologic Si-O-C pathway of carbon incorporation. These include high silica content and possible high carbon content in the parent rocks and low oxygen fugacity during impact or volcanism. The recovered natural materials should be examined in more detail by state-of-the-art analytical and structural techniques to seek evidence of carbon incorporation via mixed SiC_xO_{4-x} bonding. Silicate phases (especially glass or its devitrification/alteration products), carbon phases (including both graphite and diamond), and silicon carbide grains and their inclusions should all be examined together, preferably when they originally are in close contact in a specimen rather than as isolated, recovered grains that may have come from different environments. Given the age and likely alteration of the initial materials, this is a difficult task of mineralogical forensics and requires the collaboration of the geologic, planetary, and materials communities.

ACKNOWLEDGMENTS

This work was supported by the Deep Carbon Observatory of the Alfred P. Sloan Foundation.

REFERENCES

Chadwick, B., Claeys, P., & Simonson, B. (2001). New evidence for a large Paleoproterozoic impact: Spherules in a dolomite layer in the Ketilidian orogeny, South Greenland. *Journal of the Geological Society*, *158*, 331–340. https://doi.org/10.1144/jgs.158.2.331

Colombo, P., Mera, G., Riedel, R., & Soraru, G. D. (2010). Polymer-derived ceramics: 40 years of research and innovation in advanced ceramics. *Journal of the American Ceramic Society*, *93*(7), 1805–1837. https://doi.org/10.1111/j.1551-2916.2010.03876.x

Cuk, M., Hamilton, D. P., Lock, S. J., & Stewart, S. T. (2016). Tidal evolution of the Moon from a high-obliquity, high-angular-momentum Earth. *Nature*, *539*(7629), 402–406.

Dietz, R. S. (1964). Sudbury structure as an astrobleme. *The Journal of Geology*, *72*(4), 412–434. https://doi.org/10.1086/626999

Dobrzhinetskaya, L., Mukhin, P., Wang, Q., Wirth, R., O'Bannon, E., Zhao, W., et al. (2018). Moissanite (SiC) with metal-silicide and silicon inclusions from tuff of Israel: Raman spectroscopy and electron microscope studies. *Lithos*, *310–311*, 355–368. https://doi.org/10.1016/j.lithos.2017.04.001

Duncan, M. S., Dasgupta, R., & Tsuno, K. (2017). Experimental determination of CO_2 content at graphite saturation along a natural basalt-peridotite melt join: Implications for the fate of carbon in terrestrial magma oceans. *Earth and Planetary Science Letters*, *466*, 115–128. https://doi.org/10.1016/j.epsl.2017.03.008

Eppelbaum, L., Ben-Avraham, Z., & Katz, Y. (2004). Integrated analysis of magnetic, paleomagnetic and K-Ar data in a tectonic complex region: an example from the Sea of Galilee. *Geophysical Research Letters*, *31*, 1–4. https://doi.org/10.1029/2004GL021298

Fine, G., & Stolper, E. (1986). Dissolved carbon dioxide in basaltic glasses: concentrations and speciation. *Earth and Planetary Science Letters*, *76*, 263–278.

Gogotsi, Y., Welz, S., Ersoy, D. A., & McNallan, M. J. (2001). Conversion of silicon carbide to crystalline diamond-structured carbon at ambient pressure. *Nature*, *411*, 283–287. https://doi.org/10.1038/35077031

Grieve, R. A. F., Ames, D. E., Morgan, J. V., & Artemieva, N. (2010). The evolution of the Onaping Formation at the Sudbury impact structure. *Meteoritics & Planetary Science*, *45*(5):759–782.

Grieve, R. A. F., Stoffler, D., & Deutsch, A. (1991). The Sudbury structure: Controversial or misunderstood? *Journal of Geophysical Research: Planets*, *96*(22), 753–764. https://doi.org/10.1029/91JE02513

Haggerty, S. E. (2014). Carbonado: Physical and chemical properties, a critical evaluation of proposed origins, and a revised genetic model. *Earth-Science Reviews*, *130*, 49–72. https://doi.org/10.1016/j.earscirev.2013.12.008

Haines, J., Cambon, O., Levelut, C., Santoro, M., Gorelli, F., & Garbarino, G. (2010). Deactivation of pressure-induced amorphization in silicate SiO_2 by insertion of guest species. *Journal of the American Chemical Society*, *132*, 8860–8861. https://doi.org/10.1021/ja1034599

Huber, M. S., Črne, A. E., McDonald, I., Hecht, L., Melezhik, V. A., & Koeberl, C. (2014). Impact spherules from Karelia, Russia: Possible ejecta from the 2.02 Ga Vredefort impact event. *Geology*, *42*(5), 375–378. https://doi.org/10.1130/G35231.1

Hutchison, S. G., Richardson, L. S., & Wai, C. M. (1988). Carbothermic reduction of silicon dioxide: A thermodynamic investigation. *Metallurgical Transactions*, *19*, 249–253.

Ionescu, E., Sen, S., Mera, G., & Navrotsky, A. (2018). Structure, energetics and bioactivity of silicon oxycarbide-based amorphous ceramics with highly connected networks. *Journal of the European Ceramic Society*, *38*, 1311–1319. https://doi.org/10.1016/j.jeurceramsoc.2017.10.002

Ionescu, E., Terzioglu, C., Linck, C., Kaspar, J., Navrotsky, A., & Riedel, R. (2013). Thermodynamic control of phase composition and crystallization of metal-modified silicon oxycarbides. *Journal of the American Ceramic Society*, *96*, 1899–1903. https://doi.org/10.1111/jace.12327

Kadik, A., Pineau, F., Litvin, Y., Jendrzejewski, N., Martinez, I. & Javoy, M. (2004). Formation of carbon and hydrogen species in magmas at low oxygen fugacity. *Journal of Petrology*, *45*(7), 1297–1310. doi.org/10.1093/petrology/egh007

Kamo, S., Reimold, W. U., Krogh, T. E., & Colliston, W. P. (1996). A 2.023 Ga age for the Vredefort impact event and a first report of shock metamorphosed zircons in pseudotachylitic breccias and granophyre. *Earth and Planetary Science Letters*, *144*, 369–387. https://doi.org/10.1016/S0012-821X(96)00180-X

Kim, E. J., Fei, Y., & Lee, S. K. (2018). Effect of pressure on the short-range structure and speciation of carbon in alkali silicate and aluminosilicate glasses and melts at high pressure up to 8 GPa: ^{13}C, ^{27}Al, ^{17}O and ^{29}Si solid-state NMR study. *Geochimica et Cosmochimica Acta*, *224*, 327–343. https://doi.org/10.1016/j.gca.2018.01.006

Kinzie, C. R., Hee, S. S. Q., Stich, A., Tague, K. A., Mercer, C., Razink, J. J., et al. (2014). Nanodiamond-rich layer across three continents consistent with major cosmic impact at 12,800 Cal BP. *The Journal of Geology*, *22*, 475–506. https://doi.org/10.1086/677046

Kohn S. C., Brooker R. A., & Dupree R. (1991). 13C MAS NMR: a method for studying CO_2 speciation in glasses. *Geochimica et Cosmochimica Acta*, *55*, 3879–3884.

Le Voyer, M., Kelley, K. A., Cottrell, E., & Hauri, E. H. (2017). Heterogeneity in mantle carbon content from CO_2-undersaturated basalts. *Nature Communications*, *8*, 14062. https://doi.org/10.1038/ncomms14062

Lightfoot, P. C. (2017). *Nickel sulfide ores and impact melts: Origin of the Sudbury igneous complex*. Amsterdam: Elsevier.

Mera, G., Navrotsky, A., Sen, S., Kleebe, H.-J., & Riedel., R. (2013). Polymer-derived SiCN and SiOC ceramics: Structure and energetics at the nanoscale. *Journal of Materials Chemistry*, *1*, 3826–3836.

Mysen, B. (2017). Experimental, in-situ carbon solution mechanisms and isotope fractionation in and between (C-O-H)-saturated silicate melt and silicate-saturated (C-O-H) fluid to upper mantle temperatures and pressures. *Earth and Planetary Science Letters*, *459*, 352–361. https://doi.org/10.1016/j.epsl.2016.11.051

Nguyen, A. N., & Messenger, S. (2011). Presolar history recorded in extraterrestrial materials. *Elements*, *7*(1), 17–22. https://doi.org/10.2113/gselements.7.1.17

Oganov, A. R., Ono, S., Ma, Y., Glass, C.W., & Garcia, A. (2008). Novel high-pressure structures of $MgCO_3$, $CaCO_3$ and CO_2 and their role in Earth's lower mantle. *Earth and Planetary Science Letters*, *273*, 38–47. https://doi.org/10.1016/j.epsl.2008.06.005

Pantano, C. G., Singh, A. K., & Zhang, H. (1999). Silicon oxycarbide glasses. *Journal of Sol-Gel Science and Technology*, *14*, 7–25.

Park, J. N., Yoo, C. S., Iota, V., Cynn, H., Nicol, M. F., & Le Bihan, T. (2003). Crystal structure of bent carbon dioxide phase IV. *Physical Review B*, *68*, 14107. https://doi.org/10.1103/PhysRevB.68.014107

Pratesi, G., Viti, C., Cipriani, C., & Mellini, M. (2002). Silicate-silicate liquid immiscibility and graphic ribbons in Libyan Desert glass. *Geochimica et Cosmochimica Acta*, *66*(5), 903–11. https://doi.org/0016-7037/02

Sen, S., Widgeon, S. J., Navrotsky, A., Mera, G., Tavakoli, A., Ionescu, E., et al. (2013). Carbon substitution for oxygen in silicates in planetary interiors. *Proceedings of the National Academies of Science*, *110*, 15904–15907. https://doi.org/10.1073/pnas.1312771110

Seto, Y., Nishio-Hamane, D., Nagai, T., Sata, N., & Fujino, K. (2010). Synchrotron X-ray diffraction study for crystal structure of solid carbon dioxide CO_2-V. *Journal of Physics: Conference Series*, *215*, 012015. https://doi.org/10.1088/1742-6596/215/1/012015

Tavakoli, A. H., Armentrout, M. M., Sen, S., & Navrotsky, A. (2015). Hydrogenated Si-O-C nanoparticles: Synthesis, structure and thermodynamic stability. *Journal of Materials Research*, *30*, 295–303. https://doi.org/10.1557/jmr.2014.376

Ubide, T., Guyett, P. C., Kenny, G. G., O'Sullivan, E. M., Ames, D. E., Petrus, J. A., et al. (2017). Protracted volcanism after large impacts: Evidence from the Sudbury impact basin. *Journal of Geophysical Research: Planets*, *122*, 701–728. https://doi.org/10.1002/2016JE005085

Varga, T., Navrotsky, A., Moats, J. L., Morcos, R. M., Poli, F., Müller, K., et al. (2007). Thermodynamically stable $Si_xO_yC_z$ polymer-like amorphous ceramics. *Journal of the American Ceramic Society*, *90*, 3213–3219. https://doi.org/10.1111/j.1551-2916.2007.01874.x

Widgeon, S., Sen, S., Mera, G., Riedel, R., & Navrotsky, A. (2010). ^{9}Si and ^{13}C solid-state NMR spectroscopic study of nanometer-scale structure and mass fractal characteristics of amorphous polymer derived silicon oxycarbide ceramics. *Chemistry of Materials*, *22*, 6221–6228.

Yong, X., Tse, J. S., & Chen, J. (2018). Mechanism of chemical reactions between SiO_2 and CO_2 under mantle conditions. *ACS Earth and Space Chemistry*. https://doi.org/10.1021/acsearthspacechem.7b00144.

6

Structural and Chemical Modifications of Carbon Dioxide on Transport to the Deep Earth

Mario Santoro[1,2], Federico A. Gorelli[1,2], Kamil Dziubek[2,3], Demetrio Scelta[2,3], and Roberto Bini[2,3,4]

ABSTRACT

The structural and chemical changes to which carbon dioxide is subjected with increasing pressure and temperature are discussed here with the purpose of following the modifications of this important geochemical material on proceeding from the Earth's surface down to the core-mantle boundary. The relevance of metastabilities, and then of kinetic controlled transformations, is evidenced in the P-T ranges characteristic of both molecular phases and extended covalently bonded structures. From a chemical point of view, this analysis highlights how the characterization of the melting of the extended structures would represent an important step to understand the role of this compound in the chemistry of the Earth's mantle.

6.1. INTRODUCTION

Carbon dioxide is central in our lives, as an atmospheric component, for its contribution to global warming (Jenkinson et al., 1991), and also for its pivotal role in the physics and chemistry of the Earth's interior. In fact, carbon entering the Earth's mantle through the subduction of carbon-bearing minerals at the ocean floors not only contributes to determining the physical properties of the mantle, it also affects the Earth's interior dynamics being involved in the processes that rule the heat transfer to the surface. The presence of CO_2 has been indirectly revealed by inclusions within diamonds (Schrauder & Navon, 1993) or mantle rock-forming minerals (Frezzotti & Touret, 2014), and it has been concluded that CO_2-rich phases characterize the mantle (Dasgupta et al., 2011). Even though the distribution is rather inhomogeneous (Le Voyer et al., 2017) and the origin unclear, CO_2 release from the most stable carbonates, such as magnesite at the core-mantle interface, could be one of the possible sources (Isshiki et al., 2004). This presence is relevant in many respects, such as favoring the silicate melting and the consequent cycling to the crust of different elements and volatiles (Dasgupta & Hirschmann, 2006), or as an intermediate in the formation of diamonds (Maeda et al., 2017). In addition, the possibility that CO_2 may react with silica under high pressure and temperature conditions, leading to the formation of new compounds, is also intriguing because of the different coordination of C and Si solid oxides under any given P-T conditions. A considerable computational (Aravindh et al., 2007; Morales-García et al., 2013; Qu et al., 2016; Yong et al., 2018; Zhou et al., 2014) and experimental (Santoro et al., 2011; Santoro et al., 2014) effort has been made to solve this specific issue. Knowledge of the structural and chemical properties of CO_2 with increasing pressure and temperature is therefore essential for understanding its role in the Earth's deep carbon cycle.

The phase diagram of CO_2 is, among the simplest molecular systems, one of the most extensively studied

[1] *CNR-INO, Istituto Nazionale di Ottica, Firenze, Italy*
[2] *LENS, European Laboratory for Non-linear Spectroscopy, Firenze, Italy*
[3] *CNR-ICCOM, Istituto di Chimica dei Composti OrganoMetallici, Firenze, Italy*
[4] *Dipartimento di Chimica "Ugo Schiff", Università di Firenze, Firenze, Italy*

Carbon in Earth's Interior, Geophysical Monograph 249, First Edition. Edited by Craig E. Manning, Jung-Fu Lin, and Wendy L. Mao.

(see Datchi & Weck, 2014; Santoro & Gorelli, 2006; Yoo, 2013; and references therein) but also one of the most debated, especially for pressures in excess of 60 GPa where the formation of extended covalent solids is reported. Multiple controversies surround the nature and the stability region of the different CO_2 crystalline polymorphs all lying below 45 GPa, the number and the differences among the covalent extended phases, the existence of intermediate bonding structures, and finally the stability of the high-pressure phases against dissociation. The reasons for disagreement among different authors can be attributed to two main factors. The first is the pronounced metastability of most of the CO_2 phases, which implies the existence of large kinetic barriers to be overcome for accomplishing the structural transitions or the chemical transformations. In a simplified view, this process can be seen as a potential energy landscape where the pathway followed along a specific coordinate is characterized, from energetic point of view, by one or more local minima, each one related to an intermediate state. The height of the activation barriers separating these minima becomes the key parameter ruling the transformation towards thermodynamically stable species. The P-T path followed in the experiment therefore becomes extremely important. The second issue is experimental in nature and is related to the challenging pressures and temperatures required for exploring the most controversial P-T regions. Experimental results also appear influenced by the sample environment, by hydrostatic or nonhydrostatic conditions, and by the way in which high temperatures are achieved.

In an attempt to provide a clear description of the present knowledge on these issues, we have organized this manuscript in the following manner. The first section will discuss the crystalline molecular phases; the second section will be concerned with the extended covalent structures. The knowledge of the melting line of both molecular and extended solid phases is rather limited, but the topic deserves a careful discussion, and it will be reviewed in the third section. The high-pressure high-temperature stability of carbon dioxide versus dissociation will be the last topic discussed.

6.2. MOLECULAR CRYSTAL PHASES

CO_2 crystallizes at ambient temperature around 0.5 GPa into a primitive cubic cell ($Pa\bar{3}$, Z = 4) with the molecules aligned with the cell body diagonal. The evolution with pressure of the structural and vibrational properties of this phase, known as *dry ice* or phase I, have been recently extensively investigated (Giordano et al., 2010). Phase I transforms to the orthorhombic phase III (*Cmca*, Z = 4) with a sluggish transition starting just above 10 GPa and extending for several GPa (Aoki et al., 1994; Hanson, 1985; Iota & Yoo, 2001). Almost no volume change exists between the two crystal structures that appear strictly related: in both phases the carbon atoms are located at the face-centered positions, differing only in the orientation of the molecules, which break the F-lattice symmetry (Aoki et al., 1994). A pairing of the molecules was speculated for this phase, leading to an anomalous high-strength molecular crystal (Yoo, 2013; Iota & Yoo, 2001). Ab initio calculations challenged this conclusion; in fact, phase III resulted in a typical van der Waals crystal with a bulk modulus one order of magnitude smaller than that derived from the experimental data. This difference was ascribed to a disputable extrapolation to ambient pressure of the experimental data all acquired above 20 GPa (Bonev et al., 2003).

A new phase (II) was observed by heating phase III above 16 GPa and 500 K (Iota & Yoo, 2001). This transition was not reversible on cooling and phase II was recovered at ambient temperature, thus suggesting the metastable nature of phase III. Raman and powder X-ray diffraction (XRD) data were interpreted also in this case as suggestive of a molecular pairing leading to a very large bulk modulus. These conclusions were demonstrated to be wrong by a synchrotron XRD study where the structure was solved using samples in quasihydrostatic conditions (Datchi et al., 2014). The structure was identified as tetragonal ($P4_2/mnm$; Z = 2) with molecular and intermolecular parameters typical of a molecular crystal in perfect agreement with density functional theory (DFT) predictions (Bonev et al., 2003). Interestingly, the computed thermodynamic domain of the $P4_2/mnm$ structure matched the entire P-T region of both phases II and III, thus supporting the suggested metastability of phase III.

Phase II transforms above 13 GPa into a new phase, IV, once heated above 500 K. The transformation temperature does not depend on pressure at least within the accuracy of the experimental data (Gorelli et al., 2004). This phase could be quenched to ambient temperature being stable upon compression and transforming to phase I below 10 GPa, thus suggesting that phase II is also metastable. Spectroscopic (Yoo et al., 2001) and XRD (Park et al., 2003) data were interpreted as suggestive of a bent molecule with elongated C=O bonds, a structure intermediate between a molecular crystal and an extended covalent solid. Also in this case, these conclusions have been disputed by computational studies that found phase IV to be a typical molecular phase, with the CO_2 molecule maintaining the bond length and the linearity of the isolated molecule (Bonev et al., 2003). IR and Raman data also support a typical molecular crystal structure (Gorelli et al., 2004). Recent structural determination based on synchrotron single crystal XRD data of phase IV indicated rhombohedral space group $R\bar{3}c$, with two linear inequivalent molecules in the unit cell and bond length close to that of the free molecule (Datchi et al., 2009). These findings were confirmed by first-principles calculations, which were also capable of perfectly reproducing the structure and the vibrational spectra, and by neutron diffraction data that confirmed the structure (Palaich et al., 2014; Palaich, 2016).

A new phase, indicated as phase VII, was discovered by compressing phase I above 640 K and 12 GPa with a stability pressure range increasing with temperature (Giordano & Datchi, 2007). The transition was found to be fully reversible. High-quality synchrotron XRD patterns acquired along the 726 K isotherm provided evidence of an orthorhombic unit cell (*Cmca*, Z = 4). Such structure was predicted to be stable at high pressure by different computational studies (Bonev et al., 2003; Kuchta & Etters, 1988). The C=O bond length determined for this phase is only slightly shorter than in phase I (<2%), thus also supporting the molecular character of phase VII. Interestingly, the *Cmca* space group of phase VII is the same as that of the high-pressure phase III (Aoki et al., 1994; Yoo et al., 1999); symptomatic is also the fact that both phases form only from phase I. Indeed, whereas phase VII transforms upon compression to phase IV, it was never reobtained by decompressing phase IV. All these aspects led to the reasonable question if phase VII is thermodynamically stable. Confirmation of its thermodynamic stability was provided by the direct crystallization of this phase from the liquid at 827 K and 12.1 GPa (Giordano et al., 2007); however, when phase IV was produced by compressing phase VII, it could be isothermally decompressed until melting. In this way, the melting line of phase IV was also measured between 800 and 950 K, becoming almost indistinguishable from that of phase VII. The energy difference between these phases is therefore small, and a large energy barrier prevents the IV–VII transformation.

Very recently, quasi-harmonic electronic structure calculations were able to accurately reproduce the structural and spectroscopic properties of the molecular crystal phases I, II, IV, and VII of carbon dioxide in a broad pressure range (Sontising et al., 2017). On the contrary, the structure proposed for phase III directly relaxes to phase VII and no other structure is found by crystal structure prediction. Phases III and VII are therefore suggested to be identical and kinetically accessible at ambient temperature and thermodynamically stable above 640 K.

As a final comment, it should be remarked that the pressure evolution of the experimental volume data relative to phases I, II, III, IV, and VII is very similar and markedly different from that of phase V (see Figure 6.1), thus unequivocally pointing to the analogous (molecular) nature of all these phases and ruling out the misleading concept, both from a chemical and structural point of view, of the existence of intermediate structures between molecular and extended phases.

6.3. EXTENDED COVALENT PHASES

At pressures >40 GPa carbon dioxide transforms into silica-like extended covalent phases, thus removing, as occurs also for nitrogen (Eremets et al., 2004), the apparent ambient pressure anomalies with respect to the heavier elements of the group. The first evidence of an extended phase (V) was the appearance in the Raman spectrum of a characteristic peak of the C-O-C stretching mode after laser heating phase III to 1800 K above 40 GPa (Iota et al., 1999). This new phase could be quenched at ambient temperature down to approximately 10 GPa where it transformed to phase I (Datchi & Weck, 2014).

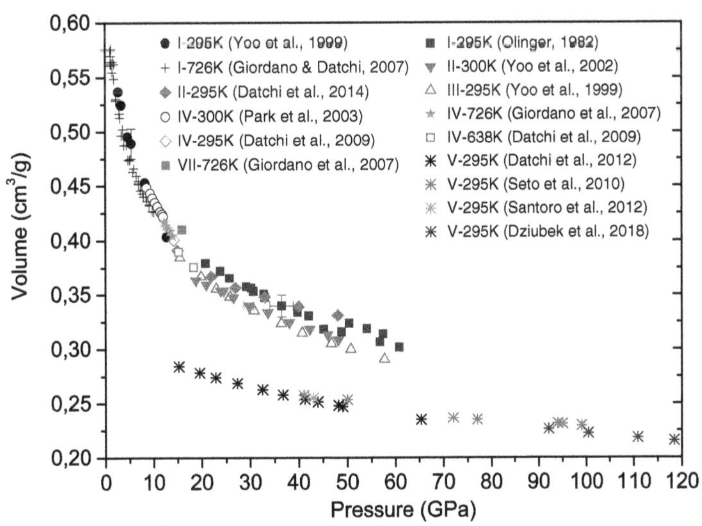

Figure 6.1 Experimental volume data of the extended phase V and of the stable and metastable molecular phases of crystalline CO_2. See electronic version for color representation of the figures in this book.

Lower pressure values have also been reported with phase V transforming directly to the fluid (Iota et al., 1999). Recently, phase V was claimed to be quenchable at ambient pressure and low temperature (Yong et al., 2016), but this conclusion was questioned because residual pressure was inferred by the frequency of the Raman modes of coexisting phase I (Datchi et al., 2017). The P-T conditions necessary for the attainment and its large stability range evidence the existence of a remarkably high energy barrier, which must be overcome to accomplish the chemical transformation from the molecular phases to phase V. This was recently demonstrated by synthesizing this phase just above 20 GPa by laser heating phase IV using Ti as a catalyst (Sengupta et al., 2012). The temperature necessary for the formation of phase V found in two follow-up studies was considerably lower (T < 1000 K) (Santoro et al., 2004; Tschauner et al., 2001).

The crystal structure of phase V has been debated for more than one decade. The structure proposed on the basis of the first synchrotron XRD studies resembled those of the tridymite family with an orthorhombic $P2_12_12_1$ unit cell and a bulk modulus of 365 GPa, therefore larger than stishovite (Yoo et al., 1999). These findings were not supported by a series of DFT computational works that predicted as stable phases α-quartz (Serra et al., 1999) or a distorted β-cristobalite structure (Dong, Tomfohr, & Sankey, 2000; Dong, Tomfohr, Sankey, Leinenweber, et al., 2000; Holm et al., 2000), with bulk modulus two to three times lower. The stability of the β-cristobalite structure was ascribed to the rigidity of the $\widehat{COC}$ angles between connected tetrahedra (120°–125°), to be compared with the wider analogous angles in silica (145°–150°), thus speculating that the extended covalent phases of CO_2 will not show the same abundant polymorphism as silica (Dong, Tomfohr, & Sankey, 2000). The structure of phase V was first suggested (Seto et al., 2010) and then independently solved by two different groups (Datchi et al., 2012; Santoro et al., 2012). As predicted by DFT calculations, a partially collapsed β-cristobalite structure (space group $I\bar{4}2d$) was found, with carbon and oxygen atoms arranged in slightly distorted CO_4 tetrahedra (tilted by ~38° about the *c* axis) linked by oxygen atoms at the corners (Figure 6.2). Computed values of the bulk modulus were definitely smaller than the one reported for the tridymite structure ranging from 115 to 154 GPa (Dong, Tomfohr, Sankey, Leinenweber, et al., 2000; Gohr et al., 2013; Holm et al., 2000; Oganov et al., 2008; Santoro et al., 2012). Experimentally, the first study provided a higher bulk modulus of 236 GPa (Seto et al., 2010), but following XRD studies performed on decompression in both pure CO_2-V samples and powder hydrostatically compressed with He allowed an accurate equation of state (EOS) from 8 to 65 GPa providing a bulk modulus of 136 GPa (Datchi et al., 2012). Higher pressure experimental volume data up to 100 GPa (Seto et al., 2010) exceed this EOS by ≤5%. Very recently, the synthesis of CO_2-V above one megabar allowed the EOS extension up to 120 GPa (Dziubek et al., 2018). The new volume data and those obtained with CO_2-V compressed in He (Datchi et al., 2012) could be fitted with an EOS that further lowers the bulk modulus to a value (114 GPa) very close to that reported in Santoro et al. (2012). Interestingly, while the *a* lattice parameter nicely fits the evolution of the lower pressure data, the *c* parameter is considerably longer with the *c/a* ratio increasing from 1.655 at 43 GPa (Santoro et al., 2012) to 1.871 at 118 GPa (Dziubek et al., 2018), thus getting closer to the fully collapsed structure, which expects a *c/a* ratio equal to 2 (Figure 6.2).

The formation of an extended amorphous phase was evidenced by FTIR spectroscopy compressing phase III at ambient temperature (Santoro et al., 2006). This material, called *a-carbonia* in analogy with silica glass, forms when pressure exceeds 45 GPa with a sluggish transformation that can be efficiently speeded up by moderate heating. The formation of an amorphous phase in the same pressure range was predicted by Serra et al. (1999). The XRD pattern and the vibrational spectra of *a-carbonia* are characterized by broad features that cannot be confused with those of phase V. First-principles molecular dynamics simulations have been capable to describe all the experimental signatures of this amorphous phase, providing insight into its bonding arrangement, which results in a mixture of three- and four-fold coordinated carbon atoms (Sun et al., 2009; Montoya et al., 2008). Interestingly, *a-carbonia* represents an excellent intermediate step for producing phase V. In fact, its strongest and characteristic IR absorption band is resonant with the CO_2 laser lines so that it can be efficiently and cleanly heated without the need of absorbing materials (Santoro et al., 2012).

Other silica-like extended solids have been claimed but not confirmed by various authors. Although their Raman spectra are significantly different (Yoo et al., 2011), the common occurrence of metastability in the phase diagram of CO_2, due to the high energy barriers characterizing most of the phase transitions, requires extreme caution in view of defining a reliable P-T path independent phase diagram. A stishovite-like phase VI (space group $P4_2/mnm$), was claimed by the compression of molecular phase II above 50 GPa and temperatures ranging between 530 and 650 K (Iota et al., 2007), with the carbon atoms having a six-fold coordination. Molecular dynamics simulations were not able to reproduce this phase, which was interpreted as due to an incomplete transformation (Sun et al., 2009). Recently, this P-T region was carefully investigated by synchrotron XRD with a resistively heated DAC in a P-T domain similar to that of Iota et al. (2007), without finding evidence of its existence (Santoro, M. unpublished data). Two other phases that appear as

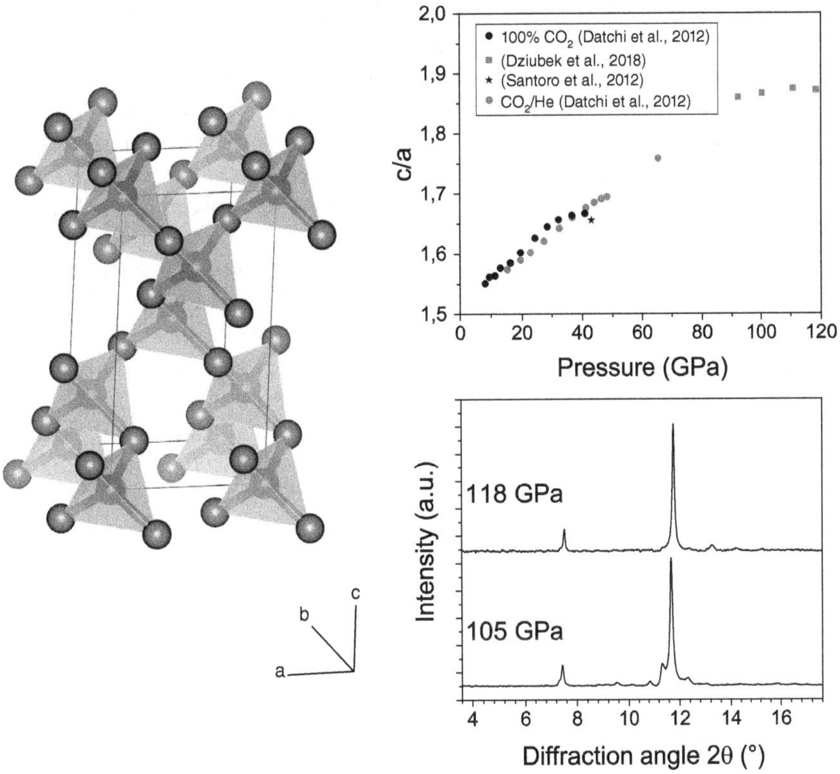

Figure 6.2 Left: refined structure of CO_2-V at 118 GPa. Right: in the lower panel the XRD patterns measured (wavelength of 0.3738 Å) after two consecutive laser heating cycles (2700 K) are reported. The annealing effects leading to the disappearance of spurious peaks are evident in the second (118 GPa) pattern. In the top panel the *c/a* ratio of phase V measured in different experiments is reported. See electronic version for color representation of the figures in this book.

kinetically frozen intermediate structures are coesite-like carbon dioxide (c-CO_2), obtained only by laser-heating quenched phase VI (Sengupta & Yoo, 2010), and CO_2-VIII, claimed to coexist with phase V above 50 GPa and at undefined temperature (lower than that necessary for the synthesis of phase V) (Sengupta & Yoo, 2009).

Upon laser heating to 1700–1800 K above 85 GPa, the extended phases CO_2-V, *a-carbonia*, and the alleged phase VI are reported to transform into a new extended ionic solid (i-CO_2) that is stable during decompression down to 10 GPa, where phase I is obtained (Yoo et al., 2011). The Raman spectrum of this material is claimed to match that of orthorhombic ($P2_12_12$) carbonyl carbonate ($COCO_3$). These results have been recently challenged by the synthesis of CO_2-V above 1 Mbar (Dziubek et al., 2018). In this work, phase V was synthesized by laser heating amorphized CO_2 at 85 GPa and 2700 K; the quenched phase V was laser heated again at the same temperature and a pressure just above 1 Mbar, the same conditions that should ensure the formation of i-CO_2, but the only effect was the obtainment of a high-quality perfectly annealed phase V (Figure 6.2). According to these results, phase V seems therefore to be the only thermodynamic stable extended phase between 120 and 10 GPa, whereas all the other extended phases are likely incompletely transformed materials, i.e local minima in the energy landscape.

6.4. MELTING LINE

The presence of fluid CO_2 has been reported to be potentially important in establishing the chemico-physical properties of the mantle as well as the depth of its partial melting (Dasgupta & Hirschmann, 2006; Isshiki et al., 2004). Therefore, one of the central issues regarding the presence and the chemistry of CO_2 in the depth is the knowledge of the melting line of both molecular and extended phases.

The melting line was determined by visual observation, fluid-solid (phase I) coexistence in the 300–800 K temperature range, and pressures up to 11 GPa (Giordano et al., 2006). In almost the same P-T range, the structure of liquid CO_2 was recently investigated by both synchrotron XRD and classical molecular dynamics simulations showing an anisotropic short-range structure (Datchi et al., 2016). This is essentially related to a change in the orientation of some molecules in the first shell from the characteristic quadrupole-driven T-shape arrangement to a slipped parallel configuration. The amount of molecules assuming the latter configuration increases with pressure in close analogy to the structural arrangement evolution in the crystal phases, since the two relative orientations are those characterizing phase I (T-shape) and phases III and VII. The melting line has been further extended up to 950 K and 15 GPa, also looking for the fluid-solid coexistence (Giordano & Datchi, 2007). The melting of phase I was observed up to 808 K (11.7 GPa), and further compression of phase I produced phase IV, whose decompression in the fluid allowed to determine the triple point between the liquid and phases I and IV at 800 K and 11.2 GPa. At 810 K (11.8 GPa), the crystal in equilibrium with the fluid was CO_2-VII, whose melting line could be followed up to 920 K. Compression of phase VII produced phase IV, which could be decompressed down to the melting so that the melting curve of phase IV could be extended up to 946 K, becoming practically indistinguishable from that of phase VII. Extrapolation of the melting line and of the IV–VII phase boundary leads to a liquid-IV-VII triple point at about 1030 K and 18.4 GPa. A good agreement with these results was shown by a more recent laser heating study that extended the melting line up to 29.5 GPa and 1600 K (Litasov et al., 2011). No additional extension of the melting line was possible because both phases II and III transformed to the extended phase V when heated above 35 GPa, and further heating of phase V between 35 and 70 GPa caused dissociation of carbon dioxide in ϵ-oxygen and elemental carbon. A triple point between the fluid, phase IV, and the extended phase V was identified at 33 GPa and 1720 K. Another triple point between phase V, fluid, and dissociated CO_2 is located around 35 GPa and 1900 K, thus making the stability region of extended CO_2 extremely limited at the P-T conditions typical of the mantle. As already stated, this claim was recently confuted by producing CO_2-V at pressures and temperatures largely exceeding these conditions (Dziubek et al., 2018).

Just as there is little available experimental data regarding the melting line at P-T conditions relevant for the Earth's interior, so too are computational studies rather limited. Ab initio calculations determined a melting line exhibiting a monotonic evolution with a positive slope up to 140 GPa (Boates et al., 2012). In addition, by evaluating the coordination number of carbon atoms with respect to oxygen, it was possible to identify the stability regions of molecular and associated (extended) liquids. A first-order molecular to extended liquid-liquid phase transition is located slightly below 50 GPa at temperatures up to 3000 K. The important consequence of this result is the possible presence of an extended solid in equilibrium with its fluid at the lowermost mantle conditions. Interestingly, CO_2 was found to be stable against dissociation up to 140 GPa and temperatures of the order of 7000 K. In a following report, the same group computed the melting curve up to 71 GPa and the molecular to extended phase transition using a phase coexistence approach from free energy calculations (Teweldeberhan et al., 2013). The triple point between phase IV, extended phase V, and the liquid was computed at 31.8 GPa and 1636 K, in excellent agreement with the experimental data (Litasov et al., 2011). The melting line computed in this work lies at higher temperatures with respect to that computed in previous single-phase simulations (Boates et al., 2012). The intersection of the melting line with the Earth's geotherm occurs at about 40 GPa and 2160 K but with a much-pronounced positive slope, so that the melting line is about 2000 K higher than the Earth's geotherm at 70 GPa. This is a relevant finding because it means that if deposits of pure carbon dioxide in the lower mantle exist, they occur in the form of extended phase V. Interestingly, the liquid in equilibrium with the extended phase V appears also as an extended liquid with a coordination continuously shifting from three to four in analogy with the amorphous extended solid phase (Montoya et al., 2008).

6.5. CARBON DIOXIDE STABILITY VERSUS DISSOCIATION

While it has been long established that, after water, carbon dioxide is the second most important volatile in the deep Earth, its role in carbon fluxes and mantle carbon content is far from being fully understood. As a matter of fact, all the analyses concerning storage of CO_2 in deep reservoirs can be misleading if its chemical form is not taken into account. It is noteworthy that the existence of CO_2-rich fluids, pivotal for major geodynamic processes such as carbon recycling, does not necessarily signify that deposits of pure carbon dioxide are found in the deep mantle. Carbon may be most likely fractionated between the free CO_2 phases, CO_2 dissolved as a stoichiometric component in silicate or carbonate melts, and minerals (carbonates, carbides or diamond). Indeed, decarbonation processes of subducted carbonates involving diamond formation from the mixtures of MORB and $CaCO_3$, or SiO_2 and $MgCO_3$, at pressures above 30 GPa and temperatures in excess of 3000 K were observed by Seto et al. (2008). Moreover, the reactivity of SiO_2 + $MgCO_3$ mixtures was characterized in a large pressure range, evidencing the diamond formation above 80 GPa and considerably lower temperatures (T ≥ 1700 K), as found by Maeda et al. (2017). Neither study provides direct evidence of CO_2 formation and its successive dissociation. Actually, the two-step mechanism of the CO_2 liberation from carbonates and the subsequent dissociation of CO_2 was questioned as a plausible theory explaining the origin of ultra-deep diamonds (Kaminsky, 2017). What is more, recent theoretical studies indicate that direct decomposition of $CaCO_3$ and $MgCO_3$ is unfavorable over the lower mantle, putting into doubt the possibility of free CO_2 existing at these depths (Pickard & Needs, 2015; Santos et al., 2019). It should be also emphasized that possible chemical processes, involving either reactions of CO_2 or carbonates with elemental iron of the Earth's outer core close to the core-mantle boundary (Dorfman et al., 2018; Martirosyan et al., 2019), or with hydrous iron minerals like goethite (Boulard et al., 2018), can dramatically change carbon speciation in the lower mantle. In light of these findings, drawing conclusions concerning the deep carbon cycle on the basis of high pressure-high temperature behavior of pure CO_2 requires extreme care.

Having said that, the stability of CO_2 with respect to dissociation was extensively studied as a major issue in the Earth's mantle chemistry. Dissociation into diamond and fluid oxygen would have important consequences on the oxidation state of the lower mantle because the O_2 formation would consistently increase the oxygen fugacity with the consequent oxidation of surrounding rocks and metals (Litasov et al., 2011). Shock compression experiments report the dissociation of carbon dioxide above 4000 K in the 0–34 GPa pressure range (Nellis et al., 1991; Oehlschlaeger et al., 2005). On the other hand, the first experiments on pure carbon dioxide reporting dissociation in DAC were performed by Tschauner et al. (2001) using a CO_2 laser as heating source without any absorber below 40 GPa, whereas at higher pressures ruby or quartz powders were necessary to reach temperatures between 2500 and 3000 K. The formation of oxygen was evidenced in all the runs by the appearance of the characteristic signatures of the ϵ phase in the Raman spectra, with and without absorbers, when the temperature overcame 2500 K. For temperatures not exceeding 1500 K, the peaks of oxygen were absent and other peaks assigned to an unidentified phase of CO_2 were observed. The instability boundary identified in this study is characterized by a negative slope and exhibits a linear evolution with pressure extending from 28 GPa and 2800 K to 69 GPa and 1700 K. This boundary, although maintaining a very similar negative slope, was shifted to lower temperatures in Litasov et al. (2011). Here, the sample heating was realized through a near IR laser absorbed by iridium couplers. CO_2 dissociated always from the extended phase V for pressures in excess of 35 GPa and temperatures below 1900 K. Both of these studies reporting the dissociation were limited to pressures ≤70 GPa.

These results contrast with the stability of phase V at P ≤ 40 GPa and temperatures up to 3000 K (Park et al., 2003) and with later studies that reported the transformation at 85 GPa and 1700 K from the extended phase V to the ionic compound identified as carbonyl carbonate (Yoo et al., 2011). In all these cases, the sample was indirectly laser heated using Pt, B, or ruby as absorbers. Both dissociation and the formation of an ionic compound are challenged by the most recent studies, where CO_2-V was synthesized with a remarkable quality by laser heating carbonia between 85 GPa, the same pressure as in Yoo et al. (2011), and 115 GPa at temperatures in the order of 2500 K (Dziubek et al., 2018). Remarkably, repeated heating cycles have the only effect of producing a better crystallinity of the extended phase V, as expected in a normal thermal annealing process, and suggest that phase V is the thermodynamic stable phase in this portion of the phase diagram. Annealing effects could explain the difficulty to reproduce the synthesis of many of the claimed extended phases (VI, VIII, i-CO_2) that are likely due to an incomplete transformation of the starting material.

As far as the dissociation is concerned, it is evident that in spite of the different P-T ranges covered, the extrapolation of the proposed instability boundaries (Litasov et al., 2011; Tschauner et al., 2001) to higher pressures suggests dissociation just above 1500 K for pressures in excess of 80 GPa. Two possible explanations are proposed to account for this discrepancy. The first one is related to the occurrence of a redox reaction where the carbon in CO_2 is reduced to diamond or graphite while the metal, and specifically rhenium, is oxidized (Dziubek et al., 2018; Santamaria Perez, 2016a; Santamaria Perez, 2016b). Obviously, this does not explain the results obtained below 40 GPa when CO_2 was laser heated without any absorber (Tschauner et al., 2001). However, in this case the P-T conditions of the dissociation fall within the fluid stability range, where dissociation requires a substantial lower energy to occur. Another possible explanation is related to the different P-T range investigated. Dissociation could be prevented at higher pressures because it would require a molecularity increase, which is disfavoured with increasing pressure. It is also worth mentioning that one of the main arguments employed for supporting the dissociation is the lower molar volume of diamond plus ϵ-O_2 with respect to CO_2-V above 22 GPa (a difference of ~6% around 100 GPa) (Litasov et al., 2011; Tschauner et al., 2001), thus indicating that in the absence of other transformation CO_2 should decompose. However, these data were related to the EOS obtained by using the data from Yoo et al. (1999) based on the wrong tridymite structure, whereas the EOS related to the partially collapsed β-cristobalite structure, determined independently by different groups (Datchi et al., 2012; Dziubek et al., 2018), lies below the one of C+ ϵ-O_2 up to the Mbar range. DFT molecular dynamics simulations do not expect dissociation below temperatures of 6000–7000 K (Boates et al., 2012), and in fact at P-T conditions well above the proposed dissociation experimental boundary (70 GPa and 3000 K), phase V is thermodynamically preferred, with differences in the free energy largely exceeding uncertainties and approximations of the method (Teweldeberhan et al., 2013).

6.6. CONCLUSIONS

We have reviewed here the present knowledge about carbon dioxide under high pressure and high temperature up to the lowermost mantle conditions (Figure 6.3). The P-T region where carbon dioxide exists as a molecular crystal ($P < 40$ GPa and $T < 1000$ K), although not as relevant for the deep Earth carbon cycle, is extremely important from a fundamental point of view. Among the five molecular phases that have been identified, the experimental data suggest that the only thermodynamic stable phases are likely I, IV, and VII. It should be remarked that in spite of being stable or metastable, all these phases are "normal" molecular crystals, in the sense that the molecular structure and the intermolecular interactions are substantially unaltered across the different phases. The observation of metastable phases over large P-T ranges evidences the presence of large energetic barriers and of several local minima in the ground state energy surface, remarking the importance of kinetic control over the transformation processes. Free-energy calculations of all the known experimental phases, particularly if temperature effects are taken into account, would greatly help to clarify the puzzling stability of CO_2 molecular phases.

The kinetic control is even more important in the formation of the new silica-like extended phases. This chemical transformation occurs, depending on the temperature value, between 35 and 60 GPa. The stoichiometry of molecular and extended phases is the same so that the same energy surface is common to all these phases. Also for the extended covalent phases, an important dependence on the P-T path followed is evidenced, leading to a wealth of local minima and corresponding metastable structures. Among the proposed extended polymorphs, phase V appears as the only one thermodynamically stable extending from pressures characteristic of the molecular phases to at least 120 GPa. Phase V is characterized by a β-cristobalite structure composed by CO_4 tetrahedra, linked by oxygen atoms at the corners, whose degree of distortion increases with pressure leading to a more compact crystal. The amorphous counterpart of phase V, *carbonia*, is composed by a mixture of three-fold and four-fold coordinated carbon atoms and forms at temperatures close to ambient and $P \geq 45$ GPa, representing a typical kinetically driven transformation with incomplete realization of the four-fold coordination of phase V.

From the geochemical point of view, the main questions related to carbon dioxide concern the form in which CO_2 exists along the geotherm and its chemical stability with respect to dissociation or to interaction with minerals.

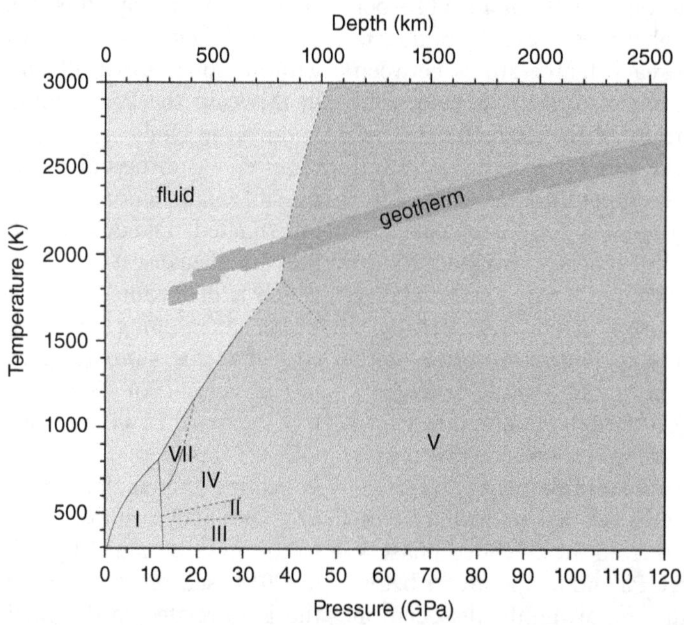

Figure 6.3 Overlap of the phase diagrams of molecular (yellow) and extended (cyan) CO_2. The two colors have been chosen to mark the different chemical nature of the two systems. In both phase diagrams, only the thermodynamically stable phases (with the exception of phases II and III) have been reported. The blended area represents the P-T region where phase V can be quenched on decompression. The melting line is designed using experimental data up to 30 GPa (Giordano et al., 2006; Giordano & Datchi, 2007; Litasov et al., 2011) and the computed data from Teweldeberhan (2013) at higher pressures. See electronic version for color representation of the figures in this book.

Recent reports extended the stability range of phase V down to the lowermost mantle conditions (~2500 km) where CO_2 was supposed to dissociate into diamond and oxygen or possibly transform into carbonyl carbonate. The stability of this phase at these depths was already predicted by computational studies that expect the geotherm to fall within the range of stability of the extended phase V above 40 GPa. The presence of pure CO_2 in the form of a covalently bonded extended solid down to the mantle-core boundary is extremely important from a chemical point of view. In fact, unless melting is induced by interaction with minerals, it is difficult to envisage an active role of CO_2 in modifying the rheological properties and also the partial melting of mantle rocks at these depths. A very important issue not yet experimentally accessed is the evolution of the melting, or alternatively decomposition, line of phase V. The melting is expected, according to computational studies, at remarkably higher temperatures than the geotherm for pressures in excess of 50 GPa; in addition, the nature of the melt could be crucial in determining the chemical activity of this species.

ACKNOWLEDGMENTS

This work was supported by the Deep Carbon Observatory.

REFERENCES

Aoki, K., Yamawaki, H., Sakashita, M., Gotoh, Y. & Takemura, K. (1994). Crystal structure of the high-pressure phase of solid CO_2. *Science*, *263*, 356–358. https://doi.org/10.1126/science.263.5145.356

Aravindh, A. et al. (2007) $Si_xC_{1-x}O_2$ alloys: A possible route to stabilize carbon-based silica-like solids? *Solid State Commun.*, *144*, 273276. https://doi.org/10.1016/j.ssc.2007.09.011

Boates, B., Teweldeberhan, A. M., & Bonev, S. A. (2012). Stability of dense liquid carbon dioxide. *Proc. Natl. Acad. Sci. USA*, *109*, 14808–14812. https://doi.org/10.1073/pnas.1120243109

Bonev, S. A., Gygi, F., Ogitsu, T., & Galli, G. (2003). High-pressure molecular phases of solid carbon dioxide. *Phys. Rev. Lett.*, *91*, 065501. https://doi.org/10.1103/PhysRevLett.91.065501

Boulard, E., Guyot, F., Menguy, N., Corgne, A., Auzende, A. L., Perrillat, J.-P., & Fiquet, G. (2018). CO_2-induced destabilization of pyrite-structured FeO_2Hx in the lower mantle. *National Science Review*, *5*, 870–877. https://doi.org/10.1093/nsr/nwy032

Dasgupta, R., & Hirschmann, M. M. (2006) Melting in the Earth's deep upper mantle caused by carbon dioxide. *Nature*, *440*, 659–662, https://doi.org/10.1038/nature04612

Dasgupta, R., Mallik, A., Tsuno, K., Withers, A. C., Hirth, G., & Hirschmann, M. M. (2011). Carbon-dioxide-rich silicate melt in the Earths upper mantle. *Nature*, *493*, 211–215. https://doi.org/10.1038/nature11731

Datchi, F., Giordano, V. M., Munsch, P., & Saitta, A. M. (2009). Structure of carbon dioxide phase IV: Breakdown of the intermediate bonding state scenario. *Phys. Rev. Lett.*, *103*, 185701. https://doi.org/10.1103/PhysRevLett.103.185701

Datchi, F., Mallick, B., Salamat, A., & Ninet, S. (2012). Structure of polymeric carbon dioxide CO_2-V. *Phys. Rev. Lett.*, *108*, 125701. https://doi.org/10.1103/PhysRevLett.108.125701

Datchi, F., Mallick, B., Salamat, A., Rousse, G., Ninet, S., Garbarino, G., et al. (2014). Structure and compressibility of the high-pressure molecular phase II of carbon dioxide. *Phys. Rev. B*, *89*, 144101. https://doi.org/10.1103/PhysRevB.89.144101

Datchi, F., Moog, M., Pietrucci, F., & Saitta, A. M.(2017). Polymeric phase V of carbon dioxide has not been recovered at ambient pressure and has a unique structure. *Proc. Natl. Acad. Sci. USA*, *114*, E656–E657. https://doi.org/10.1073/pnas.1619276114

Datchi, F., & Weck, G. (2014). X-ray crystallography of simple molecular solids up to megabar pressures: Application to solid oxygen and carbon dioxide *Z. Kristallogr.*, *229*, 135–157. DOI 10.1515/zkri-2013-1669

Datchi, F., Weck, G., Saitta, A. M., Raza, Z., Garbarino, G., Ninet, S. et al.(2016). Structure of liquid carbon dioxide at pressures up to 10 GPa. *Phys. Rev. B: Condens. Matter Mater. Phys.*, *94*, 014201. https://doi.org/10.1103/PhysRevB.94.014201

Dong, J., Tomfohr, J. K., & Sankey, O. F. (2000). Rigid intertetrahedron angular interaction of nonmolecular carbon dioxide solids. *Phys. Rev. B: Condens. Matter Mater. Phys.*, *61*, 5967–5971. https://doi.org/10.1103/PhysRevB.61.5967

Dong, J., Tomfohr, J. K., Sankey, O. F., Leinenweber, K., Somayazulu, M., & McMillan, P. F. (2000). Investigation of hardness in tetrahedrally bonded nonmolecular CO_2 solids by density-functional theory. *Phys. Rev. B: Condens. Matter Mater. Phys.*, *62*, 14685. https://doi.org/10.1103/PhysRevB.62.14685

Dorfman, S. M., Badro, J., Nabiei, F., Prakapenka, V. B, Cantoni, M., & Gillet, P. (2018). Carbonate stability in the reduced lower mantle. *Earth Planet. Sci. Lett.*, *489*, 84–91. https://doi.org/10.1016/j.epsl.2018.02.035

Dziubek, K. F., Ende, M., Scelta, D., Bini, R., Mezouar, M., Garbarino, G., & Miletich, R. (2018). Crystalline polymeric carbon dioxide stable at megabar pressures. *Nat. Comm.*, *9*, 3148. https://doi.org/10.1038/s41467-018-05593-8

Eremets, M. I., Gavriliuk, A. G., Trojan, J. A., Dzivenko, D. A., & Boeler, R. (2004). Single-bonded cubic form of nitrogen. *Nat. Mater.*, *3*, 558–563. https://doi.org/10.1038/nmat1146

Frezzotti, M. L., & Touret, J. L. R. (2014). CO_2, carbonate-rich melts, and brines in the mantle. *Geoscience Frontiers*, *5*, 697–710. https://doi.org/10.1016/j.gsf.2014.03.014

Giordano, V. M., & Datchi, F. (2007). Molecular carbon dioxide at high pressure and high temperature. *EPL*, *77*, 46002. https://doi.org/10.1209/0295-5075/77/46002

Giordano, V. M., Datchi, F., & Dewaele, A. (2006). Melting curve and fluid equation of state of carbon dioxide at high pressure and high temperature. *J. Chem. Phys.*, *125*, 054504. https://doi.org/10.1063/1.2215609

Giordano, V. M., Datchi, F., Gorelli, F. A., & Bini, R. (2010). Equation of state and anharmonicity of carbon dioxide phase I up to 12 GPa and 800 K. *J. Chem. Phys.*, *133*, 144501. https://doi.org/10.1063/1.3495951

Gohr, S., Grimme, S., Shnel, T., Paulus, B., & Schwerdtfeger, P. (2013). Pressure dependent stability and structure of carbon dioxide: A density functional study including long-range corrections. *J. Chem. Phys.*, *139*, 174501. https://doi.org/10.1063/1.4826929

Gorelli, F. A., Giordano, V. M., Salvi, P. R., & Bini, R. (2004). Linear carbon dioxide in the high-pressure high-temperature crystalline phase IV. *Phys. Rev. Lett.*, *93*, 205503. https://doi.org/10.1103/PhysRevLett.93.205503

Hanson, R. C. (1985). A new high-pressure phase of solid carbon dioxide. *J. Phys. Chem.*, *89* (21), 4499–4501. https://doi.org/10.1021/j100267a019

Holm, B., Ahuja, R., Belonoshko, A., & Johansson, B. (2000). Theoretical investigation of high pressure phases of carbon dioxide. *Phys. Rev. Lett.*, *85*, 1258–1261. https://doi.org/10.1103/PhysRevLett.85.1258

Iota, V., & Yoo, C. S. (2001). Phase diagram of carbon dioxide: Evidence for a new associated phase. *Phys. Rev. Lett.*, *86*, 5922–5925. https://doi.org/10.1103/PhysRevLett.86.5922

Iota, V., Yoo, C. S., & Cynn, H. (1999). Quartzlike carbon dioxide: An optically nonlinear extended solid at high pressures and temperatures. *Science*, *283*, 1510–1513. https://doi.org/10.1126/science.283.5407.1510

Iota, V., Yoo, C.-S., Klepeis, J.-H., Jenei, Z., Evans, W., & Cynn, H. (2007). Six-fold coordinated carbon dioxide VI. *Nat. Mater.*, *6*, 34–38. https://doi.org/10.1038/nmat1800

Isshiki, M., Irifune, T., Hirose, K., Ono, S., Ohishi, Y., Watanuki, T., et al. (2004). Stability of magnesite and its high-pressure form in the lowermost mantle. *Nature*, *427*, 60–62. https://doi.org/10.1038/nature02181

Jenkinson, D. S., Adams, D. E., & Wild, A. (1991). Model estimates of CO_2 emissions from soil in response to global warming. *Nature*, *351*, 304–306. https://doi.org/10.1038/351304a0

Kaminsky, F. V. (2017). The Earth's Lower Mantle: Composition and Structure. Springer Geology, doi 10.17/978-3-319-55684-0.

Kuchta, B. & Etters, R. D. (1988). Prediction of a high-pressure phase transition and other properties of solid CO_2 at low temperatures. *Phys. Rev. B*, *38*, 6265–6269. https://doi.org/10.1103/PhysRevB.38.6265

Le Voyer, M., Kelley, K. A., Cottrell, E., & Haur, E. H (2017). Heterogeneity in mantle carbon content from CO_2-undersaturated basalts. *Nat. Comm.*, *8*, 14062. https://doi.org/10.1038/ncomms14062

Litasov, K. D., Goncharov, A. F., & Hemley, R. J. (2011). Crossover from melting to dissociation of CO_2 under pressure: Implications for the lower mantle. *Earth and Planetary Science Letters*, *309*, 318–323. https://doi.org/10.1016/j.epsl.2011.07.006

Maeda, F. A., Ohtani, E., Kamada, S., Sakamaki, T., Hirao, N., & Ohishi, Y. (2017). Diamond formation in the deep lower mantle: a high-pressure reaction of $MgCO_3$ and SiO_2. *Sci. Rep.*, *7*, 40602. https://doi.org/10.1038/srep40602

Martirosyan, N. S., Litasov, K. D., Lobanov, S. S., Goncharov, A. F., Shatskiy, A., Ohfuji, H., & Prakapenka, V. (2019). The Mg-carbonate–Fe interaction: Implication for the fate of subducted carbonates and formation of diamond in the lower mantle. *Geoscience Frontiers*, *10*, 1449–1458. https://doi.org/10.1016/j.gsf.2018.10.003

Montoya, J. A., Rousseau, R., Santoro, M., Gorelli, F. A., & Scandolo, S. (2008). Mixed threefold and fourfold carbon coordination in compressed CO_2. *Phys. Rev. Lett.*, *100*, 163002. https://doi.org/10.1103/PhysRevLett.100.163002

Morales-García, A., Marqués, M., Menéndez, J. M., Santamaría-Pérez, D., Baonza, V. G., & Recio J. M. (2013). First-principles study of structure and stability in SiCO-based materials. *Theor. Chem. Acc.*, *132*, 1308. https://doi.org/10.1007/s00214-012-1308-6

Nellis, W. J., Mitchell, A. C., Ree, F. H., Ross, M., Holmes, N. C., Trainor, R. J., & Erskine, D. J. (1991). Equation of state of shock-compressed liquids: Carbon dioxide and air. *J. Chem. Phys.*, *95*, 5268–5272. https://doi.org/10.1063/1.461665

Oehlschlaeger, M. A., Davidson, D. F., Jeffries, J. B., & Hanson, R. K. (2005). Carbon dioxide thermal decomposition: Observation of incubation. *Z. Phys. Chem.*, *219*, 555–567. https://doi.org/10.1524/zpch.219.5.555.64325

Oganov, A. R., Ono, S., Ma, Y., Glass, C. W., & Garcia, A. (2008). Novel high-pressure structures of $MgCO_3$, $CaCO_3$ and CO_2 and their role in Earth's lower mantle. *Earth Planet. Sci. Lett.*, *273*, 38–47. https://doi.org/10.1016/j.epsl.2008.06.005

Olinger, B. (1982). The compression of solid CO_2 at 296 K to 10 GPa. *J. Chem. Phys.*, *77*, 6255–6258. https://doi.org/10.1063/1.443828

Palaich, S. Carbon in the Deep Earth: A Mineral Physics Perspective. PhD diss., University of California–Los Angeles, 2016.

Palaich, S., Tulk, C. A., Molaison, J. J.,Makhluf, A. R., Guthrie, M., Kavner, A., & Manning, C. E. (2014) Structure and thermal behavior of CO_2-IV at 18 GPa from 300–625 K. *AGU Fall Meeting Abstracts*. http://adsabs.harvard.edu/abs/2014AGUFMMR33A4332P

Park, J.-H., Yoo, C. S., Iota, V., Cynn, H., Nicol, M. F., & Le Bihan, T. (2003). Crystal structure of bent carbon dioxide phase IV. *Phys. Rev. B: Condens. Matter Mater. Phys.*, *68*, 014107. https://doi.org/10.1103/PhysRevB.68.014107

Pickard, C. J., & Needs, R. J. (2015). Structures and stability of calcium and magnesium carbonates at mantle pressures. *Phys. Rev. B*, *91*, 104101. https://doi.org/10.1103/PhysRevB.91.104101

Qu, B., Li, D., Wang, L., Wu, J., Zhou, R., Zhang, B., & Zeng, X. C. (2016). Mechanistic study of pressure and temperature dependent structural changes in reactive formation of silicon carbonate. *RSC Adv.*, *6*, 26650. https://doi.org/10.1039/c5ra21981g

Santamaria Perez, D., McGuire, C., Makhluf, A., Kavner, A., Chuliá Jordan, R., Forda, J. L. et al. (2016a). Strongly-driven $Re+CO_2$ redox reaction at high-pressure and high-temperature. *Nat. Comm.*, *7*, 13647. https://doi.org/10.1038/ncomms13647

Santamaria Perez, D., McGuire, C., Makhluf, A., Kavner, A., Chuliá Jordan, R., Forda, J. L. et al. (2016b). Exploring the chemical reactivity between carbon dioxide and three transition metal (Au, Pt and Re) at high-pressure, high-temperature conditions. *Inorg. Chem.*, *55*, 10793–10799. https://doi.org/10.1021/acs.inorgchem.6b01858

Santoro, M., & Gorelli, F. A. (2006). High pressure solid state chemistry of carbon dioxide *Chem. Soc. Rev.*, *35*, 918–931. https://doi.org/10.1039/B604306M

Santoro, M., Gorelli, F. A., Bini, R., Haines, J., Cambon, O., Levelut, C. et al. (2012). Partially collapsed cristobalite structure in the non molecular phase V in CO_2. *Proc. Natl. Acad. Sci. USA*, *109*, 5176–5179. https://doi.org/10.1073/pnas.1118791109

Santoro, M., Gorelli, F. A., Bini, R., Ruocco, G., Scandolo, S., & Crichton, W. A. (2006). Amorphous silica-like carbon dioxide. *Nature*, *441*, 857–860. https://doi.org/10.1038/nature04879

Santoro, M., Gorelli, F. A., Bini, R., Salamat, A., Garbarino, G., Levelut, C., et al. (2014). Carbon enters silica forming a cristobalite-type CO_2SiO_2 solid solution. *Nat. Comm.*, *5*, 3761. https://doi.org/10.1038/ncomms4761

Santoro, M., Gorelli, F. A., Haines, J., Cambon, O., Levelut, C., & Garbarino, G. (2011). Silicon carbonate phase formed from carbon dioxide and silica under pressure. *Proc. Natl. Acad. Sci. USA*, *108*, 7689–7692. https://doi.org/10.1073/pnas.1019691108

Santoro, M., Lin, J. F., Mao, H. K., & Hemley, R. J. (2004). In situ high P-T Raman spectroscopy and laser heating of carbon dioxide. *J. Chem. Phys.*, *121*, 2780–2787. https://doi.org/10.1063/1.1758936

Santos, S. S. M., Marcondes, M. L., Justo, J. F., & Assali, L. V. C. (2019). Stability of calcium and magnesium carbonates at Earth's lower mantle thermodynamic conditions. *Earth Planet. Sci. Lett.*, *506*, 1–7. https://doi.org/10.1016/j.epsl.2018.10.030

Schrauder, M., & Navon, O. (1993). Solid carbon dioxide in a natural diamond. *Nature*, *365*, 42–44. https://doi.org/10.1038/365042a0

Sengupta, A., Kim, M., Yoo, C.-S., & Tse, J. S. (2012). Polymerization of carbon dioxide: A chemistry view of molecular-to nonmolecular phase transitions. *J. Phys. Chem. C*, *116*, 2061–2067. https://doi.org/10.1021/jp204373t

Sengupta, A., & Yoo, C. S. (2009). Raman studies of molecular-to-nonmolecular transitions in carbon dioxide at high pressures and temperatures. *Phys. Rev. B: Condens. Matter Mater. Phys.*, *80*, 014118. https://doi.org/10.1103/PhysRevB.80.014118

Sengupta, A., & Yoo, C. S. (2010). Coesite-like CO_2: An analog to SiO_2. *Phys. Rev. B: Condens. Matter Mater. Phys.*, *82*, 012105. https://doi.org/10.1103/PhysRevB.82.012105

Serra, S., Cavazzoni, C., Chiarotti, G. L., Scandolo, S., & Tosatti, E. (1999). Pressure-induced solid carbonates from molecular CO_2 by computer simulation. *Science*, *284*, 788–790. https://doi.org/10.1126/science.284.5415.788

Seto, Y., Hamane, D., Nagai, T., & Fujino, K. (2008). Fate of carbonates within oceanic plates subducted to the lower mantle, and a possible mechanism of diamond formation. *Phys. Chem. Miner.*, *35*, 223–229. https://doi.org/10.1007/s00269-008-0215-9

Seto, Y., Nishio-Hamane, D., Nagai, T., Sata, N., & Fujino, K. (2010) Synchrotron X-ray diffraction study for crystal structure of solid carbon dioxide CO_2-V. *J. Phys. Conf. Ser.*, *215*, 012015. https://doi.org/10.1088/1742-6596/215/1/012015

Sontising, W., Heit, Y. N., McKinley, J. L., & Beran, G. J. O. (2017). Theoretical predictions suggest carbon dioxide phases III and VII are identical. *Chem. Sci.*, *8*, 7374–7382. https://doi.org/10.1039/C7SC03267F

Sun, J., Klug, D. D., Martonak, R., Montoya, J. A., Lee, M. S., Scandolo, S., & Tosatti, E. (2009). High-pressure polymeric phases of carbon dioxide. *Proc. Natl. Acad. Sci. USA*, *106*, 6077–6081. https://doi.org/10.1073/pnas.0812624106

Teweldeberhan, A. M., Boates, B., & Bonev, S. A. (2013). CO_2 in the mantle: Melting and solid-solid phase boundaries. *Earth and Planetary Science Letters*, *373*, 228–232. https://doi.org/10.1016/j.epsl.2013.05.008

Tschauner, O. Mao, H. K., & Hemley, R. J. (2001). New transformations of CO_2 at high pressures and temperatures. *Phys. Rev. Lett.*, *87*, 075701. https://doi.org/10.1103/PhysRevLett.87.075701

Yong, X., Liu, H., Wu, M., Yao, Y., Tse, J. S., Dias, R., & Yoo, C.-S. (2016). Crystal structures and dynamical properties of dense CO_2. *Proc. Natl. Acad. Sci. USA*, *113*, 11110–11115. https://doi.org/10.1073/pnas.1601254113

Yong, X., Tse, J. S., & Chen, J. (2018) Mechanism of chemical reactions between SiO_2 and CO_2 under mantle conditions. *ACS Earth Space Chem.*, *2*, 548555. https://doi.org/10.1021/acsearthspacechem.7b00144

Yoo, C.-S. (2013). Physical and chemical transformations of highly compressed carbon dioxide at bond energies. *Phys. Chem. Chem. Phys.*, *15*, 7949–7966. https://doi.org/10.1039/c3cp50761k

Yoo, C. S., Cynn, H., Gygi, F., Galli, G., Iota, V., Nicol, M. F. et al. (1999). Crystal structure of carbon dioxide at high pressure: Superhard polymeric carbon dioxide. *Phys. Rev. Lett.*, *83*, 5527–5530. https://doi.org/10.1103/PhysRevLett.83.5527

Yoo, C. S., Iota, V., & Cynn, H. (2001). C.S. Yoo, V. Iota, and H. Cynn nonlinear carbon dioxide at high pressures and temperatures. *Phys. Rev. Lett.*, *86*, 444–447. https://doi.org/10.1103/PhysRevLett.86.444

Yoo, C. S., Kohlmann, H., Cynn, H., Nicol, M. F., Iota, V., & LeBihan, T. (2002). Crystal structure of pseudo-six-fold carbon dioxide phase II at high pressures and temperatures. *Phys. Rev. B: Condens. Matter Mater. Phys.*, *65*, 104103. https://doi.org/10.1103/PhysRevB.65.104103

Yoo, C.-S., Sengupta, A., & Kim, M. (2011). Carbon dioxide carbonates in the Earth's mantle: Implications to the deep carbon cycle. *Angew. Chem. Int. Ed.*, *50*, 11219–11222. https://doi.org/10.1002/anie.201104689

Zhou, R., Qu, B., Dai, J., & Zeng, X. C. (2014) Unraveling crystalline structure of high-pressure phase of silicon carbonate. *Phys. Rev. X*, *4*, 011030. https://doi.org/10.1103/PhysRevX.4.011030

7

Carbon Redox Chemistry: Deep Carbon Dioxide and Carbonates

Choong-Shik Yoo

ABSTRACT

Carbon dioxide is an important terrestrial volatile, often considered to exist in deep Earth interior, and it plays a critical role in the deep carbon cycle. Yet it is not well understood whether or how such volatile carbon species are incorporated into the Earth's lower mantle and core. Here, we present the phase/chemical diagram of carbon dioxide that signifies the transformation of molecular CO_2 to extended solids, ionic carbonates, and ultimately its elemental decomposition products of C and O_2. We also discuss fundamental barochemical concepts governing carbon redox processes that may occur to recycle carbonaceous species in deep Earth's interior.

7.1. INTRODUCTION

Carbon dioxide exists in the atmosphere (as a major greenhouse gas), the ocean, and near the surface of Earth's crust, primarily as carbonates such as calcite and dolomite (Zhu & Ogasawara, 2002). It is also considered to exist deep in the Earth's mantle (Saal et al., 2002) and play a critical role in the deep carbon cycle (Hazen et al., 2013). The presence of carbon dioxide deep in Earth's interior is apparent from its abundance in magma plumes (Schrauder & Navon, 1993) and from assemblages of carbonate minerals such as aragonite and magnesite (Isshiki et al., 2004). In various thermal and chemical conditions, carbon dioxide converts to a wide variety of carbonaceous species, including diamond, graphite, carbon monoxide, and carbonates. Thus, the transformation of carbon, carbon dioxide, and carbonates at high pressures and temperatures is critical to understanding the origin of deep carbon, carbon budget, and carbon cycle (Hazen et al., 2013; Mao et al., 2003; Sleep et al., 2001).

Chemistries and properties of carbonaceous volatiles such as C, CO, CO_2, and carbonates are not well understood at high pressures and temperatures (P-T). Yet, it is not well understood whether or how such volatile species are incorporated into the Earth's lower mantle and core, resulting in a major uncertainty in the deep carbon cycle. The presence of CO_2 deep in the Earth can alter the properties of minerals and rocks, including the melting temperature and mechanical strength and, thereby, strongly affect the chemical and physical stability of deeply subducting plates (Dasgupta & Hirschmann, 2006; Kelemen & Manning, 2015). The mass and thermal diffusion rates of carbon dioxide into rocks and minerals are important considerations for carbon dioxide sequestration (Kampman et al., 2014; Leckner, 2003), but they are largely unknown at elevated P-T conditions, leaving questions about long-term storages in depleted oil and gas reservoirs. The relative stabilities of carbonate minerals (such as calcite $CaCO_3$, magnesite $MgCO_3$, and siderite $FeCO_3$) with respect to the mixtures of high-density extended CO_2 (polymeric CO_2 or simply XCO_2) and minerals (such as CaO, MgO, and Fe) are not well understood. Yet they are important in understanding both the deep mantle presence of volatile species like CO_2 and their geochemical transportation mechanisms. Importantly, the chemistry of CO_2 with other volatiles like H_2, H_2O, or hydrous minerals is a primitive redox process, which can produce hydrocarbons and other simple organics and drive their evolutions.

Because of the significance of carbon dioxide in geochemistry, geological processes, and the deep carbon cycle, it is important to understand the phase and chemical behaviors of carbon dioxide at high P-T. Hence, we review

Department of Chemistry and Institute of Shock Physics, Washington State University, Pullman, Washington, USA

Carbon in Earth's Interior, Geophysical Monograph 249, First Edition. Edited by Craig E. Manning, Jung-Fu Lin, and Wendy L. Mao.

the phase/chemical diagram of carbon dioxide and present four governing, fundamental transformations: (A) carbon oxidation, (B) carbonate formation, (C) ionization of extended CO_2, and (D) pressure-induced amorphization. These results indicate a large P-T stability domain of extended CO_2 and ionic carbonates, providing a geochemical mechanism for the origin of deep carbon species in the Earth's core and mantle boundary.

7.2. EXPERIMENTAL METHOD

The experiments were performed using laser-heated diamond anvil cells (DACs) coupled with confocal micro-Raman spectroscopy. Carbon dioxide samples were loaded into DACs from a liquid by condensing CO_2 gas to –35 °C and 15 atmospheres. Type IA diamond anvils were used with a flat culet size of 0.3 mm (or 0.1 mm) for higher pressures). A rhenium gasket was preindented to 0.04–0.05 mm in thickness, and a small hole of 0.12 (or 0.06) mm was drilled using an electric-discharge micro-drilling machine. A thin metal foil (Fe, Pt or B) (~0.01 mm) or a small piece of graphite was placed in the sample chamber to absorb the CW Nd: YLF laser (λ = 1054 nm) light and heat the CO_2 sample. The temperature of the sample was determined by measuring thermal emission and then fitting it for a gray-body Planck radiation equation. The pressure of the sample was measured using the Ruby luminescence method (Mao et al., 1986). Existences and transitions of different carbon dioxide phases were determined based on their characteristic Raman spectra.

7.3. RESULTS

The phase diagram of carbon dioxide is rather intricate, with an array of polymorphs exhibiting great diversity in crystal structure, chemical bonding, and collective interaction (Figure 7.1) (Yoo et al., 2011). At pressures above 40 to 60 GPa and temperatures of 300 to 1000 K, carbon dioxide transforms into a wide range of silicate-like extended solids (or XCO_2): four-fold CO_2-V, pseudo-six-fold CO_2-VI, coesite-like c-CO_2, amorphous *a*-carbonia, and ionic carbonate solid *i*-CO_2, all in high density between 3.5 and 3.8 g/cm^3. These are fundamentally new solids, consisting of monolithic 3D network structures of carbon atoms covalently bonded with oxygen atoms, largely in CO_4 tetrahedra (or orthocarbonates) similar to those of silicate minerals. These solids exhibit novel properties such as extremely low compressibility (Datchi et al., 2012; Yoo et al., 1999), strong optical nonlinearity (Iota et al., 1999), and presumably high melting temperatures (as in the case of diamond). The large disparity in chemical bonding between the extended network and molecular CO_2, on the other hand, allows these extended solids to exist over a large pressure domain (down to a few GPa), which provides opportunities for synthesis of novel materials.

To a first approximation, the crystallographic similarities between these extended CO_2 phases and SiO_2 polymorphs seem to indicate a periodic analogy between CO_2 and SiO_2 (Sengupta et al., 2010). Yet, unlike SiO_2, the polymerization of CO_2 accompanies a great level of modification in chemical bonding from weak van der Waals in

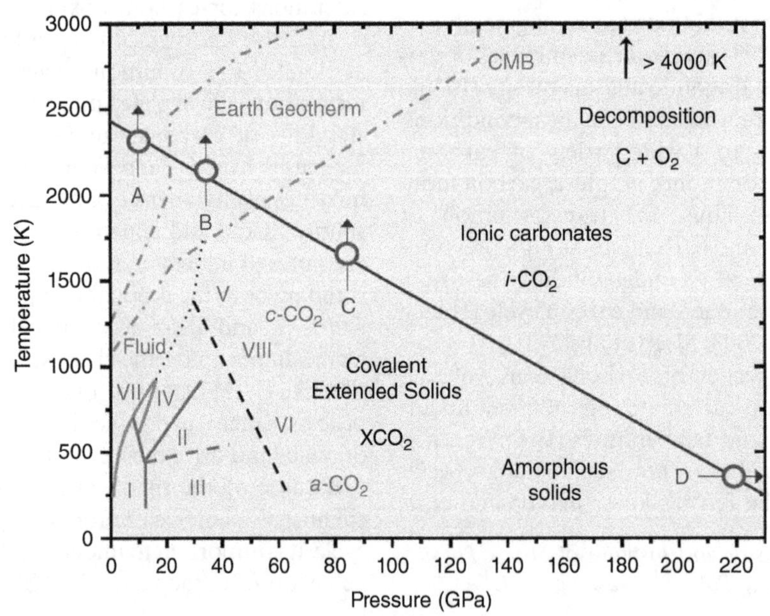

Figure 7.1 Phase and chemical transformation diagram of carbon dioxide, signifying the deep carbon cycle from molecular CO_2 phases to extended CO_2 solids (noted as XCO_2), ionic solids (i-CO_2) and, ultimately, decomposition products of C+O_2 at elevated temperatures and pressures. The blue line signifies the pressure (and temperature)–induced ionization of CO_2 phases, established by the specific transformations along the four blue arrows and described in sections 7.3.1 through 7.3.4. See electronic version for color representation of the figures in this book.

molecular solids (CO_2-I, CO_2 -III, and CO_2 -VII) to strong covalent bonds in extended CO_2. As a result, the phase diagram of carbon dioxide shows several thermal-path dependent phases (e.g., phase VI, c-coesite, and to an extent phase II and IV) and phase boundaries, underscoring the metastability and kinetics over a large P-T domain, making it difficult to determine the exact location of phase boundaries either experimentally or theoretically (Yoo, 2013).

Extended CO_2-V exhibits a stability over a large P-T domain: to 2000 K at 60 GPa and to 200 GPa at ambient temperature, above which it transforms to extended ionic solids (*i*-CO_2 or $COCO_3$) in the structure similar to post-aragonite $CaCO_3$ (Ono et al., 2005) and, ultimately, decomposes to elemental solids of carbon and oxygen at extreme temperatures above 4000 K (according to the previous shock wave studies (Nellis et al. 1991). Recently, the stability of CO_2-V has been reported even for a large P-T space well into the ionization line in Figure 7.1 (Dziubek et al., 2018), which may suggest a strong kinetic barrier associated with the transition. The presence of extended covalent (XCO_2) and ionic (*i*-CO_2) solids over a large P-T regime that overlaps with the Earth geotherm has significant implications to understanding the geochemistry of carbonaceous volatiles (C, CO, CO_2, and carbonates) and deep Earth's carbon cycle. Hence, described in sections 7.3.1 through 7.3.4 are governing carbon transformations along the specific thermal paths marked with arrows in Figure 7.1.

7.3.1. Carbon Oxidation

Carbon dioxide molecules are extremely stable at low pressures below 20 GPa, even to high temperatures above 3000 K. This is in contrast to the polymerization of CO_2 above 40 GPa. It clearly indicates the presence of a strong kinetic barrier for breaking C=O bonds at low pressures than at high pressures. In fact, CO_2-V can be made at a substantially lower pressure than 20 GPa, by heating phase IV in the presence of Ti catalyst (Sengupta et al., 2012). The polymerization of CO_2 is, on the other hand, an oxidation process converting sp^2 hybridized carbon in C=O to sp^3 carbon in C-O. As such, it is thought that the presence of oxygen can assist carbon oxidation and the formation of new CO_2 phases, especially at high temperatures. Figure 7.2 shows the catalytic reaction between hot carbon particles and oxygen molecules, producing CO_2 and $CO^{2+}CO_3^{2-}$, an ionic form of $(CO_2)_2$ at 2300 K and 8 GPa. The spectral evidence includes the characteristic vibrational peaks of CO_3^{2-} at 740 and 1075 cm^{-1} and CO^{2+} at 2230 cm^{-1} for $COCO_3$, and the Fermi resonance ν_- and ν_+ peaks of CO_2 on either sides of the diamond Raman peak at ~1335 cm^{-1}. Similar ionization reactions have previously been observed in nitrogen dioxide dimer (N_2O_4) both at high P-T (Agnew et al., 1983) and at ambient pressure and low temperature (Bolduan et al., 1984). It transforms to $NO^+NO_3^-$, whose vibrational spectrum is analogous to that of $COCO_3$. Figure 7.3 shows a striking similarity in lattice vibrations between $COCO_3$ and $NONO_3$ as an example.

These results clearly indicate the occurrence of carbon oxidation: $C + O_2 \rightarrow CO_2 \rightarrow COCO_3$ at relatively low pressures and high temperatures, thus overcoming the kinetic barrier. Note that $COCO_3$ is still a molecular ionic phase in contrast to extended ionic phase *i*-CO_3 formed at substantially higher pressure above 80 GPa (see Figure 7.1 and discussion in section 7.3.3). Nevertheless, the transformation occurs at the onset of the same phase line (the blue line in Figure 7.1), underscoring high stability of ionic solids over a large P-T region of the Earth mantle

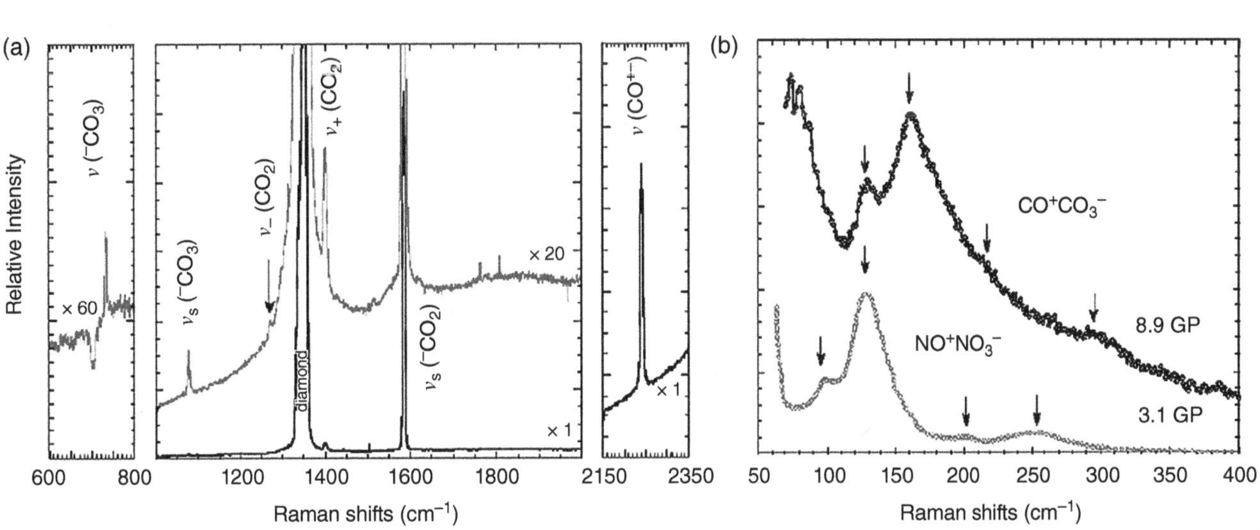

Figure 7.2 Raman spectra of laser-heated C and O_2 mixtures to 2300 K at 10 GPa in the spectral regions for (a) internal vibrations and (b) external lattice modes. It shows the evidences for formation of CO_2 and $COCO_3$ (an ionic form of CO_2 dimer). See electronic version for color representation of the figures in this book.

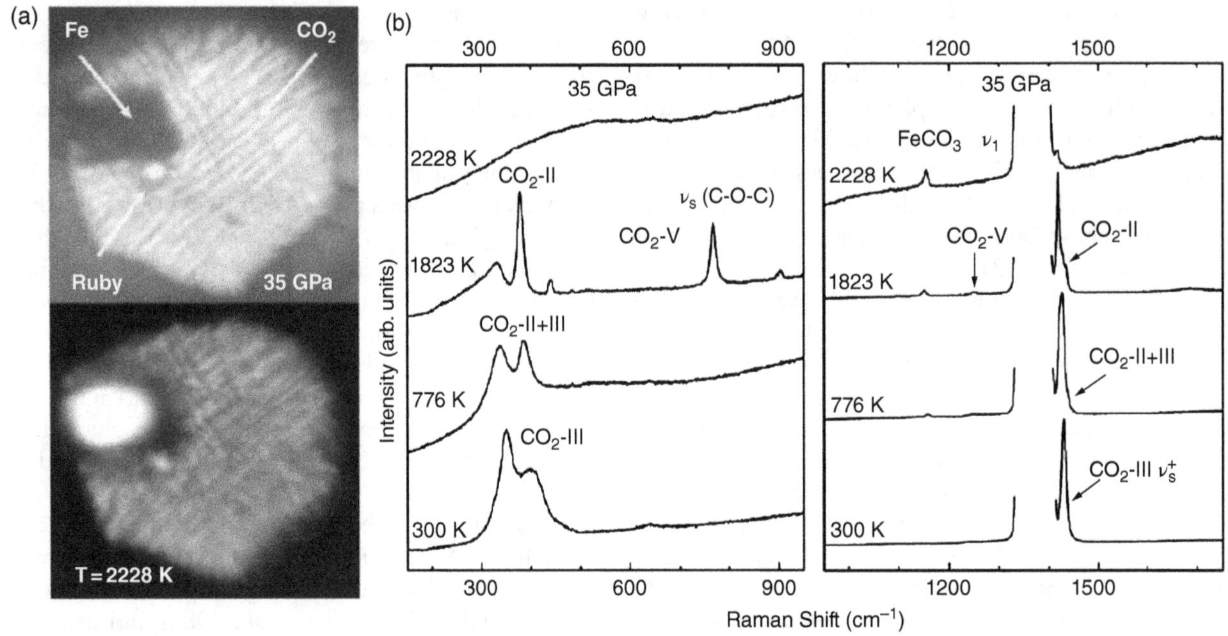

Figure 7.3 (a) Microphotograph of CO_2-Fe samples before (top, at ambient temperature) and during (bottom, at 2228 K) laser heating at 35 GPa in DAC. (b) Raman spectra of laser-heated CO_2 through Fe to 2230 K at 35 GPa, showing the transformation of CO_2-III to CO_2-II, CO_2-V, and $FeCO_3$ as temperature increases. See electronic version for color representation of the figures in this book.

and core. Thus, this result highlights the significant role of extended carbonates in the deep carbon cycle.

7.3.2. Carbonate Formation

The formation of carbonates at high temperatures and low pressures (Figure 7.2) advocates a potential geochemical mechanism of transporting carbonaceous volatiles into deep Earth interiors (Manning et al., 2013). In fact, carbonate minerals (especially of Ca and Mg) are abundant mostly in the Earth's crust (Oganov et al., 2013). To examine the formation of carbonate in deep Earth's mantle, we have investigated the transformation of CO_2 on the presence of Fe abundant in Earth's deeper mantle and core. Figure 7.3a shows the microphotograph of highly strained CO_2-III and Fe samples in DAC, before (top at ambient temperature) and during (bottom at 2228 K) laser heating at 35 GPa. Raman spectra of CO_2 laser heated through Fe were taken at several temperatures at the same pressure, as noted on Figure 7.3b. The results indicate the transformation of CO_2-III to initially CO_2-II at 776 K and then CO_2-V at 1823 K, as well as the formation of siderite $FeCO_3$ at 2228 K. The formation of $FeCO_3$ is apparent from the characteristic vibrational ν_1 peak of carbonate at ~1100 cm^{-1} (Farsang, 2018). Note that the formation of siderite occurs at the onset of the ionization line of CO_2 in Figure 7.1, underscoring the significant role of siderite in the delivery of carbonaceous volatiles deep into the Earth's mantle.

7.3.3. Ionization of XCO_2

The ionic character of the C-O bonds in XCO_2 increases further at high temperatures. Upon laser heating to ~1700–1800 K at 85 GPa, CO_2-V transforms into an extended ionic solid (i-CO_2), characterized by four characteristic Raman bands at 2000, 1200, 850, and 400 cm^{-1}, as shown in Figure 7.4a. Note that the systematic of this Raman spectrum is remarkably similar to those of ionic carbonate $COCO_3$ and previously observed ionic nitrate $NONO_3$ (Agnew et al., 1983; Bolduan et al., 1984). However, there are some important differences. For example, both the ν_1 (CO_3) at 980 cm^{-1} and particularly the ν_s (CO) at 1950 cm^{-1} appear at substantially lower frequencies than those of ionic solids at 1070 cm^{-1} and 2220 cm^{-1}, respectively, at ~10 GPa. Nevertheless, extrapolating the pressure dependence of the ν_1 (NO_3) mode to 85 GPa greatly reduces the difference to within 20–30 cm^{-1}, which can easily account for the mass difference. The substantially greater difference in the ν_s(CO) mode cannot be explained in terms of pressure dependence or mass difference, but it may indicate a local structural difference of carbonyl ions in i-CO_2 from $COCO_3$.

A plausible model for the large softening of ν_s (CO) mode is an incorporation of CO ions to adjacent carbonate layers. This would then lead to a structure analogous to the theoretically predicted mixed coordinated carbonate structures (Montoya et al., 2008; Sun et al., 2009). Hybridization of this type would certainly soften

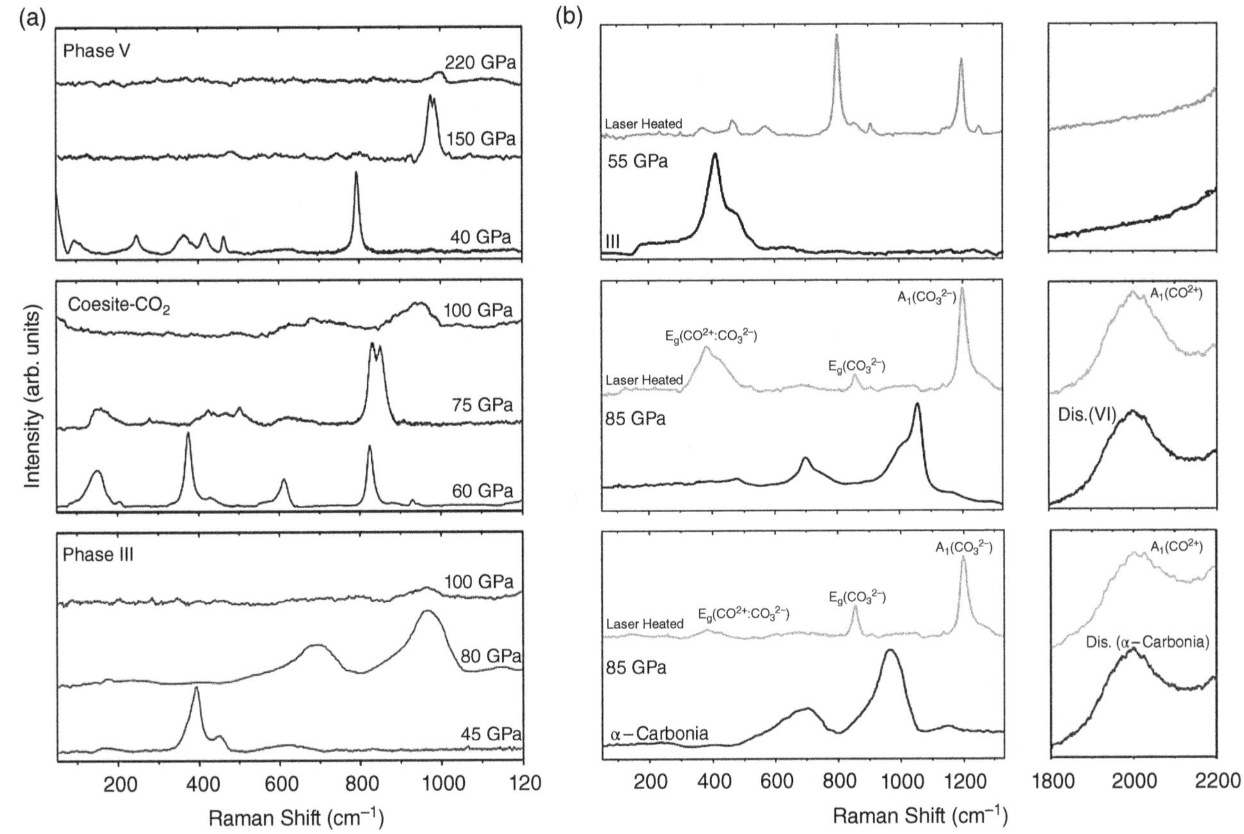

Figure 7.4 (a) Raman spectra of extended carbon dioxide phases before and after laser heating, showing the transformation to extended ionic CO_2 carbonates (*i*-CO_2) at ~85 GPa and 1700–1800 K. (b) Raman spectra of extended carbon dioxide phases to 220 GPa, showing the pressure-induced amorphization of (top) four-fold CO_2-V via new high-pressure phase V′ above 150 GPa, (middle) coesite-CO_2 (*c*-I) via a high pressure-form (*c*-II), and (bottom) phase III via *a*-carbonia, at ambient temperature. See electronic version for color representation of the figures in this book.

ν_s (CO) and ν_1 (CO_3) modes, as observed. In fact, the calculated vibration spectrum shows CO stretching in the range of 1800 to 1900 cm^{-1}, depending on the degree of mixed coordination (Lee et al., 2009; Montoya et al., 2008). On the other hand, it is important to recognize that all calculated structures produce several strong extra bands between 700 and 400 cm^{-1}, which are absent in *i*-CO_2 but present in disordered *a*-carbonia and phase VI. Therefore, the absence of such disordered peaks suggests a fully extended and more ordered *i*-CO_2 structure.

Note that *i*-CO_2 is formed only by heating extended CO_2 solids above 85 GPa; for example, heating phase III at 55 GPa produces only phase VIII (Sengupta & Yoo, 2009), whose Raman spectrum consists of two sharp peaks at 800 and 1200 cm^{-1}, without the presence of the 2000 cm^{-1} peak for CO.

7.3.4. Pressure-Induced Amorphization

Above 40 GPa, carbon dioxide polymerizes to a wide range of covalently bonded extended solids: four-fold CO_2-V (Datchi et al., 2012; Iota et al., 1999; Santoro et al., 2012; Sera et al., 1999; Yoo et al., 2013) and coesite CO_2 (*c*-CO_2) (Sengupta & Yoo, 2010), pseudo-six-fold CO_2-VI (Iota et al., 2007), and mixed three- and four-fold coordinated *a*-carbonia (Santoro et al., 2006), each with a characteristic Raman-active ν_b (C-O-C) bending vibron at around 700–1000 cm^{-1}. Upon further compression to 100–220 GPa, these XCO_2 phases become nonmetallic amorphous solids, as evident from complete loss of their vibrons and optical transparency. Figure 7.4b shows the corresponding spectral change of CO_2-V to 220 GPa as an example.

Note that pressure-induced amorphization occurs at greatly diverse pressures, depending on extended phases: for example, CO_2-VI at 80 GPa, *c*-CO_2 at 100 GPa, and CO_2-V at 220 GPa. In fact, intermediate phases CO_2-III and CO_2-IV become amorphous solids at substantially lower pressures of 60 and 80 GPa, respectively (Santoro et al., 2006; Yoo et al., 2001). Yet it is remarkable that the pressure-induced amorphization occurs when the ν_b vibron reaches about the same Raman frequency, ~1000 cm^{-1}, well above the ν_b (C-O-C) mode of six-fold configuration (CO_6) at ~900 cm^{-1}. Therefore, it is conceivable that the amorphization represents a frustration of four-fold coordinated CO4 to increase either the packing density or the coordination number as pressure increases above 100 GPa.

A similar structural frustration or disorder was observed in CO_2-VI, where carbon atoms are surrounded by an average of six oxygen atoms in a highly distorted octahedral with an average C-O distance of 1.45–1.71 Å (Iota et al., 2007), representing a substantial degree of ionic character in C-O bonds. Considering an approximately 10% in Si-O bond length between four-fold quartz (1.61 Å) and six-fold stishovite (1.76–1.81 Å) (Andrault et al., 1998), we speculate that the C-O bond length in CO_2-VI must increase even larger to ~1.65–1.75 Å in order to accommodate six oxygen atoms around relatively small carbon atoms. With further compression, a separation of this size would eventually lead to a structural destruction forming amorphous solid, in which carbon atoms are in highly mixed coordination with six or more nearest neighbor atoms. The driving force is then to increase the packing density, as is apparent from the ν_b mode shifted to ~1050 cm^{-1} (near that of carbonate). Therefore, it seems that the observed pressure-induced amorphization is driven by enhanced ionicity in carbon-oxygen bonds and topological densification.

7.4. DISCUSSION

The stability of CO_2 carbonates (both CO_2-V and *i*-CO_2) over a large pressure-temperature region relevant to the Earth's mantle and core strongly advocates a possibility that volatile carbon dioxide and carbonates are incorporated deep in the Earth's interior. Figure 7.5 supports this conjecture. It provides, for example, some constraints to evaluate the stability of carbonate minerals in descending slabs. It is also relevant to calculate the thermodynamics and phase assemblies associated with chemical reactions such as $MgCO_3 \rightarrow MgO + CO_2$ and $CaCO_3 \rightarrow CaO + CO_2$ (Dasgupta & Hirschmann, 2006). Using the equation of state of extended CO_2-V, for example, it can be readily shown that the mixture of extended CO_2 and MgO is substantially denser than $MgCO_3$ and other silicate minerals and is better matched to the Preliminary Reference Earth Model (PREM) in the upper mantle below 600 km (see Figure 7.5 inset). Therefore, even if $MgCO_3$ dissociates, CO_2 would remain either as extended CO_2-V or *i*-CO_2 in descending slabs, delivering CO_2 deep into the Earth's mantle. A small contribution of iron in $MgSiO_3$ perovskite or postperovskite (Shim, 2008) can compensate for the density difference resulting from the presence of

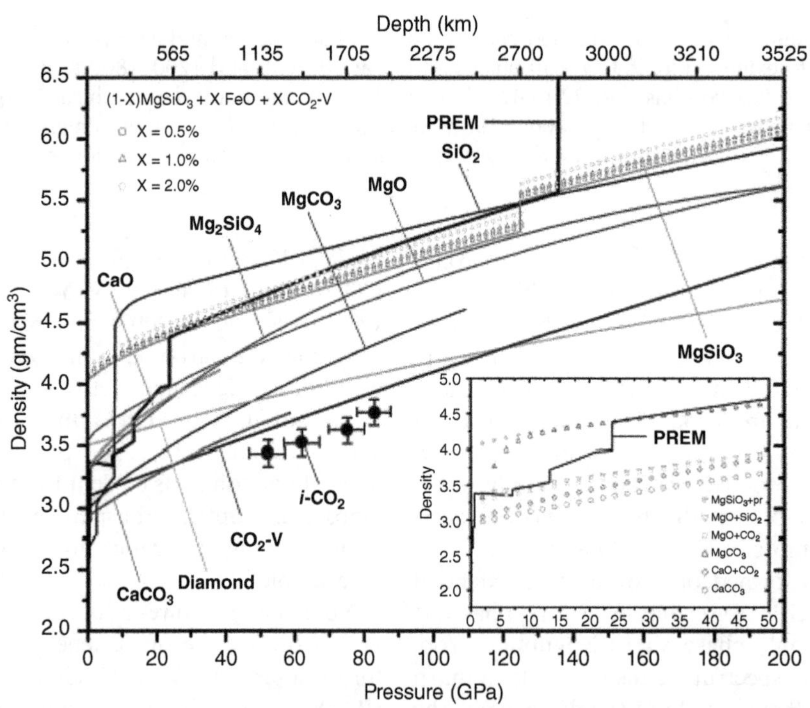

Figure 7.5 The pressure-density plots of various phases of XCO_2 and their analogs (C, $NONO_3$), the major constituents of Earth's mantle ($CaCO_3$, SiO_2, $MgCO_3$, and $MgSiO_3$), and the Preliminary Reference Earth Model (PREM). Also plotted are the calculated densities of a ternary compound of two of the most common minerals in the Earth's core mantle boundary (perovskite and wustite) and extended CO_2: $(1 - x)MgSiO_3 + xFeO + xCO_2$-V, where x = 0.005, 0.01, and 0.02. The inset also shows the similar density variation of two binary compounds of extended CO_2 with MgO and CaO compared with two common minerals from the Earth's mantle, calcite and magnesite, and highlights the distinct possibility that $MgCO_3$ and $CaCO_3$ exist down to the 560 km transition zone of the Earth's mantle. See electronic version for color representation of the figures in this book.

extended CO_2: $(Mg_{1-x}, Fe_x)(Si_{1-x}, C_x)O_3 = (1 - x)MgSiO_3 + xCO_2 + xFeO$, with $x = 0.5\%$ to 2.0% as it moves deeper into the core-mantle boundary from the 670 km transition zone. The decomposition of XCO_2 phases in the outer core region, then, provides a geochemical mechanism for the presence of deep carbon species (such as Fe_3C) (Li et al., 1996; Nakajima et al., 2009) in the Earth's outer core, originally delivered from CO_2 in the atmosphere.

The P-T-induced transformations of CO_2 can be understood in terms of carbon redox processes that may have important implications to the deep carbon cycle (see Figure 7.6a). For example, the pressure-induced transformations of molecular CO_2 to extended CO_2, covalent or ionic alike, are simply a carbon oxidation process, converting the sp^2 C=O bonds in CO_2 to the sp^3 C-O bonds in orthocarbonates (CO_4) or carbonates (CO_3) of XCO_2 phases. The pressure-induced ionization process, on the other hand, is a reduction process that converts the sp^3 C-O bonds back to the sp^2 C=O in *i*-CO_2 and further in C+O_2. Such a reduction process is, then, assisted by high temperature. The rapid temperature rise at the Earth's mantle-core boundary can further accelerate the ionization and decomposition of extended CO_2 or carbonate minerals to form deep carbons and carbonaceous minerals. As such, it gives a rebirth of fully reduced carbon species deep in the Earth's mantle–outer core boundary region, where carbon oxidation can start again as it floats up over a geological time frame.

The pressure-induced transformation in dense CO_2 also highlights several fundamental barochemistry (or pressure-induced solid-state chemistry) concepts, as illustrated in a hypothetical 2D lattice in Figure 7.6b (Yoo, 2017). At low pressures, molecular solids are relatively soft, primarily held by van der Waals interactions. Upon compression, molecular solids transform into extended polymeric structures to soften highly repulsive intermolecular interaction by delocalizing valence electrons or making bonds between nearby molecules (i.e., the pressure-induced electron delocalization). These high-symmetry polymeric structures can further undergo symmetry-breaking distortions (such as Jahn-Teller or Peierls) to enhance their packing efficiencies, which result in low-symmetry distorted or even amorphous structures with highly polarized chemical bonds. Upon further compression, the electrostatic packing energy dominates over the electron hybridization energy, converting these covalent network structures into ionic solids (i.e., pressure-induced ionization) or extended mixtures of constituting elements (phase separation), depending on the chemical/bond/structural miscibility of dense solid interfaces of multiphase and multicomponents. These transformations are strongly controlled by kinetics, giving rise to the path dependence, metastable intermediate states, and near ground-state final products, as observed in CO_2 and many other molecular systems. The collective behaviors of lattice phonons, electrons, and charges in these extended structures, on the other hand, can give rise to novel electro-optical properties such as second harmonic generation (Iota et al., 1999) and high Tc superconductivity (Drozdov et al., 2015).

7.5. CONCLUSION

We have described the transformations of carbon dioxide to extended solids, both covalent and ionic, at high pressures and temperatures. These results suggest the presence of volatile carbon dioxide and carbonates over a wide range of high P-T conditions in deep Earth's mantle, thus providing a geochemical mechanism explaining the origin of deep carbon species in the Earth's core

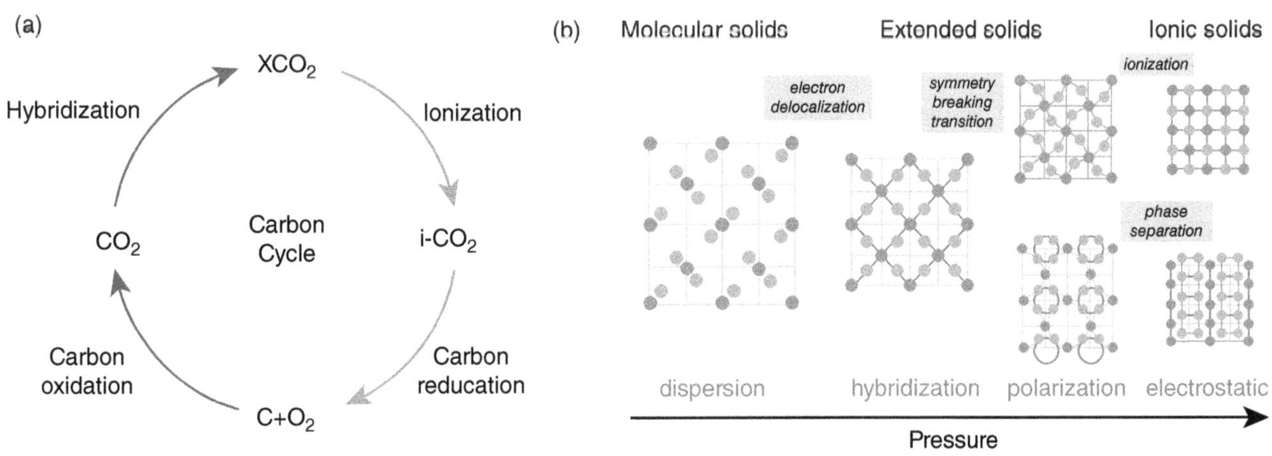

Figure 7.6 (a) A schematic of pressure-induced transformation of solid CO_2 in two dimensions, showing the fundamental concepts of pressure-induced chemistry (or barochemistry), such as the pressure-induced electron hybridization to extended solids and the pressure-induced ionization to ionic solids and decomposition products. (b) A schematic of deep carbon cycle governed by carbon Redox processes at combined high pressures and temperatures. See electronic version for color representation of the figures in this book.

and mantle boundary. In a context of carbon redox chemistry, we have conjectured a deep carbon cycle that may give a rebirth of fully reduced carbon species deep in the Earth's mantle–outer core boundary. We have also discussed fundamental barochemical concepts to explain the pressure-induced transformations in CO_2 and other molecular solids.

ACKNOWLEDGMENTS

This work has been performed in support of the National Science Foundation, Division of Materials Research (Grant No. 1701360), the Department of Energy, National Nuclear Security Administration (DE-NA0003342), the Army Research Office (W911NF-17-1-0468), the donors of the Petroleum Research Fund administered by the American Chemical Society (No. 54806-ND10), and the Sloan Foundation through the Deep Carbon Observatory—Extreme Physics and Chemistry.

REFERENCES

Agnew, S. F., Swanson, B. I., Jones, L. H., Mills, R. L., & Schiferl, D. (1983). Chemistry of nitrogen oxide (N_2O_4) at high pressure: Observation of a reversible transformation between molecular and ionic crystalline forms. *J. Phys. Chem.*, *87*, 5065–5068.

Andrault, D., Fiquet, G., Guyot, F., & Hanfland, M. (1998). Pressure-induced Landau-type transition in stishovite. *Science*, *282*, 720–724.

Bolduan, F., Jodl, H. L., & Lowenschuss, A. (1984). Raman study of solid N_2O_4: Temperature induced autoionization, *J. Chem. Phys. 80*, 1739–1743.

Dasgupta, R., & Hirschmann, M. M. (2006). Melting in the Earth's deep upper mantle caused by carbon dioxide. *Nature*, *440*, 659–662.

Datchi, F., Mallick, B., Salamat, A., & Ninet, S. (2012). Structure of polymeric carbon dioxide CO_2-V. *Phys. Rev. Lett.*, *108*, 125701-1-5.

Drozdov, A. P., Eremets, M. I., Troyan, I. A., Ksenofontov, V., & Shylin, S. I. (2015). Conventional superconductivity at 203 kelvin at high pressures in the sulfur hydride system. *Nature*, *525*, 73–76.

Dziubek, K. F., Ende, M., Scelta, D., Bini, R., Mezouar, M., Harbarino, G., & Miletich, R. (2018). Crystalline polymeric carbon dioxide stable at megabar pressures. *Nat. Comm.*, *9*, 3148.

Farsang, S., Facq, S., & Redfern, S.A.T. (2018). Raman modes of carbonate minerals as pressure and temperature gauges up to 6 GPa and 500 °C. *Am. Mineral.*, *103*, 1988–1998.

Hazen, R. M., Jones, A. P., & Baross, J. A. (2013). Carbon in Earth. *Rev. Mineral. Geochem.*, *75*, 1–698.

Iota, V., Yoo C. S., & Cynn, H. (1999). Quartzlike carbon dioxide: An optically nonlinear extended solid at high pres sures and temperatures. *Science*, *283*, 1510–1513.

Iota, V., Yoo, C. S., Klepeis, J.-H., Jenei, Z., Evans, W., & Cynn, H. (2007). Six-fold coordinated carbon dioxide VI. *Nature Mater.*, *6*, 34–38.

Isshiki, M., Irifune, T., Hirose, K., Ono, S., Ohishi, Y., Watanuki, T. (2004). Stability of magnesite and its high-pressure form in the lowermost mantle, *Nature*, *427*, 60.

Kampman, N., Bickle, M., Wigley, M., & Dubacq, B. (2014). Fluid flow and CO_2-fluid-mineral interactions during CO_2-storage in sedimentary basins *Chem. Geology*, *369*, 22–50.

Kelemen, P. B., & Manning, C. E. (2015). Reevaluating carbon fluxes in subduction zones: What goes down, mostly comes up. *Proc. Nat. Acad. Sci.*, *112*, E3997–E4006.

Leckner, K. S., 2003. A guide to CO_2 sequestration. *Science*, *300*, 1677–1678.

Lee, M. S., Montoya, J. A., & Scandolo, S. (2009). Thermodynamic stability of layered structures in compressed CO_2. *Phys. Rev. B*, *79*, 144102-1-4.

Li, J., & Agee, C. B. (1996). Geochemistry of mantle-core differentiation at high pressure. *Nature*, *381*, 686–689.

Manning, C. E., Shock, E. L., & Sverjensky, D. A. (2013). The chemistry of carbon in aqueous fluids at crustal and upper-mantle conditions. *Rev. Mineral. Geochem.*, *75* 109–148.

Mao, W. L., Mao, H. K., Eng, P., Trainor, T., Newville, M., Kao, C. C., et al. (2003). Bonding changes in compressed superhard graphite. *Science*, *302*, 425–427.

Mao, H. K., Xu J., & Bell, P. M. (1986). Calibration of the ruby gauge to 800 kbar under quasi-hydrostatic conditions. *J. Geophys. Res.*, *91*, 4673–4676.

Montoya, J. A., Rousseau, R., Santoro, M., Gorelli, F., & Scandolo, S. (2008). Mixed threefold and fourfold carbon coordination in compressed CO_2. *Phys. Rev. Lett.*, *100*, 163002-1-4.

Nakajima, Y., Takahashi, E., Suzuki, T., & Funakoshi, K.-I. (2009). "Carbon in the core" revisited. *Phys. Earth and Planet. Inter.*, *174*, 202–211.

Nellis, W. J., Mitchell, A. C., Ree, F. H., Ross, M., Holmes, N. C., Trainor, R. J., & Erskine, D. J. (1991). Equation of state of shock-compressed liquids: Carbon dioxide and air. *J. Chem. Phys.*, *95*, 5268–5272.

Oganov, A. R., Hemley, R. J., Hazen, R. M., & Jones, A. P. (2013). Structure, bonding, and mineralogy of carbon at extreme conditions. *Rev. Mineral. Geochem.*, *75*, 47–77.

Ono, S., Kikegawa, T., Ohishi, Y., & Tsuchiya, J. (2005). Post-aragonite phase transformation in CaCO3 at 40 GPa. *Am. Mineral.*, *90*, 667–671.

Saal, A. E., Hauri, E. H., Langmuir, C. H., & Perfit, M. R. (2002). Vapour undersaturation in primitive mid-ocean-ridge basalt and the volatile content of Earth's upper mantle. *Nature*, *419*, 451–455.

Santoro, M., Gorelli, F. A., Bini, R., Haines, J., Cambon, O., Levelut, C., et al. (2012). Partially collapsed cristobalite structure in the non molecular phase V in CO_2. *Proc. Nat. Acad. Sci.*, *109*, 5176–5179.

Santoro, M., Gorelli, F. A., Bini, R., Ruocco, G., Scandolo, S., & Crichton, W. A. (2006). Amorphous silica-like carbon dioxide. *Nature*, *441*, 857–860.

Schrauder, M., & Navon, O. (1993). Solid carbon dioxide in a natural diamond. *Nature*, *365*, 42–44.

Sengupta, A., Kim, M., Yoo, C. S., & Tse, J. S. (2012). Polymerization of carbon dioxide: a chemistry view of molecular-to-nonmolecular phase transitions. *J. Phys. Chem., C, 116*, 2061–2067.

Sengupta, A., & Yoo, C. S. (2009). Raman studies of molecular-to-nonmolecular transitions in carbon dioxide at high pressures and temperatures. *Phys. Rev. B, 80*, 014118-1-6.

Sengupta, A., & Yoo, C. S. (2010). Coesite-like CO_2: an analog to SiO_2. *Phys. Rev. B, 82*, 012105-1-4.

Sera, S., Corazon, C., Chiarotti, G.L., Scandolo, S., & Tossatti, E. (1999). Pressure-induced solid carbonates from molecular CO_2 by computer simulation. *Science, 284*, 788–790.

Shim, S.-H., Catalli, K., Hustoft, J., Kubo, A., Prakapenka, V. B., Caldwell, W. A., & Kunz, M. (2008). Crystal structure and thermoelastic properties of $(Mg_{0.91}Fe_{0.09})SiO_3$ postperovskite up to 135 GPa and 2700 K. *Proc. Nat. Acad. Sci., 105*, 7382–7386.

Sleep, N. H., & Zahnie, K. (2001). Carbon dioxide cycling and implications for climate on ancient Earth. *J. Geophys. Res,. 106*, 1373–1399.

Sun, J., Klug, D. D., Mortonak, R., Montoya, J. A., Lee, M. S., Scandolo, S., & Tosatti, E. (2009). High-pressure polymeric phases of carbon dioxide. *Proc. Natl. Acad. Sci., 106*, 6077–6081.

Yoo, C. S. (2013). Physical and chemical transformations of highly compressed carbon dioxide at bond energies. *Phys. Chem. Chem. Phys., 15*, 7949–7966.

Yoo, C. S. (2017). New states of matter and chemistry at extreme pressures: Low-Z extended solid. *MRS Bull., 42*, 724–728.

Yoo, C. S., Cynn, H., Gygi, F., Galli, G., Iota, V., Nicol, M. F., et al. (1999). Crystal structure of carbon dioxide at high pressure: "Superhard" polymeric carbon dioxide. *Phys. Rev. Lett., 83*, 5527–5530.

Yoo, C. S., Iota, V., & Cynn, H. (2001). Nonlinear carbon dioxide at high pressures and temperatures. *Phys. Rev. Lett., 86*, 444–447.

Yoo, C. S., Kim, M., Morgenroth, W., & Liermann, P. (2013). Transformation and structure of silicatelike CO_2-V. *Phys. Rev. B, 87*, 214103-1-9.

Yoo, C. S., Sengupta A., & Kim, M. (2011). Carbon dioxide carbonates in the Earth's mantle: Implications to the deep carbon cycle. *Angew. Chem. Int. Ed., 50*, 11415–11418.

Zhu, Y., & Ogasawara, Y. (2002). Carbon recycled into deep Earth: Evidence from dolomite dissociation in subduction-zone rocks. *Geology Soc. Am., 30*, 947–950.

8

Crystallization of Water Mediated by Carbon

Tianshu Li, Yuanfei Bi, and Boxiao Cao

ABSTRACT

Carbon is known to influence the crystallization of water in different ways. By combining forward flux sampling method and molecular dynamics simulation, we successfully investigate the formation of ice and gas hydrate induced by carbon. Our study shows that although ice nucleation on graphitic carbon follows the pathway described by the classical nucleation theory, the molecular nucleation mechanisms are complex and strongly dependent on the characters of carbon surface, namely, chemistry, crystallinity, and topography. This behavior is thermodynamically simple but complex at the molecular scale, and is rationalized by the role of water ordering induced by carbon. On the other hand, small hydrocarbon molecules such as methane are known to induce the formation of clathrate hydrate. To this end, we find the nucleation of gas hydrate follows a nonclassical pathway in that an average hydrate nucleus is amorphous-like, but exhibits a free energy landscape that can be reasonably well described by classical nucleation theory. We attempt to rationalize this seeming contradictory by making an analogy between stacking disordering in ice nucleation and the random packing of cages in hydrate nucleation. This viewpoint thus allows potentially connecting the nucleation of ice and gas hydrate through a unified picture.

8.1. INTRODUCTION

Although crystallization of water appears quite ordinary in daily life, pure water is surprisingly difficult to freeze. For example, distilled water in a freezer can easily maintain its liquid state without transforming into ice for a very long time. In fact, pure water can be supercooled down to –35 °C, a temperature usually referred to as the homogeneous nucleation temperature, below which ice crystallization becomes inevitable. This seemingly counterintuitive phenomenon was first rationalized by Gibbs about 140 years ago (Gibbs, 1876, 1877): When a new phase becomes thermodynamically more stable than the parent phase, the formation of the new phase is driven by the chemical potential difference but hindered by the surface tension between the two phases. Because the resistance scales with the interfacial area while the driving force is proportional to the volume of the new phase, a barrier must be overcome before the growth of the new phase becomes energetically favorable. The insight was further developed into the well-known homogeneous classical nucleation theory (CNT) (Volmer & Weber, 1926), which has been frequently invoked to interpret both experimental and simulation results in nucleation study.

The fundamental reason that ice freezes near its equilibrium melting point is attributed to heterogeneous nucleation, an activation process that is mediated by impurities. These impurities, often termed ice nucleators (INs), generically exist in nature and encompass a vast variety of materials, spanning from mineral dusts to organic species (Murray et al., 2012). In fact, the crystallization of nearly all materials proceeds via heterogeneous nucleation. The prevalence of heterogeneous nucleation in nature can be explained qualitatively by the extension of the CNT, which was established more than 60 years ago (Turnbull, 1950). The theory attributes the

Department of Civil and Environmental Engineering, George Washington University, Washington, DC, USA

Carbon in Earth's Interior, Geophysical Monograph 249, First Edition. Edited by Craig E. Manning, Jung-Fu Lin, and Wendy L. Mao.

enhanced nucleation rate to the reduced nucleation barrier due to the presence of a heterogeneous interface between liquid and substrate. Although the theory is well known, its quantitative validity and key conclusions have remained unconfirmed. In fact, criticisms have often been drawn towards its assumptions and quantitative validity (Gebauer & Cölfen, 2011; Sear, 2012). In complex systems, e.g., minerals, proteins, organic crystals, and hydrates, the crystallization process is often found to proceed in a multistep, nonclassical fashion (Chen et al., 2011; Erdemir et al., 2009; Gebauer & Cölfen, 2011; Gebauer et al., 2008; Jacobson et al., 2010; Vekilov, 2004). Even in simple systems such as colloids, the nucleation behaviors are often found to deviate from CNT (Auer & Frenkel, 2003).

It may seem rather unusual to relate carbon to the liquid-to-solid transformation of water, as in its first look, there appears to be a lack of apparent structural similarity between the two. Nevertheless, the crystallization of water is known to be strongly affected by the presence of carbon through a variety of forms. Many carbon-bearing materials are known to be effective INs in atmosphere, e.g., soot particles and organic compounds. In particular, graphitic carbon has been long conjectured to influence ice formation, but its role has not been clearly elucidated. In the deep ocean, small hydrocarbon molecules such as methane are known to trigger the crystallization of water into clathrate hydrate, another form of solid water.

In this chapter, we review our recent studies on the crystallization of water induced by carbon by using advanced molecular simulations. In particular, we attempt to answer the following fundamental questions: How is the crystallization of water initiated and controlled by the fundamental level interaction between carbon and water? Can we combine the two distinct processes within a unified framework? To address these questions, we employed advanced sampling method combined with molecular simulations to investigate the nucleation of ice and gas hydrate in the presence of carbon-bearing materials.

8.2. METHODS

The main challenge in modeling nucleation is to overcome the long induction time of the event, which significantly exceeds the accessible time scale of direct molecular simulation, i.e., nanosecond to microsecond. For example, a typically measured ice nucleation rate in experiment (e.g., 10^9 $cm^{-3}s^{-1}$) requires a trajectory as long as ~32 years to reproduce in a direct molecular dynamics (MD) simulation employing a simulation volume of 1000 nm^3. To overcome this shortcoming, we employed forward flux sampling (FFS) method (Allen et al., 2006) to accelerate the exploration of transition pathways between disordered (liquid) and ordered (crystal) states.

In this approach, the nucleation trajectory is decomposed into a series of consecutive transition segments through an appropriate order parameter λ. The rate constant is obtained using the "effective positive flux" expression (van Erp et al., 2003) $R_{AB} = \Phi_{\lambda_0} p(\lambda_B|\lambda_0)$, where Φ_{λ_0} is the flux rate reaching the first interface λ_0 from basin A (liquid), and $P(\lambda_B|\lambda_0)$ is the probability for a trajectory that starts from λ_0 and eventually reaches B (solid). The typically small $P(\lambda_B|\lambda_0)$ can be calculated through $P(\lambda_B|\lambda_0) = \prod_{i=1}^{n} P(\lambda_i|\lambda_{i-1})$, where $P(\lambda_i|\lambda_{i-1})$ is the crossing probability between the two adjacent interfaces λ_{i-1} and λ_i. Although FFS ensures R_{AB} is independent of the exact positions of interfaces, we find it optimal to choose λ_is that yield $P(\lambda_i|\lambda_{i-1})$ within the range of 0.01–0.2, to balance between computational efficiency and statistical variance (Li et al., 2009). Through combining FFS and backward flux sampling (BFS), one can also obtain the free energy profile along a sequence of order parameter (Bi, Porras, et al., 2016; Valeriani et al., 2007). For ice nucleation, we use the number of ice-like water molecules, which are characterized by local bond-order parameter q_6 (Li et al., 2011), in the largest crystallite as an effective order parameter; for gas hydrate, we have developed the half-cage order parameter on the basis of structural signature of clathrate (Bi & Li, 2014). The choice of both order parameters can be justified by p_B-histogram analyses (Bi, Porras, et al., 2016; Cabriolu & Li, 2015; DeFever & Sarupria, 2017; Lupi et al., 2016) that show both order parameters describe well the nucleation pathways of ice and hydrate.

8.3. HETEROGENEOUS ICE NUCLEATION FACILITATED BY GRAPHITIC CARBON

8.3.1. Verification of Heterogeneous Classical Nucleation Theory

We began our investigation by first examining ice nucleation on graphitic carbon. Water is represented by the monoatomic water model mW (Molinero & Moore, 2009), while the water-carbon interaction is described by the two-body term of the mW, where ε and σ are reparametrized to reproduce the experimental contact angle (86°) of water on graphite Lupi et al., 2014). The foremost question to address is whether CNT can describe heterogeneous ice nucleation, because although CNT has been routinely invoked to explain both experimental and simulation results, its quantitative applicability in ice nucleation remains unknown. A verification of CNT requires obtaining the key quantities (e.g., nucleation rate, critical nucleus size) independent of theory, which is challenging for both experiment and simulation. The capability of obtaining an ice nucleation rate independent of CNT by FFS allows testing the quantitative validity of the theory in heterogeneous ice nucleation for the first time.

The nucleation rates of ice forming homogeneously from supercooled water (Li et al., 2011) and heterogeneously on a graphene surface (Cabriolu & Li, 2015) are computed as a function of temperature. As shown in Figure 8.1, the computed ice nucleation rates were found to fit well according to CNT through the expression $\ln R = \ln A + C/((T - T_m)^2 T)$. Here, R, T, and T_m are the rate constant, nucleation temperature, and melting temperature of ice, respectively, while A and C are fitting constants. The fitting procedure yields the estimate of the potency factor, which measures the ratio of the barrier of heterogeneous nucleation to that of homogeneous nucleation, *i.e.*, $f(\theta) \equiv \Delta G^*_{het}/\Delta G^*_{\text{hom}} = C_{het}/C_{\text{hom}}$, directly through the fitting constant C.

On the other hand, the critical volume (or the critical size λ^*) can also be independently obtained from the ensemble of nucleation trajectories obtained in FFS calculation, using the definition of committor (Bolhuis et al., 2002) $p_B = 0.5$, *i.e.*, the critical nucleus should have the equal probabilities of dissolving and growing completely. On the basis of CNT, we can fit the obtained critical size as a function of temperature through $\lambda = B/(T_m/T - 1)^3$. Again, an excellent fitting is obtained (Figure 8.1c), from which one can obtain the volumetric ratio $\lambda^*_{het}/\lambda^*_{\text{hom}} = B_{het}/B_{\text{hom}}$.

Remarkably, the obtained volumetric ratio $B_{het}/B_{hom} = 0.480 \pm 0.011$ agrees quantitatively with the potency factor $C_{het}/C_{hom} = 0.456 \pm 0.019$. To ensure the agreement is not incidental, we also tested our conclusion at a different potency factor $f(\theta)$, through reducing the water-carbon interaction strength ε by one half so that the carbon surface now becomes more hydrophobic. We repeated the above procedures, and again obtained a quantitative agreement (Figure 8.1). To the best of our knowledge, this is the first time that the quantitative power of heterogeneous CNT has been directly supported from the independent rate constant calculation.

It should be mentioned that the quantitative equivalency between the volumetric factor and potency factor was again identified in our very recent study of ice nucleation on kaolinite surface (Sosso et al., 2016). The agreement with CNT, identified in different systems, is surely a strong support to the theory developed many decades ago, but it also raises an important question: behind this seemingly "simple" process, what determines the potency factor at the molecular level? Indeed, this has been one of the daunting questions in atmospheric chemistry for decades (Murray et al., 2012), as no strong correlation seems to exist between any of the proposed empirical criteria and the observed ice nucleation efficiency.

8.3.2. Molecular Insight Into the Complex Nature of Heterogeneous Ice Nucleation

To illustrate the molecular origin for the lack of such correlation, we examine the roles of surface crystallinity, surface hydrophilicity (Bi, Cabriolu, et al., 2016), and surface geometry (Bi et al., 2017) on ice nucleation based on the carbon-water system. We introduce Stone-Wales defects in crystalline graphene through applying the Wooten-Weaire-Winer bond-switching Monte-Carlo method (Wooten et al., 1985) to mimic the change of surface crystallinity. The resulting amorphous graphene is shown in Figure 8.2. To tune surface hydrophilicity, we modify water-carbon interaction strength ε in the wide range of $[\varepsilon_0, 10\varepsilon_0]$, where ε_0 is the original strength (Lupi et al., 2014) reproducing the water contact angle (86°) on graphite (Li & Zeng, 2012). Increasing water-carbon interaction strength makes the surface more hydrophilic (Lupi & Molinero, 2014), as carbon atoms bind water more strongly. To understand the role of surface roughness, we also create an atomically sharp wedge composed of two graphene planes with a varying wedge angle β.

We found that heterogeneous ice nucleation on a flat graphitic surface exhibits a rich spectrum of nucleation behaviors when both surface crystallinity and surface hydrophilicity were allowed to vary (Bi, Cabriolu, et al., 2016). As shown in Figure 8.2, a radical change in surface chemistry not only yields a nonmonotonic change in ice nucleation rates, but it also leads to a complex, alternate nucleation mechanism: At a low water-carbon strength (i.e., low hydrophilicity), ice nucleation is found to be mainly controlled by the hydrophilicity of carbon alone, thus independent of the crystallinity of the graphene substrate. A gradual increase of hydrophilicity is then found to differentiate between crystalline and amorphous graphene, with crystalline graphene being distinguished as a more efficient IN. Interestingly, a further increase of hydrophilicity not only leads to a sudden decrease of the nucleation rate, but also eliminates the role of crystallinity in nucleation, making ice nucleation behavior similar to what occurs at low hydrophilicity. Remarkably, with a very high surface hydrophilicity, the coupling-controlled behavior reappears so that the crystalline graphene again becomes a better IN. Such oscillating distinction between crystalline and amorphous graphene in their ice nucleation efficiencies clearly highlights the complexity of heterogeneous ice nucleation and suggests that ice nucleation can be controlled by the combined surface characteristics via coupling effects.

Similarly, we also found that surface geometry alone cannot be a good descriptor for ice nucleation efficiency either. Although surface roughness has been generally considered as a favorable factor for nucleation, our study (Bi et al., 2017) showed that a simple correlation cannot

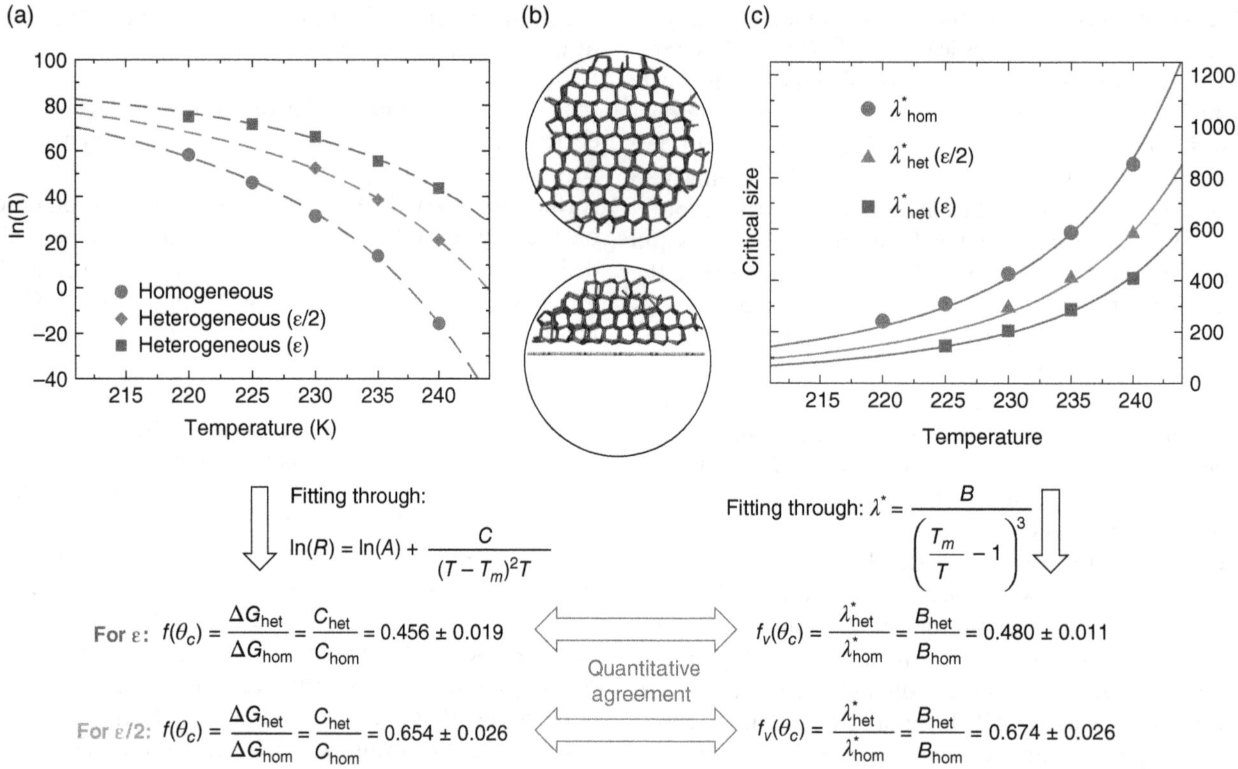

Figure 8.1 The variation of (a) the calculated ice nucleation rate (logarithm) and (c) critical size, with temperature. Data on homogeneous ice nucleation are from our previous work (Li et al., 2011). Reproduced by permission of the PCCP Owner Societies. (b) Snapshots of the critical ice nuclei for homogenous and heterogeneous nucleation at 240 K. *Source*: Cabriolu & Li (2015) Reproduced with permission of the American Physical Society. Copyright (2019) American Physical Society. See electronic version for color representation of the figures in this book.

be established between surface geometry and ice nucleation efficiency. As shown in Figure 8.3, the calculated nucleation rates of ice forming within a concave, atomically sharp wedge showed a nonmonotonic dependence on the wedge angle. In particular, the significant enhancement of ice nucleation relative to a planar surface was found to occur only under a few well-defined wedge angles: 70°, 110°, and 45°. Our studies thus suggested that it is unlikely that a single descriptor can reliably determine an IN's efficiency, and a thorough understanding of the ice nucleation capacity for an IN therefore should be achieved through a comprehensive study that explicitly considers all the necessary molecular details at the surface.

8.3.3. Role of Local Ordering of Water in Ice Nucleation

To understand this "duality" of ice nucleation, i.e., being thermodynamically "simple" but complex at molecular scales, we examine the local ordering of interfacial water. Indeed, we find the observed complex heterogenous ice nucleation behaviors can be generally interpreted based on this concept regardless of the diversity in surface character. Specifically, for ice nucleation on a flat graphene surface (Figure 8.2), we find a simple trend behind the complex ice nucleation behaviors: The better ice nucleation efficiency of crystalline graphene, whenever it occurs, is always accompanied by the appearance of *lattice registry* between ice and crystalline graphene, whether such registry occurs in the first or second contact layer of water. Further investigation shows that the occurrence of lattice registry is a direct outcome of in-plane water ordering as a result of the subtle balance between water-carbon binding strength and lattice mismatch between graphene and ice lattice.

The concept of local ordering can be also applied to explain the enhanced ice nucleation via a special surface geometry (Bi et al., 2017). Because graphene induces density layering of water that matches the density profile of ice normal to its basal plane (Lupi et al., 2014), it promotes the formation of the {0001} plane of hexagonal ice I_h or the {111} plane of cubic ice I_c. This one-dimensional density match alone leads to an enhancement of the ice nucleation rate by 25 orders of magnitude at 240 K. When two graphene planes intersect at 70°/110°, forming a wedge, they create a template that matches *two* intersecting {111} planes of cubic I_c simultaneously, which

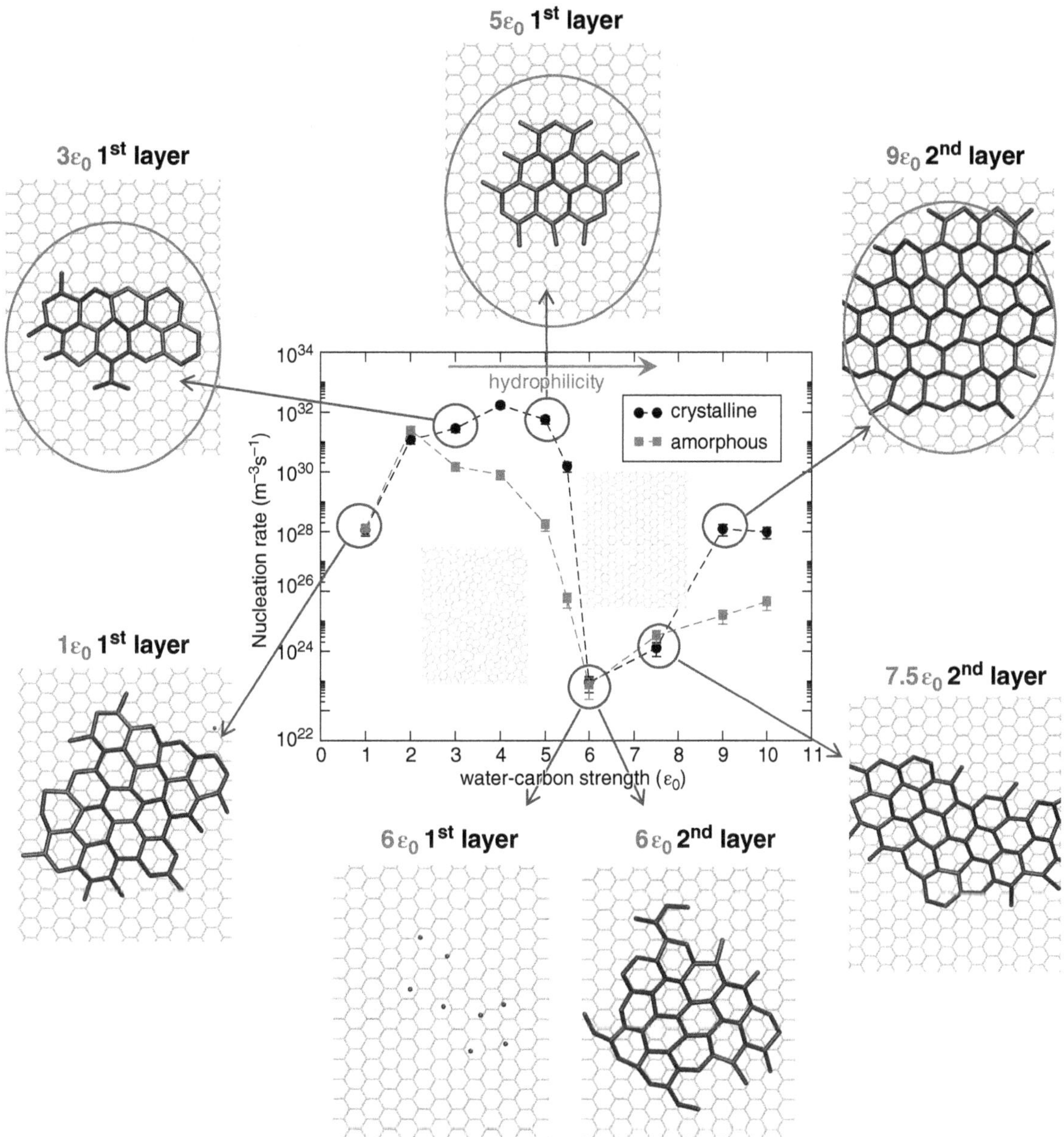

Figure 8.2 Variation of ice nucleation rates with water-carbon strength for both crystalline and amorphous graphene at 230 K. Inset: Structure of the crystalline and amorphous graphene. The panels show the structures of ice (blue) forming in the contact layers of water above graphene substrate (cyan) at different ε. Red circles highlight the ice basal plane registered with the underlying crystalline graphene lattice. *Source*: Bi, Cabriolu, et al. (2016). Reproduced with permission of the American Chemical Society. Copyright (2019) American Chemical Society. See electronic version for color representation of the figures in this book.

further enhances the ice nucleation rate by another eight orders of magnitude. Similarly, adding a third graphene sheet while creating an open tetrahedral pyramid that matches three intersecting {111} planes was found to yield spontaneous ice nucleation. It is also important to note that the increasing degree of induced ordering not only enhances the nucleation rate but also leads to an interesting enhancement of polymorph selection toward cubic ice I_c.

Importantly, we find the ordering of liquid that pertains to crystallization is not limited to apparent lattice match. In particular, noting that a 45° wedge does not accommodate any common crystalline dihedral angle, we find that the very high ice nucleation efficiency of the 45° wedge must not be explained based on the traditional wisdom of lattice match. Instead, we find that such unexpected rate enhancement is facilitated by the formation of special *topological defects* near the 45° wedge tip that

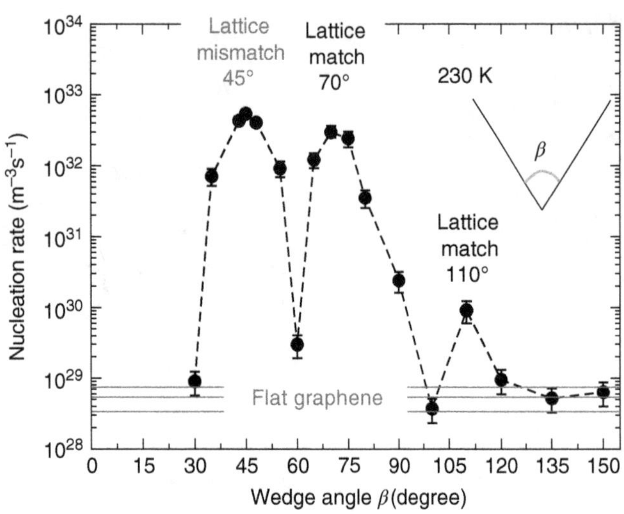

Figure 8.3 Variation of the calculated nucleation rates of ice forming within an atomically sharp, concave wedge. *Source*: Bi et al. (2017). Reproduced with permission of Nature Communications. See electronic version for color representation of the figures in this book.

consequently catalyze the growth of regular ice (Figure 8.4). The close resemblance in the molecular pathways for ice to form within the 70° wedge (Figure 8.4 g–i, nondefect initiating) and 45° wedge (Figure 8.4 c–e, defect initiating), along with their nearly degenerate ice nucleation rates (Figure 8.3), highlight the strong relevance of defects in ice nucleation. Therefore, we believe that the traditional concept of structural match or templating effect in nucleation should be extended to include a broader structural match with *noncrystalline* units.

The local ordering of liquid can naturally reconcile the duality of heterogeneous ice nucleation. From a thermodynamic viewpoint, since crystallization is driven by enthalpy gain but penalized by entropy loss, there must exist a critical point prior to which the enthalpy gain cannot be fully compensated by entropy loss. This critical point should be conceptually equivalent to the nucleation barrier in CNT. Therefore, a crystallization event can be enhanced when entropy loss is mitigated. At a molecular level, such reduction of entropy loss is reflected by the enhancement of local ordering of liquid in consistent

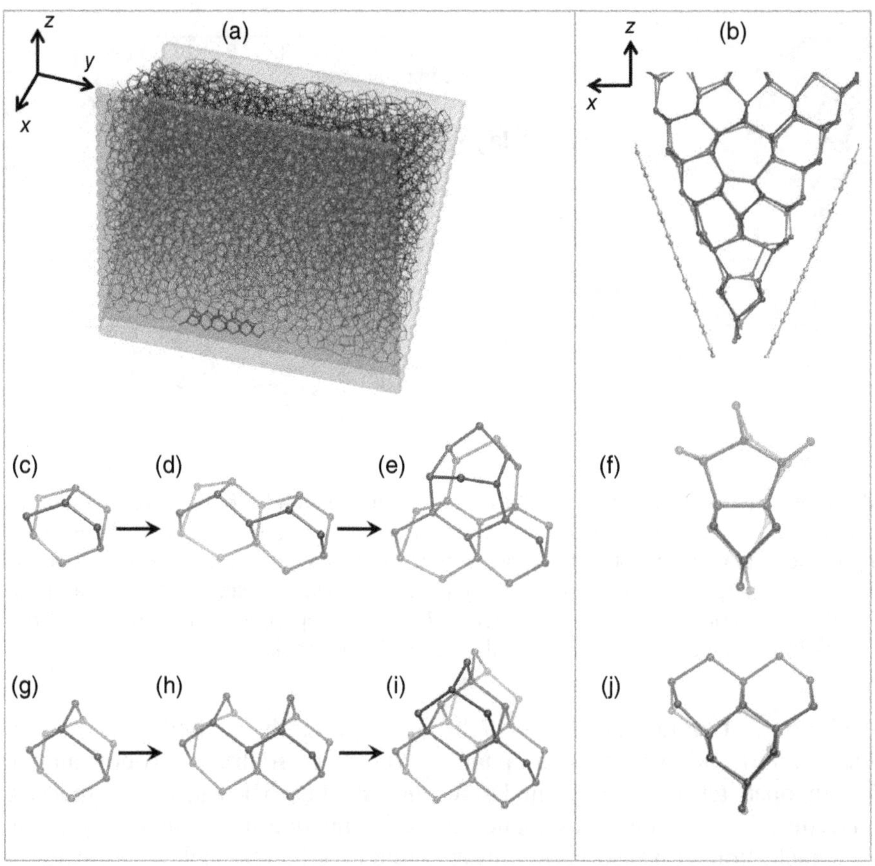

Figure 8.4 Molecular pathways of ice crystallization near tip of a 45° wedge (c–e) and a 70° wedge (g–i). (b), (f), and (j) are the side views of (a), (e), and (i), respectively. *Source*: Bi et al. (2017). Reproduced with permission of Nature Communications. See electronic version for color representation of the figures in this book.

with crystallization, which can be facilitated by certain characteristics of an IN, e.g., surface chemistry, surface crystallinity, surface geometry, etc. In this sense, the role of an IN is to "guide" water molecules to find the right entrance towards the basin of solid phase.

8.4. NUCLEATION OF GAS HYDRATE

Another major venue for water crystallization induced by carbon is the formation of gas hydrate. Small hydrocarbon molecules such as methane are known to be capable of inducing water to crystallize into clathrate structures, enclosing them as guests under high pressure. These carbon-bearing molecules, being strongly hydrophobic, have very low solubilities in water, yet their concentrations can be thousand times higher in solid gas hydrate. Although appearing chemically counterintuitive, the formation of gas hydrate can be well rationalized by both thermodynamics and kinetics. On one hand, the presence of many pairs of gas-water interaction can significantly stabilize gas hydrates, making their formation thermodynamically favorable; on the other, the close resemblance in water's structure between the hydration shell of gas molecule and clathrate cage provides a structural basis for gas hydrate nucleation. Despite these understandings, an outstanding question is how gas hydrate nucleates from liquid-gas mixture, and in particular, whether hydrate nucleation pathways can be described by CNT. Early hypotheses such as labile cluster hypothesis (LCH) (Sloan & Fleyfel, 1991) and local structuring hypothesis (LSH) (Radhakrishnan & Trout, 2002) were able to rationalize certain attributes of hydrate nucleation mechanism, but both found difficulty when compared with simulation results. The breakthrough was made when spontaneous hydrate nucleation was for the first time obtained in direct MD simulations (Walsh et al., 2009). These simulations facilitated a new view of hydrate nucleation: "blob" mechanism (Jacobson & Molinero, 2010), which reconciles labile cluster hypothesis and local structuring hypothesis. In particular, the "blob" mechanism makes an analogy between hydrate nucleation and the nucleation of minerals and protein, suggesting a "two-step" process involving the formation of amorphous nucleus followed by amorphous-to-crystal transition. The mechanism raised a fundamental question of whether hydrate nucleation can be considered nonclassical with multiple barriers, as hypothesized in the nucleation of proteins and minerals.

An unambiguous answer to this question requires a sufficient sampling of transition pathway ensemble as well as a determination of free energy profile along the transition pathway. To achieve this goal, we developed the half-cage order parameter (H-COP) (Bi & Li, 2014) on the basis of the topological hierarchy of clathrate structure (see Figure 8.5a), for driving and characterizing hydrate nucleation. H-COP is defined as the number of water molecules contained in a hydrate-like cluster assembled by half cages through certain geometrical constraints. We subsequently integrated H-COP into FFS, which allows explicitly computing hydrate nucleation rate for the first time at a condition where spontaneous hydrate nucleation is too slow to occur in direct MD simulation (Bi & Li, 2014). Combining FFS and backward flux sampling also allowed obtaining the free energy profile of hydrate nucleation along the order parameter H-COP (Bi, Porras, et al., 2016). Our study of gas hydrate nucleation was carried out based on a coarse-grained model where water is represented by the mW model and water-gas interactions are represented by the two-body interaction of the mW model (Jacobson & Molinero, 2010) that was tuned to mimic a wide range of guests with different sizes and solubilities (Jacobson et al., 2010). In this study we considered both a large (L) guest that exhibits properties comparable to oxetane and a medium (M) guest that emulates methane.

Our investigation demonstrates a few important and intriguing characteristics of gas hydrate nucleation: (1) The ensemble-averaged crystallinity of hydrate is found to be low at the early stage of nucleation but gradually increases with the size of the hydrate nucleus, thus providing a strong support to the proposed "two-step" nucleation mechanism. (2) Despite the average two-step picture, there clearly exists a structural diversity of hydrate nucleation pathway. In particular, multiple nucleation channels are shown to exist, among which hydrate is found to form directly into a crystalline structure, bypassing the amorphous stage. (3) Most surprisingly, the free energy profile of hydrate nucleation is found to follow CNT reasonably well, despite the overall "nonclassical" molecular pathways.

These observations appear contradictory, particularly (1) and (3). However, our subsequent investigation showed they may be reconciled by considering the following facts and hypotheses. First, the chemical potential difference between structure I (sI) hydrate, which is the thermodynamically stable phase of the studied gas hydrate, and the metastable phase structure II (sII) hydrate, is very small. The close proximity of structural stability in sI and sII is reminiscent of the case of ice I, where hexagonal I_h is only marginally more stable than cubic I_c. For ice, both simulations (Haji-Akbari & Debenedetti, 2015; Li et al., 2011; Moore & Molinero, 2011) and experiments (Kuhs et al., 2012; Malkin et al., 2012) have shown ice crystallites grown freshly from supercooled water are stacking disordered, rather than pure I_h or I_c. This unconventional behavior has been recently rationalized by the higher stability of stacking disordered ice crystallites,

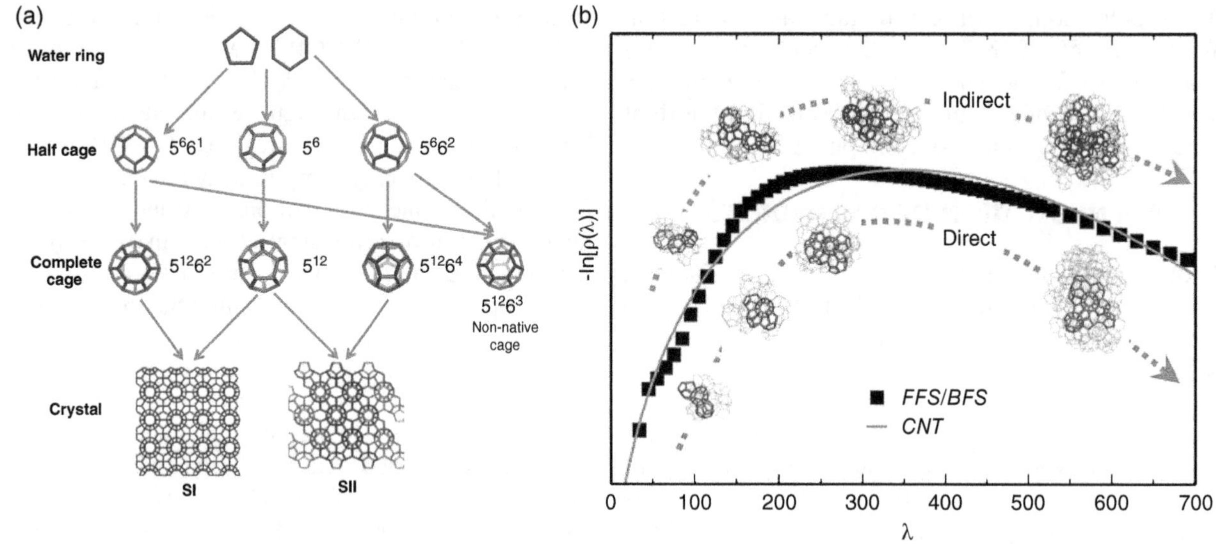

Figure 8.5 (a) Topological hierarchy of clathrate hydrate. (b) Calculated free energy profile for L-hydrate nucleation as a function of order parameter. Red line represents the best fit of classical nucleation theory. Both direct and indirect pathways are hypothesized to exhibit similar energy landscape. For indirect pathway, the color scheme for complete cages is 512 (blue), 51262 (red), 51263 (green), and 51264 (orange); for direct pathway, the sII-like hydrate cages are marked red. *Source*: Bi, Porras, et al. (2016). Reproduced with permission of AIP Publishing. See electronic version for color representation of the figures in this book.

as a result of entropy of mixing thermodynamically favoring stacking disorder (Lupi et al., 2017). From this viewpoint, we postulate that hydrate nucleation may be rationalized by a similar argument, as the close stability in the large variety of cage packing sequences guarantees a large configurational space for hydrate nucleation to explore (Hall et al., 2016). Since crystalline structures only represent a small fraction of such large configurational space (as they are outnumbered by noncrystalline packing sequences, just as stacking ordered structures are outnumbered by stacking disordered structures), the entropy of "mixing" favors a random packing of cages, thus forming noncrystalline hydrate nucleus. As a result, hydrate nuclei exhibit a wide range of crystallinity, and the ensemble-averaged crystallinity is low for small hydrate nucleus. As explained in the section above, classical nucleation theory has been shown to describe ice nucleation accurately—even when stacking disorder is considered (Lupi et al., 2017). In this sense, it is then not too surprising to find that hydrate nucleation also carries a classical-like free energy profile, albeit that small deviation is observed between the calculation and the fitting (see Figure 8.5b). One explanation for such deviation could be related to the choice of order parameter: As shown in a recent study (DeFever & Sarupria, 2017), although all the developed order parameters for hydrate nucleation, including H-COP, capture the reaction pathway well, they are in general less accurate than that for ice nucleation.

8.5. CONCLUSION

An intriguing question to address is how the crystallization of water is initiated and controlled by the fundamental level interaction between carbon and water. By developing advanced molecular modeling, we address this question at both the thermodynamic and molecular levels. We find ice crystallization, when either occurring homogeneously or heterogeneously on a graphitic carbon surface, appears to follow a pathway that can be quantitatively described by classical nucleation theory. On the other hand, at a molecular level, we also discover that ice nucleation is dictated by the subtle interplay between different molecular details of the carbon surface, including surface crystallinity, surface hydrophilicity, and surface topography. We attempt to explain this duality of ice nucleation through the role of local ordering of water. In the form of small hydrocarbons such as methane, carbon can also directly participate in the crystallization of water, leading to the formation of clathrate hydrate. To this end, by using our developed tools, we find that although on average, hydrate nucleation proceeds through a nonclassical pathway, it displays a classical-like free energy profile. We attempt to reconcile this seeming contradiction through making an analogy between the stacking disorder in ice nucleation and the random packing of cages in hydrate nucleation. This analogy potentially allows understanding the crystallization of water through a unified framework, irrespective of its end products.

ACKNOWLEDGMENTS

The work has been supported by the Sloan Foundation through the Deep Carbon Observatory, NSF through awards CMMI-1537286 and CBET-1264438, and the American Chemical Society Petroleum Research Fund.

REFERENCES

Allen, R. J., Frenkel, D., Wolde, & Ten, P. R. (2006). Simulating rare events in equilibrium or nonequilibrium stochastic systems. *J Chem Phys*, *124*(2), 024102.

Auer, S., & Frenkel, D. (2003). Line tension controls wall-induced crystal nucleation in hard-sphere colloids. *Phys Rev Lett*, *91*(1), 015703.

Bi, Y., Cabriolu, R., & Li, T. (2016). heterogeneous ice nucleation controlled by the coupling of surface crystallinity and surface hydrophilicity. *J Phys Chem C*, *120*(3), 1507–1514.

Bi, Y., Cao, B., & Li, T. (2017). Enhanced heterogeneous ice nucleation by special surface geometry. *Nat Commun*, *8*, 15372.

Bi, Y., & Li, T. (2014). Probing methane hydrate nucleation through the forward flux sampling method. *J Phys Chem B*, *118*(47), 13324–13332.

Bi, Y., Porras, A., & Li, T. (2016). Free energy landscape and molecular pathways of gas hydrate nucleation. *J Chem Phys*, *145*(21), 211909.

Bolhuis, P. G., Chandler, D., Dellago, C., & Geissler, P. L. (2002). Transition path sampling: Throwing ropes over rough mountain passes, in the dark. *Annu. Rev. Phys. Chem.*, *53*, 291–318.

Cabriolu, R., & Li, T. (2015). Ice nucleation on carbon surface supports the classical theory for heterogeneous nucleation. *Physical Review E*, *91*, 052402.

Chen, J., Sarma, B., Evans, J. M. B., & Myerson, A. S. (2011). Pharmaceutical crystallization. *Crystal Growth & Design*, *11*(4), 887–895.

DeFever, R. S., Sarupria, S. (2017). Nucleation mechanism of clathrate hydrates of water-soluble guest molecules. *J Chem Phys*, *147*(20), 204503.

Erdemir, D., Lee, A. Y., & Myerson, A. S. (2009). Nucleation of crystals from solution: Classical and two-step models. *Acc. Chem. Res.*, *42*(5), 621–629.

Gebauer, D., & Cölfen, H. (2011). Prenucleation clusters and non-classical nucleation. *Nano Today*, *6*(6), 564–584.

Gebauer, D., Volkel, A., & Cölfen, H. (2008). Stable prenucleation calcium carbonate clusters. *Science*, *322*(5909), 1819–1822.

Gibbs, J. W. (1876). On the equilibrium of heterogeneous substances. *Transactions of the Connecticut Academy of Arts and Sciences*, *3*, 108–248.

Gibbs, J. W. (1877). On the equilibrium of heterogeneous substances. *Transactions of the Connecticut Academy of Arts and Sciences*, *3*, 343–524.

Haji-Akbari, A., & Debenedetti, P. G. (2015). Direct calculation of ice homogeneous nucleation rate for a molecular model of water. *Proc Natl Acad Sci USA*, *112*(34), 10582–10588.

Hall, K. W., Carpendale, S., & Kusalik, P. G. (2016). Evidence from mixed hydrate nucleation for a funnel model of crystallization. *Proceedings of the National Academy of Sciences*, 201610437.

Jacobson, L. C., Hujo, W., & Molinero, V. (2010a). Amorphous precursors in the nucleation of clathrate hydrates. *J Am Chem Soc*, *132*(33), 11806–11811.

Jacobson, L. C., Hujo, W., & Molinero, V. (2010b). Nucleation pathways of clathrate hydrates: Effect of guest size and solubility. *J Phys Chem B*, *114*(43), 13796–13807.

Jacobson, L. C., & Molinero, V. A (2010). Methane-water model for coarse-grained simulations of solutions and clathrate hydrates. *J Phys Chem B*, *114* (21), 7302–7311.

Kuhs, W. F., Sippel, C., Falenty, A., & Hansen, T. C. (2012). Extent and relevance of stacking disorder in "ice Ic." *Proceedings of the National Academy of Sciences*, *109*(52), 21259–21264.

Li, H., & Zeng, X. C. (2012). Wetting and interfacial properties of water nanodroplets in contact with graphene and monolayer boron-nitride sheets. *ACS Nano*, *6*(3), 2401–2409.

Li, T., Donadio, D., & Galli, G. (2009). Nucleation of tetrahedral solids: A molecular dynamics study of supercooled liquid silicon. *J. Chem. Phys.*, *131*(22), 224519.

Li, T., Donadio, D., Russo, G., & Galli, G. (2011). Homogeneous ice nucleation from supercooled water. *Phys Chem Chem Phys*, *13*(44), 19807–19813.

Lupi, L., Hudait, A., & Molinero, V. (2014). Heterogeneous nucleation of ice on carbon surfaces. *J Am Chem Soc*, *136*(8), 3156–3164.

Lupi, L., Hudait, A., Peters, B., Grünwald, M., Mullen, R. G., Nguyen, A. H., & Molinero, V. (2017). Role of stacking disorder in ice nucleation. *Nature*, *551*(7679), 218–222.

Lupi, L., & Molinero, V. (2014). Does hydrophilicity of carbon particles improve their ice nucleation ability? *J Phys Chem A*, *118*(35), 7330–7337.

Lupi, L., Peters, B., & Molinero, V. (2016). Pre-ordering of interfacial water in the pathway of heterogeneous ice nucleation does not lead to a two-step crystallization mechanism. *J Chem Phys*, *145*(21), 211910.

Malkin, T. L., Murray, B. J., Brukhno, A. V., Anwar, J., & Salzmann, C. G. (2012). Structure of ice crystallized from supercooled water. *Proceedings of the National Academy of Sciences*, *109*(4), 1041–1045.

Molinero, V., & Moore, E. B. (2009). Water modeled as an intermediate element between carbon and silicon. *J Phys Chem B*, *113*(13), 4008–4016.

Moore, E. B., & Molinero, V. (2011). Is it cubic? Ice crystallization from deeply supercooled water. *Phys Chem Chem Phys*, *13*(44), 20008–20016.

Murray, B. J., O'Sullivan, D., Atkinson, J. D., & Webb, M. E. (2012). Ice nucleation by particles immersed in supercooled cloud droplets. *Chem. Soc. Rev.*, *41*(19), 6519–6554.

Radhakrishnan, R., & Trout, B. (2002). A new approach for studying nucleation phenomena using molecular simulations: Application to CO2 hydrate clathrates. *J Chem Phys*, *117*(4), 1786–1796.

Sear, R. P. (2012). The non-classical nucleation of crystals: Microscopic mechanisms and applications to molecular crystals, ice and calcium carbonate. *International Materials Reviews*, *57*(6), 328–356.

Sloan, E., & Fleyfel, F. A (1991). Molecular mechanism for gas hydrate nucleation from ice. *AIChE Journal*, *37*, 1281–1292.

Sosso, G. C., Li, T., Donadio, D., Tribello, G. A., & Michaelides, A. (2016). Microscopic mechanism and kinetics of ice formation at complex interfaces: Zooming in on kaolinite. *J Phys Chem Lett*, *7*, 2350–2355.

Turnbull, D. (1950). Kinetics of heterogeneous nucleation. *J Chem Phys*, *18*(2), 198.

Valeriani, C., Allen, R. J., Morelli, M. J., Frenkel, D., & Rein ten Wolde, P. (2007). Computing stationary distributions in equilibrium and nonequilibrium systems with forward flux sampling. *J Chem Phys*, *127*(11), 114109.

van Erp, T., Moroni, D., & Bolhuis, P. (2003). A novel path sampling method for the calculation of rate constants. *J Chem Phys*, *118*(17), 7762–7774.

Vekilov, P. G. (2004). Dense liquid precursor for the nucleation of ordered solid phases from solution. *Crystal Growth & Design*, *4*, 671–685.

Volmer, M., & Weber, A. (1926). Keimbildung in Übersättigten Gebilden. *Z. Phys. Chem. (Leipzig)*, *119*(227).

Walsh, M. R., Koh, C. A., Sloan, E. D., Sum, A. K., & Wu, D. T. (2009). Microsecond simulations of spontaneous methane hydrate nucleation and growth. *Science*, *326* (5956), 1095–1098.

Wooten, F., Winer, K., & Weaire, D. (1985). Computer generation of structural models of amorphous Si and Ge. *Phys Rev Lett*, *54*(13), 1392–1395.

9

Structures and Crystal Chemistry of Carbonate at Earth's Mantle Conditions

Marco Merlini, Sula Milani, and Juliette Maurice

ABSTRACT

We report an overview of the crystal structures of carbonates determined ab-initio with X-ray single crystal diffraction techniques at mantle conditions. The determined crystal structures of high-pressure polymorphs of $CaCO_3$ have revealed that structures denser than aragonite can exist at upper and lower mantle pressures. These results have stimulated the computational and experimental research of thermodynamically stable polymorphs. At lower mantle conditions, the carbonates transform into new structures featuring tetrahedrally coordinated carbon. The identification of a systematic class of carbonates, nesocarbonates, cyclocarbonates, and inocarbonates reveals a complex crystal chemistry, with analogies to silicates. They provide fundamental input for the understanding of deep carbonatite melt physical properties. The possible polymerization of carbonate units will affect viscosity, and the reduced polymerization in crystal structures as a function of oxidation state could suggest that also oxidation state may affect the mobility of deep carbonatitic magmas. Finally, we report two high-pressure structures of mixed alkali carbonates, which reveal that these compounds may form wide solid solutions and incorporate a sensible amount of vacancies, which would allow incorporation of high-strength elements and therefore play an important role for geochemical element partitioning in the mantle.

9.1. INTRODUCTION

Carbon, the sixth element of the periodic table, together with oxygen and hydrogen, has one of the leading roles in the oldest and still ongoing process on the planet Earth: the evolution of the geological processes that led to the creation of life on the Earth's surface (Hazen, Jones, et al., 2013). Indeed, unraveling all the steps of the deep carbon cycle is one of the main topics in the study of the dynamic history of the Earth. C seems to be a perfect element to describe the dynamism of the Earth' history, because it is a versatile element: it is present from the Earth's crust to the core, it is present in many different geological environments, and it can bond to more than 80 other elements, changing its speciation and structural environment. To date, ca. 25% of the C-bearing species must be discovered in the Earth's crust (Hazen et al., 2016), and if we consider only the C O bonding, ca. 91% of these are carbonates. The study of carbonate crystal-chemistry at deep planetary conditions can be a key point in determining the processes that characterize the storage and the importance of carbonates as C-carriers in subduction zones.

The fundamental chemical unit of the carbonates is the CO_3^{2-} anion. It is characterized by a planar trigonal geometry, where three oxygens surround the central carbon. Despite the huge chemical variety of carbonates at Earth's surface, the CO_3 group stays uniform. It is at HP-HT conditions that it will undergo many distortions that will lead to phase transitions in all the main carbonate minerals. At surface conditions, the most important and ubiquitous carbonate minerals are calcite, aragonite ($CaCO_3$), and dolomite, $CaMg(CO_3)_2$ (Hazen, Downs, et al., 2013). Iron

Dipartimento di Scienze della Terra, Università degli Studi di Milano, Milano, Italy

Carbon in Earth's Interior, Geophysical Monograph 249, First Edition. Edited by Craig E. Manning, Jung-Fu Lin, and Wendy L. Mao.

carbonate, siderite $FeCO_3$, and magnesium carbonate, magnesite $MgCO_3$, isostructural with calcite, are also important carbonates occurring especially in hydrothermal systems and constituting important ore and industrial minerals (Martin et al., 2017). Solid solution between calcite-magnesite-siderite is favored at the Earth's upper mantle condition (Franzolin et al., 2011). Among crustal minerals, carbonates of divalent cations (e.g. Sr, Ba, Pb, etc.) are also important. Alkali carbonates are also present, despite their high aqueous solubility, which prevents a massive preservation at the Earth's surface (Hazen, Downs, et al., 2013).

Calcium carbonates have two main structural types, the rhombohedral-$CaCO_3$ calcite-type structure, and the aragonite structure that is orthorhombic-$CaCO_3$. Another major rock-forming mineral on the Earth's surface is dolomite, characterized by a calcite-type structure and an ordered distribution between Mg and Ca. In the past, experimental data on carbonates were scarce, but in the last decades both experimental and computational studies (e.g. Oganov et al., 2013) have made many important discoveries on the phase transitions undertaken by carbonates at different P-T and redox conditions. These new data suggested that aragonite was not the only potentially stable carbonate at deep mantle conditions, but other polymorphic structures of dolomite and calcite were as well. One of the biggest outcomes of research in the last decades is the change of the C coordination number at HP-HT conditions. C transforms from a trigonal arrangement to tetrahedral carbonate units, which can polymerize, as silicates, in 3-fold ring carbonates (Merlini et al., 2017) or chain carbonate structures (Merlini et al., 2015; Cerantola et al., 2017), not yet found in nature. This has been both proven experimentally and predicted by computational studies (Oganov et al., 2006, 2013; Pickard & Needs, 2015). To date, the different possible structures adopted by these tetrahedrally coordinated carbonates are still unclear, because of their unquenchable nature. Notably, these structures surprisingly match 3-fold ring silicates, e.g. Ca-walstromite ($CaSiO_3$ silicate) found as inclusion in diamonds, and it has been seen that this phase undergoes polymorphic transition (Anzolini et al., 2016). By analogy with experiments performed in the 1960s on germanates as analogues of HP behavior of silicates, the input of polymorphic transition observed in Ca-walstromite suggested to expand the investigation at nonambient behavior of tetrahedrally coordinated carbonates.

In addition, the study of iron-bearing carbonates (e.g. $[Mg,Fe]CO_3$) at lower-mantle conditions has helped the scientific community to better understand the deep mantle cycle and the different redox states. With these experiments, it was observed that the spin transition, from high- to low-state, was not only occurring in candidate mantle minerals (perovskite, $[Mg,Fe]SiO_3$, and ferropericlase, $[Mg,Fe]O$) but also in iron-bearing carbonates (e.g. Liu et al., 2015). The effect of the high- to low-spin transition on the $(Mg,Fe)CO_3$ structure was that the rhomboedral phase (Phase I) was transforming to an orthorombic one (Phase II) at much lower pressure with respect to the Mg-rich analogue.

The preferential incorporation of iron in low-spin-state Fe-bearing mantle phases could support the idea that Fe partitioning is favored in Fe-bearing carbonates, because of the shallower depth at which the Fe^{2+} high-spin low-spin transition happens, with respect to other mantle minerals (Kantor et al., 2006; Kupenko et al., 2015). Recently, experiments performed on pure synthetic single crystals of $FeCO_3$ to ~ 100 GPa and 2500 K brought evidence that even iron carbonates change their structure to CO_4-based carbonates at ~ 70 GPa, proving that the CO_4-based carbonates are the most favourable C-carrier in the lower mantle (Cerantola et al., 2017; Merlini et al., 2015).

Moreover, carbonates of other elements, in particular alkali elements, could play an important role in the deep carbon cycle and the injection of alkali elements at extreme mantle conditions, as suggested by the finding of nyerereite, $(Na,K)_2(Ca,Sr,Ba)(CO_3)_2$, and nahcolite, $NaHCO_3$ as diamond inclusions (Kaminsky et al., 2009). However, a detailed study of different structures adopted by these carbonates at mantle conditions is still unknown. Recent published data on shortite ($Na_2Ca_2[CO_3]_3$) (Vennari et al., 2018) show how at pressures above 15–17 GPa the structure of shortite I transforms to shortite II, with a substantial density increase of about 2.5%. But the more relevant result is the evidence of the progressive displacement of the C atom from a trigonal planar geometry to a 3+1 coordination, resulting in a dimerization of carbonate ions units. This effect is visible on half of the carbonate groups in the shortite II phase (Vennari et al., 2018).

Given the fundamental role played by stagnant slabs in paramount petrological and geochemical processes (Rohrbach & Schmidt, 2011; Thomson et al. 2016; Walter et al., 2011), understanding the different possible structures adopted by carbonates during their polymorphic phase transitions will be of fundamental importance to reveal the past, present, and future evolution of the deep carbon cycle. In the following paragraphs, we present a synthetic overview of recent selected experimental investigations of carbonates at mantle conditions. We focus our review on in-situ single crystal X-ray diffraction experiments, because it is the most reliable technique for structural determination.

9.2. EXPERIMENTAL STRUCTURAL INVESTIGATION ON CARBONATES: METHODS

In-situ X-ray diffraction using diamond anvil cells (DACs) is the primary tool for structural determination of crystals at nonambient conditions. In the last 15 years, an upgrade in synchrotron diffraction protocols for single crystal diffraction at extreme conditions has allowed experimental ab-initio determination of structures after multiple phase transitions (Dera et al., 2005; Dubrovinsky et al., 2010; Merlini & Hanfland, 2013). The possibility to perform accurate structural refinements on tiny crystals, and especially on a small single crystal domain from a multidomain sample (e.g. original single crystal samples stabilized at HP-HT conditions), has opened a new window on Earth's material structural behavior. The experimental data, in fact, may validate computational structure prediction, as well as reveal new unpredicted structural types or crystal chemistry features, which in turn may drive novel computational works.

Historically, it is noticeable that miniaturized DAC for single-crystal X-ray diffraction purposes (i.e. the Merrill-Bassett DAC; Merrill & Bassett, 1974) was developed specifically to solve the structure of the $CaCO_3$-II polymorph (Bassett, 2009). Calcite is an interesting mineral for high-pressure crystallography. It undergoes multiple phase transitions in a relatively moderate pressure range (Bridgman, 1939). Calcite-II is a distortion of the calcite structure (Merrill & Bassett, 1975), and it is described in a symmetry group-subgroup relationship. The higher pressure polymorphs are more complex; the $CaCO_3$-III polymorph forms after a first-order phase transition from an original calcite single crystal that often promotes the formation of multiple domain grain. Several models were proposed based on powder-like data (Davis, 1964; Hagiya et al., 2005), but any attempt to correctly index the diffraction pattern failed. Recently, the use of new detectors for synchrotron X-ray diffraction with zero point spread functions allowed the correct determination of unit cell, diffraction intensities, and finally structure solution (Merlini et al., 2013). Nowadays, the use of single-crystal diffraction for structural determination after multiple phase transition has become a routine methodology in most synchrotron beamlines.

9.3. EXPERIMENTAL DETERMINATION OF CARBONATE STRUCTURES: CASE STUDIES

Experimental investigations on carbonates at nonambient conditions have revealed an unexpected complexity in structural behavior, with several phase transitions identified and new structural types determined. In the following sections, three case studies are summarized to demonstrate carbonates' rich polymorphism and implications these structures have on high-pressure carbonate mineralogy.

9.3.1. Calcium Carbonates With CO_3 Units

Calcite, $CaCO_3$, represents the most relevant carbonate at Earth's surface. Fifteen years ago, only two polymorphs were considered relevant for the Earth's mantle mineralogy of carbonates, calcite, and aragonite. $CaCO_3$-II structure was considered a metastable structure and $CaCO_3$-III polymorph an unresolved crystallographic problem. The successful structural solution of $CaCO_3$-III polymorph and the identification of higher pressure modifications, however, have suggested a more complex crystal chemistry of these phases. The calcite to $CaCO_3$-II phase involves a rotation of carbonate groups, with consequent loss in three-fold symmetry and description of the structure in a symmetry subgroub of calcite (Merrill & Bassett, 1975). $CaCO_3$-III, on the contrary, presents a structure based on the same topology as calcite, considering the arrangement of Ca atoms; however, the trigonal carbonate units are no longer parallel to each other but tilted. This structure can generate multiple superstructures, with variable coordination sites for Ca atoms, with an average increase in coordination numbers, [7–8], compared to calcite, [6]. This arrangement explains the sensible density increase of $CaCO_3$-III compared to extrapolated volume of calcite. Experimentally, in the pressure range 3–15 GPa, two structures were encountered experimentally. Both polymorphs are very likely metastable structures compared to aragonite. They confirm the theoretical prediction of multiple metastable structures of $CaCO_3$ polymorphs at high pressure with competitive energy compared to aragonite (Oganov et al., 2006). Further compression of $CaCO_3$-III phase induces a phase transition at 15 GPa towards the $CaCO_3$-VI structure. The interest of $CaCO_3$-VI, triclinic, with 10 atoms in the unit cell, arises since it has a higher density compared to aragonite (Merlini, Hanfland, et al., 2012). These results therefore indicate that dense structures of carbonates at mantle pressures based on calcite topology and different from aragonite are possible. The successful identification of the thermodynamic stable polymorphs will allow a precise phase equilibrium modeling in carbonate-bearing systems. Recently, computational works and X-ray powder diffraction experiments have provided new important information about the stable structures of $CaCO_3$ at mantle conditions, with the identification of aragonite-II and $CaCO_3$-VII polymorphs, which is potentially the predominant form in the lower mantle (Gavryuskin et al., 2017; Li et al., 2018; Pickard & Needs, 2015). For the sake of completeness, we also mention the single crystal determination of $CaCO_3$-V structure, which represents a disordered calcite structure in terms of carbonate unit position, which in turn rotates and librates around carbon position (Ishizawa et al., 2013). $CaCO_3$-V polymorph is stable at high-temperature conditions (e.g. Zhao et al., 2019).

9.3.2. Carbonates with [4]-Coordination Carbon

In the past years, the experimental studies on carbonates at lower mantle conditions were mainly focused on magnesite, which is considered an important carbonate end-member for lower mantle mineralogy. The pressure-induced sp^2 to sp^3 orbital hybridization of the carbon-oxygen bond is expected to promote new structures, based on tetrahedrally coordinated CO_4 carbonate units (Arapan et al., 2007). High-pressure phase transitions were observed in magnesite and Fe-magnesite (Boulard et al., 2011; Isshiki et al., 2004). The quality of experimental X-ray powder diffraction patterns is not suitable for ab-initio structure determinations, and the exact structure of the observed phases is still an open question, but a similarity to predicted structures based on tetrahedrally coordinated carbon is apparent in X-ray powder diffraction patterns. A fundamental discovery in these high-pressure studies of carbonates is the identification of a new class of carbonates, featuring oxidized trivalent iron (Boulard et al., 2012). An unexpected phase with chemical formula $Fe_3C_4O_{12}$ was clearly identified on quenched product from synthesis above 80 GPa. The implications for this discovery are far reaching. At ambient conditions, no anhydrous carbonate salts of trivalent common cations, such as iron and aluminum, exist among carbonate minerals. The Fe^{3+} carbonate in experiments of Boulard et al. (2012) originates from a redox reaction in the close chemical system constituted by Mg-siderite. The Fe^{3+} carbonate coexists with the high-pressure form of magnetite, hp-Fe_3O_4, and reduced carbon (diamond). The possibility to laser heat single crystals with sensible iron content for IR laser absorption has successively allowed the stabilization of single crystals suitable for ab-initio experimental phase determination of $Mg_2Fe_2(C_4O_{13})$ carbonate (Merlini et al., 2015). It belongs to a new structural class of carbonates, and it is based on truncated chains constituted by four tetrahedral groups, described by the crystal-chemical formula $[C_4O_{13}]^{10-}$. The unique structural analogue is the rare Earth element (REE) silicate $Ba_2Gd_2[Si_4O_{13}]$ (Wierzibcka-Wieczorec et al., 2010). Following this single crystal determination, other high-pressure carbonates with tetrahedral carbonate units have been determined (Table 9.1), which are $Fe_4(CO_3)_4$, $Fe_4(C_4O_{13})$, and $(Ca,Mg,Fe)_3(C_3O_9)$ (Cerantola et al., 2017; Merlini et al., 2017). These are carbonates based on tetrahedrally coordinated carbonate anionic groups. The interatomic bond lengths in $[CO_4]^{4-}$ units agree with predicted bond lengths, and the C-O-C bond angles in polymerized units are in the predicted ranges (Oganov et al., 2013). This confirms the prediction of a possible limited flexibility in polymerized CO_4 units if compared to silicates. Yet the four different experimental structures available so far indicate an interesting richness. By analogy with silicates, HP carbonate forms nesocarbonates, cyclocarbonates, and inocarbonates (Figure 9.1). The different structures are also based on variable cationic coordination polyhedra, which can accommodate elements with variable interatomic sizes and valences. The inocarbonate $Fe_4(CO_4)_3$ incorporates trivalent iron. The coexistence of this phase with elemental carbon and HP-magnetite (Boulard et al., 2012; Cerantola et al., 2017) indicates that it can be stable over a variable oxidation environment. The possible presence of iron in low-spin state (Cerantola et al., 2017) and the reduced size of low-spin iron suggests that other trivalent cation carbonates may form. There is no evidence at ambient conditions of anhydrous aluminum carbonates, but HP carbonates based on tetrahedral CO_4 groups could likely incorporate aluminum (Al being the fourth element by abundance in Earth and the planets). Cyclocarbonates, such as dolomite-IV, are based on three-fold C_3O_9 tetrahedral ring units. There is no exact structural analogue for these carbonates, but the same topology is observed in silicates and borates. The silicate walstromite, $Ba_2CaSi_3O_9$, or the pure $CaSiO_3$ cyclosilicate end-member have a closely related structure. The experimental dolomite-IV is orthorhombic and feature two different cation sites with [8] and [10] coordination, suggesting the possible existence of a wide solid solution between cations with variable sizes, Ca, Mg, and Fe in particular.

The observed polymerization of carbonate units in crystalline carbonate is also relevant for the understanding of carbonate melt structures in the lower mantle. It is a known fact that carbonatite melts in the upper mantle have a sensible low viscosity (e.g. Dobson et al., 1996; Jones et al., 2013) because of the ionic nature of these melts and the impossibility of having polymers based on trigonal carbonate units. This scenario significantly changes at lower mantle conditions, when polymerization of carbonate is possible. The observation of structural variability of carbonates as a function of Fe^{3+}/Fe^{2+} ratio, with isolated carbonate units observed only in fully oxidized Fe-carbonate, could suggest that oxygen fugacity may also have an important role in determining the viscosity and the mobility of ultradeep carbonate melts.

9.3.3. Alkali and Mixed-Alkali Carbonates

Pure alkali carbonates and carbonates containing alkali elements constitute an interesting group of carbonates and are found at variable crustal settings (Hazen, Downs, et al., 2013). Their interest in Earth's mantle mineralogy arises by the identification of alkali carbonates in diamond inclusions (Kaminsky et al., 2009), revealing that these minerals exists in the Earth's mantle and are present as chemical components in mantle fluids or melts.

Mixed-alkali carbonates, i.e. carbonates of alkali metals and other cations, represent very interesting candidate accessory minerals of the Earth's mantle (Table 9.2). They can originate from the crystallization of carbonatite melts. Among these mixed-alkali carbonates,

Table 9.1 Crystallographic information on high-pressure determination of carbonates in the Ca-Mg-Fe-C-O system based on single crystal X-ray diffraction structure solution and refinement.

Carbonate	Experimental P,T	Unit Cell	Space Group	Reference
$CaCO_3$-II	P = 1.5 GPa	a = 6.334(20) Å b = 4.948(15) Å c = 8.033(25) Å α = γ = 90° β = 107.9° Vol = 239.57 Å³	$P2_1/c$	Merrill and Bassett, 1975
$CaCO_3$-III	P = 2.8 GPa	a = 6.2811(13) Å b = 7.5073(13) Å c = 12.516(3) Å α = 93.76(2)° β = 98.95(2)° γ = 106.49(2)° Vol = 555.26(20) Å³	P-1	Merlini, Hanfland, et al., 2012
$CaCO_3$-IIIb	P = 3.1 GPa	a = 6.144(3) Å b = 6.3715(14) Å c = 6.3759(15) Å α = 93.84(2)° β = 107.34(3)° γ = 107.16(3)° Vol = 224.33(13) Å³	P-1	Merlini, Hanfland, et al., 2012
$CaCO_3$-VI	P = 30.4 GPa	a = 3.319(2) Å b = 4.883(3) Å c = 5.590(2) Å α = 103.30(5)° β = 94.73(5)° γ = 89.21(6)° Vol = 87.86 Å³	P-1	Merlini, Hanfland, et al., 2012
$Ca(Mg,Fe)(CO_3)_2$-II	P = 34.4 GPa	a = 4.7407(10) Å b = 5.3085(10) Å c = 6.7430(10) Å α = 101.42(1)° β = 89.27(1)° γ = 95.72(1)° Vol = 168.01(5) Å³	P-1	Merlini, Crichton, et al., 2012
$Ca(Mg,Fe)(CO_3)_2$-III	P = 50.6 GPa	a = 6.2346(9) Å b = 9.3025(11) Å c = 10.9893(12) Å α = 75.89(1)° β = 81.05(1)° γ = 89.48(1)° Vol = 610.32(14) Å³	P-1	Merlini, Crichton, et al., 2012
$Ca(Mg,Fe)(CO_3)_2$-IIIb	P = 36.8 GPa	a = b = 12.010(3) Å c = 13.700(5) Å α = β = 90° γ = 120° Vol = 1711.34(5) Å	R3	Merlini et al., 2017
$(Ca,Mg,Fe)_3(C_3O_9)$	P = 115 GPa (quenched from 2500 K)	a = 10.091(3) Å b = 8.090(7) Å c = 4.533(3) Å α = β = γ = 90 90 90 Vol = 370.1(4) Å³	P*nma*	Merlini et al., 2017
$Mg_2Fe_2C_4O_{13}$	P = 135 GPa (quenched from 2650 K)	a = 9.822(3) Å b = 3.9023(13) Å c = 13.154(5) Å, α = γ = 90° β = 108.02(3)° Vol = 479.4(3) Å³	C2/*c*	Merlini et al., 2015

(Continued)

Table 9.1 (Continued)

Carbonate	Experimental P,T	Unit Cell	Space Group	Reference
$Fe_4C_4O_{13}$	P = 97 GPa (quenched from 3088 K)	a = 10.261(3) Å b = 3.985(3) Å c = 13.455(5) Å α = γ = 90° β = 107.85(4)° Vol = 523.76(28) Å³	C2/c	Cerantola et al., 2017
$Fe_4C_3O_{12}$	P = 74 GPa (quenched from 1750 K)	a = b = 12.762(2) Å c = 5.332(1) Å α = γ = 90° β = 120° Vol = 752.0(3) Å³	R3c	Cerantola et al., 2017

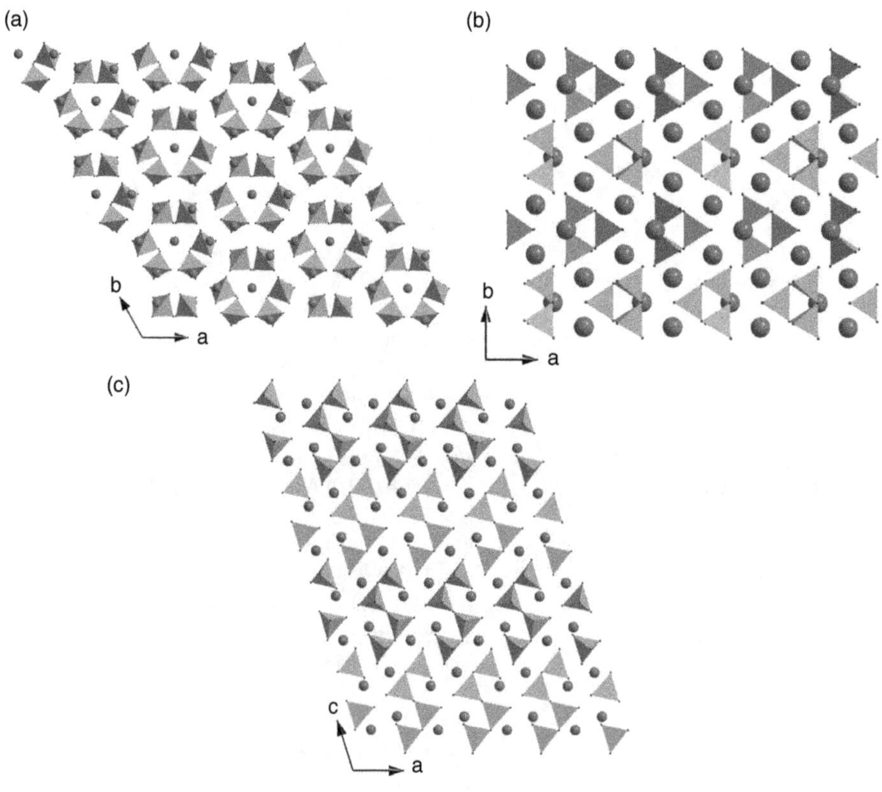

Figure 9.1 Crystal structure of (a) $Fe_4C_3O_{12}$, (b) $(Ca,Mg,Fe)_3[C_3O_9]$, and (c) $Fe_4C_4O_{13}$ as examples of nesocarbonate, cyclocarbonate, and inocarbonate stabilized at deep mantle conditions. C tetrahedra are cyan, other cations are in green (Ca, Mg, Fe), and oxygen atoms are in blue. See electronic version for color representation of the figures in this book.

it is worth mentioning the recent discovery of new high-pressure phases in the Na_2CO_3-$CaCO_3$ systems, in particular $Na_2Ca_3(CO_3)_4$ and $Na_2Ca_4(CO_3)_5$ (Gavryushkin et al., 2014; Shatskiy et al., 2013). $Na_2Ca_3(CO_3)_4$ is a liquidus phase in Na-bearing and Ca-bearing carbonatitic melt. Its structure is based on trigonal CO_3 units and large cation sites with variable coordination. We synthesized these phases in K_2CO_3-Na_2CO_3-$CaCO_3$ systems at 5 GPa and 900 °C (Figure 9.2). The results indicate the possible existence of solid solution between K and Na end-member. Chemical analyses indicate the presence of vacancies in these structures. A slight excess of Ca and depletion of alkali achieve the electrostatic balance. This aspect is very important if we consider the negligible vacancies that are possible in the calcite or aragonite structures. The possible existence of vacancies in $(Na,K)_2Ca_3(CO_3)_4$ indicates also the possible incorporation of trivalent cations in these structures, like REEs. A stoichiometric REE-bearing carbonate with formula $Na_3Ca_2La(CO_3)_5$ was synthesized in the system Na-La-Ca-C-O at 5 GPa and 900 °C (Table 9.2 and Figure 9.2). It is isostructural with the mineral burbankite

Table 9.2 Single crystal structure determination of alkali carbonate high-pressure polymorphs.

Carbonate	Experimental P,T	Unit cell	Space group	Reference
$Na_2Ca_3(CO_3)_4$	6 GPa, 1000 °C, quenched	a = 31.4421(8) Å b = 8.1960(2)Å c = 7.4360(2) Å α = γ = 90° β = 89.923(2)°	*Pn*	Gavryushkin et al., 2014
$Ca_3Na_{1,4}K_{0,6}(CO_3)_4$	5 GPa, 800 °C, quenched	a = 7.3877 Å b = 8.4967 Å c = 15.8228 Å α = β = γ = 90° V = 993.22 Å³	*Pnma*	This work
$Ca_3Na_{0,6}K_{1,4}(CO_3)_4$	5 GPa, 800 °C, quenched	a = 7.4086 Å b = 8.6407 Å c = 15.9271 Å α = β = γ = 90° V = 1019.59 Å³	*Pnma*	This work
$Na_2Ca_4(CO_3)_5$	6 GPa, 1050 °C, quenched	a = b = 10.37402(14) Å c = 6.25935(9) Å α = β = 90° γ = 120° V = 583.38(1) Å³	$P6_3mc$	Rashchenko et al., 2017
$Na_3Ca_2La(CO_3)_5$	5 GPa, 900 °C, quenched	a = b = 10.443(3) Å c = 6.3044(10) Å α = β = 90° γ = 120° V = 595.44(18) Å³	$P6_3mc$	This work

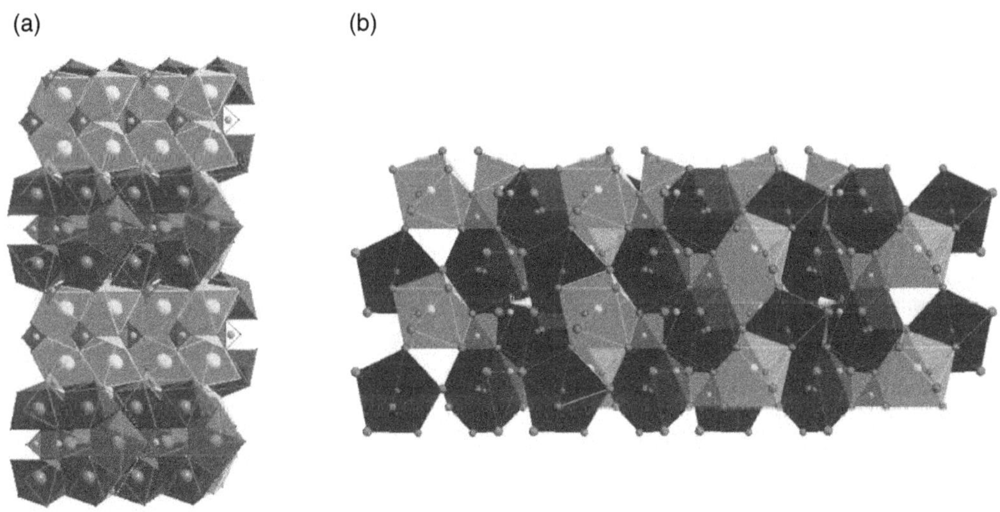

Figure 9.2 Crystal structure of (a) $Ca_3Na_{0,6}K_{1,4}(CO_3)_4$ and (b) $Na_3Ca_2La(CO_3)_5$, synthesized at 5 GPa. See electronic version for color representation of the figures in this book.

(Effenberger et al., 1985) and with the synthetic $Na_2Ca_4(CO_3)_5$ (Rashchenko et al., 2017) and demonstrates the potential chemical variation of these minerals at upper mantle pressures. The experimental evidence (i.e. Poli, 2015) of mobilization of carbon through carbonatitic melts in the upper mantle as well the natural findings of alkali carbonates in deep processes recorded by diamonds (e.g. Kaminsky et al., 2009) shows that mixed alkali carbonates are indeed important in deep processes. Their crystallographic features account for the possibility of geochemical fractionation of minor and trace elements with large size and valence state, different from other carbonates, such as calcite, aragonite, magnesite and dolomite.

9.4. CONCLUDING REMARKS

The occurrence of carbonates in kimberlites and as diamond inclusions (Berg, 1986; Kaminsky et al., 2009) and experimental studies (e.g. Poli, 2015) demonstrates the stability of carbonates in the upper mantle. Currently, there is still no direct evidence of carbonate minerals from the lower mantle. The experimental results on the structural transformation of carbonates at ultra-high pressures have, however, revealed a potential richness of new deep carbonate minerals with novel crystal chemical features, which may match the complex redox environments (Cerantola et al., 2017) and thermodynamic conditions existing in the planetary interiors.

ACKNOWLEDGMENTS

The authors acknowledge DCO support for a research grant. Diego Gatta, Stefano Poli, Nicola Rotiroti, and Andrea Risplendente are acknowledged for help and useful discussion in experimental activities at the Earth Science Department, University of Milano. ESRF and Elettra Facilities are acknowledged for synchrotron diffraction experiments.

REFERENCES

Anzolini, C., Angel, R. J., Merlini, M., Derszi, M., Tokar, K., et al. (2016). Depth of formation of $CaSiO_3$-walstromite included in super-deep diamonds. *Lithos*, *265*, 138–147.

Arapan, S., Souza De Almeida, J., Ahuja, R. (2007). Formation of sp3 hybridized bonds and stability of $CaCO_3$ at very high pressure. *Physical Review Letters*, *98*(26), 268501.

Bassett, W.A. (2009). Diamond anvil cell, 50th birthday. *High Pressure Research*, *29*, 163–186.

Berg, G.W. (1986). Evidence for carbonate in the mantle. *Nature*, *324*, 50–51.

Boulard, E., Gloter, A., Corgne, A., Antonangeli, D., Auzende, A.L., Perrillat, J.P., et al. (2011). New host for carbon in the deep Earth. *Proocedings of the National Academy of Sciences of the United States of America*, *108*, 5184–5187.

Boulard, E., Menguy, N., Auzende, A.L., Benzerara, K., Bureau, H., Antonangeli, D., et al. (2012). Experimental investigation of the stability of Fe-rich carbonates in the lower mantle. *J. Geophys. Res.*, *117*, B02208.

Bridgman, P. W. (1939). The high-pressure behavior of miscellaneous minerals. *Amer. J. Sci. 237*, 7–18.

Cerantola, V., Bykova, E., Kupenko, I., Merlini, M., Ismailova, L., McCammon, C., et al. (2017). Stability of iron-bearing carbonates in the deep Earth's interior. *Nature Communications*, *8*, 15960.

Davis, B. L. (1964). X-ray diffraction data on two high-pressure phases of calcium carbonate. *Science*, *145*(3631), 489–491.

Dera, P., Prewitt, C. T., & Jacobsen, S. D. (2005). Structure determination by single-crystal X-ray diffraction (SXD) at megabar pressures. *Journal of Synchrotron Radiation*, *12*(5), 547–548.

Dobson D., Jones A., Rabe R., Sekine T., Kurita K., Taniguchi T., et al. (1996). In-situ measurement of viscosity and density of carbonate melts at high pressure. *Earth and Planetary Science Letter*, *143*, 207–215.

Dubrovinsky, L., Boffa-Ballaran, T., Glazyrin, K., Kurnosov, A., Frost, D. Merlini, M., et al. (2010). Single-crystal X-ray diffraction at megabar pressures and temperatures of thousands of degrees. *High Pressure Research*, *30*(4), 620–633.

Effenberger, H., Kluger, F., Paulus, H., & Wölfel, E. R. (1985). Crystal structure refinement of burbankite. *Neues Jahrb. Mineral., Monatsh.*, 161–170.

Franzolin, E., Schmidt, M. W., &Poli, S. (2011). Ternary Ca-Fe-Mg carbonates: Subsolidus phase relations at 3.5 GPa and a thermodynamic solid solution model including order/disorder. *Contribution to Mineralogy and Petrology*, *161*, 213–227. https://doi: 10.1007/s00410-010-0527-x

Gavryushkin, P., Bakanin, V., Bolotina, N., Shatskiy, A., Seryotkin, Y., Litasov, K. (2014). Synthesis and crystal structure of new carbonate Ca3Na2(CO3)4 homeotypic with orthoborates M3Ln2(BO3)4 (M = Ca, Sr, and Ba). *Crystal Growth & Design*, *14*, 4610–4616.

Gavryushkin, P. N., Martirosyan, N. S., Inerbaev, T. M., Popov, Z. I., Rashchenko, S. V., Likhacheva, A. Y., et al. (2017). Aragonite-II and CaCO3-VII: New high-pressure high-temperature polymorphs of $CaCO_3$. *Crystal Growth & Design*, *17*(12), 6291–6296. https://doi.org/10.1021/acs.cgd.7b00977

Hagiya, K., Matsui, M., Kimura, Y., & Akahama, Y. (2005). The crystal data and stability of calcite III at high pressures based on single crystal X ray experiments. *Journal of Mineralogical and Petrological Sciences*, *100*(1), 31–36.

Hazen, R. M., Downs, R.T., Jones, A.P., & Kah, L. (2013). Carbon mineralogy and crystal chemistry. *Reviews in Mineralogy and Geochemistry*, *75*, 7–46.

Hazen, R. M., Hummer, D. R., Hystad, G., Downs, R. T., & Golden, J. J. (2016). Carbon mineral ecology: Predicting the undiscovered minerals of carbon. *American Mineralogist*, *101*(4), 889–906.

Hazen, R. M., Jones, A. P., & Baross, J. A. ed. (2013). Carbon in earth: Reviews in mineralogy and geochemistry, Vol. *75*.

Ishizawa, N., Setoguchi, H., &Yanagisawa, K. (2013). Structural evolution of calcite at high temperatures: Phase V unveiled. *Scientific Report*, *3*, 2832.

Isshiki, M., Irifune, T., Hirose, K., Ono, S., Ohishi, Y., Watanuki, T., et al. (2004). Stability of magnesite and its high-pressure form in the lowermost mantle. *Nature*, *427*, 60–63.

Jones, A., Genge, M., & Carmody, L. (2013). Carbonate melts and carbonatites. *Reviews in Mineralogy & Geochemistry*, *75*, 289–322.

Kaminsky, F., Wirth, R., Matsyuk, S., Schreiber, A., & Thomas, R. (2009). Nyerereite and nahcolite inclusions in diamond: evidence for lower-mantle carbonatitic magmas. *Mineralogical Magazine*, *73*(4), 521–540.

Kantor, I. Y., Dubrovisnky, L. S., & McCammon, C.A. (2006). Spin crossover in (Mg,Fe)O: A Mössbauer effect study with an alternative interpretation of X-ray emission spectroscopy data. *Physical Review B*, *73*, 100101.

Kupenko, I., McCammon, C., Sinmyo, R., Cerantola, V., Potapkin, V., Chumakov, A.I., et al. (2015). Oxidation state of the lower mantle: In situ observations of the iron electronic configuration in bridgmanite at extreme conditions. *Earth and Planetary Science Letters*, *423*, 78–86.

Li, X., Zhang, Z., Lin, J.-F., Ni, H., Prakapenka, V. B., & Mao, Z. (2018). New high-pressure phase of $CaCO_3$ at the topmost lower mantle: Implication for the deep-mantle carbon transportation. *Geophysical Research Letters*, 10.1002/2017GL076536.

Liu, J., Lin, J.-F., & Prakapenka, V. (2015). High-pressure orthorhombic ferromagnesite as a potential deep-mantle carbon carrier. *Scientific Reports*, *5*, 7640.

Martin, S., Toffolo, L., Moroni, M., Montorfano, C., Secco, L., Agnini, C., et al. (2017). Siderite deposits in northern Italy: Early Permian to Early Triassic hydrothermalism in the Southern Alps. *Lithos*, *284–285*, 276–295.

Merlini, M., Cerantola, V., Gatta, G. D., Gemmi, M., Hanfland, M., Kupenko, et al. (2017). Dolomite-IV: Candidate structure for a carbonate in the Earth's lower mantle. *American Mineralogist*, *102*(8), 1763–1766.

Merlini, M., Crichton, W., Hanfland, M., Gemmi, M., Mueller, H., Kupenko, I., & Dubrovinsky, L. (2012). Dolomite-II and dolomite-III: crystal structures and stability in the Earth's lower mantle. *PNAS*, *109*(34), 13509–13514.

Merlini, M., & Hanfland, M. (2013). Single-crystal diffraction at megabar conditions by synchrotron radiation. *High Pressure Research*, *33*(3), 511–522.

Merlini, M., Hanfland, M., & Crichton, W. A. (2012). $CaCO_3$-III and $CaCO_3$-VI, high-pressure polymorphs of calcite: Possible host structures for carbon in the Earth's mantle. *Earth and Planetary Science Letters*, *333–334*, 265–271.

Merlini, M., Hanfland, M., Salamat, A., Petitgirard, S., & Müller, H. (2015). The crystal structures of $Mg_2Fe_2C_4O_{13}$, with tetrahedrally coordinated carbon, and $Fe_{13}O_{19}$, synthesized at deep mantle condition. *American Mineralogist*, *100*(8–9), 2001–2004.

Merrill, L., & Bassett, W. A. (1974). *Review of Scientific Instruments*, *45*, 290–294.

Merrill, L., & Bassett, W. A. (1975). The crystal structure of CaCO3(II), a high-pressure metastable phase of calcium carbonate. *Acta Crystallographica*, *B31*, 343–349. https://doi.org/10.1107/S0567740875002774

Oganov, A. R., Glass, C.W., & Ono, S. (2006). High-pressure phases of $CaCO_3$: Crystal structure prediction and experiment. *Earth and Planetary Science Letters*, *241*(1–2), 95–103.

Oganov, A. R., Hemley, R. J., Hazen, R. M., & Jones, A. P. (2013). Structure, bonding, and mineralogy of carbon at extreme conditions. *Reviews in Mineralogy and Geochemistry*, *75*, 47–77.

Pickard, C. J., & Needs, R. J. (2015). Structures and stability of calcium and magnesium carbonates at mantle pressures. *Physical Review B*, *91*(10), 104101. https://doi.org/10.1103/Physrevb.91.104101

Poli, S. (2015). Carbon mobilized at shallow depths in subduction zones by carbonatitic liquids. *Nature Geoscience*, *8*, 633–636.

Rashchenko, S. V., Bakakin, V. V., Shatskiy, A. F., Gavryushkin, Seryotkin, Y. V., Konstantin, D., & Litasov, K. D. (2017). Noncentrosymmetric Na2Ca4(CO3)5 carbonate of "M13M23XY3Z" structural type and affinity between borate and carbonate structures for design of new optical materials. *Crystal Growth & Design*, *17*(11), 6079–6084.

Rohrbach, A., & Schmidt, M. W. (2011). Redox freezing and melting in the Earth's deep mantle resulting from carbon-iron redox coupling. *Nature*, *472*, 209–212.

Shatskiy, A., Sharygin, I.G., Litasov, K.D., Borzdov, Y. M., Palyanov, Y. N., & Ohtani, E. (2013). New experimental data on phase relations for the system Na2CO3-CaCO3 at 6 GPa and 900–1400 °C. *American Mineralogist*, *98*(11–12): 2164–2171.

Thomson, A. R., Walter, M. J., Kohn, S. C., & Brooker, R. A. (2016). Slab melting as a barrier to deep carbon subduction. *Nature*, *529*, 76–79.

Vennari, C. A., Beavers, C. M., & Williams, Q. (2018). High-pressure/temperature behavior of alkali/calcium carbonate shortite ($Na_2Ca_2(CO_3)_3$): Implications for carbon sequestration in Earth's transition zone. *Journal of Geophysical Research: Solid Earth*, *123*, 6574–6591.

Walter, M. J., Kohn, S. C., Araujo, D., Bulanova, G. P., Smith, C. B., Gaillou, E., et al. (2011). Deep mantle cycling of oceanic crust: Evidence from diamonds and their mineral inclusions. *Science*, *334*, 54–57.

Wierzibcka-Wieczorec, M., Kolitsch, U., & Tillmans, E. (2010). Ba2Gd2(Si4O13): A silicate with finite Si4O13 chains. *Acta Crystallographica*, *C66*, i29–i32.

Zhao, S., Schettino, E., Merlini, M., & Poli, S. (2019). The stability and melting of aragonite: An experimental and thermodynamic model for carbonated eclogites in the mantle. *Lithos*, *324–325*, 105–114.

10

Nitrogen Diffusion in Calcite

Daniele Cherniak, Morgan Schaller, and Bruce Watson

ABSTRACT

Diffusion of N has been measured in natural calcite. For diffusion normal to the {10$\bar{1}$4} cleavage surface, an activation energy for diffusion of 222 ± 17 kJ mol^{-1} and pre-exponential factor of 1.92×10^{-9} m^2sec^{-1} are obtained. Diffusion parallel to c is slightly faster than diffusion normal to {10$\bar{1}$4}. These data indicate that N contained in inclusions in calcite will diffuse through the calcite lattice at Earth's upper crustal and surface conditions slowly enough so that calcite may faithfully retain a passive signature of its formation environment for billions of years, providing much needed insight into the cycling of nitrogen between atmosphere, crust, and mantle.

10.1. INTRODUCTION

Nano to microscale mineral inclusions offer a wealth of potential information about the environments of precipitation, and in many cases can be extrapolated broadly to further our understanding large-scale Earth processes such as elemental fluxes and the history of volatile degassing (e.g., see Marty, 2012, for a review). The shallow crustal and atmospheric component of deep nitrogen and carbon cycles is surprisingly not well determined, due in large part to the paucity of minerals that serve as high-fidelity passive recorders of these reservoirs. Calcite-hosted inclusions can contain all of the major atmospheric gases (e.g., Blamey, 2012; Newman et al., 1996; Parry & Blamey, 2010), and are thus a potentially promising source for information on ancient atmospheres. However, the extent to which these records can be quantitatively interpreted depends on the capacity of the host mineral to retain these signatures over geologically relevant timescales. Unlike carbon and oxygen, nitrogen is not an essential constituent of calcite, nor is it a favorable ionic substitution upon crystallization. As such, it is unlikely to be partitioned into the calcite crystal lattice during growth. However, this incompatibility in regular lattice sites does not preclude trapping of N from the aqueous phase during precipitation and subsequent residence within inclusions or occupation of defect sites or interstices of the crystal lattice. Regardless of the mode of incorporation, if the nitrogen in calcite can be extracted and measured, it is vital to understand how retentive the calcite crystal lattice is to nitrogen. This chapter evaluates the kinetics of nitrogen retention by the calcite crystal lattice.

10.2. METHODS

10.2.1. Experimental Procedure

The experiments were performed on natural calcite from Mexico. This material has been used in our previous studies of Sr, Pb, REE, and He diffusion (Cherniak, 1997, 1998; Cherniak et al., 2015). Natural {10$\bar{1}$4} cleavage faces were used in the majority of experiments. To evaluate the potential for diffusional anisotropy, samples were also oriented to investigate diffusion normal and parallel to c.

Specimens were cut into slabs measuring ~2–3 mm on a side and 0.5–1 mm thick for use in individual experiments. The slabs were hand polished to a smooth finish using standard SiC polishing paper, followed by polishing with an automated system in a slurry of 1 μm alumina powder in water, with a final chemical polish using a 0.06 μm colloidal silica suspension.

Department of Earth and Environmental Sciences, Rensselaer Polytechnic Institute, Troy, New York, USA

Carbon in Earth's Interior, Geophysical Monograph 249, First Edition. Edited by Craig E. Manning, Jung-Fu Lin, and Wendy L. Mao.

Nitrogen was introduced into the prepared calcite samples by ion implantation. This approach has been used previously in diffusion studies, including the noble gases He (Cherniak et al., 2009, 2015; Cherniak & Watson, 2011, 2012) and Ne (Cherniak et al., 2014), as well as studies of nitrogen diffusion in other minerals (Watson et al., 2019).

The cut and polished samples were mounted on an aluminum plate with carbon paint and implanted with 100 keV ^{15}N, produced in the Extrion ion implanter at the Ion Beam Laboratory at the University at Albany. Implant doses were 1×10^{15} $^{15}N/cm^2$. ^{15}N was chosen for the ion implants over the more naturally abundant ^{14}N to facilitate analyses with nuclear reaction analysis. The $^{15}N(p, \alpha\ \gamma)^{12}C$ reaction used in these analyses permits greater sensitivity because of the comparatively large cross-section of the reaction. The gamma rays produced in the reaction are also relatively energetic, resulting in fewer interferences with gamma rays that may result from nuclear reactions with other light elements in the samples, as well as those from natural background radiation. After implantation, samples were removed from the aluminum plate and cleaned with ethanol in an ultrasonic bath.

Diffusion anneals of the implanted calcite samples were conducted at near-atmospheric pressure in sealed silica glass ampoules. The implanted samples were placed in silica glass ampoules with a small amount of $CaCO_3$ powder to ensure calcite stability during the diffusion anneals, with silica glass chips to physically separate the samples from the powder; the ampoules were then sealed under vacuum. All experiments were run in Kanthal-wound 1-atm furnaces for times ranging from 2 hours to 7 weeks at temperatures between 600 and 800 °C. Temperatures in furnaces were monitored with chromel-alumel (type K) thermocouples, with temperature uncertainties ~±2 °C.

10.2.2. Nuclear Reaction Analysis (NRA) of Nitrogen

Nitrogen distributions in samples were profiled with NRA using the reaction $^{15}N(p, \alpha\ \gamma)^{12}C$. The strong resonance at 897 keV proton energy was used (e.g. Hirvonen & Lappalainen, 1995; Kumar et al., 2005). The 4.43 MeV gamma rays from the reaction were detected with a high-efficiency bismuth germanate detector. Gamma yield as a function of energy was collected in a multichannel analyzer. Spectra were calibrated using an ^{15}N implanted olivine as a reference sample. Analytical spot sizes for incident proton beams used to induce the reactions are typically 1 mm^2. Beam energies were increased above the resonance energy of the reaction in energy steps of 1 keV in the first few hundred nanometers of depth in the sample, with steps of 5–10 keV at greater depth, to probe ^{15}N distributions at increasing depth in the samples. Depth scales for ^{15}N distributions were determined by the difference between the resonance energy and the incident beam energies, and the energy loss rates for protons in calcite. Energy loss rates were determined through calculations using the SRIM (Stopping and Range of Ions in Matter) software (Ziegler & Biersack, 2006) and standard stoichiometry and density for calcite. During each analytical session, background gamma yields in the energy region of interest were determined by analyzing unimplanted specimens of calcite.

The resultant ^{15}N profiles were fit with a model to determine the diffusion coefficient D. Uncertainties in concentration are primarily a function of counting statistics and backgrounds in the NRA spectra. Uncertainties in depth determination are dominantly a function of the statistical energy spread of individual ions comprising the beam as they travel through the sample (straggle) and the width of the resonance used for the NRA depth profiling, with the former making the largest contribution. Typical depth resolutions are ~10 nm near surface, to a few tens of nanometers several hundred nanometers into the sample (e.g., Cherniak & Lanford, 2001; Hirvonen & Lappalainen, 1995).

Diffusivities were determined using a model with an initial gaussian distribution of diffusant. For a semi-infinite medium with the concentration of diffusant equal to zero at $x = 0$, the distribution of the implanted species can be described as a function of depth x and time t as (Ryssel & Ruge, 1986):

$$N(x,t) = \frac{N_m/2}{\sqrt{1+\dfrac{2Dt}{\Delta R^2}}}\left[\exp\left[-\frac{(x-R)^2}{2\Delta R^2+4Dt}\right]\times\left[1+erf\left[\frac{\dfrac{R\sqrt{4Dt}}{\sqrt{2}\Delta R}+\dfrac{x\sqrt{2}\Delta R}{\sqrt{4Dt}}}{\sqrt{2\Delta R^2+4Dt}}\right]\right]-\exp\left[-\frac{(x+R)^2}{2\Delta R^2+4Dt}\right]\times\left[1+erf\left[\frac{\dfrac{R\sqrt{4Dt}}{\sqrt{2}\Delta R}-\dfrac{x\sqrt{2}\Delta R}{\sqrt{4Dt}}}{\sqrt{2\Delta R^2+4Dt}}\right]\right]\right] \quad (10.1)$$

where D is the diffusion coefficient, N_m is the maximum concentration of the implanted species (in an unannealed sample), R is the range (depth in the material) of the implanted species, and ΔR is the range straggle of the initial implanted distribution, which describes the width of the gaussian. Values of R and ΔR for ^{15}N from the Monte-Carlo simulation program SRIM 2006 (Ziegler & Biersack, 2006) are 2370 and 640 Å, and the ^{15}N concentration at the maximum height of the implanted distribution is ~850 ppm for the implant dose of 1×10^{15}/cm^2 for 100 keV ^{15}N in calcite.

Diffusivities were obtained with this model through nonlinear fitting routines to Eq. (10.1) using an approach similar to that outlined in previous work (e.g. Cherniak, 1991; Cherniak et al., 2014; Watson & Dohmen, 2010). Fitting was also done on spectra of implanted, unannealed samples to obtain values for R, ΔR, and N_{imp}, with D determined by fitting the profiles from annealed, implanted samples. The values for R obtained through the fitting procedure of the implanted, unannealed samples were found to agree well with those calculated by the ion implantation simulation program SRIM (Ziegler & Biersack, 2006). The values of ΔR were also found to be consistent with the range straggle predicted by SRIM. In addition to these fitting approaches, simple fits were performed using a gaussian functional form (a broadened gaussian would be the solution to the diffusion equation if the implants were sufficiently deep and/or broadening of the implanted distribution was sufficiently small such that the effects of the surface boundary condition has little influence on N profiles); this approach yielded diffusivities that agreed within uncertainties with those obtained from the other fits. Typical ^{15}N profiles are shown in Figure 10.1.

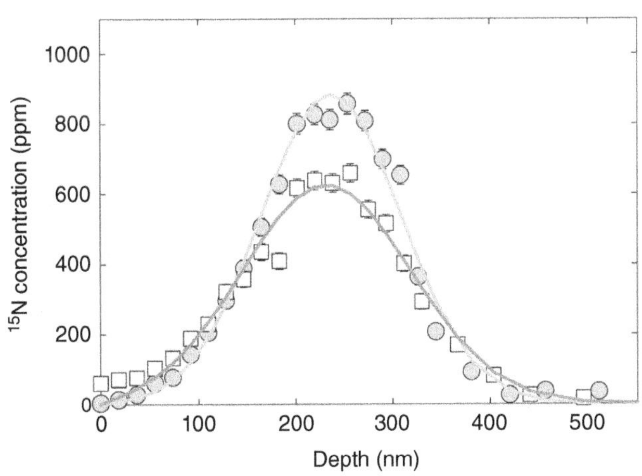

Figure 10.1 Typical ^{15}N profiles in calcite measured by NRA for experiments run at 700 °C for 73 h (grey circles) and 704 °C for 336 h (white squares). See electronic version for color representation of the figures in this book.

10.3. RESULTS

The results from N diffusion experiments on calcite are plotted in Figure 10.2 and presented in Table 10.1. For diffusion normal to {10$\bar{1}$4}, an activation energy of 222 ± 17 kJ/mol and pre-exponential factor of 1.92×10^{-9} m^2/s (log Do = –8.717 ± 0.915) are obtained. There is slight diffusional anisotropy, with diffusion parallel to c faster than diffusion normal to {10$\bar{1}$4}. For this orientation, an activation energy of 218 ± 31 kJ/mol and pre-exponential factor of 4.75×10^{-9} m^2/s (log Do = –8.323 ± 1.634) are obtained. Diffusivities obtained for samples cut parallel to c (for diffusion normal to c) agree within uncertainty with those for diffusion parallel to c.

Results for a time series at 700 °C for diffusion normal to the {10$\bar{1}$4} cleavage surface are plotted in Figure 10.3, illustrating the similarities in diffusivities for anneal times differing by more than a factor of 4, suggesting that what is being measured is volume diffusion of N. The time evolution and conformity of the profiles to the model (Eq. [1]) fit with concentration-independent diffusion coefficients indicate little evidence of concentration-dependence of N diffusion down to concentrations on order of the detection limit for ^{15}N.

10.3.1. O, C, and Noble Gas Diffusion in Carbonates

Diffusivities of other elements in calcite, including C, O, and the noble gases He and Ar, are plotted along with our

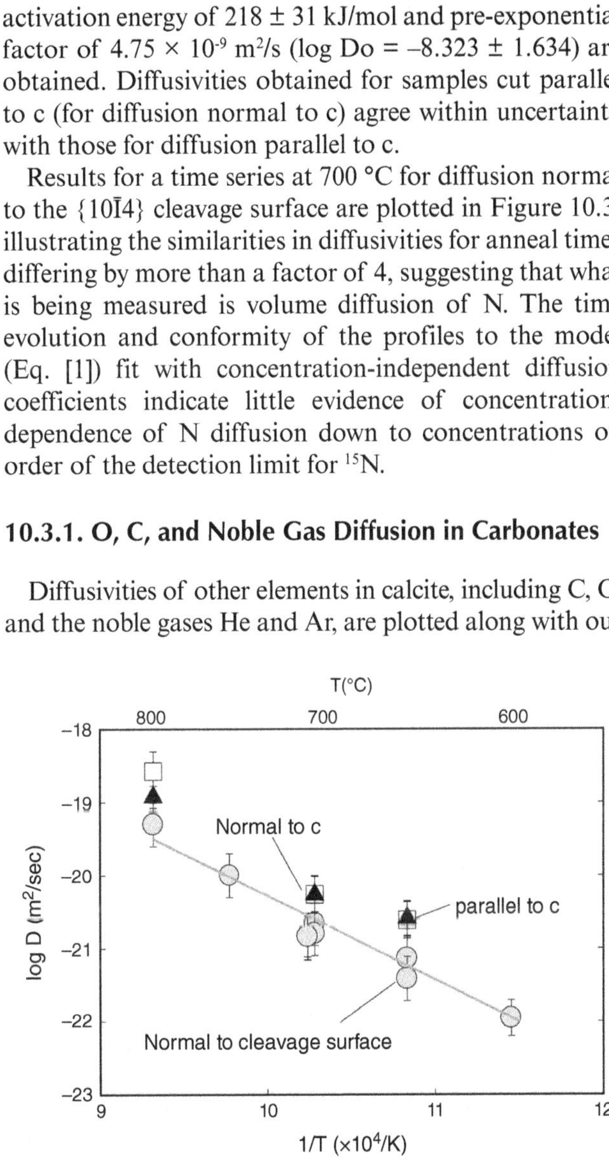

Figure 10.2 Arrhenius plot of N diffusion in calcite. An activation energy of 222 ± 17 kJ/mol and pre-exponential factor of 1.92×10^{-9} m^2/s (log D_o = –8.717 ± 0.915) are obtained from a fit to the data for diffusion normal to {10$\bar{1}$4}. For diffusion parallel to c, we determine an activation energy of 218 ± 31 kJ/mol and pre-exponential factor of 4.75×10^{-9} m^2/s (log D_o = –8.323 ± 1.634). Diffusion normal to c agrees within uncertainty with values for diffusion parallel to c. See electronic version for color representation of the figures in this book.

Table 10.1 N diffusion in calcite.

	T(°C)	time(sec)	D(m^2sec^{-1})	log D	+/–
Diffusion normal to {10$\bar{1}$4}					
NimpCC-12	600	3.79×10^6	1.10×10^{-22}	–21.96	0.18
NimpCC-6	650	7.52×10^5	7.18×10^{-22}	–21.14	0.10
NimpCC-11	650	1.24×10^6	3.82×10^{-22}	–21.42	0.16
NimpCC-2	700	2.63×10^5	2.21×10^{-21}	–20.66	0.12
NimpCC-7	700	4.21×10^5	2.19×10^{-21}	–20.66	0.11
NimpCC-8	700	6.73×10^5	1.64×10^{-21}	–20.79	0.10
NimpCC-13	704	1.21×10^6	1.44×10^{-21}	–20.84	0.28
NimpCC-9	750	8.64×10^4	9.38×10^{-21}	–20.03	0.11
NimpCC-10	800	7.20×10^3	4.96×10^{-20}	–19.30	0.19
Diffusion parallel to c					
NimpCC-19	800	1.08×10^4	1.18×10^{-19}	–18.93	0.15
NimpCC-17	700	3.46×10^5	5.41×10^{-21}	–20.27	0.25
NimpCC-15	650	7.70×10^5	2.56×10^{-21}	–20.59	0.21
Diffusion normal to c					
NimpCC-18	800	1.08×10^4	2.64×10^{-19}	–18.58	0.17
NimpCC-16	700	3.46×10^5	5.47×10^{-21}	–20.26	0.22
NimpCC-14	650	7.70×10^5	2.43×10^{-21}	–20.61	0.27

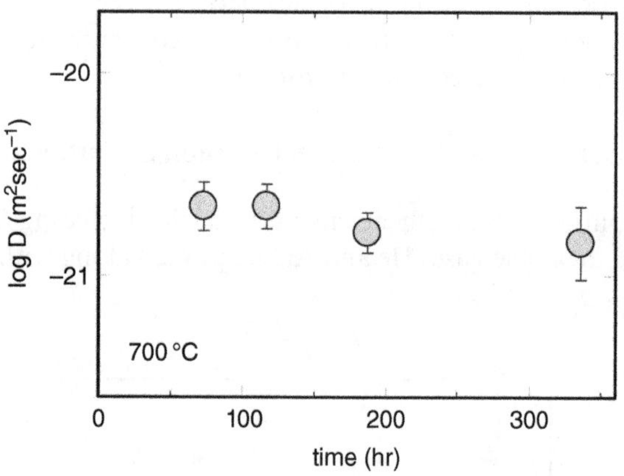

Figure 10.3 Time series for N diffusion in calcite at 700 °C. Similar diffusivities are measured over anneal times differing by up to a factor of 4.6. See electronic version for color representation of the figures in this book.

N diffusion data in Figure 10.4. The Arrhenius relation for N diffusion is bracketed by the C diffusion results for 1 atm and 100 MPa (Labotka et al., 2000; Labotka et al., 2004), with the activation energy for N diffusion falling between the values for C diffusion (291 and 166 kJ/mol for 1 atm and 100 MPa, respectively). Diffusion of N is considerably slower than oxygen diffusion, with N diffusing about two orders of magnitude slower than oxygen in a CO_2 atmosphere (Labotka et al., 2000) and three orders of magnitude slower than oxygen in the presence of H_2O (Farver, 1994). We can draw no clear conclusions regarding the mechanism of nitrogen diffusion in the calcite lattice, but recent studies of nitrogen diffusion in silicates (Watson et al., 2019) suggest it may be present in these structures as N_2.

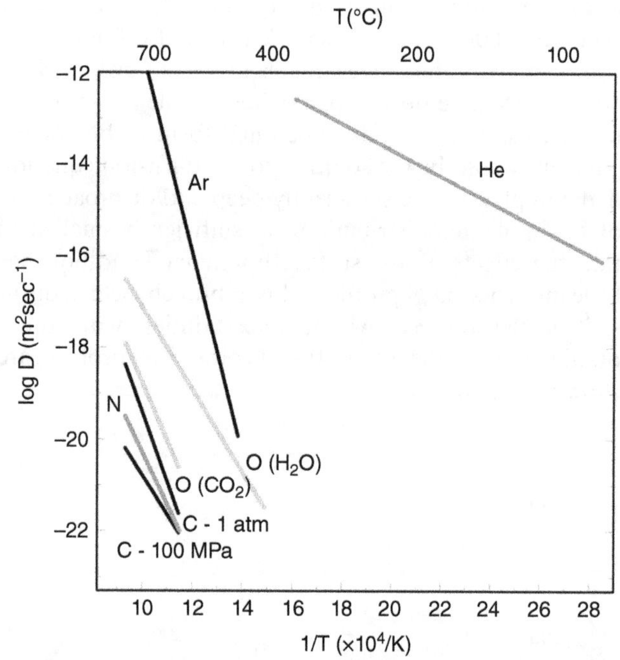

Figure 10.4 Comparison of N diffusion results with C, O, and noble gas diffusion data for calcite. Sources of data: Carbon – Labotka et al. (2000, 2004); Oxygen – Farver (1994), Labotka et al. (2000); He – Cherniak et al. (2015), diffusion normal to {10$\bar{1}$4}; Ar – Cassata and Renne (2013), calculated using an effective diffusion radius of 50 μm. See electronic version for color representation of the figures in this book.

In comparison to the noble gases, N diffuses much more slowly than He and has a much higher activation energy for diffusion (compared with 55 kJ/mol for He in calcite; Cherniak et al., 2015). For example, N will diffuse more than 10 orders of magnitude slower than He at

400 °C, with this variance increasing with decreasing temperature due to the differences in activation energies for diffusion. Ar has a much higher activation energy for diffusion (~420 kJ/mol; Cassata & Renne, 2013) than N, but diffusivities are relatively rapid at higher temperatures; for example, at 400 °C, Ar would diffuse about four orders of magnitude faster than N (with values for Ar diffusivities estimated using the values of D_0/a^2 reported by Cassata and Renne (2013) and an estimated effective diffusion radius a of 50 μm). Because of the differences in activation energies for diffusion of N and Ar, their diffusivities will converge at ~250 °C, and Ar will diffuse more slowly at low temperatures. However, diffusivities at this crossover point are slow (~1×10^{-30} m²/sec), which would yield diffusion distances of less than 400 nm over times on order of the age of the earth.

10.4. GEOLOGICAL IMPLICATIONS

The results we present here are most relevant to the retention of volatile signatures by calcite over long timescales, regardless of the original mode of incorporation. The fidelity of nitrogen hosted in the interstitial spaces within the crystal lattice (at the molecular scale) or within lattice defects and inclusions (at the nano to micro scale), is important for our interpretation of both low and high temperature calcite fluid inclusion data. The aggregate Arrhenius behavior shown in Figure 10.3 indicates that diffusion of all species (excluding He) from inclusions in calcite is slow enough at Earth's upper crustal and surface conditions that the calcite lattice may be expected to faithfully retain a passive signature of its formation environment for billions of years. This makes calcite a considerable and relatively underexploited source of information on Earth's atmospheric chemistry and volatile degassing history, and may provide much needed information on the cycling of nitrogen between atmosphere, crust, and mantle.

The few tie points we have to constrain the degassing history of the mantle come from N, $\delta^{15}N$, and Ar isotope measurements from chalcedony, chert, and other cryptocrystalline phases (Cadogan, 1977; Marty et al., 2013; Pujol et al., 2013; Sano & Pillinger, 1990). These data are relatively sparse because this information must be gleaned from rare low temperature, upper crustal phases that were originally equilibrated with the atmosphere and have survived the rock cycle at shallow depths of burial. However, fluid inclusions in calcite contain N and Ar (e.g., Newman et al., 1996; Norman et al., 2002; Norman et al., 1996) but have as yet only been treated as a record of local surface or shallow crustal processes, in part due to concerns over the fidelity of calcite as an archive. Our results, when compared to the Ar data of Renne et al. (2009), indicate that if a given sample remains below 250 °C (nominally 8 km of burial) and remains undeformed (Goldstein, 1986, 2001), the diffusive length scale of these species in the lattice will be less than 0.5 μm and the original signature is likely to be retained.

Calcite is omnipresent in surface environments as a primary biogenic and abiogenic precipitate (e.g., skeletal hard parts, speleothems, and soil carbonates, etc.), opening up a new realm of low-temperature archives with a passive record of changes in the concentration of the major atmospheric gases (and their isotope ratios). For example, the ancient atmospheric O_2 and N_2 content may be determined from fluid inclusions in soil carbonates (Schaller et al., 2017; Schaller et al., 2018), and interpretation of these data from ancient samples relies critically on the retention of N by the calcite lattice. Speleothems (Blamey et al., 2016) and biogenic carbonates like brachipod shells (Brand et al., 2016) promise a wealth of information on past gas concentrations, but these results have been treated with a healthy dose of skepticism in part due to doubts about the long-term stability of inclusions in calcite. The evidence presented and compiled here showing the slow diffusion of major atmospheric volatiles in calcite should assuage these concerns.

The calcite archive is potentially very useful in closing the crustal nitrogen cycle. The nitrogen solubility of deep crust and upper mantle minerals has been evaluated (Li & Keppler, 2014; Li et al., 2013), and are important steps in closing the deep crustal nitrogen cycle (Busigny & Bebout, 2013), but we still lack information on N cycling in the shallow crust. Calcite is a common vein-forming mineral from hydrothermal fluids; if volatiles are incorporated into calcite at depth, in the form of veins and other hydrothermal precipitates, will these signatures be retained upon exhumation? To answer this question, we examined several temperature-time path scenarios incorporating the kinetics of N diffusion in calcite. Using a finite-difference computational approach, we allowed nitrogen-bearing calcite crystals of 1 mm radius to cool exponentially from 700 °, 600 °, and 500 °C at three different initial cooling rates (500 °, 100 °, and 10 °C/MYr) and evaluated the fractional loss of nitrogen (F) for each of the nine scenarios. The results of these calculations are shown in Figure 10.5, from which is clear that for any initial temperature below ~500 °C, the loss of nitrogen during exhumation is minimal. For initial temperatures above ~500 °C, N loss can be significant, rising to ~100% for slow cooling at 10 °C/MYr from 700 °C. We emphasize that the curves in Figure 10.4b are not intended to be accurate representations of any natural system but depictions of the general effectiveness of N diffusion in calcite for a range of cooling scenarios. The main shortcomings are the implicit assumptions that (1) N is initially distributed uniformly throughout the calcite crystal and (2) that the crystal remains intact, with no growth, dissolution, or fracturing, during exhumation. These are not realistic assumptions, but the qualitative implications of the

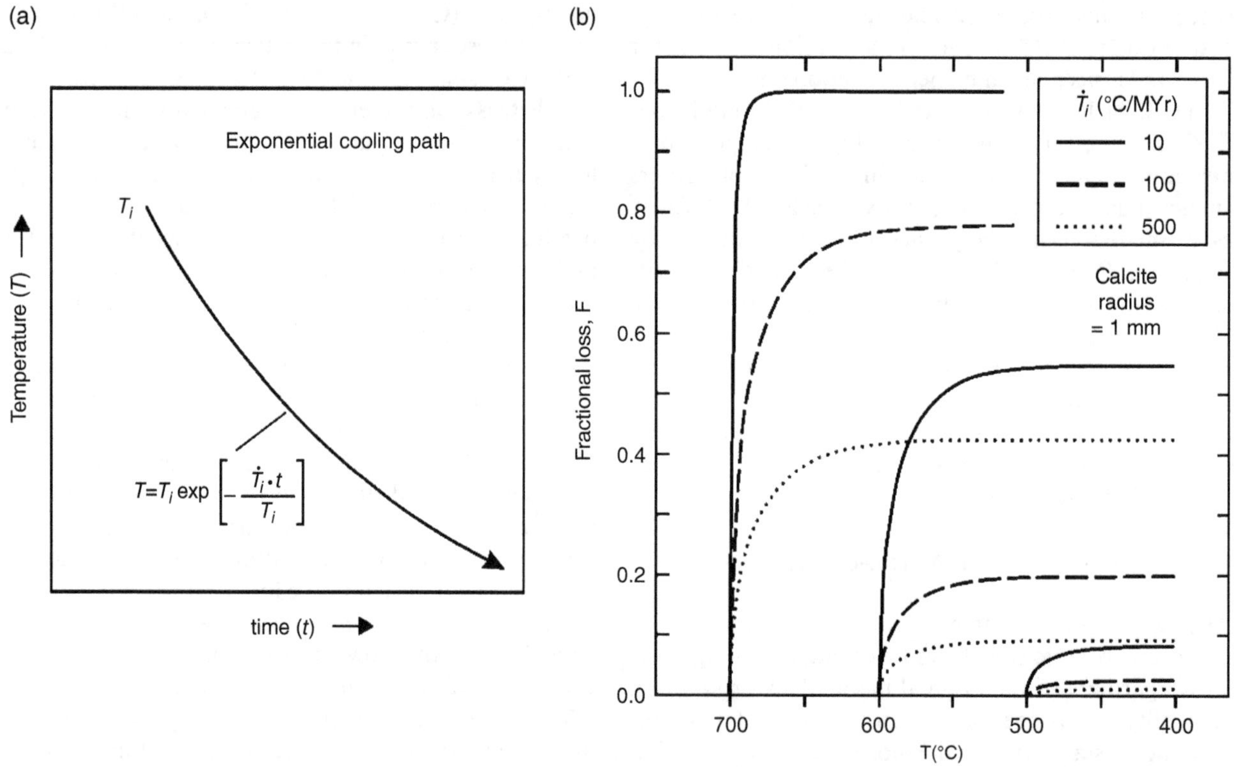

Figure 10.5 (a) Schematic diagram depicting the cooling paths and governing equation used in finite-difference simulations of nitrogen loss from calcite during exhumation and cooling. (b) Model results expressed as fractional N loss (F) vs. temperature for a spherical calcite grain (1 mm radius) cooling from three different initial temperatures (700 °, 600 °, and 500 °C) at three different rates (10 °, 100 °, and 500 °C/Myr). Note the minimal loss resulting from cooling initiated at T < ~500 °C. The results are expressed as fractional loss of N acquired at depth, but the results are equally relevant to fractional gain or isotopic exchange with the surroundings during exhumation. The faster c-parallel diffusion law was used for these calculations. See electronic version for color representation of the figures in this book.

models in terms of calcite retentiveness are nevertheless valid. It should also be borne in mind that the model results apply not only to diffusive loss of N, but also diffusive gain or isotopic exchange. Interestingly, there are some data to suggest a shallow crustal component is retained by calcite following exhumation along faults (Parry & Blamey, 2010), in which case the N_2/CO_2 ratio of calcite inclusions can be used as an indicator of the thermal histories of fluids and to track boiling/condensing processes before inclusion closure (e.g., Norman et al., 2002; Norman et al., 1996).

Busigny et al. (2003) showed that the N flux to the mantle by cool subduction zone processes is highly positively correlated with K content. However, the subducted marine sediments carry with them a substantial load of carbonate minerals, with a potential N content of ~10 ppm. Our results imply that nitrogen incorporated into calcite at the surface or near surface (e.g. from the organic components of marine calcifiers) will be retained long enough to reach a depth where it could become a nontrivial source of N to the lower crust and mantle. Indeed, calcite will survive to a much greater depth in a subduction zone before some amount of it is volatilized and returned to the atmosphere (Kerrick & Connolly, 2001; Tera et al., 1986), compared to sedimentary organic matter that is volatilized at a very shallow depth (see Fischer, 2008, for a review). The volume of this flux is not the focus of the current chapter, but the isotope ratio of this N would be almost certainly atmospheric or biogenic and thus different than the nominal upper mantle or crustal value (see Bebout et al., 2013 for a primer). Though the N in calcite is likely at very low absolute concentrations, the flux of marine biogenic calcite to the mantle is significant over geologic timescales. For a linear heating scenario of 100 °C/MYr, the diffusive opening equations of Watson and Cherniak (2013) reveal that a 2 mm diameter calcite crystal would retain 90% of hosted nitrogen at ~563 °C, 50% at 662 °C, and 10% at 723 °C, i.e., well into the mantle in the cool environment of a subduction zone.

ACKNOWLEDGMENTS

Thanks to Wayne Skala and Hassaram Bahkru for assistance with ion implantation, and to Yuan Li and an anonymous reviewer for their reviews of the manuscript. This work was supported by the Deep Carbon Observatory and the U.S. Department of Energy through grant no. DE-SC0016449.

REFERENCES

Bebout, G. E., Fogel, M. L., & Cartigny, P. (2013). Nitrogen: Highly volatile yet surprisingly compatible. *Elements*, *9*, 333–338.

Blamey, N. J. (2012). Composition and evolution of crustal, geothermal and hydrothermal fluids interpreted using quantitative fluid inclusion gas analysis. *Journal of Geochemical Exploration*, *116*, 17–27.

Blamey, N. J., Boston, P. J., & Rosales-Lagarde, L. (2016). High-resolution signatures of oxygenation and microbiological activity in speleothem fluid inclusions. *International Journal of Speleology*, *45*, 231–241.

Brand, U., Blamey, N., Garbelli, C., Griesshaber, E., Posenato, R., Angiolini, L., et al. (2016). Methane hydrate: Killer cause of Earth's greatest mass extinction. *Palaeoworld*, *25*, 496–507.

Busigny, V., Cartigny, P., Philippot, P., Ader, M., & Javoy, M. (2003). Massive recycling of nitrogen and other fluid-mobile elements (K, Rb, Cs, H) in a cold slab environment: Evidence from HP to UHP oceanic metasediments of the Schistes Lustrés nappe (western Alps, Europe). *Earth and Planetary Science Letters*, *215*, 27–42.

Busigny, V., & Bebout, G. E. (2013). Nitrogen in the silicate Earth: Speciation and isotopic behavior during mineral-fluid interactions. *Elements*, *9*, 353–358.

Cadogan, P. H. (1977). Palaeoatmospheric argon in Rhynie chert. In *Proceedings: Lunar and Planetary Science Conference 1977* (Vol. 8, p. 157).

Cassata, W. S., & Renne, P. (2013). Systematic variations of argon diffusion in feldspars and implications for thermochronometry. *Geochimica et Cosmochimica Acta*, *112*, 251–287.

Cherniak, D. J. (1997). An experimental study of Sr and Pb diffusion in calcite, and implications for carbonate diagenesis and metamorphism. *Geochimica et Cosmochimica Acta*, *61*, 4173–4179.

Cherniak, D.J. (1998). REE diffusion in calcite. *Earth and Planetary Science Letters*, *160*, 273–287.

Cherniak, D. J., Amidon, W., Hobbs, D., &Watson, E. B. (2015). Diffusion of helium in carbonates: Effects of mineral structure and composition. *Geochimica et Cosmochimica Acta*, *165*, 449–465.

Cherniak, D. J., Lanford, W. A. (2001). Nuclear reaction analysis. In Z. Alfassi (Ed.), *Non-destructive Elemental Analysis* (pp. 308–338). Blackwell Science.

Cherniak, D. J., Lanford, W. A., & Ryerson, F. J. (1991). Lead diffusion in apatite and zircon using ion implantation and Rutherford backscattering techniques. *Geochimica et Cosmochimica Acta*, *55*, 1663–1673.

Cherniak, D. J., Thomas, J. B., &Watson, E. B. (2014). Neon diffusion in olivine and quartz. *Chemical Geology*, *371*, 68–82.

Cherniak, D. J., & Watson, E. B. (2011). Helium diffusion in rutile and titanite, and consideration of the origin and implications of diffusional anisotropy. *Chemical Geology*, *288*, 149–161.

Cherniak, D. J., & Watson, E. B. (2012). Helium diffusion in olivine at 1 atm and 2.7 GPa. *Geochimica et Cosmochimica Acta*, *84*, 269–279.

Cherniak, D. J., Watson, E. B., & Thomas, J. B. (2009). He diffusion in zircon and apatite. *Chemical Geology*, *268*, 155–166.

Farver, J. R. (1994). Oxygen self-diffusion in calcite: Dependence on temperature and water fugacity. *Earth and Planetary Science Letters*, *121*, 575–587.

Fischer, T. P. (2008). Fluxes of volatiles (H_2O, CO_2, N_2, Cl, F) from arc volcanoes. *Geochemical Journal*, *42*, 21–38.

Goldstein, R. H. (1986). Reequilibration of fluid inclusions in low-temperature calcium-carbonate cement. *Geology*, *14*, 792–795.

Goldstein, R. H. (2001). Fluid inclusions in sedimentary and diagenetic systems. *Lithos*, *55*, 159–193.

Hirvonen, J. P., & Lappalainen, R. (1995). Nuclear reaction analysis: Particle-gamma reactions. In J. R. Tesmer, M. Nastasi, J. C. Barbour, C. J. Maggiore, & J. W. Mayer (Eds.), *Handbook of modern ion beam materials analysis* (pp. 167–192). Pittsburgh, PA: Materials Research Society.

Kerrick, D., & Connolly, J. (2001). Metamorphic devolatilization of subducted marine sediments and the transport of volatiles into the Earth's mantle. *Nature*, *411*, 293–296.

Kumar, S., Kumar, S. V., Reddy, G.L.N., Kain, V., Ramana, J. V., & Raju V. S. (2005). Depth profiling of nitrogen using 429 keV and 897 keV resonances in the $^{15}N(p,\alpha\gamma)^{12}C$ reaction. *Nuclear Instruments and Methods in Physics Research B*, *240*, 704–710.

Labotka, T. C., Cole, D. R., & Riciputi, L. R. (2000). Diffusion of C and O in calcite at 100 MPa. *American Mineralogist*, *85*, 488–494.

Labotka, T. C., Cole, D. R., Riciputi, L. R., & Fayek, M. (2004). Diffusion of C and O in calcite from 0.1 to 200 MPa. *American Mineralogist*, *89*,799–806.

Li, Y., & Keppler, H. (2014). Nitrogen speciation in mantle and crustal fluids. *Geochimica et Cosmochimica Acta*, *129*, 13–32.

Li, Y., Wiedenbeck, M., Shcheka, S., & Keppler, H. (2013). Nitrogen solubility in upper mantle minerals. *Earth and Planetary Science Letters*, *377*, 311–323.

Marty, B. (2012). The origins and concentrations of water, carbon, nitrogen and noble gases on Earth. *Earth and Planetary Science Letters*, *313–314*, 56–66.

Marty, B., Zimmermann, L., Pujol, M., Burgess, R., & Philippot, P. (2013). Nitrogen isotopic composition and density of the archean atmosphere. *Science*, *342*, 101–104.

Newman, B., Norman, D., Gundimeda, M., & Levy, S. (1996). Understanding the genesis of nonmarine calcite deposits through quadrupole mass spectrometric analysis of fluid inclusion gases. *Chemical Geology*, *132*, 205–213.

Norman, D. I., Blamey, N., & Moore, J. N. (2002). Interpreting geothermal processes and fluid sources from fluid inclusion organic compounds and CO_2/N_2 ratios. *Proceedings: Twenty-seventh Workshop on Geothermal Reservoir Engineering*, Stanford University, Stanford, California.

Norman, D. I., Moore, J. N., Yonaka, B., & Musgrave, J. (1996). Gaseous species in fluid inclusions: A tracer of fluids and indicator of fluid processes. *Proceedings: Twenty-first Workshop of Geothermal Reservoir Engineering*, Stanford University, Stanford, California.

Parry, W., & Blamey, N. J. (2010). Fault fluid composition from fluid inclusion measurements, Laramide age Uinta thrust fault, Utah. *Chemical Geology*, *278*, 105–119.

Pujol, M., Marty, B., Burgess, R., Turner, G., & Philippot, P. (2013). Argon isotopic composition of Archaean atmosphere probes early Earth geodynamics. *Nature*, *498*, 87–90.

Renne, P. R., Cassata, W. S., & Morgan, L. E. (2009). The isotopic composition of atmospheric argon and $^{40}Ar/^{39}Ar$ geochronology: Time for a change? *Quaternary Geochronology*, *4*, 288–298.

Ryssel, H., & Ruge, I. (1986). *Ion implantation*. New York: John Wiley.

Sano, Y., & Pillinger, C. (1990). Nitrogen isotopes and N_2/Ar ratios in cherts: An attempt to measure time evolution of atmospheric $\delta^{15}N$ value. *Geochemical Journal*, *24*, 315–325.

Schaller, M. F., Pettitt, E., & Knobbe, T. (2017). A potential new proxy for paleo-atmospheric pO_2 from soil carbonate-hosted fluid inclusions applied to pristine Chinle soils from the Petrified Forest 1A core. *Proceedings 2017 AGU Fall Meeting Abstracts*.

Schaller, M. F., Pettitt, E., Knobbe, T., & Breecker, D. (2018). A potential new proxy for paleo-atmospheric pO_2 from soil carbonate-hosted fluid inclusions. *Goldschmidt Conference Abstracts*, *2243*. Boston, MA.

Tera, F., Brown, L., Morris, J., Sacks, I. S., Klein, J., & Middleton, R. (1986). Sediment incorporation in island-arc magmas: Inferences from ^{10}Be. *Geochimica et Cosmochimica Acta*, *50*, 535–550.

Watson, E. B., & Cherniak, D. J. (2013) Simple equations for diffusion in response to heating. *Chemical Geology*, *335*, 93–104.

Watson, E. B., Cherniak, D. J., Drexler, M., Hervig, R. L., & Schaller, M. F. (2019). Nitrogen diffusion in silicate minerals, with implications for nitrogen transport and cycling in the lithosphere. *Chemical Geology*, *516*, 42–58.

Watson, E. B., & Dohmen, R. (2010). Non-traditional and emerging methods for characterizing diffusion in minerals and mineral aggregates. In Y. Zhang & D.J. Cherniak (Eds.), Diffusion in Minerals and Melts, *Reviews in Mineralogy and Geochemistry*, *72*, (pp. 61–105). Chantilly, VA: Mineralogical Society of America.

Ziegler, J. F., & Biersack, J. P. (2006). The stopping and range of ions in matter. *Computer Code SRIM 2006*, http://www.srim.org.

11

High-Pressure Transformations and Stability of Ferromagnesite in the Earth's Mantle

Eglantine Boulard, François Guyot, and Guillaume Fiquet

ABSTRACT

Ferromagnesite (Mg,Fe)CO_3 plays a key role in the transport and storage of carbon in the deep Earth. Experimental and theoretical studies demonstrated its high stability at high pressure and temperature against melting or decomposition. Several pressure-induced transformations of ferromagnesite have been reported at conditions corresponding to depths greater than ~1030 km in the Earth's lower mantle. Although there is still no consensus on their exact crystallographic structures, evidence is strong for a change in carbon environment from the low-pressure planar CO_3^{2-} ion into carbon atoms tetrahedrally coordinated by four oxygens. High-pressure iron-bearing phases concentrate a large amount of Fe^{3+} as a result of intra-crystalline self-redox reactions. These crystallographic particularities may have significant implications on carbon reservoirs and fluxes in the deep Earth.

11.1. INTRODUCTION

Carbon exchange between the Earth's interior and its surface occurs over time scales of hundreds millions of years, constituting the geodynamical carbon cycle. Superficial carbon is recycled into the deep earth by means of subduction. Estimations of this carbon influx ranges from 0.0001 to 52 megatons of carbon annually (Kelemen & Manning, 2015). This huge uncertainty hinges on poor constraints on the amount of carbon retained by subducting slabs. Carbonate inclusions in diamonds suggest that carbon is transported down to the transition zone depths (Brenker et al., 2007; Kaminsky, 2012; Wang et al., 1996); however, whether any carbon reaches the lower mantle is still controversial. Most subducted carbon is expected to melt and/or break down and return to the Earth's surface via volcanism (Kelemen & Manning, 2015; Thomson et al., 2016). However, relatively oxidizing conditions and low slab temperatures may result in the transportation of carbon to greater depths that could feed the core-mantle boundary (CMB) (Martirosyan et al., 2015). Yet, a quantitative estimation of carbon or CO_2 released at the CMB remains unconstrained, as are precise mechanisms for transportation of volatiles to the very deep mantle. In particular, the behaviors of such phases in the presence of deep mantle minerals such as silicates, iron oxides, or metallic iron remain to be evaluated.

Carbon is recycled into the deep mantle chiefly as carbonates, which mainly occur as calcite $CaCO_3$, dolomite $CaMg(CO_3)_2$, and magnesite $MgCO_3$ at the Earth's surface. Due to chemical reactions with silicates such as pyroxenes and bridgmanite, ferromagnesite (Mg-Fe)CO_3 is considered the dominant carbonate phase in the deep mantle (e.g. Biellmann et al., 1993; Kushiro et al., 1975; Wood et al., 1996). The behavior of ferromagnesite at depth is therefore critical for evaluating the storage capacity and fluxes of carbon. Because of the scarcity of natural samples coming from the lower mantle (e.g. Brenker et al., 2007; Kaminsky, 2012), knowledge of ferromagnesite's stability and behavior at depth mainly results from theoretical and experimental studies. The latter requires the ability to reach high pressure and

Sorbonne Université, UMR CNRS 7590, Muséum National d'Histoire Naturelle, IRD, Institut de Minéralogie, Physique des Matériaux et Cosmochimie-IMPMC, Paris, France

Carbon in Earth's Interior, Geophysical Monograph 249, First Edition. Edited by Craig E. Manning, Jung-Fu Lin, and Wendy L. Mao.

temperature (P-T) conditions of the Earth's mantle and to use microscale and nanoscale probes to characterize samples. Laser-heated diamond anvil cell is the main static pressure device for studying carbon-bearing phases at the Earth's mantle and core conditions. This technique permits the attainment of pressures above 300 GPa and temperatures up to 5000 K (Tateno et al., 2010) by heating with double-sided high-powered infrared lasers available in house and at synchrotron X-ray beamlines. The excellent transparency of single-crystal diamond to a wide range of electromagnetic radiation is compatible with numerous analytical probes for more comprehensive in situ characterization of high P-T behavior. This is critical in the case of nonquenchable phases or dynamic studies where "squeeze, cook, and look" experiments are not sufficient. A detailed review of the different techniques can be found in Mao and Boulard (2013).

In this chapter, we present a review of recent studies dealing with the high-pressure behavior of carbonates on the solid solution joining magnesite and siderite ($FeCO_3$). We first present ferromagnesite high-pressure behaviors and structures. We then discuss evidence of particular processes such as the self-redox reactions of Fe-bearing carbonates. Finally, we discuss potential implications for the Earth's system.

11.2. COMPRESSION OF MG-FE RHOMBOHEDRAL CARBONATE

Siderite and magnesite are isomorphous with calcite and crystallize in rhombohedral symmetry with the R-3c space group (Graf, 1961) (Figure 11.1a). Represented with a hexagonal unit cell, these carbonates contain six formula units per unit cell. They consist of an alternation of layers along the c-axis of cations (Fe^{2+}, Mg^{2+}) in six-fold oxygen coordination and carbon in trigonal planar $(CO_3)^{2-}$ groups. The orientations of two consecutive carbonate ions are staggered relative to each other with the cation at the center of symmetry.

In situ X-ray diffraction (XRD) studies at high pressure show a high stability of rhombohedral $MgCO_3$ up to ~80 GPa–2500 K (Fiquet et al., 2002). No evidence of

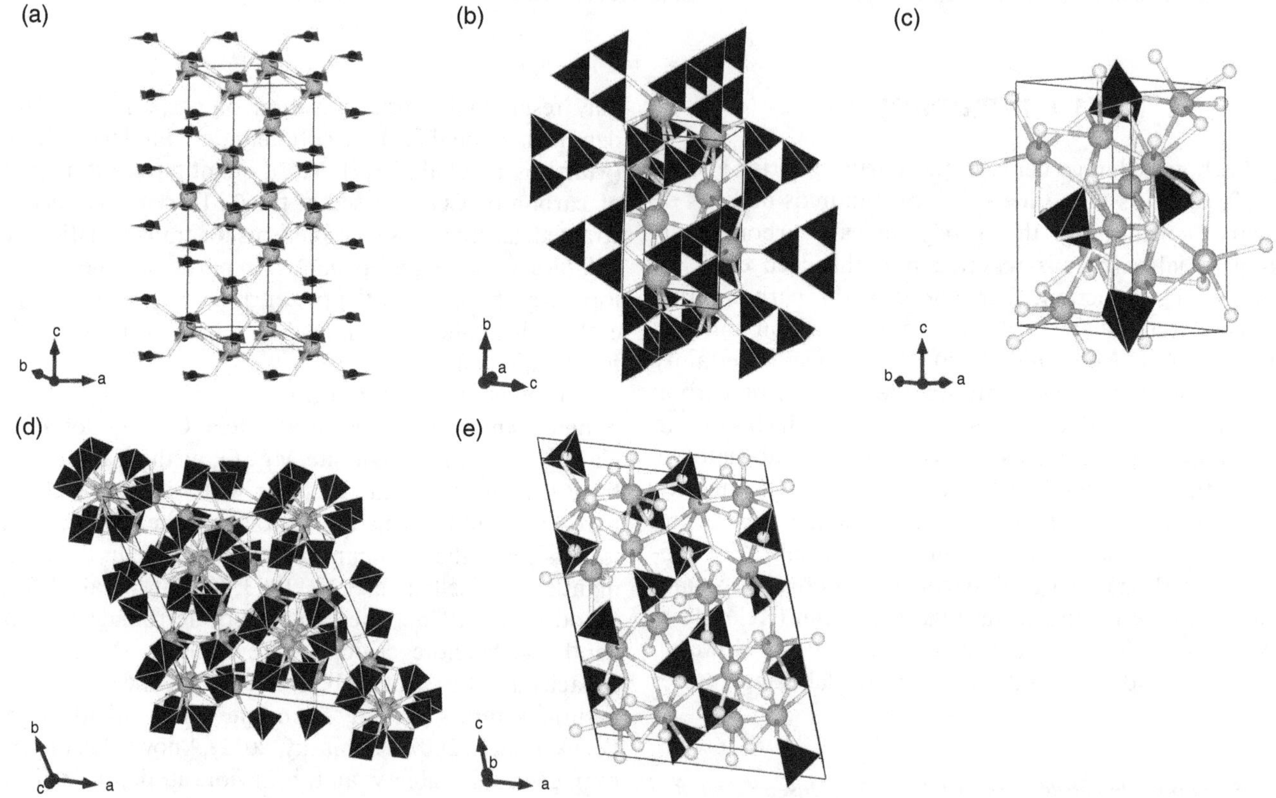

Figure 11.1 Crystallographic structures of Mg-Fe carbonates as reported in the literature. (a) The calcite-type rhombohedral structure in which Mg-Fe carbonates crystallize at ambient conditions, (b) high-pressure structure of magnesite and ferromagnesite ($MgCO_3$ and $Mg_{0.25}Fe_{0.3}[C_3O_9]_{0.233}$) (Boulard et al., 2011), (c) high-pressure phase of siderite: $Fe_4C_3O_{12}$ (Boulard et al., 2012), (d) $Fe_4C_3O_{12}$ (Cerantola et al., 2017), (e) $Fe_4C_4O_{13}$, and $Mg_2Fe_2(C_4O_{13})$ (Cerantola et al., 2017; Merlini et al., 2015). White and grey spheres are oxygens and Fe/Mg cations respectively, and black triangle or black tetrahedra are carbon polyhedrals. See electronic version for color representation of the figures in this book.

decomposition or melting under P-T conditions down to the CMB is observed (Dorogokupets, 2007; Fiquet et al., 2002; Gillet, 1993; Solopova et al., 2015). $MgCO_3$ can also be synthesized from the recombination of oxides MgO and CO_2 at mantle P-T conditions (Boulard et al., 2012; Scott et al., 2013). The c-axis is significantly more compressible than the a-axis, attributable to the tight bonding of C-O in CO_3 groups (Katsura et al., 1991). However, in situ XRD refinements and infrared (IR) spectroscopic analyses showed that from 20 to 50 GPa, C-O bonds lengthen before contracting (Fiquet et al., 2002; Santillán et al., 2005). This particular behavior, produced by the rotation of MgO_6 octahedra, likely contributes to the remarkable stability of the R-3c structure in carbonates at high pressure (Fiquet et al., 2002; Santillán et al., 2005).

The substitution of Fe^{2+} for Mg^{2+} increases the compressibility of ferromagnesite due to the Fe-O bond's length (2.141 Å) being longer than that of Mg-O (2.097 Å) (Liang et al., 2018). $FeCO_3$ and $MgCO_3$ present bulk moduli of 97 GPa and 103 GPa, respectively (Liang et al., 2018; J. Zhang et al., 1998). C-O bond length monotonically decreases upon compression up to ~40–45 GPa (Lavina et al., 2010; Santillán & Williams, 2004), the pressure at which Fe in ferromagnesite undergoes a spin transition (Liu et al., 2014; Mattila et al., 2007). It results in a lengthening of C-O bonds due to the shrinkage of Fe-O bonds, and an increase in density and incompressibility (Cerantola et al., 2015; Lavina et al., 2010; Lin et al., 2012). The Fe spin transition is expected to affect the partition coefficient of Fe between $(Mg,Fe)CO_3$ and $(Mg,Fe)SiO_3$ (Lobanov et al., 2015; Weis et al., 2017), leading to carbonate composition in equilibrium with Mg-Fe bridgmanite closer to siderite above 40–45 GPa.

Compared to magnesite, siderite decomposes at lower temperatures (~500°lower at 2 GPa) (Tao et al., 2013). Although the decarbonation boundary of siderite is very close to the average mantle geotherm at about 3 GPa (Tao et al., 2013), typical cold and hot subduction paths are well within the stability fields of both siderite and magnesite (Syracuse et al., 2010).

11.3. HIGH-PRESSURE POLYMORPHISM OF FERROMAGNESITE

High P-T phase transition in $MgCO_3$ was first demonstrated experimentally by Isshiki et al. (2004) who observed diffraction peaks at 115 GPa–2200 K that could not be assigned to any decomposition products (MgO or CO_2). Transmission electron microscopy analyses on the recovered sample showed a homogeneous amorphous sample area rich in Mg, C, and O (Irifune et al., 2005). This discovery inspired significant interest in both experimental and theoretical mineral physics. Systematic searches through databases of known crystal structures combined with energy minimization first indicated that a pyroxene structure (space group C2/c) becomes energetically more favorable than magnesite above ~100 GPa (Skorodumova, 2005). Later, Oganov et al. (2006) reported that a $C222_1$ pyroxene-type structure, also predicted in $CaCO_3$, was even more stable. Both structures contain zigzag chains of corner-sharing CO_4^{4-} tetrahedra.

Pressure-temperature conditions at which experimental studies reported phase transitions of ferromagnesite are reported in Figure 11.2. There is no consensus about those high-pressure crystallographic structures (Figure 11.1 and Table 11.1) (Boulard et al., 2011, 2012; Cerantola et al., 2017; Isshiki et al., 2004; Liu et al., 2015; Merlini et al., 2015). Those differences might be due to the existence of multiple low-enthalpy structures (metastable) that are competitive over a wide pressure range (Oganov et al., 2008) and/or to differences in compositions of the starting material.

Concerning the magnesian end-member, Isshiki et al. (2004) proposed an orthorhombic structure above 115 GPa–2200 K (noted Mag-II in Figure 11.2). No atom positions were proposed because no structural refinement could be performed. In 2011, Boulard et al. reported a transition of magnesite into a monoclinic structure above 80 GPa at 2300 K. Rietveld refinement was not possible, but through comparison with theoretical studies (Oganov et al., 2008), a crystalline structure with a P21/c space group made of groups of three $(CO_4)^{4-}$ tetrahedra sharing one corner that constitute $(C_3O_9)^{6-}$ rings, was proposed (Figure 11.1b). Mg-Fe composition was refined with the same structure (Fe-Mag-II). Nonhydrostatic conditions might have favored metastable phases in those experiments as no pressure medium was used. However, reversal reactions using oxides as starting materials (e.g. MgO + CO_2) to maximize synthesis of thermodynamically stable phases yielded same structure.

From recombination of FeO and CO_2 oxides, Boulard et al. (2012) show that rhombohedral siderite coexists with a new structure (Sid-II) from ~40 GPa–1400 K (Figure 11.1c) to ~70 GPa–2200 K, above which siderite fully disappears. The chemical composition, $Fe_4C_3O_{12}$, was deduced from electron energy-loss spectroscopy (EELS) analyses on the recovered samples following the method developed by Egerton (1996). Due to the similarity in chemical composition and unit cell parameters with the olivine-structured Laihunite silicate $(Fe^{3+},Fe^{2+})_2SiO_4$, Boulard et al. (2012) proposed a monoclinic structure ($P2_1/b$ space-group) based on isolated $(CO_4)^{4-}$ groups.

A single high-pressure phase, Sid-II/Fe-Mag-II, with an orthorhombic unit cell was proposed by Liu et al. (2015). It is still unclear whether this structure is based on CO_4 groups. More recently, single crystal XRD studies were performed on $(Mg,Fe)CO_3$ (Merlini et al., 2015) and $FeCO_3$ (Cerantola et al., 2017), allowing

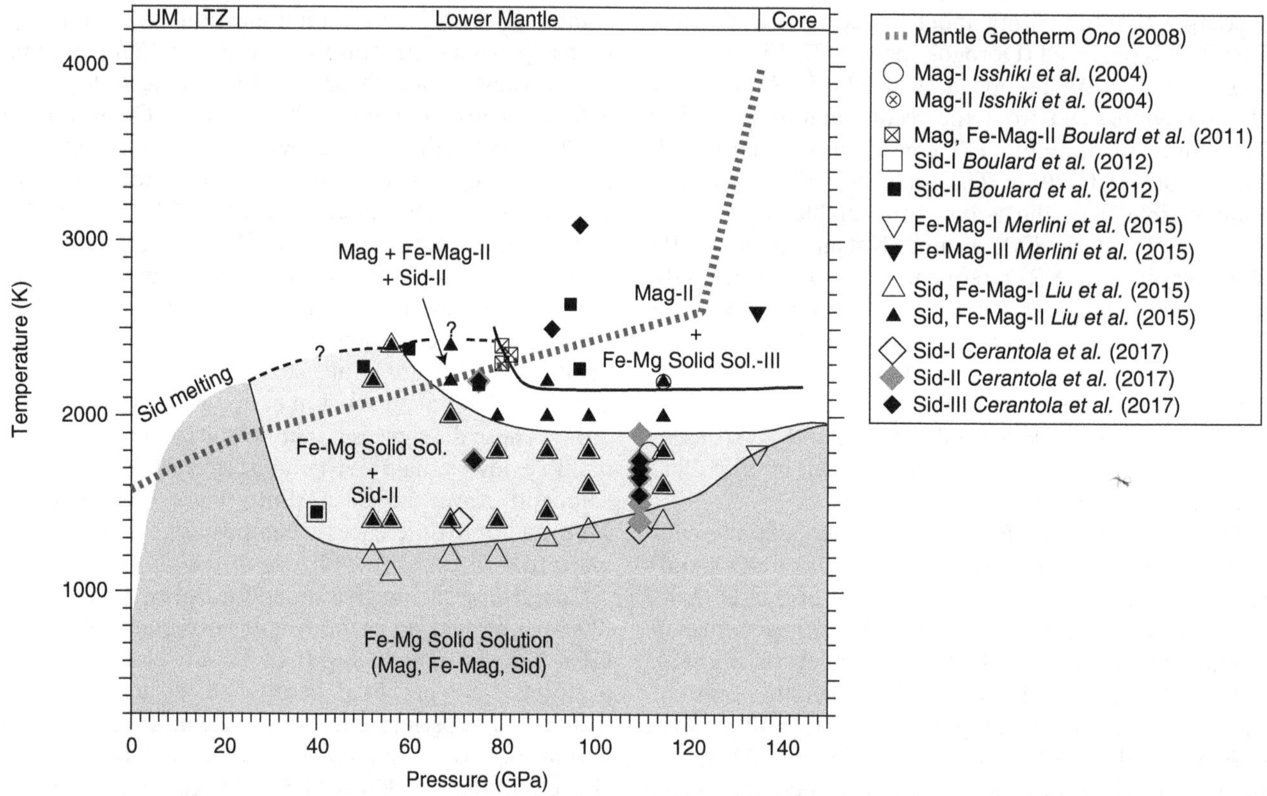

Figure 11.2 Experimental ferromagnesite phase diagram. Mag-I, Sid-I, and Fe-Mag-I refer to ambient magnesite, siderite, and ferromagnesite structures; phases II and III to the associated high-pressure polymorphs. See electronic version for color representation of the figures in this book.

structural refinements of crystallographic structures with atom position determinations. At pressure above 74 GPa and temperature between 1400 K and 1650 K, Cerantola et al. (2017) observed transformation of $FeCO_3$ into an $Fe_4C_3O_{12}$ high-pressure phase (Sid-II), an hexagonal structure with R-3c space group formed by isolated CO_4 groups. They used the same hexagonal cell to fit the diffraction pattern from Liu et al. (2015). Up to 2500 K, $Fe_4C_3O_{12}$ coexists with a second high-pressure phase, $Fe_4C_4O_{13}$ (Sid-III), a monoclinic structure with zigzag-shaped $(C_4O_{13})^{10-}$ chains formed by four corner-shared CO_4 groups, a phase previously described by Merlini et al. (2015) for ferromagnesite at 135 GPa–2600 K (2900 km).

11.3.1. Evidence for Tetrahedrally Coordinated Carbon

Identification of CO_4 groups in high-pressure structures based solely on XRD is difficult as it requires precise structural refinements. Moreover, localizing light elements such as carbon is not easy. Only Merlini et al. (2015) and Cerantola et al. (2017) could demonstrate tetrahedrally coordinated carbon from direct single-crystal-XRD measurements in the stoichiometries: $Mg_2Fe_2C_4O_{13}$, $Fe_4C_4O_{13}$, and $Fe_4C_3O_{12}$.

Vibrational spectroscopies, Raman and IR, are also particularly sensitive to carbon chemical environment and directly probe C-O bonds. Boulard et al. (2015) reported the first in situ characterization of C-O bonds in Fe-Mag-II. They found that its IR spectrum exhibits unique features not present in the low-pressure spectrum. The band assignment relied on first-principles calculations of the IR spectrum of tetrahedrally coordinated carbon in $MgCO_3$ ($P2_1/a$ space group). A mode at ~1,304 cm^{-1} at ~80 GPa, characteristic of the C-O asymmetric stretching vibration in CO_4 groups, could be used as a fingerprint of CO_4 groups in high-pressure mineral phases. An intense Raman band at ~1,025cm^{-1} (at 105 GPa) with a pressure dependence of ~1.8cm^{-1}/GPa in $P2_1/c$ $CaCO_3$ was recently proposed as characteristic of the symmetrical stretching vibration in its CO_4 groups (Lobanov et al., 2017).

While the high-pressure structures amorphized upon decompression, electron and X-ray spectroscopies at the carbon K-edge performed on recovered samples show that they preserved spectroscopic signatures in the amorphous phase, which could be associated to carbon-oxygen tetrahedral polyhedral. Analyses were either collected by EELS using transmission electron microscopy or by synchrotron radiation–based scanning transmission X-ray

Table 11.1 Crystallographic structures of high-pressure Mg-Fe carbonate polymorphs from experimental studies.

Composition	Theoretical Calculations	P Range (GPa)	Space Group	a	b	c	β
$MgCO_3$	Skorodumova et al., 2005	>113	C2/c				
	Oganov et al., 2006	>107	$C222_1$	5.552	7.201	2.880	
	Oganov et al., 2008	82–138	C2/m	8.094	6.488	6.879	103.98
		138–160	$P2_1$	4.534	7.792	5.086	104.54
	Experimental Studies						
	Starting Material						
$MgCO_3$	Isshiki et al., 2004	>115	Ortho.	7.18	5.03	4.47	
	$Mg_{0.994}Ca_{0.006}CO_3$						
	Boulard et al., 2011	>80	P21/c	8.37	6.37	6.80	104.57
	$Mg_{0.997}Fe_{0.003}CO_3$						
$Mg_2Fe_2(C_4O_{13})$	Merlini et al., 2015	135	Mono.	9.822	3.902	13.154	108.02
	$Mg_{0.26}Fe_{0.7}Mn_{0.025}Ca_{0.015}CO_3$						
$Mg_{0.25}Fe_{0.3}(C_3O_9)_{0.233}$	Boulard et al., 2011	>80	P21/c	7.83	6.37	6.73	101.97
	$Mg_{0.249}Fe_{0.748}Mn_{0.005}Ca_{0.006}CO_3$						
$Fe_4C_4O_{13}$	Cerantola et al., 2017	>74	C12/c1	10.261	3.985	13.455	107.85
	$FeCO_3$						
$Fe_4C_3O_{12}$	Cerantola et al., 2017	>74	R3c	12.762	12.762	5.332	
	$FeCO_3$						
$Fe_4C_3O_{12}$	Boulard et al., 2012 $FeO+CO_2$	>50	Mono.	10.16	6.66	6.15	93.04
$FeCO_3$	Liu et al., 2015	>50	Pmm2	10.99	6.34	5.27	
	$Fe_{0.998}Mn_{0.002}CO_3$						
	$Mg_{0.33}Fe_{0.65}Mn_{0.2}CO_3$						

Table 11.2 Self-oxidation reactions experimentally observed.

Reaction #	Reference	Chemical Reaction
R1	Merlini et al., 2015	$93Fe_{0.7}Mg_{0.3}CO_3 = 20Mg_{1.395}Fe_{2.605}(C_4O_{13}) + Fe_{13}O_{19} + 13C$
R2	Boulard et al., 2011	$20Mg_{0.25}Fe_{0.75}CO_3 = 20Mg_{0.25}Fe_{0.3}(C_3O_9)_{0.233} + 3Fe_3O_4 + 6CO$ (or $3C + CO_2$)
R3	Boulard et al., 2012	$4FeO + 4CO_2 \rightarrow Fe_4C_3O_{12} + C$
R4	Boulard et al., 2012	$4FeO + 5CO_2 = Fe_4C_3O_{12} + 2CO$
R5	Boulard et al., 2012	$2Fe_2O_3 + 3CO_2 = Fe_4C_3O_{12}$
R6	Cerantola et al., 2017	$4FeCO_3 = Fe_4C_3O_{12} + C$
R7	Cerantola et al., 2017	$7Fe_4C_3O_{12} + 3C = 6Fe_4C_4O_{13} + 2Fe_2O_3$
R8	Cerantola et al., 2017	$8Fe_4C_3O_{12} = 6Fe_4C_4O_{13} + 4Fe_2O_3 + 3O_2$

microscopy coupled to the acquisition of X-ray absorption spectra. C K-edge on an ambient pressure rhombohedral (R-3c) carbonate sample display peaks at 290.3 and 298.3 eV assigned to 1s → p* electronic transition and one peak at 300.5 eV assigned to 1s → s*, within carbonate CO_3 groups (Hofer & Golob, 1987; Zhou et al., 2008). Spectra collected on recovered samples transformed into the high-pressure phases of the two compositions $FeCO_3$ and $(Mg,Fe)CO_3$ show different spectroscopic signatures. The main peak is broader and slightly shifted to higher energy (290.47 eV in Fe-Mag-II and 290.67 eV in Sid-II), and a second peak is observed at 287.35 eV in both compositions (Boulard et al., 2012). These spectroscopic signatures are interpreted as a fingerprint of CO_4 groups, and the slight energy shift of the main peak between the two compositions may reflect different degrees of polymerization of CO_4 groups. Fe content in the $(Mg,Fe)CO_3$ solid solution likely affects polymerization of CO_4 groups. While isolated $(CO_4)^{4-}$ tetrahedra are reported in pure Fe composition, high-pressure polymorphs of Mg-rich carbonates are based on polymerized CO_4 groups, i.e. $(C_3O_9)^{6-}$ or $(C_4O_{13})^{10-}$ chains (Arapan et al., 2007; Boulard et al., 2011; Cerantola et al., 2017; Isshiki et al., 2004; Merlini et al., 2015; Oganov et al., 2008; Panero & Kabbes, 2008).

11.3.2. Self-Redox Reactions in Fe^{2+}-Bearing Carbonates

Another particularity of the crystal chemistry of Fe-bearing high-pressure structures is the preferential association of the CO_4 tetrahedral groups with trivalent iron. Incorporation of trivalent iron or of mixed 3+/2+ valences with high Fe^{3+} contents in these phases was inferred from the stoichiometries (Boulard et al. 2011, 2012; Cerantola et al., 2017; Merlini et al., 2015). Fe $L_{2,3}$-edges spectra collected ex situ by EELS or scanning transmission X-ray microscopy on the recovered samples from Sid-II or Fe-Mag-II confirmed high Fe^{3+} contents in the products of transformation at high pressure (Boulard et al. 2011, 2012). The redox counterpart for Fe^{3+} formation could eventually be the stabilization of Fe^0 as observed in the disproportionation reaction of bridgmanite (Frost & McCammon, 2008). However, Fe^0 has never been identified in transformation products of ferromagnesite. Starting from exclusively Fe^{2+}-bearing carbonates, formation of Fe^{3+} is instead balanced by partial reduction of carbon-bearing molecular groups (CO_3^{2-} or CO_2) (Table 11.2). We call this a self-redox process, since Fe^{2+} and CO_3^{2-} initially present in the low-pressure compound react with each other to yield Fe^{3+} and reduced carbon species (C or CO). Diamond coexisting with high-pressure transformation products of Fe^{2+}-bearing carbonates or Fe^{2+}-bearing oxides in presence of CO_2 was reported by Boulard et al. (2011, 2012). Coexisting Fe^{3+}-bearing iron oxides have also been reported, such as magnetite or hematite and their associated high-pressure structures (Boulard et al., 2011, 2012; Cerantola et al., 2017), as well as newly described iron oxides Fe_5O_7 (Cerantola et al., 2017) and $Fe_{13}O_{19}$ (Merlini et al., 2015). Decomposition of $FeCO_3$ into Fe_3O_4 + C was also reported in the stability field of classical CO_3^{2-}-bearing carbonates (<50 GPa) (Boulard et al., 2012; Cerantola et al., 2017). In these low-pressure experiments, only partial decomposition took place, as carbonate remained present even after heating up to one hour. The possible existence of a thermodynamic boundary of siderite decomposition remains to be further investigated. Overall, current available data suggest that the stability of high-pressure phases containing CO_4 groups enhances the disproportionation of Fe^{2+}-bearing carbonates into Fe^{3+}-bearing phases and reduced carbon species such as diamond.

11.4. CONCLUSIONS AND OUTLOOKS

Ferromagnesite $(Mg,Fe)CO_3$, or close stoichiometries containing oxidized carbon species, are very stable under extreme P-T conditions. The phase diagram of the Mg-Fe-C-O system is very rich and yields several compounds containing CO_4 groups.

Transformation from CO_3 to CO_4 groups may take place at pressure as low as 40 GPa for Fe-rich compositions, and 80 GPa for Mg-rich compositions. As SiO_4

groups in silicates, CO_4 groups may be isolated or polymerized. Impacting the thermodynamic and physical properties of carbonates and associated melts, this new carbon environment may have significant implications on carbon reservoirs and fluxes.

Fe-rich compositions of carbonate are favored at pressure above 40 GPa due to the Fe^{2+} spin transition. The observed systematic presence of trivalent iron in the Fe-rich CO_4-bearing high-pressure structures suggests that other compositions could be stabilized, such as aluminum-rich compositions that do not exist at ambient conditions (Merlini et al., 2015).

These newly described high-pressure structures represent potential oxidized carbon carriers into the lowermost mantle. Whether these phases remain stable in subducting slabs or in regular mantle lithologies is still uncertain. In a recent study, Boulard et al. (2018) reported that the deep carbon and hydrogen cycles may be more interconnected than previously thought, as $Fe_4C_3O_{12}$ replaces pyrite-structured FeO_2H_x in presence of CO_2, providing a new mechanism for hydrogen release as H_2O within the deep mantle. However, ferromagnesite and the associated high-pressure structures are sensitive to redox breakdown (Dorfman et al., 2018; Rohrbach & Schmidt, 2011; Stagno et al., 2011). Ferromagnesite reacts with metallic iron and nickel in the mantle to form either diamond or carbide, depending on the availability of metal. If this is the case, calcite, which is less sensitive to redox breakdown, could be revived as an interesting oxidized carbon carrier in the deep mantle. More oxidizing conditions like those prevailing in subducting slabs may still stabilize ferromagnesite and related stoichiometries. The redox stabilities of calcite, ferromagnesite, and of their high-pressure transformation products remain to be extensively tested as a function of T, P, and fO_2.

A next step will be to consider the Fe-Mg-C-O high-pressure phase diagram in the context of a silicate-rich lithology. Recent studies on $CaCO_3$ and $MgCO_3$ in the presence of an excess of SiO_2 or $MgSiO_3$ show that $CaCO_3$ is likely to undergo decomposition into CO_2 and Ca-perovskite under any slab P-T conditions, while $MgCO_3$ may be preserved under very cold slab P-T conditions (Kakizawa et al., 2015; Maeda et al., 2017; Seto et al., 2008; Takafuji et al., 2006; Z. Zhang et al., 2018). However, none of these studies have considered iron-rich compositions, which deserve special attention due to the effect of the iron spin transition and of self-redox processes with oxidized carbon species. There is increasing evidence that such self-redox processes might be significant. For example, self-redox reaction of ferromagnesite was recently observed in a natural sample of shocked carbonate at the Xiuyan impact crater (Chen et al., 2018). Whether it is related to impact-induced formation of CO_4 groups or not will deserve further studies. Self-redox processes might also provide relevant explanation for some of the carbonate inclusions in deep diamonds (e.g. Boulard et al., 2011; Brenker et al., 2007; Kaminsky et al., 2012).

ACKNOWLEDGMENTS

E. Boulard acknowledges DCO support during her postdoctoral work. G. Fiquet has received funding from the European Research Council (ERC) under the European Union's Horizon 2020 Research and Innovation Program (grant agreement 670787).

REFERENCES

Arapan, S., Souza de Almeida, J., & Ahuja, R. (2007). Formation of sp3 hybridized bonds and stability of CaCO3 at very high pressure. *Physical Review Letters*, *98*, 268501. https://doi.org/10.1103/PhysRevLett.98.268501

Biellmann, C., Gillet, P., Guyot, F., Peyronneau, J., & Reynard, B. (1993). Experimental evidence for carbonate stability in the Earth's lower mantle. *Earth and Planetary Science Letters*, *118*(1–4), 31–41. https://doi.org/10.1016/0012-821X(93)90157-5

Boulard, E., Gloter, A., Corgne, A., Antonangeli, D., Auzende, A., Perrillat, J.-P., et al. (2011). New host for carbon in the deep Earth. *Proceedings of the National Academy of Sciences of the United States of America*, *108*(13), 5184–5187. https://doi.org/10.1073/pnas.1016934108

Boulard, E., Guyot, F., Menguy, N., Corgne, A., Auzende, A., Perrillat, J., & Fiquet, G. (2018). CO2-induced destabilization of pyrite-structured FeO2Hx in the lower mantle. *National Science Review*, *5*(6), 870–877. https://doi.org/10.1093/nsr/nwy032

Boulard, E., Menguy, N., Auzende, a.-L., Benzerara, K., Bureau, H., Antonangeli, D., et al. (2012). Experimental investigation of the stability of Fe-rich carbonates in the lower mantle. *Journal of Geophysical Research*, *117*(B2), B02208. https://doi.org/10.1029/2011JB008733

Boulard, E., Pan, D., Galli, G., Liu, Z., & Mao, W. L. (2015). Tetrahedrally coordinated carbonates in Earth's lower mantle. *Nature Communications*, *6*, 6311. https://doi.org/10.1038/ncomms7311

Brenker, F. E., Vollmer, C., Vincze, L., Vekemans, B., Szymanski, A., Janssens, K., et al. (2007). Carbonates from the lower part of transition zone or even the lower mantle. *Earth and Planetary Science Letters*, *260*(1–2), 1–9. https://doi.org/10.1016/j.epsl.2007.02.038

Cerantola, V., Bykova, E., Kupenko, I., Merlini, M., Ismailova, L., McCammon, C., et al. (2017). Stability of iron-bearing carbonates in the deep Earth's interior. *Nature Communications*, *8*(May), 15960. https://doi.org/10.1038/ncomms15960

Cerantola, V., Mccammon, C., Kupenko, I., Kantor, I., Marini, C., Wilke, M., et al. (2015). High-pressure spectroscopic study of siderite (FeCO3) with a focus on spin crossover. *American Mineralogist*, *100*, 2670–2681.

Chen, M., Shu, J., Xie, X., Tan, D., & Mao, H. (2018). Natural diamond formation by self-redox of ferromagnesian carbonate. *Proceedings of the National Academy of Sciences*, *115*(11), 2676–2680. https://doi.org/10.1073/pnas.1720619115

Dorfman, S. M., Badro, J., Nabiei, F., Prakapenka, V. B., Cantoni, M., & Gillet, P. (2018). Carbonate stability in the reduced lower mantle. *Earth and Planetary Science Letters*, *489*, 84–91. https://doi.org/10.1016/j.epsl.2018.02.035

Dorogokupets, P. I. (2007). Equation of state of magnesite for the conditions of the Earth's lower mantle. *Geochemistry International*, *45*(6), 561–568. https://doi.org/10.1134/S0016702907060043

Egerton, R. F. (1996). *Electron energy-loss spectroscopy in the electron microscope* (2nd ed.). New York: Press Plenum.

Fiquet, G., Guyot, F., Kunz, M., Matas, J., Andrault, D., & Hanfland, M. (2002). Structural refinements of magnesite at very high pressure. *American Mineralogist*, *87*, 1261–1265. https://doi.org/10.2138/am-2002-8-927

Frost, D. J., & McCammon, C. A. (2008). The redox state of Earth's mantle. *Annual Review of Earth and Planetary Sciences*, *36*, 389–420. https://doi.org/10.1146/annurev.earth.36.031207.124322

Gillet, P. (1993). Stability of magnesite (MgCO3) at mantle pressure and temperature conditions: A Raman spectroscopic study. *American Mineralogist*, *78*, 1328–1331.

Graf, D. L. (1961). Crystallographic tables for the rhombohedral carbonates. *American Mineralogist*, *46*, 1283–1316.

Hofer, F., & Golob, P. (1987). New examples for near-edge fine structures in electron energy loss spectroscopy. *Ultramicroscopy*, *21*, 379–383.

Irifune, T., Isshiki, M., & Sakamoto, S. (2005). Transmission electron microscope observation of the high-pressure form of magnesite retrieved from laser heated diamond anvil cell. *Earth and Planetary Science Letters*, *239*(1–2), 98–105. https://doi.org/10.1016/j.epsl.2005.05.043

Isshiki, M., Irifune, T., Hirose, K., Ono, S., Ohishi, Y., Watanuld, T., et al. (2004). Stability of magnesite and its high-pressure form in the lowermost mantle. *Nature*, *427*(January), 60–63. https://doi.org/10.1029/2000C000736

Kakizawa, S., Inoue, T., Suenami, H., & Kikegawa, T. (2015). Decarbonation and melting in $MgCO_3$-SiO_2 system at high temperature and high pressure. *Journal of Mineralogical and Petrological Sciences*, *110*(4), 179–188. https://doi.org/10.2465/jmps.150124

Kaminsky, F. (2012). Mineralogy of the lower mantle: A review of "super-deep" mineral inclusions in diamond. *Earth-Science Reviews*, *110*, 127–147. https://doi.org/10.1016/j.earscirev.2011.10.005

Katsura, T., Tsuchida, Y., Ito, E., Yagi, T., Utsumi, W., & Akimoto, S. (1991). Stability of magnesite under the lower mantle conditions. *Proceedings of the Japan Academy. Ser. B: Physical and Biological Sciences*, *67*(4), 57–60. https://doi.org/10.2183/pjab.67.57

Kelemen, P. B., & Manning, C. E. (2015). Reevaluating carbon fluxes in subduction zones: What goes down, mostly comes up. *Proceedings of the National Academy of Sciences*, *112*(30), E3997–E4006. https://doi.org/10.1073/pnas.1507889112

Kushiro, I., Satake, H., Akimoto, S. (1975). Carbonate-silicate reactions at high pressures and possible presence of dolomite and magnesite in the upper mantle. *Earth and Planetary Science Letters*, *28*, 116–120. https://doi.org/10.1016/0012-821X(75)90218-6

Lavina, B., Dera, P., Downs, R., Yang, W., Sinogeikin, S., Meng, Y., et al. (2010). Structure of siderite FeCO3 to 56 GPa and hysteresis of its spin-pairing transition. *Physical Review B*, *82*(6), 1–7. https://doi.org/10.1103/PhysRevB.82.064110

Liang, W., Yin, Y., Li, Z., Li, R., Li, L., He, Y., et al. (2018). Single crystal growth, crystalline structure investigation and high-pressure behavior of impurity-free siderite (FeCO3). *Physics and Chemistry of Minerals*, *45*(9), 831–842. https://doi.org/10.1007/s00269-018-0965-y

Liu, J., Lin, J. F., Mao, Z., & Prakapenka, V. B. (2014). Thermal equation of state and spin transition of magnesiosiderite at high pressure and temperature. *American Mineralogist*, *99*(1), 84–93. https://doi.org/10.2138/am.2014.4553

Liu, J., Lin, J.-F., & Prakapenka, V. B. (2015). High-pressure orthorhombic ferromagnesite as a potential deep-mantle carbon carrier. *Scientific Reports*, *5*, 7640. https://doi.org/10.1038/srep07640

Lin, J. F., Liu, J., Jacobs, C., & Prakapenka, V. B. (2012). Vibrational and elastic properties of ferromagnesite across the electronic spin-pairing transition of iron. *American Mineralogist*, *97*(4), 583–591. https://doi.org/10.2138/am.2012.3961

Lobanov, S. S., Dong, X., Martirosyan, N. S., Samtsevich, A. I., Stevanovic, V., Gavryushkin, P. N., et al. (2017). Raman spectroscopy and x-ray diffraction of sp3-CaCO3 at lower mantle pressures. *Physical Review B*, *96*, 104101. https://doi.org/10.1103/PhysRevB.96.104101

Lobanov, S. S., Goncharov, A. F., & Litasov, K. D. (2015). Optical properties of siderite (FeCO3) across the spin transition: Crossover to iron-rich carbonates in the lower mantle. *American Mineralogist*, *100*(5–6), 1059–1064. https://doi.org/10.2138/am-2015-5053

Maeda, F., Ohtani, E., Kamada, S., Sakamaki, T., Hirao, N., & Ohishi, Y. (2017). Diamond formation in the deep lower mantle: A high-pressure reaction of MgCO3 and SiO2. *Scientific Reports*, *7*(January), 40602. https://doi.org/10.1038/srep40602

Mao, W. L., & Boulard, E. (2013). Nanoprobes for deep carbon. *Reviews in Mineralogy and Geochemistry*, *75*(1), 423–448. https://doi.org/10.2138/rmg.2013.75.13

Martirosyan, N. S., Litasov, K. D., Shatskiy, A. F., & Ohtani, E. (2015). Reactions of iron with calcium carbonate at 6 GPa and 1273-1873 K: Implications for carbonate reduction in the deep mantle. *Russian Geology and Geophysics*, *56*(9), 1322–1331. https://doi.org/10.1016/j.rgg.2015.08.008

Mattila, a, Pylkkänen, T., Rueff, J.-P., Huotari, S., Vankó, G., Hanfland, M., et al. (2007). Pressure induced magnetic transition in siderite FeCO3 studied by x-ray emission spectroscopy. *Journal of Physics: Condensed Matter*, *19*(38), 386206. https://doi.org/10.1088/0953-8984/19/38/386206

Merlini, M., Hanfland, M., Salamat, A., Petitgirard, S., & Müller, H. (2015). The crystal structures of Mg2Fe2C4O13 with tetrahedrally coordinated carbon, and Fe13O19, synthesized at deep mantle conditions. *American Mineralogist*, *100*, 2001–2004. https://doi.org/10.1524/zkri.219.10.621.50817

Oganov, A. R., Glass, C. W., & Ono, S. (2006). High-pressure phases of CaCO3: Crystal structure prediction and experiment. *Earth and Planetary Science Letters*, *241*(1–2), 95–103. https://doi.org/10.1016/j.epsl.2005.10.014

Oganov, A. R., Ono, S., Ma, Y., Glass, C. W., & Garcia, A. (2008). Novel high-pressure structures of MgCO3, CaCO3 and CO2 and their role in Earth's lower mantle. *Earth and Planetary Science Letters*, *273*, 38–47. https://doi.org/10.1016/j.epsl.2008.06.005

Ono, S. (2008). Experimental constraints on the temperature profile in the lower mantle. *Physics of the Earth and Planetary Interiors*, *170*, 267–273. doi10.1016/j.pepi.2008.06.033

Panero, W. R., & Kabbes, J. E. (2008). Mantle-wide sequestration of carbon in silicates and the structure of magnesite II. *Geophysical Research Letters*, *35*(14), 1–5. https://doi.org/10.1029/2008GL034442

Rohrbach, A., & Schmidt, M. W. (2011). Redox freezing and melting in the Earth's deep mantle resulting from carbon-iron redox coupling. *Nature*, *472*(7342), 209–12. https://doi.org/10.1038/nature09899

Santillán, J., Catalli, K., & Williams, Q. (2005). An infrared study of carbon-oxygen bonding in magnesite to 60 GPa. *American Mineralogist*, *90*(10), 1669–1673. https://doi.org/10.2138/am.2005.1703

Santillán, J., & Williams, Q. (2004). A high-pressure infrared and X-ray study of FeCO3 and MnCO3: Comparison with CaMg(CO3)2-dolomite. *Physics of the Earth and Planetary Interiors*, *143*(1–2), 291–304. https://doi.org/10.1016/j.pepi.2003.06.007

Scott, H. P., Doczy, V. M., Frank, M. R., Hasan, M., Lin, J.-F., & Yang, J. (2013). Magnesite formation from MgO and CO2 at the pressures and temperatures of Earth's mantle. *American Mineralogist*, *98*(7), 1211–1218. https://doi.org/10.2138/am.2013.4260

Seto, Y., Hamane, D., Nagai, T., & Fujino, K. (2008). Fate of carbonates within oceanic plates subducted to the lower mantle, and a possible mechanism of diamond formation. *Physics and Chemistry of Minerals*, *35*, 223–229. https://doi.org/10.1007/s00269-008-0215-9

Skorodumova, N. V. (2005). Stability of the MgCO3 structures under lower mantle conditions. *American Mineralogist*, *90*, 1008–1011. https://doi.org/10.2138/am.2005.1685

Solopova, N. A., Dubrovinsky, L., Spivak, A. V., Litvin, Y. A., & Dubrovinskaia, N. (2015). Melting and decomposition of MgCO3 at pressures up to 84 GPa. *Physics and Chemistry of Minerals*, *42*(1), 73–81. https://doi.org/10.1007/s00269-014-0701-1

Stagno, V., Tange, Y., Miyajima, N., Mccammon, C. A., Irifune, T., & Frost, D. J. (2011). The stability of magnesite in the transition zone and the lower mantle as function of oxygen fugacity. *Geophysical Research Letters*, *38*, L19309. https://doi.org/10.1029/2011GL049560

Syracuse, E. M., Van Keken, P. E., Abers, G. A., Suetsugu, D., Editor, G., Bina, C., et al. (2010). The global range of subduction zone thermal models. *Physics of the Earth and Planetary Interiors*. https://doi.org/10.1016/j.pepi.2010.02.004

Takafuji, N., Fujino, K., Nagai, T., Seto, Y., & Hamane, D. (2006). Decarbonation reaction of magnesite in subducting slabs at the lower mantle. *Physics and Chemistry of Minerals*, *33*(10), 651–654. https://doi.org/10.1007/s00269-006-0119-5

Tao, R., Fei, Y., & Zhang, L. (2013). Experimental determination of siderite stability at high pressure. *American Mineralogist*, *98*, 1565–1572. https://doi.org/10.2138/am.2013.4351

Tateno, S., Hirose, K., Ohishi, Y., & Tatsumi, Y. (2010). The structure of iron in Earth's inner core. *Science*, *330*, 359–361. https://doi.org/10.1126/science.1194662

Thomson, A. R., Walter, M. J., Kohn, S. C., & Brooker, R. A. (2016). Slab melting as a barrier to deep carbon subduction. *Nature*, *529*(7584), 76–79. https://doi.org/10.1038/nature16174

Wang, A., Pasteris, J. D., Meyer, H. O. A., & Dele-duboi, M. L. (1996). Magnesite-bearing inclusion assemblage in natural diamond, *141*, 293–306.

Weis, C., Sternemann, C., Cerantola, V., Sahle, C. J., Spiekermann, G., Harder, M., et al. (2017). Pressure driven spin transition in siderite and magnesiosiderite single crystals. *Scientific Reports*, *7*(1), 1–10. https://doi.org/10.1038/s41598-017-16733-3

Wood, B. J., Pawley, A., & Frost, D. R. (1996). Water and carbon in the Earth's mantle. *Philosophical Transactions of the Royal Society of London*, *354*, 1495 1511.

Zhang, J., Martinez, I., Guyot, F., & Reeder, R. J. (1998). Effects of Mg-Fe 2+ substitution in calcite-structure carbonates: Thermoelastic properties. *American Mineralogist*, *83*, 280–287.

Zhang, Z., Mao, Z., Liu, X., Zhang, Y., & Brodholt, J. (2018). Stability and reactions of CaCO 3 polymorphs in the Earth's deep mantle. *Journal of Geophysical Research: Solid Earth*, *123*(8), 6491–6500. https://doi.org/10.1029/2018JB015654

Zhou, D., Metzler, R. A., Tyliszczak, T., Guo, J., Abrecht, M., Coppersmith, S. N., & Gilbert, P.U.P.A. (2008). Assignment of polarization-dependent peaks in carbon K-edge spectra from biogenic and geologic aragonite. *Journal of Physical Chemistry B*, *112*(41), 13128–13135. https://doi.org/10.1021/jp803176z

12

Spin Transition of Iron in Deep-Mantle Ferromagnesite

Jiachao Liu, Suyu Fu, and Jung-Fu Lin

ABSTRACT

Iron-bearing magnesite, or ferromagnesite ([Mg,Fe]CO_3), is an abundant component of Earth's deep-mantle carbonates. Under Earth's lower mantle pressure-temperature conditions, Fe^{2+} in (Mg,Fe)CO_3 undergoes an electronic high-spin (HS) to low-spin (LS) transition, which has been reported to influence a broad spectrum of the physical, chemical, and transport properties of the host ferromagnesite. These observations in turn help us better understand the effects of the Fe spin transition on other lower-mantle minerals. In this chapter, we review recent high-pressure experimental and theoretical results in the magnesite ($MgCO_3$) and siderite ($FeCO_3$) solid solution system. We first introduce the thermodynamic background of the Fe^{2+} spin transition in (Mg,Fe)CO_3 and summarize the characterizations of this transition using various high-pressure techniques. Using literature results, we focus on elucidating the effects of the Fe^{2+} spin transition on the equation of state, sound velocities, elastic moduli, and seismic radial anisotropies of (Mg,Fe)CO_3 in the Earth's mantle. Along an expected lower mantle geotherm, the Fe^{2+} spin crossover in (Mg,Fe)CO_3 occurs at ~ 50–80 GPa. Within this pressure range, the mixed-spin state of ferromagnesite exhibits significant increase in density but softening in V_P. Most significantly, (Mg,Fe)CO_3 exhibits a much higher V_S splitting anisotropy of ~45%–65% at lower mantle pressures compared with major candidate lower mantle minerals (e.g. ferropericlase: 3%–24%, bridgmanite: 11%–12% at 24–60 GPa). The strong anisotropy in textured (Mg,Fe)CO_3 developed due to its low strength could be used to explain high seismic Vs anisotropy observed in mantle subducting slabs at depths greater than 300 km. These observations suggest that (Mg,Fe)CO_3 as a deep-mantle carbon carrier can display unique physical and chemical properties.

12.1. INTRODUCTION

Carbonates are believed to be major deep-mantle carbon carriers that transport carbon stored in deep-sea sediments and oceanic mantle lithosphere (serpentinized and carbonated depleted peridotite) to the Earth's deep interior through subduction. The subduction of slabs with carbon-rich materials thus serves as a major influx source for the deep-Earth carbon cycle (e.g. Dasgupta & Hirschmann, 2010; Kelemen & Manning, 2015). At the same time, reduced carbon species such as diamond, Fe-C alloys, and carbon-bearing melts can return back to shallower depths along mantle upwellings and be oxidized to carbonates or carbonatite melts (Rohrbach & Schmidt, 2011). The presence of carbonates in subducting slabs may trigger deep-focus earthquakes and could be manifested in strong V_S radial anisotropy ($\xi = [V_{SH}/V_{SV}]^2$, where V_{SH} and V_{SV} are horizontally and vertically polarized shear waves, respectively) (J. Li et al., 2018). Therefore, studying the physical and chemical properties of carbonates under mantle pressure-temperature (P-T) conditions is critical for a better understanding of the deep carbon storage and cycling as well as its role in mantle geophysics and geochemistry.

Carbonates in the magnesite-siderite-calcite ($MgCO_3$-$FeCO_3$-$CaCO_3$) ternary system are the major forms of crustal and mantle carbonates. In most parts of the mantle, magnesite ($MgCO_3$) is the most abundant carbonate among these end members because of its

Department of Geological Sciences, Jackson School of Geosciences, University of Texas at Austin, Austin, Texas, USA

Carbon in Earth's Interior, Geophysical Monograph 249, First Edition. Edited by Craig E. Manning, Jung-Fu Lin, and Wendy L. Mao.

extensive P-T stability (Fiquet et al., 2002; Isshiki et al., 2004; Katsura & Ito, 1990; Z. Li et al., 2017; Tao et al., 2013). $MgCO_3$ and $FeCO_3$ form a complete solid solution series (Shatskiy et al., 2014), with the Mg-rich (Mg,Fe)CO_3 named as ferromagnesite and the Fe-rich (Mg,Fe)CO_3 named as magnesiosiderite. Previous experimental studies show that the coexisting ferromagnesite and olivine have comparable Fe/Mg ratios at 2.0-3.5 GPa and 1273–1373 K (Dalton & Wood, 1993), while Fe/Mg ratios in ferropericlase are 9–19 times higher than that of coexisting ferromagnesite at 6.5–7.5 GPa, 1273–1873 K (Palyanov et al., 2013). These partitioning data indicate that ferromagensite would be one of the most abundant and stable form of carbonates in the Earth's deep mantle. Siderite, on the other hand, may be unstable under most mantle P-T conditions (Cerantola et al., 2017; Tao et al., 2013). Subducted $CaCO_3$ and $CaMg(CO_3)_2$, on the other hand, could react with surrounding (Mg,Fe)SiO_3 and $(Mg,Fe)_2SiO_4$ to form $MgCO_3$ and $CaMg(SiO_3)_2$ (e.g. Biellmann et al., 1993; Kushiro, 1975; Jie Li et al., this monograph). Such chemical reactions would shift the compositions of deep mantle carbonates toward the (Mg,Fe)CO_3 compositional joint.

Fe is the most abundant 3d transition element in mantle minerals, and the variations in Fe electronic spin and valence states at mantle P-T conditions have been documented to play a significant role in controlling the geophysical and geochemical properties of the lower mantle ferropericlase (Fp) and bridgmanite (Bdg) (reviewed by Badro, 2014; Lin et al., 2013). This chapter focuses on the Fe^{2+} spin transition in (Mg,Fe)CO_3 and the corresponding effects on its physical and chemical properties, which include the density, sound velocity, elastic moduli, and V_S splitting anisotropy and their associated physical relevance to the spin transitions of Fp and Bdg in the lower mantle.

12.2. FE^{2+} SPIN TRANSITION IN (MG,FE)CO_3

12.2.1. Crystal Field Theory and Parameters of the Spin Transition in Ferromagnesite

Ferromagnesite is stable with a rhombohedral structure in the $R\bar{3}c$ space group (Figure 12.1a). Mg^{2+} and Fe^{2+} occupy the octahedral sites composed of O^{2-} atoms, while the planar CO_3^{2-} units are perpendicular to the c-axis and the C-O bonds are parallel to the a-axis (Figure 12.1a). High-pressure single-crystal X-ray diffraction (XRD) measurements show that the octahedra are fairly isometric (Lavina, Dera, Downs, Yang, et al., 2010), which makes ferromagnesite an ideal case study for the spin transition behavior of Fe^{2+} in six-fold coordinated octahedral sites under high pressures. According to the symmetry of the charge density distribution and crystal field theory (Burns, 1993), the five $3d$

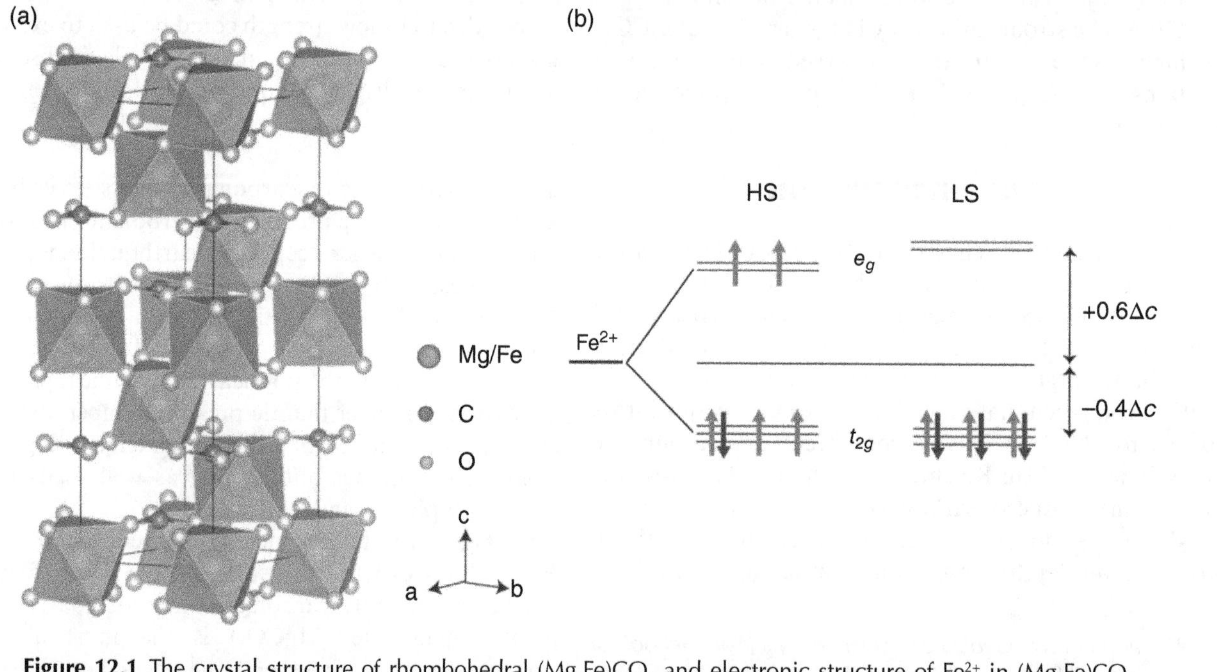

Figure 12.1 The crystal structure of rhombohedral (Mg,Fe)CO_3 and electronic structure of Fe^{2+} in (Mg,Fe)CO_3. (a) Rhombohedral (Mg,Fe)CO_3 is in $R\bar{3}c$ space group (Lavina et al., 2009). Mg^{2+} and Fe^{2+} occupy the octahedral sites shown as orange boxes. (b) Crystal field splitting diagrams of Fe^{2+} in the high-spin and low-spin states in the octahedral site. Δc is the crystal field splitting parameter; t_{2g} orbitals localize charge density between the coordinate axes; e_g orbitals localize charge density along the coordinate axes. See electronic version for color representation of the figures in this book.

orbitals of Fe^{2+} in the octahedral site of ferromagnesite are divided into two groups in a Cartesian coordination system: a set of three orbitals (d_{xy}, d_{xz}, d_{yz}) localize charge density between the coordinate axes, named as t_{2g} orbitals; the other two orbitals ($d_{x^2-y^2}$, d_{z^2}) with charge density concentrating along the coordinate axes are termed e_g orbitals. With the presence of oxygen ligands in the crystal structure, the degeneracy of the five *d*-orbitals of Fe^{2+} is lifted because of the different extent of the interaction between 3*d* electrons and ligands' electrons. As Fe^{2+} in $(Mg,Fe)CO_3$ occupies the six-fold coordinated octahedral site, e_g orbitals have higher energy than t_{2g} orbitals because of the proximity of e_g orbitals to the ligands (Burns, 1993). This results in an energy difference between e_g and t_{2g} orbitals, which is termed crystal-field splitting energy (Δc). The competition between Δc and the spin-paring energy Λ determines the electronic spin configuration of Fe^{2+} in $(Mg,Fe)CO_3$ (Figure 12.1). At ambient conditions where $\Delta c < \Lambda$, the number of paired electrons is minimized; therefore, Fe^{2+} in $(Mg,Fe)CO_3$ adopts a high-spin configuration with two paired electrons in one t_{2g} orbital, and the other four electrons are unpaired (Figure 12.1b). At higher pressures, the Fe-O bonds in $(Mg,Fe)CO_3$ are shortened (Lavina et al., 2009; Lavina, Dera, Downs, Tschauner, et al., 2010; Lavina, Dera, Downs, Yang, et al., 2010), resulting in a significant increase of Δc from ~14238 cm^{-1} (1.765 eV) to ~17517 cm^{-1} (2.172 eV) across the Fe^{2+} spin transition at 45.5 GPa and 300 K (Lobanov et al., 2015). In contrast, Λ is not sensitive to pressure increase as it is an intraionic quantity (Ohnishi, 1978), which is ~14193 cm^{-1} (1.760 eV) for siderite (Lobanov et al., 2015). Considering that the Fe-O length in $FeCO_3$ changes from 1.971 Å to 1.892 Å across the Fe^{2+} spin transition at ~45 GPa and 300 K (Lavina et al., 2010b), $\Delta c^{LS}/\Delta c^{HS}$ is comparable to $(r^{LS}/r^{HS})^{-5}_{Fe\text{-}O}$, consistent with theoretical prediction by Burns (1993). Once $\Delta c > \Lambda$ at ~ 45 GPa for $(Mg,Fe)CO_3$, all 3*d* electrons pair up in t_{2g} orbitals to form the LS state (Figure 12.1b). Optical absorption measurements have also been conducted for Fp and Bdg samples (e.g. Goncharov et al., 2006; Goncharov et al., 2008; Keppler et al., 2007, 2008), but the experimental results are still controversial due to Fe^{2+}-Fe^{3+} interaction and the complexity of crystal chemistry. Therefore, the change of Δc measured across the Fe^{2+} spin transition in $FeCO_3$ (Lobanov et al., 2015) provides the most solid experimental basis for understanding how the Fe^{2+} spin transition influences the thermal transport properties of lower-mantle minerals.

The change of Gibbs free energy $\Delta G_{LS\text{-}HS}$ across the Fe^{2+} spin transition can be expressed as follows (Sherman, 1988):

$$\Delta G_{LS-HS} = \Delta U_{LS-HS} + P\Delta V_{LS-HS} - T\Delta S_{LS-HS}, \quad (12.1)$$

where $\Delta U_{LS\text{-}HS}$, $\Delta V_{LS\text{-}HS}$, and $\Delta S_{LS\text{-}HS}$ are the differences in internal energy, volume, and entropy between LS and HS states, respectively; *P* is pressure; and *T* is temperature. Based on the crystal field theory (Burns, 1993), $\Delta U_{LS\text{-}HS}$ depends on the magnitudes of Δc and Λ. $\Delta V_{LS\text{-}HS}$ is negative because of the volume collapse across the spin transition (Lavina et al., 2009; Lavina, Dera, Downs, Tschauner, et al., 2010; Lavina, Dera, Downs, Yang, et al., 2010). Increasing pressure would promote the spin transition by lowering $\Delta G_{LS\text{-}HS}$ from a combined contribution of Δc and $\Delta V_{LS\text{-}HS}$. Increasing temperature, on the other hand, increases $\Delta G_{LS\text{-}HS}$ through the $-T\Delta S_{LS\text{-}HS}$ term in equation (12.1) and stabilizes the HS state with respect to the LS state. Two major components of $\Delta S_{LS\text{-}HS}$ are magnetic and configuration terms (Tsuchiya et al., 2006): The magnetic entropy for each spin state is expressed as $S_{mag} = k_B \ln[n\bullet(2S+1)]$ (Burns, 1993), where k_B is the Boltzmann constant, *S* is the total spin momentum of the 3*d* electrons in iron ion, and *n* is the electronic configuration degeneracy. Across the Fe^{2+} spin transition, S_{mag} decreases from $k_B \ln(15)$ to 0 ($S = 2$ and $n = 3$ for HS Fe^{2+}; $S = 0$, $n = 1$ for LS Fe^{2+} in $[Mg,Fe]CO_3$, respectively). Compared with spin transition pressure at 300 K, higher pressure is therefore needed to compensate the increases of $\Delta G_{LS\text{-}HS}$ from the $-T\Delta S_{LS\text{-}HS}$ term in equation (12.1) at high temperature condition. This is consistent with literature findings that pressures of the spin transition in $(Mg_{0.35}Fe_{0.65})CO_3$ increasing at a rate of ~10 MPa/K (Liu et al., 2014), indicating a generally positive Clapeyron slope. In addition, the configuration entropy of the coexisting HS and LS Fe^{2+} is defined as $S_{conf} = -k_B[n\ln n + (1-n)\ln(1-n)]$, where *n* is the LS fraction (Tsuchiya et al., 2006). High temperature condition would favor the mixed-spin state by decreasing ΔG_{mix} through the $-T\Delta S_{conf}$ term, which broadens the pressure range of the mixed-spin state. For example, the pressure interval of the mixed-spin state is ~4 GPa for $(Mg,Fe)CO_3$ at 300 K, but it broadens to ~10 GPa at 1200 K (Liu et al., 2014).

12.2.2. Characterizations of the Fe^{2+} Spin Transition in $(Mg,Fe)CO_3$ at High Pressure

The spin transition of Fe^{2+} in $(Mg,Fe)CO_3$ has been well constrained to occur at 40–50 GPa and 300 K using synchrotron X-ray and laser spectroscopies coupled with high-pressure diamond anvil cells (DAC) (Figure 12.2). The Fe^{2+} spin transition can cause significant changes in a range of physical properties of $(Mg,Fe)CO_3$, which in turn have been used to determine the occurrence of the Fe^{2+} spin transition. Specifically, the Fe^{2+} spin transition induces shortening of Fe-O bonds and lengthening of C-O bonds in $(Mg,Fe)CO_3$ (e.g. Lavina et al., 2010; Liu et al., 2019), which results in a reduction in the unit-cell volume and significant changes in vibrational properties

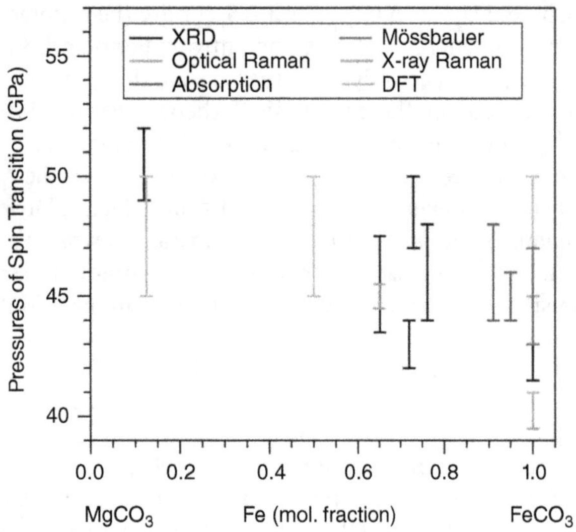

Figure 12.2 A summary of the reported pressure ranges for the Fe^{2+} spin transition in (Mg, Fe)CO_3 samples observed using various techniques at 300 K. Each vertical line with bars marks the pressure range of mixed-spin state for the corresponding composition. Black lines: XRD (Lavina et al., 2009; Lavina et al. 2010; Lavina et al., 2010; Nagai et al., 2010; Lin et al., 2012; Farfan et al., 2012; Liu et al., 2014, 2015); red line: Mössbauer (Cerantola et al., 2015, Liu et al., 2019); pink line: optical Raman (Lin et al. 2012); orange line: X-ray Raman (Weis et al., 2017); blue lines: optical absorption spectroscopy (Lobanov et al., 2015, 2016; Taran et al., 2017); green lines: DFT (Hsu & Huang, 2016). For simplicity, some previous DAC results obtained at relatively nonhydrostatic conditions, which can influence the spin transition pressure, are not shown here. See electronic version for color representation of the figures in this book.

(Cerantola et al., 2015; Lin et al., 2012; Spivak et al., 2014; Weis et al., 2017). Thus far, the Fe^{2+} spin transition in (Mg,Fe)CO_3 has been diagnosed based on the abrupt reduction of the unit-cell volume in XRD measurements (Farfan et al., 2012; Lavina et al., 2009; Lavina et al., 2010; Lavina et al., 2010; Lin et al., 2012; Liu et al., 2014, 2015; Nagai et al., 2010) and significant changes in the Raman shifts of vibrational modes in both optical and X-ray Raman spectroscopic studies (Cerantola et al., 2015; Fu et al., 2017; Lin et al., 2012; Spivak et al., 2014; Weis et al., 2017). Across the Fe^{2+} spin transition, the reduction of total spin momentum in Fe^{2+} can be captured by X-ray emission spectroscopy (Mattila et al., 2007). Conventional and synchrotron Mössbauer spectroscopies have also been used to study the Fe^{2+} spin transition in (Mg,Fe)CO_3 (Cerantola et al., 2015; Liu et al., 2019). Both of the fitted hyperfine parameters, center shift (CS) and quadrupole splitting (QS), show significant changes across the Fe spin transition, which have been used to determine spin transition pressures for not only (Mg,Fe)CO_3 (Cerantola et al., 2015) but also Fp and Bdg (e.g. Lin et al., 2013; Liu et al., 2018). In addition, the optical and near IR absorption spectra of (Mg,Fe)CO_3 also change significantly across the Fe^{2+} spin transition due to the dramatic increase in Δc (Lobanov et al., 2015, 2016; Taran et al., 2017). Theoretically, density function theory calculations predict the transition pressures that are in agreement with aforementioned experimental results (Hsu & Huang, 2016) (Figure 12.2). However, an early density function theory study underestimates the spin transition by ~15–30 GPa (Shi et al., 2008).

The starting pressure and pressure ranges of the Fe^{2+} spin transition in (Mg,Fe)CO_3 are sensitive to the hydrostatic condition in high-pressure DAC experiments: When relatively hydrostatic helium pressure medium is used in DAC experiments, the spin transition in $FeCO_3$ occurs at 40.4 ± 0.1 GPa with a transition width of 0.7 GPa at 300 K (Weis et al., 2017). In contrast, the spin transition pressure changes to 44.3 ± 0.4 GPa and the transition width broadens to 4.4 GPa when argon is used as the pressure medium (Weis et al., 2017). The Earth's mantle is mostly under relatively hydrostatic compression condition, except regions near subducting slabs where stress could be ~100–300 MPa (e.g. Houseman & Gubbins, 1997; Běhounková & Čížková, 2008). Therefore, hydrostatic results on the spin transition and its associated effects are certainly more applicable to the deep mantle. Specifically, it now becomes clearer that much of the inconsistencies in the study of the Fe spin transition in the lower mantle Fp is likely due to experimental conditions, together with possible iron clustering in the lattice (Kantor et al., 2009). Therefore, the broad spin transition pressures of $MgCO_3$-$FeCO_3$ observed in some DAC studies may be caused by artifacts from nonhydrostatic conditions. In addition, negligible compositional dependence of the spin transition pressures is found in (Mg,Fe)CO_3 (Figure 12.2). This is explained by weak Fe^{2+}-Fe^{2+} exchange interactions in (Mg,Fe)CO_3 due to the long distance between neighboring Fe^{2+} ions (3.458 Å at 1 bar, Lavina et al., 2010), which are separated by $(CO_3)^{2-}$ units along the c-axis (Figure 12.1a). In comparison, spin transition pressures of Fe^{2+} in Fp strongly depend on Fe content (Lin et al., 2006) because of the strong Fe^{2+}-Fe^{2+} exchange interactions, especially for FeO-rich magnesiowüstite.

12.2.3. The Effects of the Fe^{2+} Spin Transition on Physical Properties of (Mg,Fe)CO_3

12.2.3.1. Equation of State Anomaly Across the Fe^{2+} Spin Transition

The spin transition in (Mg,Fe)CO_3 is associated with a volume collapse due to the reconfiguration of the electronic orbitals and thus shortening of the Fe-O bonds (e.g. Lavina et al., 2009; Lavina et al., 2010; Lavina et al., 2010; Lin et al., 2012; Liu et al., 2014). The relationship of pressure-volume-temperature, called the equation of state (EOS), is derived from high P-T XRD experiments in DACs and therefore can be used to infer the spin transition as well as the fraction of low spin Fe^{2+} (n_{LS}) at given P-T conditions. These XRD results have been used to model the EOS parameters across the spin transition and to construct the spin-crossover diagram at lower-mantle P-T conditions (e.g. Lavina et al., 2009; Lavina et al., 2010; Lavina et al., 2010; Lin et al., 2012; Liu et al., 2014).

A summary of the 300-K pressure-volume relationship for a range of compositions is shown in Figure 12.3a. The 1-bar unit cell volume of $(Mg,Fe)CO_3$ increases with Fe^{2+} content, consistent with the larger radius of HS Fe^{2+} (r ~ 0.92 Å; Shannon, 1976) than Mg^{2+} (r ~ 0.86 Å; Shannon, 1976) in the octahedral site (Figure 12.3b). In contrast, the fitted 1-bar unit cell volume of LS $(Mg,Fe)CO_3$ decreases with increasing Fe content (Figure 12.3b), which can be explained by the smaller radius of LS-Fe^{2+} (r ~ 0.75 Å; Shannon, 1976) in comparison with Mg^{2+}. These represent 8%–10% collapse of the 1-bar unit cell volume for Fe^{2+} ion from the HS to LS state in $(Mg,Fe)CO_3$ (Figure 12.3b). Further analysis of these results shows that the Fe^{2+} spin transition in $(Mg,Fe)CO_3$ induces a significant reduction in the unit-cell volume. For example, the unit-cell volumes of $(Mg_{0.35}Fe_{0.65})CO_3$ (Lin et al., 2012; Liu et al., 2014) and $FeCO_3$ (Lavina, Dera, Downs, Yang, et al., 2010) shrink by 6% and 10%, respectively, at around 45 GPa and 300 K (Figure 12.3a). The significant volume reduction caused by the Fe^{2+} spin transition makes the unit-cell volume of LS $(Mg,Fe)CO_3$ smaller than that of $MgCO_3$ (Figure 12.3a). The interatomic distances in $(Mg,Fe)CO_3$ are also significantly influenced by the Fe^{2+} spin transition: The shorter radius of LS Fe^{2+} decreases the Fe-O bond length by 0.078 Å at 45 GPa and 300 K for $FeCO_3$ (Lavina et al., 2010). The O-O bond lengths also shrink by 0.135–0.074 Å (Lavina et al., 2010). In contrast, the C-O bond exhibits lengthening across the Fe^{2+} spin transition, which may reflect a stress release of the C-O bond due to the shrinkage of the Fe-O bond (Lavina et al., 2010; Liu et al., 2019). The corresponding *a* and *c* axes shrink by 3% and 4%, respectively, for $FeCO_3$ (Lavina, Dera, Downs, Yang, et al., 2010) and 2% and 3%, respectively, for $(Mg_{0.35}Fe_{0.65})CO_3$ (Lin et al., 2012), which results in ~1% decrease in the *c*/*a* ratio (Lin et al., 2012). The higher compressibility of the *c*-axis compared with the *a*-axis can be explained by the rigidity of planar $(CO_3)^{2-}$ units (Lin et al., 2012), which are along the *a*-axis but perpendicular to the *c*-axis (Figure 12.1a). Based on the EOS modeling, the isothermal bulk moduli (K_T) of both HS and LS $(Mg,Fe)CO_3$ are not sensitive to its Fe content (Figure 12.3c), which can also be explained by the weak Fe^{2+}-Fe^{2+} exchange interactions (Lin et al., 2012). On the other

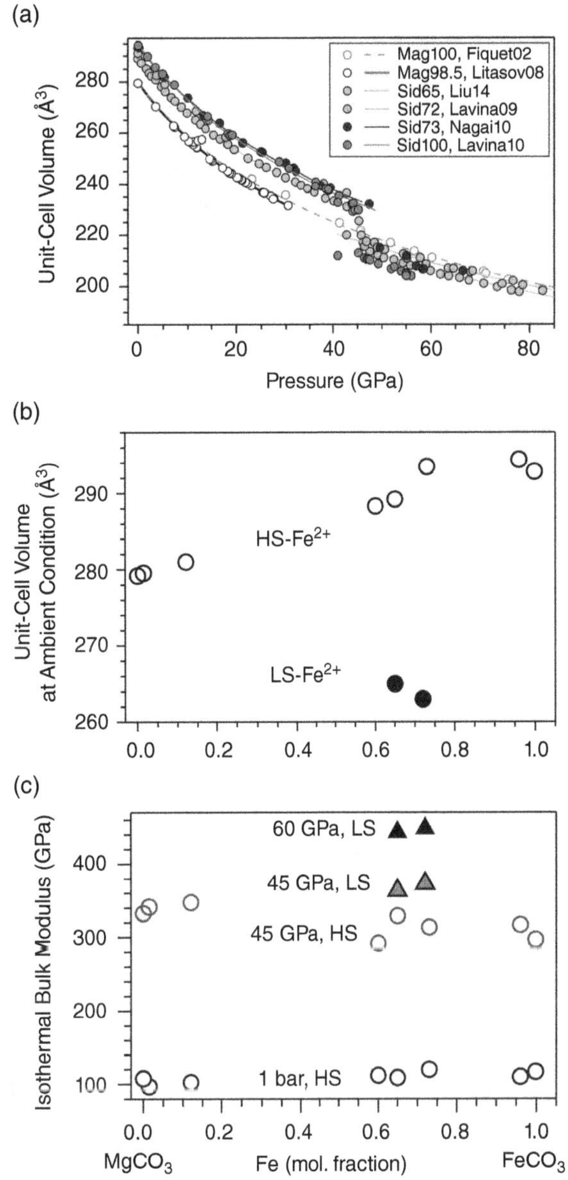

Figure 12.3 Spin transition of iron in $(Mg,Fe)CO_3$ at high pressure and 300 K. (a) A summary of compression curves for $(Mg,Fe)CO_3$ with different compositions at 300 K. (b) 1-bar and 300-K unit-cell volume (V_0) for high-spin Fe^{2+} (open circles) and fitted V_0 for low-spin Fe^{2+} (black circles) of $(Mg,Fe)CO_3$. (c) 300-K bulk moduli of $(Mg,Fe)CO_3$ at 1 bar (high-spin Fe^{2+}, open black circles), 45 GPa (high-spin Fe^{2+}, open blue circles), 45 GPa (low-spin Fe^{2+}, red triangles), and 60 GPa (low-spin Fe^{2+}, black triangles). Data sources are listed as follows: Fiquet02: (Fiquet et al., 2002); Litasov08: (Litasov et al., 2008); Liu14: (Liu et al., 2014); Lavina09: (Lavina et al., 2009); Nagai10: (Nagai et al., 2010); Lavina10: (Lavina et al., 2010). See electronic version for color representation of the figures in this book.

hand, the Fe^{2+} spin transition significantly influences the isothermal bulk modulus K_T (Figure 12.3c): K_T of LS $(Mg,Fe)CO_3$ is ~10%–19% larger than that of HS $(Mg,Fe)CO_3$ at 45 GPa and 300 K (Lavina et al., 2009; Lin et al., 2012; Liu et al., 2014). Within the pressure range of the mixed-spin state, the reduction of K_T is as high as ~86% for $(Mg_{0.35}Fe_{0.65})CO_3$ at 300 K (Liu et al., 2014), which is also associated with a significant decrease in elasticity and V_P (discussed in section 12.2.3.2). Such significant reduction in K_T results from the significant change in unit-cell volume ($\Delta V_{\text{LS-HS}}$), as dictated by the following equation from Wentzcovitch et al. (2009) based on an ideal mixing model:

$$\frac{V}{K} = (1-n_{\text{LS}})\frac{V_{\text{HS}}}{K_{\text{HS}}} + n_{\text{LS}}\frac{V_{\text{LS}}}{K_{\text{LS}}} - (V_{\text{LS}} - V_{\text{HS}})\left(\frac{\partial n_{\text{LS}}}{\partial P}\right)_T \quad (12.2)$$

The thermal EOS results can be used to model the relevant thermoelastic parameters of $(Mg,Fe)CO_3$ along an expected mantle geotherm (Brown & Shankland, 1981). By using resistive and laser-heated DACs coupled with synchrotron XRD techniques, Liu et al. (2014) reported the P-T dependence of isothermal bulk modulus and thermal expansion coefficient. In addition, Liu et al. (2014, 2015) found that increasing temperature broadens the spin transition pressures of $(Mg,Fe)CO_3$ and the mixed-spin state would exist between ~53 and ~80 GPa along an expected geotherm (Brown & Shankland, 1981). Within the spin crossover, the magnitude of the unit-cell volume reduction relative to HS $(Mg_{0.35}Fe_{0.65})CO_3$ decreases from 6.0% at 300 K to 5.5% at 1200 K (Liu et al., 2014). Such P-T effects on the unit-cell volume translate into strong anomaly in the thermal expansion coefficient across the spin crossover (Liu et al., 2014). Compared with the HS counterpart, LS $(Mg_{0.35}Fe_{0.65})CO_3$ has ~10% higher bulk modulus and ~ 20% lower thermal expansion coefficient, making it 5%–7% denser than the HS phase at lower mantle P-T conditions. Therefore, dense LS ferromagnesite becomes thermodynamically more stable than its HS counterpart because of the negative $P\Delta V_{\text{LS-HS}}$ term in calculating $\Delta G_{\text{LS-HS}}$ (eq. [1]).

12.2.3.2. Anomalous Elastic Properties Across the Spin Transition

The concentration, valence, and spin states of Fe in mantle minerals could significantly influence the mantle seismic velocities, which could help explain some seismically observed heterogeneities (Fu et al., 2017; Liu et al., 2018; Wu & Wentzcovitch, 2014). The definitions of V_P and V_S splitting anisotropies are as follows:

$$A_P = 2(V_{P,\max} - V_{P,\min})/(V_{P,\max} + V_{P,\min}) \times 100\% \quad (12.3)$$

$$A_S = 2(V_{S1} - V_{S2})/(V_{S1} + V_{S2}) \times 100\%, \quad (12.4)$$

where $V_{P,max}$ and $V_{P,min}$ are the azimuthal maximum and minimum V_P, respectively, and V_{S1} and V_{S2} are two orthogonally polarized V_S velocities for a given propagation direction (Mainprice, 2015). Under the P-T conditions of the Earth's upper mantle, $MgCO_3$ exhibits 37%–41% V_P splitting anisotropy and 37%–41% V_S splitting anisotropy (Yang et al., 2014). The presence of Fe in $(Mg,Fe)CO_3$ together with its spin transition–induced elastic anomaly could further enhance the signatures of seismic velocity anisotropies.

The effect of the Fe^{2+} spin transition on the single-crystal elasticity of $(Mg,Fe)CO_3$ can be modeled using the following equation:

$$S^{ij}V(n) = nS^{ij}_{LS}V_{LS} + (1-n) S^{ij}_{HS}V_{HS} - \left(\frac{\partial G_{LS}}{\partial \sigma_j} - \frac{\partial G_{HS}}{\partial \sigma_j}\right)\frac{\partial n_{LS}}{\partial \sigma_i}, \quad (12.5)$$

where S^{ij} is the elastic compliance; σ_i and σ_j are the *ith* and *jth* stress components, respectively, in the Voigt notation; and G_{HS}, G_{LS} are the Gibbs free energies of the HS and LS states, respectively (Wu et al., 2013; Fu et al., 2017). The corresponding elastic constants C_{ij} can be calculated based on the relationship between C_{ij} and S^{ij} (Fu et al., 2017). The third term on the right-hand side of equation (12.5) determines the C_{ij} anomalies across the spin transition: $\frac{\partial n_{LS}}{\partial \sigma_i} \neq 0$ for $i = 1,2,3$ and $\frac{\partial n_{LS}}{\partial \sigma_i} = 0$ for $i = 4,5,6$. This is because n_{LS} is an even function of the shear stress components (Fu et al., 2017). As a result, drastic softening occurs on C_{11}, C_{12}, C_{13} and C_{33}, which are the elastic constants related to the compressional stress component. On the other hand, slight stiffening occurs on C_{14} and C_{44}, which are the elastic constants related to the shear stress component. Furthermore, calculating the aggregate elastic moduli (K_S and G) of ferromagnesite using Voigt-Reuss-Hill averages (Hill, 1952) shows that the calculation of K_S would involve all the softened elastic constants, C_{11}, C_{12}, C_{13}, and C_{33}, whereas the softening parts cancel out for the calculation of G (Fu et al., 2017). As a result, drastic softening in K_S and V_P and slight stiffening in G and V_S would be expected across the Fe^{2+} spin transition in $(Mg,Fe)CO_3$. Notably, equation (12.5) is also applicable to Fp and Bdg (Fu et al., 2018; Yang et al., 2015), both of which show significant softening in K_S and V_P but almost smooth profiles for G and V_S across the spin transition pressures of Fe^{2+} and Fe^{3+} in the octahedral sites of Fp and Bdg, respectively.

A recent experimental study using Brillouin light scattering (BLS) and impulsive stimulated light scattering (ISS) techniques has determined the effects of the Fe^{2+}

spin transition on the single-crystal elastic constants (C_{ij}) of $(Mg_{0.35}Fe_{0.65})CO_3$ (Fu et al., 2017). Within the pressure range of the Fe^{2+} spin transition, V_P and V_S of $(Mg_{0.35}Fe_{0.65})CO_3$ are reduced by as much as ~40% and ~21%, respectively, at 300 K (Figure 12.4), which are consistent with the thermodynamic predictions discussed above. However, the elastic moduli of $FeCO_3$ derived from the phonon dispersion frequency probed by the inelastic X-ray scattering technique do not show softening across the pressure range of the Fe^{2+} spin transition (Stekiel et al., 2017). Such an inconsistence has also been seen in sound velocity measurements of Fp samples (Antonangeli et al., 2011; Crowhurst et al., 2008; Yang et al., 2015), which could be explained by the different frequencies and scattering vectors used in inelastic X-ray scattering compared with BLS and ISS (Lin et al., 2013). Because BLS and ISS directly measure the acoustic wave velocities of samples, the effects of Fe^{2+} spin transition on the sound velocity of $(Mg,Fe)CO_3$ determined by these techniques are more applicable to seismic implications.

To quantitatively understand the effects of Fe on the elasticity of $(Mg,Fe)CO_3$, the velocities of $(Mg,Fe)CO_3$ at high pressures are compared with those of end-member $MgCO_3$ (Figure 12.4). At 1 bar to ~ 40 GPa, V_P and V_S of HS $(Mg_{0.35}Fe_{0.65})CO_3$ are ~17% and 22% lower than that of $MgCO_3$, respectively (Figure 12.4 b, c). Within the pressure range of the Fe^{2+} spin transition, V_P of $(Mg_{0.35}Fe_{0.65})CO_3$ is reduced by as much as 56% with respect to $MgCO_3$ (Figure 12.4 b, c). At higher pressures, there is ~9% reduction in both V_P and V_S for LS $(Mg_{0.35}Fe_{0.65})CO_3$, up to 70 GPa with respect to $MgCO_3$ (Figure 12.4 b, c). Assuming a linear relationship between velocities and Fe content, V_P would decrease with an average dV_P/dFe of about –0.026 km/s/mol.% for the HS state at 0–40 GPa and dV_P/dFe of about –0.0016 km/s/mol.% for the LS state at 40–60 GPa. V_S decreases with dV_S/dFe of about –0.019 km/s/mol.% for the HS state and dV_S/dFe of about –0.007 km/s/mol.% for the LS state. Compared with $MgCO_3$, the shear-wave splitting anisotropy factor A_S of $(Mg,Fe)CO_3$ increases, particularly across the pressure range of the Fe^{2+} spin transition (Figure 12.5). For $MgCO_3$, A_S increases from ~37% to ~48% with pressure increasing from 0 to 60 GPa (Yang et a., 2014; Stekiel et al., 2017). In comparison, A_S of $(Mg_{0.35}Fe_{0.65})CO_3$ increases from ~37% to ~58% between 1 bar and 40 GPa and reaches a peak at ~64% across the spin transition, but it decreases to ~53% for LS state at 47 GPa and 300 K (Figure 12.5). The V_S anisotropy factor of ferromagnesite across the spin transition is much higher than that of major lower-mantle minerals, such as Bdg with ~10%–20% A_S (Karki et al., 1997) and Fp with ~20% A_S under lower-mantle pressures (Marquardt et al., 2009; Yang et al., 2015). Here we interpolated the seismic properties of $(Mg_{0.90}Fe_{0.10})CO_3$ from those of $MgCO_3$ and magnesiosiderite compositions (Figures 12.4 and 12.5). The

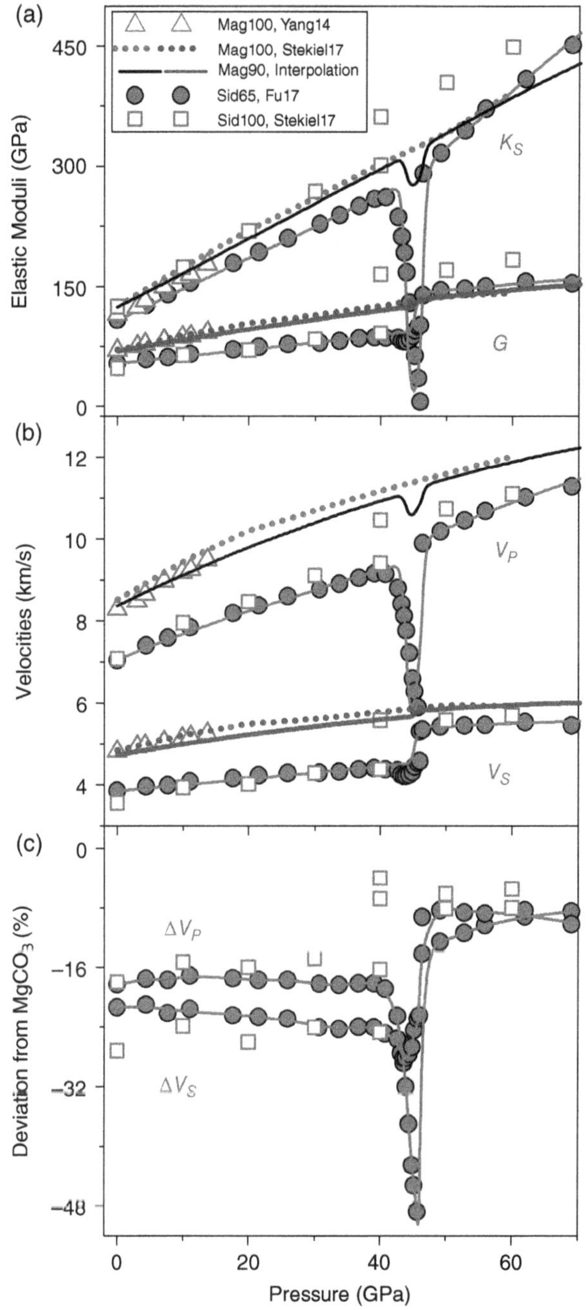

Figure 12.4 The pressure dependence of the elastic properties of $(Mg,Fe)CO_3$ at high pressure and 300 K. (a) and (b) Aggregate elastic properties (K_S, G, V_P, V_S) of $(Mg,Fe)CO_3$ with different iron contents; (c) Deviations of sound velocities in Fe-bearing $(Mg,Fe)CO_3$ with respect to $MgCO_3$ magnesite. Data source: open squares, $FeCO_3$ from inelastic X-ray scattering measurements (Stekiel et al., 2017); open triangles, $MgCO_3$ from BLS measurements (Yang et al., 2014); solid circles, $(Mg_{0.35}Fe_{0.65})CO_3$ using BLS and ISS measurements (Fu et al., 2017); solid curves, fittings to results of Fu et al. (2017); dotted curves, $MgCO_3$ using DFT calculation (Stekiel et al., 2017); black and blue curves, interpolated results for $(Mg_{0.90}Fe_{0.10})CO_3$. See electronic version for color representation of the figures in this book.

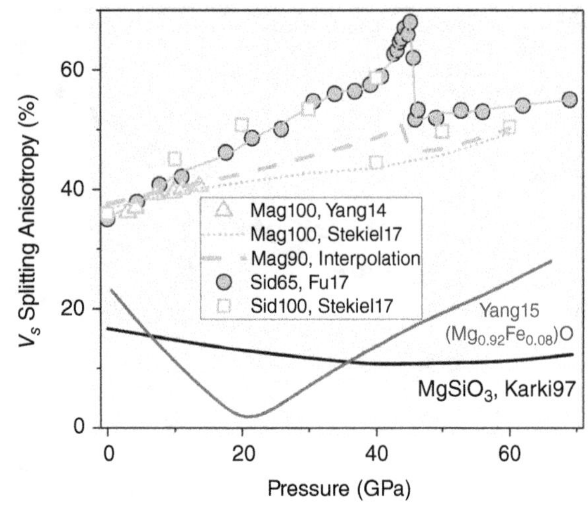

Figure 12.5 The pressure dependence of the V_S anisotropy factor (A_S) of (Mg,Fe)CO_3 in comparison with those of $MgSiO_3$ bridgmanite and $(Mg_{0.92}Fe_{0.08})O$ ferropericlase at high pressure and 300 K. The data sources for (Mg,Fe)CO_3 are the same as for Figure 12.4. The anisotropy factor of $MgSiO_3$ bridgmanite and $(Mg_{0.92}Fe_{0.08})O$ Fp are from Karki et al. (1997) and Yang et al. (2015), respectively. See electronic version for color representation of the figures in this book.

modeled results show ~5% decrease in V_P and ~2% increase in V_S would occur in $(Mg_{0.90}Fe_{0.10})CO_3$ across the spin transition region. A_S of $(Mg_{0.90}Fe_{0.10})CO_3$ is comparable with that of $MgCO_3$ in the upper mantle, but 5% and 13% higher at the bottom of the transition zone and the middle mantle, respectively (Figure 12.5). At deeper depths, A_S of LS $(Mg_{0.90}Fe_{0.10})CO_3$ becomes comparable with that of $MgCO_3$ again (Figure 12.5). In addition, A_S of HS $(Mg_{0.90}Fe_{0.10})CO_3$ is about 4–13 times higher than those of the two major lower-mantle minerals, Bdg and Fp, at the uppermost lower mantle depths; in the middle mantle depths, A_S of LS $(Mg_{0.90}Fe_{0.10})CO_3$ is about 2–4 times higher than those of Bdg and Fp (Figure 12.5).

In the mantle, especially near the subduction regions, deformation-induced lattice-preferred orientations of minerals together with their A_S could generate seismic V_S radial anisotropy (e.g. Panning & Romanowicz, 2004; Wookey and Kendall, 2004). Previous deformation experiments show that the strength of $MgCO_3$ is several orders of magnitude lower than peridotite at mantle P-T conditions (Holyoke et al., 2014). By analyzing the micro-structures and lattice orientations of recovered $MgCO_3$ samples from deformation experiments, Ulrich (2015) found that the *c*-axis tends to parallel to the maximum compressive stress under high P-T conditions. These experimental results indicate that ferromagnesite conveyed by subducting slabs would localize strain and develop a lattice-preferred orientation with the *c*-axis parallel to the maximum compressive stress. Together with its strong anisotropy, ferromagenesite would be a leading candidate for explaining the strong seismic anisotropy. A recent seismic study observed strong seismic anisotropy of ~25% below ~300 km (J. Li et al., 2018), which is attributed to the presence of either magnesite or carbonatite melt. The addition of Fe^{2+} into magnesite and the associated increase of A_S due to Fe^{2+} spin transition from the bottom of the transition zone to the middle lower mantle (Figure 12.5) would further promote ferromagnesite as a leading candidate to explain seismic anisotropy near the subduction regions (J. Li et al., 2018).

12.3. CONCLUSIONS

The pressure-temperature conditions of the Fe^{2+} spin transition in ferromagnesite (Mg,Fe)CO_3 and its influence on the geophysical properties of this mantle major carbonate have been extensively studied by both experimental and theoretical methods. Along an expected mantle geotherm, the Fe^{2+} spin transition in ferromagnesite is expected to occur at ~50–80 GPa. Within this pressure range, significant density increase, V_P softening and slight V_S stiffening are expected. At deeper depths, ferromagnesite with low-spin Fe^{2+} would have higher density, incompressibility, and sound velocities than its high-spin Fe^{2+} counterpart. The strong anisotropy and low strength of ferromagnesite make it a potential candidate to explain the strong seismic radial anisotropy in the subduction regions beneath 300 km. Moreover, the robust results on the Fe^{2+} spin transition and its associated changes in the physical properties of ferromagnesite would provide the baseline for understanding the iron spin transition in the lower-mantle ferropericlase and bridgmanite.

Future work could focus on determining how spin states of Fe^{2+} influence iron partitioning, especially across melting temperature, as well as the rheological and transport properties of (Mg,Fe)CO_3 at lower-mantle P-T conditions. For example, based on the change of the crystal-field splitting energy across the spin transition in (Mg,Fe)CO_3, Lobanov et al. (2015) infer that Fe would strongly partition into low-spin (Mg,Fe)CO_3 compared with coexisting bridgmanite. Upon melting, Fe^{2+} in carbonated melt can also experience a spin transition, which may also significantly influence Fe partitioning (Fu et al., 2018). In addition, the relationship between the spin states of Fe^{2+} and the crystal structures/chemical compositions of carbonates under the lower-mantle conditions would shed light on the fate of these oxidized components in the reduced lower mantle.

ACKNOWLEDGMENTS

The authors acknowledge the support of the Deep Carbon Observatory and Geophysics Program of the National Science Foundation.

REFERENCES

Antonangeli, D., Siebert, J., Aracne, C. M., Farber, D. L., Bosak, A., Hoesch, M., et al. (2011). Spin crossover in ferropericlase at high pressure: A seismologically transparent transition? *Science*, *331*, 64–67.

Badro, J. (2014). Spin transitions in mantle minerals. *Annu Rev Earth Pl Sc*, *42*, 231–248.

Běhounková, M., & Čížková, H. (2008). Long-wavelength character of subducted slabs in the lower mantle. *Earth and Planetary Science Letters*, *275*, 43–53.

Biellmann, C., Gillet, P., Guyot, F., Peyronneau, J., & Reynard, B. (1993). Experimental evidence for carbonate stability in the Earth's lower mantle. *Earth and Planetary Science Letters*, *118*, 31–41.

Brown, J. M., & Shankland, T. J. (1981). Thermodynamic parameters in the Earth as determined from seismic profiles. *Geophysical Journal of the Royal Astronomical Society 66*, 579–596.

Burns, R. G. (1993). *Mineralogical applications of crystal field theory*. Cambridge, UK: Cambridge University Press.

Cerantola, V., Bykova, E., Kupenko, I., Merlini, M., Ismailova, L., McCammon, C., et al. (2017). Stability of iron-bearing carbonates in the deep Earth's interior. *Nature Communications*, *8*, 15960.

Cerantola, V., McCammon, C., Kupenko, I., Kantor, I., Marini, C., Wilke, M., et al. (2015). High-pressure spectroscopic study of siderite ($FeCO_3$) with a focus on spin crossover. *American Mineralogist*, *100*, 2670–2681.

Crowhurst, J. C., Brown, J. M., Goncharov, A. F., & Jacobsen, S. D. (2008). Elasticity of (Mg,Fe)O through the spin transition of iron in the lower mantle. *Science*, *319*, 451–453.

Dalton, J. A., & Wood, B. J. (1993). The partitioning of Fe and Mg between olivine and carbonate and the stability of carbonate under mantle conditions. *Contrib Mineral Petr*, *114*, 501–509.

Dasgupta, R., & Hirschmann, M. M. (2010). The deep carbon cycle and melting in Earth's interior. *Earth Planet Sci Lett*, *298*, 1–13.

Farfan, G., Wang, S., Ma, H., Caracas, R., & Mao, W.L. (2012). Bonding and structural changes in siderite at high pressure. *American Mineralogist*, *97*, 1421–1426.

Fiquet, G., Guyot, F., Kunz, M., Matas, J., Andrault, D., & Hanfland, M. (2002). Structural refinements of magnesite at very high pressure. *American Mineralogist*, *87*, 1261–1265.

Fu, S., Yang, J., & Lin, J.-F. (2017). Abnormal elasticity of single-crystal magnesiosiderite across the spin transition in Earth's lower mantle. *Phys. Rev. Lett.*, *118*, 036402.

Fu, S., Yang, J., Zhang, Y., Liu, J., Greenberg, E., Prakapenka, V.B., et al. (2018). Melting behavior of the lower-mantle ferropericlase across the spin crossover: Implication for the ultra-low velocity zones at the lowermost mantle. *Earth and Planetary Science Letters*, *503*, 1–9.

Fu, S., Yang, J., Zhang, Y., Okuchi, T., McCammon, C., Kim, H.-I., et al. (2018). Abnormal elasticity of Fe-bearing bridgmanite in the Earth's lower mantle. *Geophysical Research Letters*, *45*, 4725–4732.

Goncharov, A. F., Haugen, B. D., Struzhkin, V. V., Beck, P., & Jacobsen, S. D. (2008). Radiative conductivity in the Earth's lower mantle. *Nature*, *456*, 231–234.

Goncharov, A. F., Struzhkin, V. V., & Jacobsen, S. D. (2006). Reduced radiative conductivity of low-spin (Mg,Fe)O in the lower mantle. *Science*, *312*, 1205–1208.

Hill, R. (1952). The elastic behaviour of a crystalline aggregate. *Proceedings of the Physical Society. Section A*, *65*, 349.

Holyoke, C. W., Kronenberg, A. K., Newman, J., & Ulrich, C. (2014). Rheology of magnesite. *Journal of Geophysical Research: Solid Earth*, *119*, 6534–6557.

Houseman, G. A., & Gubbins, D. (1997). Deformation of subducted oceanic lithosphere. *Geophys J Int*, *131*, 535–551.

Hsu, H., & Huang, S.-C. (2016). Spin crossover and hyperfine interactions of iron in $(Mg,Fe)CO_3$ ferromagnesite. *Phys. Rev. B*, *94*, 060404.

Isshiki, M., Irifune, T., Hirose, K., Ono, S., Ohishi, Y., Watanuki, T., et al. (2004). Stability of magnesite and its high-pressure form in the lowermost mantle. *Natur,e 427*, 60–63.

Kantor, I., Dubrovinsky, L., McCammon, C., Steinle-Neumann, G., Kantor, A., Skorodumova, N., et al. (2009). Short-range order and Fe clustering in $Mg_{1-x}Fe_xO$ under high pressure. *Phys. Rev. B*, *80*, 014204.

Karki, B. et al. (1997). Elastic properties of orthorhombic MgSiO3 perovskite at lower mantle pressures. *Am Mineral*, *82*, 635–638.

Katsura, T., & Ito, E. (1990). Melting and subsolidus phase relations in the $MgSiO_3$-$MgCO_3$ system at high pressures: implications to evolution of the Earth's atmosphere. *Earth and Planetary Science Letters*, *99*, 110–117.

Kelemen, P. B., & Manning, C. E. (2015). Reevaluating carbon fluxes in subduction zones, what goes down, mostly comes up. *P Natl Acad Sci USA 112*, E3997–E4006.

Keppler, H., Dubrovinsky, L. S., Narygina, O., & Kantor, I. (2008). Optical absorption and radiative thermal conductivity of silicate perovskite to 125 gigapascals. *Science*, *322*, 1529–1532.

Keppler, H., Kantor, I., & Dubrovinsky, L. S. (2007). Optical absorption spectra of ferropericlase to 84 GPa. *Am Mineral*, *92*, 433–436.

Kushiro, I. (1975). Carbonate-silicate reactions at high pressures and possible presence of dolomite and magnesite in the upper mantle. *Earth and Planetary Science Letters*, *28*, 116–120.

Lavina, B., Dera, P., Downs, R. T., Prakapenka, V., Rivers, M., Sutton, S., & Nicol, M. (2009). Siderite at lower mantle conditions and the effects of the pressure-induced spin-pairing transition. *Geophys. Res. Lett.*, *36*, L23306.

Lavina, B., Dera, P., Downs, R. T., Tschauner, O., Yang, W., Shebanova, O., & Shen, G. (2010). Effect of dilution on the spin pairing transition in rhombohedral carbonates. *High Pressure Research*, *30*, 224–229.

Lavina, B., Dera, P., Downs, R. T., Yang, W., Sinogeikin, S., Meng, Y., et al. (2010). Structure of siderite $FeCO_3$ to 56 GPa and hysteresis of its spin-pairing transition. *Phys. Rev. B*, *82*, 064110.

Li, J., Zheng, Y., Thomsen, L., Lapen, T. J., & Fang, X. (2018). Deep earthquakes in subducting slabs hosted in highly anisotropic rock fabric. *Nature Geoscience*, *11*, 696.

Li, Z., Li, J., Lange, R., Liu, J., & Militzer, B. (2017). Determination of calcium carbonate and sodium carbonate melting curves up to Earth's transition zone pressures with implications for the deep carbon cycle. *Earth and Planetary Science Letters*, *457*, 395–402.

Lin, J.-F., Gavriliuk, A. G., Struzhkin, V. V., Jacobsen, S. D., Sturhahn, W., Hu, M. Y., et al. (2006). Pressure-induced electronic spin transition of iron in magnesiowustite-(Mg,Fe)O. *Phys. Rev. B*, *73*, 113107.

Lin, J.-F., Liu, J., Jacobs, C., & Prakapenka, V. B. (2012). Vibrational and elastic properties of ferromagnesite across the electronic spin-pairing transition of iron. *American Mineralogist*, *97*, 583–591.

Lin, J.-F., Speziale, S., Mao, Z., & Marquardt, H. (2013). Effects of the electronic spin transitions of iron in lower mantle minerals: implications for deep mantle geophysics and geochemistry. *rev. geophys.* *51*, 244–275.

Litasov, K. D., Fei, Y., Ohtani, E., Kuribayashi, T., & Funakoshi, K. (2008). Thermal equation of state of magnesite to 32 GPa and 2073 K. *Physics of the Earth and Planetery Interiors*, *168*, 191–203.

Liu, J., Dorfman, S.M., Zhu, F., Li, J., Wang, Y., Zhang, D., et al. (2018). Valence and spin states of iron are invisible in Earth's lower mantle. *Nature Communications*, *9*, Article 1284.

Liu, J., Lin, J.-F., Mao, Z., & Prakapenka, V. B. (2014). Thermal equation of state and spin transition of magnesiosiderite at high pressure and temperature. *American Mineralogist*, *99*, 84–93.

Liu, J., Lin, J.-F., & Prakapenka, V. B. (2015). High-pressure orthorhombic ferromagnesite as a potential deep-mantle carbon carrier. *Scientific Reports* *5*, srep07640.

Liu, J., Wang, W., Yang, H., Wu, Z., Hu, M., Zhao, J., et al. (2019). Carbon isotopic signatures of super-deep diamonds mediated by iron redox chemistry. *Geochemical Perspectives Letters.* doi: 10.7185/geochemlet.1915

Lobanov, S. S., Goncharov, A. F., & Litasov, K. D. (2015). Optical properties of siderite ($FeCO_3$) across the spin transition: Crossover to iron-rich carbonates in the lower mantle. *American Mineralogist*, *100*, 1059–1064.

Lobanov, S. S., Holtgrewe, N., & Goncharov, A. F. (2016). Reduced radiative conductivity of low spin FeO_6-octahedra in $FeCO_3$ at high pressure and temperature. *Earth and Planetary Science Letters*, *449*, 20–25.

Mainprice, D. (2015). Seismic anisotropy of the deep Earth from a mineral and rock physics perspective. In G. Schubert (Ed.), *Treatise on geophysics* (2nd ed., pp. 487–538). Oxford: Elsevier.

Marquardt, H., Speziale, S., Reichmann, H. J., Frost, D. J., & Schilling, F. R. (2009). Single-crystal elasticity of $(Mg_{0.9}Fe_{0.1})O$ to 81 GPa. *Earth and Planetary Science Letters*, *287*, 345–352.

Mattila, A., Pylkkänen, T., Rueff, J.-P., Huotari, S., Vankó, G., Hanfland, M., et al. (2007). Pressure induced magnetic transition in siderite $FeCO_3$ studied by x-ray emission spectroscopy. *J. Phys.: Condens. Matter*, *19*, 386206.

Nagai, T., Ishido, T., Seto, Y., Nishio-Hamane, D., Sata, N., & Fujino, K. (2010). Pressure-induced spin transition in $FeCO_3$-siderite studied by X-ray diffraction measurements. *J. Phys.: Conf. Ser.*, *215*, 012002.

Ohnishi, S. (1978). A theory of the pressure-induced high-spin-low-spin transition of transition-metal oxides. *Physics of the Earth and Planetary Interiors*, *17*, 130–139.

Palyanov, Y. N., Bataleva, Y. V., Sokol, A. G., Borzdov, Y. M., Kupriyanov, I. N., Reutsky, V. N., & Sobolev, N. V. (2013). Mantle-slab interaction and redox maechanism of diamond formation. *PNAS*, *110*, 20408–20413.

Panning, M., & Romanowicz, B. (2004). Inferences on flow at the base of Earth's mantle based on seismic anisotropy. *Science*, *303*, 351–353.

Rohrbach, A., & Schmidt, M. W. (2011). Redox freezing and melting in the Earth's deep mantle resulting from carbon-iron redox coupling. *Nature*, *472*, 209–212.

Shannon, R. D. (1976). Revised effective ionic radii and systematic studies of interatomic distances in halides and chalcogenides. *Acta Crystallographica Section A: Crystal Physics, Diffraction, Theoretical and General Crystallography*, *32*, 751–767.

Shatskiy, A., Borzdov, Y. M., Litasov, K. D., Kupriyanov, I. N., Ohtani, E., & Palyanov, Y. N. (2014). Phase relations in the system $FeCO_3$-$CaCO_3$ at 6 GPa and 900–1700 °C and its relation to the system $CaCO_3$-$FeCO_3$-$MgCO_3$. *American Mineralogist*, *99*, 773–785.

Sherman, D. M. (1988). High-spin to low-spin transition of iron(II) oxides at high pressures: Possible effects on the physics and chemistry of the lower mantle. In S. Ghose, J.M.D. Coey, & E. Salje (Eds.), *Structural and magnetic phase transitions in minerals: Advances in physical geochemistry* (pp. 113–128). New York: Springer.

Shi, H., Luo, W., Johansson, B., & Ahuja, R. (2008). First-principles calculations of the electronic structure and pressure-induced magnetic transition in siderite $FeCO_3$. *Phys. Rev. B*, *78*, 155119.

Spivak, A., Solopova, N., Cerantola, V., Bykova, E., Zakharchenko, E., Dubrovinsky, L., & Litvin, Y. (2014). Raman study of $MgCO_3$–$FeCO_3$ carbonate solid solution at high pressures up to 55 GPa. *Phys Chem Minerals*, *41*, 633–638.

Stekiel, M., Nguyen-Thanh, T., Chariton, S., McCammon, C., Bosak, A., & Morgenroth, W. (2017). High pressure elasticity of $FeCO_3$-$MgCO_3$ carbonates. *Physics of the Earth and Planetary Interiors*, *271*, 57–63.

Tao, R., Fei, Y., & Zhang, L. (2013). Experimental determination of siderite stability at high pressure. *American Mineralogist*, *98*, 1565–1572.

Taran, M. N., Müller, J., Friedrich, A., & Koch-Müller, M. (2017). High-pressure optical spectroscopy study of natural siderite. *Phys Chem Minerals*, *44*, 537–546.

Tsuchiya, T., Wentzcovitch, R. M., da Silva, C.R.S., & de Gironcoli, S. (2006). Spin transition in magnesiowüstite in Earth's lower mantle. *Phys. Rev. Lett.* *96*, 198501.

Ulrich, C. A, (2015). *Deformation mechanisms and microstructures of experimentally deformed magnesite.* Master's thesis, Texas A & M University.

Weis, C., Sternemann, C., Cerantola, V., Sahle, C. J., Spiekermann, G., Harder, M., et al. (2017). Pressure driven spin transition in siderite and magnesiosiderite single crystals. *Scientific Reports*, *7*, 16526.

Wentzcovitch, R. M., Justo, J. F., Wu, Z., da Silva, C.R.S., Yuen, D. A., & Kohlstedt, D. (2009). Anomalous compressibility of ferropericlase throughout the iron spin cross-over. *P Natl Acad Sci USA*, *106*, 8447–8452.

Wookey, J., & Kendall, J.-M. (2004). Evidence of midmantle anisotropy from shear wave splitting and the influence of shear-coupled P waves. *Journal of Geophysical Research: Solid Earth, 109*.

Wu, Z., Justo, J. F., & Wentzcovitch, R M. (2013). Elastic anomalies in a spin-crossover system: ferropericlase at lower mantle conditions. *Phys. Rev. Lett.*, *110*, 228501.

Wu, Z., & Wentzcovitch, R. M. (2014). Spin crossover in ferropericlase and velocity heterogeneities in the lower mantle. *P Natl Acad Sci USA*, *111*, 10468–10472.

Yang, J., Mao, Z., Lin, J.-F., & Prakapenka, V. B. (2014). Single-crystal elasticity of the deep-mantle magnesite at high pressure and temperature. *Earth and Planetary Science Letters*, *392*, 292–299.

Yang, J., Tong, X. Y., Lin, J. F., Okuchi, T., & Tomioka, N. (2015). Elasticity of ferropericlase across the spin crossover in the Earth's lower mantle. *Sci Rep-Uk 5*.

13

High-Pressure Na-Ca Carbonates in the Deep Carbon Cycle

Sergey Rashchenko[1,2], Anton Shatskiy[1,2,3], and Konstantin Litasov[1,2,3]

ABSTRACT

Carbonates play a central role in the subduction transport of oxidized carbon from Earth's surface to the deep mantle. The melting of carbonated rocks also leads to the release of carbonate melts, regarded as important oxidizing and metasomatizing agents, from the subducting slab. Although at upper mantle conditions the chemistry of carbonates within the subducting slab is restricted to the thoroughly studied ternary $CaCO_3$-$MgCO_3$-$FeCO_3$ system, recent experimental and natural evidences strongly suggest that at conditions of the mantle transition zone, Na-Ca carbonates become the main host of oxidized carbon and define solidus temperatures and chemistry of deep melts. In this chapter, we discuss transport of carbonates to the mantle transition zone within the subducting slab, evidences of sodium migration from silicates into carbonates under high pressure, and crystal chemistry of currently known high-pressure Na-Ca carbonates.

13.1. CARBONATES IN SUBDUCTION INFLUX OF CARBON

Although deep reservoirs can host carbon in both reduced (diamond, carbides, reduced C-O-H fluid) and oxidized (carbonates, CO_2) forms, the transport of carbon from Earth's surface to mantle depths is indeed dominated by its oxidized form: carbonates.

According to data from the Ocean Drilling Program (ODP), the most carbon-rich part of the subducting lithosphere is its sedimentary layer, which hosts 2–3 wt.% of CO_2, predominantly in the form of carbonates (Plank & Langmuir, 1998; Staudigel, 2014). The latter corresponds to the subduction influx of 8.9×10^{11} mol C/year. Hydrothermal alteration also leads to the accumulation of carbon in basaltic oceanic crusts, up to the average CO_2 content established by the ocean drilling program of 0.2 wt.% (Shilobreeva et al., 2011). Of these, ~80% correspond to carbonates, while the remaining 20% represent hydrocarbons and other organic carbon species. The smaller CO_2 content compared to overlying sedimentary layer is compensated by the ~7 km thickness of basaltic oceanic crust, which provides a carbon flux of 2.9×10^{12} mol C/year (i.e. more than three times higher than that from sediments).

The most controversial reservoir of subducting carbon is serpentinized peridotites localized in the lower part of the oceanic lithosphere. The results of ODP demonstrate that significant amounts of carbonates may appear during hydrothermal serpentinization of peridotite, so that up to 13 vol.% of the resulting serpentinite may be represented by carbonate veins (Kodolányi et al., 2011). An estimated degree of serpentinization of lithospheric peridotites entering the subduction zone, however, varies from ~10% in models considering bending faults-related serpentinization (Faccenda, 2014) to the order of magnitude lower in models that reject the latter phenomenon (Korenaga, 2017). An average carbonate content of related serpentinites also remains unclear. Moreover, a well-known process of CO_2 and carbonate reduction into abiogenic methane and graphite during serpentinization of peridotite (Brovarone et al., 2017, and references therein) raises additional question about the abundance

[1] *Sobolev Institute of Geology and Mineralogy, Novosibirsk, Russia*

[2] *Novosibirsk State University, Novosibirsk, Russia*

[3] *Vereshchagin Institute for High Pressure Physics, Moscow, Russia*

Carbon in Earth's Interior, Geophysical Monograph 249, First Edition. Edited by Craig E. Manning, Jung-Fu Lin, and Wendy L. Mao.

of carbonates in serpentinized peridotites. On the other hand, Debret et al. (2015) recently pointed out that inevitable deserpentinization of subducted peridotites enriched in Fe^{3+}, will lead to the release of oxidizing fluids (between FMQ + 1 and FMQ + 4), definitely incompatible with any hosts of reduced carbon in the subducting slab.

Even if abiogenic methane from the serpentinized peridotites and organic compounds from the overlying basaltic oceanic crust and sedimentary layer represent a significant part of subduction carbon influx, they readily leave the slab during devolatilization. In contrast, stabilization of aragonite, dolomite, and magnesite inhibits the carbon escape (Kerrick & Connolly, 2001; Litasov, 2011), making carbonates the main reservoir transporting carbon from Earth's surface to mantle depth.

13.2. SUBDUCTED CARBONATES IN THE UPPER MANTLE

The mineralogy of carbonates entering subduction zones within subducting slabs is represented by calcite, biogenic aragonite, dolomite, magnesite (occurring in serpentinites), and sedimentary Mg-calcite, often with admixture of Fe^{2+}. During subduction-related high-pressure metamorphism, calcite transforms into aragonite, and dolomite decomposes into aragonite + magnesite (Luth, 2001; Sato & Katsura, 2001), so the latter two become the main carbon reservoir in the slab sinking through the upper mantle. The fate of carbon in the slab depends on the position of subduction geotherm in relation to the solidus of carbonated pelite (for carbonates delivered within sedimentary layer) or eclogite (for those from basaltic oceanic crust).

The exact composition of subducting eclogite and pelite, however, is poorly constrained, and a range of solidi is reported for corresponding systems depending on composition and H_2O content (see reviews: Hammouda & Keshav, 2015; Litasov & Shatskiy, 2018; Stagno, 2019, and references therein). According to Grassi and Schmidt (2011), warm subduction geotherms cross the solidus of carbonated pelite between 6 and 9 GPa depending on H_2O content (~190–270 km), where decarbonation melting produces potassium-rich carbonate melts (Table 13.1). In the case of colder geotherms (especially in dry conditions), aragonite and magnesite can escape melting, being transported further to the mantle transition zone (Grassi & Schmidt, 2011). The solidus of carbonated eclogite is a few hundred degrees higher, so the hot subduction geotherm crosses it only near 13 GPa (~390 km) with production of calcium-rich carbonate melts (Table 13.1), whereas average and cold geotherms allow the transport of magnesite directly to the mantle transition zone (Thomson et al., 2016).

13.3. SUBDUCTED CARBONATES IN THE MANTLE TRANSITION ZONE

As we mentioned above, subducted carbonates in the form of aragonite and magnesite can be transported to the mantle transition zone within carbonated pelites along the cold subduction geotherm, and within carbonated eclogites along average and cold ones. The latter agrees with the data of Kiseeva et al. (2018), who observed an anomalous Fe^{3+} enrichment in garnets originated from the mantle transition zone, and suggested that corresponding oxidizing agents are carbonates delivered to the mantle transition zone by subduction. The oxidizing nature of subducted carbonates results from the fact that carbonate ↔ carbon equilibria buffer oxygen fugacity several log units higher than typical for deep mantle values close to iron ↔ wustite buffer (Stagno, 2019; Stagno et al., 2015).

Although $CaCO_3$ and $MgCO_3$ polymorphs stable at lower mantle pressures were extensively studied in the last decade (Gavryushkin et al., 2017; Merlini et al., 2017; Oganov et al., 2008; Pickard & Needs, 2015; etc.), a widespread treatment of such phases as lower mantle carbon hosts should be justified by investigation of whether the transport of carbonates to the lower mantle via mantle transition zone is possible in real lithologies. Related data were reported by Grassi and Schmidt (2011) for carbonated pelite, Litasov et al. (2013) for alkali-bearing carbonatite compositions, and Thomson et al. (2016) for carbonated eclogite, who observed two important phenomena: (1) at pressures of the mantle transition zone, the solidus of carbonated rocks dramatically drops over a narrow pressure interval by 100–200 °C, and (2) at the same pressures, new Na-Ca-carbonate substitute aragonite and magnesite in the subsolidus assemblage.

According to Grassi and Schmidt (2011), carbonated pelite, delivered to the mantle transition zone along colder subduction geotherms, will melt near 20 GPa (~570 km) during the heating of deflected slab towards mantle adiabat. The resulting carbonatite melt, however, drastically differs from that produced during the melting of carbonated pelite along warm subduction geotherms in the upper mantle (Table 13.1). Whereas potassium behaves as an incompatible element during melting at upper mantle pressures, giving 24 wt.% K_2O in the K-Ca carbonate melt, it completely switches behavior in the mantle transition zone, so that corresponding carbonate melt possesses 3 wt.% of K_2O and is not significantly enriched in potassium compared with the bulk rock (Table 13.1). In contrast, sodium, being perfectly compatible during melting at upper mantle pressure (just 1 wt.% Na_2O in the melt, Table 13.1), becomes highly incompatible as soon as melting occurs in the mantle transition zone, so that the Na_2O content of the corresponding Na-Ca carbonate melt reaches 23 wt.%. (Table 13.1). A similar phenomenon

Table 13.1 Compositions of near-solidus carbonate melts derived from dry carbonated pelite (Grassi & Schmidt, 2011) and eclogite (Thomson et al., 2016) in the upper mantle (UM) and mantle transition zone (MTZ), and subsolidus Na-Ca-carbonates stable in the MTZ.

	Carbonated Pelite				Carbonated Eclogite			
	Starting Material	UM Melt	MTZ Melt	Subsolidus Carbonate*	Starting Material	UM Melt	MTZ Melt	Subsolidus Carbonate*
wt. %		8 GPa 1100 °C	22 GPa 1500 °C	22 GPa 1400 °C		13.1 GPa 1450 °C	20.7 GPa 1200 °C	20.7 GPa 1100 °C
CO_2	4.5	43.67	43.12	41.91	2.52	41.88	45.78	43.55
SiO_2	54.63	0.26	0.05	–	50.35	0.36	1.10	–
CaO	5.88	16.3	15.3	34.84	10.80	33.61	26.03	38.39
MgO	2.92	2.95	7.36	5.72	7.15	6.67	6.66	3.54
FeO	4.86	7.55	7.32	6.15	11.35	9.50	8.25	1.61
Na_2O	3.20	0.95	22.7	11.04	2.48	4.63	10.04	12.10
K_2O	2.21	24.4	3.27	0.33	0.06	0.76	0.39	0.56
CaO/Na_2O	1.84	17.16	0.57	3.16	4.35	7.26	2.59	3.17
K_2O/Na_2O	0.69	25.68	0.14	0.03	0.02	0.16	0.04	0.05

* Coexisting with magnesite.

was reported by Thomson et al. (2016) for carbonated eclogite that also melts near 20 GPa (~570 km) being subducted along an average subduction geotherm: the Na_2O content of the mantle transition zone Ca-Na carbonate melt (10 wt.%) is twice than that of the melt produced in the upper mantle.

The reason for the change in alkali speciation is the appearance of a new Na-Ca carbonate mineral that is stable in the subsolidus assemblages (Table 13.1), which (1) effectively extracts Na_2O from silicates into fusible carbonate phase, (2) drastically lowers the solidus temperatures, and (3) provides specific Ca-Na carbonate melts when the solidus is crossed by a geotherm. The idealized formula of this Na-Ca carbonate is $(Na,K)_2(Ca,Mg,Fe)_4(CO_3)_5$, with pronounced dominance of Na and Ca cations (Table 13.1). The recent findings of Na-Ca carbonates in inclusions in "superdeep" Juina diamonds (Kaminsky, 2017) that allowed to distinguish "deep natrocarbonatite association" strongly support the experimental data on the existence of Na-Ca-carbonate and related melts in the mantle transition zone.

The reported Na-Ca carbonate formation is closely related to the strong enrichment of subducting pyroxene in $NaAlSi_2O_6$ end-member due to preferential dissolution of sodium-poor end-members into coexisting majoritic garnet, so that above 13 GPa, pyroxene in both carbonated pelite and eclogite is represented by nearly pure jadeite (Grassi & Schmidt, 2011; Thomson et al., 2016). In carbonate-free lithologies, the further pressure increase would lead to the dissolution of jadeite in majoritic garnet in the form of Na-majorite end-member. In contrast, in carbonated lithologies above 13 GPa (~390 km), jadeite reacts with Ca-Mg carbonates, producing Na-Ca carbonate.

The corresponding reaction in carbonated pelite takes place between 13 and 16 GPa (Grassi & Schmidt, 2011) and may be written for end-members as follows:

$$15CaCO_3\,(\text{aragonite}) + 6NaAlSi_2O_6\,(\text{jadeite}) \rightarrow 3Na_2Ca_4(CO_3)_5 + Ca_3Al_2Si_3O_{12}\,(\text{Ca-garnet}) + 2Al_2O_3\,(\text{corundum}) + 9SiO_2\,(\text{stishovite}). \tag{13.1}$$

In the case of carbonated eclogite, where magnesite is the only carbonate entering the mantle transition zone, the corresponding reaction, taking place between 15 and 20 GPa (Thomson et al., 2016) should be written in a different way:

$$15MgCO_3\,(\text{magnesite}) + 6NaAlSi_2O_6\,(\text{jadeite}) + 4Ca_3Al_2Si_3O_{12}\,(\text{Ca-garnet}) \rightarrow 3Na_2Ca_4(CO_3)_5 + 5Mg_3Al_2Si_3O_{12}\,(\text{Mg-garnet}) + 2Al_2O_3\,(\text{corundum}) + 9SiO_2\,(\text{stishovite}). \tag{13.2}$$

The latter reaction demonstrates an interesting ability of carbonates to extract not only sodium but also calcium from coexisting silicates at pressures of the mantle transition zone. Partitioning of calcium into Na-Ca carbonate also leads to the disappearance of aragonite from the subsolidus assemblage of carbonated pelite (and, in general, any Na-containing silicate lithology). The latter suggests that the only reliable solid-state hosts of oxidized carbon in the lower part of the mantle transition zone are Na-Ca carbonate and/or magnesite. Although the data of Grassi and Schmidt (2011) and Thomson et al. (2016) suggest that cold subduction geotherms pass well below the solidi of carbonated pelite and eclogite down to 660 km discontinuity, the possibility of carbonates transport from the mantle transition zone to the lower mantle requires additional studies.

13.4. CRYSTAL CHEMISTRY OF HIGH-PRESSURE NA-CA CARBONATES

In contrast to carbonates of calcium, magnesium, and iron, extensively studied during the last decade up to megabar pressures both theoretically and experimentally, the high-pressure chemistry of Na-Ca carbonates that hosts oxidized carbon in the mantle transition zone remains terra incognita.

The first systematic study of the Na_2CO_3-$CaCO_3$ system was done by Cooper et al. (1975) who reported its phase diagram at 1 kbar with two orthorhombic intermediate compounds: $Na_2Ca(CO_3)_2$ and $Na_2Ca_2(CO_3)_3$. Both of these compounds exist in natural forms, known as nyerereite and shortite minerals, respectively, although natural nyerereite always contains certain admixture of potassium, causing incommensurate modulation of its structure (Bolotina et al., 2017). Recently Song et al. (2017) reported hydrothermal synthesis at 220 °C of another intermediate compound, $Na_6Ca_5(CO_3)_8$, probably stable in the low-temperature field of Na_2CO_3-$CaCO_3$ phase diagram.

Room-temperature high-pressure behavior of nyerereite and shortite was recently studied by in situ Raman spectroscopy and X-ray diffraction (Borodina et al., 2018; Rashchenko, Goryainov, et al., 2017; Vennari et al., 2018). The equilibrium high-pressure investigation of the Na_2CO_3-$CaCO_3$ system was initiated by studies of its phase diagram at 3 and 6 GPa (Podborodnikov et al., 2018; Shatskiy et al., 2013, 2015; Litasov et al., this volume). The chemistry of intermediate Na-Ca carbonates turned out to be very sensitive to pressure (Figure 13.1): upon pressure increase from 0.1 to 3 GPa, the $Na_2Ca_2(CO_3)_3$ phase becomes unstable, while a new high-pressure $Na_2Ca_3(CO_3)_4$ phase appears on the phase diagram. A further pressure increase to 6 GPa leads to even more drastic changes: disappearance of $Na_2Ca(CO_3)_2$

and emergence of two new high-pressure phases, $Na_4Ca(CO_3)_3$ and $Na_2Ca_4(CO_3)_5$ on the phase diagram.

$Na_2Ca_3(CO_3)_4$. The crystal structure of $Na_2Ca_3(CO_3)_4$, stable at least between 3 and 6 GPa, was reported by Gavryushkin et al. (2014). The low symmetry (monoclinic) and long unit cell (a = 31.4421 Å, Table 13.2) of the compound are compensated by its strong pseudosymmetry, which allows reliable description of the structure using an orthorhombic cell with halved a parameter. Since the common approach of using cation-centered anion polyhedra could not provide a reliable structure description, Gavryushkin et al. proposed an alternative description based on stacking of cationic nets perpendicular to the a axis, with $(CO_3)^{2-}$ anions filling large cationic voids between adjacent nets. In the structure of $Na_2Ca_3(CO_3)_4$, cationic nets form two types of anion-filled layers, denoted A and B in Figure 13.2. In the A-type layers, $(CO_3)^{2-}$ anions are parallel to the a axis, whereas in the B-type layers they are oblique. The cationic voids hosting $(CO_3)^{2-}$ anions have the same geometry in both types of layers, namely a two-capped trigonal prism (Figure 13.3a). The Raman spectrum of the $Na_2Ca_3(CO_3)_4$ (Figures 13.4, 13.5c) is characterized by splitting of ν_4, ν_1, and $2\nu_2$ $(CO_3)^{2-}$ bands into up to three components due to the presence of several symmetrically independent carbonate anions in the structure.

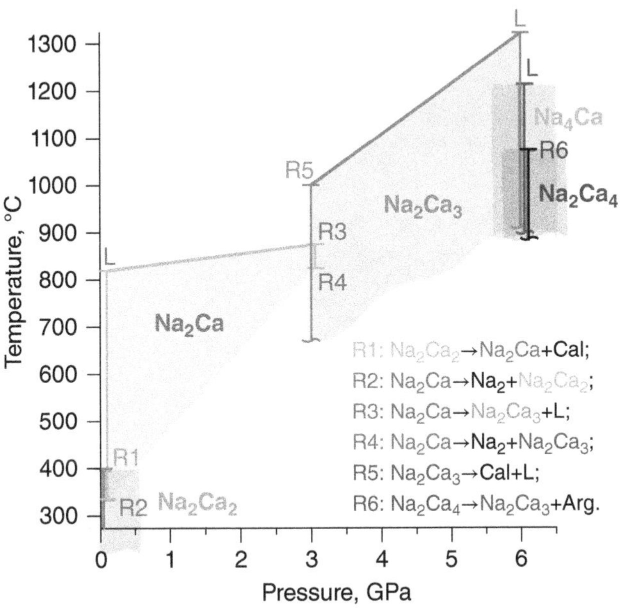

Figure 13.1 Variety of Na-Ca double carbonates established in the system Na_2CO_3–$CaCO_3$ at 0.1 GPa (Cooper et al., 1975), at 3 GPa (Podborodnikov et al., 2018), and at 6 GPa (Shatskiy et al., 2013). Na_2 = Na_2CO_3, Na_4Ca = $Na_4Ca(CO_3)_3$, Na_2Ca = $Na_2Ca(CO_3)_2$, Na_2Ca_2 = $Na_2Ca_2(CO_3)_3$, Na_2Ca_3 = $Na_2Ca_3(CO_3)_4$, Na_2Ca_4 = $Na_2Ca_4(CO_3)_5$, Arg = aragonite, Cal = calcite, L = carbonate melt. See electronic version for color representation of the figures in this book.

$Na_4Ca(CO_3)_3$. The crystal structure of $Na_4Ca(CO_3)_3$, stable at least at 6 GPa, was reported by Rashchenko et al. (2018). Rather unusual for carbonates, it has a cubic *Ia–3d* space group and can be described in a very elegant way using the same approach of cationic sublattice with anion-filled cavities. The cationic sublattice is represented by primitive cubic lattice with period of 0.25·a, where 25% of cubic voids are centered by Na^+ anions, and the remaining 75% are filled by $(CO_3)^{2-}$ anions (Figure 13.2). The resulting coordination of $(CO_3)^{2-}$ anion is a two-capped cube (Figure 13.3b). Although there is only one $(CO_3)^{2-}$ site in the structure, the Raman spectrum of the $Na_4Ca(CO_3)_3$

Table 13.2 Known Na-Ca carbonates (high-pressure phases are given in bold).

Formula and Synthesis Conditions	Space Group	Unit Cell, Å	Natural Analogue	Reference
$Na_2Ca(CO_3)_2$ 0.1 GPa/450 °C	*P2$_1$ca* *Z* = 8	*a* = 10.0713(5) *b* = 8.7220(2) *c* = 12.2460(4)	*nyerereite* *(Na,K)$_2$Ca(CO$_3$)$_2$*	(Gavryushkin et al., 2016)
$Na_2Ca_2(CO_3)_3$ 220 °C (hydrothermal)	*Amm2* *Z* = 2	*a* = 4.9720(9) *b* = 11.068(3) *c* = 7.1271(14)	*shortite* *Na$_2$Ca$_2$(CO$_3$)$_3$*	(Song et al., 2017)
$Na_6Ca_5(CO_3)_8$ 220 °C (hydrothermal)	*P6$_3$mc* *Z* = 2	*a* = 10.0708(19) *c* = 12.604(4)	–	(Song et al., 2017)
$Na_2Ca_3(CO_3)_4$ 6 GPa/1400 °C	*P1n1* *Z* = 8	*a* = 31.4421(8) *b* = 8.1960(2) *c* = 7.4360(2) β = 89.923(2)	–	(Gavryushkin et al., 2014)
$Na_4Ca(CO_3)_3$ 6 GPa/1200 °C	*Ia–3d* *Z* = 16	*a* = 14.5770(5)	–	(Rashchenko et al., 2018)
$Na_2Ca_4(CO_3)_5$ 6 GPa/1050 °C	*P6$_3$mc* *Z* = 2	*a* = 10.37402(14) *c* = 6.25935(9)	*calcioburbankite* *(Na,Ca,REE)$_3$(Ca, REE,Sr)$_3$(CO$_3$)$_5$*	(Rashchenko, Bakakin, et al., 2017)

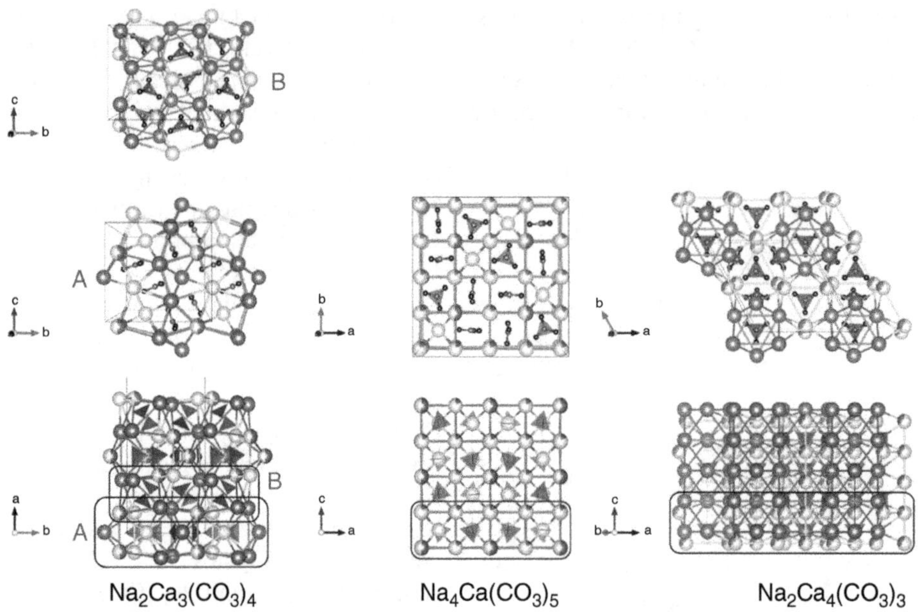

Figure 13.2 Crystal structures of high-pressure Na-Ca carbonates. Cationic layers filled with anions are shown in the top rows; overall view of structures with outlined layers is in the bottom one. Na^+ and Ca^{2+} cations are shown as yellow and blue, respectively; combined colors correspond to mixed sites. The drawing was prepared using VESTA software (Momma & Izumi, 2011). See electronic version for color representation of the figures in this book.

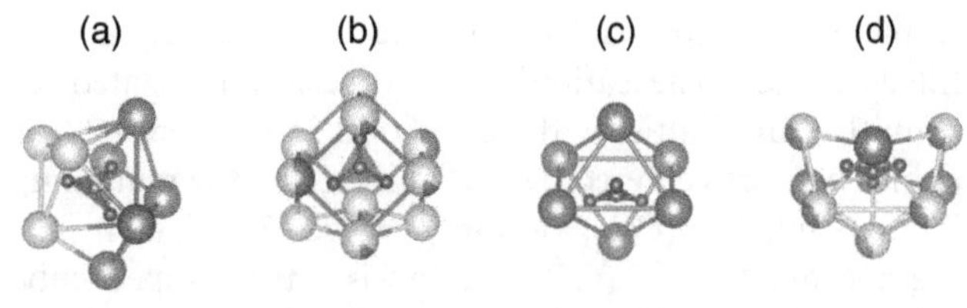

Figure 13.3 $(CO_3)^{2-}$ filled voids in cationic sublattice of high-pressure Na-Ca carbonates: (a) two-capped trigonal prism ($Na_2Ca_3(CO_3)_4$ and $Na_2Ca_4(CO_3)_5$), (b) two-capped cube ($Na_4Ca(CO_3)_5$), (c) octahedron ($Na_2Ca_4(CO_3)_5$), (d) three-capped trigonal prism ($Na_2Ca_4(CO_3)_5$). Na^+ and Ca^{2+} cations are shown as yellow and blue, respectively; combined colors correspond to mixed sites. See electronic version for color representation of the figures in this book.

exhibits clear splitting of ν_4, ν_1, and $2\nu_2$ $(CO_3)^{2-}$ bands (Figures 13.4, 13.5b), which can be explained by the lower (C_{2v}) local symmetry of the $(CO_3)^{2-}$ site.

$Na_2Ca_4(CO_3)_5$. The most interesting high-pressure Na-Ca carbonate is hexagonal ($P6_3mc$) $Na_2Ca_4(CO_3)_5$, synthesized at 6 GPa and strongly resembling the $(Na,K)_2(Ca,Mg,Fe)_4(CO_3)_5$ stoichiometry stable in carbonated pelite and eclogite under mantle transition zone conditions (Litasov et al., 2013; Thomson et al., 2016) (Table 13.1). Its crystal structure, reported by Rashchenko, Bakakin, et al., (2017), is also based on cationic nets perpendicular to the *c* axis (Figure 13.2), and contains three types of voids: two-capped trigonal prisms, also typical for $Na_2Ca_3(CO_3)_4$ (Figure 13.3a), octahedra (Figure 13.3c), and three-capped trigonal prisms (Figure 13.3d). The abundance of differently coordinated carbonate anions in the structure leads to the pronounced splitting of ν_4, ν_1, and $2\nu_2$ $(CO_3)^{2-}$ Raman bands into up to four components (Figures 13.4, 13.5a).

Importantly, among all reported high-pressure Na-Ca carbonates, the $Na_2Ca_4(CO_3)_5$ is the only one that has a natural analogue, *calcioburbankite* mineral, $(Na,Ca,REE)_3(Ca,REE,Sr)_3(CO_3)_5$ (Belovitskaya & Pekov, 2004). The apparent difference in formulae disappears if one rewrites the chemical $Na_2Ca_4(CO_3)_5$ formula into a structural $^{M1}(Na_2Ca)^{M2}Ca_3(CO_3)_5$ one, considering that the M1 cation site is occupied by ⅔Na^+ and ⅓Ca^{2+}, whereas M2 one is occupied solely by Ca^{2+} (Figure 13.2). A heterovalent $2Ca^{2+} \leftrightarrow Na^+ + REE^{3+}$ substitution herewith leads to *REE*-rich compositions, corresponding to natural calcioburbankites. Such a substitution (as well as incorporation of large Sr^{2+} and Ba^{2+} cations) very efficiently stabilizes the resulting structures at moderate pressures, so that natural calcioburbankites occur in carbonatites and alkali-rich pegmatites far from gigapascal pressures. A possibility of $Na_2Ca_4(CO_3)_5$ structure to incorporate large cations may have very important geochemical implications in the case of existence of this phase in the mantle transition zone. In contrast to $CaCO_3$ and $MgCO_3$ with limited capability to host REE and LILE, the appearance of the Na-Ca carbonate in equilibria with silicates will strongly deplete them in REE and LILE, leading to new patterns of trace element partitioning in the Earth's mantle.

An important crystal-chemical feature that unites all these high-pressure Na-Ca carbonates is their relation to

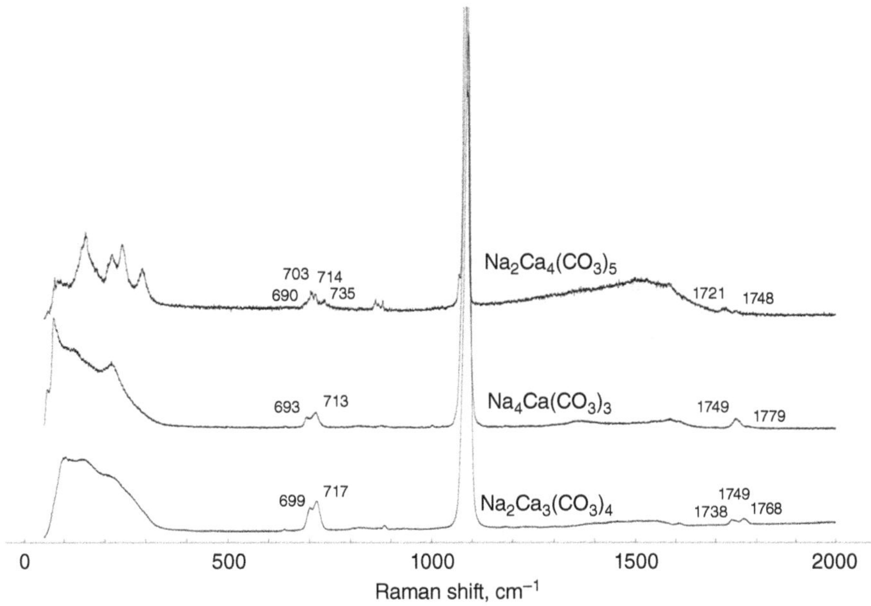

Figure 13.4 Raman spectra of high-pressure Na-Ca carbonates (deconvolutions of the main ν_1 $(CO_3)^{2-}$ band into individual components are shown in Figure 13.5). See electronic version for color representation of the figures in this book.

isostructural *borates*. Although crystal structures of ambient pressure Na-Ca carbonates have nothing in common with the latter class of compounds, their high-pressure counterparts adopt exactly the same crystal structures that were described years ago for borate materials (Table 13.3). Another crystal-chemical feature of high-pressure Na-Ca carbonates is a presence of mixed Na-Ca sites in their crystal structures (Figure 13.2), unusual for nyerereite and shortite.

Besides the high-pressure Na-Ca carbonates discussed above, we should mention the following:

1. In two subsolidus experiments (at 22 GPa/1350 °C and 23.5 GPa/1400 °C) Grassi and Schmidt (2011) also reported the existence of nearly pure $Na_2Ca_2(CO_3)_3$ phase together with $(Na,K)_2(Ca,Mg,Fe)_4(CO_3)_5$ discussed above, in the carbonated pelite.

2. In a subsolidus experiment at 21 GPa/1200 °C, Kiseeva et al. (2013) reported $(Na,K)_2(Ca,Mg,Fe)(CO_3)_2$ phase coexisting with "Na-aragonite" (Na-Ca carbonate with Na_2O up to 8 wt.%) in a carbonated eclogite. A similar assemblage was reported by Litasov et al. (2013) for subsolidus conditions of Na-bearing carbonatite composition at 15.5 GPa and 1100 °C. Unfortunately, only the chemical composition was reported for Na-Ca carbonates in all listed studies.

Although at ambient pressure $Na_2Ca_2(CO_3)_3$ and $Na_2Ca(CO_3)_2$ stoichiometries correspond to shortite and nyerereite, which decompose below 3 and 6 GPa, respectively, one cannot exclude that high-pressure polymorphs of these minerals may exist at pressures of the mantle transition zone. "Na-aragonite" may represent an unknown Ca-rich high-pressure Na-Ca carbonate, because no evidence of sodium incorporation into aragonite was reported by Shatskiy et al. (2013) or Podborodnikov et al. (2018), at least up to 6 GPa. On the other hand, taking into account a known problem of Na_2O underestimation in carbonates by WDS and EDS (Shatskiy et al., 2013), one can suggest that real Na_2O content of "Na-aragonite" may exceed 10 wt.%, actually corresponding to the discussed $(Na,K)_2(Ca,Mg,Fe)_4(CO_3)_5$ phase.

13.5. CONCLUSIONS

High-pressure experiments in the carbonate-silicate systems replicating eclogite (Kiseeva et al., 2013; Thomson et al., 2016), pelite (Grassi & Schmidt, 2011), and model carbonatite (Litasov et al., 2013) revealed a number of Na-Ca carbonates resembling stoichiometries of $Na_2Ca(CO_3)_2$ nyerereite, $Na_2Ca_2(CO_3)_3$ shortite, and $Na_2Ca_4(CO_3)_5$ burbankite. Stabilization of these carbonates within oceanic crust subducted to the mantle transition zone can decrease drastically the solidus temperatures and cause formation of sodic dolomitic carbonatite melts.

Considering the phase relations in the Na_2CO_3–$CaCO_3$ system (Cooper et al., 1975; Podborodnikov et al., 2018; Shatskiy et al., 2013, 2015), a range of Na-Ca double carbonates changes in the following sequence upon pressure and temperature increase: $Na_2Ca(CO_3)_2$, $Na_2Ca_2(CO_3)_3$, (0.1 GPa) → $Na_2Ca(CO_3)_2$, $Na_2Ca_3(CO_3)_4$ (3 GPa) → $Na_4Ca(CO_3)_3$, $Na_2Ca_3(CO_3)_4$,

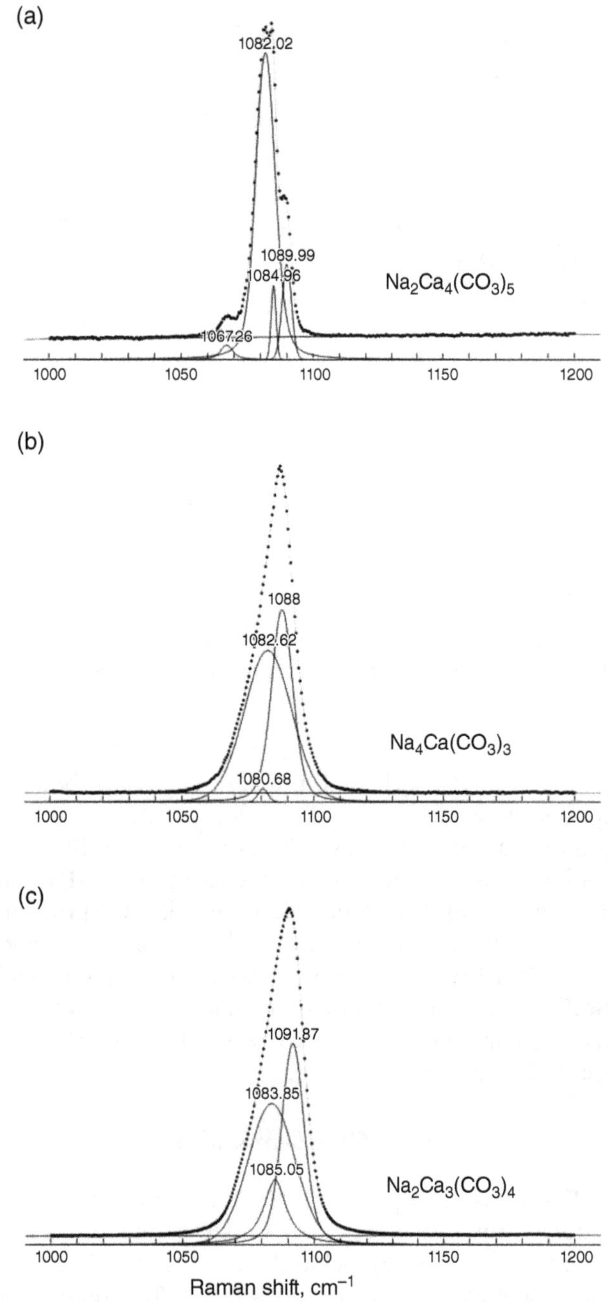

Figure 13.5 Deconvolutions of the main ν_1 $(CO_3)^{2-}$ band into individual components for the Raman spectra of high-pressure Na-Ca carbonates. See electronic version for color representation of the figures in this book.

$Na_2Ca_4(CO_3)_5$ (6 GPa). The Na-Ca carbonates recovered from experiments at 0.1, 3, and 6 GPa were characterized by Raman spectroscopy and their crystal structures were solved: $Na_4Ca(CO_3)_3$ (*Ia*−3*d*) (Rashchenko et al., 2018), $Na_2Ca(CO_3)_2$ ($P2_1ca$) nyerereite (Gavryushkin et al., 2016), $Na_2Ca_3(CO_3)_4$ (P1n1) (Gavryushkin et al., 2014; Shatskiy et al., 2015), $Na_2Ca_4(CO_3)_5$ ($P6_3mc$) burbankite (Rashchenko, Bakakin, et al., 2017).

Table 13.3 Borate analogs of high-pressure Na-Ca carbonates.

High-Pressure Na-Ca Carbonates	Space Group	Ambient-Pressure Borates	Reference
$Na_2Ca_3(CO_3)_4$ a = 7.4360 Å b = 15.72 Å c = 8.1960 Å	*Pnma**	$Nd_2Ba_3(BO_3)_4$ a = 7.7143(11) Å b = 16.7790(34) Å c = 8.9478(14) Å	(Yan & Hong, 1987)
$Na_4Ca(CO_3)_3$ a = 14.5770(5) Å	*Ia*–3*d*	$NaBa_4(BO_3)_3$ a = 15.783(4) Å	(Kokh et al., 2004)
$Na_2Ca_4(CO_3)_5$ a = 10.37402(14) Å c = 6.25935(9) Å	$P6_3mc$	$Ca_3La_3(BO_3)_5$ a = 10.530(3) Å c = 6.398(2) Å	(Zhou & Ye, 2008)

* Pseudosymmetry for $Na_2Ca_3(CO_3)_4$.

Although in situ high-pressure studies of Na-Ca carbonates have just begun (Borodina et al., 2018; Vennari et al., 2018), they already suggest an existence of unquenchable high-pressure polymorphs of $Na_2Ca_2(CO_3)_3$ and probably $Na_2Ca(CO_3)_2$ at pressures exceeding 15 GPa. The compounds with similar stoichiometries were also observed in quench experiments. Grassi and Schmidt (2011) detected carbonate resembling shortite composition at 22 and 23.5 GPa in carbonated pelite, whereas Kiseeva et al. (2013) found carbonate resembling nyerereite stoichiometry at 21 GPa in carbonated eclogite. Both compounds were observed in subsolidus assemblages of model Na-bearing carbonatite (Litasov et al., 2013).

ACKNOWLEDGMENTS

The work was supported by the Russian Science Foundation (No 14-17-00609-P) and performed under the Deep Carbon Observatory program.

REFERENCES

Belovitskaya, Y. V., & Pekov, I. V. (2004). Genetic mineralogy of the burbankite group. *New Data on Minerals*, *39*, 50–64.

Bolotina, N. B., Gavryushkin, P. N., Korsakov, A. V., Rashchenko, S. V., Seryotkin, Y. V., Golovin, A. V., et al. (2017). Incommensurately modulated twin structure of nyerereite $Na_{1.64}K_{0.36}Ca(CO_3)_2$. *Acta Crystallographica B*, *73*, 276–284.

Borodina, U., Likhacheva, A., Golovin, A., Goryainov, S., Rashchenko, S., & Korsakov, A. (2018). Raman spectra of shortite $Na_2Ca_2(CO_3)_3$ compressed up to 8 GPa. *High Pressure Research*, *38*, 293–302.

Brovarone, A. V., Martinez, I., Elmaleh, A., Compagnoni, R., Chaduteau, C., Ferraris, C., & Esteve, I. (2017). Massive production of abiotic methane during subduction evidenced in metamorphosed ophicarbonates from the Italian Alps. *Nature Communications*, *8*, 14134.

Cooper, A. F., Gittins, J., & Tuttle, O. F. (1975). The system Na_2CO_3-K_2CO_3-$CaCO_3$ at 1 kilobar and its significance in carbonatite petrogenesis. *American Journal of Science*, *275*, 534–560.

Debret, B., Bolfan-Casanova, N., Padrón-Navarta, J. A., Martin-Hernandez, F., Andreani, M., Garrido, C. J., et al. (2015). Redox state of iron during high-pressure serpentinite dehydration. *Contributions to Mineralogy and Petrology*, *169*, 313.

Faccenda, M. (2014). Water in the slab: A trilogy. *Tectonophysics*, *614*, 1–30.

Gavryushkin, P. N., Bakakin, V. V., Bolotina, N. B., Shatskiy, A. F., Seryotkin, Y. V., & Litasov, K. D. (2014). Synthesis and crystal structure of new carbonate $Ca_3Na_2(CO_3)_4$ homeotypic with orthoborates $M_3Ln_2(BO_3)_4$ (M = Ca, Sr, and Ba). *Crystal Growth & Design*, *14*, 4610–4616.

Gavryushkin, P. N., Martirosyan, N. S., Inerbaev, T. M., Popov, Z. I., Rashchenko, S. V., Likhacheva, A. Y., et al. (2017). Aragonite-II and $CaCO_3$-VII: New high-pressure, high-temperature polymorphs of $CaCO_3$. *Crystal Growth & Design*, *17*, 6291–6296.

Gavryushkin, P. N., Thomas, V. G., Bolotina, N. B., Bakakin, V. V., Golovin, A. V., Seryotkin, Y. V., et al. (2016). Hydrothermal synthesis and structure solution of $Na_2Ca(CO_3)_2$: "Synthetic analogue" of mineral nyerereite. *Crystal Growth & Design*, *16*, 1893–1902.

Grassi, D., & Schmidt, M. W. (2011). The melting of carbonated pelites from 70 to 700 km depth. *Journal of Petrology*, *52*, 765–789.

Hammouda, T., & Keshav, S. (2015). Melting in the mantle in the presence of carbon: Review of experiments and discussion on the origin of carbonatites. *Chemical Geology*, *418*, 171–188.

Kaminsky, F. V. (2017). *The Earth's lower mantle: Composition and structure*. New York: Springer.

Kerrick, D. M., & Connolly, J.A.D. (2001). Metamorphic devolatilization of subducted oceanic metabasalts: Implications for seismicity, arc magmatism and volatile recycling. *Earth and Planetary Science Letters*, *189*, 19–29.

Kiseeva, E. S., Litasov, K. D., Yaxley, G. M., Ohtani, E., & Kamenetsky, V. S. (2013). Melting and phase relations of carbonated eclogite at 9–21 GPa and the petrogenesis of alkali-rich melts in the deep mantle. *Journal of Petrology*, *54*, 1555–1583.

Kiseeva, E. S., Vasiukov, D. M., Wood, B. J., McCammon, C., Stachel, T., Bykov, M., et al. (2018). Oxidized iron in garnets from the mantle transition zone. *Nature Geoscience*, *11*, 144–147.

Kodolányi, J., Pettke, T., Spandler, C., Kamber, B. S., & Gméling, K. (2011). Geochemistry of ocean floor and fore-arc serpentinites: Constraints on the ultramafic input to subduction zones. *Journal of Petrology*, *53*, 235–270.

Kokh, A. E., Kononova, N. G., Bekker, T. B., Furmanova, N. G., Maximov, B. A., Bolotina, N. B., et al. (2004). New sodium barium orthoborate $NaBa_4(BO_3)_3$. *Russian Journal of Inorganic Chemistry*, *49*, 984–988.

Korenaga, J. (2017). On the extent of mantle hydration caused by plate bending. *Earth and Planetary Science Letters*, *457*, 1–9.

Litasov, K. D. (2011). Physicochemical conditions for melting in the Earth's mantle containing a C–O–H fluid (from experimental data). *Russian Geology and Geophysics*, *51*, 475–492.

Litasov, K. D., & Shatskiy, A. (2018). Carbon-bearing magmas in the Earth's deep interior. In Y. Kono & C. Sanloup (Eds.), *Magmas Under Pressure* (pp. 43–82). Amsterdam: Elsevier.

Litasov, K. D., Shatskiy, A., Ohtani, E., & Yaxley, G. M. (2013). Solidus of alkaline carbonatite in the deep mantle. *Geology*, *41*, 79–82.

Luth, R.W. (2001). Experimental determination of the reaction aragonite plus magnesite = dolomite at 5 to 9 GPa. *Contributions to Mineralogy and Petrology, 141*, 222–232.

Merlini, M., Cerantola, V., Gatta, G. D., Gemmi, M., Hanfland, M., Kupenko, I., et al. (2017). Dolomite-IV: Candidate structure for a carbonate in the Earth's lower mantle. *American Mineralogist*, *102*, 1763–1766.

Momma, K., & Izumi, F. (2011). VESTA 3 for three-dimensional visualization of crystal, volumetric and morphology data. *Journal of Applied Crystallography*, *44*, 1272–1276.

Oganov, A. R., Ono, S., Ma, Y. M., Glass, C. W., & Garcia, A. (2008). Novel high-pressure structures of $MgCO_3$, $CaCO_3$ and CO_2 and their role in Earth's lower mantle. *Earth and Planetary Science Letters*, *273*, 38–47.

Pickard, C. J., & Needs, R. J. (2015). Structures and stability of calcium and magnesium carbonates at mantle pressures. *Physical Review B*, *91*, 104101.

Plank, T., & Langmuir, C. H. (1998). The chemical composition of subducting sediment and its consequences for the crust and mantle. *Chemical Geology*, *145*, 325–394.

Podborodnikov, I. V., Shatskiy, A., Arefiev, A. V., Rashchenko, S. V., Chanyshev, A. D., & K. D. Litasov (2018). The system Na_2CO_3–$CaCO_3$ at 3 GPa. *Physics and Chemistry of Minerals*, *45*, 773–787.

Rashchenko, S. V., Bakakin, V. V., Shatskiy, A. F., Gavryushkin, P. N., Seryotkin, Y. V., & Litasov K. D. (2017). Noncentrosymmetric $Na_2Ca_4(CO_3)_5$ carbonate of "$M1_3M2_3XY_3Z$" structural type and affinity between borate and carbonate structures for design of new optical materials. *Crystal Growth & Design*, *17*, 6079–6084.

Rashchenko, S. V., Goryainov, S. V., Romanenko, A. V., Golovin, A. V., Korsakov, A. V., Moine, B. N., & Mikhno A. O. (2017). High-pressure Raman study of nyerereite from Oldoinyo Lengai. *Journal of Raman Spectroscopy*, *48*, 1438–1442.

Rashchenko, S., Shatskiy, A., Arefiev, A., Seryotkin Y., & Litasov, K. (2018). $Na_4Ca(CO_3)_3$: A novel carbonate analog of borate optical materials, *CrystEngComm*, *20*, 5228–5232,.

Sato, K., & Katsura, T. (2001). Experimental investigation on dolomite dissociation into aragonite+magnesite up to 8.5 GPa. *Earth and Planetary Science Letters*, *184*, 529–534.

Shatskiy, A., Gavryushkin, P. N., Litasov, K. D., Koroleva, O. N., Kupriyanov, I. N., Borzdov, Y. M., et al. (2015). Na-Ca carbonates synthesized under upper-mantle conditions: Raman spectroscopic and X-ray diffraction studies. *European Journal of Mineralogy*, *27*, 175–184.

Shatskiy, A., Sharygin, I. S., Litasov, K. D., Borzdov, Y. M., Palyanov, Y. N., & Ohtani, E. (2013). New experimental data on phase relations for the system Na_2CO_3-$CaCO_3$ at 6 GPa and 900-1400 °C. *American Mineralogist*, *98*, 2164–2171.

Shilobreeva, S., Martinez, I., Busigny, V., Agrinier, P., & Laverne, C. (2011). Insights into C and H storage in the altered oceanic crust: Results from ODP/IODP Hole 1256D. *Geochimica et Cosmochimica Acta*, *75*, 2237–2255.

Song, Y., Luo, M., Zhao, D., Peng, G., Lin, C., & Ye, N. (2017). Explorations of new UV nonlinear optical materials in the Na_2CO_3–$CaCO_3$ system. *Journal of Materials Chemistry C*, *5*, 8758–8764.

Stagno, V. (2019). Carbon, carbides, carbonates and carbonatitic melts in the Earth's interior. *Journal of Geological Society*, doi.org/10.1144/jgs2018-1095 (in press).

Stagno, V., Frost, D., McCammon, C., Mohseni, H., & Fei, Y. (2015). The oxygen fugacity at which graphite or diamond forms from carbonate-bearing melts in eclogitic rocks. *Contributions to Mineralogy and Petrology*, *169*(2), 1–18.

Staudigel, H. (2014). Chemical fluxes from hydrothermal alteration of the oceanic crust. In K. K. Turekian & H. D. Holland (Eds.), *Treatise on Geochemistry* (2nd ed., pp. 583–606) Oxford: Elsevier.

Thomson, A. R., Walter, M. J., Kohn, S. C., & Brooker, R. A. (2016). Slab melting as a barrier to deep carbon subduction. *Nature*, *529*, 76–79.

Vennari, C. E., Beavers, C. M., & Williams, Q. (2018). High-pressure/temperature behavior of the alkali/calcium carbonate shortite ($Na_2Ca_2(CO_3)_3$): Implications for carbon sequestration in Earth's transition zone. *Journal of Geophysical Research: Solid Earth*, *123*, 6574–6591.

Yan, J. F., & Hong, H.Y.-P. (1987). Crystal structure of a new mini-laser material. $Nd_2Ba_3(BO_3)_4$. *Materials Research Bulletin*, *22*, 1347–1353.

Zhou, T., & Ye, N. (2008). Redetermination of tricalcium trilanthanum pentakis (orthoborate) from single-crystal data. *Acta Crystallographica E*, *64*, i37.

14

Phase Diagrams of Carbonate Materials at High Pressures, with Implications for Melting and Carbon Cycling in the Deep Earth

Konstantin Litasov[1,2,3], Anton Shatskiy[1,2,3], Ivan Podborodnikov[1,2,3], and Anton Arefiev[1,2,3]

ABSTRACT

In this chapter, we review phase diagrams of alkali and alkaline earth carbonates at high pressures, particularly simple, binary, and ternary systems, which were recently constrained at pressures of 3 and 6 GPa. These studies revealed a number of new alkali-alkaline earth double carbonates. Major transformations of high-pressure carbonates, including changes in carbon coordination, spin transition, and valence state in Fe-bearing carbonates up to the lower mantle levels, were also discussed. We emphasize the importance of carbonate systems for understanding the low-degree partial melting of carbonated mantle rocks and explaining carbonate inclusions in diamond and other deep-seated minerals. The question of carbonate stability versus the presumably reduced nature of the deep Earth's mantle provides significant impact on the further study of material transport and deep volatile cycle through the history of our planet.

14.1. INTRODUCTION

Carbonates and carbonatite melt can play a key role in the dynamics of the upper mantle and transition zone and may be more important than water or reduced C-O-H compounds (Dasgupta, 2013; Litasov & Shatskiy, 2018). Although total carbon content in the deep Earth is expected to be very low, in the range of hundreds of parts per million, it can be concentrated locally, for example, by the melting of subducted carbonates in the transition zone to produce a significant amount of melt, which can migrate through the mantle towards the lithosphere-asthenosphere boundary. Then, carbonate can segregate below this boundary to form the source region for carbonatite-kimberlite and other alkaline magmas. Even the generation of mid-ocean ridge basalt can be enhanced by carbonate activity at greater depths below ridge axes (Stagno et al., 2013). However, the role of carbonates in the lower mantle is difficult to constrain due to presumably reduced conditions corresponding to iron-wüstite equilibria or even lower fO_2 (Frost & McCammon, 2008) and variable degree of oxidation by subducted slabs penetrating into the lower mantle over time. Overall, carbonates can be involved in all notable mantle processes, including (a) subduction (Dasgupta & Hirschmann, 2010; Plank & Langmuir, 1998; Staudigel, 2014), (b) melting and material transport (Litasov, 2011; Wyllie & Huang, 1976), (c) generation of deep-seated magma (Sharygin et al., 2015; Shatskiy, Litasov, et al., 2017), (d) mantle metasomatism (Agashev et al., 2013; Shu & Brey, 2015; Yaxley & Green, 1996), and (e) diamond formation (Kopylova et al., 2010; Navon et al., 1988).

The origin of carbonates in the deep mantle is difficult to constrain. Subduction of carbonated materials is one of the efficient ways to deliver carbonates as deep as the core-mantle boundary (CMB). However, for Archean-Proterozoic time, the volumes of this delivery could be limited. An important issue is a survival of carbonates in subducted crust and lithosphere through the hydrous melting in the devolatilization region beneath the island

[1] *Sobolev Institute of Geology and Mineralogy, Novosibirsk, Russia*

[2] *Vereshchagin Institute for High Pressure Physics, Moscow, Russia*

[3] *Novosibirsk State University, Novosibirsk,, Russia*

Carbon in Earth's Interior, Geophysical Monograph 249, First Edition. Edited by Craig E. Manning, Jung-Fu Lin, and Wendy L. Mao.

arcs. The existing data indicate almost entire removal of carbonates in the presence of hydrous melt/fluid (Foley et al., 2009; Kelemen & Manning, 2015; Litasov et al., 2011). On the other hand, there are other experimental observations, as well as petrological and numerical modeling data, showing that 20%–80% of subducted carbonates can penetrate beneath the arc devolatilization depths and can be transported to the transition zone or even to the lower mantle (Johnston et al., 2011; Kerrick & Connolly, 2001; Litasov, 2011; Molina & Poli, 2000). Furthermore, the evidences for the deep subduction of carbonates are abundant from the recovered ultra-high-pressure metamorphic rocks worldwide containing diamond and Ca-Mg carbonates (Dobrzhinetskaya et al., 2006).

Carbonates are rare in mantle igneous rocks, xenoliths, and diamonds due to poor solubility of CO_2 in basaltic and alkaline magmas, significant degassing and decarbonation at the subsurface conditions, and postmagmatic alteration. This is particularly applicable for the alkali-bearing carbonates and carbonatites. An example of ultrafast alteration of sodium carbonatites at Oldoinyo Lengai with formation of calcium carbonatites in a matter of days to months (Zaitsev & Keller, 2006) indicates that some of the calcitic and dolomitic carbonatites found in different geological sequences could be initially enriched in Na and K. The scale of carbonate activity in the mantle can be indirectly estimated from the inclusions in diamond and other mantle minerals. Currently, we have various new information about the presence of carbonates in the upper mantle, transition zone, and even the lower mantle. Primary Ca-Mg-carbonates appear in melt pockets in spinel peridotite and eclogite xenoliths (Rudnick et al., 1993; Yaxley et al., 1998) and also in the inclusions in garnet and diamonds (Brenker et al., 2007; Stachel et al., 1998; Zedgenizov et al., 2014).

Recently, various alkali-bearing carbonates were established as inclusions in kimberlite groundmass minerals from Siberia (Golovin et al., 2017). Canada, and Greenland (Kamenetsky et al., 2009, 2013). as well as in minerals from sheared peridotites originated from 180–230 km depths (Golovin et al., 2018; Sharygin et al., 2013). The findings of eitelite- and nyerereite-bearing assemblages in carbonatitic inclusions in diamonds from the Juina (Mato Grosso state, Brazil) (Kaminsky et al., 2016) indicate the possible presence of Na-Ca-Mg carbonatite melts at depths corresponding to the lower mantle. Alkali-bearing fluid and carbonatite melts coexist with peridotite and eclogite minerals in microinclusions in fibrous diamonds, which crystallize prior to the kimberlite eruption (Klein-BenDavid et al., 2009; Logvinova et al., 2011; Navon et al., 1988; Weiss et al., 2009; Zedgenizov et al., 2007, 2011). Similar microinclusions were found along the twinning plane in ancient gem-quality diamonds (Jablon & Navon, 2016). This suggests that alkali-rich carbonate melts have been introduced into the reduced lithospheric mantle since the Archaean and that these melts are responsible for the formation of most diamonds. The important findings of Na-K-Ca-rich carbonatitic inclusions from spinel harzburgites originating from 110–115 km depths indicate that alkali-carbonatite melts can appear in the shallow lithospheric mantle, above the diamond stability field (Giuliani et al., 2012). This is consistent with the interpretation of electrical conductivity anomalies observed in the continental lithospheric mantle 80–120 km beneath the Slave craton (NE Canada) (Jones et al., 2001, 2003) and >120 km beneath the São Francisco craton (SE Brazil) (Pinto et al., 2010) as carbonate-bearing regions based on experimental measurements (Yoshino et al., 2018).

Considering the abundance and importance of carbonates in the shallow Earth's mantle, extensive experimental and theoretical studies have been performed on various carbonate systems at high P-T conditions. The major effect of carbonate on melting of silicate mantle is drastically different from that of hydrous or hydrocarbon fluid and is manifested by the formation of carbonatite magmas, which are stable in equilibrium with peridotite and eclogite in a wide temperature range (Falloon & Green, 1989; Gudfinnsson & Presnall, 2005; Litasov, 2011; Litasov & Shatskiy, 2018; Wyllie & Huang, 1975). Yet Ca-Mg carbonates may be too refractory to initiate melting at geothermal conditions relevant to the continental lithosphere, even in the presence of $FeCO_3$ or H_2O (Irving & Wyllie, 1975; Müller et al., 2017; Shatskiy et al., 2014, 2018). In contrast, minor amounts of alkalis can drastically reduce the solidus temperature of carbonated silicate mantle (Brey et al., 2011; Dasgupta & Hirschmann, 2007a; Grassi & Schmidt, 2011b; Kiseeva et al., 2013; Litasov, 2011; Litasov, Shatskiy, & Ohtani, 2013; Shatskiy, Litasov, & Palyanov, 2015). The decrease in melting temperatures can exceed 400–500 °C relative to alkali-free systems, resulting in a solidus ~100–200 °C below the cratonic geotherm (Litasov, Shatsky, Ohtani, et al., 2013).

Low-degree partial melts in the carbonated peridotite and eclogite systems show strong enrichments in Na_2O and K_2O (Dasgupta & Hirschmann, 2007a; Dasgupta et al., 2004, 2005; Ghosh et al., 2009; Kiseeva et al., 2013; Litasov & Ohtani, 2009, 2010; Sweeney et al., 1995; Wallace & Green, 1988). However, precise determination of the composition of this melt in the natural-like systems with realistic bulk compositions, including very small proportions of CO_2, is problematic due to their trace amounts (Dasgupta & Hirschmann, 2007b; Dasgupta et al., 2004; Sweeney, 1994; Sweeney et al., 1995). In addition, the mass balance calculations for the samples obtained below the apparent solidi often produce clear deficits of alkalis, suggesting the presence of minor alkali-bearing liquid or solid carbonate phases. In contrast, the study of carbonatite or pure carbonate systems allows careful determination of subsolidus carbonate phases, minimum melting temperatures, and composition of incipient melt, which facilitate interpretation of data obtained in the natural-like carbonate-bearing silicate systems (Litasov, Shatsky, & Ohtani, 2013; Shatskiy, Litasov, & Palyanov, 2015).

Recently, we have thoroughly studied the K-Na-Ca-Mg-Fe carbonate systems at 3–6 GPa (see references in section 14.3) and revealed a number of new alkali-alkaline earth double carbonates (Rashchenko et al., this volume). Here, we review available data on the melting, phase transitions, and phase diagrams of the pure carbonate systems and show their importance for understanding the low-degree partial melting of carbonated mantle rocks. We emphasize the importance of study of pure Na_2CO_3 and K_2CO_3 carbonates for geological purposes, since these endmembers are mandatory for the thermodynamic modeling of alkali-alkaline earth carbonate melts and solids.

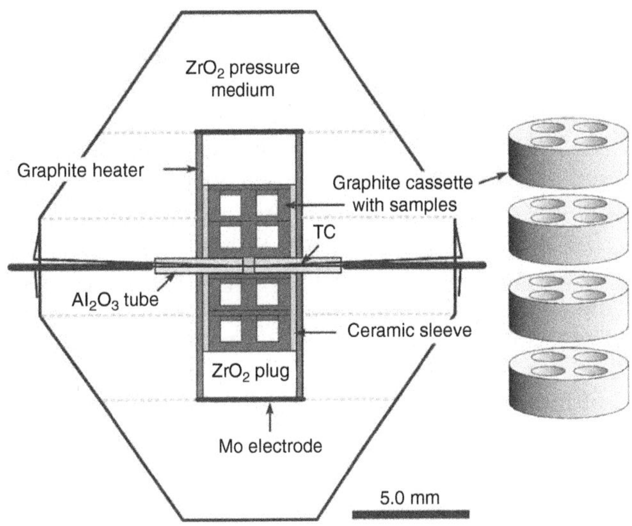

Figure 14.1 High-pressure cell assembly employed to study phase relations in the carbonate systems at 3–6 GPa. The cell contains four graphite cassettes. Each cassette contains four samples. TC – W-$Re_{3\%/25\%}$ thermocouple. See electronic version for color representation of the figures in this book.

14.2. EXPERIMENTAL METHODS

Here we briefly describe several important experimental techniques developed for the study of carbonate and carbonate-silicate systems including particular aspects for sample preparation and electron microprobe measurements. Most experiments were carried out using the DIA-type multianvil apparatuses Discoverer-1500 at the Sobolev Institute of Geology and Mineralogy SB RAS (Novosibirsk, Russia) and similar configuration 3000-ton press at the Department of Earth Sciences (Tohoku University, Sendai, Japan).

Fujilloy N-05 26-mm tungsten carbide cubes with truncation edge length of 12 mm were employed as anvils for Kawai-type cell. Pressure media were made of semi-sintered ZrO_2 ceramics shaped as a 20.5 mm octahedron with ground edges and corners. Pyrophyllite gaskets, 4.0 mm in both width and thickness were used to seal the compressed volume and support the anvil flanks (Shatskiy et al., 2010, 2011). The cell assembly contains four graphite cassettes (multiple sample holders), allowing simultaneous study of 16 samples in a single run (Figure 14.1). The heating was achieved using a tubular graphite heater, 4.5/4.0 mm outer/inner diameter and 11 mm length. The heater and capsules were separated by ceramic sleeve made of talc fired at 1000 °C for 1 hour. The sample temperature was monitored by a $W_{97}Re_3$-$W_{75}Re_{25}$ thermocouple inserted in the heater center via walls and electrically insulated by Al_2O_3 tubes.

The temperature gradients in the cell were examined using thermal modeling software (Hernlund et al., 2006) and verified experimentally (Shatskiy, Sharygin, Gavryushkin et al., 2013) using a two-pyroxene equilibrium (Brey & Kohler, 1990). The maximum temperature gradients within individual sample and across the sample charge were less than 5° and 30 °C/mm at 1300 °C, respectively. We also found no differences in the phase composition of identical samples loaded in different cassettes. The pressure calibration was reported by Shatskiy et al. (2018). Deviation of pressure from the desired value during heating from room temperature in the given cell did not exceed ±0.2 GPa, as confirmed by comparison of pressure calibration at room temperature and 900 °C. In all experiments, no correction for the effect of pressure on the thermocouple electromotive force was applied.

Starting materials were prepared by blending reagent grade Na_2CO_3, K_2CO_3, $CaCO_3$, natural magnesite (<0.1% impurity), dolomite (<0.3% impurity), and siderite (Sd-1: $Fe_{0.95}Mn_{0.05}CO_3$ and Sd-2: $Fe_{0.87}Mn_{0.06}Mg_{0.07}CO_3$) in an agate mortar with acetone and loaded as a powder into graphite cassettes. Since Na_2CO_3 and K_2CO_3 are highly hygroscopic materials, special attention was paid to minimize the amount of moisture in the sample absorbed from the atmosphere. For this purpose, the loaded cassettes were dried at 300 °C for 1–2 h and then prepared assemblies were stored at 200 °C in a vacuum for ≥12 h prior to the experiment. In general, experiments were conducted at low (<25%) indoor humidity.

After completion of the experiments, the recovered graphite cassettes were immediately filled with epoxy in vacuum. Then cassettes were cut using a low-speed diamond saw to get vertical cross-sections of the samples. The obtained specimens were mounted in a Plexiglas holder with epoxy and polished in low-viscosity oil using a sandpaper. The sample surface was cleaned using an oil spray between each step of polishing. The final polishing was done on a satin cloth with 3 μm diamond paste and oil spray. We used petroleum benzene to remove the oil after polishing. The clean samples were stored in petroleum benzene prior to carbon coating and loading into a scanning electron microscope.

Samples were studied using a MIRA 3 LMU scanning electron microscope (Tescan Orsay Holding) coupled with an INCA-450 energy-dispersive X-ray microanalysis system equipped with the liquid nitrogen-free Large Area EDS X-Max-80 Silicon Drift Detector (Oxford Instruments Nanoanalysis Ltd) at IGM SB RAS (Lavrent'ev et al., 2015). Energy-dispersive X-ray spectra (EDS) were collected by using an electron beam-rastering method, in which the stage is stationary while the electron beam moves over the surface area, with dimensions 5–50 μm (for mineral phases) and 50–500 μm (for a quenched melt) at 20 kV accelerating voltage and 1.5 nA beam current. Live counting time for X-ray spectra was 20–30 s. Sample measurements were always combined with standard carbonate calibrations. This technique allows careful determination of the Na_2O and K_2O contents of carbonates, which can be missing during conventional electron microprobe analyses.

Diamond anvil cell (DAC) experiments on the carbonated systems are usually performed using powdered or single crystal samples with an addition of the internal laser-absorber for sample heating. The typical loading includes carbonate powder mixed with Pt metal and sandwiched between the salt (NaCl, KCl) insulators or pressure medium. Argon or neon gases can also serve as a pressure medium, whereas CO_2 may be placed as a reactant to synthesize carbonates from oxide mixtures. In case of Fe-carbonate, an additional absorber is not needed (Cerantola et al., 2017). The sample is placed between diamond anvils into the hole in the preindented Re gasket. The pressure is calibrated using ruby, metal, neon, or diamond equation of state. The most powerful tool for measurements is an in situ synchrotron X-ray diffraction (Dorfman et al., 2018; Maeda et al., 2017; Martirosyan et al., 2019) with further transmission electron microscopy of recovered sample films. Reliable data can also be obtained by in situ Raman spectroscopy at high temperature or at room temperature on the quenched sample (Bayarjargal et al., 2018). However, DAC has serious limitations related to the achievement of chemical equilibrium, thermal gradient, and temperature measurements in the sample (Dewaele et al., 1998; Kavner & Nugent, 2008; Rainey et al., 2013). Uncertainties in pressure scales of calibrants are essential at pressures above 50–100 GPa. We recommend using recent pressure scales from Sokolova et al. (2013, 2016). where internally consistent equations of state were calculated for most of the important pressure calibrants and can be used for pressure recalculations. In spite of these limitations, the data from DAC experiments are the only available source of information for phase relations at pressures exceeding 30–50 GPa.

The common methods of ab initio computations include density functional theory approach implemented into the VASP or other computer codes using the plane wave basis set and the projector augmented wave method. Exchange-correlation effects have been taken into account in the generalized gradient or local density approximation. Phase diagrams at high temperatures can be calculated using modeling of the Helmholtz free energy within lattice dynamic and the quasi-harmonic approximation or within molecular dynamic simulations. The latter accounts for the effect of anharmonicity at high temperatures more precisely (Gavryushkin et al., 2017; Zhang et al., 2018).

14.3. PHASE DIAGRAMS OF CARBONATES

14.3.1. Melting and Phase Transitions in Simple Carbonate Systems

$CaCO_3$. Recent experimental and theoretical studies indicate that $CaCO_3$ has extremely complex phase diagram at high pressure and many phases are still poorly characterized (Bayarjargal et al., 2018; Gavryushkin et al., 2017; X. Li et al., 2018; Litasov et al., 2017; Lobanov et al., 2017; Ono et al., 2005, 2007; Zhang et al., 2018). The major sequence of phase transitions established from ab initio computations include $R\bar{3}c$ calcite (0–2 GPa), *Pmcn* aragonite (2–42 GPa), *Pmmn* post-aragonite (42–37 GPa), $C222_1$ pyroxene-type structure (137–240 GPa), and *Pmcn*-$CaCO_3$ at higher pressures (Arapan et al., 2007; Oganov et al., 2008). However, near 30 GPa this pattern is complicated by the appearance of several candidate structures with close enthalpies. Their crystal structures vary in different studies. Pickard and Needs (2015) computed the lowest enthalpy for the $P2_1/$c-l phase at 32–48 GPa and $P2_1$c-h for pressures above 67 GPa. Gavryushkin et al. (2017) argued that the most stable phase in the pressure range of 32–46 GPa is $CaCO_3$-VII, which is a polymorph with unit cell parameters of $P2_1/$c-h and atomic arrangement of $P2_1/$c-l. In their study, $P2_1/c$ aragonite II and $P2_1$c-l have very close enthalpy to $CaCO_3$-VII. Smith et al. (2018) came to similar conclusions but named $CaCO_3$-VII as $P2_1$c-II. The computation of equation of state from quasi-harmonic approximation and molecular dynamics revealed high-temperature stability of $P2_1$c-l phase and limited stability of $CaCO_3$-VII (Gavryushkin et al., 2017; Smith et al., 2018; Zhang et al., 2018) (Figures 14.2 and 14.3). which is roughly consistent with the experimental data for the pressure range of 25–45 GPa.

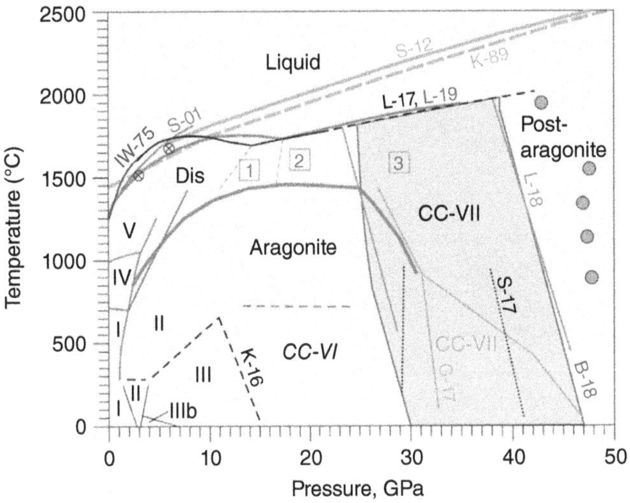

Figure 14.2 Phase diagram of $CaCO_3$ in the pressure range 0–50 GPa. Green lines show melting line and aragonite stability field from present work (L-19). Dis = disordered aragonite. Numbers 1, 2, and 3 indicate observation of new phases. Phase 3 is consistent with $CaCO_3$-VII/$P2_1$/c-l. Other melting lines are from Irving and Wyllie, 1975 (IW-75); Kerley, 1989 (K-89); Suito et al., 2001 (S-01); Spivak et al., 2012 (S-12); and Z. Li et al., 2017 (L-17). Crossed circles show melting points of $CaCO_3$ at 3 and 6 GPa (Shatskiy et al., 2018). Transition boundaries between phases I, II, III, and IIIb are from Pippinger et al. (2015). K-16 = boundary between $CaCO_3$-III and VI (CC-VI) (Koch-Müller et al., 2016). Gray shaded area is the stability field of $CaCO_3$-VII (CC-VII) according to experimental data by X. Li et al., 2018 (L-18), and Bayarjargal et al., 2018 (B-18); and ab initio computations by Gavryushkin et al., 2017 (G-17), and Smith et al., 2018 (S-17). Orange circles indicate observation of CC-VII and post-aragonite in the experiments (Gavryushkin et al., 2017). See electronic version for color representation of the figures in this book.

Experimental observations of the phase transitions in $CaCO_3$ are also highly controversial. At low pressures, calcite transforms to disordered $R\overline{3}m$ modifications IV and V at 715 and 970 °C, respectively (Ishizawa et al., 2013). At low temperatures, several modifications $CaCO_3$-II, III, and IIIb were described. The transition boundaries at 25 °C correspond to the pressures of 1.7 GPa (I → II). 2.0 GPa (II → IIIb). and 3.4 GPa (IIIb → III) (Pippinger et al., 2015). The upper temperature limit of phase II is not well established (Figure 14.2). Phase III transforms to phase VI at about 15 GPa and room temperature (Merlini, Hanfland, et al., 2012). Koch-Müller et al. (2016) observed this transition at higher temperatures with a negative Clapeyron slope (Figure 14.2). The temperature stability of phase VI is poorly resolved (Bayarjargal et al., 2018). The phases II, III, IIIb, and VI were not confirmed by thermodynamic calculations or even clearly observed in the multianvil experiments with better equilibration of the sample at high temperatures. Therefore, these phases can be metastable.

Calcite-aragonite transition was determined in several studies by quench experiments (Irving & Wyllie, 1973), differential thermal analysis (DTA) (Mirwald, 1976), thermodynamic calculations (Salje & Viswanathan, 1976), and electrical impedance measurements (Bagdassarov & Slutskii, 2003). All these data are consistent with later in situ determinations (Suito et al., 2001) and revealed transition at 2 GPa and 700 °C (Figure 14.2). At pressures of 5–8 GPa and 1200–1300 °C, aragonite transforms to a presumably disordered phase observed by the disappearance of most of the diffraction peaks (Figure 14.2). It is not clear whether this phase is similar to $CaCO_3$-V or not. Most likely, it is different from calcite and corresponds to disordered aragonite based on the observation that minor peaks are consistent with aragonite structure. In situ X-ray diffraction in a multianvil apparatus indicates that heating of aragonite at 14 and 19 GPa revealed transition to new phases at temperatures near 1400 °C (Figure 14.2). These phases are not yet characterized. At 29 GPa and 1100 °C, aragonite is transformed to a new phase that has clear similarities with $CaCO_3$-VII. This is in agreement with the data obtained from DAC experiments using both in situ X-ray diffraction (X. Li et al., 2018) and Raman spectroscopy (Bayarjargal et al., 2018). The pressure of the transition is different for 3–4 GPa (Figure 14.2); however, this may be connected with differences in the pressure scales used for the experiments. Bayarjargal et al. (2018) believe that $CaCO_3$-VII and $P2_1$c-l should be the same phase.

Original phase transitions to post-aragonite and to a pyroxene structure were discovered by Ono et al. (2005, 2007). However, phase boundaries were not clearly identified. Phase $P2_1$c-h was synthesized and characterized by Lobanov et al. (2017). Figure 14.3 shows phase boundaries in $CaCO_3$ from ab initio computation (Zhang et al., 2018). These boundaries are consistent with available experimental data except for aragonite to $P2_1$c-l ($CaCO_3$-VII) transition, which was placed at lower pressures.

The melting line of $CaCO_3$ (Figure 14.2) was determined in several works at pressures to 3.5 GPa by quench experiments (Irving & Wyllie, 1975), to about 8 GPa by in situ multianvil experiments (Suito et al., 2001) by electrical conductivity measurements (ECM) and falling sphere method to 21 GPa (Z. Li et al., 2017). DAC experiments at 10–40 GPa (Spivak et al., 2012) and thermodynamic calculations based on properties of calcite, aragonite, and melt (Kerley, 1989). The DAC data are scarce and may overestimate melting temperatures of $CaCO_3$. The results of our new in situ X-ray diffraction multianvil experiments for melting line of $CaCO_3$ obtained at pressures up to 30 GPa with press oscillation procedure are generally in agreement with the results

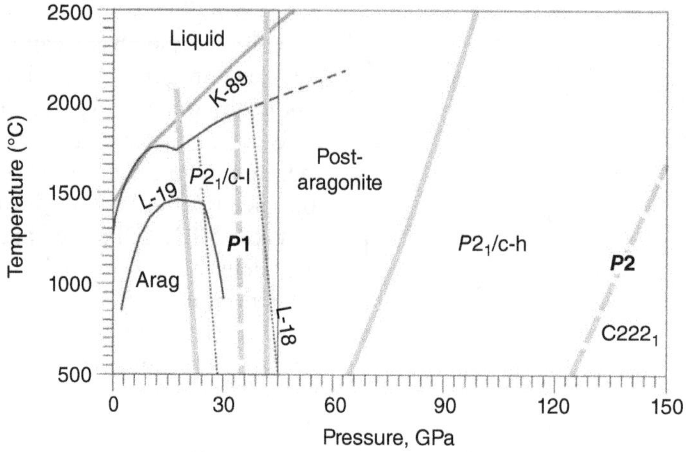

Figure 14.3 Phase transitions in $CaCO_3$ to 150 GPa (Zhang et al., 2018) shown by gray lines. Some data from Figure 14.2 are shown for comparison. P1 and P2 show boundaries of direct transition from aragonite to post-aragonite and to $C222_1$ phase. $P2_1/c$-I is equal to $CaCO_3$-VII. See electronic version for color representation of the figures in this book.

obtained by Z. Li et al. (2017). We may emphasize uncertain position of temperature minimum at solidus connected with phase transition from disordered phase to new high-pressure phase. Z. Li et al. (2017) placed this minimum at 14 GPa, whereas we observed minor minimum near 17 GPa (Figure 14.2).

At temperatures near 3000 °C, which is about 1000 °C above the melting line, $CaCO_3$ melt decomposes to CaO, diamond/graphite and O_2 fluid (Spivak et al., 2012). This temperature is consistent with decomposition parameters revealed by shock compression experiments (Bobrovsky et al., 1976; Ivanov & Deutsch, 2002; Martinez et al., 1995).

$MgCO_3$. Ab initio evolutionary search in the $MgCO_3$ system revealed stability of magnesite at pressures up to 82 GPa, where it transforms to C2/m phase II with three-membered rings of carbonate tetrahedra. Phase II transforms to $P2_1$ phase III with similar structural features. Finally, at ~160 GPa phase III transforms to $Pna2_1$ phase with chains of carbonate tetrahedra (Oganov et al., 2008). In the later study, Pickard and Needs (2015) proposed an additional phase with $P\bar{1}$ symmetry, which was energetically favorable at 85–101 GPa, and $P2_12_12_1$ phase, which was marginally stable along with $P2_1$ structure near 144 GPa. Calculation of high-temperature equations of state indicates limited temperature stability of $P\bar{1}$ phase (<1000 °C) and location of magnesite to phase II transition at 73 GPa with near-zero Clapeyron slope (Zhang et al., 2018). In DAC experiments, transition of magnesite to high-pressure phase was observed at pressures above 100 GPa (Isshiki et al., 2004). In the study of the $MgCO_3$+SiO_2 reaction at lower mantle pressures, magnesite to phase II transition was observed at 76 GPa and 1500 °C (Maeda et al., 2017), which is perfectly consistent with theoretical prediction (Zhang et al., 2018).

Dissociation and melting line of magnesite was determined by quench experiments using piston-cylinder and multianvil technique at pressures to 15 GPa (Byrnes & Wyllie, 1981; Harker & Tuttle, 1955a; Irving & Wyllie, 1975; Katsura & Ito, 1990; Müller et al., 2017; Shatskiy et al., 2018) and in DAC experiments to 84 GPa (Solopova et al., 2015). At pressures below 2.6 GPa, magnesite dissociates to periclase and CO_2 fluid with a narrow pressure range near 3 GPa; it melts incongruently to produce $MgCO_3$–MgO liquid and CO_2 fluid (Figure 14.4). At higher pressures, magnesite melts congruently and its melting line locates slightly above that of $CaCO_3$ (Figure 14.4). The melting line from DAC and large-volume press experiments are consistent with each other. Similar to $CaCO_3$, $MgCO_3$ melt dissociates to MgO, diamond, and O_2 at about 500 °C above the melting line (at 2500–3000 °C) (Solopova et al., 2015).

$FeCO_3$. Phase transitions in $FeCO_3$ were intensively studied using DAC owing to particular interest to spin transitions, valence changes in Fe, and coordination changes in carbon at high pressures (Boulard et al., 2012; Cerantola et al., 2017; Lavina et al., 2009; J. Liu et al., 2015; Lobanov et al., 2015; Mattila et al., 2007). Accordingly, a range of novel carbonate-like phases were discovered upon $FeCO_3$ decomposition at lower mantle pressures. Thermodynamic stability fields of these phases are not yet constrained, just as a detailed ab initio computation on $FeCO_3$ and related compositions has not yet been performed. Figure 14.5 summarizes the schematic phase diagram for $FeCO_3$ (Cerantola et al., 2017). The high-spin to low-spin transition occurs at about 45 GPa (Figure 14.5), whereas two new compounds with tetrahedrally coordinated carbon $R\bar{3}c$ $Fe^{3+}{}_4C_3O_{12}$ and C2/c $Fe^{2+}{}_2Fe^{3+}{}_2C_4O_{13}$ appear at higher pressures of 75–120 GPa and 1000–2200 °C. They coexist with Fe-oxides. Similarly, Fe-oxides and diamond were observed as quench products of siderite decomposition at pressures of 20–60 GPa (Figure 14.5). Mossbauer spectra of the samples quenched from the molten state at temperatures above 1800 °C show unambiguously the presence of $FeCO_3$ as well as iron

oxides: α-Fe_2O_3 (hematite) at pressures below ~25 GPa and HP-Fe_3O_4 (Bykova et al., 2016) at pressures above ~31 GPa. Thus, the location of the melting/dissociation line at pressures above 25 GPa is not clear.

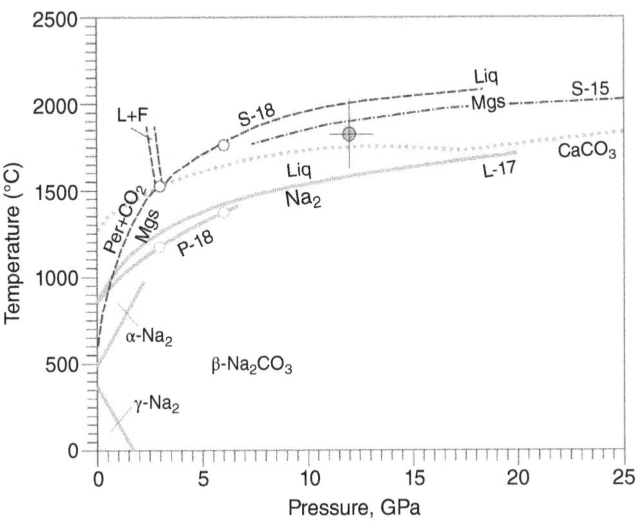

Figure 14.4 Melting and phase transitions in $MgCO_3$ and Na_2CO_3 at pressures to 25 GPa. S-18 = magnesite (Mgs) dissociation and melting line after Shatskiy et al. (2018), including data from Byrnes and Wyllie (1981), Harker and Tuttle (1955a), Irving and Wyllie (1975), and Katsura and Ito (1990). S-15 and grey circle with error bar are results from DAC experiments on magnesite melting (Solopova et al., 2015). Stability of low-pressure Na_2CO_3 phases is after Shatskiy, Litasov, and Palyanov (2015). Melting lines of Na_2CO_3 (Na_2) are after Z. Li et al., 2017 (L-17), and Podborodnikov, Shatskiy, Arefiev, Rashchenko, et al., 2018 (P-18). The $CaCO_3$ melting line (Z. Li et al., 2017) is shown for comparison. See electronic version for color representation of the figures in this book.

The melting line of siderite is not well constrained at lower pressures either. It may strongly depend on the oxidation state during experiment. At pressures below 1 GPa, dissociation lines $FeCO_3 = Fe_3O_4+C+CO_2$ and $Fe_3O_4+C = L+CO_2$ were determined by Weidner (1972, 1982). Tao et al. (2013) performed experiments on dissociation and melting of siderite to 10 GPa and determined the invariant point of transition from dissociation to melting at 7.5 GPa and 1500 °C, whereas Kang et al. (2015) reported the melting line of siderite to 22 GPa. The results of these works do not coincide with our results for decomposition/melting of siderite at 3 and 6 GPa. According to our data, the invariant point should be located near 5.7 GPa and 1570 °C (Figure 14.6). The differences can be explained by peculiarities of experimental technique used, i.e. run duration, details of phase identification, and capsule material, which may affect the oxidation state of the sample. We performed experiments in graphite capsules, which maintain fO_2 near the C–CO_2 (CCO) oxygen buffer and prevent sample oxidation due to large graphite/sample volume ratio. Tao et al. (2013) performed experiments in Pt capsule, which can absorb Fe and cause sample oxidation:

$$4FeCO_3(\text{siderite}) = Fe^0(\text{Fe}-\text{Pt alloy}) + Fe_3O_4(\text{magnetite}) + 4CO_2(\text{fluid}). \quad (14.1)$$

As a result, they observed the $FeCO_3$–Fe_3O_4 eutectics near 1425 °C at 6 GPa, which is 180 °C lower than the siderite melting point established in our experiments. In this case, oxidation state was likely controlled by the FeO–Fe_3O_4 or Fe_3O_4–Fe_2O_3 buffer, which are 1–3 orders of magnitude higher than CCO at 6 GPa and 1400 °C.

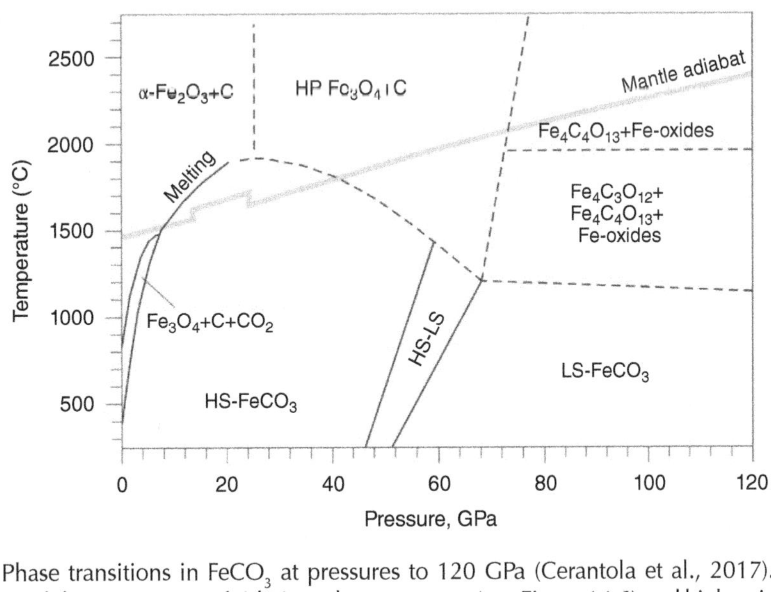

Figure 14.5 Phase transitions in $FeCO_3$ at pressures to 120 GPa (Cerantola et al., 2017). The black solid lines show melting and decomposition of siderite at low pressures (see Figure 14.6) and high-spin to low-spin transition (HS-LS) for magnesiosiderite (J. Liu et al., 2015). The gray line is the mantle adiabat (Litasov & Shatskiy, 2018). See electronic version for color representation of the figures in this book.

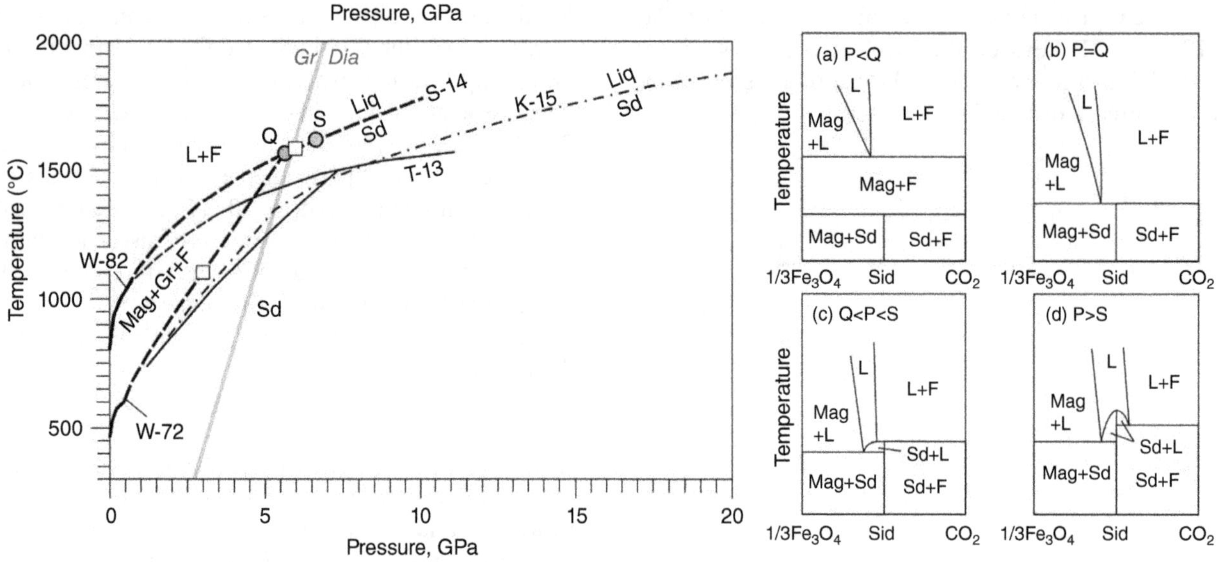

Figure 14.6 The sequence of siderite decomposition reactions, incongruent and congruent melting with increasing pressure based on the experimental data: W-72 = Weidner, 1972; W-82 = Weidner, 1982; T-13 = Tao et al., 2013; S-14 = Shatskiy et al., 2014; K-15 = Kang et al., 2015. Squares show decomposition and melting points of siderite at 3 and 6 GPa. Mag = magnetite; Sd = siderite; Gr = graphite; Q = invariant point, where incongruent melting begins; S = singular point above which congruent melting occurs. Gr/Dia is graphite-to-diamond transition (Kennedy & Kennedy, 1976). Schematic isobaric projections on the right (a–d) illustrate siderite decomposition and melting pathways expected in the Fe–C–3O system. We assume that graphite is present in excess and skip it on these diagrams. See electronic version for color representation of the figures in this book.

Kang et al. (2015) used a double Pt-graphite capsule at 6 GPa and a Au-Pd capsule at 10–20 GPa. The run duration was quite short and did not exceed 1.5 hours at 6 GPa and 20 min at higher pressures. In many subsolidus runs at 10–20 GPa, they observed siderite coexisting with Fe-oxide. This indicates oxidation of the sample and, therefore, lower melting line for siderite. The oxidation can be caused by absorption of Fe by the Au capsule (this may be significant for Fe-rich compositions; Ratajeski & Sisson, 1999) or penetration of oxygen or moisture from cell assembly parts during experiment or from atmosphere prior to loading (this possibility was noticed in the original work). We also suggest oxidation of the sample in Kang et al. (2015) at 6 GPa. Although they used Pt-graphite capsules, their solidus position coincides with that by Tao et al. (2013). In the experiments at 6 GPa we used Sd-2 as starting material, which contains a significant amount of impurities (Shatskiy et al., 2014). At 3 Gpa, we tested both Sd-1 (nearly pure siderite) and Sd-2 and obtained the same results for dissociation point.

Summarizing the data for dissociation and melting of siderite, we can emphasize their dependence on the oxidation state. For the fO_2-conditions close to CCO=0 to CCO–1 (1 log unit below the CCO buffer) at 6 GPa, melting line and invariant point may correspond to our data (Shatskiy et al., 2014); however at more oxidized conditions, when fO_2 is buffered by FeO–Fe_3O_4 (which should be close to quartz–fayalite–magnetite, QFM buffer) the melting line of siderite follows that proposed by Kang et al. (2015) (Figure 14.6).

Na_2CO_3. High-pressure behavior of alkali carbonates is complicated due to the ferroelastic phase transition and lattice melting (Harris & Dove, 1995; Swainson et al., 1995), change of carbon coordination number (CO_3^{2-} to CO_4^{4-}) (Al-Shemali & Boldyrev, 2002; Cancarevic et al., 2007; Mellot-Draznieks et al., 2002), and ability to form disordered (Depater, 1979) and incommensurate (Dusek et al., 2003; Harris & Salje, 1992) phases.

Natrite, γ-Na_2CO_3 (*C2/m*), is stable at ambient conditions (Dusek et al., 2003) and transforms to a monoclinic *C2/m* β-modification at 330–360 °C, and a hexagonal *$P6_3/mmc$* α-phase at 481 °C (Swainson et al., 1995). In situ X-ray diffraction experiments (Shatskiy, Gavryushkin, et al., 2013) revealed the stability of β-Na_2CO_3 at least to 1200 °C at 15.2 GPa (Figure 14.4). Melting curves of Na_2CO_3 obtained by ECM to 18 GPa and falling sphere method at 3 and 6 GPa are marginally consistent with each other and indicate that Na_2CO_3 is the most fusible among major carbonate phases under consideration (Figure 14.4).

Ab initio evolutionary search for stable crystal structures of Na_2CO_3 showed phase transition from γ-Na_2CO_3 to *$P6_3/mcm$* phase at 5 GPa and to *$P2_1/m$* phase at 35 GPa (Gavryushkin et al., 2016). At the same time, at high pressures orthocarbonate phases with tetrahedrally coordinated carbon can preferentially be stable instead of carbonates. The phase transition from Na_2CO_3 to Na_4CO_4 (+CO_2) with $I\bar{4}2m$ symmetry was suggested at 40–60 GPa (Cancarevic et al., 2007).

K_2CO_3. Potassium carbonate is monoclinic *C2/c* β-K_2CO_3 at ambient conditions (Figure 14.7) and transforms to a hexagonal α-phase ($P6_3/mmc$) at 420 °C (Becht & Struikmans, 1976; Dinnebier et al., 2005; Schneide & Levin, 1973). The α–β transition was observed using DTA to 0.5 GPa (Klement & Cohen, 1975). The melting temperature of K_2CO_3 was determined by DTA at pressures to 0.5 GPa (Klement & Cohen, 1975); by falling sphere method at 1.8–3.2 GPa (Q. Liu et al., 2007), at 3 and 6 GPa (Arefiev et al., 2018; Shatskiy, Sharygin, Gavryushkin, et al., 2013), and at 4–11.5 GPa; and by ECM to 20 GPa (Z. Li, 2015) (Figure 14.7). The resulting melting lines vary considerably: The differences between three data sets exceeds 250 °C at 6 GPa. This indeed may be connected with the highly hygroscopic nature of K_2CO_3 and possible differences in data obtained by ECM and falling spheres. ECM data indicate a significant positive slope of the K_2CO_3 melting line, which intersects that of $CaCO_3$ at pressures of 10–11 GPa.

In situ X-ray diffraction experiments (Z. Li, 2015) revealed appearance of several new high-pressure modifications of K_2CO_3, which, however, are not characterized and for which phase boundaries are not established (Figure 14.7). There is a definite phase transition at ~4.5 GPa with a solid-solid-liquid triple point bracketed at 1150–1300 °C. This phase transition is either from α to β phase, or from new-1 to new-2.

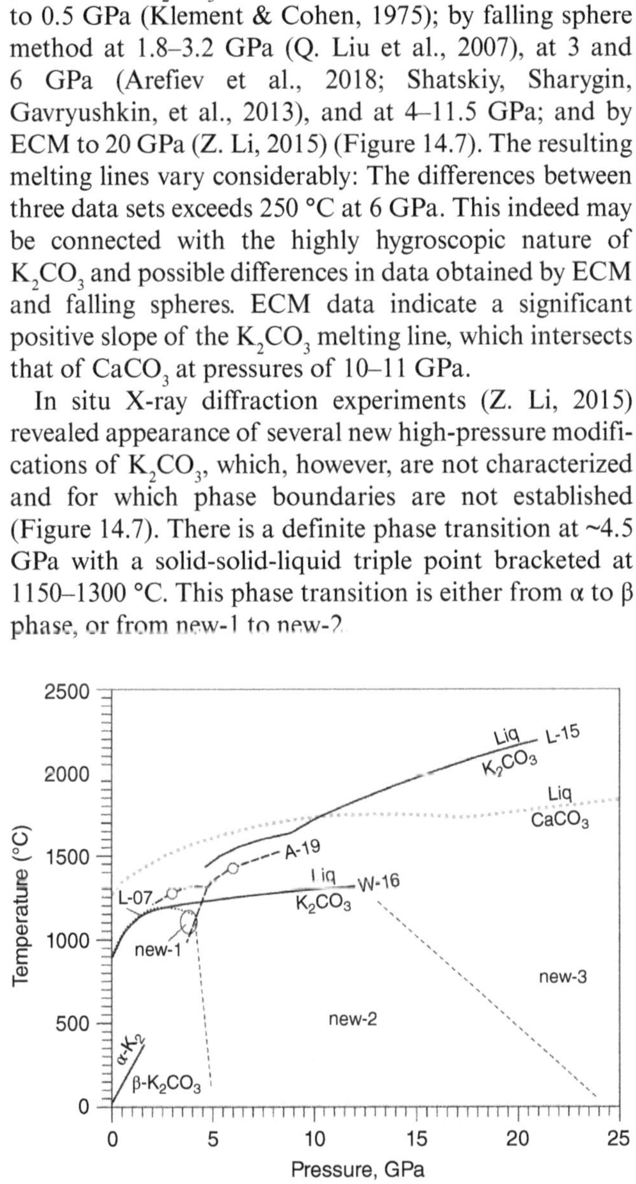

Figure 14.7 Melting lines and phase relations in K_2CO_3. L-07 (dotted curve) = J. Liu et al., 2007; L-15 = Z. Li, 2015; W-16 = Wang et al., 2016; A-19 = Arefiev et al., 2018; Shatskiy, Sharygin, Gavryushkin, et al., 2013b. α-β transition (Klement & Cohen, 1975). New phases 1 (oval field), 2 and 3 and approximate transition lines (Z. Li, 2015). The $CaCO_3$ melting line (Z. Li et al., 2017) is shown for comparison. See electronic version for color representation of the figures in this book.

An ab initio evolutionary search indicates stability of C2/c β-K_2CO_3 (in some works γ-K_2CO_3, as it is isostructural with γ-Na_2CO_3) at pressures to 12 GPa, where it transforms to $P\bar{1}$ structure, which is very similar with $P2_1/m$ Na_2CO_3, and then to *C2/c* phase at 53.5 GPa (Gavryushkin et al., 2016). It should be noted that phases predicted by ab initio computations and those observed in in situ X-ray diffraction experiments were not compared with each other. In turn, stability of K_4CO_4 orthocarbonates with $I\bar{4}$ and $P4_2/n$ symmetry was suggested at 25–40 GPa and above 40–45 GPa, respectively (Cancarevic et al., 2007).

14.3.2. Binary Carbonate Systems at High Pressures

Figures 14.8–14.11 summarize data for the binary carbonate systems at pressures of 3 and 6 GPa, which were systematically studied by our group in recent years and compared with low-pressure or 1 atm data. Some of the phases observed in these systems were studied at higher pressures (Golubkova et al., 2015). Crystal structures of the novel high-pressure Na-Ca carbonates are reviewed by Rashchenko et al. (this volume). The studied binary systems include $CaCO_3$-$MgCO_3$ (Shatskiy et al., 2018). Na_2CO_3-$CaCO_3$ at 3 GPa (Podborodnikov, Shatskiy, Arefiev, Rashchenko, et al., 2018) and 6 GPa (Shatskiy, Sharygin, Litasov, et al., 2013). Na_2CO_3–$MgCO_3$ at 3 GPa (Podborodnikov, Shatskiy, Arefiev, Chanyshev, et al., 2018) and 6 GPa (Shatskiy, Gavryushkin, et al., 2013) K_2CO_3–$CaCO_3$ at 3 GPa (Arefiev et al., 2019) and 6 GPa (Shatskiy, Borzdov, et al., 2015). K_2CO_3–$MgCO_3$ at 3 GPa (Arefiev et al., 2018) and 6 GPa (Shatskiy et al., 2013), and Fe-bearing systems at 6 GPa: $FeCO_3$–$MgCO_3$ and K_2CO_3–$FeCO_3$ (Shatskiy, Litasov, Ohtani, et al., 2015). $FeCO_3$–$CaCO_3$ (Shatskiy et al., 2014). and Na_2CO_3–$FeCO_3$ (Shatskiy, Rashchenko, et al., 2015). Previous review of the data available before 2015 with a wider variety of carbonates (such as Li, Ba, Sr-carbonates) was presented by Shatskiy, Litasov, & Palyanov (2015).

$CaCO_3$–$MgCO_3$. This is one of the most important systems in petrology; along with the carbonate-silicate reaction, it determines major features of melting and phase relations in the carbonated systems. Phase relations in the $CaCO_3$–$MgCO_3$ system were studied at $P(CO_2)$ = 0.1 GPa and 625–850 °C (Graf & Goldsmith, 1955), $P(CO_2)$ < 0.1–0.3 GPa and 500–900 °C (Harker & Tuttle, 1955b), $P(CO_2)$ < 1.0 GPa and 700–1200 °C (Goldsmith & Heard, 1961), P = 1.0 GPa and 900–1400 °C (Byrnes & Wyllie, 1981), and P = 3.0 GPa and 1100–1600 °C (Irving & Wyllie, 1975). The studies revealed a two-phase stability field (Mgs + Arg) at low temperatures and limited solid-solutions fields (Mgs + Dol and Dol + Cal/Arg) at higher temperatures (Figure 14.8a, d). The *T-X* diagrams have temperature minimums on solidus and liquidus curves that shift from

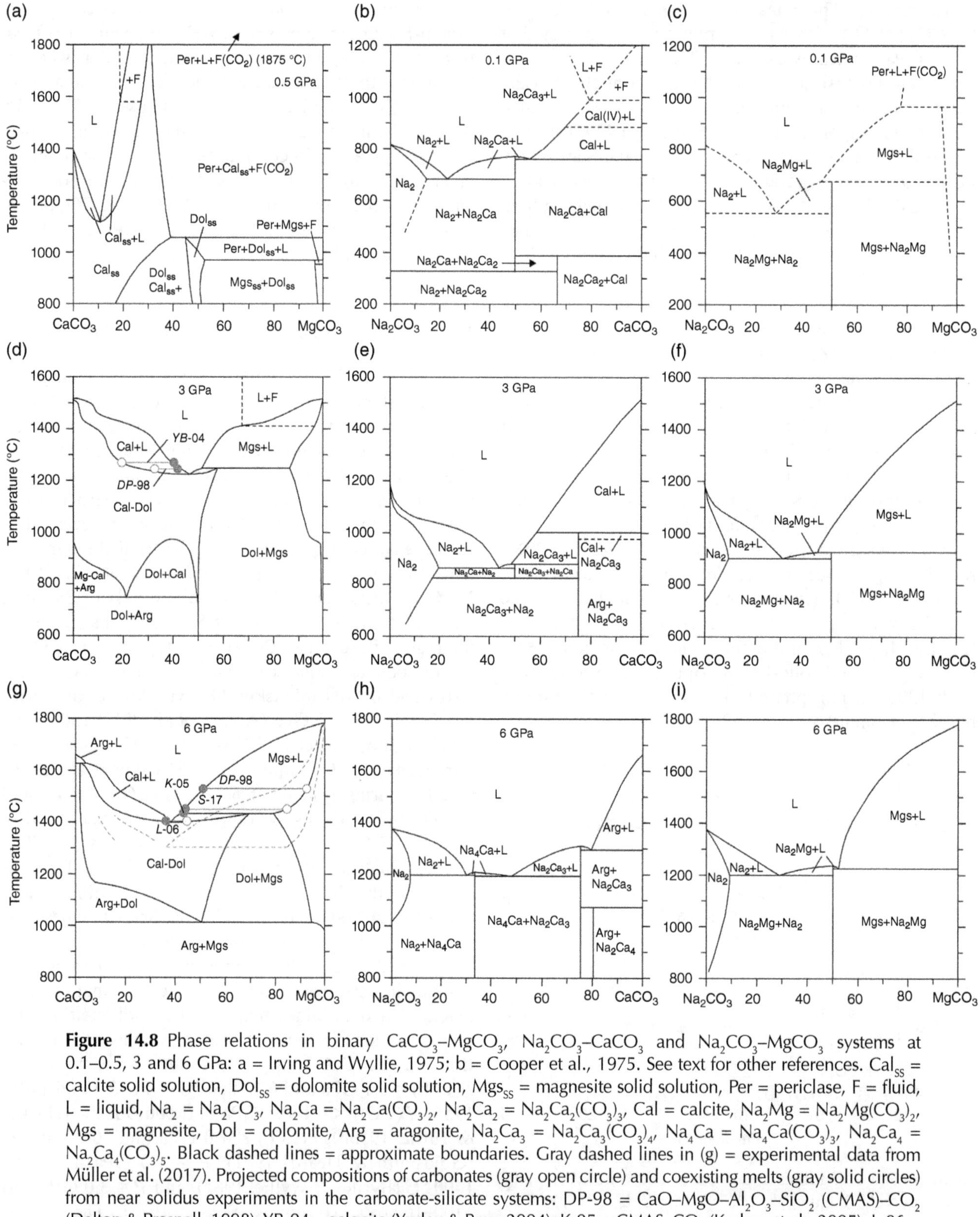

Figure 14.8 Phase relations in binary $CaCO_3$–$MgCO_3$, Na_2CO_3–$CaCO_3$ and Na_2CO_3–$MgCO_3$ systems at 0.1–0.5, 3 and 6 GPa: a = Irving and Wyllie, 1975; b = Cooper et al., 1975. See text for other references. Cal_{SS} = calcite solid solution, Dol_{SS} = dolomite solid solution, Mgs_{SS} = magnesite solid solution, Per = periclase, F = fluid, L = liquid, Na_2 = Na_2CO_3, Na_2Ca = $Na_2Ca(CO_3)_2$, Na_2Ca_2 = $Na_2Ca_2(CO_3)_3$, Cal = calcite, Na_2Mg = $Na_2Mg(CO_3)_2$, Mgs = magnesite, Dol = dolomite, Arg = aragonite, Na_2Ca_3 = $Na_2Ca_3(CO_3)_4$, Na_4Ca = $Na_4Ca(CO_3)_3$, Na_2Ca_4 = $Na_2Ca_4(CO_3)_5$. Black dashed lines = approximate boundaries. Gray dashed lines in (g) = experimental data from Müller et al. (2017). Projected compositions of carbonates (gray open circle) and coexisting melts (gray solid circles) from near solidus experiments in the carbonate-silicate systems: DP-98 = CaO–MgO–Al_2O_3–SiO_2 (CMAS)–CO_2 (Dalton & Presnall, 1998); YB-04 = eclogite (Yaxley & Brey, 2004); K-05 = CMAS–CO_2 (Keshav et al., 2005); L-06 = diopside–CO_2 system (Luth, 2006); S-17 = $CaMgSi_2O_6$+2$MgCO_3$ system (Shatskiy, Podborodnikov, et al., 2017). See electronic version for color representation of the figures in this book.

Ca to Mg side with pressure from Ca# = 58 (Ca# = 100 Ca/(Ca+Mg+Fe) and 1100 °C at 0.5 GPa to Ca# = 58 and 1300 °C at 3 GPa (Figure 14.8a). Our new data at 3 GPa (Shatskiy et al., 2018) indicate that this join is located at slightly lower temperature 1225 °C and Ca# = 53. More significant shift was detected for temperature of the intersection of the Dol-Mgs solvus and the melting loop, which is about 1250 °C in Shatskiy et al. (2018) and 1385 °C in Irving and Wyllie (1975) (Figure 14.8d). The obvious differences are explained by very short duration of the experiments in Irving and Wyllie (1975) and poor identification of chemical compositions in their work. At temperatures above 1425 °C and Ca# ≤ 30, the liquid quenches to dendritic carbonate and periclase and contains rounded voids, indicating an incongruent melting of magnesite to MgO in liquid and CO_2 in fluid and/or liquid (Figure 14.8d).

At 6 GPa the phase diagram was studied by Buob et al. (2006). Müller et al. (2017). and Shatskiy et al. (2018) (Figure 14.8g). Arg + Mgs assemblage is stable below 1000 °C. An appearance of dolomite between 1000 and 1050 °C splits the system into two partial binaries: Arg + Dol and Dol + Mgs. The dolomite-magnesite solvus intersects the melting loop between 1400 and 1450 °C, and forms the isothermal three-phase line, which represents the peritectic reaction: Dol (Ca# 31) = Mgs (Ca# 21) + liquid (Ca# 57). The melting loop for the $CaCO_3$–$MgCO_3$ join has a liquidus minimum at 1400 °C and Ca# = 62. The robustness of the obtained data for the $CaCO_3$–$MgCO_3$ system is confirmed by the excellent consistency of the geometry of the $CaCO_3$–$MgCO_3$ melting loop at 3 and 6 GPa with phase relations in the carbonated eclogite (Yaxley & Brey, 2004) and peridotite (Dalton & Presnall, 1998) systems, where Ca-dolomite melt coexists with Mg-calcite in eclogite and peridotite at 3 GPa, and dolomite melt coexists with magnesite in peridotite at 6 GPa (Figure 14.8d, g). In Müller et al. (2017). the liquidus minimum and dolomite-magnesite solvus are located at 100–150 °C lower, relative to the data by Shatskiy et al. (2018) (Figure 14.8g). It is difficult to explain these uncertainties. They may be related to a limited amount of experiments in Müller et al. (2017), calibration of the temperature measurements, and unavoidable hydration of the sample by, for example, parts of the cell assembly.

The $CaCO_3$–$MgCO_3$ system contains the important intermediate compound dolomite, which was intensively studied at high pressures. Original experiments on dolomite stability showed its decomposition to magnesite and aragonite at pressures near 5–6 GPa (Figure 14.9). There are some uncertainties between the data in different works; however, major consistency is observed near 6 GPa and 1000 °C (Hermann et al., 2016; Luth, 2001; Martinez et al., 1996; Morlidge et al., 2006; Shirasaka et al., 2002; Tao et al., 2014). Later dolomite composition

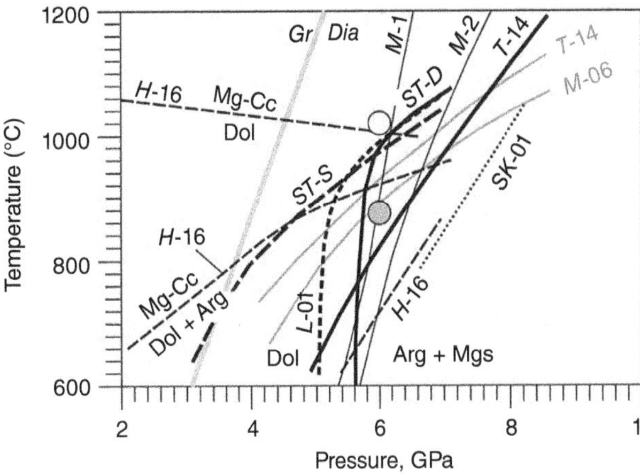

Figure 14.9 Decomposition boundary of $CaMg(CO_3)_2$ dolomite (black lines) and $CaFe(CO_3)_2$ ankerite solid solution (gray lines). SK-01 = Sato & Katsura, 2001; L-01 = (Luth, 2001); ST-D = dolomite decomposition, and ST-S = dolomite synthesis experiments (Shirasaka et al., 2002); M-1 and M-2 = thermodynamic estimations of phase transition (Martinez et al., 1996); T-14 = Tao et al., 2014; M-06 = Morlidge et al., 2006. The reactions from Hermann et al. (2016) (H-16) correspond to Dol = Mg-Cc (Mg-rich Ca-carbonate); Dol+Arg = Mg-Cc and Dol = Arg+Mgs. Open circle = our data for dolomite at 6 GPa (Shatskiy et al., 2018). Gray circle = our data for ankerite solid solution at 6 GPa (Shatskiy et al., 2014). Thick gray line is graphite to diamond transition (Kennedy & Kennedy, 1976). See electronic version for color representation of the figures in this book.

was reinvestigated at very high pressures, where it can reappear as a stable phase with different crystal structures if one adds >10 mol.% $FeCO_3$ to dolomite (see section 14.3.3). In pure $CaMg(CO_3)_2$ composition, dolomite transforms to high-pressure dolomite-II with $CaCO_3$-III-like structure upon compression to 20 GPa (Santillán et al., 2003). Merlini et al. (2017) showed that dolomite II transforms further to dolomite IIIc with $P\bar{1}$ symmetry (Z = 8). This crystal structure is different from Fe-dolomite III and IIIb (see below). Stability of high-pressure polymorphs of dolomite upon heating is not reported.

Na_2CO_3–$CaCO_3$. The phase relations in the Na_2CO_3–$CaCO_3$ reveal several intermediate compounds, which are different at 0.1, 3, and 6 GPa (Figure 14.8). At 0.1 GPa, the system has one intermediate phase, nyerereite $Na_2Ca(CO_3)_2$, near the solidus at 400–700 °C. It melts congruently at 817 °C and 0.1 GPa. $Na_2Ca(CO_3)_2$ is unstable below 300 °C and breaks down into calcite and shortite $Na_2Ca_2(CO_3)_3$ (Cooper et al., 1975) (Figure 14.8b). At 3 GPa, the system has two intermediate compounds, $Na_2Ca(CO_3)_2$ and $Na_2Ca_3(CO_3)_4$, at 850 °C. The maximum solubility of $CaCO_3$ in Na_2CO_3 is 20 mol.% at 850 °C. The Na_2CO_3–$Na_2Ca(CO_3)_2$ eutectic locates near 860 °C and 56 mol.% Na_2CO_3. $Na_2Ca(CO_3)_2$ melts incongruently to $Na_2Ca_3(CO_3)_4$ and a liquid containing about 51 mol.% Na_2CO_3 at ~880 °C.

$Na_2Ca_3(CO_3)_4$ disappears above 1000 °C via incongruent melting to calcite and a liquid containing about 43 mol.% Na_2CO_3 (Figure 14.8e) (Podborodnikov, Shatskiy, Arefiev, Rashchenko, et al., 2018).

At 6 GPa, the system has three eutectics at 1200 °C and 70 mol.% Na_2CO_3, at 1200 °C and 52 mol.% Na_2CO_3, and at 1300 °C and 21 mol.% Na_2CO_3, and three intermediate compounds. Two of them, $Na_4Ca(CO_3)_3$ and $Na_2Ca_3(CO_3)_4$, melt congruently at 1210 and 1310 °C, respectively, and one, $Na_2Ca_4(CO_3)_5$, decomposes to the $Na_2Ca_3(CO_3)_4$ + aragonite at 1100 °C (Figure 14.8h) (Shatskiy, Sharygin, Litasov, et al., 2013; Shatskiy, Rashchenko, et al., 2015). The maximum solubility of $CaCO_3$ in Na_2CO_3 is 6–8 mol.% at 1100–1300 °C, whereas Na_2CO_3 solubility in aragonite does not exceed the detection limit (<0.5 mol.%). Thus, a range of the intermediate compounds along the $CaCO_3$–Na_2CO_3 join changes with increasing pressure in the following sequence: $Na_2Ca(CO_3)_2$, $Na_2Ca_2(CO_3)_3$ (0.1 GPa) → $Na_2Ca(CO_3)_2$, $Na_2Ca_3(CO_3)_4$ (3 GPa) → $Na_4Ca(CO_3)_3$, $Na_2Ca_3(CO_3)_4$, $Na_2Ca_4(CO_3)_5$ (6 GPa). Thus, the nyerereite stability field extends to the shallow mantle pressures, whereas the shortite stability field terminates somewhere between 0.1 and 3 GPa. Consequently, findings of nyerereite and shortite among daughter phases in the melt inclusions in olivine from the sheared garnet peridotites are consistent with their mantle origin.

Na_2CO_3–$MgCO_3$. Phase relations in the system Na_2CO_3–$MgCO_3$ at ambient or low pressures have not been studied so far. Schematic diagram is presented in Figure 14.8c. The system has likely one intermediate compound, $Na_2Mg(CO_3)_2$ eitelite. It was constrained using available data on eitelite melting at 0.12 GPa and 677 °C (Eitel & Skaliks, 1929) and eitelite stability in the related systems $K_2Ca(CO_3)_2$–$Na_2Mg(CO_3)_2$ (McKie, 1990) and Na_2CO_3–MgF_2 at 0.1 GPa (Mitchell & Kjarsgaard, 2011). Eitelite is stable in the subsolidus assemblages below 550 °C in both systems.

At 3 and 6 GPa, the subsolidus system topology is not changed (Figure 14.8f, i). The subsolidus assemblages comprise two fields of Na_2CO_3 + eitelite and eitelite + magnesite. At 3 GPa, the Na-carbonate–eitelite eutectic was established at 900 °C and 69 mol.% Na_2CO_3. Eitelite melts incongruently to magnesite and a liquid containing about 55 mol.% Na_2CO_3 at 925 ± 25 °C. At 6 GPa, the system has two eutectics at 1200 °C and 71 mol.% Na_2CO_3 and at 1250 °C and 48 mol.% Na_2CO_3. Eitelite melts congruently at 1225 °C. The $MgCO_3$ solubility in Na_2CO_3 approaches 9 mol.% at 900 °C and 3 GPa and 10 mol.% at 1200 °C and 6 GPa, whereas Na_2CO_3 contents in $MgCO_3$ are below detection limit (Podborodnikov, Shatskiy, Arefiev, Chanyshev et al., 2018; Shatskiy, Gavryushkin, et al., 2013).

Compression of $Na_2Mg(CO_3)_2$ at room temperature reveals a phase transition to high-pressure phase at 14 GPa; however, its crystal structure is not yet characterized (Golubkova et al., 2015).

K_2CO_3–$CaCO_3$. The system K_2CO_3–$CaCO_3$ was studied at ambient pressure within 50–100 mol.% K_2CO_3 (Niggli, 1916) and 0–50 mol.% K_2CO_3 (Kröger et al., 1943). Similar to ambient pressures at 0.1 GPa, the system has two intermediate compounds near the solidus at 600–700 °C $K_2Ca(CO_3)_2$ (fairchildite) and $K_2Ca_2(CO_3)_3$ (Figure 14.10a) (Cooper et al., 1975). Below 512 °C, $K_2Ca_2(CO_3)_3$ becomes unstable and breaks down into calcite and $K_2Ca(CO_3)_2$ (buetschliite). Fairchildite melts congruently at 809 °C, whereas $K_2Ca_2(CO_3)_3$ melts incongruently to calcite + liquid at 810 °C. Mutual solubility of the endmembers was not well studied. It was suggested that $CaCO_3$ solubility in K_2CO_3 is up to 18 mol.%.

At 3 GPa and 800 °C (Figure 14.10b). the system has also two intermediate compounds: $K_2Ca(CO_3)_2$ buetschliite, and $K_2Ca_2(CO_3)_3$. However, at 850 °C, a third compound, $K_2Ca_3(CO_3)_4$, appears. The K-carbonate–$K_2Ca(CO_3)_2$ eutectic is established near 970 °C and 56 mol.% K_2CO_3. $K_2Ca(CO_3)_2$ melts incongruently at 990 °C to $K_2Ca_2(CO_3)_3$ and a liquid containing 53 mol.% K_2CO_3. In turn, $K_2Ca_2(CO_3)_3$ melts congruently just above 1100 °C. The $K_2Ca_2(CO_3)_3$–$K_2Ca_3(CO_3)_4$ eutectic locates near 1085 °C and 29 mol.% K_2CO_3. $K_2Ca_3(CO_3)_4$ melts incongruently at 1100 °C to calcite and a liquid containing 28 mol.% K_2CO_3. The maximum solubility of $CaCO_3$ in K_2CO_3 is 18 mol.% at 950 °C (Arefiev et al., 2019).

At 6 GPa, the system has three eutectics at 1180 °C and 63 mol.% K_2CO_3, at 1180 °C and 44 mol.% K_2CO_3, and at 1300 °C and 23 mol.% K_2CO_3, as well as three intermediate compounds. Two intermediate phases, $K_8Ca_3(CO_3)_7$ and $K_2Ca_3(CO_3)_4$, melt congruently at 1225 °C and 1350 °C, respectively (Figure 14.10c). The $K_2Ca(CO_3)_2$ compound decomposes to $K_8Ca_3(CO_3)_7$ + $K_2Ca_3(CO_3)_4$ above 950 °C. $CaCO_3$ solubility in K_2CO_3 and K_2CO_3 solubility in aragonite are below the detection limit (<0.5 mol.%) (Shatskiy, Borzdov, et al., 2015). This indicates significant differences in crystal structure of K_2CO_3 at 3 and 6 GPa, which is consistent with the phase transition near 4.5 GPa (Figure 14.7).

The crystal structures of different K-Ca phases obtained at 3–6 GPa have not been determined yet. Preliminary single crystal X-ray diffraction data for $K_2Ca_3(CO_3)_4$ show an existence of two quenchable polymorphs at 3 and 6 GPa. Both phases are orthorhombic and have very similar unit cell parameters: a = 7.38749(17) Å, b = 8.8128(2) Å, c = 16.4873(4) Å for the 3 GPa polymorph with space group $P2_12_12_1$; and a = 7.53915(18) Å, b = 8.7799(2) Å, c = 16.1811(4) Å for the 6 GPa polymorph with space group *Pnam*. Higher symmetry of the 6 GPa polymorph together with broadening of its Raman bands suggests that it may be a disordered variety of the ordered 3 GPa counterpart (Arefiev et al., 2019).

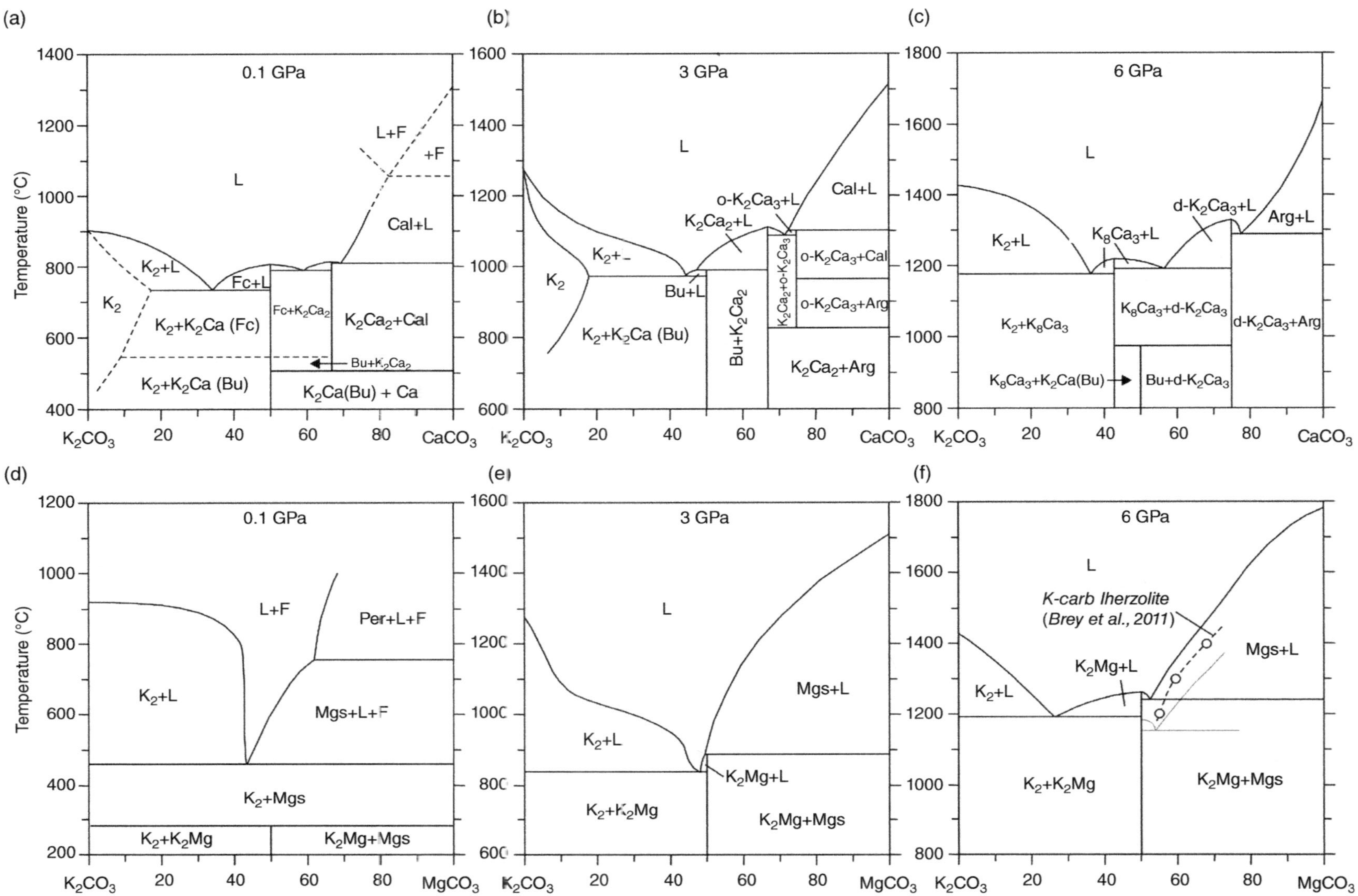

Figure 14.10 Phase relations in binary K_2CO_3–$CaCO_3$ and K_2CO_3–$MgCO_3$ systems at 0.1, 3, and 6 GPa: a = Cooper et al., 1975; d = Ragone et al., 1966. See text for other references. K_2 = K_2CO_3, K_2Ca = $K_2Ca(CO_3)_2$, Bu = buetschliite, Fc = fairchildite, K_2Ca_2 = $K_2Ca_2(CO_3)_3$, Cal = calcite, F = fluid, L = liquid, o-K_2Ca_3 = ordered $K_2Ca_3(CO_3)_4$, Arg = aragonite, K_8Ca_3 = $K_8Ca_3(CO_3)_7$, d-K_2Ca_3 = disordered $K_2Ca_3(CO_3)_4$, Per = periclase, K_2Mg = $K_2Mg(CO_3)_2$, Mgs = magnesite. Dashed lines = approximate boundaries. Melt compositions from K-bearing carbonated peridotite (Brey et al., 2011) is shown at f along with part of K_2CO_3–$FeCO_3$ diagram (gray lines) (see Figure 14.11d). See electronic version for color representation of the figures in this book.

Thus, a range of K-Ca double carbonates changes upon pressure and temperature increase in the following sequence: $K_2Ca(CO_3)_2$ (buetschliite), $K_2Ca_2(CO_3)_3$ (≤0.1 GPa; < 547 °C) → $K_2Ca(CO_3)_2$ (fairchildite), $K_2Ca_2(CO_3)_3$ (≤0.1 GPa; 547–835 °C) → $K_2Ca(CO_3)_2$ (buetschliite), $K_2Ca_2(CO_3)_3$, $K_2Ca_3(CO_3)_4$ (ordered) (3 GPa; 800–1100 °C) → $K_8Ca_3(CO_3)_7$, $K_2Ca(CO_3)_2$ (buetschliite), $K_2Ca_3(CO_3)_4$ (disordered) (6 GPa; 900–1300 °C) (Arefiev et al., 2019).

K_2CO_3–$MgCO_3$. The system K_2CO_3–$MgCO_3$ at $P(CO_2)$ = 3.4 MPa (Eitel & Skaliks, 1929) and at 0.1 GPa (Figure 14.10d) (Ragone et al., 1966) contains one intermediate phase $K_2Mg(CO_3)_2$ below 300 °C. It has an *R3m* space group at 25 MPa (Hesse & Simons, 1982) and breaks down into K_2CO_3 and $MgCO_3$ at higher temperatures. The eutectic is located at 460 °C and 56 mol.% K_2CO_3. Magnesite melts incongruently at 755 °C to periclase + liquid containing 37 mol.% K_2CO_3.

At 3 GPa, two fields appear at subsolidus conditions: K_2CO_3 + $K_2Mg(CO_3)_2$ and $K_2Mg(CO_3)_2$ + $MgCO_3$ with the boundary near 50 mol.% K_2CO_3. The K_2CO_3–$K_2Mg(CO_3)_2$ eutectic is located at 840 °C and 52 mol.% K_2CO_3. $K_2Mg(CO_3)_2$ melts incongruently at 890 °C to magnesite and a liquid containing 51 mol.% K_2CO_3. (Figure 14.10e) (Arefiev et al., 2018). At 6 GPa, the system has two eutectics at 1200 °C and 75 mol.% K_2CO_3 and at 1300 °C and 48 mol.% K_2CO_3, and the same stable intermediate phase, $K_2Mg(CO_3)_2$, which melts congruently near 1250 °C (Shatskiy, Sharygin, Gavryushkin, et al., 2013).

The near-solidus melt of K-bearing carbonated peridotite at 6 GPa and 1200 °C is dominated by K_2CO_3 (40.8 mol.%) and $MgCO_3$ (45.5 mol.%). It also contains 6.0 mol.% $FeCO_3$ and 5.5 mol.% $CaCO_3$. With increasing temperature from 1200 to 1400 °C, the melt evolves towards Mg-Fe-rich compositions. The SiO_2 content of the melt in this temperature range increases from 1.1 to 4.8 mol.% (Brey et al., 2011). The compositions of partial melts are broadly consistent with those in the K_2CO_3–$MgCO_3$ (Figure 14.10f) and K_2CO_3–$FeCO_3$ binaries (Figure 14.11d). Consequently, the partial melt composition in K-rich carbonated lherzolite at 6 GPa and 1200–1400 °C is determined by melting phase relations along the K_2CO_3–(Mg,Fe)CO_3 join. Golubkova et al. (2015) reported a phase transition of $K_2Mg(CO_3)_2$ to monoclinic phase upon compression to 8.1 GPa.

$FeCO_3$–$MgCO_3$. Similar to the $FeCO_3$ phase diagram, there are some uncertainties in phase relations in the binary $FeCO_3$–$MgCO_3$ systems at 3–6 GPa. At 3 GPa, we observed dissociation of siderite at 1100 °C. Accordingly, we constrained an approximate phase diagram (Figure 14.11a). At 6 GPa, a continuous solid solution phase diagram was established with experimental points up to 1700 °C (Shatskiy, Litasov, Ohtani, et al., 2015b). Alternative phase diagrams with asymmetric melting loop were proposed by Kang et al. (2016). They constrained thermodynamic model and calculated phase diagrams in the Fe-rich part of the system with liquidus and solidus minimum located at 10 GPa, 18 mol.% $MgCO_3$, and 1580 °C; and at 20 GPa, 28 mol.% $MgCO_3$, and 1830 °C. At present, this model is too speculative and is not sufficiently supported by experimental data; therefore, we only mention it here.

$FeCO_3$–$CaCO_3$. Subsolidus phase relations in the system $FeCO_3$–$CaCO_3$ were investigated at $P(CO_2)$ = 0.2–0.4 GPa and 300–550 °C (Rosenberg, 1963), 1.5 GPa and 600–800 °C (Goldsmith et al., 1962), 2–3 GPa and 650–800 °C (Davidson, 1994), and at 3.5 GPa and 600–900 °C (Franzolin et al., 2011). The experiments show a large miscibility gap of calcium and iron carbonates at low temperatures, which narrows with increasing temperature and terminates near 1000 °C. Pressure does not affect the topology of diagrams within the studied range. Piston-cylinder experiments in double Pt-graphite capsules (Franzolin et al., 2011) resulted in decomposition of $Fe_{0.75}Ca_{0.25}CO_3$ in the Fe-rich part of the system at 1100 °C and 3.5 GPa:

$$12\,Fe_{0.75}Ca_{0.25}CO_3 = 6\,Fe_{0.5}Ca_{0.5}CO_3 + 2\,Fe_3O_4 + C + 5\,CO_2. \qquad (14.2)$$

At 6 GPa, the $FeCO_3$–$CaCO_3$ system shows melting loop with a minimum on the liquidus/solidus curves established at 1280 °C and 56 mol.% $CaCO_3$ (Figure 14.10b) (Shatskiy et al., 2014). At low temperatures, the existence of solid solutions is limited by the appearance of siderite + ankerite assemblage at about 900 °C and ankerite + aragonite below 1000 °C. Below 880 °C, ankerite is unstable and decomposes into siderite and aragonite (Figure 14.9) (Morlidge et al., 2006; Tao et al., 2014). Aragonite solid solution can be transformed to disordered phase at temperatures above 1200 °C (Figure 14.2). The formation of CO_2 fluid and presence of iron oxide among quenched products of carbonate melt within 0–30 mol.% $CaCO_3$ and 1500–1700 °C indicate incongruent dissolution of siderite in the melt above 1500 °C.

Na_2CO_3–$FeCO_3$. At 6 GPa, the system Na_2CO_3–$FeCO_3$ has eutectic at 1000 °C and 66 mol.% Na_2CO_3 and one intermediate compound $Na_2Fe(CO_3)_2$, which melts incongruently at 1050 °C to siderite and melt with 55 mol.% Na_2CO_3 (Figure 14.11c) (Shatskiy, Rashchenko, et al., 2015).

K_2CO_3–$FeCO_3$. At 6 GPa, the system K_2CO_3–$FeCO_3$ has two eutectics at 1200 °C and 74 mol.% K_2CO_3 and at 1250 °C and 48 mol.% K_2CO_3, and one intermediate compound $K_2Fe(CO_3)_2$, which melts congruently at 1210 °C (Figure 14.11d) (Shatskiy, Litasov, Ohtani, et al., 2015).

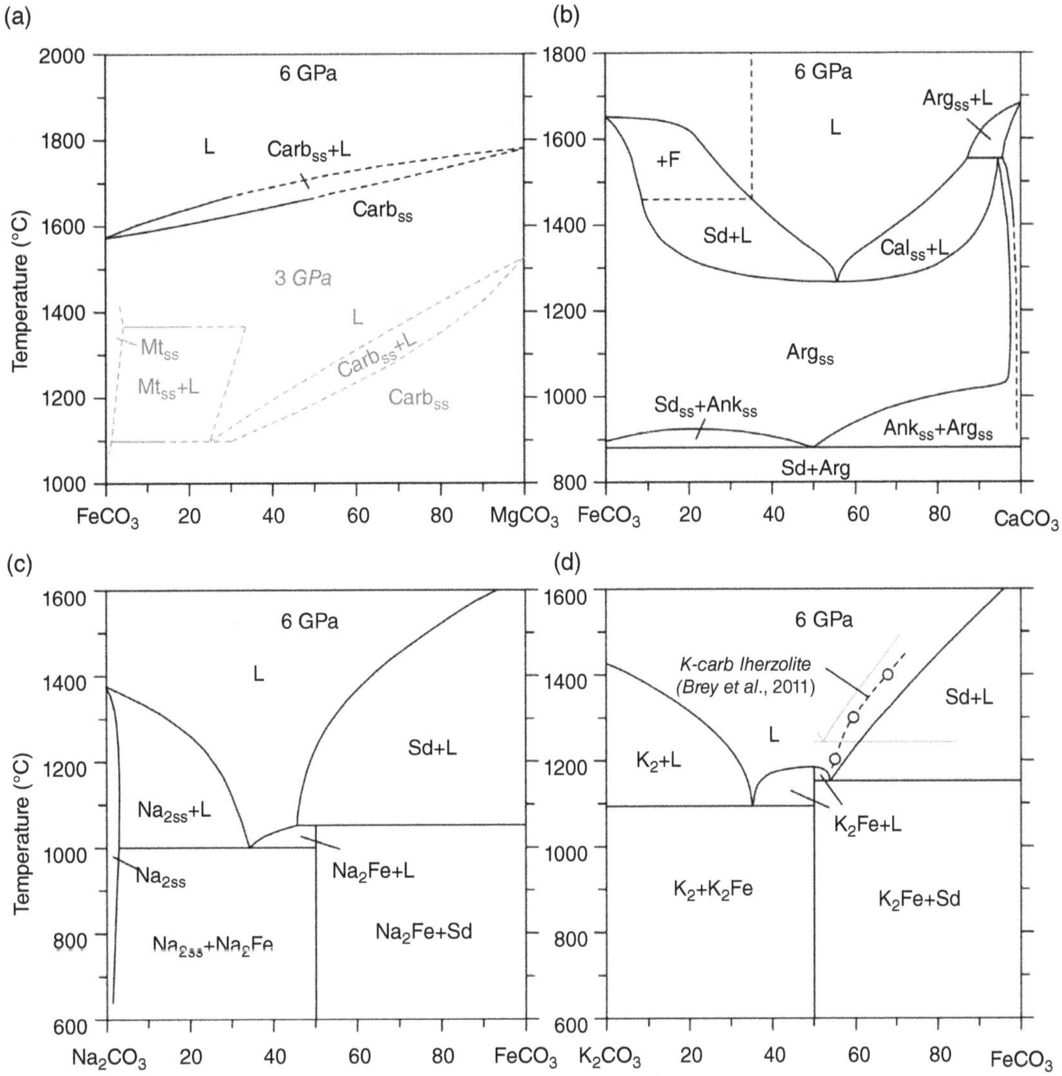

Figure 14.11 Phase relations in the binary siderite-bearing systems at 6 GPa (Shatskiy et al., 2014; Shatskiy, Litasov, Ohtani, et al., 2015; Shatskiy, Rashchenko, et al., 2015). Mt_{SS} = magnetite solid solution, $Carb_{SS}$ = Mg-Fe carbonate solid solution, L = liquid, F = fluid, Sd_{SS} = siderite solid solution, Ank_{SS} = ankerite solid solution, Arg_{SS} = aragonite solid solution, Cal_{SS} = calcite solid solution, Na_{2SS} = Na-carbonate solid solution, Na_2Fe = $Na_2Fe(CO_3)_2$, Sd = siderite, K_2 = K_2CO_3, K_2Fe = $K_2Fe(CO_3)_2$. Dashed lines = estimated boundaries. Gray lines at (a) show phase diagram at 3 GPa (our data in preparation). Melt compositions from K-bearing carbonated peridotite (Brey et al., 2011) is shown at (d) along with part of K_2CO_3–$MgCO_3$ diagram (gray lines) (see Figure 14.10f). See electronic version for color representation of the figures in this book.

14.3.3. Ternary Carbonate Phase Diagrams

$CaCO_3$–$FeCO_3$–$MgCO_3$. Subsolidus phase relations in the ternary system $CaCO_3$–$MgCO_3$–$FeCO_3$ were studied under the same conditions as $CaCO_3$–$FeCO_3$ system (see references in previous section). The resulting diagrams show limited solid solutions of $FeCO_3$ in $CaCO_3$ and a complete series of $FeCO_3$ and $MgCO_3$ solid solutions at 3.5 GPa (Figure 14.12) (Franzolin et al., 2011). The solubility of siderite in calcite and calcite in ankerite increases with temperature. At 6 GPa, incorporation of $FeCO_3$ extends the zone of dolomite–ankerite solid solutions into the low-temperature portion of phase diagram (Shatskiy et al., 2014). At the same time, obtained data suggest a minor decrease in the minimum melting temperature with adding geologically relevant amounts of siderite or ankerite in the $MgCO_3$–$CaCO_3$ system, because the minimum melting temperature of the $FeCO_3$–$CaCO_3$ system is less than 100 °C lower than that of $MgCO_3$–$CaCO_3$ (Figure 14.12). Indeed, no melting

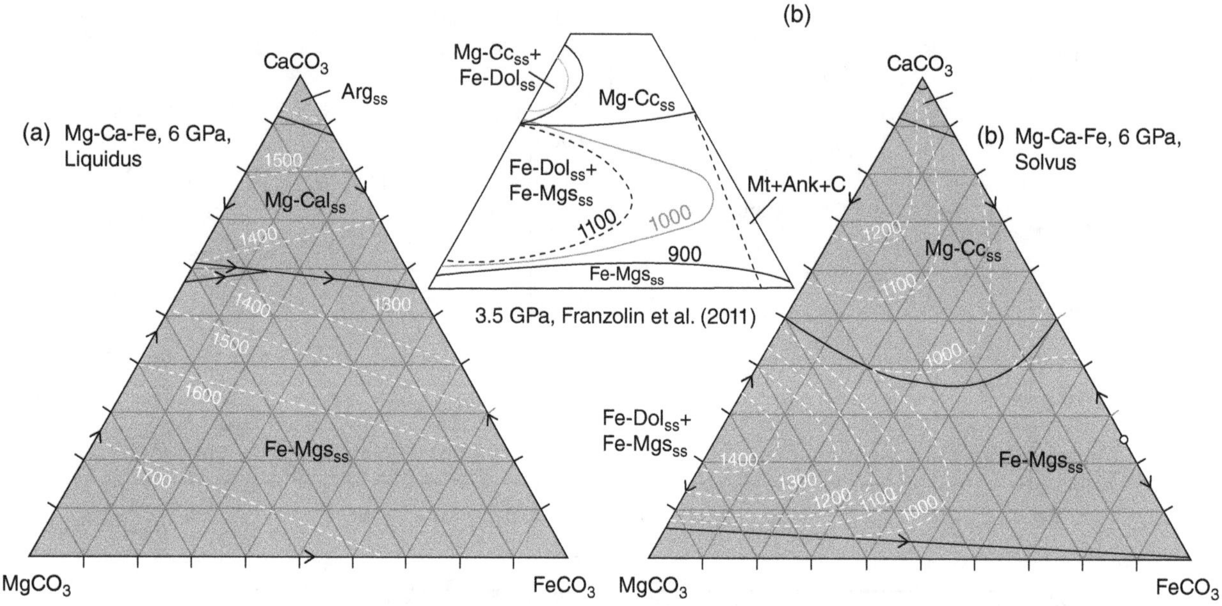

Figure 14.12 Phase relations in the system $CaCO_3$–$FeCO_3$–$MgCO_3$ at 6 GPa (Shatskiy et al., 2014). Inset shows solvus projections at 3.5 GPa and 900–1100 °C (Franzolin et al., 2011). Arg = aragonite; Mg-Cal = Mg-calcite; Mg-Cc = Mg-Ca-carbonate; Fe-Mgs = Fe-Mg carbonate; Fe-Dol = Fe-dolomite; SS = solid solution; Mt = magnetite; Ank = ankerite. See electronic version for color representation of the figures in this book.

occurs along the $FeCO_3$–$Ca_{0.5}Mg_{0.5}CO_3$ join at 1300 °C and 6 GPa.

An addition of Fe-carbonate to dolomite stabilizes this compound at pressures exceeding 20 GPa. Two compositions have been studied. Mao et al. (2011) investigated phase transitions in natural $Ca_{0.99}Mg_{0.92}Fe_{0.08}Mn_{0.01}(CO_3)_2$ dolomite. They observed transition to dolomite II at 17 GPa. Laser heating of the sample in DAC at 27–30 GPa revealed stability of dolomite II to 1400 °C and decomposition to $CaCO_3$ and magnesite at a higher temperature. Heating of the sample at 36–40 GPa indicates appearance of new phase dolomite III, which was stable near 1250 °C.

Merlini, Crichton, et al. (2012) refined crystal structures of phases synthesized from $Ca(Mg_{0.6}Fe_{0.4})(CO_3)_2$ composition using DAC. Dolomite II appears at 17 GPa and crystallizes in triclinic $P\bar{1}$ symmetry with Z = 4. Dolomite III appears at 36 GPa and has the same $P\bar{1}$ space group with Z = 8, but with significantly different structure. Laser heating indicates stability of dolomite III to 2300 °C at 43 GPa, i.e. to temperatures close to the melting point. X-ray diffraction of dolomite III is not consistent with that of dolomite III from Mao et al. (2011). This indicates variety of the stable phases of dolomitic compositions in this pressure range. Moreover, Merlini et al. (2017) reported further transition of dolomite to IIIb phase observed at 36–115 GPa. Its structure corresponds to the *R*3 space group. An appearance of different structures in the pressure range for dolomite III stability indicates the possible existence of several structural modifications with similar energy. The structural transitions in dolomite show similarities with calcite I–II–III transitions. Indeed, $CaCO_3$ has several metastable polymorphs with close energy at high pressures.

At 115 GPa and 2200 °C, new phase dolomite IV was synthesized (Merlini et al., 2017). It is orthorhombic, space group *Pnma*, $a = 10.091(3)$. $b = 8.090(7)$. $c = 4.533(3)$ Å, $V = 370.1(4)$ Å ^{3}, at 115.2 GPa and ambient temperature. The structure is based on the presence of threefold C_3O_9 carbonate rings, with carbon in tetrahedral coordination. The structure of dolomite-IV presents similarities with the structural models proposed for the high-pressure polymorphs of $MgCO_3$.

Ab initio computations indicate that monoclinic *C2/c* Fe-bearing dolomite is more stable than single-cation carbonate mixtures at lower mantle pressures (Solomatova & Asimow, 2018); however, other structures were not considered, and an evolutionary search has not been performed.

Na_2CO_3–$CaCO_3$–$MgCO_3$. Phase relations in the Na_2CO_3–$CaCO_3$–$MgCO_3$ system have been studied in details at 3 GPa and 6 GPa (Figure 14.13a, b) (Podborodnikov et al., 2019; Shatskiy, Litasov, Sharygin, et al., 2016).

At 3 GPa and 700 °C, the system has five intermediate compounds: dolomite, Mg-bearing $Na_2Ca_4(CO_3)_5$ burbankite, $Na_2Ca_3(CO_3)_4$, $Na_4Ca(CO_3)_3$, and eitelite. As temperature increases to 800 °C, the system is complicated by an appearance of Ca-dolomite and Mg-bearing shortite, while $Na_2Ca_4(CO_3)_5$ disappears. $Na_4Ca(CO_3)_3$ decomposes to produce Na carbonate and nyerereite at 850 °C. The latter melts incongruently at 875 °C to form $Na_2Ca_3(CO_3)_4$. Incongruent melting of eitelite, producing magnesite and a liquid, occurs at 925 ± 25 °C. Mg-bearing shortite melts

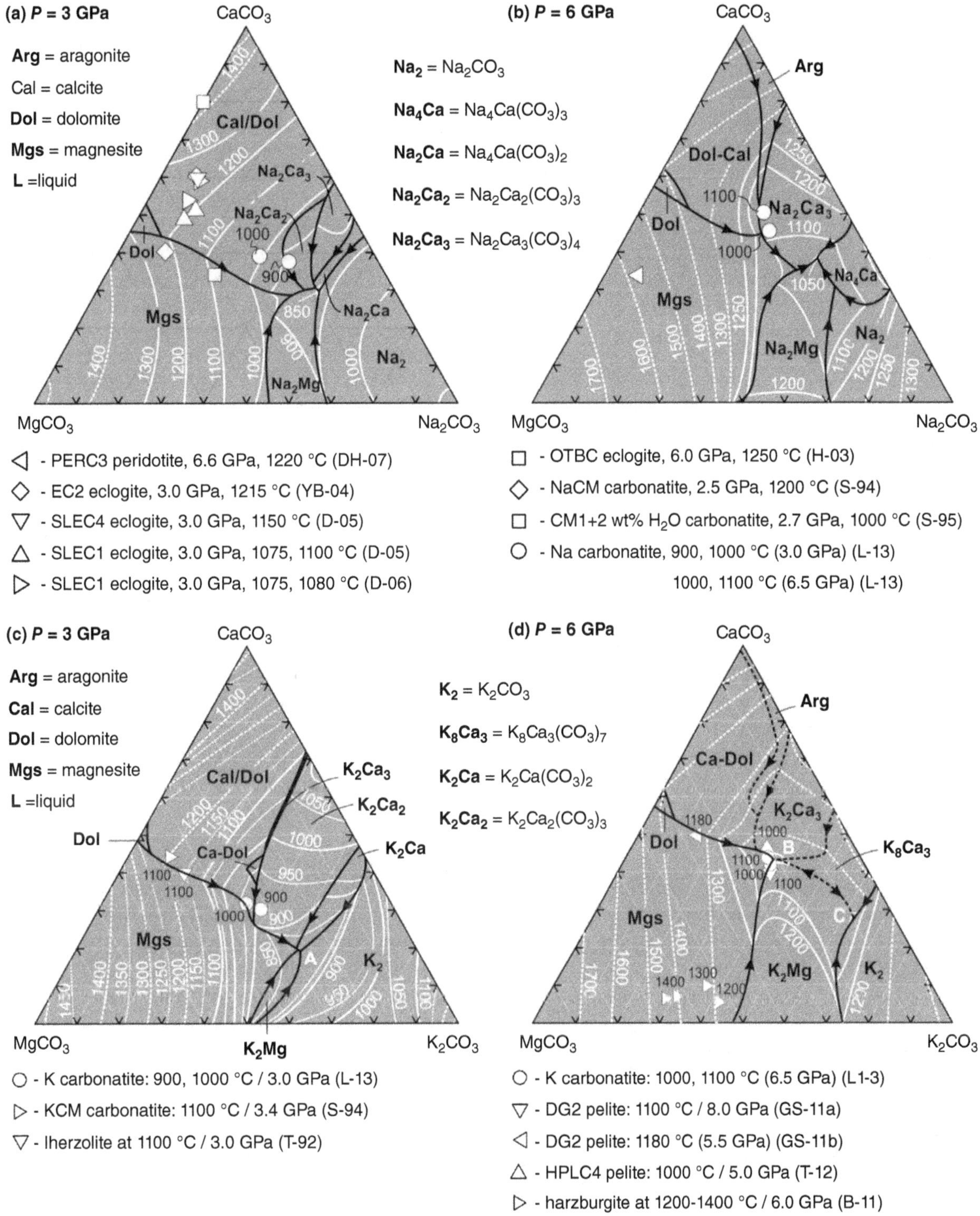

Figure 14.13 Melting phase relations in the Na_2CO_3–$MgCO_3$–$CaCO_3$ (a–b) and K_2CO_3–$MgCO_3$–$CaCO_3$ (c–d) ternary systems at 3 and 6 GPa (Podborodnikov et al., 2019; Shatskiy, Litasov, Palyanov, et al., 2016; Shatskiy, Litasov, Sharygin, et al., 2016). Black lines (dashed where inferred) are boundary lines and white lines (dashed where inferred) are liquidus isotherms with temperatures in °C. Compositions of carbonatite melts revealed from partial melting experiments in silicate–CO_2 systems are shown: DH-07 = (Dasgupta & Hirschmann, 2007a), YB-04 = (Yaxley & Brey, 2004), D-05 = (Dasgupta et al., 2005), D-06 = (Dasgupta et al., 2006), H-03 = (Hammouda, 2003), S-94 = (Sweeney, 1994), S-95 = (Sweeney et al., 1995), L-13 = (Litasov, Shatskiy, Ohtani, & Yaxley, 2013), T-92 = (Thibault et al., 1992), GS-11a and GS-11b = (Grassi & Schmidt, 2011a, 2011b), T-12 = (Tsuno et al., 2012), B-11 = (Brey et al., 2011). See electronic version for color representation of the figures in this book.

incongruently at about 950 °C producing $Na_2Ca_3(CO_3)_4$ and liquid. $Na_2Ca_3(CO_3)_4$ disappears at 1000 °C via incongruent melting to calcite + liquid. The liquidus projection of the studied ternary system has seven primary solidification phase regions for magnesite, dolomite-calcite solid solutions, $Na_2Ca_3(CO_3)_4$, Mg-bearing shortite, nyerereite, eitelite, and Na carbonate (Figure 14.13a). The system has six ternary peritectic points and one minimum on the liquidus at 825 °C and 52 $Na_2CO_3 \cdot 48\ (Ca_{0.62}Mg_{0.38})CO_3$. Since the system has a single eutectic at 3 GPa, there is no thermal barrier preventing continuous liquid fractionation from Na-poor to Na-rich dolomitic compositions more alkaline than eitelite and nyerereite.

At 6 GPa and 900–1000 °C, the system has four intermediate compounds: $Na_2Ca_4(CO_3)_5$ burbankite, $Na_2Ca_3(CO_3)_4$, $Na_4Ca(CO_3)_3$, and $Na_2Mg(CO_3)_2$ eitelite. The Na-Ca compounds dissolve noticeable amounts of Mg component, whereas eitelite dissolves a few percent of Ca component: $Na_2(Ca_{\geq 0.91}Mg_{\leq 0.09})_4(CO_3)_5$, $Na_2(Ca_{\geq 0.94}Mg_{\leq 0.06})_3(CO_3)_4$, $Na_4(Ca_{\geq 0.67}Mg_{\leq 0.33})(CO_3)_3$, and $Na_2(Mg_{\geq .93}Ca_{\leq 0.07})(CO_3)_2$. At 1050 °C, the system is complicated by the appearance of dolomite. Na-Ca burbankite decomposes at 1075 ± 25 °C via congruent melting between 1200 and 1250 °C producing aragonite plus $Na_2Ca_3(CO_3)_4$, $Na_4Ca(CO_3)_3$ and eitelite disappear. $Na_2Ca_3(CO_3)_4$ remains stable over the entire studied temperature range (900–1250 °C). The liquidus projection of the studied ternary system has eight primary solidification phase regions for magnesite, dolomite, calcite-dolomite solid solutions, aragonite, $Na_2Ca_3(CO_3)_4$, $Na_4Ca(CO_3)_3$, and Na_2CO_3 solid solutions (Figure 14.13b). The system has five ternary peritectic reaction points and one minimum on the liquidus at 1050 °C and $48Na_2CO_3 \cdot 5$ $2(Ca_{0.75}Mg_{0.25})CO_3$. The minimum point resembles a eutectic controlled by a four-phase reaction, by which a liquid transforms into three solid phases upon cooling: $Na_2(Ca_{0.94}Mg_{0.06})_3(CO_3)_4$, $Na_4(Ca_{0.67}Mg_{0.33})(CO_3)_3$, and $Na_2(Mg_{0.93}Ca_{0.07})(CO_3)_2$ eitelite.

Taking into account the data from the binary systems at low pressures, we can constrain Na-Ca-Mg double carbonate sequences upon pressure and temperature increase: $Na_2Ca_2(CO_3)_3$ (A*mm*2) shortite, $Na_2Ca(CO_3)_2$ ($P2_1ca$) nyerereite, $Na_2Mg(CO_3)_2$ ($R\bar{3}$) eitelite (0.1 GPa) → $Na_2(Ca_{0.97–0.98}Mg_{0.02–0.03})_4(CO_3)_5$ ($P6_3mc$). $Na_2(Ca_{\geq 0.91}Mg_{\leq 0.09})_3(CO_3)_4$ ($P1n1$). $Na_2(Ca\geq_{0.81}Mg_{0\leq 0.19})(CO_3)_2$ ($R\bar{3}$) nyerereite, $Na_2(Ca_{0.77–0.93}Mg_{0.07–0.23})_2(CO_3)_3$ (A*mm*2) shortite, $Na_4(Ca_{0.90–0.98}Mg_{0.02–0.10})(CO_3)_3$ ($Ia3d$). $Na_2(Mg_{\geq 0.9}Ca_{0\leq 0.1})(CO_3)_2$ ($P2_1ca$) eitelite (3 GPa) → $Na_2(Ca_{\geq 0.87}Mg_{0\leq 0.13})_4(CO_3)_5$ ($P6_3mc$). $Na_2(Ca_{\geq 0.89}Mg_{\leq 0.11})_3(CO_3)_4$ ($P1n1$). $Na_4(Ca\geq_{0.7}Mg_{0\leq 0.3})(CO_3)_3$ ($Ia3d$). $Na_2(Mg_{\geq 0.92}Ca_{0\leq 0.08})(CO_3)_2$ ($P2_1ca$) eitelite (6 GPa).

K_2CO_3–$CaCO_3$–$MgCO_3$. At 3 GPa, the liquidus projection of this ternary system has nine primary solidification phase regions for magnesite, dolomite, Ca-dolomite, calcite-dolomite solid solutions, $K_2Ca_3(CO_3)_4$, $K_2Ca_2(CO_3)_3$, $K_2Ca(CO_3)_2$ buetschliite, $K_2Mg(CO_3)_2$, and K_2CO_3 solid solutions containing up to 24 mol% $CaCO_3$ and less than 2 mol% $MgCO_3$ (Figure 14.13c). The system has six ternary peritectic reaction points and one eutectic at 825 ± 25 °C and $53K_2CO_3 \cdot 47Ca_{0.4}Mg_{0.6}CO_3$ (A) controlled by a four-phase reaction, where a liquid transforms into three solid phases on cooling: $K_2(Mg_{0.78}Ca_{0.22})(CO_3)_2$, $K_2(Ca_{0.70}Mg_{0.30})(CO_3)_2$ buetschliite, and $K_{1.70}Ca_{0.23}Mg_{0.07}CO_3$ solid solution.

At 6 GPa, the diagram has eight primary phase fields for magnesite, dolomite, calcite-dolomite solid solutions, aragonite, $K_2Ca_3(CO_3)_4$, $K_8Ca_3(CO_3)_7$, K_2CO_3, and $K_2Mg(CO_3)_2$. The system has four peritectic points and two eutectics near 1000 °C with compositions $36K_2CO_3 \cdot 64(Ca_{0.65}Mg_{0.35})CO_3$ (B) and $62K_2CO_3 \cdot 38Ca_{0.72}Mg_{0.28}CO_3$ (C) (Figure 14.13d). The eutectic (B) is controlled by three-phase assemblage of $K_2Ca_3(CO_3)_4$ + $K_8Ca_3(CO_3)_7$ + $K_2Mg(CO_3)_2$, whereas the eutectic (C) is controlled by three-phase assemblage of $K_8Ca_3(CO_3)_7$ + K_2CO_3 + $K_2Mg(CO_3)_2$.

Na_2CO_3–$FeCO_3$–$MgCO_3$. The ternary Na_2CO_3–$FeCO_3$–$MgCO_3$ system can be constrained at 6 GPa combining data from the corresponding binary systems: two systems with the intermediate $Na_2(Mg,Fe)(CO_3)_2$ phase, which melts congruently at the Mg-rich side and incongruently at the Fe-rich side, and the $(Mg,Fe)CO_3$ system with complete solid solutions. With decreasing $MgCO_3$ content, the eutectic shifts from 1225 °C and Na# = 48% toward lower temperatures and Na# =50%, where it changes to peritectic, which shifts to 1050 °C and Na# = 55%. The measurable amounts of $MgCO_3$ (up to 9 mol.%) and $FeCO_3$ (up to 3 mol.%) in Na_2CO_3 suggest an existence of the limited range of sodium carbonate solid solutions (Shatskiy, Rashchenko, et al., 2015).

K_2CO_3–$FeCO_3$–$MgCO_3$. Similarly, the ternary K_2CO_3–$FeCO_3$–$MgCO_3$ system at 6 GPa has two cotectic curves. One joining the two binary eutectics at 1250 °C and 48 mol.% K_2CO_3 at the Mg-side and at 1150 °C and 46 mol.% K_2CO_3 at the Fe-side. Another cotectic joins two eutectics at 1200 °C and 74 mol.% K_2CO_3 at the Mg-side and at 1100 °C and 65 mol.% K_2CO_3 at the Fe-side (Shatskiy, Litasov, Ohtani, et al., 2015).

14.4. IMPLICATIONS FOR MELTING IN THE DEEP EARTH'S MANTLE

14.4.1. Carbonates in the Complex Experimental Systems

Alkali-bearing carbonates were detected in several carbonated systems at P-T conditions relevant to the Earth's mantle: carbonated eclogite (Kiseeva et al., 2013; Thomson et al., 2016), pelite (Grassi & Schmidt, 2011b), and model Na- and K-bearing carbonatite (Litasov, Shatskiy, Ohtani, & Yaxley, 2013). Litasov, Shatskiy,

Ohtani, and Yaxley (2013) studied phase relations in model Na-rich (K# = 16, NaK# = 13, and Ca# = 54) and K-rich (K# = 70, NaK# = 9, and Ca# = 33) carbonatites at 3–21 GPa, where K# = 100·K/(K+Na) and NaK# = 100·(K_2O+Na_2O)/(K_2O+Na_2O+CaO+MgO+FeO). It has been found that Na is mainly hosted by Ca-rich crystalline carbonate in both systems over the entire pressure range, whereas K is mostly stored in $K_2Mg(CO_3)_2$. Using a new approach for carbonate analysis (see section 14.2) we re-analyzed and reinterpreted the data from Litasov, Shatskiy, Ohtani, and Yaxley (2013). Thus, unlike the original work, in the "Na-carbonatite" system, $Na_2Ca_4(CO_3)_5$ coexists with dolomite and eitelite at 3 GPa and 750 °C. This is consistent with the subsolidus phase assemblage, dolomite + $Na_2Ca_4(CO_3)_5$ + $Na_2Mg(CO_3)_2$, established in the ternary Na_2CO_3–$CaCO_3$–$MgCO_3$ system at 3 GPa and 700 °C (Podborodnikov et al., 2019). Similarly, at 6 GPa, the $Na_2Ca_3(CO_3)_4$ phase is stable instead of Na-aragonite reported in the original work.

Kiseeva et al. (2013) examined phase relations in carbonated eclogite (GA1-cc and Volga-cc) from 9 to 21 GPa. They found that the subsolidus carbonates are represented by magnesite and nearly pure aragonite to 9 GPa in Volga-cc and to 13 GPa in GA1-cc. At higher pressures, the subsolidus carbonates were represented by magnesite and Na-Ca carbonate with composition similar to that reported by Litasov, Shatskiy, Ohtani, and Yaxley (2013) (10–15 mol.% Na_2CO_3 + K_2CO_3, K# = 3–13, Ca# = 76–81). This carbonate was also identified as aragonite based on the Raman spectra. According to the binary and ternary carbonate systems at 3 and 6 GPa, $CaCO_3$ aragonite does not dissolve Na as well as K, Mg, and Fe. Instead, incorporation of Na into Ca-bearing carbonates form $Na_2Ca_4(CO_3)_5$, and $Na_2Mg(CO_3)_2$. However, the Raman spectra of $Na_2Ca_4(CO_3)_5$ exhibit the main bands in the same spectral region as aragonite, which can lead to misinterpretation. The Na-deficit relative to that required by the $Na_2Ca_4(CO_3)_5$ and $Na_2Mg(CO_3)_2$ stoichiometries is presumably due to the migration of Na under a stationary electron beam for both WDS and EDS analyses discussed in section 14.2.

Grassi and Schmidt (2011b) determined the phase relations in the carbonated pelite systems (DG2 and AM) from 5.5 to 23.5 GPa. They found that the decomposition of clinopyroxene at ≥ 16 GPa causes Na-Ca carbonate (16–20 mol.% Na_2CO_3 + K_2CO_3, K# = 1–7, Ca# = 73–84) to replace aragonite. Similar tendency was established in carbonated MORB composition. It was found that dissolution of Na-poor pyroxene components into coexisting garnet at pressures just above 13 GPa results in redistribution of Na from silicate (clinopyroxene) to Na-Ca carbonate $(Na_{0.97}K_{0.03})_2(Ca_{0.86}Mg_{0.11}Fe_{0.03})_4(CO_3)_5$, lowering the solidus by ~200 °C and yielding a Na-rich carbonatite melt (Thomson et al., 2016). The composition of this carbonate is very similar to the Na-Ca carbonate in Grassi and Schmidt (2011b) and to the $Na_2(Ca_{\geq 0.87}Mg_{\leq 0.13})_4(CO_3)_5$ compound established in the Na_2CO_3–$CaCO_3$–$MgCO_3$ system at 6 GPa and 900–1000 °C and at 3 GPa and 700 °C. The Na-Ca carbonate similar to $Na_2Ca_4(CO_3)_5$ was also detected among the run products from the experiment in natural Udachnaya-East kimberlite system at 6.5 GPa and 900 °C (Sharygin et al., 2015).

In the DG2 composition (dry pelite) at 22 and 23.5 GPa and temperature ≤ 1400 °C, carbonate with an approximate stoichiometry $Na_2Ca_2(CO_3)_3$ and almost no Mg, Fe, K has been identified (Grassi & Schmidt, 2011b). This carbonate resembles stoichiometry of Mg-free shortite, which is not stable at 3 GPa and ≥ 800 °C (Podborodnikov, Shatskiy, Arefiev, Rashchenko, et al., 2018) and at 6 GPa and ≥ 900 °C (Shatskiy et al., 2016b). In the Volga-cc composition (eclogite) at 21 GPa and 1200 °C, a carbonate with an empirical formula $(Na_{0.89}K_{0.11})_2(Ca_{0.68}Mg_{0.24}Fe_{0.08})(CO_3)_2$ was detected (Kiseeva et al., 2013). This carbonate resembles stoichiometry of nyerereite, which is stable at 3 GPa and 850 °C (Figure 14.7d). but disappears at 6 GPa and ≥ 900 °C (Figure 14.10) (Shatskiy et al., 2016b) according to the reaction: $Na_2Ca(CO_3)_2$ = $Na_4Ca(CO_3)_3$ + $Na_2Ca_3(CO_3)_4$ (Podborodnikov, Shatskiy, Arefiev, Rashchenko, et al., 2018). The appearance of carbonates with the same stoichiometries as shortite and neyrereite at pressures ≥ 21–22 GPa may indicate the existence of high-pressure polymorphs of these phases.

14.4.2. The Lower Temperature Stability Limits of Carbonatite Melts at 100–200 km Depths

Partial fusion of mantle peridotite/eclogite with CO_2 at depths shallower than 80 km produces silicate melts, whereas at greater depths the first incipient liquids, produced at much lower temperatures, are essentially carbonatitic (Wyllie & Huang, 1976; Yaxley & Brey, 2004; Yaxley & Green, 1994). Carbonatite melts, which are extremely effective in transporting incompatible elements, cause metasomatic modification of the shallow subcontinental mantle (Green & Wallace, 1988). The geochemical signatures of carbonatite metasomatism are commonly recognized in eclogite and peridotite xenoliths from alkaline basalts and kimberlites (Pokhilenko et al., 2015; Yaxley et al., 1991, 1998). It is therefore important to discuss the temperature range of carbonatite melt stability and their compositional trends at depths 100–200 km.

The lower temperature stability limits of dolomitic melt inferred from the liquidus-solidus minima on the join $CaCO_3$–$MgCO_3$ (CM–CO_2) are 1225 ± 25 °C at 3 GPa and 1400 °C at 6 GPa (Shatskiy et al., 2018). These temperatures coincide, within experimental uncertainty, with the solidi of the CaO–MgO–SiO_2 ± Al_2O_3–CO_2 (CMS–CO_2 and CMAS–CO_2) systems modeling carbonated peridotite (Gudfinnsson & Presnall, 2005; Wyllie & Huang, 1976) (Figure 14.14). Similar to CM–CO_2, low-degree partial melting of CMS–CO_2 and CMAS–CO_2 yields silica-poor dolomitic melt. Sodium-bearing (NCMAS–CO_2) system and natural compositions (PERC and Hawaiian pyrolite) modeling carbonated peridotites at 3 GPa, revealed lower solidi temperatures, 1150 ± 50, 1090 ± 15, and ~1040 °C, respectively (Dasgupta & Hirschmann, 2006; Falloon & Green, 1989; Moore, 2012). The solidi of carbonated eclogites fall into the same temperature range: 1140 ± 40 °C (Yaxley & Brey, 2004), 1063–1113 °C (Dasgupta et al., 2005). As it can be seen in Figure 14.14, the experimentally inferred solidi of carbonated peridotites and eclogites cross the cratonic geotherms at depths ranging from 100 to 200 km and greater, yielding formation of essentially dolomitic melts. It was observed that near-solidus melts in the model peridotite–CO_2 and eclogite–CO_2 systems replicating natural compositions show strong enrichments in alkalis (Figure 14.13a) (Dasgupta & Hirschmann, 2006, 2007a; Dasgupta et al., 2004; Sweeney et al., 1995; Wallace & Green, 1988; Yaxley & Brey, 2004).

In the Na_2CO_3–$CaCO_3$–$MgCO_3$ system (NCM–CO_2) at 3 GPa, the compositions of melt in equilibrium with Ca-Mg carbonates show systematic variation with temperature from alkali-poor dolomitic at about 1225 °C, near the CM–CO_2 eutectic, toward sodic dolomitic (with 52 mol.% Na_2CO_3) at 825 °C, near the NCM–CO_2 minimum melting (Figure 14.14b). Sweeney (1994) has reported carbonatite melt with NaK# = 11, K# = 22, and Ca# = 45 established in equilibrium with phlogopite lherzolite at 1200 °C and 2.5 GPa in the NaCM system (Figure 14.13a). Later, Sweeney et al. (1995) have published new results on the phase relations of a "primary" sodic dolomitic carbonatite (CM1). determined by Wallace and Green (1988) to be in equilibrium with an amphibole lherzolite assemblage. They found that in the CM1 + 2 wt.% H_2O starting composition annealed at 2.7 GPa and 1000 °C, the carbonatite melt has NaK# = 26, K# = 2, and Ca# = 46. Litasov, Shatskiy, Ohtani, & Yaxley (2013) have reported experimental data on model sodic carbonatite, which is compositionally similar to a low-degree partial melt of carbonated eclogite at 3–10 GPa. We have reanalyzed the melt composition obtained by Litasov, Shatskiy, Ohtani, & Yaxley (2013) at 3 GPa and 900 °C and found that the melt has NaK# = 41, K# = 13, and Ca# = 64 (Figure 14.13a). These data show close agreement with the melt compositions observed in the related ternary carbonate systems.

The P-T plot in Figure 14.14b shows isopleths of Na_2CO_3 in the melt inferred from the carbonate phase

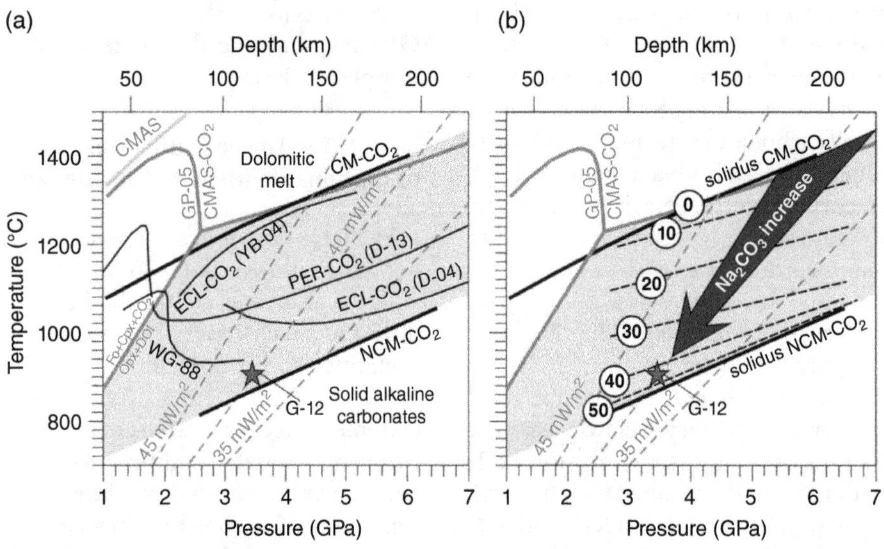

Figure 14.14 (a) Solidi of the carbonate and carbonate-silicate systems, compared with 35–45 mW/m² cratonic geotherms (gray dashed lines) (Hasterok & Chapman, 2011). G-12 (star) = P-T estimations for spinel harzburgite with primary Na-K-Ca-Mg carbonatite inclusions in metasomatic ilmenite (Giuliani et al., 2012). CM–CO_2 = $CaCO_3$–$MgCO_3$; NCM–CO_2 = Na_2CO_3–$CaCO_3$–$MgCO_3$; CMAS = CaO–MgO–SiO_2–Al_2O_3 (Gudfinnsson & Presnall, 1996); CMAS–CO_2 = CaO–MgO–SiO_2–Al_2O_3–CO_2 (Gudfinnsson & Presnall, 2005); D-13 = peridotite–CO_2 (Dasgupta et al., 2013); WG-88 = carbonate- and amphibole-bearing peridotite (0.3% H_2O and 0.5%–2.5% CO_2) (Wallace & Green, 1988); YB-04 = eclogite–CO_2 (Yaxley & Brey, 2004); D-04 = eclogite–CO_2 (Dasgupta et al., 2004). (b) P-T plot illustrating enrichment of magnesiocarbonatite melt with sodium during its upward percolation in subcratonic mantle lithosphere constrained. Black dashed lines are Na_2CO_3 isopleths with numbers indicating the Na_2CO_3 content in the melt in mol.%. See electronic version for color representation of the figures in this book.

diagrams at 3 and 6 GPa. The cratonic geotherms cross the isopleths, so that the carbonatite melt percolating upward via the continental mantle lithosphere should become progressively enriched by Na, evolving from alkali-poor dolomitic composition toward sodic dolomitic with Na_2CO_3 content 52 mol.% at the NCM–CO_2 solidus. These melts are buoyant and mobile, owing their low viscosity (Kono et al., 2014; Stagno et al., 2018). Upward percolation of such melts could yield metasomatic enrichment and wehrlitization in the shallower levels of subcratonic mantle lithosphere, documented in mantle xenoliths (Rudnick et al., 1993; Yaxley et al., 1998).

14.4.3. Implication of Carbonate Phase Diagrams to Mantle-Derived Carbonatite Inclusions

Kimberlites, lamproites, and alkaline basalts carry debris of the underlying lithospheric mantle. Many of these xenoliths show chemical, textural, and mineralogical evidences for metasomatism by fluids or melts within the mantle (Green & Wallace, 1988; Menzies & Hawkesworth, 1986). Giuliani et al. (2012) reported Na-K-Ca-Mg carbonatite melts preserved in primary multiphase inclusions hosted by metasomatic ilmenite grains in a spinel harzburgite from Bultfontein kimberlite (Kimberley, South Africa). Mineral thermometry indicates that the spinel harzburgite crystallized at ~860 °C, which corresponds to pressures near 3.5 GPa (~115 km) on a 40 mW/m^2 geotherm. This temperature corresponds to the low stability limit of Na-Ca-Mg carbonate melt according to the phase relations in the Na_2CO_3–$CaCO_3$–$MgCO_3$ system (Figure 14.14). The melts, detected by Giuliani et al. (2012). should be extremely enriched in alkalis, containing about 50 mol.% $(Na,K)_2CO_3$ and have Ca# ~60.

The sequence of Na-Ca and K-Ca carbonates established by phase diagram studies makes it possible to correctly interpret alkali carbonate inclusions in kimberlitic minerals. According to the phase relations at 6 GPa, shortite $Na_2Ca_2(CO_3)_3$ and nyerereite $Na_2Ca(CO_3)_2$, the double carbonates observed at 0.1 GPa (Cooper et al., 1975), are not stable in the deep mantle (Shatskiy, Sharygin, Litasov, et al., 2013). However, nyerereite is stable at the shallow mantle pressures at least up to 3 GPa and can be of the deep origin. Pure $Na_2Ca_2(CO_3)_3$ shortite is not stable at 3 GPa and ≥ 800 °C, however partial Mg substitution for Ca stabilizes shortite as a liquidus phase at 3 GPa and 800–900 °C. Consequently, the findings of nyerereite and shortite among daughter phases in melt inclusions in olivine from the sheared garnet peridotites (Golovin et al., 2017) are consistent with their mantle origin at least at pressures of kimberlite magma ascent to the surface.

14.4.4. Carbonates in the Lower Mantle

The highly reduced nature of the lower mantle and, possibly, part of the upper mantle and transition zone indicates the instability of carbonates in the deep Earth's interior. At the conditions of the iron-wüstite buffer, which are claimed for most of the mantle (Frost & McCammon, 2008), diamond or Fe-carbide should be the major host for carbon.

Ca-Mg carbonate cannot coexist with metallic iron even at pressures of the D" layer at CMB and react to form Fe-carbide or diamond (Dorfman et al., 2018; Martirosyan et al., 2019). However, this may not be the case for Fe-carbonate and its related compounds stable under the lower mantle conditions. Moreover, oxidation state of the mantle can be strongly heterogenous. Oxidized subducting slabs or mantle wedge materials can be entrained in convective mixing but can remain heterogenous for a long time due to sluggish kinetics of exchange reactions, even in partially molten matrices. In this case, carbonate or carbonatite melt can survive in the deep mantle, even when surrounded by the reduced material of the ambient mantle (Martirosyan et al., 2016) and be submerged as deep as CMB. This possibility is clearly indicated by carbonate inclusions in superdeep diamonds originating from the uppermost lower mantle (Brenker et al., 2007; Kaminsky, 2012).

There are several possible scenarios for the interaction of subducted material with rocks in the D" layer and the CMB. In the first model, subducted rocks will never appear in contact with the CMB and metallic iron, but will interact only with the uppermost layer of dense D" rocks before being incorporated in mantle convection or plumes. In this case, subducted carbonates will be mostly transformed to diamond (Maeda et al., 2017). A second scenario assumes active reaction of subduction materials with the D" and even with metal from the core. In this case, we may expect formation of diamond in the top portions of the slab/D′′ contact layer and Fe-carbide if carbonate or diamond appears in the direct contact with metallic Fe (Dorfman et al., 2018; Martirosyan et al., 2019) (Figure 14.15).

If both hydrogen and carbon penetrate to the D" depths with subduction slabs they can be remobilized. The reactions of Fe with mixed carbon and hydrogen have been modeled only theoretically. Belonoshko et al. (2015) showed formation of hydrocarbons from the interaction of Fe with C–O–H or H_2O–CO_2 fluids. These hydrocarbons can serve as fusible components for the formation and movement of mantle plumes. It is important to note that Fe is preferentially bonded to carbon and oxygen so that these elements may be consumed by the core. In contrast, hydrogen does not form bonds with Fe and is mostly involved into plume recycling (Litasov & Shatskiy, 2018). This observation has not yet been verified experimentally.

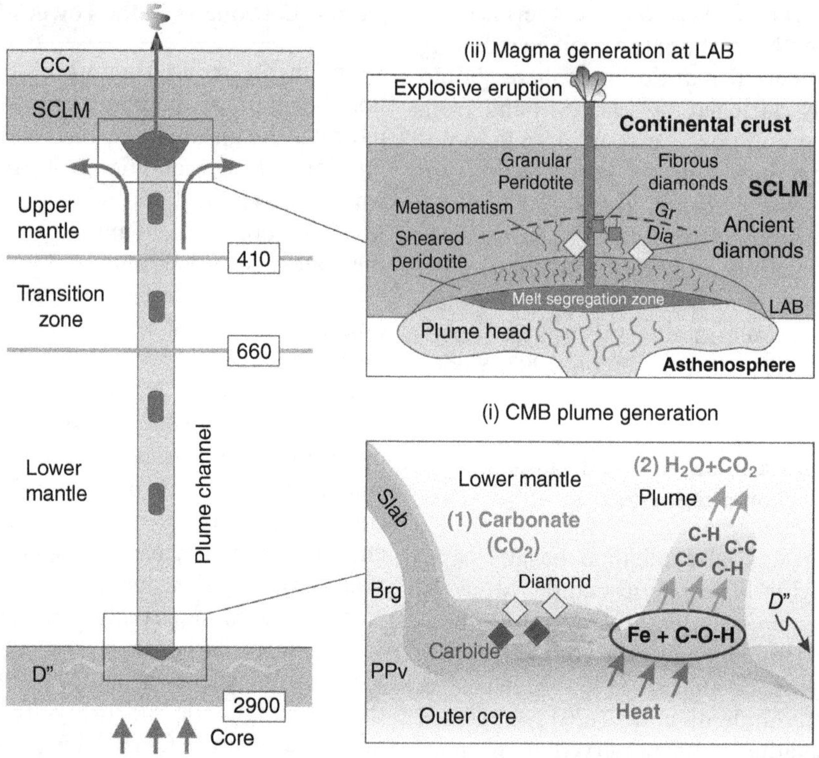

Figure 14.15 Schematic model for mantle plume generation at the core-mantle boundary (CMB) associated with burial and remobilization of carbon-bearing compounds. Right panels, bottom to top: (i) Carbonate reduction at the CMB and generation of the CMB plume. In hydrogen-free conditions carbonates are reduced to diamond in contact with reduced lithologies of the D″ layer and to carbide if they encounter metallic Fe. In this case, remobilization of carbon from the CMB is difficult. In hydrous conditions (H_2O+CO_2), carbon and hydrogen react with metallic iron or reduced rocks to form hydrocarbons (shown as C–C and C–H bonds), which can enhance formation of mantle plumes (Belonoshko et al., 2015; Martirosyan et al., 2019). (ii) Carbonatite or hydrocarbon-bearing hydrous melts approach the lithosphere-asthenosphere boundary (LAB), where they can solidify to form the source for later magmatism or, in case of high enough capacity, immediately cause lithosphere erosion, deformation, and metasomatism with formation of carbonatite, kimberlite, or other alkaline magmas. The examples for kimberlite magmas are shown in Pokhilenko et al., 2015; Sharygin et al., 2015; and Shatskiy, Litasov, et al., 2017). Abbreviations: OC = oceanic crust, CC = continental crust, SCLM = subcontinental lithospheric mantle, Gr = graphite, Dia = diamond, Brg = bridgmanite, Ppv = postperovskite (Mg,Fe)SiO_3. See electronic version for color representation of the figures in this book.

ACKNOWLEDGMENTS

This work was supported by Russian Science Foundation (No 14-17-00609-P) and performed under the Deep Carbon Observatory program.

REFERENCES

Agashev, A. M., D. A. Ionov, N. P. Pokhilenko, A. V. Golovin, Y. Cherepanova, & I. S. Sharygin (2013). Metasomatism in lithospheric mantle roots: Constraints from whole-rock and mineral chemical composition of deformed peridotite xenoliths from kimberlite pipe Udachnaya, *Lithos*, *160–161*, 201–215.

Al-Shemali, M., & A. I. Boldyrev (2002). Search for ionic orthocarbonates: Ab initio study of Na_4CO_4. *The Journal of Physical Chemistry A*, *106*, 8951–8954.

Arapan, S., J. S. De Almeida, & R. Ahuja (2007). Formation of sp^3 hybridized bonds and stability of $CaCO_3$ at very high pressure. *Physical Review Letters*, *98*, 268501.

Arefiev, A. V., A. Shatskiy, I. V. Podborodnikov, S. V. Rashchenko, A. D. Chanyshev, & K. D. Litasov (2019). The system K_2CO_3–$CaCO_3$ at 3 GPa: link between phase relations and variety of K–Ca double carbonates at ≤ 0.1 and 6 GPa. *Physics and Chemistry of Minerals*, *46*, 229–244.

Arefiev, A. V., A. Shatskiy, I. V. Podborodnikov, S. V. Rashchenko, & K. D. Litasov (2018). The system K_2CO_3–$MgCO_3$ at 3 GPa. *High Pressure Research*, *38*, 422–439.

Bagdassarov, N. S., & A. B. Slutskii (2003). Phase transformations in calcite from electrical impedance measurements, *Phase Transitions*, *76*, 1015–1028.

Bayarjargal, L., C.-J. Fruhner, N. Schrodt, & B. Winkler (2018). $CaCO_3$ phase diagram studied with Raman spectroscopy at pressures up to 50 GPa and high temperatures and DFT modeling. *Physics of the Earth and Planetary Interiors*, *281*, 31–45.

Becht, H. Y., & R. Struikmans (1976). A monoclinic high-temperature modification of potassium carbonate. *Acta Crystallographica B, 32*, 3344–3346.

Belonoshko, A. B., T. Lukinov, A. Rosengren, T. Bryk, & K. D. Litasov (2015). Synthesis of heavy hydrocarbons at the core-mantle boundary, *Scientific Reports, 5*, 18382.

Bobrovsky, S. V., V. M. Gogolev, B. V. Zamyshlyaev, V. P. Lozhkina, and V. V. Rasskazov (1976). The study of thermal decomposition influence on the spallation velocity for strong shock waves in solids. *Soviet Mining Science, 3*, 49–57.

Boulard, E., N. Menguy, A. L. Auzende, K. Benzerara, H. Bureau, D. Antonangeli, A. Corgne, et al. (2012). Experimental investigation of the stability of Fe-rich carbonates in the lower mantle. *Journal of Geophysical Research-Solid Earth, 117*, B02208, doi: 10.1029/2011jb008733.

Brenker, F. E., C. Vollmer, L. Vincze, B. Vekemans, A. Szymanski, K. Janssens, I. Szaloki, et al. (2007). Carbonates from the lower part of transition zone or even the lower mantle. *Earth and Planetary Science Letters, 260*, 1–9.

Brey, G. P., V. K. Bulatov, & A. V. Girnis (2011). Melting of K-rich carbonated peridotite at 6–10 GPa and the stability of K-phases in the upper mantle. *Chemical Geology, 281*, 333–342.

Brey, G. P., & T. Kohler (1990). Geothermobarometry in four-phase lherzolites II. New thermobarometers, and practical assessment of existing thermobarometers. *Journal of Petrology, 31*, 1353–1378.

Buob, A., R. W. Luth, M. W. Schmidt, & P. Ulmer (2006). Experiments on $CaCO_3$–$MgCO_3$ solid solutions at high pressure and temperature, *American Mineralogist, 91*, 435–440.

Bykova, E., L. Dubrovinsky, N. Dubrovinskaia, M. Bykov, C. McCammon, S. Ovsyannikov, H.-P. Liermann, et al. (2016). Structural complexity of simple Fe_2O_3 at high pressures and temperatures. *Nature Communications, 7*, 10661.

Byrnes, A. P., & P. J. Wyllie (1981). Subsolidus and melting relations for the join $CaCO_3$–$MgCO_3$ at 10 kbar, *Geochimica et Cosmochimica Acta, 45*, 321–328.

Cancarevic, Z. P., J. C. Schon, & M. Jansen (2007). Possible existence off alkali metal orthocarbonates at high pressure. *Chemistry-a European Journal, 13*, 7330–7348.

Cerantola, V., E. Bykova, I. Kupenko, M. Merlini, L. Ismailova, C. McCammon, M. Bykov, et al. (2017). Stability of iron-bearing carbonates in the deep Earth's interior. *Nature Communications, 8*, 15960.

Cooper, A. F., J. Gittins, & O. F. Tuttle (1975). The system Na_2CO_3–K_2CO_3–$CaCO_3$ at 1 kilobar and its significance in carbonatite petrogenesis. *American Journal of Science, 275*, 534–560.

Dalton, J. A., & D. C. Presnall (1998). Carbonatitic melts along the solidus of model lherzolite in the system CaO–MgO–Al_2O_3–SiO_2–CO_2 from 3 to 7 GPa. *Contributions to Mineralogy and Petrology, 131*, 123–135.

Dasgupta, R. (2013). Ingassing, storage, and outgassing of terrestrial carbon through geologic time. *Reviews in Mineralogy and Geochemistry, 75*, 183–229.

Dasgupta, R., & M. M. Hirschmann (2006). Melting in the Earth's deep upper mantle caused by carbon dioxide. *Nature, 440*, 659–662.

Dasgupta, R., & M. M. Hirschmann (2007a). Effect of variable carbonate concentration on the solidus of mantle peridotite. *American Mineralogist, 92*, 370–379.

Dasgupta, R., & M. M. Hirschmann (2007b). A modified iterative sandwich method for determination of near-solidus partial melt compositions. II. Application to determination of near-solidus melt compositions of carbonated peridotite. *Contributions to Mineralogy and Petrology, 154*, 647–661.

Dasgupta, R., & M. M. Hirschmann (2010). The deep carbon cycle and melting in Earth's interior. *Earth and Planetary Science Letters, 298*, 1–13.

Dasgupta, R., M. M. Hirschmann, & N. Dellas (2005). The effect of bulk composition on the solidus of carbonated eclogite from partial melting experiments at 3 GPa, *Contributions to Mineralogy and Petrology, 149*, 288–305.

Dasgupta, R., M. M. Hirschmann, & K. Stalker (2006). Immiscible transition from carbonate-rich to silicate-rich melts in the 3 GPa melting interval of eclogite plus CO_2 and genesis of silica-undersaturated ocean island lavas. *Journal of Petrology, 47*, 647–671.

Dasgupta, R., M. M. Hirschmann, & A. C. Withers (2004). Deep global cycling of carbon constrained by the solidus of anhydrous, carbonated eclogite under upper mantle conditions, *Earth and Planetary Science Letters. 227*, 73–85.

Dasgupta, R., A. Mallik, K. Tsuno, A. C. Withers, G. Hirth, & M. M. Hirschmann (2013). Carbon-dioxide-rich silicate melt in the Earth's upper mantle. *Nature, 493*, 211–215.

Davidson, P. M. (1994). Ternary iron, magnesium, calcium carbonates: A thermodynamic model for dolomite as an ordered derivative of calcite-structure solutions. *American Mineralogist, 79*, 332–339.

Depater, C. J. (1979). Disordered structure of Na_2CO_3 at 400 °C, *Physica B & C, 96*, 89–95.

Dewaele, A., G. Fiquet, & P. Gillet (1998). Temperature and pressure distribution in the laser-heated diamond-anvil cell. *Review of Scientific Instruments, 69*, 2421–2426.

Dinnebier, R. E., S. Vensky, M. Jansen, & J. C. Hanson (2005). Crystal structures and topological aspects of the high-temperature phases and decomposition products of the alkali-metal oxalates $M_2(C_2O_4)$ (M = K, Rb, Cs). *Chemistry-a European Journal, 11*, 1119–1129.

Dobrzhinetskaya, L. F., R. Wirth, & H. W. Green (2006). Nanometric inclusions of carbonates in Kokchetav diamonds from Kazakhstan: A new constraint for the depth of metamorphic diamond crystallization. *Earth and Planetary Science Letters, 243*, 85–93.

Dorfman, S. M., J. Badro, F. Nabiei, V. B. Prakapenka, M. Cantoni, & P. Gillet (2018). Carbonate stability in the reduced lower mantle. *Earth and Planetary Science Letters, 489*, 84–91.

Dusek, M., G. Chapuis, M. Meyer, & V. Petricek (2003). Sodium carbonate revisited. *Acta Crystallographica B, 59*, 337–352.

Eitel, W., & W. Skaliks (1929). Some double carbonates of alkali and earth alkali. *Zeitschrift Fur Anorganische Und Allgemeine Chemie, 183*, 263–286.

Falloon, T. J., & D. H. Green (1989). Solidus of carbonated fertile peridotite. *Earth and Planetary Science Letters, 94*, 364–370.

Foley, S. F., G. M. Yaxley, A. Rosenthal, S. Buhre, E. S. Kiseeva, R. P. Rapp, & D. E. Jacob (2009). The composition of near-solidus melts of peridotite in the presence of CO_2 and H_2O between 40 and 60 kbar. *Lithos*, *112*(S1). 274–283.

Franzolin, E., M. W. Schmidt, & S. Poli (2011). Ternary Ca–Fe–Mg carbonates: Subsolidus phase relations at 3.5 GPa and a thermodynamic solid solution model including order/disorder. *Contributions to Mineralogy and Petrology*, *161*, 213–227.

Frost, D. J., & C. A. McCammon (2008). The redox state of Earth's mantle. *Annual Review of Earth and Planetary Sciences*, *36*, 389–420.

Gavryushkin, P. N., A. Behtenova, Z. I. Popov, V. V. Bakakin, A. Y. Likhacheva, K. D. Litasov, & A. Gavryushkin (2016). Toward analysis of structural changes common for alkaline carbonates and binary compounds: Prediction of high-pressure structures of Li_2CO_3, Na_2CO_3, and K_2CO_3. *Crystal Growth & Design*, *16*, 5612–5617.

Gavryushkin, P. N., N. S. Martirosyan, T. M. Inerbaev, Z. I. Popov, S. V. Rashchenko, A. Y. Likhacheva, S. S. Lobanov, et al. (2017). Aragonite-II and $CaCO_3$-VII: new high-pressure, high-temperature polymorphs of $CaCO_3$. *Crystal Growth & Design*, *17*, 6291–6296.

Ghosh, S., E. Ohtani, K. D. Litasov, & H. Terasaki (2009). Solidus of carbonated peridotite from 10 to 20 GPa and origin of magnesiocarbonatite melt in the Earth's deep mantle. *Chemical Geology*, *262*, 17–28.

Giuliani, A., V. S. Kamenetsky, D. Phillips, M. A. Kendrick, B. A. Wyatt, & K. Goemann (2012). Nature of alkali-carbonate fluids in the sub-continental lithospheric mantle. *Geology*, *40*, 967–970.

Goldsmith, J. R., D. L. Graf, J. Witters, & D. A. Northrop (1962). Studies in the system $CaCO_3$–$MgCO_3$–$FeCO_3$: 1. Phase relations. 2. A method for major-element spectrochemical analysis. 3. Compositions of some Ferroan dolomites. *Journal of Geology*, *70*, 659–688.

Goldsmith, J. R., & H. C. Heard (1961). Subsolidus phase relations in the system $CaCO_3$-$MgCO_3$. *Journal of Geology*, *69*, 45–74.

Golovin, A. V., I. S. Sharygin, V. S. Kamenetsky, A. V. Korsakov, & G. M. Yaxley (2018). Alkali-carbonate melts from the base of cratonic lithospheric mantle: links to kimberlites. *Chemical Geology*, *483*, 261–274.

Golovin, A. V., I. S. Sharygin, & A. V. Korsakov (2017). Origin of alkaline carbonates in kimberlites of the Siberian craton: Evidence from melt inclusions in mantle olivine of the Udachnaya-East pipe. *Chemical Geology*, *455*, 357–375.

Golubkova, A., M. Merlini, & M. W. Schmidt (2015). Crystal structure, high-pressure, and high-temperature behavior of carbonates in the $K_2Mg(CO_3)_2$–$Na_2Mg(CO_3)_2$ join. *American Mineralogist*, *100*, 2458–2467.

Graf, D. L., & J. R. Goldsmith (1955). Dolomite-magnesian calcite relations at elevated temperatures and CO_2 pressures. *Geochimica et Cosmochimica Acta*, *7*, 109–128.

Grassi, D., & M. W. Schmidt (2011a). Melting of carbonated pelites at 8–13 GPa: generating K-rich carbonatites for mantle metasomatism. *Contributions to Mineralogy and Petrology*, *162*, 169–191.

Grassi, D., & M. W. Schmidt (2011b). The melting of carbonated pelites from 70 to 700 km depth. *Journal of Petrology*, *52*, 765–789.

Green, D. H., & M. E. Wallace (1988). Mantle metasomatism by ephemeral carbonatite melts. *Nature*, *336*, 459–462.

Gudfinnsson, G. H., & D. C. Presnall (1996). Melting relations of model lherzolite in the system CaO-MgO-Al_2O_3-SiO_2 at 2.4–3.4 GPa and the generation of komatiites. *Journal of Geophysical Research: Solid Earth*, *101*, 27701–27709.

Gudfinnsson, G. H., & D. C. Presnall (2005). Continuous gradations among primary carbonatitic, kimberlitic, melilititic, basaltic, picritic, and komatiitic melts in equilibrium with garnet lherzolite at 3–8 GPa. *Journal of Petrology*, *46*, 1645–1659.

Hammouda, T. (2003). High-pressure melting of carbonated eclogite and experimental constraints on carbon recycling and storage in the mantle. *Earth and Planetary Science Letters*, *214*, 357–368.

Harker, R. I., & O. F. Tuttle (1955a). Studies in the system CaO–MgO–CO_2. 1. The thermal dissociation of calcite, dolomite and magnesite. *American Journal of Science*, *253*, 209–224.

Harker, R. I., & O. F. Tuttle (1955b). Studies in the system CaO–MgO–CO_2: 2. Limits of solid solution along the binary join, $CaCO_3$–$MgCO_3$. *American Journal of Science*, *253*, 274–282.

Harris, M. J., & M. T. Dove (1995). Lattice melting at structural phase-transitions. *Modern Physics Letters B*, *9*, 67–85.

Harris, M. J., & E. K. H. Salje (1992). The incommensurate phase of sodium carbonate: An infrared absorption study. *Journal of Physics-Condensed Matter*, *4*, 4399–4408.

Hasterok, D., & D. Chapman (2011). Heat production and geotherms for the continental lithosphere. *Earth and Planetary Science Letters*, *307*, 59–70.

Hermann, J., U. Troitzsch, & D. Scott (2016). Experimental subsolidus phase relations in the system $CaCO_3$–$CaMg(CO_3)_2$ up to 6.5 GPa and implications for subducted marbles. *Contributions to Mineralogy and Petrology*, *171*, 84.

Hernlund, J., K. Leinenweber, D. Locke, & J. A. Tyburczy (2006). A numerical model for steady-state temperature distributions in solid-medium high-pressure cell assemblies. *American Mineralogist*, *91*, 295–305.

Hesse, K.-F., & B. Simons (1982). Crystalstructure of synthetic $K_2Mg(CO_3)_2$. *Zeitschrift fur Kristallographie*, *161*, 289–292.

Irving, A., & P. Wyllie (1973). Melting relationships in CaO–CO_2 and MgO–CO_2 to 36 kilobars with comments on CO_2 in the mantle. *Earth and Planetary Science Letters*, *20*, 220–225.

Irving, A. J., & P. J. Wyllie (1975). Subsolidus and melting relationships for calcite, magnesite and the join $CaCO_3$–$MgCO_3$ to 36 kb. *Geochimica et Cosmochimica Acta*, *39*, 35–53.

Ishizawa, N., H. Setoguchi, & K. Yanagisawa (2013). Structural evolution of calcite at high temperatures: Phase V unveiled. *Scientific Reports*, *3*, 02832.

Isshiki, M., T. Irifune, K. Hirose, S. Ono, Y. Ohishi, T. Watanuki, E. Nishibori, et al. (2004). Stability of magnesite and its high-pressure form in the lowermost mantle. *Nature*, *427*, 60–63.

Ivanov, B. A., & A. Deutsch (2002). The phase diagram of $CaCO_3$ in relation to shock compression and decomposition. *Physics of the Earth and Planetary Interiors*, *129*, 131–143.

Jablon, B. M., & O. Navon (2016). Most diamonds were created equal. *Earth and Planetary Science Letters*, *443*, 41–47.

Johnston, F. K. B., A. V. Turchyn, & M. Edmonds (2011). Decarbonation efficiency in subduction zones: Implications for warm Cretaceous climates. *Earth and Planetary Science Letters*, *303*, 143–152.

Jones, A. G., I. J. Ferguson, A. D. Chave, R. L. Evans, & G. W. McNeice (2001). Electric lithosphere of the Slave craton. *Geology*, *29*, 423–426.

Jones, A. G., P. Lezaeta, I. J. Ferguson, A. D. Chave, R. L. Evans, X. Garcia, & J. Spratt (2003). The electrical structure of the Slave craton. *Lithos*, *71*, 505–527.

Kamenetsky, V. S., H. Grütter, M. B. Kamenetsky, & K. Gömann (2013). Parental carbonatitic melt of the Koala kimberlite (Canada): Constraints from melt inclusions in olivine and Cr-spinel, and groundmass carbonate. *Chemical Geology*, *353*, 96–111.

Kamenetsky, V. S., M. B. Kamenetsky, Y. Weiss, O. Navon, T. F. D. Nielsen, & T. P. Mernagh (2009). How unique is the Udachnaya-East kimberlite? Comparison with kimberlites from the Slave Craton (Canada) and SW Greenland. *Lithos*, *112*(S1). 334–346.

Kaminsky, F. (2012). Mineralogy of the lower mantle: A review of 'super-deep' mineral inclusions in diamond. *Earth-Science Reviews*, *110*, 127–147.

Kaminsky, F. V., I. D. Ryabchikov, & R. Wirth (2016). A primary natrocarbonatitic association in the Deep Earth. *Mineralogy and Petrology*, *110*, 387–398.

Kang, N., M. W. Schmidt, S. Poli, J. A. Connolly, & E. Franzolin (2016). Melting relations in the system $FeCO_3$–$MgCO_3$ and thermodynamic modelling of Fe–Mg carbonate melts. *Contributions to Mineralogy and Petrology*, *171*, 74.

Kang, N., M. W. Schmidt, S. Poli, E. Franzolin, & J.A.D. Connolly (2015). Melting of siderite to 20 GPa and thermodynamic properties of $FeCO_3$-melt. *Chemical Geology*, *400*, 34–43.

Katsura, T., & E. Ito (1990). Melting and subsolidus phase relations in the $MgSiO_3$–$MgCO_3$ system at high pressures: Implications to evolution of the Earth's atmosphere. *Earth and Planetary Science Letters*, *99*, 110–117.

Kavner, A., & C. Nugent (2008). Precise measurements of radial temperature gradients in the laser-heated diamond anvil cell. *Review of Scientific Instruments*, *79*, 024902.

Kelemen, P. B., & C. E. Manning (2015). Reevaluating carbon fluxes in subduction zones, what goes down, mostly comes up. *Proceedings of the National Academy of Sciences*, *112*, E3997–E4006.

Kennedy, C. S., & G. C. Kennedy (1976). The equilibrium boundary between graphite and diamond. *Journal of Geophysical Research*, *81*, 2467–2470.

Kerley, G. I. (1989). Equations of state for calcite minerals: I. Theoretical model for dry calcium carbonate. *High Pressure Research*, *2*, 29–47.

Kerrick, D. M., & J. A. D. Connolly (2001). Metamorphic devolatilization of subducted marine sediments and the transport of volatiles into the Earth's mantle. *Nature*, *411*, 293–296.

Keshav, S., A. Corgne, G. H. Gudfinnsson, M. Bizimis, W. F. McDonough, & Y. W. Fei (2005). Kimberlite petrogenesis: Insights from clinopyroxene–melt partitioning experiments at 6 GPa in the CaO–MgO–Al_2O_3–SiO_2–CO_2 system. *Geochimica et Cosmochimica Acta*, *69*, 2829–2845.

Kiseeva, E. S., K. D. Litasov, G. M. Yaxley, E. Ohtani, & V. S. Kamenetsky (2013). Melting and phase relations of carbonated eclogite at 9–21 GPa and the petrogenesis of alkali-rich melts in the deep mantle. *Journal of Petrology*, *54*(8). 1555–1583.

Klein-BenDavid, O., A. M. Logvinova, M. Schrauder, Z. V. Spetius, Y. Weiss, E. H. Hauri, F. V. Kaminsky, et al. (2009). High-Mg carbonatitic microinclusions in some Yakutian diamonds: a new type of diamond-forming fluid. *Lithos*, *112*(S2). 648–659.

Klement, W., & L. H. Cohen (1975). Solid-solid and solid-liquid transitions in K_2CO_3, Na_2CO_3 and Li_2CO_3: Investigations to ≥ 5 kbar by differential thermal analysis; thermodynamics and structural correlations. *Berichte der Bunsengesellschaft für physikalische Chemie*, *79*, 327–334.

Koch-Müller, M., S. Jahn, N. Birkholz, E. Ritter, & U. Schade (2016). Phase transitions in the system $CaCO_3$ at high P and T determined by in situ vibrational spectroscopy in diamond anvil cells and first-principles simulations. *Physics and Chemistry of Minerals*, *43*, 545–561.

Kono, Y., C. Kenney-Benson, D. Hummer, H. Ohfuji, C. Park, G. Shen, Y. Wang, et al. (2014). Ultralow viscosity of carbonate melts at high pressures. *Nature Communications*, *5*, 5091.

Kopylova, M., O. Navon, L. Dubrovinsky, & G. Khachatryan (2010). Carbonatitic mineralogy of natural diamond-forming fluids. *Earth and Planetary Science Letters*, *291*, 126–137.

Kröger, C., K. W. Illner, & W. Graeser (1943). Über die systeme alkalioxyd CaO–Al_2O_3–SiO_2–CO_2. XI. die reaktionsdrucke im system K_2O–CaO–SiO_2–CO_2. *Zeitschrift Fur Anorganische Und Allgemeine Chemie*, *251*, 270–284.

Lavina, B., P. Dera, R. T. Downs, V. Prakapenka, M. Rivers, S. Sutton, & M. Nicol (2009). Siderite at lower mantle conditions and the effects of the pressure-induced spin-pairing transition. *Geophysical Research Letters*, *36*, L23306, doi: 10.1029/2009gl039652.

Lavrent'ev, Y. G., N. S. Karmanov, & L. V. Usova (2015). Electron probe microanalysis of minerals: Microanalyzer or scanning electron microscope? *Russian Geology and Geophysics*, *56*, 1154–1161.

Li, X., Z. Zhang, J. F. Lin, H. Ni, V. B. Prakapenka, & Z. Mao (2018). New high-pressure phase of $CaCO_3$ at the topmost lower mantle: Implication for the deep-mantle carbon transportation. *Geophysical Research Letters*, *45*, 1355–1360.

Li, Z. (2015). Melting and structural transformations of carbonates and hydrous phases in Earth's mantle (PhD thesis, 126 pp). The University of Michigan.

Li, Z., J. Li, R. Lange, J. Liu, & B. Militzer (2017). Determination of calcium carbonate and sodium carbonate melting curves up to Earth's transition zone pressures with implications for the deep carbon cycle. *Earth and Planetary Science Letters*, *457*, 395–402.

Litasov, K. D. (2011). Physicochemical conditions for melting in the Earth's mantle containing a C–O–H fluid (from experimental data). *Russian Geology and Geophysics*, *51*, 475–492.

Litasov, K. D., & E. Ohtani (2009). Solidus and phase relations of carbonated peridotite in the system CaO–Al_2O_3–MgO–SiO_2–Na_2O–CO_2 to the lower mantle depths. *Physics of the Earth and Planetary Interiors*, *177*, 46–58.

Litasov, K. D., & E. Ohtani (2010). The solidus of carbonated eclogite in the system $CaO–Al_2O_3–MgO–SiO_2–Na_2O–CO_2$ to 32 GPa and carbonatite liquid in the deep mantle. *Earth and Planetary Science Letters, 295*, 115–126.

Litasov, K. D., & A. Shatskiy (2018). Carbon-bearing magmas in the Earth's deep interior. In Y. Kono and C. Sanloup (Eds.), *Magmas Under Pressure* (pp. 43–82), Elsevier.

Litasov, K. D., A. Shatskiy, P. N. Gavryushkin, A. E. Bekhtenova, P. I. Dorogokupets, B. S. Danilov, Y. Higo, et al. (2017). PVT equation of state of $CaCO_3$ aragonite to 29 GPa and 1673 K: In situ X-ray diffraction study. *Physics of the Earth and Planetary Interiors, 265*, 82–91.

Litasov, K. D., A. Shatskiy, & E. Ohtani (2013). Earth's mantle melting in the presence of C-O-H-bearing fluid. In S. Karato (Ed.), *Physics and Chemistry of the Deep Earth* (pp. 38–65). New York: Wiley.

Litasov, K. D., A. Shatskiy, E. Ohtani, & G. M. Yaxley (2013). Solidus of alkaline carbonatite in the deep mantle. *Geology, 41*, 79–82.

Litasov, K. D., A. F. Shatskiy, & N. P. Pokhilenko (2011). Phase relations and melting in the systems of peridotite–$H_2O–CO_2$ and eclogite–$H_2O–CO_2$ at pressures up to 27 GPa. *Doklady Earth Sciences, 437*, 498–502.

Liu, J., J.-F. Lin, & V. B. Prakapenka (2015). High-pressure orthorhombic ferromagnesite as a potential deep-mantle carbon carrier. *Scientific Reports, 5*, 7640.

Liu, Q., T. J. Tenner, & R. A. Lange (2007). Do carbonate liquids become denser than silicate liquids at pressure? Constraints from the fusion curve of K_2CO_3 to 3.2 GPa. *Contributions to Mineralogy and Petrology, 153*, 55–66.

Lobanov, S. S., X. Dong, N. S. Martirosyan, A. I. Samtsevich, V. Stevanovic, P. N. Gavryushkin, K. D. Litasov, et al. (2017). Raman spectroscopy and x-ray diffraction of sp^3 $CaCO_3$ at lower mantle pressures. *Physical Review B, 96*, 104101.

Lobanov, S. S., A. F. Goncharov, & K. D. Litasov (2015). Optical properties of siderite ($FeCO_3$) across the spin transition: Crossover to iron-rich carbonates in the lower mantle. *American Mineralogist, 100*, 1059–1064.

Logvinova, A. M., R. Wirth, A. A. Tomilenko, V. P. Afanas'ev, & N. V. Sobolev (2011). The phase composition of crystal-fluid nanoinclusions in alluvial diamonds in the northeastern Siberian Platform. *Russian Geology and Geophysics, 52*, 1286–1297.

Luth, R. W. (2001). Experimental determination of the reaction aragonite+magnesite = dolomite at 5 to 9 GPa. *Contributions to Mineralogy and Petrology, 141*, 222–232.

Luth, R. W. (2006). Experimental study of the $CaMgSi_2O_6–CO_2$ system at 3–8 GPa. *Contributions to Mineralogy and Petrology, 151*, 141–157.

Maeda, F., E. Ohtani, S. Kamada, T. Sakamaki, N. Hirao, & Y. Ohishi (2017). Diamond formation in the deep lower mantle: a high-pressure reaction of $MgCO_3$ and SiO_2. *Scientific Reports, 7*, 40602.

Mao, Z., M. Armentrout, E. Rainey, C. E. Manning, P. Dera, V. B. Prakapenka, & A. Kavner (2011). Dolomite III: A new candidate lower mantle carbonate. *Geophysical Research Letters, 38*, L22303, doi: 22310.21029/22011GL049519.

Martinez, I., A. Deutsch, U. Schärer, P. Ildefonse, F. Guyot, & P. Agrinier (1995). Shock recovery experiments on dolomite and thermodynamical calculations of impact induced decarbonation. *Journal of Geophysical Research: Solid Earth, 100*, 15465–15476.

Martinez, I., J. Z. Zhang, & R. J. Reeder (1996). In situ X-ray diffraction of aragonite and dolomite at high pressure and high temperature: Evidence for dolomite breakdown to aragonite and magnesite. *American Mineralogist, 81*, 611–624.

Martirosyan, N. S., K. D. Litasov, S. S. Lobanov, A. F. Goncharov, A. Shatskiy, H. Ohfuji, & V. B. Prakapenka (2019). The Mg-carbonate–Fe interaction: Implication for the fate of subducted carbonates and formation of diamond in the lower mantle. *Geoscience Frontiers, 10*, 1449–1458.

Martirosyan, N. S., T. Yoshino, A. Shatskiy, A. D. Chanyshev, & K. D. Litasov (2016). The $CaCO_3$–Fe interaction: Kinetic approach for carbonate subduction to the deep Earth's mantle. *Physics of the Earth and Planetary Interiors, 259*, 1–9.

Mattila, A., T. Pylkkänen, J. Rueff, S. Huotari, G. Vanko, M. Hanfland, M. Lehtinen, et al. (2007). Pressure induced magnetic transition in siderite $FeCO_3$ studied by X-ray emission spectroscopy, *Journal of Physics: Condensed Matter, 19*, 386206.

McKie, D. (1990). Subsolidus phase relations in the system $K_2Ca(CO_3)_2–Na_2Mg(CO_3)_2$ at 1 kbar; the fairchildite ss-buetschliite-eitelite eutectoid. *American Mineralogist, 75*, 1147–1150.

Mellot-Draznieks, C., S. Girard, G. Ferey, J. C. Schon, Z. Cancarevic, & M. Jansen (2002). Computational design and prediction of interesting not-yet-synthesized structures of inorganic materials by using building unit concepts. *Chemistry-a European Journal, 8*, 4103–4113.

Menzies, M., & C. Hawkesworth (1986). *Mantle metasomatism*, Academic Press, London.

Merlini, M., V. Cerantola, G. D. Gatta, M. Gemmi, M. Hanfland, I. Kupenko, P. Lotti, et al. (2017). Dolomite-IV: Candidate structure for a carbonate in the Earth's lower mantle. *American Mineralogist, 102*, 1763–1766.

Merlini, M., W. A. Crichton, M. Hanfland, M. Gemmi, H. Müller, I. Kupenko, & L. Dubrovinsky (2012a). Structures of dolomite at ultrahigh pressure and their influence on the deep carbon cycle. *Proceedings of the National Academy of Sciences, 109*, 13509–13514.

Merlini, M., M. Hanfland, & W. A. Crichton (2012b). $CaCO_3$-III and $CaCO_3$-VI, high-pressure polymorphs of calcite: Possible host structures for carbon in the Earth's mantle, *Earth and Planetary Science Letters, 333–334*, 265–271.

Mirwald, P. W. (1976). A differential thermal analysis study of the high-temperature polymorphism of calcite at high pressure. *Contributions to Mineralogy and Petrology, 59*, 33–40.

Mitchell, R. H., & B. A. Kjarsgaard (2011). Experimental studies of the system $Na_2CO_3–CaCO_3–MgF_2$ at 0.1 GPa: Implications for the differentiation and low-temperature crystallization of natrocarbonatite. *Journal of Petrology, 52*, 1265–1280.

Molina, J. F., & S. Poli (2000). Carbonate stability and fluid composition in subducted oceanic crust: an experimental study on H_2O-CO_2-bearing basalts. *Earth and Planetary Science Letters, 176*, 295–310.

Moore, K. R. (2012). Experimental study in the $Na_2O–CaO–MgO–Al_2O_3–SiO_2–CO_2$ system at 3 GPa: the effect of sodium on mantle melting to carbonate-rich liquids and implications for the petrogenesis of silicocarbonatites. *Mineralogical Magazine*, *76*, 285–309.

Morlidge, M., A. Pawley, & G. Droop (2006). Double carbonate breakdown reactions at high pressures: an experimental study in the system CaO–MgO–FeO–MnO–CO_2. *Contributions to Mineralogy and Petrology*, *152*, 365–373.

Müller, J., M. Koch-Müller, D. Rhede, F. D. Wilke, & R. Wirth (2017). Melting relations in the system $CaCO_3$–$MgCO_3$ at 6 GPa. *American Mineralogist*, *102*, 2440–2449.

Navon, O., I. Hutcheon, G. Rossman, & G. Wasserburg (1988). Mantle-derived fluids in diamond micro-inclusions. *Nature*, *335*, 784–789.

Niggli, P. (1916). Gleichgewichte zwischen TiO_2 und CO_2, sowie SiO_2 und CO_2 in alkali-, kalk-alkali und alkali-aluminatschmelzen. *Zeitschrift Fur Anorganische Und Allgemeine Chemie*, *98*, 241–326.

Oganov, A. R., S. Ono, Y. M. Ma, C. W. Glass, & A. Garcia (2008). Novel high-pressure structures of $MgCO_3$, $CaCO_3$ and CO_2 and their role in Earth's lower mantle. *Earth and Planetary Science Letters*, *273*, 38–47.

Ono, S., T. Kikegawa, & Y. Ohishi (2007). High-pressure transition of $CaCO_3$. *American Mineralogist*, *92*, 1246–1249.

Ono, S., T. Kikegawa, Y. Ohishi, & J. Tsuchiya (2005). Post-aragonite phase transformation in $CaCO_3$ at 40 GPa. *American Mineralogist*, *90*, 667–671.

Pickard, C. J., & R. J. Needs (2015). Structures and stability of calcium and magnesium carbonates at mantle pressures. *Physical Review B*, *91*, 104101.

Pinto, L. G. R., M. B. de Pádua, N. Ussami, Í. Vitorello, A. L. Padilha, & C. Braitenberg (2010). Magnetotelluric deep soundings, gravity and geoid in the south São Francisco craton: Geophysical indicators of cratonic lithosphere rejuvenation and crustal underplating. *Earth and Planetary Science Letters*, *297*, 423–434.

Pippinger, T., R. Miletich, M. Merlini, P. Lotti, P. Schouwink, T. Yagi, W. A. Crichton, et al. (2015). Puzzling calcite-III dimorphism: crystallography, high-pressure behavior, and pathway of single-crystal transitions. *Physics and Chemistry of Minerals*, *42*, 29–43.

Plank, T., & C. H. Langmuir (1998). The chemical composition of subducting sediment and its consequences for the crust and mantle. *Chemical Geology*, *145*, 325–394.

Podborodnikov, I. V., A. Shatskiy, A. V. Arefiev, A. D. Chanyshev, & K. D. Litasov (2018). The system Na_2CO_3–$MgCO_3$ at 3 GPa. *High Pressure Research*, *38*, 281–292.

Podborodnikov, I. V., A. Shatskiy, A. V. Arefiev, & K. D. Litasov (2019). Phase relations in the system Na_2CO_3–$CaCO_3$–$MgCO_3$ at 3 GPa with implications for carbonatite genesis and evolution. *Lithos*, *330–331*, 74–89.

Podborodnikov, I. V., A. Shatskiy, A. V. Arefiev, S. V. Rashchenko, A. D. Chanyshev, & K. D. Litasov (2018). The system Na_2CO_3–$CaCO_3$ at 3 GPa. *Physics and Chemistry of Minerals*, *45*, 773–787.

Pokhilenko, N. P., A. M. Agashev, K. D. Litasov, & L. N. Pokhilenko (2015). Carbonatite metasomatism of peridotite lithospheric mantle: implications for diamond formation and carbonatite-kimberlite magmatism. *Russian Geology and Geophysics*, *56*, 280–295.

Ragone, S. E., R. K. Datta, D. M. Roy, & O. F. Tuttle (1966). The system potassium carbonate-magnesium carbonate. *Journal of Physical Chemistry*, *70*, 3360–3361.

Rainey, E. S. G., J. W. Hernlund, & A. Kavner (2013). Temperature distributions in the laser-heated diamond anvil cell from 3-D numerical modeling. *Journal of Applied Physics*, *114*, 204905.

Ratajeski, K., & T. W. Sisson (1999). Loss of iron to gold capsules in rock-melting experiments. *American Mineralogist*, *84*, 1521–1527.

Rosenberg, P. E. (1963). Subsolidus relations in the system $CaCO_3$–$FeCO_3$. *American Journal of Science*, *261*, 683–689.

Rudnick, R. L., W. F. McDonough, & B. W. Chappell (1993). Carbonatite metasomatism in the northern Tanzanian mantle: petrographic and geochemical characteristics. *Earth and Planetary Science Letters*, *114*, 463–475.

Salje, E., & K. Viswanathan (1976). The phase diagram calcite-aragonite as derived from the crystallographic properties. *Contributions to Mineralogy and Petrology*, *55*, 55–67.

Santillán, J., Q. Williams, & E. Knittle (2003). Dolomite-II: A high-pressure polymorph of $CaMg(CO_3)_2$. *Geophysical Research Letters*, *30*, 1054, doi: 1010.1029/2002GL016018

Sato, K., & T. Katsura (2001). Experimental investigation on dolomite dissociation into aragonite+magnesite up to 8.5 GPa. *Earth and Planetary Science Letters*, *184*, 529–534.

Schneide, S. J., & E. M. Levin (1973). Polymorphism of K_2CO_3. *Journal of the American Ceramic Society*, *56*, 218–219.

Sharygin, I., K. Litasov, A. Shatskiy, A. Golovin, E. Ohtani, & N. Pokhilenko (2015). Melting phase relations of the Udachnaya-East group-I kimberlite at 3.0–6.5 GPa: Experimental evidence for alkali-carbonatite composition of primary kimberlite melts and implications for mantle plumes. *Gondwana Research*, *28*, 1391–1414.

Sharygin, I. S., A. V. Golovin, A. V. Korsakov, & N. P. Pokhilenko (2013). Eitelite in sheared peridotite xenoliths from Udachnaya-East kimberlite pipe (Russia) – a new locality and host rock type. *European Journal of Mineralogy*, *25*, 825–834.

Shatskiy, A., Y. M. Borzdov, K. D. Litasov, I. N. Kupriyanov, E. Ohtani, & Y. N. Palyanov (2014). Phase relations in the system $FeCO_3$–$CaCO_3$ at 6 GPa and 900–1700 °C and its relation to the system $CaCO_3$–$FeCO_3$–$MgCO_3$. *American Mineralogist*, *99*, 773–785.

Shatskiy, A., Y. M. Borzdov, K. D. Litasov, I. S. Sharygin, Y. N. Palyanov, & E. Ohtani (2015). Phase relationships in the system K_2CO_3–$CaCO_3$ at 6 GPa and 900–1450°C. *American Mineralogist*, *100*, 223–232.

Shatskiy, A., P. N. Gavryushkin, I. S. Sharygin, K. D. Litasov, I. N. Kupriyanov, Y. Higo, Y. M. Borzdov, et al. (2013). Melting and subsolidus phase relations in the system Na_2CO_3–$MgCO_3$±H_2O at 6 GPa and the stability of $Na_2Mg(CO_3)_2$ in the upper mantle. *American Mineralogist*, *98*, 2172–2182.

Shatskiy, A., T. Katsura, K. D. Litasov, A. V. Shcherbakova, Y. M. Borzdov, D. Yamazaki, A. Yoneda, et al. (2011). High pressure generation using scaled-up Kawai-cell. *Physics of the Earth and Planetary Interiors*, *189*, 92–108.

Shatskiy, A., K. D. Litasov, E. Ohtani, Y. M. Borzdov, A. I. Khmelnicov, & Y. N. Palyanov (2015). Phase relations in the K_2CO_3–$FeCO_3$ and $MgCO_3$–$FeCO_3$ systems at 6 GPa and 900–1700 °C. *European Journal of Mineralogy*, *27*, 487–499.

Shatskiy, A. F., K. D. Litasov, & Y. N. Palyanov (2015). Phase relations in carbonate systems at pressures and temperatures of lithospheric mantle: Review of experimental data. *Russian Geology and Geophysics*, *56*, 113–142.

Shatskiy, A., K. D. Litasov, Y. N. Palyanov, & E. Ohtani (2016). Phase relations on the K_2CO_3–$CaCO_3$–$MgCO_3$ join at 6 GPa and 900–1400 °C: Implications for incipient melting in carbonated mantle domains. *American Mineralogist*, *101*, 437–447.

Shatskiy, A., K. D. Litasov, I. S. Sharygin, I. A. Egonin, A. M. Mironov, Y. N. Palyanov, & E. Ohtani (2016). The system Na_2CO_3–$CaCO_3$–$MgCO_3$ at 6 GPa and 900–1250 °C and its relation to the partial melting of carbonated mantle. *High Pressure Research*, *36*, 23–41.

Shatskiy, A., K. D. Litasov, I. S. Sharygin, & E. Ohtani (2017). Composition of primary kimberlite melt in a garnet lherzolite mantle source: constraints from melting phase relations in anhydrous Udachnaya-East kimberlite with variable CO_2 content at 6.5 GPa. *Gondwana Research*, *45*, 208–227.

Shatskiy, A., K. D. Litasov, H. Terasaki, T. Katsura, & E. Ohtani (2010). Performance of semi-sintered ceramics as pressure-transmitting media up to 30 GPa. *High Pressure Research*, *30*, 443–450.

Shatskiy, A., I. V. Podborodnikov, A. V. Arefiev, K. D. Litasov, A. D. Chanyshev, I. S. Sharygin, N. S. Karmanov, et al. (2017). Effect of alkalis on the reaction of clinopyroxene with Mg-carbonate at 6 GPa: Implications for partial melting of carbonated lherzolite. *American Mineralogist*, *102*, 1934–1946.

Shatskiy, A., I. V. Podborodnikov, A. V. Arefiev, D. A. Minin, A. D. Chanyshev, & K. D. Litasov (2018). Revision of the $CaCO_3$–$MgCO_3$ phase diagram at 3 and 6 GPa. *American Mineralogist*, *103*, 441–452.

Shatskiy, A., S. V. Rashchenko, E. Ohtani, K. D. Litasov, M. V. Khlestov, Y. M. Borzdov, I. N. Kupriyanov, et al. (2015). The system Na_2CO_3–$FeCO_3$ at 6 GPa and its relation to the system Na_2CO_3–$FeCO_3$–$MgCO_3$. *American Mineralogist*, *100*, 130–137.

Shatskiy, A., I. S. Sharygin, P. N. Gavryushkin, K. D. Litasov, Y. M. Borzdov, A. V. Shcherbakova, Y. Higo, et al. (2013). The system K_2CO_3–$MgCO_3$ at 6 GPa and 900–1450 °C. *American Mineralogist*, *98*, 1593–1603.

Shatskiy, A., I. S. Sharygin, K. D. Litasov, Y. M. Borzdov, Y. N. Palyanov, & E. Ohtani (2013). New experimental data on phase relations for the system Na_2CO_3–$CaCO_3$ at 6 GPa and 900–1400 °C. *American Mineralogist*, *98*, 2164–2171.

Shirasaka, M., E. Takahashi, Y. Nishihara, K. Matsukage, & T. Kikegawa (2002). In situ X-ray observation of the reaction dolomite = aragonite+magnesite at 900–1300 K. *American Mineralogist*, *87*, 922–930.

Shu, Q., & G. P. Brey (2015). Ancient mantle metasomatism recorded in subcalcic garnet xenocrysts: temporal links between mantle metasomatism, diamond growth and crustal tectonomagmatism. *Earth and Planetary Science Letters*, *418*, 27–39.

Smith, D., K. V. Lawler, M. Martinez-Canales, A. W. Daykin, Z. Fussell, G. A. Smith, C. Childs, et al. (2018). Postaragonite phases of $CaCO_3$ at lower mantle pressures. *Physical Review Materials*, *2*, 013605.

Sokolova, T. S., P. I. Dorogokupets, A. M. Dymshits, B. S. Danilov, & K. D. Litasov (2016). Microsoft excel spreadsheets for calculation of PVT relations and thermodynamic properties from equations of state of MgO, diamond and nine metals as pressure markers in high-pressure and high-temperature experiments. *Computers & Geosciences*, *94*, 162–169.

Sokolova, T. S., P. I. Dorogokupets, & K. D. Litasov (2013). Self-consistent pressure scales based on the equations of state for ruby, diamond, MgO, B2-NaCl, as well as Au, Pt and other metals to 4 Mbars and 3000 K. *Russian Geology and Geophysics*, *54*, 181–199.

Solomatova, N. V., & P. D. Asimow (2018). First-principles calculations of high-pressure iron-bearing monoclinic dolomite and single-cation carbonates with internally consistent Hubbard U. *Physics and Chemistry of Minerals*, *45*, 293–302.

Solopova, N. A., L. Dubrovinsky, A. V. Spivak, Y. A. Litvin, & N. Dubrovinskaia (2015). Melting and decomposition of $MgCO_3$ at pressures up to 84 GPa. *Physics and Chemistry of Minerals*, *42*, 73–81.

Spivak, A. V., Y. A. Litvin, S. V. Ovsyannikov, N. A. Dubrovinskaia, & L. S. Dubrovinsky (2012). Stability and breakdown of $Ca^{13}CO_3$ melt associated with formation of ^{13}C-diamond in static high pressure experiments up to 43 GPa and 3900 K. *Journal of Solid State Chemistry*, *191*, 102–106.

Stachel, T., J. W. Harris, & G. P. Brey (1998). Rare and unusual mineral inclusions in diamonds from Mwadui, Tanzania. *Contributions to Mineralogy and Petrology*, *132*, 34–47.

Stagno, V., D. O. Ojwang, C. A. McCammon, & D. J. Frost (2013). The oxidation state of the mantle and the extraction of carbon from Earth's interior. *Nature*, *493*, 84–88.

Stagno, V., V. Stopponi, Y. Kono, C. E. Manning, & T. Irifune (2018). Experimental determination of the viscosity of Na_2CO_3 melt between 1.7 and 4.6 GPa at 1200–1700°C: Implications for the rheology of carbonatite magmas in the Earth's upper mantle. *Chemical Geology*, *501*, 19–25.

Staudigel, H. (2014). Chemical fluxes from hydrothermal alteration of the oceanic crust, in Turekian, K. K., Holland, H. D (Eds.), *Treatise on Geochemistry* (2nd ed., pp. 583–606), Oxford: Elsevier.

Suito, K., J. Namba, T. Horikawa, Y. Taniguchi, N. Sakurai, M. Kobayashi, A. Onodera, et al. (2001). Phase relations of $CaCO_3$ at high pressure and high temperature. *American Mineralogist*, *86*, 997–1002.

Swainson, I. P., M. T. Dove, & M. J. Harris (1995). Neutron powder diffraction study of the ferroelastic phase transition and lattice melting in sodium carbonate, Na_2CO_3. *Journal of Physics-Condensed Matter*, *7*, 4395–4417.

Sweeney, R. J. (1994). Carbonatite melt compositions in the Earth's mantle. *Earth and Planetary Science Letters*, *128*, 259–270.

Sweeney, R. J., T. J. Falloon, & D. H. Green (1995). Experimental constraints on the possible mantle origin of natrocarbonatite. In Bell, K., Keller, J. (Eds.), *Carbonatite Volcanism* (pp. 191–207), Springer.

Tao, R., Y. Fei, & L. Zhang (2013). Experimental determination of siderite stability at high pressure. *American Mineralogist*, *98*, 1565–1572.

Tao, R., L. Zhang, Y. Fei, & Q. Liu (2014). The effect of Fe on the stability of dolomite at high pressure: Experimental study and petrological observation in eclogite from southwestern Tianshan, China. *Geochimica et Cosmochimica Acta*, *143*, 253–267.

Thibault, Y., A. D. Edgar, & F. E. Lloyd (1992). Experimental investigation of melts from a carbonated phlogopite lherzolite: implications for metasomatism in the continental lithospheric mantle. *American Mineralogist, 77*, 784–794.

Thomson, A. R., M. J. Walter, S. C. Kohn, & R. A. Brooker (2016). Slab melting as a barrier to deep carbon subduction. *Nature, 529*, 76–79.

Tsuno, K., R. Dasgupta, L. Danielson, & K. Righter (2012). Flux of carbonate melt from deeply subducted pelitic sediments: Geophysical and geochemical implications for the source of Central American volcanic arc. *Geophysical Research Letters, 39*, L16307, doi:10.1029/2012gl052606.

Wallace, M. E., & D. H. Green (1988). An experimental determination of primary carbonatite magma composition. *Nature, 335*, 343–346.

Wang, M., Q. Liu, T. Inoue, B. Li, S. Pottish, J. Wood, C. Yang, et al. (2016). The K_2CO_3 fusion curve revisited: New experiments at pressures up to 12 GPa. *Journal of Mineralogical and Petrological Sciences, 111*, 241–251.

Weidner, J. R. (1972). Equilibria in the system Fe–CO; Part I, Siderite-magnetite-carbon-vapor equilibrium from 500 to 10,000 bars. *American Journal of Science, 272*, 735–751.

Weidner, J. R. (1982). Iron-oxide magmas in the system Fe–CO. *Canadian Mineralogist, 20*, 555–566.

Weiss, Y., R. Kessel, W. L. Griffin, I. Kiflawi, O. Klein-BenDavid, D. R. Bell, J. W. Harris, et al. (2009). A new model for the evolution of diamond-forming fluids: Evidence from microinclusion-bearing diamonds from Kankan, Guinea. *Lithos, 112*(S2). 660–674.

Wyllie, P. J., & W. L. Huang (1975). Inflence of mantle CO_2 ingeneration of carbonatites and kimberlites. *Nature, 257*, 297–299.

Wyllie, P. J., & W. L. Huang (1976). Carbonation and melting reactions in the system CaO–MgO–SiO_2–CO_2 at mantle pressures with geophysical and petrological applications. *Contributions to Mineralogy and Petrology, 54*, 79–107.

Yaxley, G. M., & G. P. Brey (2004). Phase relations of carbonate-bearing eclogite assemblages from 2.5 to 5.5 GPa: implications for petrogenesis of carbonatites. *Contributions to Mineralogy and Petrology, 146*, 606–619.

Yaxley, G. M., A. J. Crawford, & D. H. Green (1991). Evidence for carbonatite metasomatism in spinel peridotite xenoliths from western Victoria, Australia. *Earth and Planetary Science Letters, 107*, 305–317.

Yaxley, G. M., & D. H. Green (1994). Experimental demonstration of refractory carbonate-bearing eclogite and siliceous melt in the subduction regime. *Earth and Planetary Science Letters, 128*, 313–325.

Yaxley, G. M., & D. H. Green (1996). Experimental reconstruction of sodic dolomitic carbonatite melts from metasomatised lithosphere. *Contributions to Mineralogy and Petrology, 124*, 359–369.

Yaxley, G. M., D. H. Green, & V. Kamenetsky (1998). Carbonatite metasomatism in the southeastern Australian lithosphere. *Journal of Petrology, 39*, 1917–1930.

Yoshino, T., B. Gruber, & C. Reinier (2018). Effects of pressure and water on electrical conductivity of carbonate melt with implications for conductivity anomaly in continental mantle lithosphere. *Physics of the Earth and Planetary Interiors, 281*, 8–16.

Zaitsev, A. N., & J. Keller (2006). Mineralogical and chemical transformation of Oldoinyo Lengai natrocarbonatites, Tanzania. *Lithos, 91*, 191–207.

Zedgenizov, D. A., A. L. Ragozin, V. S. Shatsky, D. Araujo, & W. L. Griffin (2011). Fibrous diamonds from the placers of the northeastern Siberian Platform: carbonate and silicate crystallization media. *Russian Geology and Geophysics, 52*, 1298–1309.

Zedgenizov, D. A., S. Rege, W. L. Griffin, H. Kagi, & V. S. Shatsky (2007). Composition of trapped fluids in cuboid fibrous diamonds from the Udachnaya kimberlite: LAM-ICPMS analysis. *Chemical Geology, 240*, 151–162.

Zedgenizov, D. A., A. Shatskiy, A. L. Ragozin, H. Kagi, & V. S. Shatsky (2014). Merwinite in diamond from São Luiz, Brazil: A new mineral of the Ca-rich mantle environment. *American Mineralogist, 99*, 547–550.

Zhang, Z., Z. Mao, X. Liu, Y. Zhang, & J. Brodholt (2018). Stability and reactions of $CaCO_3$ polymorphs in the Earth's deep mantle. *Journal of Geophysical Research: Solid Earth, 123*, 6491–6500.

15

Reactive Preservation of Carbonate in Earth's Mantle Transition Zone

Jie Li[1], Feng Zhu[2], Jiachao Liu[3], and Junjie Dong[4]

ABSTRACT

Calcium-rich carbonate may be preserved in fast-descending slabs to reach the mantle transition zone (MTZ), which is known to be at least locally hydrous. At MTZ pressures, the melting curve of $CaCO_3$ crosses the geotherm and is further depressed by water; hence, Ca-rich carbonate may be mobilized by hydrous melting and escape the MTZ. Here we show that aragonite reacts with wadsleyite to produce magnesite under the pressure and temperature conditions of cold slabs in the MTZ. Water considerably enhances conversion of Ca-rich carbonate into more refractory magnesite, helping to retain carbonate in the deep mantle.

15.1. INTRODUCTION

Carbonates are common rock-forming minerals in the Earth's crust and act as sinks of atmospheric carbon dioxide (CO_2) released from the mantle through volcanism. Subduction of hydrothermally altered oceanic lithosphere returns carbonates and organic matter from near-surface reservoirs to the interior, where more than three quarters of Earth's carbon is stored (e.g. Chen et al., 2014; Dasgupta & Hirschmann, 2010). A significant fraction of the subducted carbonates will join shallow cycles within the top 200 km of the solid Earth, whereas the rest reaches greater depths to participate in deep cycles, possibly extending to the core (Dasgupta & Hirschmann, 2010; Kelemen & Manning, 2015). The long-term evolution of CO_2 in the atmosphere is influenced by the residence time of carbonates in solid Earth, which, in turn, depends on their melting behavior and chemical stability under relevant conditions.

Calcium carbonate ($CaCO_3$), dolomite ($CaMg[CO_3]_2$), and magnesium carbonate ($MgCO_3$) are dominant ingredients in subducted slabs (e.g. Alt & Teagle, 1999). Although magnesite is expected to be the ultimately stable carbonate in the deep mantle (Biellmann et al., 1993; Keshav et al., 2011; Kushiro et al., 1975; Yaxley & Brey, 2004), Ca-carbonate in fast-descending slabs may be preserved to reach the mantle transition zone (MTZ) or lower mantle, as indicated by calcite and nyerereite (Na-K-Ca bicarbonate) inclusions in superdeep diamonds from Juina, Brazil (Brenker et al., 2007; Kaminsky et al., 2009). The MTZ is widely considered a potential water reservoir and known to be at least locally wet, as indicated by hydrous ringwoodite and ice VII inclusions in superdeep diamonds (Pearson et al., 2014; Tschauner et al., 2018) and seismic and electrical conductivity observations (Kelbert et al., 2009; Schmandt et al., 2014). The melting point of $CaCO_3$ is comparable to the mantle

[1] *Department of Earth and Environmental Sciences, University of Michigan, Ann Arbor, Michigan, USA*

[2] *Hawaii Institute of Geophysics and Planetology, University of Hawai'i at Mānoa, Honolulu, Hawaii, USA*

[3] *Department of Geological Sciences, Jackson School of Geosciences, University of Texas at Austin, Austin, Texas, USA*

[4] *Department of Earth and Planetary Sciences, Harvard University, Cambridge, Massachusetts, USA*

Carbon in Earth's Interior, Geophysical Monograph 249, First Edition. Edited by Craig E. Manning, Jung-Fu Lin, and Wendy L. Mao.

temperature beneath ocean islands (Z. Li et al., 2017), and it is further reduced by the presence of water (Wyllie & Tuttle, 1960). In contrast, magnesite remains refractory at the MTZ conditions (Katsura & Ito, 1990); thus, converting $CaCO_3$ to magnesite would help preserve carbonate in the MTZ. Previous work examined phase relations of complex carbonated silicates at the MTZ pressure, mainly under nominally anhydrous conditions (e.g. Grassi & Schmidt, 2011; Keshav & Gudfinnsson 2010; Keshav et al., 2011; Kiseeva et al., 2013; Thomson et al., 2016). It is unknown if hydrous melting can lead to loss of Ca-carbonate before it reacts with silicates to form magnesite.

In this exploratory study, we investigate the effect of water on the reaction between aragonite and wadsleyite at the conditions of cold slabs in the MTZ through high-pressure experiments under nominally anhydrous conditions and with free or structural water. The results are applied to assess the stability of carbonates in the MTZ and explore the implications for Earth's deep carbon cycle.

15.2. MATERIALS AND METHODS

15.2.1. High-Pressure Experiments

Synthesis and reaction experiments were conducted using the multianvil apparatus at the University of Michigan. All experiments used Toshiba F-grade tungsten-carbide anvils with 3 mm truncation edge length and standard COMPRES cell assemblies (Leinenweber et al., 2012). Sample pressure was calibrated using fixed-point phase transitions in SiO_2, $MgSiO_3$, and Mg_2SiO_4 and bears a precision of 5% and an estimated uncertainty of less than ±1 GPa at 10–20 GPa and 1000 °C–1500 °C (Z. Li & Li, 2015). Temperature-power relation was established by using a type C thermocouple placed along the cylindrical axis of the heater. The effect of pressure on the emf is unknown and therefore ignored (J. Li et al., 2003). The power curve is reproducible within ±100 °C, according to multiple calibration experiments.

The starting material for the nominally anhydrous and free water reaction experiments consists of one layer of $CaCO_3$ powder (Alfa Aesar 43073, 99.997%) and one layer of synthetic forsterite (Fo100) powder packed inside a gold capsule. In the free water experiments, ~0.6 microliter deionized water was added, corresponding to ~10 wt% water in the starting composition, which is well above the estimated range of water in the MTZ (e.g. Pearson et al., 2014). In the experiments with structurally bound water, an aggregate piece of hydrous wadsleyite, synthesized at 20 GPa and 1400 °C for 4 hrs in experiment M011217, was embedded in $CaCO_3$ powder and packed into a gold capsule. The water content of the synthetic wadsleyite is estimated at 0.8–1.2 wt% according to its b/a ratio of 2.018±0.001 (Chang et al., 2015). The starting material contains roughly equal volumes of carbonate and silicate in the nominally anhydrous and free water experiments, and more carbonate in the structurally bounded water experiment. The grain sizes of the starting $CaCO_3$ and forsterite are the same in all experiments, and that of wadsleyite in the structurally bound water experiment is larger and estimated at 5–10 microns. Sample was compressed at room temperature to target pressure, and then heated at the rate of 1 °C per second to target power. It was equilibrated at constant power for 120 min and then quenched by shutting off the power and decompressed overnight to ambient pressure.

15.2.2. Raman Spectroscopy Analysis

Experimental product was mounted in epoxy inside an acrylic disc. The sample was exposed using diamond or silicon carbide grinding discs and then polished with alumina powder on lapping discs. It was then cleaned ultrasonically with alcohol.

Raman analysis was performed using a Renishaw inVia confocal Raman microscope with a 532 nm continuous wave laser. The input laser power varied between 4 and 40 mW and the laser beam was focused through a 50x Mitutoyo objective lens to a diameter of 3–5 microns or through a 20x Olympus objective lens to a beam diameter of 7–12 microns. Raman signal was dispersed with an 1800 grooves/cm grating to achieve spectral resolution of 1–2 wave number (wn). The spectral window was centered at 520 or 3500 wn. The signal was recorded with a Peltier cooled CCD. Acquisition time ranged from 1 to 120 seconds per CCD window depending on the signal intensity of the sample. Replicate measurements were conducted to evaluate heterogeneity.

15.2.3. Electron Microprobe Analyses

For textural and chemical analysis, recovered sample was coated with a thin conductive film of carbon or aluminum. Back-scattered electron (BSE) and energy dispersive spectroscopy point analyses and maps were collected using the JOEL 7700 JOEL-7800FLV field-emission scanning electron microscope (FE-SEM) at the University of Michigan. Quantitative chemical analyses were conducted using the Electron Probe Micro-Analyzer (EPMA) on the CAMECA SX-100 at University of Michigan. Forsterite Mg_2SiO_4 and wollastonite $CaSiO_3$ were used as standards for Mg, Si, O, and Ca. The accelerating voltage and beam current were set to 15 kV and 10 nA, respectively, and the electron beam was focused to 1 μm.

15.3. RESULTS

In all experiments, $CaCO_3$ reacted with Mg_2SiO_4 at 15 GPa and 1200 °C to form $MgCO_3$, MgO, and $CaSiO_3$. The reaction can be described as

$$CaCO_3\,(\text{aragonite}) + Mg_2SiO_4\,(\text{wadsleyite}) = MgCO_3\,(\text{magnesite}) + CaSiO_3\,(\text{perovskite}) + MgO\,(\text{periclase}). \quad (15.1)$$

Reaction products were identified on the basis of anti-correlation between the Mg and Si and positive correlation between Ca and Si in the energy dispersive spectroscopy maps (Figures 15.1, 15.2, 15.3). $MgCO_3$ and MgO grains appear bright in the Mg map (MgO is brighter than $MgCO_3$) and dark in the Si map, whereas $CaSiO_3$ grains appeared bright in the Ca and Si maps and dark in the Mg map. The mean atomic numbers of $CaSiO_3$, MgO, and $MgCO_3$, calculated as the sum of weight percentage and atomic number of each element are 13.6, 10.4, and 8.9, respectively, and they appear as the bright, gray, and dark regions in the BSE image.

In the nominally anhydrous experiment M040316, a 50-micron-thick reaction zone formed between $CaCO_3$ and Mg_2SiO_4 layers in 120 min (Figure 15.1a). The grain size is typically 1–2 microns. With free water added to the starting material in experiment M031416, the reaction proceeded to consume all the $CaCO_3$ and formed a ~500 micron thick reaction zone in 120 min (Figure 15.3a). The reaction products may contain some hydroxide $Mg(OH)_2$ in addition to oxide MgO, as the experimental temperature of 1200 °C overlaps with the dehydration temperature of $Mg(OH)_2$ near 1250 °C at 15 GPa within uncertainties (Johnson & Walker 1993). With a weight mean atomic number of 9.1, $Mg(OH)_2$ would appear darker than MgO and similar to $MgCO_3$ in the BSE image. We did not find sufficiently large grains with the expected Raman signal and chemical composition of $Mg(OH)_2$. The grain sizes of the reaction products are 10–20 micron in some areas and 5–10 micron in others (Figure 15.2). These are larger than the nominally anhydrous reaction products. In experiment M011517, the hydrous wadsleyite grain in the starting material was fully consumed (Figure 15.3). The identification of reaction products was confirmed by additional O and C maps (Figure 15.3c). The grain size is up to 5 microns along the rim of the hydrous wadsleyite pseudomorph and finer inside.

Quantitative measurements showed limited solid-solution in the reaction products. In both the nominally

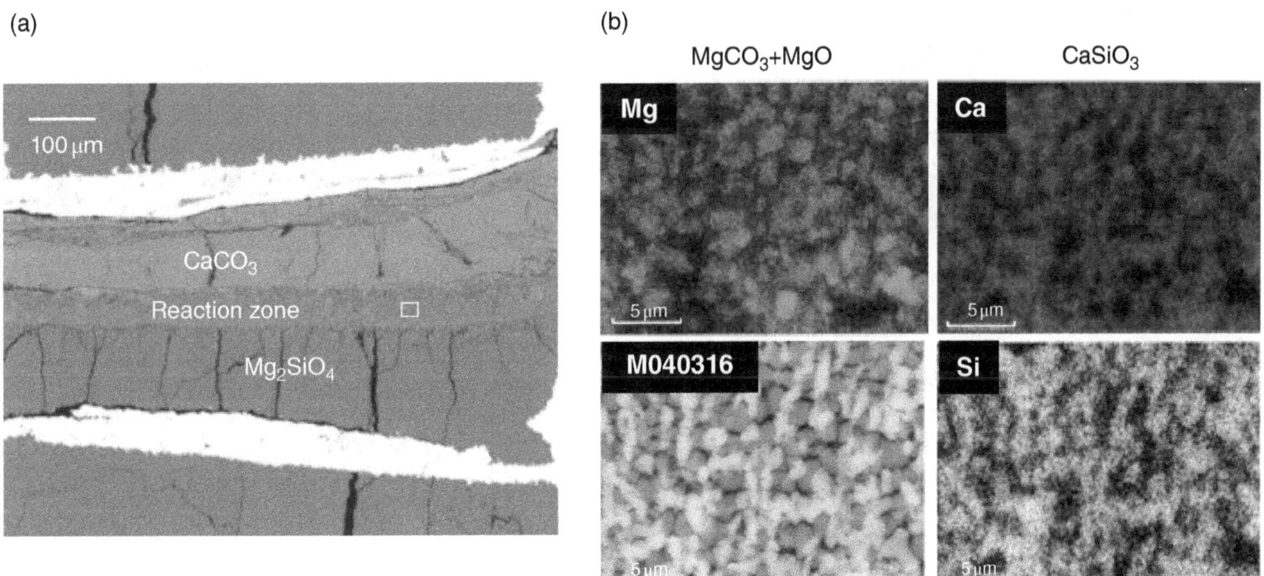

Figure 15.1 Product of reaction between nominally anhydrous $CaCO_3$ and Mg_2SiO_4 at 15 GPa and 1200 °C for 2 hrs. (a) Back-scattered electron (BSE) image of M040316 showing a ~50 micron reaction zone between a $CaCO_3$ layer and Mg_2SiO_4 layer. (b) Energy-dispersive spectroscopy (EDS) maps and BSE image of region marked by the yellow box in (a). The area consists of reaction products $CaSiO_3$, $MgCO_3$, and MgO, corresponding to the bright, dark, and gray regions in the BSE image in (b), respectively. The $CaSiO_3$ phases appear as the bright regions in the Ca and Si maps, and as the dark regions in the Mg map. The $MgCO_3$ and MgO phases appear as the bright regions in the Mg map (MgO is brighter than $MgCO_3$) and as the dark regions in the Ca and Si maps. See electronic version for color representation of the figures in this book.

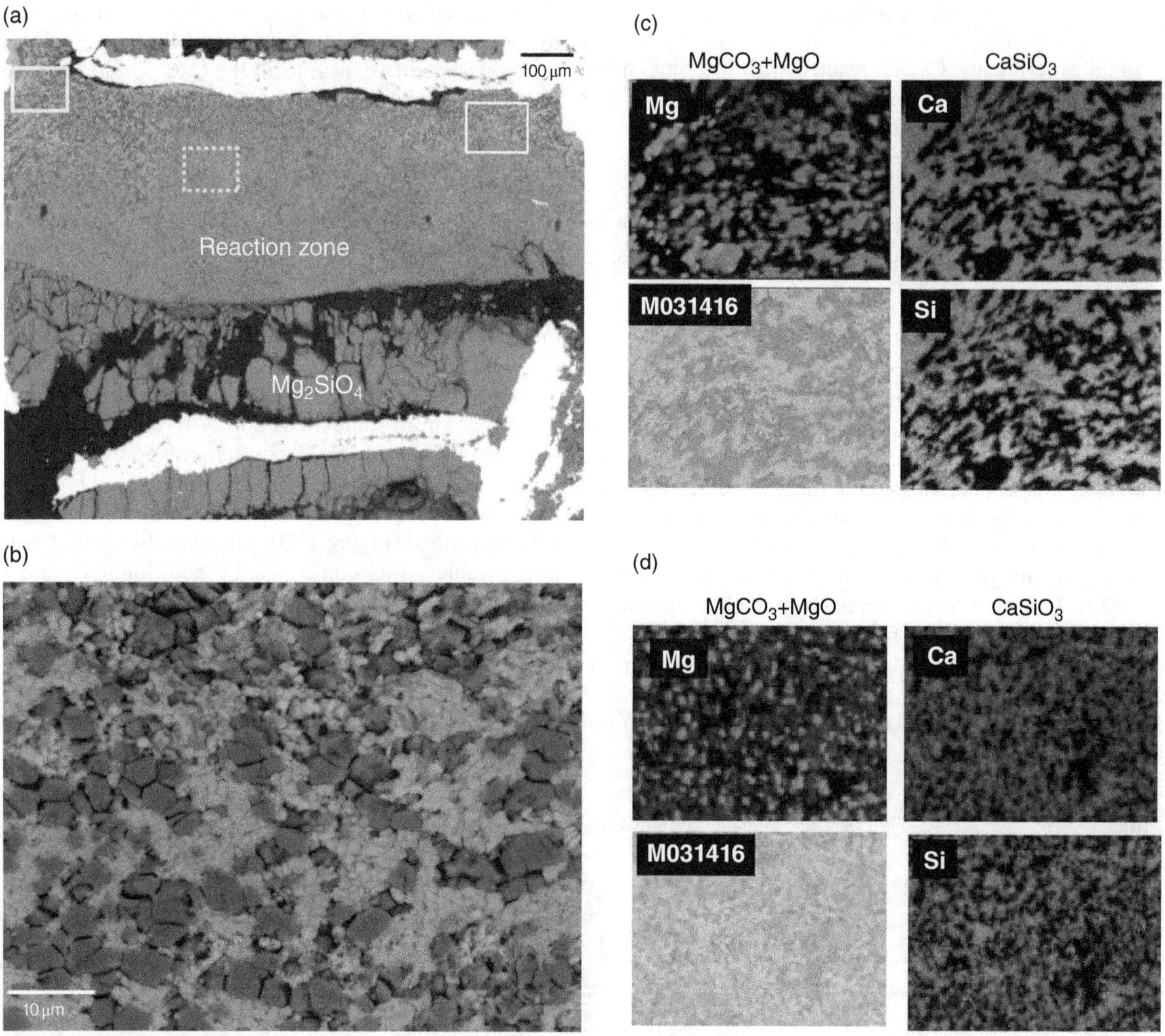

Figure 15.2 Product of reaction between $CaCO_3$ and Mg_2SiO_4 in the presence of H_2O at 15 GPa and 1200 °C for 2 hrs. (a) BSE image of M031416 showing a ~400 micron thick reaction zone above excess Mg_2SiO_4. (b) BSE image of the area marked by the white box on the upper left in (a). (c) EDS maps and BSE image of the region marked by the solid yellow boxes on the upper right in (a). (d) EDS maps and BSE image of the region marked by the dotted yellow boxes in (a). In (c) and (d), reaction products $CaSiO_3$, $MgCO_3$, and MgO appear as the bright, dark, and gray regions in the BSE image, respectively. $CaSiO_3$ appear as the bright regions in the Ca and Si maps, and as the dark regions in the Mg map. $MgCO_3$ and MgO appear as the bright regions in the Mg map (MgO is brighter than $MgCO_3$) and as dark regions in the Ca and Si maps. See electronic version for color representation of the figures in this book.

anhydrous and free water experiments, the $MgCO_3$ contained 4–5 at% Ca, and $CaSiO_3$ contained less than 1 at% Mg. In the hydrous wadsleyite experiment, the $MgCO_3$ contained 5–7 at% Ca, and $CaSiO_3$ contained 1–4 at% Mg. The extents of solid-solution are comparable in all experiments within analytical uncertainties.

Raman analyses allowed us to identify different polymorphs of phases in the recovered products (Figure 15.4). The remaining $CaCO_3$ in both the nominally anhydrous and structurally bound water experiments is aragonite, and the remaining Mg_2SiO_4 in both the anhydrous and free water experiments is wadsleyite. These are consistent with the known phase diagrams of $CaCO_3$ (Litasov et al., 2017) and Mg_2SiO_4 (Fei & Bertka 1999). The reaction product $MgCO_3$ is magnesite in all experiments and agrees with the known phase diagram (Litasov et al., 2008). The reaction product $CaSiO_3$ appears amorphous, consistent with the previous findings that $CaSiO_3$ adopts

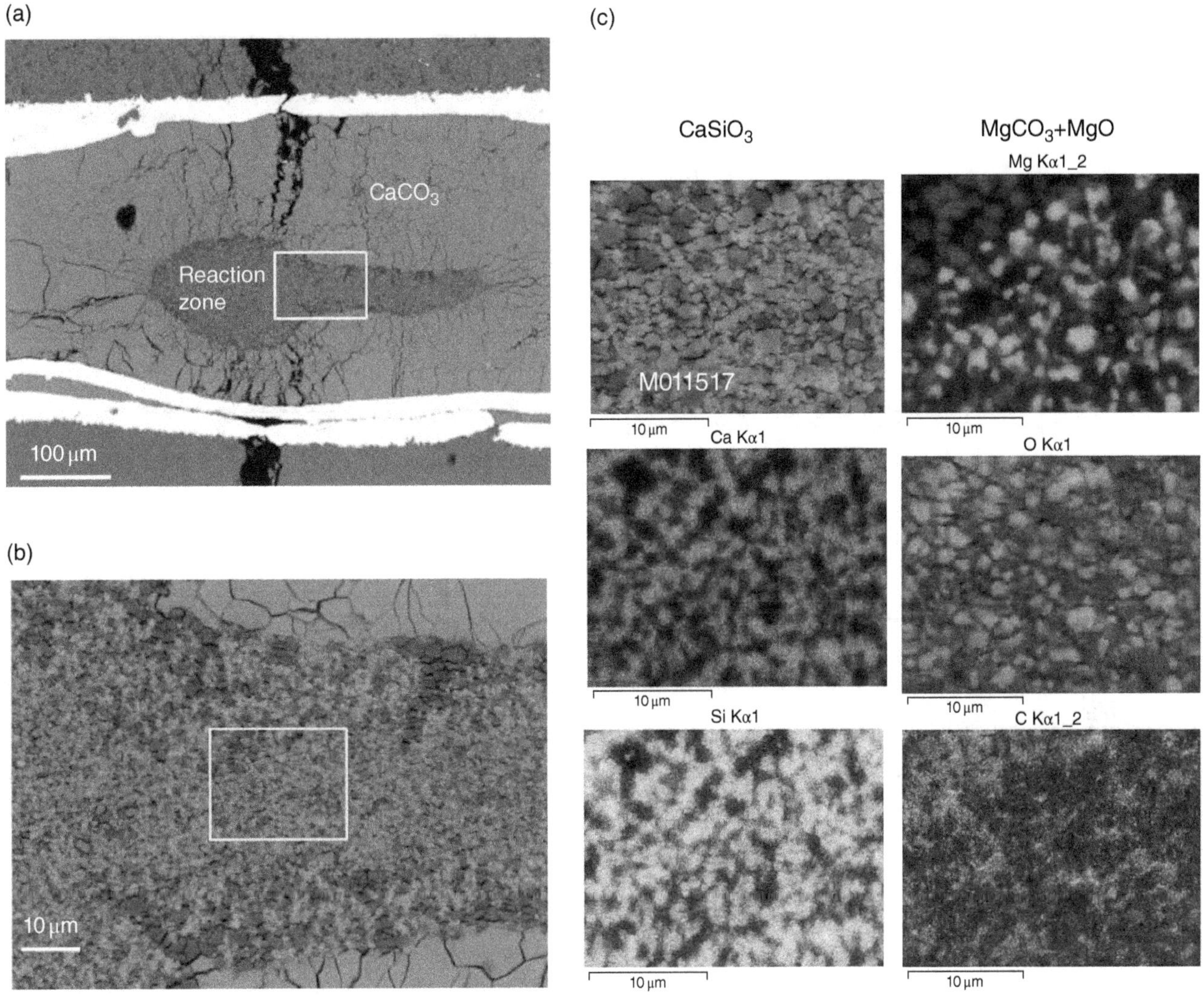

Figure 15.3 Product of reaction between $CaCO_3$ and hydrous wadsleyite (Mg_2SiO_4) at 15 GPa and 1200 °C for 2 hrs. (a) BSE image of M011517 showing hydrous wadsleyite in the starting material replaced by reaction products. (b) BSE image of the area marked by the white box in (a). (c) EDS maps and BSE image of the region marked by the yellow box in (b). Reaction products $CaSiO_3$, $MgCO_3$, and MgO appear as the bright, dark, and gray regions in the BSE image in (b), respectively. $CaSiO_3$ appears as the bright regions in the Ca and Si maps and as the dark regions in the Mg map. $MgCO_3$ and MgO appear as the bright regions in the Mg map (MgO is brighter than $MgCO_3$) and as dark regions in the Ca and Si maps. See electronic version for color representation of the figures in this book.

the perovskite structure at 15 GPa (Gasparik et al., 1994) and becomes amorphous in just a few hours upon pressure release to ambient condition (Mao et al., 1989).

The presence of structural water was inferred from the Raman data in the 3000–4000 cm^{-1} range (Figure 15.4). In the nominally anhydrous experiment, none of the remaining reactants or products shows detectable OH peak in this range, as expected. In the free water experiment, the remaining Mg_2SiO_4 is hydrous wadsleyite with an OH peak at 3410 cm^{-1}, near the broad OH peak of the synthetic hydrous wadsleyite at 3320 cm^{-1}, and the OH peak of a hydrous wadsleyite with 1.6 wt% structural water at 3380 cm^{-1} (Kleppe et al., 2001).

15.4. DISCUSSION

15.4.1. Chemical Stability of Carbonates in MTZ

This study shows that at 15 GPa and 1200 °C, aragonite is unstable and reacts with wadsleyite to form magnesite under both anhydrous and hydrous conditions. The experimental pressure corresponds to ~440 km depth, in the upper part of the MTZ. Similar reactions occur between carbonate, including calcite, aragonite, or dolomite, and enstatite in the upper mantle pressure range at 4–7 GPa (Kushiro et al., 1975). They have also been observed at the MTZ and lower mantle pressures, between dolomite and

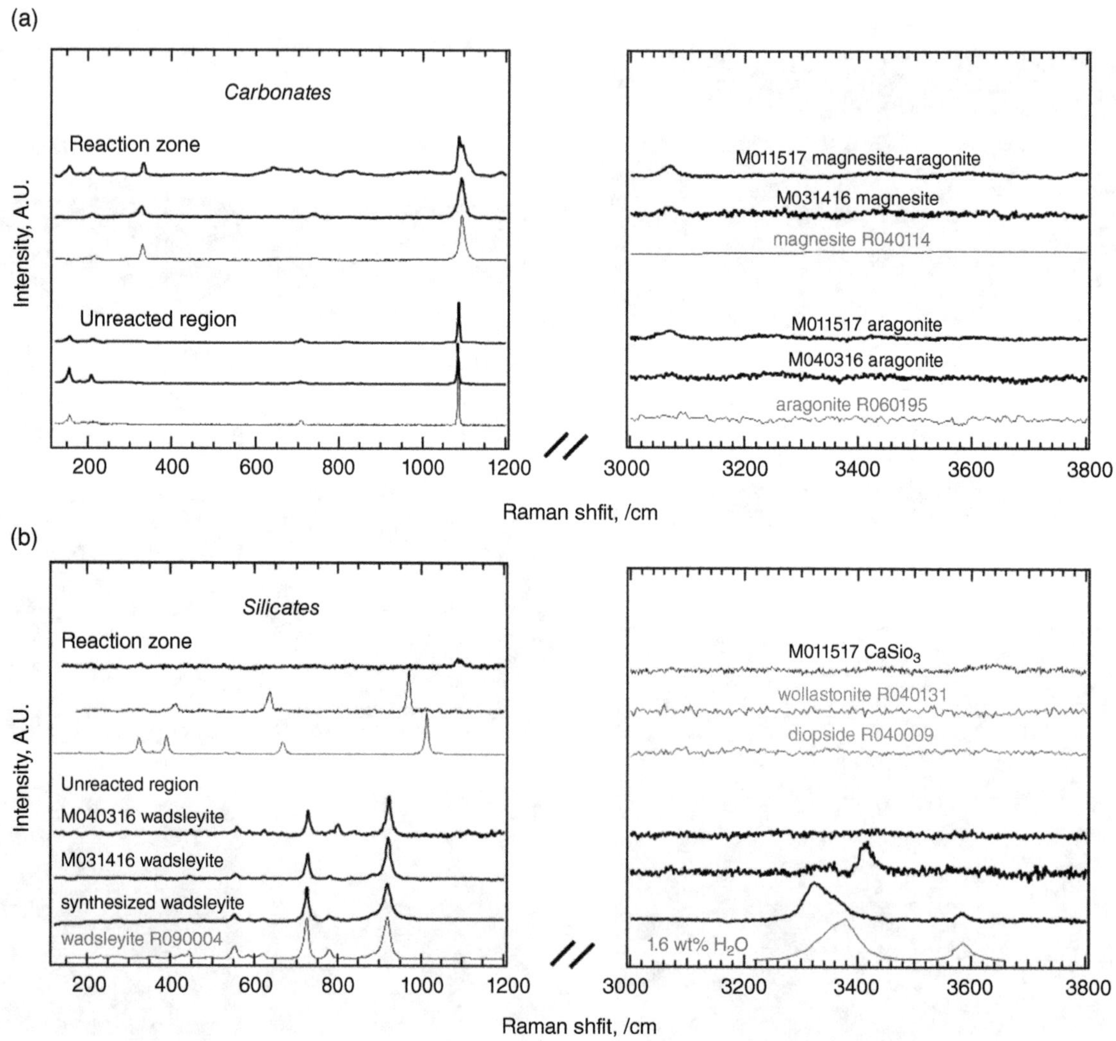

Figure 15.4 Raman spectra of samples recovered from the $CaCO_3$ and Mg_2SiO_4 reaction experiments. (a) Raman spectra of carbonates in the experimental products match the reference spectra of magnesite (RRUFF 040114 and aragonite (RRUFF 060195). (b) Raman spectra of silicates in the experimental products and the synthesized hydrous wadsleyite from experiment M062716 match the reference spectra of wadsleyite (RRUFF 090004) and hydrous wadsleyite (Kleppe et al., 2001). The spectrum of $CaSiO_3$ is mostly featureless and does not show diagnostic peaks of wollastonite (RRUFF 040131) or diopside (RRUFF 040009), walstromite (Anzolini et al., 2016), majorite (Hofmeister et al., 2004), or perovskite (Nestola et al., 2018). See electronic version for color representation of the figures in this book.

enstatite at 20–50 GPa, and between dolomite and olivine at 50 GPa (Biellmann et al., 1993). Stabilization of magnesite at high pressures in natural compositions such as carbonated eclogite has also been reported (e.g., Keshav & Gudfinnsson, 2010; Seto et al., 2008). Here, magnesite was found stable in the presence of hydrous wadsleyite, confirming its stability in the MTZ.

The strong affinity of Mg for carbonate and Ca for silicate at high pressures may be explained by volume reduction. At ambient conditions, the volume change of the cation exchange reaction from calcite and Mg-bridgmanite to magnesite and Ca-perovskite is –5.95 cc/mol (Biellmann et al., 1993), and that of reaction (1) is –2.54 cc/mol. The bulk moduli of aragonite, magnesite, wadsleyite, CaSi-perovskite, and periclase at 1 bar and 300 K are 66, 108, 173, and 232, and 160 GPa, respectively (Fiquet et al., 2002; Katsura et al., 2009; Litasov et al., 2017; Shim et al., 2000). Because the products $CaSiO_3$ perovskite and magnesite are less compressible than the reactants wadsleyite and aragonite, respectively, the reaction volume becomes less negative at higher pressures. At 20 GPa and 1600 K, the volume change of reaction (1) is estimated at –1.11 cc/mol, still favoring magnesite.

Temperature affects reaction kinetics and may influence the direction of the cation exchange. At 8 GPa and 800 °C, aragonite and forsterite did not react, possibly because the temperature was too low (Kushiro et al., 1975). At upper mantle pressures, the boundaries of relevant reactions have only a slightly positive d*T*/d*P* slope (Figure 15.5), suggesting relatively small effect of temperature on the Gibbs energy of the reaction.

15.4.2. Water Enhances Carbonate-Silicate Reaction

In this study we observed a strong effect of water on the extent of reaction between carbonate and silicate. The reaction zone in the free water experiment is nearly one order of magnitude broader than in the nominally anhydrous experiment. We speculate that an aqueous fluid formed at relatively low temperature and facilitated the

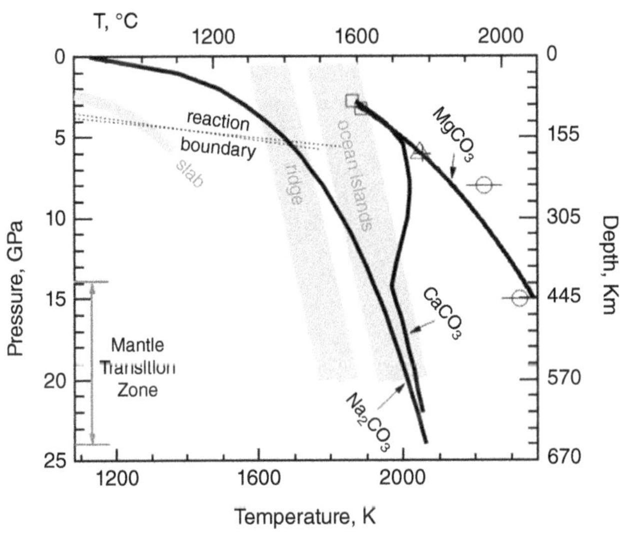

Figure 15.5 Melting curves of carbonates at the pressures of the upper mantle and transition zone. The melting curves of $CaCO_3$ and $MgCO_3$ diverge with increasing pressure and differ by by more than 400 °C at the pressures of the mantle transition zone. The melting curves of Na_2CO_3 and $CaCO_3$ converge with increasing pressure and become comparable near 14 GPa, both crossing the mantle geotherms beneath ridges (Katsura et al., 2010) in the vicinity of the mantle transition zone. Adding 1–2 wt% water further depresses the melting temperature of $CaCO_3$ (Wyllie & Tuttle, 1960). At 4–7 GPa, temperature has little influence on the boundaries of reactions from Ca-carbonate and Mg-silicate to Ca-silicate and Mg-carbonate (dotted lines; Kushiro et al., 1975). The melting curves of $CaCO_3$ and Na_2CO_3 are from Z. Li et al. (2017). That of $MgCO_3$ is a Simon's law fitting to the data from Irving and Wyllie, 1975 (open squares), Müller et al., 2017 (open triangle), Shatskiy et al., 2018 (cross), and Katsura and Ito, 1990 (open circles with error bars). The pressures in Irving and Wyllie 1975 are corrected by −10% (Byrnes & Wyllie 1981). See electronic version for color representation of the figures in this book.

cation exchange reaction. Such a fluid is expected to be stable at temperatures well below the eutectic melting point of the $CaCO_3$-H_2O binary system, and its stability field expands with increasing temperature (Wyllie & Tuttle, 1960). The fluid may circulate inside the sealed Au capsule and carry $CaCO_3$ to react with the silicate. Similar fluxing effect of a fluid with ppm level of dissolved silica is known to enhance the growth of quartz crystal in water. The presence of a fluid phase in the hydrous experiments is supported by the grain size of the reaction products. The smaller grain size in the anhydrous experiment is consistent with solid-state reaction, whereas the coarser grains in the hydrous experiments indicate fluid-mediated crystal growth.

A hydrous carbonate melt could play a similar role, but the experimental temperature is likely too low to generate such a melt. At 15 GPa, the melting temperature of $CaCO_3$ is more than 1600 °C. At 1 kbar, as much as 20 wt% water is needed to reduce the melting point of $CaCO_3$ by 300 °C (Wyllie & Tuttle, 1960). The presence of a hydrous carbonate melt cannot be excluded, however, considering that laboratory experiments at ~4 GPa showing formation of Ca-rich hydrous carbonatitic liquid at temperatures as low as ~900 °C, nearly 700 °C below the melting point of $CaCO_3$ at this pressure (Poli, 2015), and that a small amount of water was found to depress the melting point of $CaCO_3$ by 200 °C–300 °C at 8 GPa (Z. Li et al., 2017). In any case, neither an aqueous fluid nor carbonate melt was directly detected in the experimental products.

15.4.3. Reactive Preservation of Carbonate

Conversion of $CaCO_3$ into magnesite effectively raises the melting point of carbonate at pressures above ~5 GPa, where magnesite is more refractory than $CaCO_3$ (Figure 15.5). The melting slope of $CaCO_3$ becomes flat and possibly turns negative between 5 and 14 GPa (Z. Li et al., 2017), whereas that of magnesite continues to rise with increasing pressure (Katsura & Ito, 1990). At ~14 GPa, the melting point of magnesite is ~400 °C higher than that of aragonite. By taking the more refractory form, solid carbonate can remain frozen to higher temperatures, and Ca-rich carbonate melt may react with Mg-silicate and freeze up.

Redox freezing is a well-known example of reactive preservation, where carbonatitic melt is reduced by metallic iron to form immobile diamond or iron carbide (Palyanov et al., 2013; Rorhbach & Schmidt, 2011). Decomposition of siderite $FeCO_3$ melt at high temperature can also change carbon into the native element state (Kang et al., 2015). At the lower mantle pressures, carbon can also be transferred into more refractory host phases through various types of reactions. Dolomite or magnesite may turn into diamonds by reacting with iron (Dorfman et al., 2018) or SiO_2 (Maeda et al., 2017; Seto et al., 2008). A recent study showed that $FeCO_3$ transformed into $Fe_4C_3O_{12}$ and $Fe_3C_3O_{12}$ with tetrahedral-coordinated carbon, which remain solid well above the extrapolated melting curve of $FeCO_3$ (Cerantola et al., 2017). These reactions not only keep carbon in solid forms to higher temperatures and may also freeze carbon-bearing melt.

Cation exchange may also cause melting and loss of carbonate. For example, the steep depression in the solidus of carbonated eclogite at 13–21 GPa is attributed to cation exchange between carbonate and silicate (Thomson et al., 2016). In the experiments, the clinopyroxene became increasingly enriched in sodium (Na) at high pressures as the Na-poor clinopyroxene component dissolved into coexisting garnet. Eventually, the clinopyroxene became so Na-rich that a coexisting Na,Ca-rich carbonate phase stabilized in the subsolidus assemblage, causing the solidus depression. Earlier studies of carbonated eclogite, peridotite, and pelite found a wide range of melting behavior with the solidus temperature spanning 500 °C at 15 GPa (Grassi & Schmidt, 2011; Keshav & Gudfinnsson, 2010; Keshav et al., 2011; Kiseeva et al., 2013). The low solidus ledge exists for a composition containing less and perhaps more relevant CO_2 at ~2.5 wt.% (Thomson et al., 2016) but was not observed in all studies, suggesting strong dependence on the bulk composition.

15.4.4. Implications for Deep Carbon Cycle

As carbonates are subducted into the deep Earth, they may be mobilized at various depths through hydrous melting below the arcs (Kelemen & Manning, 2015), incipient melting at up to 330 km depths (Dasgupta & Hirschmann, 2006), and reactive melting in the MTZ (Grassi & Schmidt, 2011; Thomson et al., 2016). On the other hand, carbonate melt may freeze through reactions with magnesian silicates or metallic iron (e.g Rohrbach & Schmidt, 2011). While sub-arc melting releases carbon from the solid Earth, reactive preservation of solid carbonates and freezing of carbonate melt at higher pressures help retain carbon in the deep mantle (Figure 15.6).

Water has competing effects on the stability of Ca-rich carbonate in the MTZ. The presence of water reduces the melting point of $CaCO_3$ and may induce hydrous melting at the MTZ pressures (Z. Li et al., 2017). On the other hand, water enhances the conversion of Ca-rich

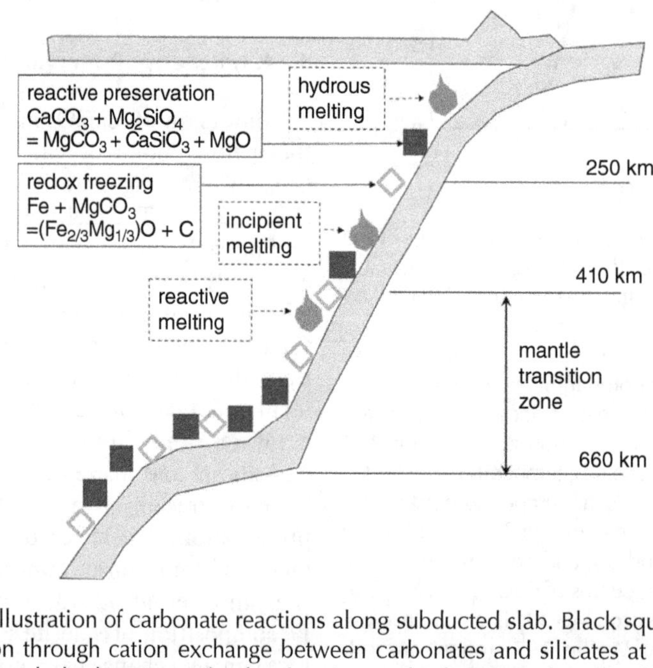

Figure 15.6 Cartoon illustration of carbonate reactions along subducted slab. Black squares mark the locations of reactive preservation through cation exchange between carbonates and silicates at pressures above 5 GPa. Blue open diamonds mark the locations of redox freezing at depth greater than 250 km. Red droplets denote hydrous melting at the mantle wedge (Kelemen & Manning, 2015), incipient melting at up to 330 km depths (Dasgupta & Hirschmann, 2006), and reactive melting at 13–17 GPa (Thomson et al., 2016). The extent of redox freezing and reactive preservation depends on kinetics and local redox state. They are likely incomplete and occur simultaneously at a given depth. See electronic version for color representation of the figures in this book.

carbonate into more refractory magnesite, thus helping to preserve carbonate. High temperature also increases reaction rates. The efficiency of carbonate retention in the MTZ thus depends on how rapidly Ca-carbonate converts to the more refractory magnesite before slabs warm up sufficiently to induce melting. The extent of carbonate retention is further complicated by other processes such as reverse conversion of magnesite to Ca-Na-rich carbonate (Grassi & Schmidt 2011; Thomson et al., 2016) and reduction of magnesite into elemental carbon or iron carbides (Palyanov et al., 2013). A reliable assessment of carbonate stability in the MTZ thus requires quantifying the effects of water and temperature on the rates of these reactions.

Preservation of carbonates in the MTZ would facilitate their subduction into the lower mantle where convection is considerably more sluggish, thus prolonging the residence time in the deep cycles. After crossing the MTZ, residual $CaCO_3$ may continue to react with Mg-bridgmanite to form magnesite (Biellmann et al., 1993), which, in turn, may be reduced by iron metal into elemental carbon or iron carbides (Dorfman et al., 2018) or may react with SiO_2 to form diamond (Maeda et al., 2017). Magnesite, carbon, or iron carbide remain solid in most of the lower mantle and may carry carbon all the way to the base of the mantle and even into the core (Liu et al., 2016).

15.5. CONCLUSIONS

This experimental study shows that aragonite is unstable in cold slabs at MTZ conditions and reacts with wadsleyite to form magnesite. The conversion of aragonite to magnesite effectively raises the melting point of carbonate at MTZ pressures and helps its retention in the deep mantle. Free or structural water enhances the reaction to preserve carbonate against hydrous melting. By inference, Ca-rich carbonate melt may freeze through cation exchange with magnesian silicate. In order to estimate the residence time of carbonates in solid Earth, further studies are needed to quantity the effects of water and temperature on the rates of relevant reactions.

ACKNOWLEDGMENTS

We thank Owen Neill for assistance with electron microprobe analyses and interpretation, and Dave Walker for providing synthetic Fo100 and valuable comments, and two reviewers for providing critical and constructive comments.

There are no conflicting interests. The data supporting the conclusion can be found in the figures. This work is partially supported by Sloan Foundation Deep Carbon Observatory Grant G-2017-9954, National Science Foundation grants EAR 1763189 and AST 1344133.

REFERENCES

Alt, J. C., & Teagle, D.A.H. (1999). Uptake of carbon during alteration of oceanic crust. *Geochim. Cosmochim. Acta*, *63*, 1527–1535.

Anzolini, C., Angel, R. J., Merlini, M., Derzsi, M., Tokár, K., Milani, S., et al. (2016). Depth of formation of $CaSiO_3$-walstromite included in super-deep diamonds. *Lithos*, *265*, 138–147.

Biellmann, C., Gillet, P., Guyot, F., Peyronneau, J., & Reynard, B. (1993). Experimental evidence for carbonate stability in the Earth's lower mantle. *Earth and Planetary Science Letters*, *118*, 31–41.

Brenker, F. E., Vollmer, C., Vincze, L., Vekemans, B., Szymanski, A., Janssens, K., et al. (2007). Carbonates from the lower part of transition zone or even the lower mantle. *Earth and Planetary Science Letters*, *260*, 1–9.

Byrnes, S.A.P., & Wyllie, P.J. (1981). Subsolidus and melting relations for the join $CaCO_3$-$MgCO_3$ at 10 kbar. *Geochim. Cosmochim. Acta 45*, 321–328.

Cerantola, V., Bykova, E., Kupenko, I., Merlini, M., Ismailova, L., McCammon, C., et al. (2017). Stability of iron-bearing carbonates in the deep Earth's interior. *Nature Communications*, *8*, 15960.

Chang, Y.-Y., Jacobsen, S.D., Bina, C.R., Thomas, S.-M., Smyth, J.R., Frost, D., Ballaran, T.B., McCammon, C., Hauri, E., Inoue, T., Yurimoto, H., Meng, Y. and Dera, P. (2015). Comparative compressibility of hydrous wadsleyite and ringwoodite: Effect of H_2O and implications for detecting water in the transition zone. *J. Geophys. Res. Solid Earth 120*, 8259–8280.

Chen, B., Z., L., Zhang, D., Liu, J., Hu, M. Y., Zhao, J., et al. (2014). Hidden carbon in Earth's inner core revealed by shear softening in dense Fe_7C_3. *Proceedings of the National Academy of Sciences of the United States of America*, *111*, 17755–17758.

Dasgupta, R. & Hirschmann, M. M. (2006). Melting in the Earth's deep upper mantle caused by carbon dioxide. *Nature*, *440*, 659–662.

Dasgupta, R., & Hirschmann, M. M. (2010). The deep carbon cycle and melting in Earth's Interior. *Earth and Planetary Science Letters*, *298*, 1–13.

Dorfman, S. M., Badro, J., Nabieia, F., Prakapenka, V. B., Cantonie, M., & Gillet, P. (2018). Carbonate stability in the reduced lower mantle. *Earth and Planetary Science Letters*, *498*, 84–91.

Fei, Y., & Bertka, C. (1999). Phase transitions in the Earth's mantle and mantle mineralogy, in: Fei, Y., Bertka, C., Mysen, B.O. (Eds.), Mantle Petrology: Field Observations and High Pressure Experimentation: a tribute to Francis R. (Joe) Boyd, pp. 189–207.

Fiquet, G., Guyot, F., Kunz, M., Matas, J., Andrault, D., & Hanfland, M. (2002). Structural refinements of magnesite at very high pressure. *American Mineralogist*, *87*, 1261–1265.

Gasparik, T., Wolf, K., & Smith, C. M. (1994). Experimental determination of phase relation in the $CaSiO_3$ system from 8 to 15 GPa. *American Mineralogist*, *79*, 1219–1222.

Grassi, D., & Schmidt, M. W. (2011). The melting of carbonated pelites from 70 to 700 km depth. *Journal of Petrology*, *52*, 765–789.

Hofmeister, A.M., Giesting, P.A., Wopenka, B., Gwanmesia, G.D. and Jolliff, B.L. (2004) Vibrational spectroscopy of pyrope-majorite garnets: Structural implications. *Am. Mineral. 889*, 132–146.

Irving, A. J., & Wyllie, P. J. (1975). Subsolidus and melting relationships for calcite, magnesite and the join $CaCO_3$–$MgCO_3$ to 36 kb. *Geochim. Cosmochim. Acta, 39*, 35–53.

Johnson, M. C., & Walker, D. (1993). Brucite [$Mg(OH)_2$] dehydration and the molar volume of H2O to 15 GPa, *American Mineralogist, 78*, 271–284.

Kaminsky, F., Wirth, R., Matsyuk, S., Schreiber, A., & Thomas, R. (2009). Nyerereite and nahcolite inclusions in diamond: evidence for lower-mantle carbonatitic magmas. *Mineralogical Magazine, 73*, 797–816.

Kang, N., Schmidt, M. W., Poli, S., Franzolin, E., & Connolly, J.A.D. (2015). Melting of siderite to 20 GPa and thermodynamic properties of $FeCO_3$-melt. *Chemical Geology, 400*, 34–43.

Katsura, T., & Ito, E. (1990). Melting and subsolidus phase relations in the $MgSiO_3$-$MgCO_3$ system at high pressures: implications to evolution of the Earth's atmosphere. *Earth and Planetary Science Letters, 99*, 110–117.

Katsura, T., Shatskiy, A., Geeth, M.A., Manthilake, M., Zhai, S., Yamazaki, D., et al. (2009). P-V-T relations of wadsleyite determined by in situ X-ray diffraction in a large-volume high-pressure apparatus. *Geophysical Research Letters, 36*, L11307.

Katsura, T., Yoneda, A., Yamazaki, D., Yoshino, T., & Ito, E. (2010). Adiabatic temperature profile in the mantle. *Physics of the Earth and Planetary Interiors, 183*, 212–218.

Kelbert, A., Schultz, A., & Egbert, G. (2009). Global electromagnetic induction constraints on transition-zone water content variations. *Nature, 460*, 1003–1006.

Kelemen, P. B., & Manning, C. E. (2015). Reevaluating carbon fluxes in subduction zones, what goes down, mostly comes up. *Proc. Natl. Acad. Sci. USA, 112*, E3997–E4006.

Keshav, S., & Gudfinnsson, G. H. (2010). Experimentally dictated stability of carbonated oceanic crust to moderately great depths in the Earth: Results from the solidus determination in the system CaO–MgO–Al_2O_3–SiO_2–CO_2. *J. Geophys. Res., 115*, B05205.

Keshav, S., Gudfinnsson, G. H., & Presnall, D. (2011). Melting phase relations of simplified carbonated peridotite at 12–26 GPa in the systems CaO-MgO-SiO_2-CO_2 and CaO-MgO-Al_2O_3-SiO_2-CO_2: highly calcic magmas in the transition zone of the Earth. *Journal of Petrology, 52*, 2265–2291.

Kiseeva, E. S., Litasov, K. D., Yaxley, G. M., Ohtani, E. & Kamenetsky, V. S. (2013). Melting and phase relations of carbonated eclogite at 9–21 GPa and the petrogenesis of alkali-rich melts in the deep mantle. *Journal of Petrology, 54*, 1555–1583.

Kleppe, A. K., Jephcoat, A. P., Olijnyk, H., Slesinger, A. E., Kohn, S. C. & Wood, B. J. (2001). Raman spectroscopic study of hydrous wadsleyite β-Mg_2SiO_4 to 50 GPa. *Physics and Chemistry of Minerals, 28*, 232–241.

Kushiro, I., Satake, H., & Akimoto, S. (1975). Carbonate-silicate reactions at high pressures and possible presence of dolomite and magnesite in the upper mantle. *Earth and Planetary Science Letters, 28*, 116–120.

Leinenweber, K., Tyburczy, J. A., Sharp, T., Soignard, E., Diedrich, T., Petuskey, W. B., Wang, Y., & Mosenfelder, J. (2012). Cell assemblies for reproducible multi-anvil experiments (the COMPRES assemblies). *American Mineralogist, 97*, 353–368.

Li, J., Hadidacos, C., Mao, H. K., Fei, Y., & Hemley, R. J. (2003). Effect of pressure on thermocouples in a multi-anvil apparatus. *High Press. Res., 23*, 389–401.

Li, Z., & Li, J. (2015). Melting curve of NaCl to 20 GPa from electrical measurements of capacitive current. *American Mineralogist, 100*, 1892–1898.

Li, Z., Li, J., Liu, J., Lange, R., & Militzer, B. (2017). Determination of calcium carbonate and sodium carbonate melting curves up to Earth's transition zone pressures with implications for the deep carbon cycle. *Earth and Planetary Science Letters, 457*, 395–402.

Litasov, K. D., Fei, Y., Ohtani, E., Kuribayashi, T, & Funakoshi, K. (2008). Thermal equation of state of magnesite to 32 GPa and 2073 K, *Physics of the Earth and Planetary Interiors, 168* 191–203.

Litasov, K. D., Shatskiy, A., Gavryushkin, P. N., Bekhtenova, A. E., Dorogokupets, P. I., Danilov, B.S., et al. (2017). P-V-T equation of state of $CaCO_3$ aragonite to 29 GPa and 1673 K: In situ X-ray diffraction study. *Physics of the Earth and Planetary Interiors, 265*, 82–91.

Liu, J., Li, J., Hrubiak, R.& J., S. (2016). Origin of ultra-low velocity zones through mantle-derived metallic melt. *Proceedings of the National Academy of Sciences of the United States of America, 113*, 5547–5551.

Maeda, F., Ohtani, E., Kamada, S., Sakamaki, T., Hirao, N., & Ohishi, Y. (2017). Diamond formation in the deep lower mantle: a high-pressure reaction of $MgCO_3$ and SiO_2. *Scientific Reports, 7*, 40602.

Mao, H. K., Chen, L. C., Hemley, R. J., Jephcoat, A. P., Wu, Y. & Bassett, W. A. (1989). Stability and equation of state of $CaSiO_3$-perovskite to 134 GPa. *Journal of Geophysical Research, 94*, 1789–17894.

Müller, J., Koch-Müller, M., Rhede, D., Wilke, F.D.H., & Wirth, R. (2017). Melting relations in the system $CaCO_3$-$MgCO_3$ at 6 GPa. *Am. Mineral., 102*, 2400–2449.

Nestola, F., Korolev, N., Kopylova, M., Rotiroti, N., Pearson, D. G., Pamato, M. G., et al. (2018). $CaSiO_3$ perovskite in diamond indicates the recycling of oceanic crust into the lower mantle. *Nature, 555*, 237–241.

Palyanov, Y. N., Bataleva, Y. V., Sokol, A. G., Borzdov, Y. M., Kupriyanov, I. N., Reutsky, V. N., & Sobolev, N. V. (2013). Mantle–slab interaction and redox mechanism of diamond formation. *Proc. Natl. Acad. Sci. USA*, 201313340.

Pearson, D. G., Brenker, F. E., Nestola, F., McNeill, J., Nasdala, L., Hutchison, M. T., et al. (2014). Hydrous mantle transition zone indicated by ringwoodite included within diamond. *Nature, 507*, 221–224.

Poli, S. (2015). Carbon mobilized at shallow depths in subduction zones by carbonatitic liquids. *Nature Geoscience, 8*, 633–637.

Rohrbach, A., & Schmidt, M. W. (2011). Redox freezing and melting in the Earth's deep mantle resulting from carbon-iron redox coupling. *Nature, 472*, 209–212.

Schmandt, B., Jacobsen, S. D., Becker, T. W., Liu, Z., & Dueker, K. G. (2014). Dehydration melting at the top of the lower mantle. *Science*, *344*, 1265–1268.

Seto, Y., Hamane, D., Nagai, T., & Fujino, K. (2008). Fate of carbonates within oceanic plates subducted to the lower mantle, and a possible mechanism of diamond formation. *Phys Chem Minerals*, *35*, 223–229.

Shatskiy, A., Podborodnikov, I. V., Arefiev, A. V., Minin, D. A., Chanyshev, A. D. & Litasov, K. D. (2018). Revision of the $CaCO_3$–$MgCO_3$ phase diagram at 3 and 6 GPa. *Am. Mineral.*, *103*, 441–452.

Shim, S., Duffy, T., & Shen, G. (2000). The stability and P-V-T equation of state of CaSiO3 perovskite in the Earth's lower mantle. *Journal of Geophysical Research*, *105*, 25955–25968.

Thomson, A. R., Walter, M. J., Kohn, S. C., & Brooker, R. A. (2016). Slab melting as a barrier to deep carbon subduction. *Nature*, *529*, 76–79.

Tschauner, O., Huang, S., Greenberg, E., Prakapenka, V. B., Ma, C., Rossman, G. R., et al. (2018). Ice-VII inclusions in diamonds: Evidence for aqueous fluid in Earth's deep mantle. *Science*, *359*, 1136–1139.

Wyllie, P. J., & Tuttle, O. F. (1960). The system CaO-CO_2-H_2O and the origin of carbonatites. *Journal of Petrology*, *1*, 1–46.

Yaxley, G. M., & Brey, G. P. (2004). Phase relations of carbonate–bearing eclogite assemblages from 2.5–5.5 GPa: implications for petrogenesis of carbonatites. *Contrib Mineral Petrol*, *146*, 606–619.

16

Carbon Speciation and Solubility in Silicate Melts

Natalia Solomatova[1], Razvan Caracas[1,2], and Ronald Cohen[3,4]

ABSTRACT

To improve our understanding of the Earth's global carbon cycle, it is critical to characterize the distribution and storage mechanisms of carbon in silicate melts. Presently, the carbon budget of the deep Earth is not well constrained and is highly model-dependent. In silicate melts of the uppermost mantle, carbon exists predominantly as molecular carbon dioxide and carbonate, whereas at greater depths, carbon forms complex polymerized species. The concentration and speciation of carbon in silicate melts is intimately linked to the melt's composition and affects its physical and dynamic properties. Here we review the results of experiments and calculations on the solubility and speciation of carbon in silicate melts as a function of pressure, temperature, composition, polymerization, water concentration, and oxygen fugacity.

16.1. INTRODUCTION

Evidence of carbon-bearing phases in the Earth's mantle includes the release of CO_2 in volcanic eruptions, dissolved CO_2 in magmatic glasses and glass inclusions (Mörner & Etiope, 2002), diamonds and carbonate minerals in mantle xenoliths (Eggler, 1987; Sobolev & Shatsky, 1990), and the existence of carbonatite and kimberlite magmas (Wyllie et al., 1990). There are two possible sources of carbon: primordial carbon and the carbon delivered by later cometary and asteroid bombardment. Primordial carbon existed in the proto-Earth and subsequently survived the moon-forming impact, and its amount is currently unknown. From all existent carbon, part of it might be locked in the core, part in some deep mantle reservoir, another fraction lies at the surface, and the remaining is resurfaced after surviving subduction. To determine how much carbon may have remained in the Earth after the giant impact, the chemistry and thermodynamics of carbon in silicate (particularly, with the bulk silicate Earth composition) must be determined as a function of pressure and temperature. Estimates of the amount of carbon exchanged between the surface and the mantle range between 30 and 130 megatons per year, and estimates of the carbon concentration stored within the core range between 0.2 and 4 wt.% (McDonough, 2003; Mookherjee et al., 2011; Wood, 1993). Carbon is mainly subducted into the Earth in the form of carbonates within metasomatically calcium-enriched basaltic rock, calcified serpentinites, and sedimentary carbonaceous ooze of the seafloor (Brenker et al.,

[1] CNRS, Ecole Normale Supérieure de Lyon, Université de Lyon, Laboratoire de Géologie de Lyon, Lyon, France
[2] CEED, The Center for Earth Evolution and Dynamics, University of Oslo, Oslo, Norway
[3] Extreme Materials Initiative, Carnegie Institution for Science, Washington, DC, USA
[4] Department für Geo- und Umweltwissenschaften, Ludwig Maximilians Universität München, München, Germany

Carbon in Earth's Interior, Geophysical Monograph 249, First Edition. Edited by Craig E. Manning, Jung-Fu Lin, and Wendy L. Mao.

2006). Some carbonates may survive subduction by undergoing transitions to denser carbonate phases (Brenker et al., 2006; Isshiki et al., 2004; Mao et al., 2011; Merlini et al., 2012; Oganov et al., 2008; Ono et al., 2005; Solomatova & Asimow, 2017). If a fraction of carbon is subducted into the lower mantle, all possible carbon sinks need to be determined. Due to carbon's low solubility in crystalline silicates (Keppler et al., 2003; Panero & Kabbes, 2008; Shcheka et al., 2006), carbon strongly partitions into silicate melts relative to silicate minerals. This chapter briefly summarizes the experiments and calculations that have been performed on carbon-bearing silicate melts and glasses, particularly the speciation and solubility of carbon as a function of pressure, with a primary goal to expose areas that remain controversial or unknown that would especially benefit from ab initio molecular dynamics simulations. There have been at least three excellent review papers on carbon in silicate melts (see Holloway & Blank, 1994; Ni & Keppler, 2013; Webster et al., 2011), as well as several papers that have done an excellent job summarizing, tabulating, and plotting together data from previous studies (e.g. Duan, 2014; Shishkina et al., 2014). Here, we provide a perspective from computational mineral physics on what is critical to investigate in the near future. Furthermore, we contribute our calculations on carbonated forsterite melts to improve our understanding of diamond formation from silicate melts.

16.2. EXPERIMENTS ON SILICATE MELTS

Experimentally, standard difficulties hinder the in situ analysis of silicate melts: absence of periodicity, which prevents refining structure and density from X-ray diffraction, and high background optical radiation, which hinders spectroscopic measurements. Glasses from quenched melts are amenable to study, but the structure may change and carbon may diffuse during quench. Nevertheless, the carbon content of quenched samples that were equilibrated with other phases (solids, liquids or gases) can be determined using a bulk analyzer, secondary-ion mass spectrometry, nuclear magnetic resonance experiments, and most commonly, Fourier-transform infra-red (FTIR) spectroscopy. FTIR is able to determine both CO_2 and H_2O concentrations with a high level of accuracy if the molar absorptivity is well calibrated against absolute techniques, such as a bulk carbon analyzer. The solubility and speciation of carbon depends on the composition, oxygen fugacity, H_2O and CO_2 concentrations, temperature, pressure, and NBO/T (the ratio between the number of Non-Bridging Oxygen atoms and Tetrahedrally-coordinated cations; this ratio is defined in slightly different ways by different authors, depending on which cations are considered tetrahedrally coordinated). Experiments have generally been conducted in large-volume presses in the pressure range of 1–4 GPa and temperature range of 1000 °C–2000 °C. Typically, the procedure is to saturate a melt at high pressure and temperature with CO_2 until equilibrium is reached and then rapidly quench the melt to produce a glass that ideally retains the structure of the melt. The examination of carbon speciation has mostly been limited on quantifying the CO_2-to-CO_3^{2-} ratio as a function of composition. Unfortunately, with a few exceptions (e.g. Sen et al., 2013), experimental studies have not pursued the formation of more complex species, such as carbon-carbon polymers, organic molecules, or even silicon-carbon and iron-carbon polymers.

16.2.1. Speciation of Carbon

Experiments have demonstrated that at 1–4 GPa and 1000 °C–2000 °C, the $CO_2/(CO_2+CO_3^{2-})$ ratio is most strongly dependent on the composition of the melt. It has been shown that carbon exists almost exclusively as CO_3^{2-} (rather than CO_2) in mafic melts, such as basalt (Fine & Stolper, 1986; Fogel & Rutherford, 1990; Shishkina et al., 2014; Thibault & Holloway, 1994), olivine melitite (Brey, 1976; Brey & Green, 1976), phono-tephrite (Behrens, 2009), nephelinite (Brooker et al., 1999; Shishkina et al., 2014), and icelandite (Jakobsson, 1997). In the sequence of basalt, andesite, rhyodacite, and rhyolite melts, Fogel and Rutherford (1990) showed a general increase in the CO_2/CO_3^{2-} ratio, such that carbon exists exclusively as CO_2 in rhyolite and as CO_3^{2-} in basalt, while in albite-anorthite melts, Brey (1976) found that the CO_2/CO_3^{2-} ratio increases with increasing albite content. King and Holloway (2002) found that highly polymerized melts with high ionic porosities (>48.3%) and low nonbridging oxygen/tetrahedral oxygen (<0.3) have a higher concentration of molecular CO_2. In melts along the $NaAlO_2$-SiO_2 join (from $NaAlSi_6O_{14}$ rhyolite rock composition to pure $NaAlSiO_4$ nepheline), Brooker et al. (1999) found that the relative and absolute abundance of CO_3^{2-} increases with increasing $NaAlO_2/SiO_2$ at the expense of CO_2, such that carbon in $NaAlSiO_4$ exists almost exclusively as CO_3^{2-}, whereas carbon in $NaAlSi_6O_{14}$ exists almost exclusively as CO_2. The total solubility of carbon decreases as $NaAlO_2$ is added to SiO_2, up to a $NaAlO_2/SiO_2$ ratio of about 0.5, after which the trend reverses and there is a sharp increase in the solubility of carbon. The CO_2/CO_3^{2-} ratio decreases with increasing pressure at constant temperature (in other words, carbon increases coordination with pressure). The effect of temperature is subtler: at constant pressure, the CO_3^{2-} concentration decreases with increasing temperature between 1450 °C and 1600 °C and increases with increasing temperature from 1600 °C to 1700 °C, in agreement with the results of Fine and Stolper (1985). Konschak and Keppler (2014) examined the speciation of carbon in dacite, phonolite, basaltic andesite, and alkali silicate melts using synchrotron infrared spectroscopy and diamond anvil cells up to 1000 °C and 200 kbar, finding a correlation between NBO/T and speciation. In alkali silicate and basaltic andesite, carbon exists exclusively as carbonate, while in dacite and phonolite, carbon exists as a mixture of CO_2 and CO_3^{2-}. With increasing temperature, the effect of composition on carbon's speciation decreases as carbon shifts from existing as CO_3^{2-} to CO_2. They find that the speciation of carbon is largely insensitive to pressure up to 200 kbar, and that composition and temperature have a much stronger effect.

The speciation of carbon in highly reduced melts is poorly understood. Whereas Morizet et al. (2010) suggested that applying reducing conditions would not affect carbon speciation in hydrous melts, Kadik (2004) observed that carbon is dissolved primarily as atomic carbon in reduced ferrobasalt. Kadik observed direct Si-C bonds in ferrobasaltic glass using Raman spectroscopy, and Sen et al. (2013) observed that carbon exclusively bonds to silicon rather than oxygen in a reduced SiLiOC melt using nuclear magnetic resonance experiments, giving rise to their contrary conclusions. Although it is typically expected that in shallow melts carbon will bond to oxygen to form CO_2 or CO_3^{2-}, it is possible that in more reduced melts and especially at higher pressures, carbon will bond directly to silicon instead of oxygen. See the computational section for theoretical evidence of Si-C bonds in pyrolitic melt. Additional experiments and specific measurements are needed to constrain the speciation of carbon in reduced silicate melts. In summary, at lithospheric conditions, i.e. within the pressure range of 0.5–4 GPa and at temperatures below 2000 °C, the $CO_2/(CO_2+CO_3^{2-})$ ratio is most strongly affected by the composition, and in particular, the silica content of the melt (Figure 16.1). Carbon exists almost exclusively as CO_2 in silica-rich felsic melts (e.g., rhyolite) and as CO_3^{2-} in silica-poor mafic melts (e.g., basalt, olivine melitite and nepheline melts). The $CO_2/(CO_2+CO_3^{2-})$ ratio is also negatively correlated with pressure and generally insensitive to temperature; there remain disagreement and insufficient data on the effect of the H_2O content on the speciation of carbon. Additional studies are needed to refine the effect of temperature, H_2O content, and NBO/T, as well as the effect of network-modifying cations within the melt.

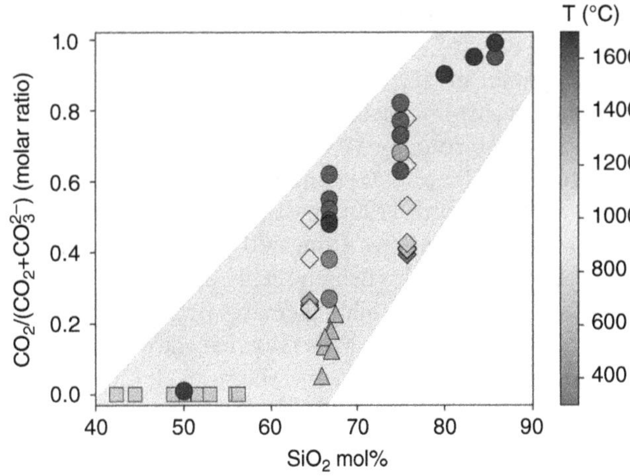

Figure 16.1 Molar ratio, $CO_2/(CO_2+CO_3^{2-})$ as a function of silica content at 300 °C–1700 °C for melts and glasses along the $NaAlO_2$-SiO_2 join from Brooker et al. (1999) (circles), andesite melts from King and Holloway (2002) (triangles), phonolite and dacite melts from Konschak and Keppler (2014) (diamonds), and mafic melts from Shishkina et al. (2014) (squares). The data from Brooker et al. (1999) and King and Holloway (2002) were obtained at 1–2 GPa and 1 GPa, respectively. Shishkina et al. (2014) did not observe any molecular CO_2 species for the range of mafic melts investigated at 1200 °C–1250 °C up to 0.5 GPa. Konschak and Keppler (2014) reported that the carbon speciation is independent of pressure in the pressure range of 3–20 GPa at temperatures of up to 1000 °C; however, Brooker et al. (1999) observed a negative correlation of $CO_2/(CO_2+CO_3^{2-})$ and pressure. In either case, the effect of pressure and temperature is secondary to the effect of silica content on carbon's speciation. See electronic version for color representation of the figures in this book.

16.2.2. Solubility of Carbon

The solubility of carbon within a silicate melt depends most strongly on pressure and composition. The effects of temperature, water content, and oxidation state are less well defined at the present time, due to discrepancies in the literature. Solubility can be difficult to measure accurately due to several factors. Hydrogen and carbon may diffuse into the silicate melt, or small amounts of melt components may dissolve into the fluid phase, affecting the measured carbon solubility. The oxidation state may change due to the capsule material, resulting in changes in carbon speciation, and thus, solubility. Furthermore, the use of different molar absorptivity standards for the CO_2 and CO_3^{2-} for the FTIR bands may be a source of inconsistencies in reported solubilities. Here we summarize the current data on carbon solubility in a variety of silicate melt compositions, highlighting the need for additional high-quality experiments to resolve discrepancies and gaps in the literature. We use the term *carbon solubility* to encompass the cumulative solubility of CO_2 and CO_3^{2-}, making the distinction when necessary.

There is good agreement within the experimental literature that carbon solubility in silicate melts increases rapidly with increasing pressure (Behrens et al., 2004; Blank,

1993; Botcharnikov et al., 2005; Botcharnikov et al., 2007; Brooker et al., 1999; Fogel & Rutherford, 1990; Jendrzejewski et al., 1997; Shishkina et al., 2014). For example, in rhyolitic melts at 1123–1323 K, CO_2 solubility increases from roughly 120 ppm at 20 MPa to 800–1000 ppm at 140 MPa to 3700 ppm at 660 MPa (Blank, 1993; Fogel & Rutherford, 1990). In dacitic melts, the maximum CO_2 solubility increases from 800 ppm at 100 MPa to 3000 ppm at 500 MPa (Behrens et al., 2004), and in andesitic melts, the CO_2 solubility increases from 175 ppm at 50 MPa to 3300 ppm at 500 MPa (Botcharnikov et al., 2005). Similarly, in shoshonitic melts, Vetere et al. (2011) found that the concentration of CO_2 (in the form of carbonate) increased from 307 ppm at 50 MPa to 2932 ppm at 400 MPa. Thus, there is a consensus that more carbon species can dissolve in silicate melts with increasing pressure, and thus with increasing depth. There is also a strong dependence of the carbon solubility on the composition of the silicate melts, most notably affected by the degree of polymerization (Brooker et al., 2001a; Brooker et al., 2001b; Mysen et al., 2009; Shishkina et al., 2014) and concentration of CaO (Brey, 1976; Holloway, 1976; Iacono-Marziano et al., 2012; King & Holloway, 2002; Moore, 2008; Papale et al., 2006; Shishkina et al., 2014). Holloway (1976) observed that carbon solubility is positively correlated with the Ca/(Ca+Mg) ratio. Brey (1976) found that the solubility of carbon in olivine melitite (8.3–9.5 wt.% CO_2) is much higher than in albite-anorthite (1.6–2.5 wt.% CO_2), for which the solubility increases slightly with increasing anorthite content. Brooker et al. (2001a, 2001b) observed a negative correlation with MgO content and a positive correlation with NBO/T. King and Holloway (2002) found that carbon solubility is negatively correlated with the Si+Al mole fraction and positively correlated with cations that have a large Gibbs free energy of decarbonation or a high charge-to-radius ratio (e.g. Ca). Alkalis have also been found to have a positive correlation with the solubility of CO_2 (Behrens et al., 2009; Lesne et al., 2011; Vetere et al., 2014), although the relationship is more complex. Shishkina et al. (2014) found that at 500 MPa and the H_2O content extrapolated to 0 wt.%, the CO_2 concentration increases from tholeiite (~3400 ppm) to alkali basalt (~5200 ppm), to basanite (~7400 ppm), to phonotephrite (~8500 ppm), to nephelinite (~14000 ppm). To describe the complex relationship between composition and carbon solubility, Shishkina et al. formulated a modified Π parameter (originally proposed by Dixon, 1997, for a limited range of silica contents) using eight compositions of silicate melt combined with previous data sets. The Π parameter in Shishkina et al. is defined as the ratio between the freely diffusing cations and the tetrahedral network, $(Ca^{2+} + 0.8K^+ + 0.7Na^+ + 0.4Mg^{2+} + 0.4Fe^{2+})/(Si^{4+} + Al^{3+})$, as reflecting the degree of polymerization of the melt. The relationship between the natural log of the solubility of CO_2 and Π was found to be $1.167*\ln P + 0.671* \Pi + 0.65$ (where P is pressure), obtained from the linear regression of 48 values of CO_2 solubilities at 50–500 MPa. The newly formulated Π parameter in Shishkina et al. is the most rigorous parameterization of the compositional effect on carbon solubility at the present time.

Although most studies demonstrate a negative correlation between total carbon solubility and H_2O content (Behrens et al., 2004; Blank et al., 1993; Botcharnikov et al., 2005; Botcharnikov et al., 2006; Fogel & Rutherford, 1990; Mattey, 1991; Moussallam et al., 2016; Pan et al., 1991; Pawley et al., 1992; Shishkina et al., 2010; Stolper & Holloway, 1988; Tamic et al., 2001), several studies observed a positive correlation (Behrens et al., 2004; Behrens et al., 2009; Iacono-Marziano et al., 2012; King & Holloway, 2002; Jakobssen, 1997). Below about 2×10^4 ppm H_2O, there is generally a weak correlation between the H_2O and CO_2 concentrations, while at higher H_2O concentration the CO_2 solubilities tend to drop (Ni & Keppler, 2013). For example, at 2 kbar and 1250–1300 °C, the CO_2 solubility in basaltic and rhyolitic melts plateaus around 1000 ppm CO_2 up to around 2×10^4 ppm H_2O, gradually dropping to 0 ppm CO_2 by 5×10^4 ppm H_2O in basaltic melt (Botcharnikov et al., 2005; Shishkina et al., 2010) and by 5.6×10^4 ppm H_2O in rhyolitic melt (Tamic et al., 2001). On the other hand, King and Holloway (2002) observed a total CO_2 solubility increase by about 0.06 wt.% per wt.% of H_2O in andesite glass at 10 kbar and 1300 C: as the H_2O concentration increased from 0 to 3.4 wt.%, molecular CO_2 decreased from 0.07 to 0.01 wt.% and CO_3^{2-} increased from 0.25 to 0.57 wt.%. Iacono-Marziano et al. (2012) hypothesized that a positive correlation between CO_2 and H_2O concentrations may be due to NBO formation, which promotes CO_3^{2-} formation, thus increasing the total carbon solubility. The solubility data of icelandite from Jakobssen (1997) shows no trend with H_2O when plotted against dissolved CO_2 (Ni & Keppler, 2013), whereas there is a positive trend when plotted against dissolved CO_2 normalized to the CO_2 partial pressure. Nevertheless, the majority of studies see a negative correlation and it is not clear at the present time what is responsible for the discrepancy in these studies.

The effect of temperature on the solubility of carbon is also poorly constrained. Generally, carbon solubility seems to be weakly dependent on temperature, such that the sign of the trend depends on the composition. For example, it has been shown that carbon solubility decreases weakly with increasing temperature for albitic melts (Blank & Brooker, 1994; Brooker et al., 1999; Stolper et al., 1987) and decreases strongly for rhyolitic melts at 0.1 GPa (Blank, 1993; Fogel & Rutherford, 1990; Ni & Keppler, 2013). Eitel and Weyl (1932) observed a negative dependence in sodium metasilicate melts at 750 atm, and Pearce (1964) observed pronounced negative temperature dependence in Na_2O-SiO_2 melts at 1 atm. However, Brooker et al. (1999) observed a slight increase in solubility with increasing temperature for $NaAlSiO_4$ and no dependence for melts with compositions between $NaAlSi_4O_{10}$ and $NaAlSi_6O_{14}$. In more mafic compositions, there is even less consensus. Brey and Green (1976) observed a slight negative relationship in olivine melilitite at 3 GPa with 9% CO_2 dissolved at 1450 °C, 8.5% at 1550 °C, and 8.3% at 1650 °C, while Mysen et al. (1976) observed a strong positive correlation with temperature at 1400–1700C above 1 GPa. Pan et al. (1991) demonstrated that carbon solubility is temperature-independent in tholeiitic basalt melt at 1–1.5 GPa and 1300 °C–1600 °C. It is possible that there are discrepancies in the effect of temperature due to different quenching methods and/or quality of the glass as an analog for certain compositions. In any case, additional studies are needed to constrain the effect of temperature on carbon solubility. At least at uppermost mantle pressures and temperatures, increasing oxygen fugacity increases the solubilities of carbon and hydrogen (Kadik, 2004; Morizet et al., 2010; Stanley et al., 2011). Morizet et al. (2010) found a strong positive correlation between carbon solubility and oxidation state in haploblastic glasses at 1250 °C and 200–300 MPa, finding that the CO_2 content changes from 680 to 1320 ppm between ΔFMQ – 2.6 and ΔFMQ + 2.6 independently of H_2O content changing from 1.3 to 4.0 wt.%. Stanley et al. (2011) found that the CO_2 content increases from 51 ppm at IW to 510 ppm at IW + 1 in a synthetic melt based on the Adirondack-class Humphrey basalt, at 1–2.5 GPa and 1400 °C–1625 °C. At greater depths, however, the relationship between oxygen fugacity and carbon solubility is unknown. It has been hypothesized that carbon may behave like oxygen in reduced high-pressure melts (Sen et al., 2013) and possibly result in an increase in solubility.

16.3. SIMULATIONS OF SILICATE MELTS

It remains a great challenge to experimentally measure the structural properties of melts at simultaneously high pressures and temperatures. The majority of experiments require quenching of the melt prior to analysis, which may change the local structure and coordination environment of the atoms, as well as the volatile concentration (Behrens & Nowak, 2003; Brey, 1976; Guillot & Sator, 2011). Additionally, experiments on carbon solubility and speciation in silicate melts have mostly been limited to pressures of up to 4 GPa and temperatures of up to 2000 °C, while calculations can determine the precise behavior of atoms and directly calculate thermodynamic properties at a wide range of pressures and temperatures. On the other hand, the atomistic simulations have strong size limitations, as they typically contain a few hundred atoms in a periodically repeated simulation box. This may be somehow mitigated by long simulation times and large number of configurations, in a push to cover as much as possible from the configurational space.

Molecular dynamics calculations can be performed with classical or ab initio methods. In classical molecular dynamics, the forces between atoms are calculated from empirical interatomic potentials, whereas in ab initio molecular dynamics, forces are determined from electronic structure calculations (e.g. with density functional theory). In both cases, the motions of the atoms are calculated with Newton's second law and are treated classically. Due to computational expense, ab initio molecular dynamics are limited by the number of atoms and duration of the simulation; the latter can be orders of magnitude higher than for classical molecular dynamics. The level of accuracy of classical molecular dynamics depends on the quality of the parametrization of the interatomic potentials, which depends on the accuracy of the reference data. Furthermore, extrapolation to pressure-temperature conditions beyond the conditions of the reference data decreases the accuracy of the potentials. Although ab initio molecular dynamics depends on several approximations (e.g., the exchange-correlation energy and the treatment of core electrons), it is based on first principles in the sense that it does not require parameterization with experimental data. For the case of high-pressure, high-temperature silicate melts, the reference data is scarce or nonexistent, especially within the large range of pressure and temperature conditions of planetary interiors, and so ab initio methods are presumably more reliable. For example, Vuilleumier et al. (2015) found that their classical simulations were not sufficiently viscous, as the atomic diffusion coefficients were larger than experimental values. They also found disagreement in the CO_2 concentration as a function of composition when comparing their results from classical simulations to ab initio simulations, concluding that the empirical force fields need to be improved. Furthermore, the ratio of CO_2 to CO_3 is severely overestimated with classical molecular dynamics in the calculations of Guillot and Sator (2011) (see section 3.2). Thus, even though empirical methods allow for a larger number of atoms and longer simulation duration, experimental results on carbon speciation are far better reproduced with ab initio methods. The effect of cell size (i.e., number of atoms) on carbon speciation in ab initio molecular dynamics simulations is not currently well understood.

Thus far, there have been five studies on carbonated silicate melts using molecular dynamics simulations. The two first studies were conducted at pressure and temperature conditions of the upper mantle: Guillot and Sator

(2011) implemented classical molecular dynamics to quantify the solubility of carbon in rhyolite, MORB, and kimberlite melts at 2–15 GPa and 1473–2273 K; and Vuilleumier et al. (2015) used a combination of classical and ab initio molecular dynamics to study the speciation of carbon in basaltic and kimberlitic melts with ~20 wt.% CO_2 at 2073 K and 12 GPa. Subsequently, simulations at lower mantle pressures and temperatures were conducted. Ghosh et al. (2017) used ab initio molecular dynamics to examine the speciation of carbon in $MgSiO_3$ melt with 5–16 wt.% CO_2 at 2200–6000 K and 140 GPa, while Ghosh and Karki (2017) examined the transport properties of $MgSiO_3$ melt with 16.1 wt.% CO_2 at 2200–5000 K up to 140 GPa. Most recently, Solomatova et al. (2019) used ab initio molecular dynamics to examine the speciation and polymerization of carbon in pyrolite melt with 5–10 wt.% CO_2 at 3000–5000 K up to 140 GPa. The compositions and carbon concentrations of the ab initio molecular dynamics simulations are summarized in Table 16.1.

In this chapter, we add our simulations on pure and carbonated Mg_2SiO_4 forsterite melt at 3000 K and up to 120 GPa. We have worked on several distinct compositions: Mg_2SiO_4, Mg_2SiO_4+C, $Mg_2SiO_4+CO_2$, and $Mg_2SiO_4+MgCO_3$, and study the effect of carbonation on the melt properties as well as the differences between the various carbon species. We have employed the Qbox package (Gygi, 2018). We start the calculations with solid Mg_2SiO_4 forsterite. Forsterite, the Mg-end member term of the mineral olivine, has an orthorhombic structure with 4 formula units per unit cell, namely 28 atoms. We consider a 2x1x2 supercell containing 112 atoms. After static structural relaxation, we perform a molecular-dynamics calculation on the solid, raising the temperature well above the melting point. Once we reach melting, we decrease the temperature and thermalize the melt at the desired value closer to realistic telluric conditions. Here, we performed a high-pressure study along the 3000 K isotherm. All simulations are NVT type, meaning the Number of particles, the Volume of the simulation box, and the Temperature are kept constant. This is numerically simpler than the NPT ensemble, where pressure instead of volume is fixed. Temperature is controlled via a Nose-Hoover thermostat, and the time step of the simulations is 0.75 fs. Temperature is monitored throughout the simulations to maintain fluctuations within a couple hundred degrees above and below the average 3000 K. We compute the electronic density and wave functions sampling the Brillouin zone in the reciprocal space using only the Γ point. We add carbon in the silicate melt as C atoms, CO_2 molecules, or $MgCO_3$ groups. The addition of the supplementary atoms is done in the interstitial pores of the melt at low pressure. CO_2 is added as a linear molecule, and $MgCO_3$ as a Mg cation about 2 Å apart on top of a CO_3 triangular anionic group. The new melts are first thermalized for one full picosecond in order to dissipate the interatomic forces and equilibrate the strains and then compressed along the 3000 K isotherm up to lower-mantle pressures.

16.3.1. Solubility of Carbon Dioxide in the Silicate Melt

Guillot and Sator (2011) observed a strong positive correlation of carbon solubility with pressure. For example, at 1673 K, carbon solubility in a basaltic melt increased from 2 wt.% CO_2 at 2 GPa to 30 wt.% CO_2 at 10 GPa. A weak negative correlation with temperature was observed and the effect of temperature increases with increasing pressure, such that at 2 GPa, there is almost no

Table 16.1 Carbon concentrations in the simulated melts using ab initio molecular dynamics, expressed as the formula units of carbon species added, wt.% of carbon species added, wt.% of C and mol% of C. The chosen carbon concentrations of 1.4–4.7 wt.% C reflect the range of carbon present in carbonaceous chondrites (0–6 wt.% C) (Pearson et al., 2006), which are regarded as the building blocks of our planet.

Composition, Supercell Formula	Reference	Carbon Species Added	wt.% of Carbon Species Added	wt.% C	mol% C
Enstatite, $32MgSiO_3$	Ghosh et al. (2017)	4 CO	3.37 wt.% CO	1.45	2.38
		4 CO_2	5.20 wt.% CO_2	1.42	2.33
		14 CO	10.88 wt.% CO	4.66	7.45
		14 CO_2	16.09 wt.% CO_2	4.39	6.93
Pyrolite, $NaCa_2Fe_4Mg_{30}Al_3Si_{24}O_{89}$	Solomatova et al. (2019)	4 CO	3.35 wt.% CO	1.44	2.48
		4 CO_2	5.16 wt.% CO_2	1.41	2.42
		8 CO	6.48 wt.% CO	2.78	4.97
		8 CO_2	9.82 wt.% CO_2	2.68	4.52
Forsterite, $16Mg_2SiO_4$	This study	8 C	4.09 wt.% C	4.09	6.67
		4 CO_2	7.25 wt.% CO_2	1.98	3.23
		8 CO_2	13.52 wt.% CO_2	3.69	5.88
		4 $MgCO_3$	13.03 wt.% $MgCO_3$	1.86	3.03

difference in solubility with temperature, whereas at 5 GPa, the solubility of CO_2 decreases from 10 wt.% at 1673 K to 7 wt.% at 2273 K. At low pressure, they demonstrate excellent agreement with experiments (Hammouda, 2003; Mattey, 1991; Pan et al., 1991; Thomsen & Schmidt, 2008). For example, both Guillot and Sator (2011) and Mattey (1991) predict CO_2 solubility in basaltic melt of 1.5 wt.% at 2 GPa and 1673 K. At 3.5 GPa and 1573 K, Thomsen and Schmidt (2008) report a CO_2 solubility of 5 wt.% CO_2 in a complex carbonatite melt, similar to the value of 6 wt.% CO_2 at 3.5 GPa and 1673 K predicted by Guillot and Sator (2011). The solubility is also weakly correlated with silica content, only becoming noticeable above about 10 GPa. The solubility of CO_2 at 10 GPa and 2273 K is about 25 wt.% in basalt and 24 wt.% in rhyolite, while at 15 GPa and 2273 K, the solubility of CO_2 is about 37 wt.% in basalt and 34 wt.% in rhyolite. Although Guillot and Sator (2011) predict excellent agreement of solubility with experiments, the form in which carbon exists (CO_2 vs. CO_3^{2-}) is significantly different than experimental observations (see below).

16.3.2. Carbon Coordinated by Oxygen

Guillot and Sator (2011) found that the proportion of CO_3^{2-} increases with increasing pressure and decreases with increasing temperature relative to CO_2. Although experiments on quenched mafic melts have shown that carbon exists exclusively as CO_3^{2-} (e.g. Fogel & Rutherford, 1990; Shishkina et al., 2014; Thibault & Holloway, 1994), Guillot and Sator predicted a significant CO_2 content ($CO_2/[CO_2+CO_3^{2-}]$~0.6 at 1873 K), noting that a low-temperature extrapolation to 300 K would agree with the experimental results, implying that the speciation (and thus, solubility) of carbon in the recovered silicate glasses (i.e. quenched silicate melt) is not an accurate representation of the high-temperature silicate melt. The implication is that the experimental results on carbon-bearing silicate glasses are unreliable. However, when the $CO_2/(CO_2+CO_3^{2-})$ ratio is plotted for the computational studies on carbon-bearing $MgSiO_3$ (Ghosh et al., 2017) and carbon-bearing pyrolite (Solomatova et al., 2019), the concentrations of CO_2 are substantially lower than in Guillot and Sator (2011) at comparable temperatures (Figure 16.2). For example, at 2–3 GPa, Ghosh et al. (2017) observed 1%–4% $CO_2/(CO_2+CO_3^{2-})$ at 2200 K in $MgSiO_3$, but Guillot and Sator (2011) observed 70% $CO_2/(CO_2+CO_3^{2-})$ at 2273 K in MORB. The discrepancy between the computational studies may be best explained by the difference in methodology. Guillot and Sator (2011) used classical molecular dynamics on a CO_2 saturated composition (~25 wt.% CO_2 at 10 GPa), but Ghosh et al. (2017) and Solomatova et al. (2019) used ab initio molecular dynamics on melts with 5–16 wt.% CO_2. Vuilleumier et al. (2015) used a combination of classical and ab initio molecular dynamics to study the speciation of carbon in basaltic and kimberlitic melts with ~20 wt.% CO_2 at 2073 K and 12 GPa, finding that most of the carbon exists as CO_3^{2-}. With ab initio molecular dynamics, the $CO_2/(CO_2+CO_3^{2-})$ ratio is 0.15 for the basaltic melt and 0.03

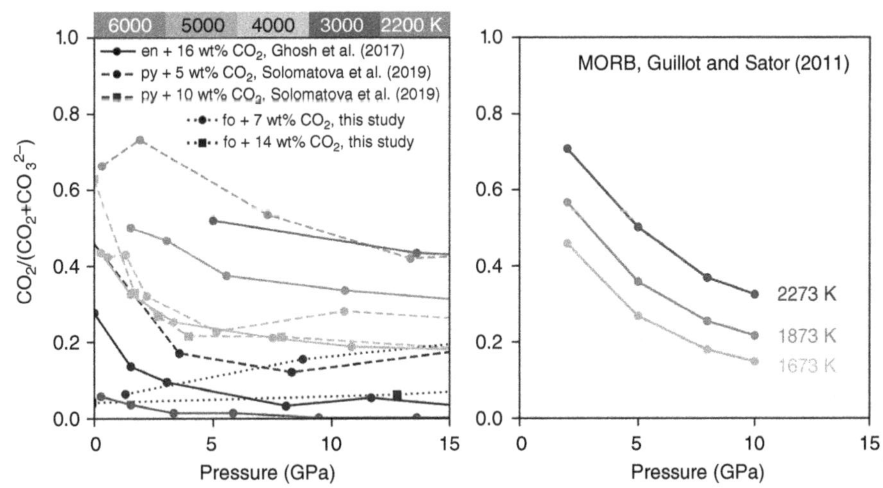

Figure 16.2 Molar ratio, $CO_2/(CO_2+CO_3^{2-})$, as a function of pressure and temperature. Guillot and Sator (2011) predict significantly more CO_2 relative to CO_3^{2-} in MORB melt than Ghosh et al. (2017), Solomatova et al. (2019), and this study in enstatite melt (en), pyrolite melt (py), and forsterite melt (fo), if an extrapolation to lower temperatures is made. In fact, the high $CO_2/(CO_2+CO_3^{2-})$ ratios predicted by Guillot and Sator (2011) at 2273 K are only achieved at 5000 K in Ghosh et al. (2017) and Solomatova et al. (2019). Experiments on basaltic melts at 1–4 GPa and 1000–2200 K observe no CO_2 (Fine & Stolper, 1986; Fogel & Rutherford, 1990; Shishkina et al., 2014; Thibault & Holloway, 1994), in agreement with Ghosh et al. (2017) and in disagreement with Guillot and Sator (2011). See electronic version for color representation of the figures in this book.

for the kimberlitic melt, but with classical molecular dynamics, the ratio is approximately 0.2 for both melts, higher than what is predicted with classical molecular dynamics. At similar conditions (~10–15 GPa and 2200 K), Ghosh et al. (2017) observed essentially no CO_2 in $MgSiO_3$ melt, similar to the results of Vuilleumier et al. (2015) for kimberlitic melt with ab initio molecular dynamics.

In carbonated forsterite melts, the fate of carbon depends on its entry state in the system and, consequently, on the oxidation state of the melt. The general tendency is that the coordination of carbon by oxygen increases with increasing pressure. In forsterite melt with carbon species added as CO_2 and $MgCO_3$, carbon exists almost exclusively in the form of CO_3^{2-} and increases in coordination to CO_4^{4-} with increasing pressure such that carbon in the melt is mostly CO_4^{4-} at core-mantle boundary pressures. In forsterite with reduced carbon added, carbon exists as a mixture of C, CO, and CO_2; with increasing pressure, carbon increases coordination to CO_3^{2-}. At lower-mantle pressures, carbon in the reduced forsterite melt exists as an approximately even mixture of CO, CO_2, and CO_3^{2-} with no CO_4^{4-}. The relative proportions of CO_2, CO_3^{2-}, and CO_4^{4-} are similar at all mantle pressures regardless of the melt chemistry (pyrolite, enstatite, or forsterite), as demonstrated in Figure 16.3.

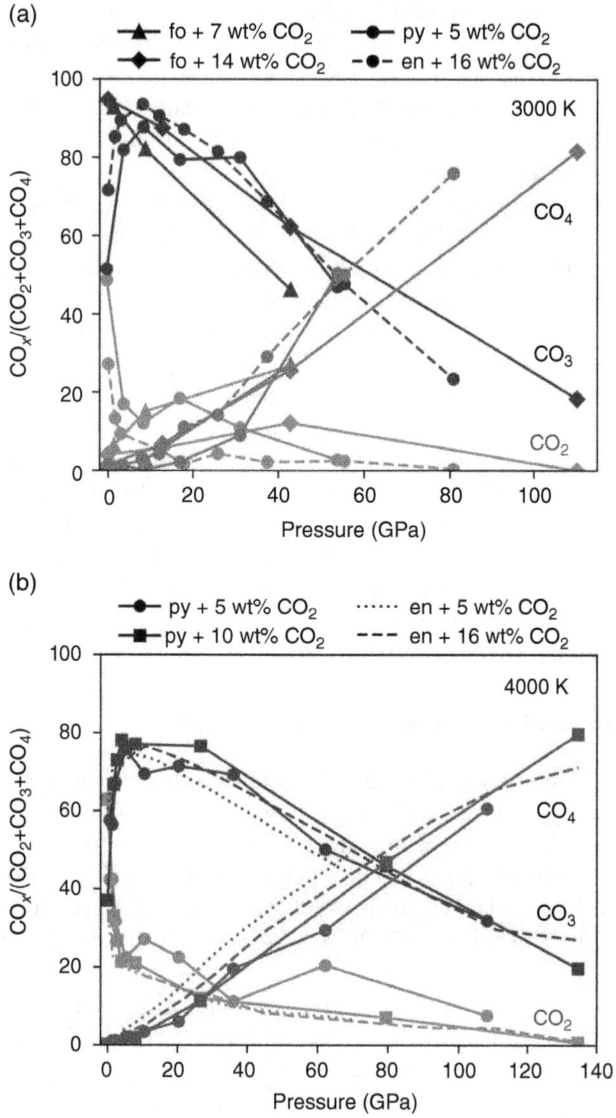

Figure 16.3 Molar ratio, $CO_x/(CO_2+CO_3^{2-}+CO_4^{4-})$ at 3000 K (a) and 4000 K (b) for enstatite (en; Ghosh et al., 2017), pyrolite (py; Solomatova et al., 2019) and forsterite (fo; this study) melts with carbon added into the melt as CO_2. See electronic version for color representation of the figures in this book.

16.3.3. Carbon Complexes

Although at ambient pressure carbon exists in silicate melts exclusively in unpolymerized forms (e.g. CO, CO_2, and CO_3^{2-}), with increasing pressure it becomes increasingly polymerized (i.e. a higher fraction of carbon atoms are bonded to other carbon atoms) and the carbon species increasingly complex (Ghosh et al., 2017; Solomatova et al., 2019). For example, in pyrolite with 3.35 wt.% CO, nearly none of the carbon is polymerized while at 0 GPa, whereas most of the polymerized carbon exists in C_2O, C_2O_2, and C_2O_3 molecules at ~60 GPa and predominantly as $C_2O_{2\text{-}6}$ and $C_3O_{4\text{-}6}$ molecules at ~115 GPa and 4000 K (Solomatova et al., 2019). At 12 GPa and 2073 K, Vuilleumier et al. (2015) observed that about 3.7% of all carbon species in basalt + 20 wt.% CO_2 are C_2O_5, which means that 7% of carbon atoms are bonded to another carbon atom, similar to pyrolite + 9.82 wt.% CO_2, in which 5% of carbon is polymerized (Solomatova et al., 2019), and $MgSiO_3$ + 16.1 wt.%, in which 7% of carbon is polymerized (Ghosh et al., 2017), at an equivalent pressure and 4000 K (Figure 16.3).

The polymerization of carbon is inversely dependent on temperature because of the increase in entropy with decreasing polymerization. At 3000 K and 20 GPa, pyrolite with 3.35 wt.% CO has 45% of its carbon polymerized, whereas at 5000 K only about 3% of its carbon is polymerized (Figure 16.4; Solomatova et al., 2019). In addition to the carbon-carbon polymerization, the degree of clustering between carbon atoms and other cations (predominantly, iron and silicon) also increases with increasing pressure (Figure 16.4). The predisposition of carbon to bond directly to iron suggests that carbon would follow iron during segregation of an iron-rich metallic liquid during core formation in the early Earth, both prior to and after the moon-forming impact.

The oxidation state of the melt has a significant effect on the types of species present (Ghosh et al., 2017; Solomatova et al., 2019). For example, the abundance of

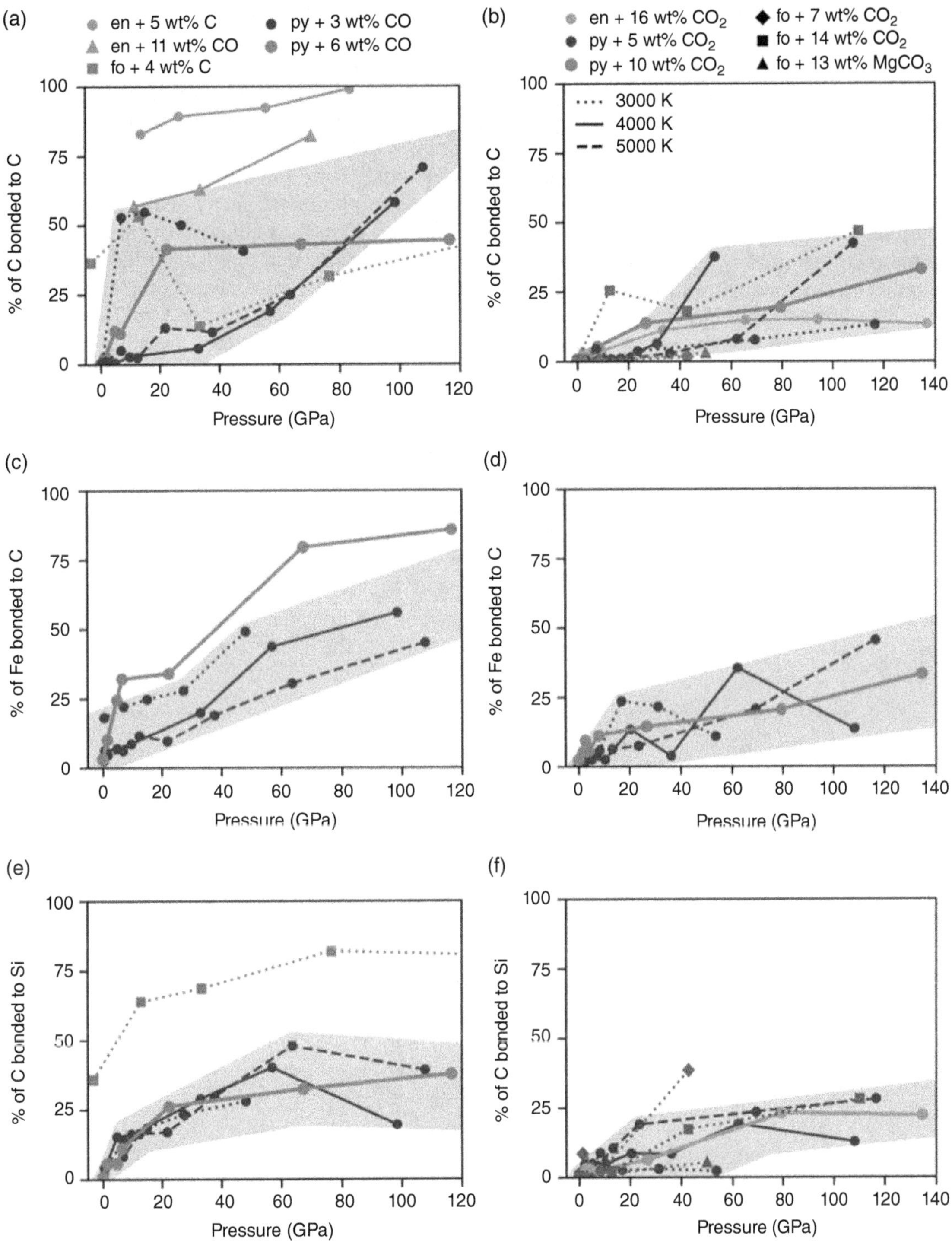

Figure 16.4 Formation of carbon complexes as a function of pressure in simulations on enstatite melt (en; Ghosh et al., 2017); pyrolite melt (py; Solomatova et al., 2019) and forsterite melt (fo; this study) at 3000 K (dotted lines), 4000 K (solid lines), and 5000 K (dashed lines). Left column (a, c, e) includes melts with more reduced carbon species (C and CO); right column (b, d, f) includes melts with more oxidized species (CO_2 and $MgCO_3$). Gray regions mark pyrolite with 3 wt.% CO (left column) and pyrolite with 5 wt.% CO_2 (right column). The degree of carbon-carbon, carbon-iron, and carbon-silicon polymerization generally increases with increasing pressure and decreasing oxidation state of the melts. See electronic version for color representation of the figures in this book.

carbonate ions relative to molecular carbon dioxide is positively correlated to oxygen fugacity. More oxidized pyrolitic melts only contain C_2O_4, C_2O_5, C_2O_6, and C_2O_7, with nearly no C_3O_x species forming while the more reduced melts have a wide range of C_xO_y species present. The oxygen fugacity of the melt is negatively correlated with the degree of carbon-carbon polymerization, as well as the degree of direct carbon-iron and carbon-silicon bonding. Thus, more carbon will bond to other carbon atoms and other cations, such as iron and silicon, in a more reducing environment. Indeed, as Figure 16.4 shows, there is a considerably larger amount of direct carbon-iron bonds for the pyrolite + CO compositions than for the pyrolite + CO_2 compositions. The same is true, and maybe even more pronounced, for the carbon-silicon bonds in case of the forsterite + C compared to forsterite + CO_2. The pyrolite + CO_2 and forsterite + CO_2 show similar behavior, suggesting once more that the oxygen fugacity and/or the state of entry of C into the system are the dominant factors determining its speciation. These findings are in agreement with experimental studies that observed direct carbon-silicon bonds in reduced ferrobasalt melt (Kadik, 2004) and reduced SiLiOC melt (Sen et al., 2013).

16.3.4. Elemental Diffusivities

Ghosh and Karki (2017) examined the transport properties of $MgSiO_3$ enstatite melt with 16.1 wt.% CO_2 at 2200–5000 K up to 140 GPa and found that the addition of carbon increases the diffusivity rates of all elements. However, the diffusivity rates of cations in more complex melts, such as basalt and kimberlite at 12 GPa and 1273 K (Vuilleumier et al., 2015) and pyrolite up to 140 GPa and 3000–5000 K, appear to be unaffected by the presence of carbon (Figure 16.5). The diffusion rates of oxygen and magnesium are nearly identical in carbonated pyrolite and carbonated enstatite, while silicon and carbon appear to have higher diffusion rates in carbonated pyrolite than in carbonated enstatite, indicating that as network formers, silicon and carbon are more mobile in the less polymerized pyrolite melt relative to the more polymerized enstatite melt. There is good agreement in the change of slope of the elemental diffusivities with pressure at around 5 GPa and 4000 K in the different melts. At 3000 K, the diffusion rates of magnesium, silicon, carbon, and oxygen increase with decreasing polymerization of the melt: enstatite < pyrolite < forsterite

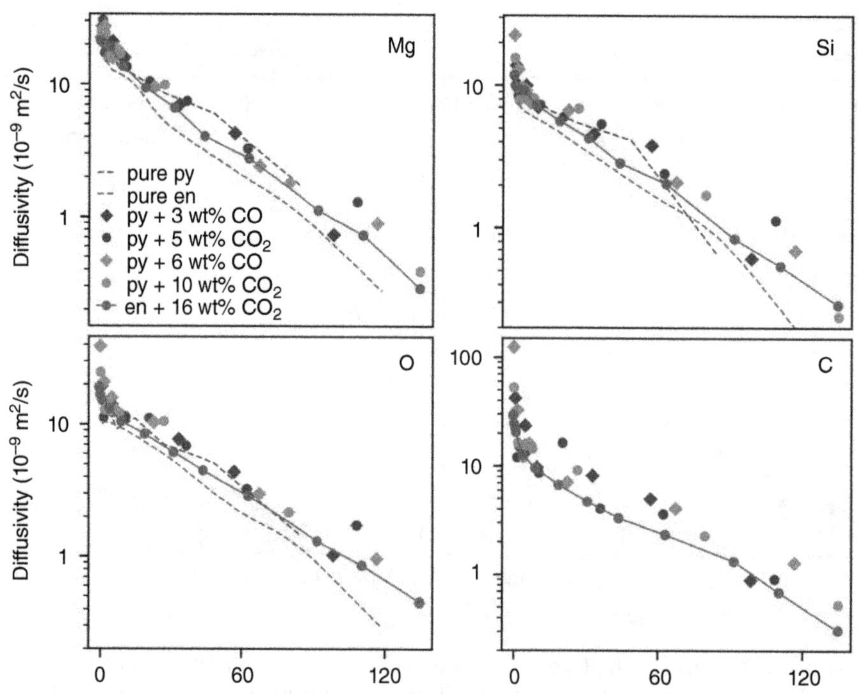

Figure 16.5 Elemental diffusivities of magnesium, silicon, oxygen and carbon at 4000 K in pure pyrolite melt ("pure py"), pure enstatite melt ("pure en"), pyrolite melt with 3.35 wt.% CO ("py + 3 wt.% CO"), pyrolite melt with 5.16 wt.% CO_2 ("pyr + 5 wt.% CO_2"), pyrolite melt with 6.48 wt.% CO ("py + 6 wt.% CO"), pyrolite melt with 9.82 wt.% CO_2 ("py + 10 wt.% CO_2"), and enstatite melt melt with 16.1 wt.% CO_2 ("en + 16 wt.% CO_2"). Pyrolite data is from Solomatova et al. (2019) and enstatite data is from Ghosh and Karki (2017). See electronic version for color representation of the figures in this book.

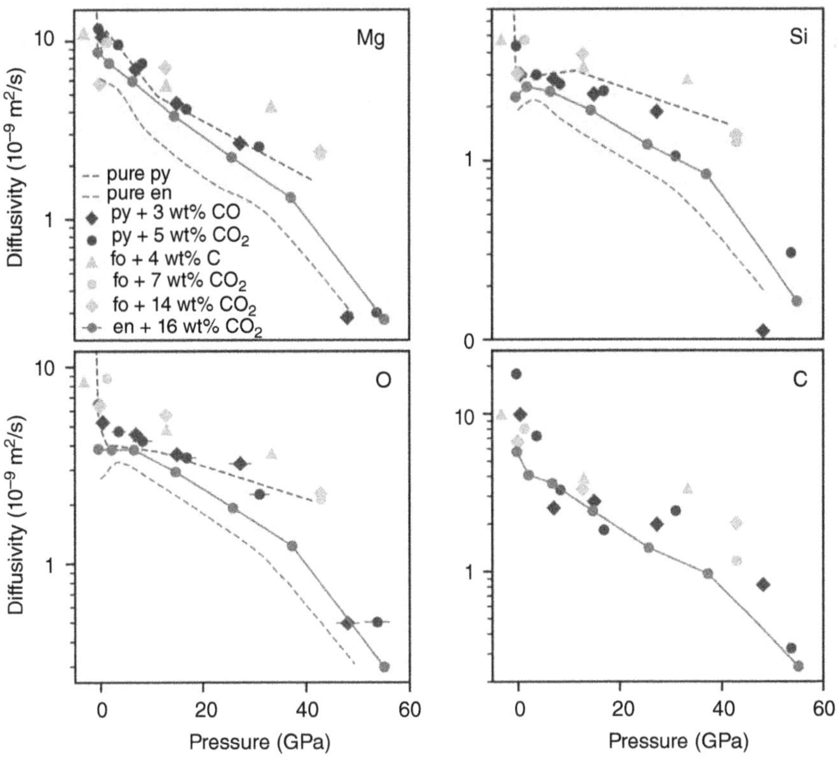

Figure 16.6 Elemental diffusivities of magnesium, silicon, oxygen, and carbon at 3000 K in pure pyrolite melt ("pure py"), pure enstatite melt ("pure en"), pyrolite melt with 3.35 wt.% CO ("py + 3 wt.% CO"), pyrolite melt with 5.16 wt.% CO_2 ("py + 5 wt.% CO_2"), forsterite melt with 4.09 wt.% C ("fo + 4 wt.% C"), forsterite melt with 7.25 wt.% CO_2 ("fo + 7 wt.% CO_2") and forsterite melt with 13.52 wt.% CO_2 ("fo + 14 wt.% CO_2"), and enstatite melt melt with 16.1 wt.% CO_2 ("en + 16 wt.% CO_2"). Pyrolite data is from Solomatova et al. (2019), forsterite data is from this study, and enstatite data is from Ghosh and Karki (2017). See electronic version for color representation of the figures in this book.

(Figure 16.6). When the atoms are more strongly bound in a silicate network, their diffusion rates are lower.

16.3.5. Equation of State

Solomatova et al. (2019) found that K_0 is generally positively correlated with the degree of oxidation and negatively correlated with temperature and carbon concentration; K'_0 is positively correlated with the degree of oxidation, temperature, and carbon concentration. Table 16.2 lists the equation of state parameters for a wide range of melts. Solomatova et al. find that the density difference between carbon-bearing and carbon-free pyrolite melts is larger at low pressure (where the density ratio of carbonated pyrolite to pure pyrolite ranges from 0.90 to 0.97 at 0 GPa), while above 10–20 GPa the densities converge, resulting in a density ratio of about 0.98–1. Consequently, carbon-bearing melts would be expected to be relatively less buoyant at lower-mantle pressures than at pressures of the upper mantle.

16.3.6. Diamond Formation

From a strict thermodynamical point of view, diamonds should not exist on the surface of the Earth. We find them naturally in rocks because they form at depth and are brought to the surface by uprising molten rock (magmas). Some of these diamonds coming from the deep mantle contain geochemical and mineralogical keys to the functioning of the entire silicate mantle. Hence, understanding their formation is fundamental to tracing the cycles of matter inside our planet. The formation of diamonds in Earth's deep mantle is poorly understood due to the limitations of experimental methods and our inability to access the diamond-forming depths of the mantle. Diamonds likely form in a wide range of pressure-temperature regimes and chemical environments (Cartigny et al., 2001; Kaminsky, 2012; Meyer, 1985). In the uppermost mantle, diamonds are thought to form during metamorphic reactions (Boyd & Finnerty, 1980), metasomatism (Shee et al., 1982; Sverjensky & Huang, 2015) and in magmatic processes (Meyer, 1985). It has

Table 16.2 Equation of state parameters for silicate melts with varying concentrations of carbon. We compare our results to the experimental studies on carbon-bearing peridotitic and basaltic melts (Sakamaki et al., 2011[not in refs]; Ghosh et al., 2007[2009 or 2017?]). An isothermal fourth-order Birch Murnaghan equation of state was fitted to the pressure-density data sets for the pyrolite and carbonated Mg_2SiO_4 melts. An isothermal third-order Birch Murnaghan equation of state was used to fit the pressure-volume data for Mg_2SiO_4 (this study), $MgSiO_3$ (Ghosh et al., 2017), peridotite (Sakamaki et al., 2011[no]) and basalt (Ghosh et al., 2007[no]).

Melt Composition	T (K)	ρ_0 (g/cm^3)	K_0(GPa)	K'_0	K''_0
pyrolite[a]	3000	2.265(2)	12.3(1)	7.8(1)	−2.4(1)
pyrolite + 3.35 wt.% CO[a]		2.19(6)	12(5)	7(3)	−2(3)
pyrolite + 5.16 wt.% CO_2[a]		2.14(1)	10.0(6)	7.6(4)	−2.5(5)
Mg_2SiO_4[b]		2.658(1)	56.3(2)	3.60(4)	—
Mg_2SiO_4 + 0.53 wt.% C[b]		2.4(1)	24(10)	6(1)	−0.5(5)
Mg_2SiO_4 + 1.92 wt.% CO_2[b]		2.13(5)	10(6)	6.0(5)	−1.0(4)
Mg_2SiO_4 + 3.61 wt.% $MgCO_3$[b]		2.684(7)	39(1)	4.75(8)	−0.13(1)
$MgSiO_3$[c]		2.56(2)	18(1)	6.9(2)	—
$MgSiO_3$ + 5.2 wt.% CO_2[c]		2.43(3)	16(2)	6.5(2)	—
peridotite + 2.5 wt.% CO_2[d]		—	23(1)	7(1)	—
basalt + 5 wt.% CO_2, exp[e]		—	16(1)	5.2(2)	—
pyrolite[a]	4000	2.03(4)	8(1)	8(1)	−4(3)
pyrolite + 3.35 wt.% CO[a]		1.92(4)	7(2)	6.7(9)	−2(1)
pyrolite + 5.16 wt.% CO_2[a]		1.96(4)	9(2)	6.2(9)	−1(1)
pyrolite + 6.48 wt.% CO[a]		1.71(4)	3.3(6)	8.2(6)	−8(3)
pyrolite + 9.82 wt.% CO_2[a]		1.78(1)	3.8(3)	8.8(4)	−9(2)
pyrolite[a]	5000	1.77(2)	4.9(6)	7.4(5)	−4(1)
pyrolite + 3.35 wt.% CO[a]		1.56(5)	2.2(8)	9(2)	−17(15)
pyrolite + 5.16 wt.% CO_2[a]		1.56(6)	2(1)	10(3)	−19(24)

[a] Ab initio molecular dynamics calculations on pyrolite melt from Solomatova et al. (2019).
[b] Ab initio molecular dynamics calculations on Mg_2SiO_4 forsterite melt, this study.
[c] Ab initio molecular dynamics calculations on $MgSiO_3$ enstatite melt from Ghosh et al. (2017).
[d] Experimental results on peridotite melt from Sakamaki et al. (2011).
[e] Experimental results on basalt melt from Ghosh et al. (2007).

been proposed that many of the diamonds interpreted to have originated in the transition zone and lower mantle formed from subducted carbon (Burnham et al., 2015; Pal'yanov et al., 2002; Walter et al., 2011; Zedgenizov et al., 2015). However, it is likely that a fraction of superdeep diamonds, many of which may have never reached the surface, formed during the crystallization of the mantle after the moon-forming impact in the Hadean and subsequently during the Archaean and Proterozoic eons (Gurney et al., 2010; Helmstaedt et al., 2010). There have been a few computational ab initio studies examining the behavior of carbon in silicate melts with a focus on carbon coordination by oxygen (Ghosh et al., 2017; Guillot & Sator, 2011; Vuilleumier et al., 2015); however, there have been no computational studies characterizing the genesis of diamonds in pyrolitic melts.

The analysis of diamonds in metamorphic rocks provides some information on the depths at which the diamonds formed and the mechanisms by which they grew. For example, octahedral diamonds in peridotitic and eclogitic rocks displayed a spiral growth mechanism, indicating that the crystal moved freely in carbon-bearing silicate melt (Bulanova, 1995). The formation and growth of diamonds has been experimentally observed in a kimberlitic melt at 1800 °C–2200 °C and 7–7.7 GPa (Arima et al., 1993), and from dolomite+iron through carbon-iron redox reactions (Dorfman et al., 2018). However, the coordination of carbon in the melt and the mechanism of subsequent polymerization of carbon could not be determined due to limitations in experimental techniques.

Although redox reactions, such as the oxidation of methane or reduction of carbon dioxide, have been considered a necessary step for the formation of diamonds, it has been recently shown that diamonds can form from a reaction of carboxylates (e.g. acetate [CH_3COO^-]) with water at constant oxygen fugacity (Sverjensky et al., 2014; Sverjensky & Huang, 2015). In pyrolitic melts, Solomatova et al. (2019) observed various oxo-carbon species, such as C_2O_2, i.e. ethylenedione, and C_2O_4, i.e. oxalate, (Figure 16.5), suggesting that in the presence of hydrogen, the formation of diamonds from acetate is possible. These polymerized carbon chains are reminiscent of carboxylates and may be considered precursors to the formation of diamonds. This suggests a mechanism for diamond formation in pyrolitic melts at large depths that does not necessitate the direct formation of oxygen-free carbon clusters; instead, diamond embryos on the nano scale could form through a pathway involving polymerized hydrocarbons, as described in Sverjensky et al. (2014) and Sverjensky and Huang (2015).

16.4. CONCLUSIONS

Experiments on silicate melts have demonstrated that at 1–4 GPa and 1000 °C–2000 °C, both the solubility and speciation of carbon depend most strongly on pressure and composition. Carbon solubility increases rapidly with small increases in pressure and is generally positively correlated with the concentration of CaO in the melt, while the $CO_2/(CO_2+CO_3^{2-})$ ratio is negatively correlated with pressure and is highly sensitive to the silica content, such that carbon exists almost exclusively as CO_2 in silica-rich felsic melts (e.g. rhyolite) and as CO_3^{2-} in silica-poor mafic melts (e.g. basalt). Meanwhile, molecular dynamics simulations of silicate melts at 0–140 GPa and 1400–5000 K have come to a general conclusion that there is a strong positive correlation of carbon solubility with pressure, a weak negative correlation of solubility with temperature, and a positive correlation of $CO_2/(CO_2+CO_3^{2-})$ with temperature. Simulations predict that the concentration of CO_2 sharply decreases at the expense of CO_3^{2-} with increasing pressure between 0 and 5 GPa, in agreement with experimental observations; between 10 and 140 GPa, pressures above what has been achieved by most experiments on carbonated silicate melts, CO_4^{4-} species appear and their concentration increases with increasing pressure at the expense of $CO_2+CO_3^{2-}$.

Here we also demonstrate a need for high-pressure high-temperature experiments on carbonated multicomponent silicate melts to confirm the computationally predicted carbon-carbon polymerization, carbon-silicon bonding, and carbon-iron complexes. Ab initio computational studies on CO_2-H_2O-bearing silicate melts would help understand the effect of water on carbon's speciation and solubility. The detailed effect of composition and polymerization in complex silicate melts needs to be determined both experimentally and computationally, to thoroughly understand the speciation and solubility of carbon in the interior of the Earth. Finally, to constrain the carbon concentration of Earth's interior, it is crucial to determine the fraction of carbon that was retained in the planet after the Giant Impact via calculations on the volatility of carbon in pyrolitic-type melt and of partitioning between the molten magma ocean and molten core.

ACKNOWLEDGMENTS

This research was supported by the European Research Council (ERC) under the European Union's Horizon 2020 research and innovation program (grant agreement No. 681818 – IMPACT to R. Caracas) and by the Deep Carbon Observatory. R. Caracas acknowledges access to the GENCI supercomputers (Occigen, Ada, and Curie) through the stl2816 series of eDARI computing grants. R. Cohen acknowledges access to the Teragrid/Xsede facilities via grant EAR080015. REC was supported by the ERC Advanced Grant ToMCaT and the Carnegie Institution. R. Cohen gratefully acknowledges the Gauss Center for Supercomputing e.V. (www.gauss-centre.eu) for funding this project by providing computing time on the GCS Supercomputer SuperMUC at Leibniz Supercomputing Centre (LRZ, www.lrz.de).

REFERENCES

Arima, M., Nakayama, K., Akaishi, M., Yamaoka, S., & Kanda, H. (1993). Crystallization of diamond from a silicate melt of kimberlite composition in high-pressure and high-temperature experiments. *Geology*, *21*(11), 968. https://doi.org/10.1130/0091-7613(1993)021<0968:CODFAS>2.3.CO;2

Behrens, H., Misiti, V., Freda, C., Vetere, F., Botcharnikov, R. E., & Scarlato, P. (2009). Solubility of H_2O and CO_2 in ultrapotassic melts at 1200 and 1250 °C and pressure from 50 to 500 MPa. *American Mineralogist*, *94*(1), 105–120. https://doi.org/10.2138/am.2009.2796

Behrens, H., & Nowak, M. (2003). Quantification of H_2O speciation in silicate glasses and melts by IR spectroscopy: In situ versus quench techniques. *Phase Transitions*, *76*(1–2), 45–61. https://doi.org/10.1080/0141159031000076048

Behrens, H., Zhang, Y., & Xu, Z. (2004). H_2O diffusion in dacitic and andesitic melts. *Geochimica et Cosmochimica Acta*, *68*(24), 5139–5150. https://doi.org/10.1016/j.gca.2004.07.008

Blank, J. G. (1993). An experimental investigation of the behavior of carbon dioxide in rhyolitic melt (PhD thesis). California Institute of Technology. Retrieved from http://resolver.caltech.edu/CaltechETD:etd-05302007-075656

Blank, J. G., & Brooker, R. A. (1994). Experimental studies of carbon dioxide in silicate melts: Solubility, speciation, and stable carbon isotope behavior. *Reviews in Mineralogy and Geochemistry*, *30*, 157–186.

Blank, J. G., Stolper, E. M., & Carroll, M. R. (1993). Solubilities of carbon dioxide and water in rhyolitic melt at 850 °C and 750 bars. *Earth and Planetary Science Letters*, *119*(1–2), 27–36. https://doi.org/10.1016/0012-821X(93)90004-S

Botcharnikov, R. E., Behrens, H., & Holtz, F. (2006). Solubility and speciation of C–O–H fluids in andesitic melt at T= 1100–1300 C and P= 200 and 500 MPa. *Chemical Geology*, *229*(1–3), 125–143. https://doi.org/10.1016/j.chemgeo.2006.01.016

Botcharnikov, R. E., Koepke, J., Holtz, F., McCammon, C., & Wilke, M. (2005). The effect of water activity on the oxidation and structural state of Fe in a ferro-basaltic melt. *Geochimica et Cosmochimica Acta*, *69*(21), 5071–5085. https://doi.org/10.1016/j.gca.2005.04.023

Botcharnikov, Roman E., Holtz, F., & Behrens, H. (2007). The effect of CO_2 on the solubility of H_2O-Cl fluids in andesitic melt. *European Journal of Mineralogy*, *19*(5), 671–680. https://doi.org/10.1127/0935-1221/2007/0019-1752

Boyd, F. R., & Finnerty, A. A. (1980). Conditions of origin of natural diamonds of peridotite affinity. *Journal of Geophysical Research: Solid Earth*, *85*(B12), 6911–6918. https://doi.org/10.1029/JB085iB12p06911

Brenker, F., Vollmer, C., Vincze, L., Vekemans, B., Szymanski, A., Janssens, K., et al. (2006). CO_2-recycling to the deep convecting mantle. *Geochimica et Cosmochimica Acta*, *70*(18), 66.

Brey, G. (1976). CO_2 solubility and solubility mechanisms in silicate melts at high pressures. *Contributions to Mineralogy and Petrology*, *57*(2), 215–221. https://doi.org/10.1007/BF00405226

Brey, G. P., & Green, D. H. (1976). Solubility of CO_2 in olivine melilitite at high pressures and role of CO_2 in the earth's upper mantle. *Contributions to Mineralogy and Petrology*, *55*(2), 217–230. https://doi.org/10.1007/BF00372228

Brooker, R. A., Kohn, S. C., Holloway, J. R., & McMillan, P. F. (2001a). Structural controls on the solubility of CO_2 in silicate melts: Part I. Bulk solubility data. *Chemical Geology*, *174*(1), 225–239. https://doi.org/10.1016/S0009-2541(00)00353-3

Brooker, R. A., Kohn, S. C., Holloway, J. R., & McMillan, P. F. (2001b). Structural controls on the solubility of CO_2 in silicate melts: Part II. IR characteristics of carbonate groups in silicate glasses. *Chemical Geology*, *174*(1), 241–254. https://doi.org/10.1016/S0009-2541(00)00318-1

Brooker, R. A., Kohn, S. C., Holloway, J. R., McMillan, P. F., & Carroll, M. R. (1999). Solubility, speciation and dissolution mechanisms for CO_2 in melts on the NaAlO2–SiO2 join. *Geochimica et Cosmochimica Acta*, *63*(21), 3549–3565. https://doi.org/10.1016/S0016-7037(99)00196-9

Bulanova, G. P. (1995). The formation of diamond. *Journal of Geochemical Exploration*, *53*(1), 1–23. https://doi.org/10.1016/0375-6742(94)00016-5

Burnham, A. D., Thomson, A. R., Bulanova, G. P., Kohn, S. C., Smith, C. B., & Walter, M. J. (2015). Stable isotope evidence for crustal recycling as recorded by superdeep diamonds. *Earth and Planetary Science Letters*, *432*, 374–380. https://doi.org/10.1016/j.epsl.2015.10.023

Cartigny, P., Harris, J. W., & Javoy, M. (2001). Diamond genesis, mantle fractionations and mantle nitrogen content: a study of δ13C–N concentrations in diamonds. *Earth and Planetary Science Letters*, *185*(1), 85–98. https://doi.org/10.1016/S0012-821X(00)00357-5

Dixon, J. E. (1997). Degassing of alkalic basalts. *American Mineralogist*, *82*(3–4), 368–378. https://doi.org/10.2138/am-1997-3-415

Dorfman, S. M., Badro, J., Nabiei, F., Prakapenka, V. B., Cantoni, M., & Gillet, P. (2018). Carbonate stability in the reduced lower mantle. *Earth and Planetary Science Letters*, *489*, 84–91. https://doi.org/10.1016/j.epsl.2018.02.035

Duan, X. (2014). A general model for predicting the solubility behavior of H_2O–CO_2 fluids in silicate melts over a wide range of pressure, temperature and compositions. *Geochimica et Cosmochimica Acta*, *125*, 582–609. https://doi.org/10.1016/j.gca.2013.10.018

Eggler, D. H. (1987). Discussion of recent papers on carbonated peridotite, bearing on mantle metasomatism and magmatism: An alternative. *Earth and Planetary Science Letters*, *82*(3), 398–400. https://doi.org/10.1016/0012-821X(87)90214-7

Eitel, W., & Weyl, W. (1932). Residuals in the melting of commercial glasses. *Journal of the American Ceramic Society*, *15*(3), 159–166. https://doi.org/10.1111/j.1151-2916.1932.tb13916.x

Fine, G., & Stolper, E. (1985). The speciation of carbon dioxide in sodium aluminosilicate glasses. *Contributions to Mineralogy and Petrology*, *91*(2), 105–121. https://doi.org/10.1007/BF00377759

Fine, G., & Stolper, E. (1986). Dissolved carbon dioxide in basaltic glasses: Concentrations and speciation. *Earth and Planetary Science Letters*, *76*(3–4), 263–278. https://doi.org/10.1016/0012-821X(86)90078-6

Fogel, R. A., & Rutherford, M. J. (1990). The solubility of carbon dioxide in rhyolitic melts: A quantitative FTIR study. *American Mineralogist*, *75*(11–12), 1311–1326.

Ghosh, D. B., Bajgain, S. K., Mookherjee, M., & Karki, B. B. (2017). Carbon-bearing silicate melt at deep mantle conditions. *Scientific Reports*, *7*(1). https://doi.org/10.1038/s41598-017-00918-x

Ghosh, D. B., & Karki, B. B. (2017). Transport properties of carbonated silicate melt at high pressure. *Science Advances*, *3*(12), e1701840. https://doi.org/10.1126/sciadv.1701840

Ghosh, S., Ohtani, E., Litasov, K., Suzuki, A., & Sakamaki, T. (2007). Stability of carbonated magmas at the base of the Earth's upper mantle. *Geophysical Research Letters*, *34*(22). https://doi.org/10.1029/2007GL031349

Guillot, B., & Sator, N. (2011). Carbon dioxide in silicate melts: A molecular dynamics simulation study. *Geochimica et Cosmochimica Acta*, *75*(7), 1829–1857. https://doi.org/10.1016/j.gca.2011.01.004

Gurney, J. J., Helmstaedt, H. H., Richardson, S. H., & Shirey, S. B. (2010). Diamonds through time. *Economic Geology*, *105*(3), 689–712. https://doi.org/10.2113/gsecongeo.105.3.689

Gygi, F. (2018). Qbox: A large-scale parallel implementation of first-principles molecular dynamics. http://www.qboxcode.org/

Hammouda, T. (2003). High-pressure melting of carbonated eclogite and experimental constraints on carbon recycling and storage in the mantle. *Earth and Planetary Science Letters*, *214*(1–2), 357–368. https://doi.org/10.1016/S0012-821X(03)00361-3

Helmstaedt, H. H., Gurney, J. J., & Richardson, S. H. (2010). Ages of cratonic diamond and lithosphere evolution: Constraints on precambrian tectonics and diamond exploration. *The Canadian Mineralogist*, *48*(6), 1385–1408. https://doi.org/10.3749/canmin.48.5.1385

Holloway, J. R. (1976). Fluids in the evolution of granitic magmas: Consequences of finite CO_2 solubility. *GSA Bulletin*, *87*(10), 1513–1518. https://doi.org/10.1130/0016-7606(1976)87<1513:FITEOG>2.0.CO;2

Holloway, J. R., & Blank, J. G. (1994). Application of experimental results to C-O-H species in natural melts. *Reviews in Mineralogy and Geochemistry*, *30*(1), 187–230.

Iacono-Marziano, G., Morizet, Y., Le Trong, E., & Gaillard, F. (2012). New experimental data and semi-empirical parameterization of H_2O–CO_2 solubility in mafic melts. *Geochimica et Cosmochimica Acta*, *97*, 1–23. https://doi.org/10.1016/j.gca.2012.08.035

Isshiki, M., Irifune, T., Hirose, K., Ono, S., Ohishi, Y., Watanuki, T., et al. (2004). Stability of magnesite and its high-pressure form in the lowermost mantle. *Nature*, *427*(6969), 60–63. https://doi.org/10.1038/nature02181

Jakobsson, S. (1997). Solubility of water and carbon dioxide in an icelandite at 1400 °C and 10 kilobars. *Contributions to Mineralogy and Petrology*, *127*(1), 129–135. https://doi.org/10.1007/s004100050270

Jendrzejewski, N., Trull, T. W., Pineau, F., & Javoy, M. (1997). Carbon solubility in mid-ocean ridge basaltic melt at low pressures (250–1950 bar). *Chemical Geology*, *138*(1–2), 81–92. https://doi.org/10.1016/S0009-2541(96)00176-3

Kadik, A. (2004). Formation of carbon and hydrogen species in magmas at low oxygen fugacity. *Journal of Petrology*, *45*(7), 1297–1310. https://doi.org/10.1093/petrology/egh007

Kaminsky, F. (2012). Mineralogy of the lower mantle: A review of 'super-deep' mineral inclusions in diamond. *Earth-Science Reviews*, *110*(1), 127–147. https://doi.org/10.1016/j.earscirev.2011.10.005

Keppler, H., Wiedenbeck, M., & Shcheka, S. S. (2003). Carbon solubility in olivine and the mode of carbon storage in the Earth's mantle. *Nature*, *424*(6947), 414–416. https://doi.org/10.1038/nature01828

King, P. L., & Holloway, J. R. (2002). CO_2 solubility and speciation in intermediate (andesitic) melts: The role of H_2O and composition. *Geochimica et Cosmochimica Acta*, *66*(9), 1627–1640. https://doi.org/10.1016/S0016-7037(01)00872-9

Konschak, A., & Keppler, H. (2014). The speciation of carbon dioxide in silicate melts. *Contributions to Mineralogy and Petrology*, *167*(5). https://doi.org/10.1007/s00410-014-0998-2

Lesne, P., Scaillet, B., Pichavant, M., & Beny, J.-M. (2011). The carbon dioxide solubility in alkali basalts: An experimental study. *Contributions to Mineralogy and Petrology*, *162*(1), 153–168. https://doi.org/10.1007/s00410-010-0585-0

Mao, Z., Armentrout, M., Rainey, E., Manning, C. E., Dera, P., Prakapenka, V. B., & Kavner, A. (2011). Dolomite III: A new candidate lower mantle carbonate: high-pressure phase of Fe-dolomite. *Geophysical Research Letters*, *38*(22), n/a-n/a. https://doi.org/10.1029/2011GL049519

Mattey, D. P. (1991). Carbon dioxide solubility and carbon isotope fractionation in basaltic melt. *Geochimica et Cosmochimica Acta*, *55*(11), 3467–3473. https://doi.org/10.1016/0016-7037(91)90508-3

McDonough, W. F. (2003). Compositional model for the Earth's core. *Treatise on Geochemistry*, 547–568.

Merlini, M., Crichton, W. A., Hanfland, M., Gemmi, M., Muller, H., Kupenko, I., & Dubrovinsky, L. (2012). Structures of dolomite at ultrahigh pressure and their influence on the deep carbon cycle. *Proceedings of the National Academy of Sciences*, *109*(34), 13509–13514. https://doi.org/10.1073/pnas.1201336109

Meyer, H. O. A. (1985). Genesis of diamond: A mantle saga. *American Mineralogist*, *70*(3–4), 344–355.

Mookherjee, M., Nakajima, Y., Steinle-Neumann, G., Glazyrin, K., Wu, X., Dubrovinsky, L., et al. (2011). High-pressure behavior of iron carbide (Fe_7C_3) at inner core conditions. *Journal of Geophysical Research: Solid Earth*, *116*(B4). https://doi.org/10.1029/2010JB007819

Moore, G. (2008). Interpreting H_2O and CO_2 contents in melt inclusions: Constraints from solubility experiments and modeling. *Reviews in Mineralogy and Geochemistry*, *69*(1), 333–362. https://doi.org/10.2138/rmg.2008.69.9

Morizet, Y., Paris, M., Gaillard, F., & Scaillet, B. (2010). C–O–H fluid solubility in haplobasalt under reducing conditions: An experimental study. *Chemical Geology*, *279*(1), 1–16. https://doi.org/10.1016/j.chemgeo.2010.09.011

Mörner, N.-A., & Etiope, G. (2002). Carbon degassing from the lithosphere. *Global and Planetary Change*, *33*(1), 185–203. https://doi.org/10.1016/S0921-8181(02)00070-X

Moussallam, Y., Morizet, Y., & Gaillard, F. (2016). H_2O–CO_2 solubility in low SiO2-melts and the unique mode of kimberlite degassing and emplacement. *Earth and Planetary Science Letters*, *447*, 151–160. https://doi.org/10.1016/j.epsl.2016.04.037

Mysen, B. O. (1976). The role of volatiles in silicate melts: Solubility of carbon dioxide and water in feldspar, pyroxene, and feldspathoid melts to 30 kb and 1625 degrees C. *American Journal of Science*, *276*(8), 969–996.

Mysen, B. O., Fogel, M. L., Morrill, P. L., & Cody, G. D. (2009). Solution behavior of reduced COH volatiles in silicate melts at high pressure and temperature. *Geochimica et Cosmochimica Acta*, *73*(6), 1696–1710. https://doi.org/10.1016/j.gca.2008.12.016

Ni, H., & Keppler, H. (2013). Carbon in silicate melts. *Reviews in Mineralogy and Geochemistry*, *75*(1), 251–287. https://doi.org/10.2138/rmg.2013.75.9

Oganov, A. R., Ono, S., Ma, Y., Glass, C. W., & Garcia, A. (2008). Novel high-pressure structures of $MgCO_3$, $CaCO_3$ and CO_2 and their role in Earth's lower mantle. *Earth and Planetary Science Letters*, *273*(1–2), 38–47. https://doi.org/10.1016/j.epsl.2008.06.005

Ono, S. (2005). Post-aragonite phase transformation in CaCO3 at 40 GPa. *American Mineralogist*, *90*(4), 667–671. https://doi.org/10.2138/am.2005.1610

Pal'yanov, Y. N., Sokol, A. G., Borzdov, Y. M., Khokhryakov, A. F., & Sobolev, N. V. (2002). Diamond formation through carbonate-silicate interaction. *American Mineralogist*, *87*(7), 1009–1013. https://doi.org/10.2138/am-2002-0726

Pan, V., Holloway, J. R., & Hervig, R. L. (1991). The pressure and temperature dependence of carbon dioxide solubility in tholeiitic basalt melts. *Geochimica et Cosmochimica Acta*, *55*(6), 1587–1595. https://doi.org/10.1016/0016-7037(91)90130-W

Panero, W. R., & Kabbes, J. E. (2008). Mantle-wide sequestration of carbon in silicates and the structure of magnesite II. *Geophysical Research Letters*, *35*(14). https://doi.org/10.1029/2008GL034442

Papale, P., Moretti, R., & Barbato, D. (2006). The compositional dependence of the saturation surface of H_2O+CO_2 fluids in silicate melts. *Chemical Geology*, *229*(1), 78–95. https://doi.org/10.1016/j.chemgeo.2006.01.013

Pawley, A. R., Holloway, J. R., & McMillan, P. F. (1992). The effect of oxygen fugacity on the solubility of carbon-oxygen fluids in basaltic melt. *Earth and Planetary Science Letters*, *110*(1), 213–225. https://doi.org/10.1016/0012-821X(92)90049-2

Pearce, M. L. (1964). Solubility of carbon dioxide and variation of oxygen ion activity in soda-silica melts. *Journal of the American Ceramic Society*, *47*(7), 342–347. https://doi.org/10.1111/j.1151-2916.1964.tb12998.x

Pearson, V. K., Sephton, M. A., Franchi, I. A., Gibson, J. M., & Gilmour, I. (2006). Carbon and nitrogen in carbonaceous chondrites: Elemental abundances and stable isotopic compositions. *Meteoritics & Planetary Science*, *41*(12), 1899–1918. https://doi.org/10.1111/j.1945-5100.2006.tb00459.x

Sakamaki, T., Ohtani, E., Urakawa, S., Terasaki, H., & Katayama, Y. (2011). Density of carbonated peridotite magma at high pressure using an X-ray absorption method. *American Mineralogist*, *96*(4), 553–557. https://doi.org/10.2138/am.2011.3577

Sen, S., Widgeon, S. J., Navrotsky, A., Mera, G., Tavakoli, A., Ionescu, E., & Riedel, R. (2013). Carbon substitution for oxygen in silicates in planetary interiors. *Proceedings of the National Academy of Sciences*, *110*(40), 15904–15907. https://doi.org/10.1073/pnas.1312771110

Shcheka, S. S., Wiedenbeck, M., Frost, D. J., & Keppler, H. (2006). Carbon solubility in mantle minerals. *Earth and Planetary Science Letters*, *245*(3), 730–742. https://doi.org/10.1016/j.epsl.2006.03.036

Shee, S. R., Gurney, J. J., & Robinson, D. N. (1982). Two diamond-bearing peridotite xenoliths from the finsch kimberlite, South Africa. *Contributions to Mineralogy and Petrology*, *81*(2), 79–87. https://doi.org/10.1007/BF00372045

Shishkina, T. A., Botcharnikov, R. E., Holtz, F., Almeev, R. R., Jazwa, A. M., & Jakubiak, A. A. (2014). Compositional and pressure effects on the solubility of H_2O and CO_2 in mafic melts. *Chemical Geology*, *388*, 112–129. https://doi.org/10.1016/j.chemgeo.2014.09.001

Shishkina, T. A., Botcharnikov, R. E., Holtz, F., Almeev, R. R., & Portnyagin, M. V. (2010). Solubility of H_2O- and CO_2-bearing fluids in tholeiitic basalts at pressures up to 500 MPa. *Chemical Geology*, *277*(1), 115–125. https://doi.org/10.1016/j.chemgeo.2010.07.014

Sobolev, N. V., & Shatsky, V. S. (1990). Diamond inclusions in garnets from metamorphic rocks: A new environment for diamond formation. *Nature*, *343*(6260), 742–746. https://doi.org/10.1038/343742a0

Solomatova, N. V., & Asimow, P. D. (2017). Ab initio study of the structure and stability of $CaMg(CO_3)_2$ at high pressure. *American Mineralogist*, *102*(1), 210–215. https://doi.org/10.2138/am-2017-5830

Solomatova, N. V., Caracas, R., & Manning, M. E. (2019). Carbon sequestration during core formation implied by complex carbon polymerization. *Nature Communications*, *10*(1), 789. https://doi.org/10.1038/s41467-019-08742-9

Stanley, B. D., Hirschmann, M. M., & Withers, A. C. (2011). CO_2 solubility in Martian basalts and Martian atmospheric evolution. *Geochimica et Cosmochimica Acta*, *75*(20), 5987–6003. https://doi.org/10.1016/j.gca.2011.07.027

Stolper, E., Fine, G., Johnson, T., & Newman, S. (1987). Solubility of carbon dioxide in albitic melt. *American Mineralogist*, *72*(11–12), 1071–1085.

Stolper, E., & Holloway, J. R. (1988). Experimental determination of the solubility of carbon dioxide in molten basalt at low pressure. *Earth and Planetary Science Letters*, *87*(4), 397–408. https://doi.org/10.1016/0012-821X(88)90004-0

Sverjensky, D. A., & Huang, F. (2015). Diamond formation due to a pH drop during fluid–rock interactions. *Nature Communications*, *6*(1). https://doi.org/10.1038/ncomms9702

Sverjensky, D. A., Stagno, V., & Huang, F. (2014). Important role for organic carbon in subduction-zone fluids in the deep carbon cycle. *Nature Geoscience*, *7*(12), 909–913. https://doi.org/10.1038/ngeo2291

Tamic, N., Behrens, H., & Holtz, F. (2001). The solubility of H_2O and CO_2 in rhyolitic melts in equilibrium with a mixed CO_2–H_2O fluid phase. *Chemical Geology*, *174*(1), 333–347. https://doi.org/10.1016/S0009-2541(00)00324-7

Tamic, N., Behrens, H., & Holtz, F. (2001). The solubility of H_2O and CO_2 in rhyolitic melts in equilibrium with a mixed CO_2–H_2O fluid phase. *Chemical Geology*, *174*(1), 333–347. https://doi.org/10.1016/S0009-2541(00)00324-7

Thibault, Y., & Holloway, J. R. (1994). Solubility of CO_2 in a Ca-rich leucitite: Effects of pressure, temperature, and oxygen fugacity. *Contributions to Mineralogy and Petrology*, *116*(1), 216–224. https://doi.org/10.1007/BF00310701

Thomsen, T. B., & Schmidt, M. W. (2008). Melting of carbonated pelites at 2.5–5.0 GPa, silicate–carbonatite liquid immiscibility, and potassium–carbon metasomatism of the mantle. *Earth and Planetary Science Letters*, *267*(1), 17–31. https://doi.org/10.1016/j.epsl.2007.11.027

Vetere, F., Botcharnikov, R. E., Holtz, F., Behrens, H., & De Rosa, R. (2011). Solubility of H_2O and CO_2 in shoshonitic melts at 1250 °C and pressures from 50 to 400MPa: Implications for Campi Flegrei magmatic systems. *Journal of Volcanology and Geothermal Research*, *202*(3), 251–261. https://doi.org/10.1016/j.jvolgeores.2011.03.002

Vetere, F., Holtz, F., Behrens, H., Botcharnikov, R. E., & Fanara, S. (2014). The effect of alkalis and polymerization on the solubility of H_2O and CO_2 in alkali-rich silicate melts. *Contributions to Mineralogy and Petrology*, *167*(5), 1014. https://doi.org/10.1007/s00410-014-1014-6

Vuilleumier, R., Seitsonen, A. P., Sator, N., & Guillot, B. (2015). Carbon dioxide in silicate melts at upper mantle conditions: Insights from atomistic simulations. *Chemical Geology*, *418*, 77–88. https://doi.org/10.1016/j.chemgeo.2015.02.027

Walter, M. J., Kohn, S. C., Araujo, D., Bulanova, G. P., Smith, C. B., Gaillou, E., et al. (2011). Deep mantle cycling of oceanic crust: Evidence from diamonds and their mineral inclusions. *Science*, *334*(6052), 54–57. https://doi.org/10.1126/science.1209300

Webster, J. D., & Botcharnikov, R. E. (2011). Distribution of sulfur between melt and fluid in S-O-H-C-Cl-bearing magmatic systems at shallow crustal pressures and temperatures. *Reviews in Mineralogy and Geochemistry*, *73*(1), 247–283. https://doi.org/10.2138/rmg.2011.73.9

Wood, B. J. (1993). Carbon in the core. *Earth and Planetary Science Letters*, *117*(3), 593–607. https://doi.org/10.1016/0012-821X(93)90105-I

Wyllie, P. J., Baker, M. B., & White, B. S. (1990). Experimental boundaries for the origin and evolution of carbonatites. *Lithos*, *26*(1), 3–19. https://doi.org/10.1016/0024-4937(90)90037-2

Zedgenizov, D. A., Shatsky, V. S., Panin, A. V., Evtushenko, O. V., Ragozin, A. L., & Kagi, H. (2015). Evidence for phase transitions in mineral inclusions in superdeep diamonds of the Sao Luiz deposit (Brazil). *Russian Geology and Geophysics*, *56*(1), 296–305. https://doi.org/10.1016/j.rgg.2015.01.021

17

The Effect of Variable Na/K on the CO_2 Content of Slab-Derived Rhyolitic Melts

Michelle Muth[1,2], Megan S. Duncan[3], and Rajdeep Dasgupta[1]

ABSTRACT

We conducted high pressure, high temperature experiments to investigate the effect of variable alkali ratio on the CO_2-rich fluid solubility in hydrous rhyolitic melts at sub-arc depths. Experiments were performed at 3.0 and 1.5 GPa, 1300 °C on rhyolitic compositions similar to low-degree partial melts of subducted slab lithologies, with fixed total alkalis (Na_2O+K_2O ~11.5 wt.%, volatile-free), but Na# (molar $Na_2O/[Na_2O+K_2O]$) varying from 0.15 to 0.88. In the experimental glasses, total dissolved CO_2 ($CO_2^{tot.}$) ranged from 2.14 ± 0.07 to 3.20 ± 0.07 wt.% at 3.0 GPa, and from 0.70 ± 0.02 to 1.19 ± 0.02 wt.% at 1.5 GPa. Experiments showed a general positive correlation between Na# and $CO_2^{tot.}$, with the exception of the highest Na# experiment at 1.5 GPa. Carbon was dissolved as molecular CO_2 ($CO_2^{mol.}$) and carbonate (CO_3^{2-}). As Na# increased, $CO_2^{mol.}/CO_2^{tot.}$ decreased from 0.94 to ~0.00 in the 1.5 GPa experiments and from 0.65 to 0.05 in the 3.0 GPa experiments. Variability in CO_2 concentration is larger and more clearly correlated with Na# at 3.0 GPa, indicating that this effect is pressure dependent. Our results show that compositional variability in silicic melts must be considered to accurately place constraints on the limit of CO_2 transfer in subduction zones.

17.1. INTRODUCTION

On million-year to billion-year timescales, the whole Earth carbon cycle is controlled by exchange between two reservoirs: the exosphere, which includes the atmosphere, crust, and ocean, and the mantle. This exchange plays an important role in long-term climate variability and in deep earth dynamics (Dasgupta & Hirschmann, 2010; Hayes & Waldbauer, 2006; Sleep & Zahnle, 2001). While the outgassing of carbon occurs through magmatic and volcanic processes at many volcanic centers, ingassing is achieved via subducting slabs that contain carbonated basaltic crust and carbon-bearing sediments (Dasgupta, 2013; Dasgupta & Hirschmann, 2010; Kelemen & Manning, 2015). As these slabs pass beneath volcanic arcs, a portion of this slab carbon is transported upward into the volcanic arc mantle source. The remaining carbon is transported to greater depths, where it may eventually be released into other areas of the mantle (e.g., Thomson et al., 2016) or become effectively sequestered within the residual slab material. The proportion of carbon released from the slab at sub-arc conditions and the exact agent that transports carbon to the arc magma mantle source remains incompletely constrained.

Measurements of carbon isotopes and $CO_2/^3He$ ratios of volcanic arcs indicate that ~71%–96% of carbon outgassed comes from the subducted slab (de Leeuw et al., 2007; Sano & Williams, 1996; Shaw et al., 2003). Most subduction zones today are likely too cold to liberate significant amounts of stored carbon via dry partial melting or decarbonation. However, increasing evidence indicates that many subduction zones likely undergo

[1] Department of Earth, Environmental and Planetary Sciences, Rice University, Houston, Texas, USA

[2] Department of Earth Sciences, University of Oregon, Eugene, Oregon, USA

[3] Department of Geosciences, Virginia Tech, Blacksburg, Virginia, USA

Carbon in Earth's Interior, Geophysical Monograph 249, First Edition. Edited by Craig E. Manning, Jung-Fu Lin, and Wendy L. Mao.

fluid-saturated slab melting (e.g., Kerrick & Connolly, 2001), and that this melt is a feasible carrier of CO_2 into arc source regions (Cooper et al., 2012; Duncan & Dasgupta, 2015; Martin & Hermann, 2018; Skora et al., 2015; Syracuse et al., 2010). Therefore, carbon carrying capacity, as CO_2, of the slab partial melt places an upper limit on the amount of carbon that reaches arc source regions from subducting slabs. Previous work that considered melting of subducted lithologies found that with the exception of carbonatitic and carbonated silicate melts generated at comparatively high pressures (e.g., Thomsen & Schmidt, 2008; Tsuno et al., 2012), the partial melts generated were generally rhyolitic in composition (Figure 17.1; e.g. Schmidt, 2015). Therefore, determining CO_2 carrying capacity of rhyolitic partial melts is crucial to understanding carbon flux in subduction settings.

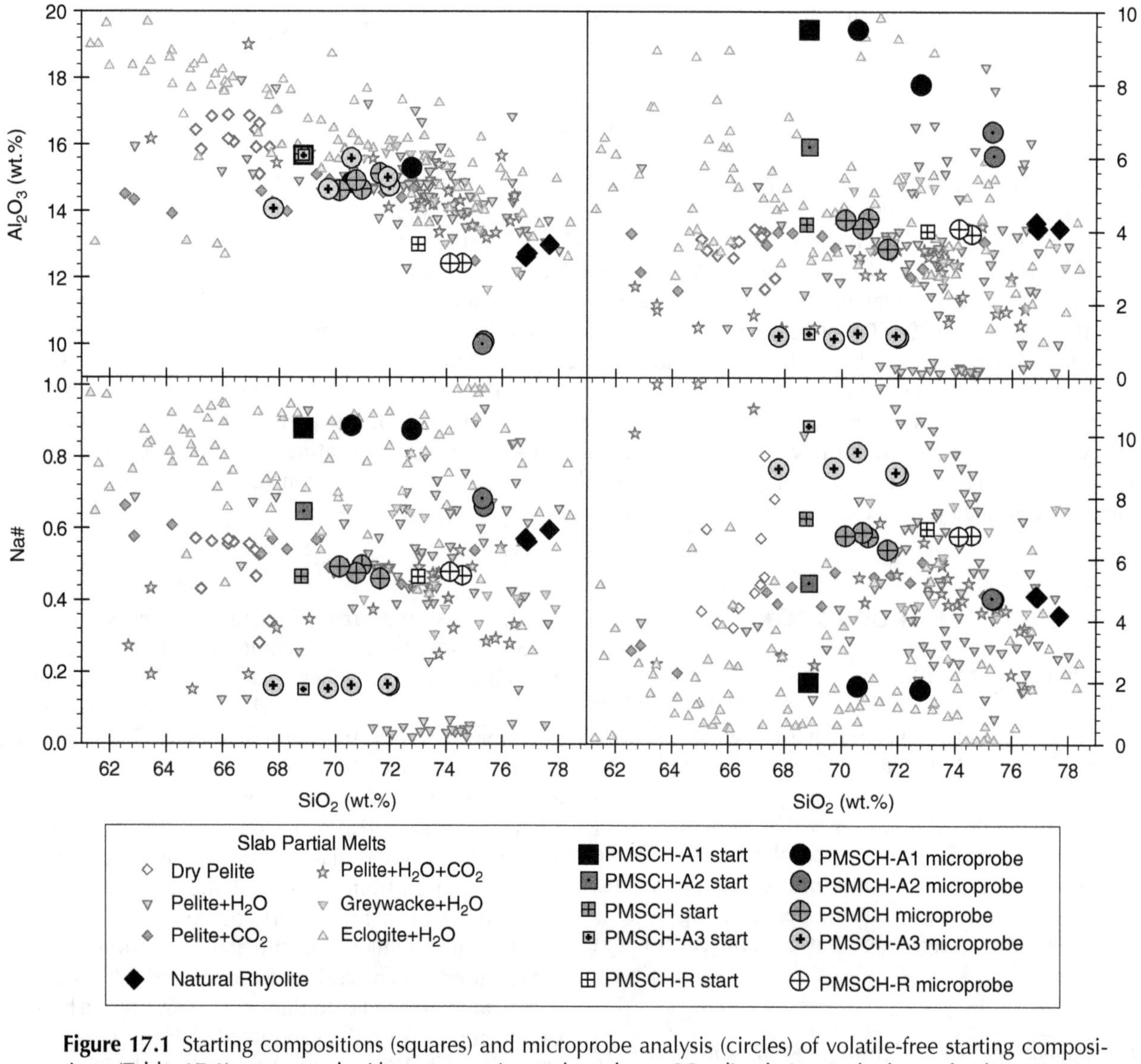

Figure 17.1 Starting compositions (squares) and microprobe analysis (circles) of volatile-free starting compositions (Table 17.1), compared with past experimental work on CO_2 dissolution in hydrous rhyolite (PMSCH; Duncan & Dasgupta, 2014), past experimental work that determined partial melt compositions of slab lithologies: dry metapelite (open diamonds; Spandler et al., 2010); fluid-present, hydrous pelite (inverted triangles; Hermann & Green, 2001; Hermann & Spandler, 2008; Johnson & Plank, 1999; Schmidt et al., 2004; Tsuno & Dasgupta, 2012); dry, carbonated pelite (gray diamonds; Tsuno & Dasgupta, 2011); fluid-present, hydrous, carbonated pelite (stars; Thomsen & Schmidt, 2008; Tsuno & Dasgupta, 2012); hydrous graywacke (inverted triangles; Auzanneau et al., 2006; Schmidt et al., 2004); hydrous basalt (triangles; Carter et al., 2015; Klimm et al., 2008; Laurie & Stevens, 2012; Mann & Schmidt, 2015; Martin & Hermann, 2018; Prouteau et al., 2001; Qian & Hermann, 2013; Rapp & Watson, 1995; Sen & Dunn, 1994; Skora et al., 2015; Wolff et al., 2013); and natural rhyolite compositions from previous CO_2 solubility studies (large black diamonds; Blank et al., 1993; Fogel & Rutherford, 1990; Tamic et al., 2001). See electronic version for color representation of the figures in this book.

Many previous studies that investigated the CO_2 solubility of rhyolitic melts were either at pressure-temperature (*P-T*) conditions not relevant to the sub-arc mantle (Blank et al., 1993; Fogel & Rutherford, 1990; Tamic et al., 2001), or used simplified melt compositions (Brey, 1976; Mysen, 1976; Mysen et al., 1976; Mysen & Virgo, 1980a). More recent work (Duncan & Dasgupta, 2014, 2015) investigated the CO_2 carrying capacity of slab-derived melts by focusing on CO_2 dissolution in a model rhyolitic partial melt composition of subducted slab lithologies over the appropriate *P-T* range. These studies found that differences in CO_2 dissolution in silicate melts were linked to the degree of melt polymerization caused by varying amounts of SiO_2 or H_2O (Fogel & Rutherford, 1990; Mysen, 1976), where less polymerized melts have a higher CO_2 capacity (e.g., Brooker et al., 2001; Eguchi & Dasgupta, 2018).

Other factors controlling melt polymerization include major cation proportions, which are a potentially important control on CO_2 capacity. Dixon (1997) presented a model that incorporates a CO_2 dependence on relative amounts of Ca^{2+}, Mg^{2+}, Na^+, and K^+ in basaltic melts, consistent with subsequent studies that found CO_2 concentrations in mafic melts increased with total alkali content (e.g. Ni & Keppler, 2013) and that for mafic-ultramafic melts both Ca^{2+} and Mg^{2+} complexes with CO_3^{2-}, with the affinity of Ca^{2+} being much stronger (Duncan et al., 2017). One study at low pressures found that melt CO_2 capacity increased with increasing Na# = $Na_2O/(Na_2O+K_2O)$ on a molar basis, in phonotephritic melts (Vetere et al., 2014), while another found the opposite trend in nephelinitic melts (Morizet et al., 2014). This contradiction may be due to other differing compositional factors between the two studies. No similar work has been done in rhyolitic partial melts, and the contradictory results of the two previous studies, albeit on more silica-poor melts, leaves a gap in our knowledge on the possible effect of Na# on CO_2 carrying capacity of slab-derived silicic melts. Given that rhyolitic melts are poor in Ca, Mg, and Fe, Na# is an important compositional variable to explore.

In combination with a wide range of *P-T* conditions, compositional heterogeneity in subducting materials produce a wide spectrum of partial melt compositions. Previous experiments that determined partial melt compositions of both silicic sediment and hydrous basalt report a wide range of Na# from 0.07 to 0.98 (Figure 17.1, see caption for references). This ratio is sensitive to the *P-T* conditions at which a melt is generated; Na becomes more compatible in residual lithologies at high pressures, driving a decrease in Na# in partial melts with increasing pressure (Poli & Schmidt, 2002).

To investigate the compositional dependence of CO_2 dissolution in slab-derived, hydrous, silicic melts, we conducted piston cylinder experiments on rhyolitic partial melts under *P-T* conditions relevant to subduction zones. In order to isolate the effects of alkalis from other cations, we focused our investigation on compositions with variable alkali ratios that were otherwise identical.

17.2. METHODS

17.2.1. Starting Material

Starting compositions were variations of the PMSCH composition from Duncan and Dasgupta (2014), which in turn was modeled from previous studies of low-degree partial melts of subducted lithologies (Figure 17.1, Table 17.1). The three main starting compositions for this study were identical to PMSCH in all major elements except for Na_2O and K_2O, which varied in concentration while maintaining a fixed total alkali content (~11.5 wt.%). Na# ranged from 0.15 to 0.88, which spans the Na# of both sediment and hydrous basalt partial melts. Synthetic starting compositions were made from powders of oxides (SiO_2, TiO_2, Al_2O_3, MnO_2, MgO, P_2O_5) and carbonates ($FeCO_3$, $CaCO_3$, Na_2CO_3, K_2CO_3) to add ~7 wt.% CO_2, and $Al(OH)_3$ to add ~3 wt.% H_2O (Table 17.1). To further assess the effects of H_2O and SiO_2 on CO_2 dissolution, two additional mixes were made: one with a Na# of 0.15 but a higher H_2O content, and a second with a Na# of 0.46 but a slightly higher concentration of SiO_2 (PMSCH-R). In the subsequent discussions, we focus primarily on experimental results from the three main compositions. Starting oxides were fired for several hours prior to mixing to avoid excess absorbed water. Components were then weighed, mixed, and ground in ethanol for 60 minutes to homogenize the starting powder, then stored in a 110 °C oven along with assembly parts to minimize moisture absorption.

17.2.2. Experimental Technique

Experiments were conducted in an end-loaded piston cylinder device in a half-inch $BaCO_3$ assembly at Rice University. $Au_{75}Pd_{25}$ capsules containing the starting mixtures were embedded in crushable MgO spacers. Pressure calibration and temperature gradient for this assembly and apparatus is reported in Tsuno and Dasgupta (2011). Temperature was controlled and monitored using a type C thermocouple, placed above the capsule, separated by a 1 mm MgO disc. Experiments were pressurized to 3.0 GPa or 1.5 GPa, then heated at 100 °C/min to 1300 °C and held for 48 hours, after which they were quenched by turning off power to the heater while maintaining pressure on the assembly. The capsule was then depressurized slowly, cut in half with a wire saw, set in PETROPOXY 154, and polished for analysis.

Table 17.1 Starting compositions used in the experiments.

Comp.	PMSCH-A1	PMSCH-A2	PMSCH-A3[a]	PMSCH-R	PMSCH[b]
SiO_2	68.85	68.87	68.86	73.01	68.77
TiO_2	0.50	0.50	0.50	0.30	0.50
Al_2O_3	15.68	15.67	15.67	13.00	15.71
FeO	0.99	0.99	0.99	0.88	0.99
MnO	0.50	0.50	0.50	0.34	0.50
MgO	0.09	0.09	0.09	0.26	0.19
CaO	1.47	1.47	1.47	1.00	1.47
Na_2O	9.51	6.30	1.20	4.00	4.19
K_2O	2.01	5.22	10.33	7.00	7.33
P_2O_5	0.41	0.38	0.39	0.21	0.35
Total	100.00	100.00	100.00	100.00	100.00
Na#	0.88	0.65	0.15	0.46	0.46
NBO/T	0.10	0.08	0.04	0.07	0.06
H_2O	8.46	7.83	6.80	6.95	7.49
$CO_2^{tot.}$	3.00	3.00	3.00	3.00	3.00

[a] This mix was made twice; the second contained a water content ~8 wt.%.
[b] Compositions re-reported from Duncan and Dasgupta (2014).

17.2.3. Analytical Techniques

Major elements were measured using a Cameca SX 100 electron microprobe at NASA Johnson Space Center, with an accelerating voltage of 15 kV, an electron beam current of 10 nA, and a 20 μm spot. Counting times for Si, Al, Ca, Ti, Fe, Mn, and Mg were 30 s on peak and 10 s on each background, and for Na, P, and K were 20 s on peak and 10 s on each background. Natural minerals and synthetic glasses were used as analytical standards. Averages of analyses on several spots of each experiment on each glass are reported in Table 17.2.

CO_2 and H_2O speciation and abundances were determined using a Thermo Nicolet Fourier Transform Infrared Spectrometer at Rice University. Experimental glasses were doubly polished to 15–50 μm, then cleaned with acetone before analysis. Conditions were set to 128 scans, at 4 cm^{-1} resolution, and a 40 μm spot size under a nitrogen purge. Peaks attributed to water were detected at 3530 cm^{-1} (OH^{-}+H_2O: O-H stretching) and at 1630 cm^{-1} (molecular water). Peaks attributed to carbon were detected at 2349 cm^{-1} (molecular CO_2, $CO_2^{mol.}$) and from 1425 cm^{-1} to 1530 cm^{-1} (carbonate doublet, CO_3^{2-}); concentration of CO_3^{2-}, calculated as CO_2, was determined using the 1425 cm^{-1} peak, and total dissolved CO_2 ($CO_2^{tot.}$) was simply $CO_2^{mol.}$ + CO_3^{2-}. Peak heights and areas were measured with linear backgrounds.

Volatile concentrations in the glass were quantified with the Beer-Lambert law:

$$C_i = 100 * M_i A_j / d \rho \varepsilon_j, \quad (17.1)$$

where C_i is the concentration of species i in weight percent, M_i is the molecular weight of the species, A_j is the absorbance of band j, unitless or in cm^{-1}, d is the thickness of the section in cm, ρ is the calculated density of the glass in g L^{-1} (Lange & Carmichael, 1987; Silver, 1988), and ε_j is the absorption coefficient for the peak j in L mol^{-1} cm^{-1} or L mol^{-1} cm^{-2} from previous studies (Duncan & Dasgupta, 2015; Silver & Stolper, 1989; Stolper et al., 1987). Sample thickness was measured using a digital micrometer after the analysis and calculated using the "interference fringe" method (Tamic et al., 2001) that uses $\lambda = 1/2dn$ with an empirical value for refractive index (n) and λ measured from each spectrum.

17.3. RESULTS

17.3.1. Texture and Major Element Composition

Quench products showed a glassy phase with large, rounded vesicles, indicating the presence of a CO_2-rich vapor phase. In each experimental glass, the electron microprobe total was less than 100%, ranging from 92.55% to 99.02%, suggesting the presence of dissolved volatiles (Table 17.2). While measured glass compositions generally match starting materials, at both pressures there were some deviations from the starting compositions. For example, experimental glasses from the three main compositions varied in SiO_2 content from ~68 to 75 wt.% and CaO content from ~1.2 to 1.6 wt.%. However, the measured analyses of a given composition were consistent.

We did attempt to estimate oxygen fugacity (fO_2) of these experiments based on the amount of iron that was dissolved in the AuPd capsule; unfortunately, the dissolved Fe concentrations were within the error of the measurement, making calculation of fO_2 highly uncertain. However, the absence of graphite in the quench products and evidence for equilibrium with a CO_2 rich

Table 17.2 Measured experimental glass compositions and run conditions.

	PMSCH-A1		PMSCH-A2		PMSCH-A3		PMSCH-A3[a]			PMSCH-R		PMSCH[b]	
Run No.	B311	B319	B316	B321	B331	B337	B308	B323	B322	B229	B231	B116	B139
P (GPa)	3.0	1.5	3.0	1.5	3.0	1.5	3.0	3.0	1.5	3.0	2.5	3.0	1.5
T (°C)	1300	1300	1300	1300	1300	1300	1300	1300	1300	1300	1300	1300	1300
T_{ae} (°C)	530	566	525	533	523	537	503	501	526	513	529	537	563
n	10	10	10	10	10	10	10	10	10	20	20	9	10
SiO_2	70.6 (1)	72.8 (3)	75.4 (8)	75.3 (2)	69.7 (2)	67.8 (3)	70.6 (4)	72.0 (3)	71.9 (2)	74.6 (3)	74.1 (4)	71.6 (3)	70.1 (2)
TiO_2	0.46 (3)	0.45 (2)	0.55 (4)	0.56 (5)	0.50 (3)	0.49 (4)	0.52 (3)	0.59 (2)	0.53 (3)	0.31 (3)	0.32 (3)	0.50 (3)	0.50 (2)
Al_2O_3	14.9 (1)	15.3 (2)	10.1 (2)	10.01 (6)	14.6 (2)	14.1 (1)	15.6 (2)	14.8 (2)	15.0 (1)	12.4 (2)	12.4 (3)	15.1 (1)	14.6 (2)
FeO	0.20 (4)	0.05 (3)	0.24 (8)	0.13 (3)	0.09 (5)	0.24 (4)	0.16 (3)	0.20 (5)	0.08 (2)	0.25 (7)	0.4 (1)	0.56 (9)	0.17 (4)
MnO	0.35 (3)	0.11 (2)	0.30 (5)	0.30 (5)	0.19 (5)	0.32 (5)	0.33 (5)	0.39 (4)	0.19 (3)	0.18 (5)	0.21 (4)	0.34 (4)	0.21 (3)
MgO	0.50 (5)	0.29 (3)	1.2 (4)	0.45 (3)	0.31 (2)	0.31 (3)	0.30 (2)	0.42 (3)	0.5 (1)	0.44 (3)	0.47 (4)	0.41 (1)	0.9 (2)
CaO	1.58 (6)	1.20 (4)	1.4 (3)	1.66 (5)	1.44 (5)	1.4 (1)	1.59 (6)	1.56 (8)	1.57 (5)	0.95 (7)	1.08 (6)	1.53 (5)	2.3 (2)
Na_2O	9.5 (1)	8.00 (9)	6.1 (2)	6.7 (1)	1.07 (5)	1.13 (7)	1.21 (6)	1.12 (5)	1.15 (4)	3.9 (2)	4.1 (1)	3.5 (1)	4.31 (8)
K_2O	1.87 (1)	1.76 (9)	4.7 (1)	4.7 (1)	9.0 (1)	8.9 (1)	9.49 (9)	8.75 (9)	8.8 (1)	6.8 (1)	6.8 (1)	6.3 (2)	6.77 (9)
P_2O_5	0.11 (5)	0.06 (6)	0.13 (6)	0.13 (8)	0.3 (1)	0.3 (1)	0.26 (8)	0.23 (1)	0.19 (5)	0.16 (8)	0.1 (1)	0.10 (7)	0.13 (7)
Total	97.62	99.02	95.13	97.03	97.23	95.07	92.55	93.49	96.94	92.96	92.88	97.52	96.35
Na#	0.89 (1)	0.87 (1)	0.66 (1)	0.68 (1)	0.15 (1)	0.16 (1)	0.16 (1)	0.16 (1)	0.17 (1)	0.47 (1)	0.48 (1)	0.459 (9)	0.492 (5)
NBO/T	0.104 (5)	0.038 (3)	0.15 (3)	0.149 (4)	0.011 (5)	0.027 (4)	0.016 (3)	0.019 (3)	0.015 (4)	0.065 (1)	0.075 (8)	0.035 (9)	0.1 (0)
H_2O[c]	2.84 (5)	1.67 (3)	3.18 (8)	2.6 (1)	3.01 (4)	2.3 (1)	4.91 (6)	5.27 (1)	2.88 (7)	3.86 (5)	2.81 (5)	2.45 (5)	1.70 (2)
$CO_2^{tot.d}$	3.20 (7)	0.70 (2)	3.10 (9)	1.19 (2)	2.14 (7)	0.81 (9)	2.01 (6)	2.02 (7)	0.77 (7)	1.80 (7)	1.84 (8)	1.95 (2)	0.82 (5)
$CO_2^{mol.}$	0.26 (2)	0.406 (1)	0.15 (2)	n.d.	1.40 (6)	0.76 (8)	1.44 (6)	1.42 (6)	0.68 (7)	0.61 (9)	0.65 (6)	1.00 (3)	0.11 (1)
CO_3^{2-}	2.94 (8)	0.30 (2)	2.9 (1)	1.18 (2)	0.74 (5)	0.05 (2)	0.58 (3)	0.60 (2)	0.099 (7)	1.2 (1)	1.19 (5)	0.95 (2)	0.71 (5)
$CO_2^{mol.}/CO_2^{tot.}$	0.081 (5)	0.58 (2)	0.048 (7)	n.a.	0.65 (2)	0.94 (2)	0.71 (1)	0.704 (9)	0.87 (1)	0.34 (5)	0.35 (2)	0.513 (1)	0.13 (1)
$CO_2^{mol.e}$	2.6 (2)	0.59 (5)	2.4 (2)	0.92 (8)	1.8 (1)	0.7 (1)	1.7 (1)	1.7 (1)	0.65 (9)	1.5 (1)	1.5 (1)	1.63 (9)	0.66 (8)
CO_3^{2-e}	0.6 (2)	0.12 (4)	0.7 (2)	0.26 (7)	0.3 (1)	0.13 (5)	0.3 (1)	0.3 (1)	0.12 (5)	0.3 (1)	0.3 (1)	0.32 (9)	0.16 (5)
$CO_2^{mol.}/CO_2^{tot.e}$	0.80 (5)	0.84 (5)	0.78 (6)	0.78 (6)	0.85 (4)	0.84 (4)	0.85 (4)	0.84 (4)	0.85 (4)	0.82 (5)	0.82 (5)	0.84 (5)	0.81 (5)

Note. Oxide concentrations are reported on a volatile-free basis with the 1σ error, in last digits cited, are given in parentheses based on spot to spot variation in EPMA and FTIR analyses for each sample; for example, 70.6 (1) wt.% should be read as 70.6 ± 0.1 wt.%. *n* = number of EMPA spots averaged.

[a] Exactly the same in all oxides as PMSCH-A3, but with a starting water content of ~8 wt.%.

[b] Compositions re-reported from Duncan and Dasgupta (2014).

[c] Water concentrations were calculated using absorption coefficients from Stolper et al. (1987) and Silver and Stolper (1989).

[d] Carbon concentrations were calculated using absorption coefficients from Duncan and Dasgupta (2015).

[e] Species concentrations recalculated using Konschak and Keppler (2014).

vapor phase indicates fO_2 conditions above the CCO (C + O_2 = CO_2) buffer.

17.3.2. Dissolved CO_2 and H_2O

Peak heights and shapes associated with $CO_2^{mol.}$ and CO_3^{2-} shifted with changing pressure and between each composition (Figure 17.2 a, b). At 3.0 GPa, with increasing Na#, the carbon species was increasingly dominated by CO_3^{2-}. The carbonate doublet also changed shape, with the doublet split decreasing with increasing Na#; in the high Na# glass, CO_3^{2-} peaks were asymmetrical and closely spaced at ~1504 cm^{-1} and 1439 cm^{-1}, while low Na# glass peaks were symmetrical and located at ~1530 cm^{-1} and ~1425 cm^{-1}. Experiments conducted at 1.5 GPa showed a similar, if less clearly delineated, trend. Unlike experiments at 3.0 GPa, the carbonate doublet did not show a progression in split or relative height, and became harder to detect, as it sits on the shoulder of Al-Si network peaks (~1600 and 1800 cm^{-1}; Newman et al., 1986) at the lowest Na#. The high Na# (Na# = 0.88) experiment at 1.5 GPa in particular had a carbonate doublet that looks distinctly different from its counterpart at 3.0 GPa.

Quantitative estimation of dissolved CO_2 and H_2O (Table 17.2) was performed from the collected FTIR spectra using previously determined absorption coefficients (Duncan & Dasgupta, 2015). In the main series of experiments, total dissolved CO_2 ranged from 2.14 ± 0.07 to 3.20 ± 0.07 wt.% at 3.0 GPa, and from 0.70 ± 0.02 to 1.19 ± 0.02 wt.% at 1.5 GPa (Figure 17.3d), with a general trend of increasing total CO_2 content with increasing Na#. The ratio of $CO_2^{mol.}/CO_2^{tot.}$ ranged from ~0.00 to 0.94 at 1.5 GPa to 0.05 to 0.65 at 3.0 GPa. Within both 1.5 GPa and 3.0 GPa experiments, there was a general

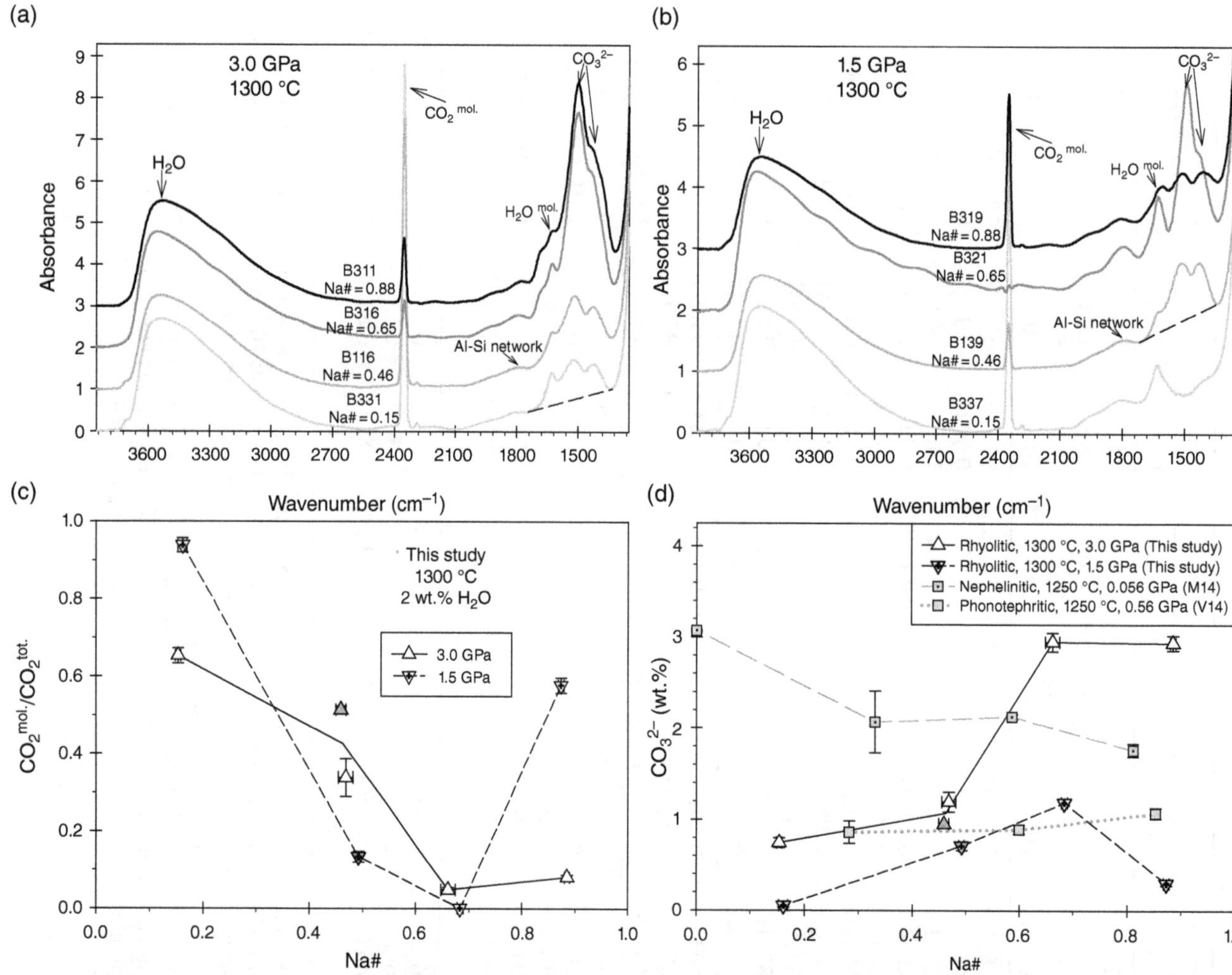

Figure 17.2 Top: FTIR spectra for each composition at (a) 3.0 GPa and (b) 1.5 GPa, marked with identified peak locations and Na# of each composition; gray Na# = 0.46 data are from Duncan and Dasgupta (2014). Bottom: Calculated (c) $CO_2^{mol.}/CO_2^{tot.}$ and (d) CO_3^{2-} estimated from the FTIR spectra, plotted versus Na#. Data shown for experiments at 1.5 GPa and 3.0 GPa. Data from previous studies on phonotephritic and nephelinitic compositions shown for comparison (squares; Morizet et al., 2014; Vetere et al., 2014). See electronic version for color representation of the figures in this book.

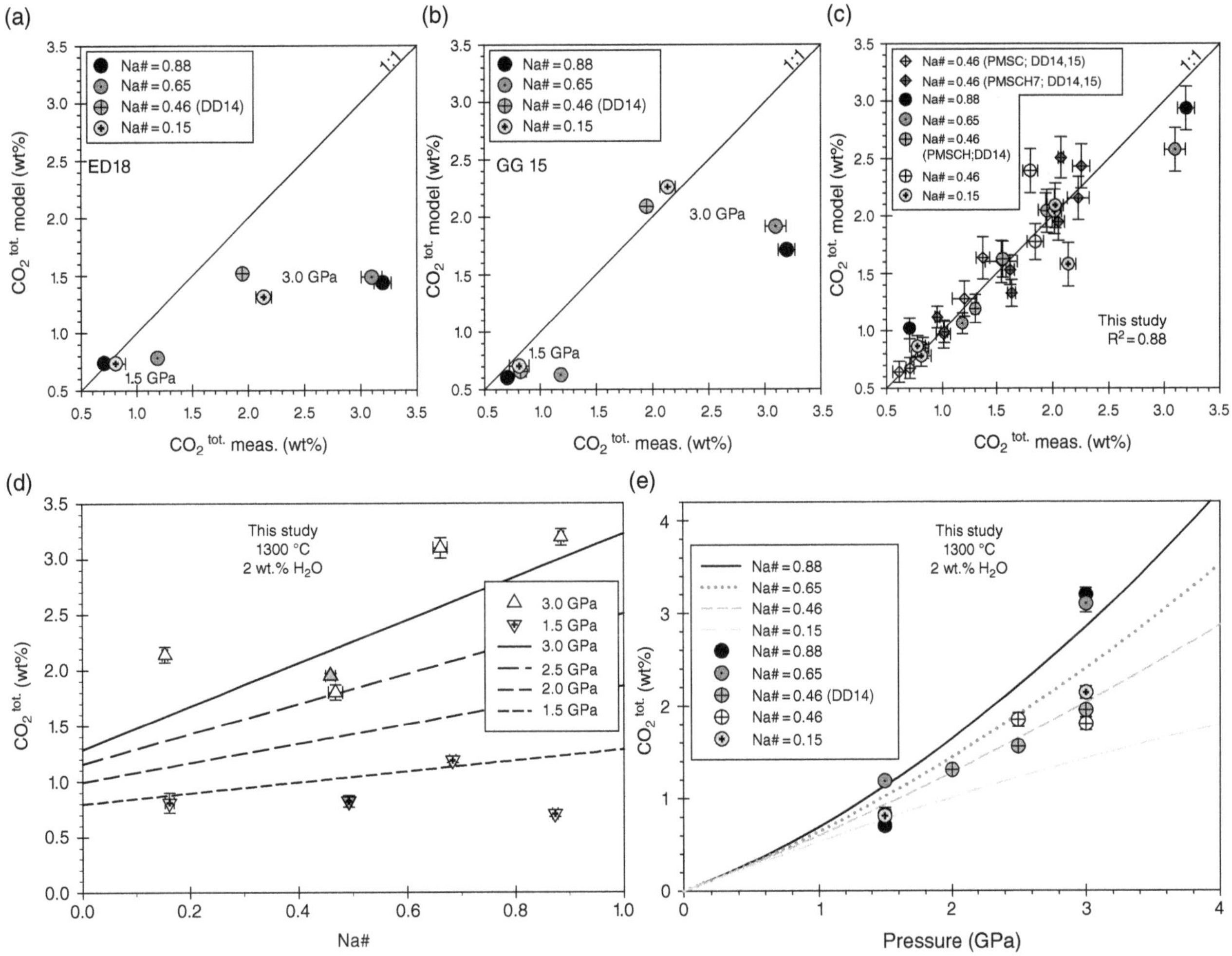

Figure 17.3 Measured CO_2 contents of rhyolitic glasses in this study compared to (a) the model of Eguchi and Dasgupta (2018), (b) the model of Ghiorso and Gualda (2015), and (c) values from our parameterization to show goodness of fit. All experimental data used for fitting are shown; see text for references. Calculated CO_2 content as a function of (d) Na#, as in Figure 17.4, and (e) pressure using equation (17.5) and an assumed H_2O content of 2 wt.%, with experimental data from the current study shown for comparison. Symbols as in Figure 17.1; DD14 = Duncan and Dasgupta (2014); DD15 = Duncan and Dasgupta (2015). See electronic version for color representation of the figures in this book.

trend of decreasing $CO_2^{mol.}$ with increasing Na#, while the behavior of CO_3^{2-} was the inverse, increasing with increasing Na# (Figure 17.2 c, d). At 3.0 GPa, experiments with slightly higher SiO_2 than the main series, composition PMSCH-R, contained 1.80 ± 0.07 wt.% total dissolved CO_2 with a $CO_2^{mol.}/CO_2^{tot.}$ value of 0.34. These values are comparable to results from experiments with the same Na#, compositions PMSCH, and are consistent with the observed patterns in CO_2 dissolution within the main series of experiments. An exception to this general behavior was the highest Na# experiment at 1.5 GPa, which had a lower CO_3^{2-} and $CO_2^{tot.}$ content than expected based on these overall trends.

It should be noted that the measured carbon speciation in the experimental glasses as determined using FTIR is likely affected during rapid cooling of the melts and that the actual magmatic speciation was different (see complete discussion in Duncan & Dasgupta, 2014). The measurement therefore represents the speciation at the glass transition temperature or apparent equilibrium temperature (T_{ae}, Table 17.2) that can be calculated for our experiments based on the dissolved water content (Zhang et al., 1995). Previous work showed that decreasing temperature toward T_{ae} will transform $CO_2^{mol.}$ to CO_3^{2-} (Morizet et al., 2001; Nowak et al., 2003), without changing the total amount of CO_2 dissolved. This effect was quantified, under limited experimental conditions (silicate melt compositions and water contents), by Konschak and Keppler (2014), who approximated the effect as:

$$\ln K = X_{CO_3^{2-}} / X_{CO_2^{mol.}} = \left[-2.69 - 21.38(\mathrm{NBO/T})\right] + \left[1480 + 38810(\mathrm{NBO/T})\right] / T \quad (17.2)$$

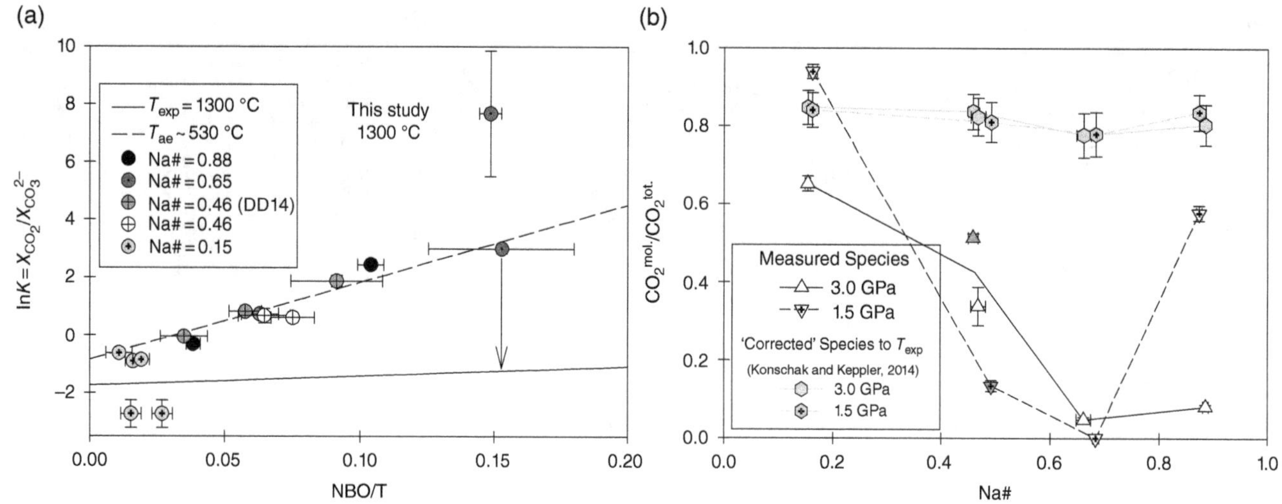

Figure 17.4 (a) Calculated ln*K* after Konschak and Keppler (2014) for experiments in this work and from Duncan and Dasgupta (2014) plotted as a function of estimated NBO/T. Various curves show the temperature dependence of ln*K* on silicate melt composition according to Konschak and Keppler (2014). Solid line: calculated ln*K* for our experimental temperatures (T_{exp}); dashed line: ln*K* determined for the apparent equilibrium temperatures (T_{ae}) calculated from Zhang et al. (1995). (b) Open triangles: $CO_2^{mol.}/CO_2^{tot.}$ values measured using FTIR on quenched glasses for experiments in this study and those from Duncan and Dasgupta (2014; DD14). Filled hexagons: "Corrected" $CO_2^{mol.}/CO_2^{tot.}$ values for the same experiments after Konschak and Keppler (2014). See electronic version for color representation of the figures in this book.

where NBO/T is a silicate compositional parameter (the ratio of nonbridging oxygens to tetrahedrally coordinated cations) and *T* is the temperature in K. When we "correct" our measured speciation from the T_{ae} to the experimental temperatures, all $CO_2^{mol.}/CO_2^{tot.}$ values are between 0.8 and 0.9 (Figure 17.4, Table 17.2). Given that NBO/T parameter does not account for differences in Na and K, only the total nonbridging oxygens, which is constant in our experiments, it is not surprising that "correcting" the data removes any trends in speciation as a function of Na#. Given the differences in CO_3^{2-} concentrations at different pressures and/or Na# but similar T_{ae}, we argue that the effect of pressure on enhanced CO_3^{2-} dissolution is real, although the absolute values of CO_2 solubility in the form of CO_3^{2-} may be influenced by quenching.

17.4. DISCUSSION

17.4.1. Composition-Dependent CO_2 Dissolution in Silicate Melt: Comparison to Previous Studies

Analysis of the experimental glasses showed that carbon was dissolved as $CO_2^{mol.}$ and CO_3^{2-} (Table 17.2). This suggests that the following carbon dissolution reactions were taking place (Mysen & Virgo, 1980b; Stolper et al., 1987):

$$\underset{\text{vapor}}{CO_2} \leftrightarrow \underset{\text{melt}}{CO_2} \tag{17.3}$$

$$\underset{\text{vapor}}{CO_2} + \underset{\text{melt}}{O^{2-}} \leftrightarrow \underset{\text{melt}}{CO_3^{2-}} \tag{17.4}$$

The previously discussed changes in carbonate doublet splitting visible in the FTIR spectra were likely caused by changes in the bonding environment of the carbonate group (e.g. Ni & Keppler, 2013). This observation combined with the positive correlation between the amount of CO_3^{2-} and Na# (Figure 17.2 c, d) suggests that at these conditions Na^+ may act as a stronger network modifier and therefore create melt carbonate complexes more effectively than K^+. Because of this, CO_3^{2-} may be accommodated more readily into the melt structure as Na# increases, explaining the simultaneous increase in CO_3^{2-} and total dissolved CO_2.

Variables besides Na# such as *P*, *T*, melt polymerization, and other cation concentrations also play a role in CO_2 dissolution and may moderate the effect of Na#. These compositional variables may explain the opposite trends seen in previous experiments (Morizet et al., 2014; Vetere et al., 2014), and variation within the general trends in the present study. For example, our high Na# experiment at 1.5 GPa, which contained less $CO_2^{tot.}$ and CO_3^{2-} than expected, contained approximately 0.4 wt.% less CaO than other compositions, which may have contributed to its anomalous behavior. However, $CO_2^{tot.}$ and CO_3^{2-} contents of experiments with a slightly more silicic composition, PMSCH-R, were consistent with the previously discussed patterns in Na#, $CO_2^{tot.}$, and $CO_2^{mol.}/CO_2^{tot.}$. This similarity suggests that small variations in melt polymerization alone are not enough to generate the observed variations in CO_2 dissolution within our experiments.

Correlations in total CO_2 and $CO_2^{mol.}/CO_2^{tot.}$ were much stronger at 3.0 GPa than at 1.5 GPa (Figures 17.2 c, d, and 17.3d), indicating that the Na# effect is sensitive to pressure. Additionally, the observed range in total CO_2 dissolution and speciation was also larger in the 3.0 GPa experiments. Within the main experimental series at 1.5 GPa, the total dissolved CO_2 varied by 0.49 wt.% and $CO_2^{mol.}/CO_2^{tot.}$ varied by 0.60. In contrast, at 3.0 GPa the total dissolved CO_2 varied by 1.06 wt.% and $CO_2^{mol.}/CO_2^{tot.}$ varied by 0.94. This is consistent with CO_3^{2-} dissolution in silicic melt becoming important at higher pressures (Duncan & Dasgupta, 2014, 2015, 2017), where at higher pressure CO_2 behavior would be more sensitive to parameters such as Na# that influence dissolution of C in the melt as CO_3^{2-}.

The observed trend of increasing CO_3^{2-} with Na# in rhyolitic compositions (Figure 17.2d) is similar to the trend observed in a previous study conducted on phonotephritic composition (Vetere et al., 2014), and in contrast with the trend observed for a nephelinitic composition (Morizet et al., 2014). However, our experiments differ from this previous work in that we observed complex speciation (i.e., variable proportions of $CO_2^{mol.}$ and CO_3^{2-} within experimental glasses at the same pressure). As both of these previous studies were conducted at lower pressures on silica-poor compositions, it is difficult to make a direct comparison between their data and the current study.

Our data trends can also be compared with recent composition-dependent CO_2 solubility models for silicate melts (Figure 17.3 a, b; e.g. Eguchi & Dasgupta, 2018; Ghiorso & Gualda, 2015). While the model of Ghiorso and Gualda (2015) captures our low to intermediate CO_2 concentrations reasonably well, it does not capture the trend of increasing CO_2 with increasing Na# and as a consequence distinctly underpredicts our high CO_2 data. Because our CO_2 solubility data are for hydrous silicate melts, they cannot be compared directly with the model of Eguchi and Dasgupta (2018), which was built for nominally anhydrous systems and is expected to mostly underpredict CO_2 contents for melts with ~1–3 wt.% H_2O, where our data are focused. However, the model of Eguchi and Dasgupta (2018) also does not capture the specific trend of increasing $CO_2^{tot.}$ with increasing Na#, even when this difference in melt H_2O content is taken into account.

17.4.2. Parameterization

Because existing models that are applicable for a wide range of silicate melt compositions do not specifically capture the Na# effect observed in our study, we used all of the data from the current study and data from previous studies on natural rhyolitic compositions at high pressure (i.e. Duncan & Dasgupta, 2014, 2015) to derive an empirical fit of total dissolved CO_2 as a function of Na#. The variation in CO_2 speciation observed in our study and the opposing Na# effects observed in previous work (Morizet et al., 2014; Vetere et al., 2014) indicate that this effect is complex and is likely moderated by other compositional variables in addition to pressure and temperature. A comprehensive model describing the effect of Na# across a broad range of *P-T* conditions and major element compositions would require an expanded data set or intensive statistical treatment of available experimental data, both of which are beyond the scope of the current study. We focus here instead on creating an empirical parameterization to describe CO_2 solubility in natural rhyolitic melts at high pressure conditions.

The data set used in the empirical fit is drawn from a total of 31 experiments on rhyolitic compositions with variable Na#. Within these experiments, pressures range from 1.5 to 3.0 GPa, temperatures range from 1150 °C to 1300 °C, initial water contents range from 0 to ~8 wt.%, and Na# ranges from 0.15 to 0.89. This data was fit to a predefined form of equation using least-squares regression. This form of equation is similar to one used in previous work (Duncan & Dasgupta, 2014), where it was demonstrated that the effect of pressure is best described as a polynomial function. The other equation terms (Na#, X_{H2O}, T) describe the other variables within the data set and are arranged to provide the best fit to the data based on tests of several equation forms. This form of equation also captures the diminishing solubility of CO_2 as pressure approaches 1 atm. The equation and empirical fit are as follows:

$$\underset{\text{melt}}{CO_2(\text{wt.\%})} = (a_1\text{Na\#} + a_2X_{H_2O} + a_3)P^2 + (b_1T + b_2X_{H_2O} + b_3)P \qquad (17.5)$$

where X_{H2O} is mole fraction of H_2O in the melt, T is temperature in °C, and P is pressure in GPa. The constants are as follows: $a_1 = 2.16 \pm 0.04 \times 10^{-1}$, $a_2 = -4.07 \pm 0.07 \times 10^{-1}$, $a_3 = -2.8 \pm 0.1 \times 10^{-1}$, $b_1 = -5.6 \pm 0.4 \times 10^{-4}$, $b_2 = 3.7 \pm 0.1 \times 10^{0}$, $b_3 = 9.9 \pm 0.3 \times 10^{-1}$, $R^2 = 0.88$ (Figure 17.3c). The fit accounts for $CO_2^{mol.}$ and CO_3^{2-} as $CO_2^{tot.}$, but as this simple parameterization does not take into account the relative concentrations of the different dissolved species, it should not be used for melt compositions that are likely to have only CO_3^{2-} (such as basalts) or only $CO_2^{mol.}$ (such as rhyolites at low pressures). However, using $CO_2^{tot.}$ does remove any potential effects of quench-modified speciation. It can be applied to rhyolitic melts at pressures, temperatures, and H_2O, SiO_2, and CaO contents close to the calibration data set as detailed above.

Comparisons of modeled and predicted values show that this empirical fit does a reasonable job of capturing CO_2 dissolution within the calibration data set (Figure 17.3c). Based on the relative difference between the modeled and predicted values, experiments within the main series that deviate most from our fit are the low Na# experiment at 3.0 GPa (~35% difference), the high Na# experiment at 1.5 GPa (~31% difference), and the intermediate Na# (Na# = 0.66) experiment at 3.0 GPa (~20% difference). This is expected, as these experiments differ from overall trends within the data set (Figures 17.2 and 17.3). In order to compare the behavior of this parameterization to experimental data, we fixed temperature (1300 °C) and water content (~2 wt.%) and calculated CO_2 as a function of Na# (Figure 17.3d) and pressure (Figure 17.3e). Calculated CO_2 contents spread farther apart between each composition with increasing pressure. This is consistent with our data, which as previously discussed indicate that pressure has an effect on the importance of Na# in CO_2 dissolution.

Although the simple empirical parameterization describes our data overall, there are other compositional factors that influenced the measured CO_2 concentrations. For example, despite our attempt to hold all other oxide compositions constant, the experimental glasses varied in SiO_2 and CaO content. Given the widely recognized roles of melt polymerization and melt CaO on CO_2 dissolution, these variations might have contributed towards the variation of measured total CO_2. However, these compositional variations are smaller than the Na# variation, and therefore, our Na#-based parameterization captures most of the dissolved CO_2 variability observed in our experiments, to first order.

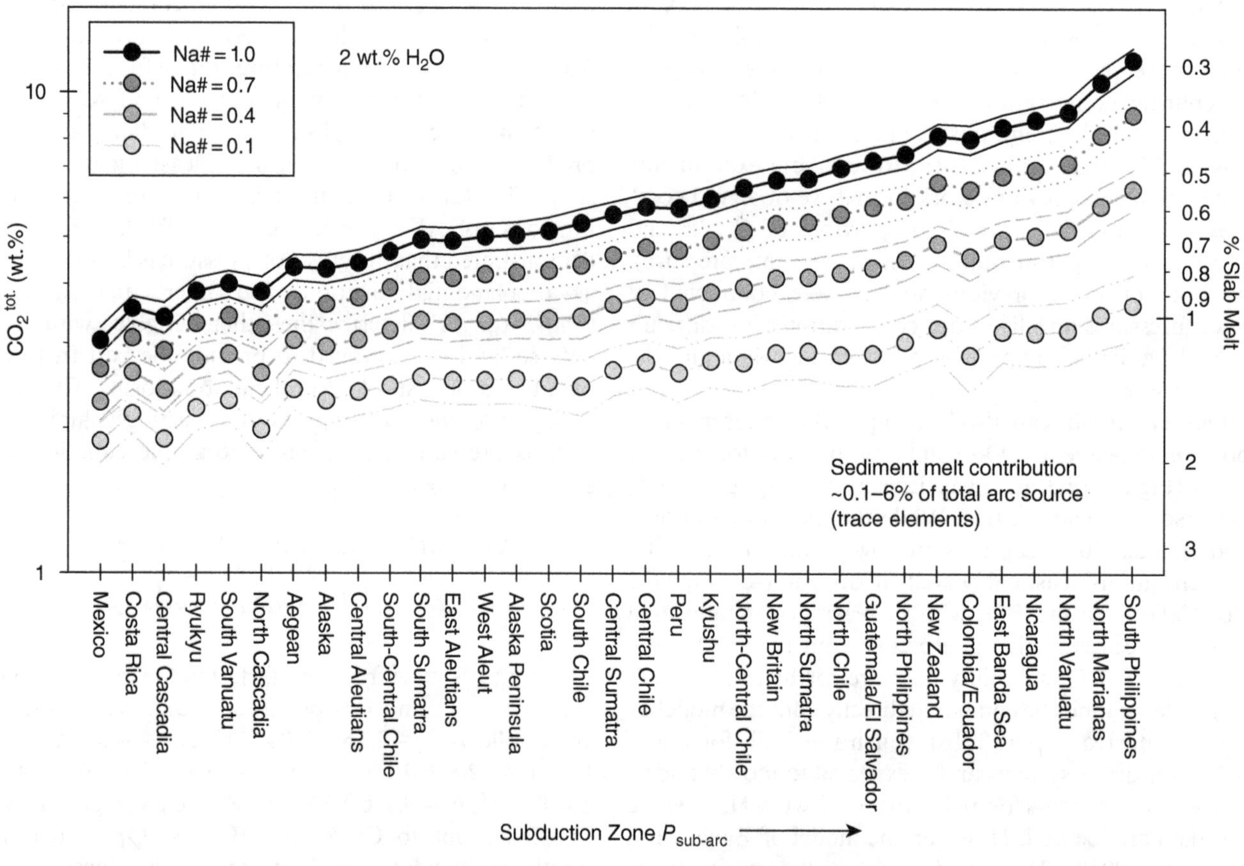

Figure 17.5 Total CO_2 content of rhyolitic slab melts (primary *y*-axis) calculated using the model for different melt Na# at fixed water content of 2 wt.% using calculated at the sub-arc *P-T* from the D80 model of Syracuse et al. (2010) for slabs above the hydrous solidus of Mann and Schmidt (2015). Subduction zones are arranged by increasing sub-arc depths. Calculated slab melt contribution to the sub-arc melt (secondary *y*-axis) assumes a primary arc basalt CO_2 content of 3000 ppm (Blundy et al., 2010; Wallace, 2005), an average mantle wedge CO_2 of 107 ppm (Cartigny et al., 2008; Dasgupta & Hirschmann, 2010; Hirschmann & Dasgupta, 2009), 15% mantle wedge melt, and mixing of mantle wedge-derived basaltic melt with a rhyolitic slab melt that is saturated in a CO_2-rich fluid (see Duncan & Dasgupta, 2015, for details). See electronic version for color representation of the figures in this book.

17.4.3. Implications for Slab-Mantle CO_2 Transfer in Subduction Zones

Our results indicate that changes in Na_2O versus K_2O concentrations in silicate melts have a significant effect on the CO_2 carrying capacities of slab-derived rhyolitic magmas at subduction zones. Because of this, variations in slab melt Na# and related variation in CO_2 carrying capacity can have significant impact on subduction zone carbon cycling efficiency. To illustrate this, we used our empirical fit to calculate the CO_2 content of slab melts and how much of such melts would be needed to mix with mantle melts to generate a primary arc basalt with 3000 ppm CO_2 (Blundy et al., 2010; Skora & Blundy, 2010; Wallace, 2005). If the extent of arc mantle source melting is 15%, the source mantle must contain 450 ppm CO_2, assuming batch melting and $D_C^{solid/melt}$ ~0. For example, based on the sub-arc conditions of North Cascadia, 3.09 GPa, 936 °C, from the D80 model of Syracuse et al. (2010), a slab melt with an Na# = 1 would have a saturated CO_2 content of 3.84 wt.%, assuming 2 wt.% H_2O is also present in the melt. For a given mantle wedge CO_2 content of 107 ppm (an average based on values from Cartigny et al., 2008; Dasgupta & Hirschmann, 2010; Hirschmann & Dasgupta, 2009; Marty, 2012; and Saal et al., 2002), the slab melt would only need to make up 0.89 wt.% of the total arc source region (Figure 17.5). We did this calculation for all slabs whose surface T at sub-arc depths fall above the hydrous solidus of Mann and Schmidt (2015).

According to this calculation, the CO_2-rich, fluid-saturated, rhyolitic slab melt contribution varies from 0.3% (Na# = 1.0) to 1.8% (Na# = 0.1), which is well within values of sediment contributions that are based on geochemistry of arc magmas (e.g., Class et al., 2000; Plank, 2005; Shimoda et al., 1998). Although we are extrapolating our empirical fit to a lower temperature range, the effect of temperature on CO_2 carrying capacity is small relative to the pressure effect (e.g., Duncan & Dasgupta, 2015) and the agreement with geochemical arguments from natural data gives us confidence that our values are reasonable as a point of discussion. Nevertheless, we emphasize that this model is intended to illustrate the importance of melt Na# on CO_2 transport from the slabs to the sub-arc mantle and is not intended to provide accurate estimates of slab melt contribution within individual arc segments.

In sub-arc settings with higher degrees of mantle wedge melting, lower water contents in the slab melts, CO_2-undersaturated silicic melt, and/or higher concentrations of CO_2 in primary arc magmas, a higher slab contribution to the sub-arc mantle would be required to generate estimated primary arc magma CO_2 concentrations. These effects have been observed and discussed in previous investigations of CO_2 cycling, but this calculation illustrates the additional importance of evaluating the impact of Na# in slab melts. According to our parametrization and calculations described here, differences in slab melt Na# could drive variations in slab contribution estimates by up to 120%.

17.5. CONCLUSIONS

We conducted high P-T experiments to determine the relationship between Na# and CO_2 in rhyolitic partial melts under pressure and temperatures relevant to subduction settings. Results indicate that there is likely a positive relationship between Na# of rhyolitic melts and carbonate dissolution, tied to changes in carbon speciation. The effect is stronger at higher pressures, where CO_3^{2-} dissolution in silicic melt becomes more significant. Our results could be interpreted to suggest greater affinity of CO_3^{2-} towards Na^+ rather than K^+, at least in silicic melts. No current CO_2 solubility models adequately capture the effect of melt Na/K ratio observed in our study. Future model regressions for CO_2 dissolution should be built differently, ensuring that effects of individual cation modifiers are reproduced. The melt compositional effect on CO_2 dissolution studied here has implications for carbon cycling in subduction settings. More specifically, our results indicate that accurately predicting carbon transport via slab melt in subduction zones requires taking the effects of network-modifying cation proportions on carbonate dissolution into consideration.

ACKNOWLEDGMENTS

We thank James Eguchi for his important discussion. Malcolm Massuyeau and an anonymous reviewer are acknowledged for their valuable comments on the original manuscript. The study was made possible by financial support from NSF grant OCE-1338842 and the Deep Carbon Observatory.

REFERENCES

Auzanneau, E., Vielzeuf, D., & Schmidt, M. W. (2006). Experimental evidence of decompression melting during exhumation of subducted continental crust. *Contrib. Mineral. Petrol.*, *152*(2), 125–148. doi:10.1007/s00410-006-0104-5

Blank, J. G., Stolper, E. M., & Carroll, M. R. (1993). Solubilities of carbon dioxide and water in rhyolitic melt at 850 °C and 750 bars. *Earth Planet. Sci. Lett.*, *119*(1–2), 27–36.

Blundy, J., Cashman, K. V., Rust, A., & Witham, F. (2010). A case for CO_2-rich arc magmas. *Earth Planet. Sci. Lett.*, *290*(3–4), 289–301. doi:10.1016/j.epsl.2009.12.013

Brey, G. (1976). CO_2 solubility and solubility mechanisms in silicate melts at high pressures. *Contrib. Mineral. Petrol.*, *57*(2), 215–221.

Brooker, R. A., Kohn, S. C., Holloway, J. R., & McMillan, P. F. (2001). Structural controls on the solubility of CO_2 in silicate melts: Part I. Bulk solubility data. *Chem. Geol.*, *174*(1–3), 225–239.

Carter, L. B., Skora, S., Blundy, J. D., De Hoog, J.C.M., & Elliott, T. (2015). An experimental study of trace element fluxes from subducted oceanic crust. *J. Petrol.*, *56*(8), 1585–1605. doi:10.1093/petrology/egv046

Cartigny, P., Pineau, F., Aubaud, C., & Javoy, M. (2008). Towards a consistent mantle carbon flux estimate: Insights from volatile systematics (H_2O/Ce, δD, CO_2/Nb) in the North Atlantic mantle (14°N and 34°N), *Earth Planet. Sci. Lett.*, *265*(3–4), 672–685. doi:10.1016/j.epsl.2007.11.011

Class, C., Miller, D. M., Goldstein, S. L., & Langmuir, C. H. (2000). Distinguishing melt and fluid subduction components in Umnak Volcanics, Aleutian Arc. *Geochem. Geophys. Geosys.*, *1*, 1004. doi:10.1029/1999gc000010

Cooper, L. B., Ruscitto, D. M., Plank, T., Wallace, P. J., Syracuse, E. M., & Manning, C. E. (2012). Global variations in H_2O/Ce: 1. Slab surface temperatures beneath volcanic arcs. *Geochem. Geophys. Geosys.*, *13*, Q03024. doi:10.1029/2011gc003902

Dasgupta, R. (2013). Ingassing, storage, and outgassing of terrestrial carbon through geologic time. *Rev. Mineral. Geochem.*, *75*, 183–229. doi:10.2138/rmg.2013.75.7

Dasgupta, R., & Hirschmann, M. M. (2010). The deep carbon cycle and melting in Earth's interior. *Earth Planet. Sci. Lett.*, *298*(1–2), 1–13. doi:10.1016/j.epsl.2010.06.039

de Leeuw, G.A.M., Hilton, D. R., Fischer, T. P., & Walker, J. A. (2007). The He-CO_2 isotope and relative abundance characteristics of geothermal fluids in El Salvador and Honduras: New constraints on volatile mass balance of the Central American Volcanic Arc. *Earth Planet. Sci. Lett.*, *258*(1–2), 132–146. doi:10.1016/j.epsl.2007.03.028

Dixon, J. E. (1997). Degassing of alkalic basalts. *Am. Mineral.*, *82*(3–4), 368–378.

Duncan, M. S., & Dasgupta, R. (2014). CO_2 solubility and speciation in rhyolitic sediment partial melts at 1.5–3.0 GPa: Implications for carbon flux in subduction zones. *Geochim. Cosmochim. Acta*, *124*, 328–347. doi:10.1016/j.gca.2013.09.026

Duncan, M. S., & Dasgupta, R. (2015). Pressure and temperature dependence of CO_2 solubility in hydrous rhyolitic melt: Implications for carbon transfer to mantle source of volcanic arcs via partial melt of subducting crustal lithologies. *Contrib. Mineral. Petrol.*, *169*(54). doi:10.1007/s00410-015-1144-5

Duncan, M. S., & Dasgupta, R. (2017). Rise of Earth's atmospheric oxygen controlled by efficient subduction of organic carbon. *Nat. Geosci.*, *10*(5), 387. doi:10.1038/Ngeo2939

Duncan, M. S., Dasgupta, R., & Tsuno, K. (2017). Experimental determination of CO_2 content at graphite saturation along a natural basalt-peridotite melt join: Implications for the fate of carbon in terrestrial magma oceans. *Earth Planet. Sci. Lett.*, *466*, 115–128. doi:10.1016/j.epsl.2017.03.008

Eguchi, J., & Dasgupta, R. (2018). A CO_2 solubility model for silicate melts from fluid saturation to graphite or diamond saturation. *Chem. Geol.*, *487*, 23–38. doi:10.1016/j.chemgeo.2018.04.012

Fogel, R. A., & Rutherford, M. J. (1990). The solubility of carbon dioxide in rhyolitic melts: A quantitative FTIR study. *Am. Mineral.*, *75*(11–12), 1311–1326.

Ghiorso, M. S., & Gualda, G.A.R. (2015). An H_2O-CO_2 mixed fluid saturation model compatible with rhyolite-MELTS. *Contrib. Mineral. Petrol.*, *169*(53). doi:10.1007/s00410-015-1141-8

Hayes, J. M., & Waldbauer, J. R. (2006). The carbon cycle and associated redox processes through time. *Philos. Trans. Roy. Soc. B*, *361*(1470). 931–950. doi:10.1098/rstb.2006.1840

Hermann, J., & Green, D. H. (2001). Experimental constraints on high pressure melting in subducted crust. *Earth Planet. Sci. Lett.*, *188*(1–2), 149–168.

Hermann, J., & Spandler, C. J. (2008). Sediment melts at sub-arc depths: An experimental study. *J. Petrol.*, *49*(4), 717–740. doi:10.1093/petrology/egm073.

Hirschmann, M. M., & Dasgupta, R. (2009). The H/C ratios of Earth's near-surface and deep reservoirs, and consequences for deep Earth volatile cycles. *Chem. Geol.*, *262*(1–2), 4–16. doi:10.1016/j.chemgeo.2009.02.008

Johnson, M. C., & Plank, T. (1999). Dehydration and melting experiments constrain the fate of subducted sediments. *Geochem. Geophys. Geosys.*, *1*, 1999GC000014.

Kelemen, P. B., & Manning, C. E. (2015). Reevaluating carbon fluxes in subduction zones, what goes down, mostly comes up. *Proc. Nat. Acad. Sci.*, *112*(30), E3997–E4006. doi:10.1073/pnas.1507889112

Kerrick, D. M., & Connolly, J.A.D. (2001). Metamorphic devolatilization of subducted oceanic metabasalts: Implications for seismicity, arc magmatism and volatile recycling. *Earth Planet. Sci. Lett.*, *189*(1–2), 19–29.

Klimm, K., Blundy, J. D., & Green, T. H. (2008). Trace element partitioning and accessory phase saturation during H_2O-saturated melting of basalt with implications for subduction zone chemical fluxes. *J. Petrol.*, *49*(3), 523–553. doi:10.1093/petrology/egn001

Konschak, A., & Keppler, H. (2014). The speciation of carbon dioxide in silicate melts. *Contrib. Mineral. Petrol.*, *167*(5), 998. doi:10.1007/S00410-014-0998-2

Lange, R. A., & Carmichael, I.S.E. (1987). Densities of Na_2O-K_2O-CaO-MgO-FeO-Fe_2O_3-Al_2O_3-TiO_2-SiO_2 liquids: New measurements and derived partial molar properties. *Geochim. Cosmochim. Acta*, *51*(11), 2931–2946.

Laurie, A., & Stevens, G. (2012). Water-present eclogite melting to produce Earth's early felsic crust. *Chem. Geol.*, *314*, 83–95. doi:10.1016/j.chemgeo.2012.05.001

Mann, U., & Schmidt, M. W. (2015). Melting of pelitic sediments at subarc depths: 1. Flux vs. fluid-absent melting and a parameterization of melt productivity. *Chem. Geol.*, *404*, 150–167. doi:10.1016/j.chemgeo.2015.02.032

Martin, L.A.J., & Hermann, J. (2018). Experimental phase relations in altered oceanic crust: Implications for carbon recycling at subduction zones. *J. Petrol.*, *59*(2), 299–320. doi:10.1093/petrology/egy031

Marty, B. (2012). The origins and concentrations of water, carbon, nitrogen and noble gases on Earth. *Earth Planet. Sci. Lett.*, *313*, 56–66. doi:10.1016/j.epsl.2011.10.040

Morizet, Y., Kohn, S. C., & Brooker, R. A. (2001). Annealing experiments on CO_2-bearing jadeite glass: An insight into the true temperature dependence of CO_2 speciation in silicate melts. *Mineral. Mag.*, *65*(6), 701–707.

Morizet, Y., Paris, M., Gaillard, F., & Scaillet, B. (2014). Carbon dioxide in silica-undersaturated melt: Part I. The effect of mixed alkalis (K and Na) on CO_2 solubility and speciation. *Geochim. Cosmochim. Acta*, *141*, 45–61. doi:10.1016/j.gca.2014.06.014

Mysen, B. O. (1976). Role of volatiles in silicate melts: Solubility of carbon dioxide and water in feldspar, pyroxene, and feldpathoid melts to 30 kb and 1625 °C. *Am. J. Sci.*, *276*(8), 969–996.

Mysen, B. O., Eggler, D. H., Seitz, M. G., & Holloway, J. R. (1976). Carbon dioxide in silicate melts and crystals: 1. Solubility measurements. *Am. J. Sci.*, *276*(4), 455–479.

Mysen, B. O., & Virgo, D. (1980a). The solubility behavior of CO_2 in melts on the join $NaAlSi_3O_8$-$CaAl_2Si_2O_8$-CO_2 at high pressures and temperatures: A Raman spectroscopic study. *Am. Mineral.*, *65*, 1166–1175.

Mysen, B. O., & Virgo, D. (1980b). Solubility mechanisms of carbon dioxide in silicate metals: A Raman spectroscopic study. *Am. Mineral.*, *65*(9–10), 885–899.

Newman, S., Stolper, E. M., & Epstein, S. (1986). Measurement of water in rhyolitic glasses: Calibration of an infrared spectroscopic technique. *Am. Mineral.*, *71*(11–12), 1527–1541.

Ni, H. W., & Keppler, H. (2013). Carbon in silicate melts. *Rev. Mineral. Geochem.*, *75*, 251–287. doi:10.2138/rmg.2013.75.9

Nowak, M., Porbatzki, D., Spickenbom, K., & Diedrich, O. (2003). Carbon dioxide speciation in silicate melts: A restart. *Earth Planet. Sci. Lett.*, *207*(1–4), 131–139. doi:10.1016/S0012-821x(02)01145-7

Plank, T. (2005). Constraints from thorium/lanthanum on sediment recycling at subduction zones and the evolution of the continents. *J. Petrol.*, *46*(5), 921–944. doi:10.1093/petrology/egi005

Poli, S., & Schmidt, M. W. (2002). Petrology of subducted slabs. *Ann. Rev. Earth Planet. Sci.*, *30*, 207–235.

Prouteau, G., Scaillet, B., Pichavant, M., & Maury, R. (2001). Evidence for mantle metasomatism by hydrous silicic melts derived from subducted oceanic crust. *Nature*, *410*(6825), 197–200.

Qian, Q., & Hermann, J. (2013). Partial melting of lower crust at 10–15 kbar: Constraints on adakite and TTG formation. *Contrib. Mineral. Petrol.*, *165*(6), 1195–1224. doi:10.1007/s00410-013-0854-9

Rapp, R. P., & Watson, E. B. (1995). Dehydration melting of metabasalt at 8–32 kbar: Implications for continental growth and crust-mantle recycling. *J. Petrol.*, *36*(4), 891–931.

Saal, A. E., Hauri, E. H., Langmuir, C. H., & Perfit, M. R. (2002). Vapour undersaturation in primitive mid-ocean-ridge basalt and the volatile content of Earth's upper mantle. *Nature*, *419*(6906), 451–455. doi:10.1038/Nature01073

Sano, Y., & Williams, S. N. (1996). Fluxes of mantle and subducted carbon along convergent plate boundaries. *Geophys. Res. Lett.*, *23*(20), 2749–2752.

Schmidt, M. W. (2015). Melting of pelitic sediments at subarc depths: 2. Melt chemistry, viscosities and a parameterization of melt composition. *Chem. Geol.*, *404*, 168–182. doi:10.1016/j.chemgeo.2015.02.013

Schmidt, M. W., Vielzeuf, D., & Auzanneau, E. (2004). Melting and dissolution of subducting crust at high pressures: The key role of white mica. *Earth Planet. Sci. Lett.*, *228*(1–2), 65–84. doi:10.1016/j.epsl.2004.09.020

Sen, C., & Dunn, T. (1994). Dehydration melting of a basaltic composition amphibolite at 1.5 and 2.0 GPa: Implications for the origin of adakites. *Contrib. Mineral. Petrol.*, *117*(4), 394–409. doi:10.1007/Bf00307273

Shaw, A. M., Hilton, D. R., Fischer, T. P., Walker, J. A., & Alvarado, G. E. (2003). Contrasting He-C relationships in Nicaragua and Costa Rica: Insights into C cycling through subduction zones. *Earth Planet. Sci. Lett.*, *214*(3–4), 499–513. doi:10.1016/S0012-821x(03)00401-1

Shimoda, G., Tatsumi, Y., Nohda, S., Ishizaka, K., & Jahn, B. M. (1998). Setouchi high-Mg andesites revisited: Geochemical evidence for melting of subducting sediments. *Earth Planet. Sci. Lett.*, *160*(3–4), 479–492.

Silver, L., & Stolper, E. (1989). Water in albitic glasses. *J. Petrol.*, *30*(3), 667–709.

Silver, L. A. (1988). Water in silicate glasses (Ph. D. thesis). California Institute of Technology, Pasadena, CA.

Skora, S., & Blundy, J. (2010). High-pressure hydrous phase relations of radiolarian clay and implications for the involvement of subducted sediment in arc magmatism. *J. Petrol.*, *51*(11), 2211–2243. doi:10.1093/petrology/egq054

Skora, S., Blundy, J. D., Brooker, R. A., Green, E.C.R., de Hoog, J.C.M., & Connolly, J.A.D. (2015). Hydrous phase relations and trace element partitioning behaviour in calcareous sediments at subduction zone conditions. *J. Petrol.*, *56*(5), 953–980. doi:10.1093/petrology/egv024

Sleep, N. H., & Zahnle, K. (2001). Carbon dioxide cycling and implications for climate on ancient Earth. *J. Geophys. Res.*, *106*(E1), 1373–1399.

Spandler, C., Yaxley, G., Green, D. H., & Scott, D. (2010). Experimental phase and melting relations of metapelite in the upper mantle: Implications for the petrogenesis of intraplate magmas. *Contrib. Mineral. Petrol.*, *160*(4), 569–589. doi:10.1007/s00410-010-0494-2

Stolper, E., Fine, G., Johnson, T., & Newman, S. (1987). Solubility of carbon dioxide in albitic melt. *Am. Mineral.*, *72*(11 12), 1071 1085.

Syracuse, E. M., van Keken, P. E., & Abers, G. A. (2010). The global range of subduction zone thermal models. *Phys. Earth Planet. Inter.*, *183*(1–2), 73–90. doi:10.1016/j.pepi.2010.02.004

Tamic, N., Behrens, H., & Holtz, F. (2001). The solubility of H_2O and CO_2 in rhyolitic melts in equilibrium with a mixed CO_2-H_2O fluid phase. *Chem. Geol.*, *174*(1–3), 333–347.

Thomsen, T. B., & Schmidt, M. W. (2008). Melting of carbonated pelites at 2.5–5.0 GPa, silicate-carbonatite liquid immiscibility, and potassium-carbon metasomatism of the mantle. *Earth Planet. Sci. Lett.*, *267*(1–2), 17–31. doi:10.1016/j.epsl.2007.11.027

Thomson, A. R., Walter, M. J., Kohn, S. C., & Brooker, R. A. (2016). Slab melting as a barrier to deep carbon subduction. *Nature*, *529*(7584), 76. doi:10.1038/nature16174

Tsuno, K., & Dasgupta, R. (2011). Melting phase relation of nominally anhydrous, carbonated pelitic-eclogite at 2.5–3.0 GPa and deep cycling of sedimentary carbon. *Contrib. Mineral. Petrol.*, *161*, 743–763. doi:10.1007/s00410-010-0560-9

Tsuno, K., & Dasgupta, R. (2012). The effect of carbonates on near-solidus melting of pelite at 3 GPa: Relative efficiency of H_2O and CO_2 subduction. *Earth Planet. Sci. Lett.*, *319*, 185–196. doi:10.1016/j.epsl.2011.12.007

Tsuno, K., Dasgupta, R., Danielson, L., & Righter, K. (2012). Flux of carbonate melt from deeply subducted pelitic sediments: Geophysical and geochemical implications for the source of Central American volcanic arc. *Geophys. Res. Lett.*, *39*, L16307. doi:10.1029/2012gl052606

Vetere, F., Holtz, F., Behrens, H., Botcharnikov, R. E., & Fanara, S. (2014). The effect of alkalis and polymerization on the solubility of H_2O and CO_2 in alkali-rich silicate melts. *Contrib. Mineral. Petrol.*, *167*(5). doi:10.1007/S00410-014-1014-6

Wallace, P. J. (2005). Volatiles in subduction zone magmas: Concentrations and fluxes based on melt inclusion and volcanic gas data. *J. Volcanol. Geotherm. Res.*, *140*(1–3), 217–240. doi:10.1016/j.jvolgeores.2004.07.023

Wolff, P. E., Koepke, J., & Feig, S. T. (2013). The reaction mechanism of fluid-induced partial melting of gabbro in the oceanic crust. *Eur. J. Mineral.*, *25*(3), 279–298. doi:10.1127/0935-1221/2013/0025-2314

Zhang, Y. X., Stolper, E. M., & Ihinger, P. D. (1995). Kinetics of the reaction $H_2O + O = 2OH$ in rhyolitic and albitic glasses: Preliminary results. *Am. Mineral.*, *80*(5–6), 593–612.

18

Hydrous Carbonatitic Liquids Drive CO_2 Recycling From Subducted Marls and Limestones

Erwin Schettino[1,2] and Stefano Poli[1]

ABSTRACT

Pelagic limestones are subducted in a variety of subduction zones worldwide. Despite the geochemical relevance of systems enriched in $CaCO_3$, previous experimental investigations mostly focused on carbonated pelites, with low Ca/(Ca+Mg+Fe) ratio. We present the compositions and the formation conditions of liquids in the model system CaO-Al_2O_3-SiO_2-H_2O-CO_2 (CASHC), building on phase relationships in the subsystems CHC and CSHC, where a second critical endpoint was suggested at temperatures as low as 515 °C, and 3.2 GPa. Multianvil experiments were performed at 4.2 and 6.0 GPa on five bulk compositions at variable Ca/Si/Al ratios. H_2O contents vary from 5.6 to 21 wt%. Aragonite + kyanite + vapor and minor lawsonite form at 700 °C, replaced by zoisite/grossular at 800 °C. Between 850 °C and 950 °C, a complex sequence of textural features is observed upon quenching of a single volatile-rich liquid phase formed at run conditions. Precipitates include dendritic $CaCO_3$, silicate glass, and Al-rich whiskers. The bulk composition of such hydrous carbonatitic liquids is retrieved by image analysis on X-ray maps, showing Ca/Si ratio increasing with pressure and temperature. Hydrous Ca-carbonatitic liquids are efficient media for scavenging volatiles from subducted crustal material and for metasomatizing the mantle wedge.

18.1. INTRODUCTION

At subduction zones, mass and heat are exchanged between different Earth reservoirs. On subduction, devolatilization of the altered oceanic crust and of serpentinized upper mantle drive the formation of mobile phases, leading to the generation of refertilized domains in the mantle wedge, capable of feeding the magmatic activity at arcs. Among subducted crustal lithologies, carbonate sedimentary rocks are heterogeneously distributed in the oceans. In modern Earth, pelagic oozes, largely constituted of calcitic nannofossils and foraminifera, ubiquitously accumulate on the seafloor above the carbonate compensation depth, except where upwelling conditions prevail and diatoms-radiolarians form siliceous oozes. Burial and plate drift are then responsible for sediment redistribution and progressive building of sedimentary covers characterized by lithological units, which are expected to maintain their compositional identity on subduction processes.

While decomposition of hydrous minerals in the slab has been thoroughly modeled to account for water-budget being recycled in arc systems (Schmidt & Poli, 2014), estimates of the magnitude of carbonate masses fed in the subduction system, and the relative efficiency of decarbonation, carbonate dissolution, and melting, driving carbon recycling, are still affected by very large uncertainties (Kelemen & Manning, 2015). Nonetheless, a global assessment of carbon dioxide and total sulphur abundances (CO_2/S_T ratio) in volcanic arc gases provides evidence for correlations between subduction of carbonate rocks and CO_2 outgassing (Aiuppa et al., 2017), thereby

[1] *Dipartimento di Scienze della Terra "Ardito Desio", Università degli Studi di Milano, Milano, Italy*

[2] *Instituto Andaluz de Ciencias de la Tierra, CSIC-Universidad de Granada, Granada, Spain*

Carbon in Earth's Interior, Geophysical Monograph 249, First Edition. Edited by Craig E. Manning, Jung-Fu Lin, and Wendy L. Mao.

suggesting the existence of processes and media capable to fractionate carbon from the uppermost geochemical reservoir of the slab. The occurrence of aragonite, and dolomite-bearing mantle xenoliths from Andean volcanoes, SW Colombia, is a direct evidence of the migration of carbon-rich metasomatic agents able to mobilize CO_2 in the upper mantle (Ferri et al., 2017). Entrainment of calcareous rocks of pelagic origin in the orogenic cycle is further documented by the occurrence of calc-schists and impure marbles, forming large tectono-stratigraphic units in mountain belts (e.g. Alps).

Despite the geochemical relevance of pelagic limestones and marls as a major carbon reservoir, experimental modeling has largely underevaluated the role of chemical systems enriched in $CaCO_3$ in sedimentary materials, and a global weighted average composition of subducted sediments (GLOSS; Plank, 2014) has been extensively used as the proxy for sedimentary contribution to the subduction factory. Experimental studies have aimed at reconstructing phase transformations on carbonated metapelites close to the composition of GLOSS (Brey et al., 2015; Grassi & Schmidt, 2011; Skora et al., 2015; Thomsen & Schmidt, 2008; Tsuno & Dasgupta, 2012). The liquid phase found on the solidus for all such compositions is silicate at P < 5.5 GPa, whereas carbonatitic liquids form at higher pressures (Grassi & Schmidt, 2011) or at higher temperatures, on the liquidus surface (Thomsen & Schmidt, 2008; see also Figure 11 in Schmidt & Poli, 2014). At subarc conditions, carbonatitic liquids, as efficient media for carbon transfer, are therefore restricted to relatively high temperatures (>1020 °C, Figure 18.1, unless bulk composition is enriched in Ca; Poli, 2015, 2016), and were therefore considered to be uncommon products at mature subduction zones (Syracuse et al., 2010).

Although melting of $CaCO_3$ at high pressures occurs at very high temperatures (T > 1500 °C at P > 2 GPa; Irving & Wyllie, 1975; Zhao et al., 2019, and references therein), it has long been known that water is responsible for a profound depression of the liquidus surface. Paterson (1958) documented melting of calcite in the presence of water and carbon dioxide at 930 °C, and 50 bar total vapor pressure. Wyllie and Tuttle (1959, 1960) and Wyllie and Boettcher (1969) provided the frame of reactions for the system $CaO-CO_2-H_2O$ and experimentally demonstrated that eutectic melting occurs at temperatures down to 524 °C at 4 GPa from $CaCO_3 + Ca(OH)_2$ + vapor. Such an extreme influence of H_2O on melting of calcium carbonate (more than a thousand degree difference between dry and wet melting) has been recently confirmed and characterized by Foustoukos and Mysen (2015). H_2O solubility is extremely high in carbonate melts even at very low pressure; Keppler (2003) measured H_2O contents of 14 wt.% in Ca, Na, Mg carbonate melt at 900 °C, 0.22 GPa, then confirmed by qualitative estimates in

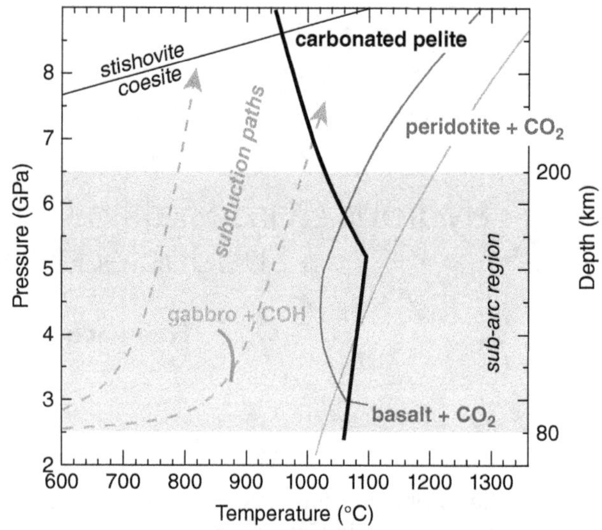

Figure 18.1 Experimentally derived reactions marking the appearance of carbonatitic liquids. The solidus on carbonated peridotite (green) is from Dasgupta & Hirschmann (2006); the solidus on carbonated basalt (red) is from Dasgupta et al. (2004). Carbonatitic liquid in carbonated pelites (brown) appears on the solidus at pressures higher than 5 GPa, whereas it exsolves on the liquidus surface at lower pressures (Thomsen & Schmidt, 2008). Hydrous carbonatitic liquids can be generated from carbonated anorthite-rich gabbros (Poli, 2015) at temperatures attained in subduction zones along slab surface (Syracuse et al., 2010). See electronic version for color representation of the figures in this book.

Foustoukos and Mysen (2015). H_2O concentrations in carbonatitic melts are four times larger than in silicate melts at conditions of diamond growth (Sokol et al., 2013; Taylor et al., 2016).

The subsolidus phase relations in the system $CaO-CO_2-H_2O$ are characterized by carbonate dissolution in the vapor phase (for the sake of clarity, we will use here the nomenclature used by Boettcher & Wyllie, 1969) marked by the increasing solubility of $CaCO_3$ in water, with increasing pressure (Caciagli & Manning, 2003; Manning et al., 2013). There is an ample enumeration of natural records pointing to carbonate dissolution, driven by infiltrating fluids released by subducting slabs, as a mechanism relevant for CO_2 recycling at the top of the subducting slab, both at blueschist (Ague & Nicolescu, 2014) and eclogite (Frezzotti et al., 2011) facies. Carbon solubility in fluids has been shown to extend to more than 1 wt.% at pressures > 4 GPa and temperatures of 800 °C. Consequently, as the solubility of solids increases in the vapor, and that of H_2O increases in the liquid, the miscibility gap tends to shrink with increasing pressure, and a "second critical end-point" marks the appearance of a region in the phase diagram where there is compositional continuity between liquid and vapor. It is worth pointing out that such continuity is restricted to the cotectic region of the phase diagram, and liquid + vapor still coexist as discrete phases at different compositional domains (see Figure 5 in Wyllie & Boettcher, 1969, and Figure S1 in Poli, 2015). Supercritical behavior has been proposed by Wyllie and Boettcher (1969), and Boettcher and Wyllie (1969) in the systems CaO-CO_2-H_2O and CaO-SiO_2-CO_2-H_2O, respectively.

In this study, we will test if the $CaCO_3$-rich component of subducted sediments, here modeled in the system CaO-Al_2O_3-SiO_2-H_2O-CO_2 (CASHC), is keen to promote the generation of carbonatitic liquids at sub-arc conditions, possibly at *super-critical* conditions; the mobilization of such liquids would represent an important pathway for carbon transfer from the slabs to the mantle wedge.

18.2. MODELING IMPURE MARBLES: THE SYSTEM CaO-Al_2O_3-SiO_2-H_2O-CO_2

Despite their limited thickness in comparison to the underlying altered crystalline basement, pelagic sediments concentrate large masses of carbon. As examples, DSDP site 495 on Cocos Plate, Middle America (Aubouin et al., 1982), site 504 on Nazca Plate (Cann et al., 1983), sites 212 and 260 on the Indo-Australian Plate (Heirtzler et al., 1974; von der Borch et al., 1974), and site 1124 on Pacific Plate (Carter et al., 1999) display stratigraphic columns where calcitic nannofossil oozes and chalks may constitute up to more than 90% of lithological units, with typical thicknesses of individual units in the order of tens or a few hundred meters, interbedded with minor clays, zeolites, and variable proportions of siliceous oozes. $CaCO_3$ and SiO_2 are therefore the major chemical components, where clays and zeolites introduce significant amount of Al_2O_3.

Impure marbles and calc-schists outcropping in collisional orogens constitute proxies of pelagic sediments, exhumed after high-pressure metamorphism and deformation. A large dataset of the compositions of “Schistes lustrés” and “Bündnerschiefer,” representing marine sediments accumulated in alpine Tethys Ocean, is available from Garofalo (2012) and da Mommio (2018). Figure 18.2 represents the composition of marbles and calc-schists from Alpine ophiolites, and of carbonate-rich sediments from Plank and Langmuir (1998) compared to the bulk compositions adopted in experimental studies devoted to reconstruct the fate of subducted sediments at high pressure. Figure 18.2 is a chemographic projection built on the basis of phase assemblages predicted by thermodynamic calculations. Pseudosections (not shown) were calculated following the approach of Connolly (1990), using the internally consistent thermodynamic dataset and the CORK equation of state for fluids of Holland and Powell (2011), implemented by the carbonate solution model by Franzolin et al. (2011) and aragonite properties by Zhao et al. (2019). Calculated

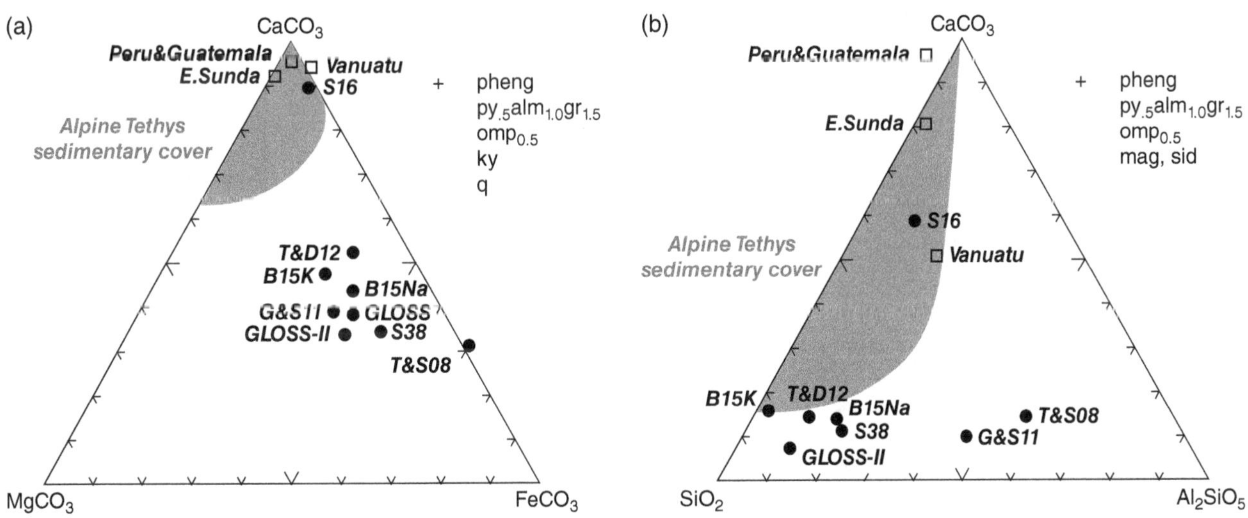

Figure 18.2 Chemographic representation of carbonated sedimentary materials in the system $CaCO_3$-$MgCO_3$-$FeCO_3$ (a) and $CaCO_3$-SiO_2-Al_2SiO_5 (b), projected from garnet, omphacite, phengite, and quartz + kyanite in (a), and magnesite + siderite in (b). The green field stands for the compositional variation of impure marbles, carbonate-silicate rocks, and calc-schists from the ophiolitic suites in the Alps (Garofalo, 2012; da Mommio, 2018). Squares are the average compositions of calcite-rich sedimentary units being subducted at Sunda, Vanuatu, Peru, and Guatemala (Plank & Langmuir, 1998). Filled circles show the bulk compositions of starting materials used in experimental studies: T&S08 = Thomsen & Schmidt (2008); G&S11 = Grassi & Schmidt (2011); T&D11 = Tsuno & Dasgupta (2012); B15K and B15Na = Brey et al. (2015), K- and Na-enriched, respectively; S16 and S38 = Skora et al. (2015), bulk composition 144-16 and 144-38, respectively. GLOSS and GLOSS II from Plank (2014). See electronic version for color representation of the figures in this book.

phase proportions are dominated by aragonite, ranging from 40 wt.%, in calc-schists, and 70 wt.% in impure marbles. The abundance of Na_2O and K_2O controls the amount of omphacite$_{s.s.}$ and white mica, respectively, over a wide range of P-T conditions. The main hydrous minerals that occur are lawsonite and zoisite, their breakdown responsible to form garnet + kyanite + coesite + H_2O. At pressures higher than ~4 GPa, aragonite coexists with minor magnesite. On the basis of the general features of calculated pseudosections, Figure 18.2 is therefore constructed projecting from garnet ($Mg_{0.5}Fe_{1.0}Ca_{1.5}Al_2Si_3O_{12}$), omphacite ($Na_{0.5}Ca_{0.5}[Mg,Fe]_{0.5}Al_{0.5}Si_2O_6$), and phengitic mica ($K[M,Fe]_{0.3}Al_{1.7}Al_{0.7}Si_{3.3}[OH]_2O_{12}$), plus Al_2SiO_5 and silica in Figure 18.2a, and $(Mg,Fe)CO_3$ in Figure 18.2b. Figure 18.2 shows that bulk composition used in the experiments of Thomsen and Schmidt (2008), Tsuno and Dasgupta (2012), Grassi and Schmidt (2011), Skora et al. (2015) (bulk 144-38), and Brey et al. (2015) are characterized by low Ca/(Ca+Mg+Fe) ratio and high $SiO_2+Al_2O_3$, similarly to the model GLOSS (Plank & Langmuir, 1998). Only bulk composition 144-16 in Skora et al. (2015) is representative of pelagic sediments discussed here.

Although the composition of garnet, omphacite, mica, and (Mg,Fe)-carbonate will vary at different P-T, a common feature of this family of marine sediments is the predominance of $CaCO_3$, and a more or less pronounced saturation in silica and Al_2SiO_5. Phase relationships explored in this experimental work are then modeled in the five-components system, $CaO-Al_2O_3-SiO_2-CO_2-H_2O$, on five different bulk compositions with three variable Al_2O_3/SiO_2 ratios, in order to stabilize, at high pressure, aragonite + coesite + an alumino-silicate phase such as garnet, zoisite or kyanite, depending on pressure and temperature conditions. Silica saturation and addition of alumina are expected to prevent the crystallization of dellaite and portlandite, experimentally recorded on the eutectic of the system $CaO-SiO_2-CO_2-H_2O$ (Boettcher & Wyllie, 1969). The amount of volatile components were defined over different $CO_2:H_2O$ ratios, so that phase relationships between a low-density vapor, solute-rich vapor, volatile-rich liquid, and, possibly, supercritical liquid were explored, at different P-T conditions, over a wide compositional range.

18.3. EXPERIMENTAL AND ANALYTICAL METHODS

Experiments were performed with mixtures of crystalline powders plus distilled water, when required. Reagent grade $CaCO_3$ calcite, amorphous silica, $Al(OH)_3$, and synthetic zoisite were used as starting materials. The $Al(OH)_3$ was employed as a source of alumina, and of H_2O for bulk 1 and 1b (Table 18.1), in order to favor reaction kinetics at relatively low temperature. Bulk compositions were constructed to lie close to the compositional pseudosection $CaCO_3$–H_2O as projected from Al_2SiO_5 and SiO_2, and were therefore at a constant $CaO:CO_2$ ratio but variable $CaCO_3:H_2O$ ratio (molar fractions of components in Table 18.1). Bulk compositions were then obtained by mixtures of calcite + $Al(OH)_3$ + amorphous silica (bulk 1 and 1b), calcite + zoisite + water (bulk 2), and calcite + zoisite + silica + water (bulk 3 and bulk 3b). Capsules were made of Au tubing, 3.0 mm outer diameter, and loaded with ~7 mg of solids and up to 1.5 microliter of water, if any; accuracy in H_2O addition by water is in the order of 7%, whereas <1% when $Al(OH)_3$ is used. All capsules were welded after loading the starting materials, previously permanently stored in a vacuum oven at 110 °C. Distilled water was added at the bottom of the run charge by using a high-precision microsyringe. In order to prevent water loss, capsules were held into a frozen steel support during welding.

Table 18.1 Bulk composition of starting materials.

	1	1b	2	3	3b
CaO	40.02	35.08	38.46	35.70	42.15
Al_2O_3	13.00	10.41	4.81	4.47	3.91
SiO_2	8.58	21.38	6.08	12.82	16.09
CO_2	31.41	27.53	27.17	25.22	30.64
H_2O	7.00	5.60	23.48	21.80	7.22
Total	100.00	100.00	100.00	100.00	100.00
X_{CaO}	0.274	0.257	0.182	0.162	0.290
X_{CO2}	0.274	0.257	0.166	0.146	0.268
X_{H2O}	0.299	0.256	0.606	0.618	0.309
$X_{Al2SiO5}$	0.147	0.126	0.033	0.033	0.045
X_{SiO2}	0.006	0.104	0.013	0.041	0.088

Note. Oxides in wt%; components in molar fractions: $X_{CaO} = CaO/(CaO+CO_2+H_2O+Al_2SiO_5+SiO_2)$.

Experiments were performed using a Walker-type multianvil apparatus on a 1000 ton press at the Dipartimento di Scienze della Terra, Università degli Studi di Milano. WC cubes had edge lengths of 32 mm and truncation edge lengths of 17 mm. Assemblies were composed by pyrophyllite gaskets, a prefabricated MgO-octahedron (containing 5 wt% Cr_2O_3) with 25 mm edge length, and cylindrical graphite heater. A MgO sleeve was placed between capsules and graphite. The axial thermocouple (Pt- $Pt_{90}Rh_{10}$, S-type) was placed in direct contact with the capsules, and a mullite thermocouple ceramic was used. Thermal gradients over the length of the capsule in this relatively large assembly are less than 20 °C–30 °C at experimental temperatures of 700 °C–1000 °C. Pressure calibration for the 25 M assembly was performed at 1000 °C by using the coesite-stishovite and $CaGeO_3$ garnet-perovskite phase transitions, occurring respectively at 8.7 GPa (Zhang et al., 1996) and 6.1 GPa (Susaki et al., 1985), with uncertainties of about ±3%.

Our study assumes that redox reactions are negligible, on the basis of relatively low permeability of gold to hydrogen (Chou, 1986), and of absence of precipitation of graphite, as verified by XRPD on selected run charges. Therefore, vapor composition is expected to adjust internally in order to minimize the free energy of the reactive system (see also Poli & Schmidt, 1998).

At the end of the experiments, recovered capsules were longitudinally embedded in epoxy resins. Sample mounts were ground by using ethanol, instead of water, in order to avoid dissolution of the quench phases. After exposing the very first portion of the run charge, the capsules were impregnated under vacuum with low-viscosity epoxy resin. The samples were later polished with a 1 μm diamond paste, and finally carbon coated for scanning electron microprobe (SEM) analysis. Textural and compositional characterizations were performed using a WDS-equipped JEOL JXA8200 electron microprobe analyzer (EMPA). Analytical conditions were set to 15 kV acceleration voltage, and 5 nA current for a less than ~1 μm spot-size beam. Defocused beam was used whenever possible. X-ray elemental maps on experimental products were acquired with a 0.3 μm step-size and 20 ms time acquisition for each element. Standards used are anorthite USNM137041 for aluminum, and wollastonite for calcium and silicon.

18.4. RESULTS

18.4.1. Microstructural Analysis

A total of 13 successful experiments were performed between 700 °C and 950 °C at two pressures, 4.2 and 6 GPa. Run conditions, starting materials, duration of the experiments, mineralogical assemblages, and microstructural features of quench products are reported in Table 18.2. The attainment of textural equilibrium at run conditions was recognized by the development of polygonal grain boundaries, and 120° triple junctions between the solid phases.

The existence of a mobile, volatile-rich, phase at the P-T conditions of the experiments was determined based on microstructural criteria. The occurrence of vapor and solute-rich vapor are expected to produce a diffuse porosity, provided the relatively high proportions of H_2O available. Carbonatitic liquids are characterized by low dihedral angle (<30°; Minarik & Watson, 1995; Watson

Table 18.2 Summary of the experiments and of the run products.

Run	Bulk	P (GPa)	T (°C)	t (h)	Starting Materials	Min. Assemblage	Quench Products
SPCC1	1	4.2	700	81	cc + $Al(OH)_3$ + si	ky + law + ar	—
SPCC2	1	4.2	800	72	cc + $Al(OH)_3$ + si	ky + ar	cc-drop si-drop al-whisk
SPCC4	1	4.2	850	62	cc + $Al(OH)_3$ + si	cor + zo (ky) + ar	cc-dend al-whisk
SPCC5	1b	4.2	850	110	cc + $Al(OH)_3$ + si	coe + zo (ky) + ar	cc-film si-drop al-whisk
SPCC9	3	4.2	850	122	cc + zo + si + w	coe + gr (zo) + ar	cc-dend si-drop al-whisk
SPCC14	3b	4.2	850	120	cc + zo + si + w	coe + gr (zo) + ar	glass cc-drop
SPCC3	1	4.2	900	83	cc + $Al(OH)_3$ + si	cor + zo (ky) + ar	cc-dend si-drop al-whisk
SPCC6	1b	4.2	900	49	cc + $Al(OH)_3$ + si	coe + zo + ar	cc-dend si-drop al-whisk
SPCC7	2	4.2	900	120	cc + zo + w	cor + gr (zo) + ar	cc-dend si-drop al-whisk
SPCC8	3	4.2	900	79	cc + zo + si + w	gr (zo) + ar	cc-dend glass
SPCC13	3b	4.2	900	113	cc + zo + si + w	coe + gr + ar	cc-dend si-drop glass
SPCC10	3	6	900	118	cc + zo + si + w	coe + gr (zo) + ar	cc-dend si-drop al-whisk
SPCC11	3	6	950	89	cc + zo + si + w	coe + gr (zo) + ar	cc-dend si-glob glass

Note. Phase abbreviations: cc = calcite; si = amorphous silica; zo = zoisite; w = water; ky = kyanite; law = lawsonite; ar = aragonite; gr = grossular; coe = coesite; cc-drop = droplets of $CaCO_3$; si-drop = droplets of SiO_2; al-whisk = Al-Si-rich whiskers; cc-dend = dendrites of $CaCO_3$.

et al., 1990), and they are expected to fractionate at the hot portion of the capsule, by Soret thermal diffusion, related to thermal gradient in the multianvil assembly. Carbonatitic liquids cannot be quenched to a glass, because of extremely low viscosities (Genge et al., 1995; Kono et al., 2014), and melting of the system can be inferred by the development of crystal morphologies that are indicative of textural disequilibrium due to rapid cooling rates, such as tabular, skeletal, branching, chain, feather-like, dendritic, spherulitic textures (at increasing cooling rates, or undercooling steps; Donaldson, 1976; Lofgren, 1974). Spherulites, and spherical droplets ("fish eggs" in Skora et al., 2015), can be the result of precipitation of amorphous solid material exsolved from the hydrous vapor into insoluble colloidal particles.

In the occurrence of coexisting vapor and liquid at run conditions, vapor bubbles are expected to coalesce into very few, large, subspherical bubbles, again as a result of low viscosity of the liquid. The presence of diffuse microvesicles within quench material can be attributed to the exsolution of volatile components during cooling, from a volatile-rich liquid present at high pressure/high temperature (e.g. see Figure 2 in Thomsen & Schmidt, 2008).

At 700 °C, 4.2 GPa, in bulk composition 1, the mineralogical assemblage is defined by aragonite + lawsonite + kyanite. Aragonite is organized in an equigranular polygonal texture, characterized by 120° triple junctions, while kyanite occurs in aggregates of tabular crystals. A diffuse interstitial porosity in the capsule testifies to the presence of a vapor phase coexisting with the solid assemblage. At 800 °C, 4.2 GPa, bulk 1, pores contain spherical droplets, whiskers, and spherulites of radially arranged fibers. All such disequilibrium textures are attained during rapid crystal growth for large steps of undercooling, therefore testifying to the presence of a solute-rich vapor at run conditions. Spherical droplets are both silicatic (Figure 18.3a) and calcitic (Figure 18.3b), while whiskers are relatively alumina rich (see section 18.4.2). Interestingly, silica droplets occur at the tip end of whiskers, suggesting the former precipitates late in the sequence. The solid assemblage (Figure 18.4a) shows that lawsonite is no longer present, and the assemblage aragonite + zoisite + kyanite is stable. Phase assemblages shown in Figure 18.4, calculated from thermodynamic data, should be regarded as schematic because the projection assumes a free binary vapor, for simplicity.

At 850 °C and 4.2 GPa (Figure 18.3 c, d, and e), run products display a larger proportion of silicate and carbonate material precipitated on quenching, suggesting that the vapor phase becomes increasingly solute enriched with increasing temperature. At 850 °C, bulk compositions 1, 1b, and 3, we observe the precipitation of dendritic carbonates, of intergranular films composed of $CaCO_3$ between aragonite grain boundaries, and of $CaCO_3$ whiskery overgrowths. Precipitates also include Al-enriched whiskers and a large proportion of silica droplets and spherulites. Since the observed textures do not provide any evidence for a biphasic coexistence of vapor and melt, the transition from a low-density to a solute-rich liquid should be regarded as the result of the progressive increase in solubility of both carbonates and silicates in the "fluid" phase, with increasing temperature. In the frame of a continuous transition from vapor to liquid, the microstructural criterion used to highlight the appearance of a liquid-like phase is based on the assumption that crystal habits (from bladed, to chain, to feathery, see Donaldson 1976) develop when the mobile phase, quenched at rates typical of multianvil assemblies (ΔT ~300 °C–400 °C/sec), is characterized by a framework favoring growth. Dendrites of calcite were observed to form on quenching hydrous $CaCO_3$ melt (Paterson, 1958). In bulk composition 1, the mobile phase coexists with a solid assemblage defined by aragonite + zoisite + corundum (Figure 18.4b) organized in polygonal texture, while kyanite only appears as inclusions in zoisite crystals. In bulk composition 1b (derived from silica addition), the solid assemblage is aragonite + zoisite + coesite. In bulk composition 3, aragonite develops polygonal textures and planar grain boundaries along with grossular and coesite, while zoisite only occurs as inclusions in poikilitic grossular. In bulk 3b, characterized by a low H_2O/SiO_2 ratio, a silicate glass forms. Subspherical blebs of $CaCO_3$ precipitate embedded in the silicate glass (Figure 18.3e), suggesting that immiscibility between a silicate and carbonatitic liquid might have occurred at run conditions. However, immiscibility in alkali free systems is questionable, and it will not be discussed further in this work.

At 900 °C and 4.2 GPa, run products in all experiments are characterized by the presence of a *meniscus* at the top of the capsule marked by a sharp boundary between a complex intergrowth of dendrites, droplets, spherulites, and glass, and a polygonal aggregate of coarse grains (Figure 18.3 f, g). This texture is interpreted as evidence of a hydrous carbonatitic liquid segregating at the hot top of the capsule. Large, spherical bubbles were never observed, suggesting that volatiles are entirely dissolved in the liquid and no vapor coexists with such liquid at run conditions. In bulk composition 1, the solid assemblage is aragonite + zoisite + corundum (Figure 18.4c), while kyanite occurs as inclusions in zoisite. A large proportion of calcium carbonates precipitated in a dendritic texture, characterized by rhombohedral intergrowths of acicular and chain-like crystals of $CaCO_3$ (Figure 18.3 g, h, i), with Al-rich whiskers radially arranged to form spherulite-like morphologies. The outer shell of the spherules is built of silica droplets precipitated at the tips of the whiskers (Figure 18.3h). In bulk composition 2, solid assemblage includes aragonite + grossular + corundum, and zoisite occurs as inclusions in poikilitic grossular. In bulk 3 (run SPCC8), the solid assemblage is defined by aragonite + grossular. Despite the large fraction of H_2O in the starting material for bulk 3 (~21 wt.%), there is no textural evidence for a vapor phase coexisting with the liquid at run conditions. On the contrary, glass is

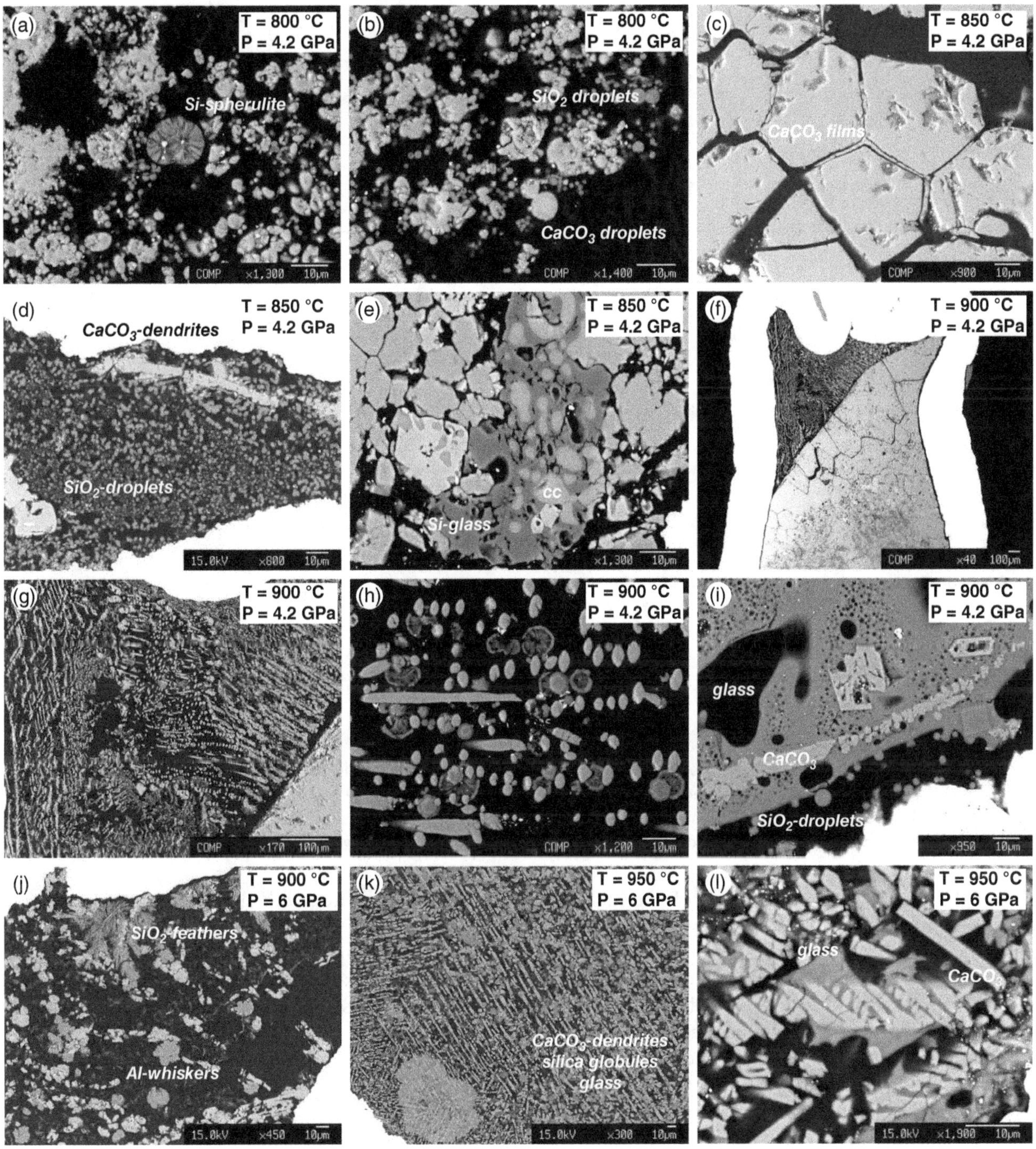

Figure 18.3 Back-scattered electron images on representatives run products: (a) and (b) 4.2 GPa, 800 °C, bulk 1 run SPCC2 (see Table 18.2); (c) 4.2 GPa, 850 °C, bulk 1b, run SPCC5; (d) 4.2 GPa, 850 °C, bulk 3, run SPCC9; (e) 4.2 GPa, 850 °C, bulk 3b, run SPCC14; (f), (g), and (h) 4.2 GPa, 900 °C, bulk 1, run SPCC3; (i) 4.2 GPa, 900 °C, bulk 3b, run SPCC13; (j) 6 GPa, 900 °C, bulk 3, run SPCC 10; (k) and (l) 6 GPa, 950 °C, bulk 3, run SPCC11. See electronic version for color representation of the figures in this book.

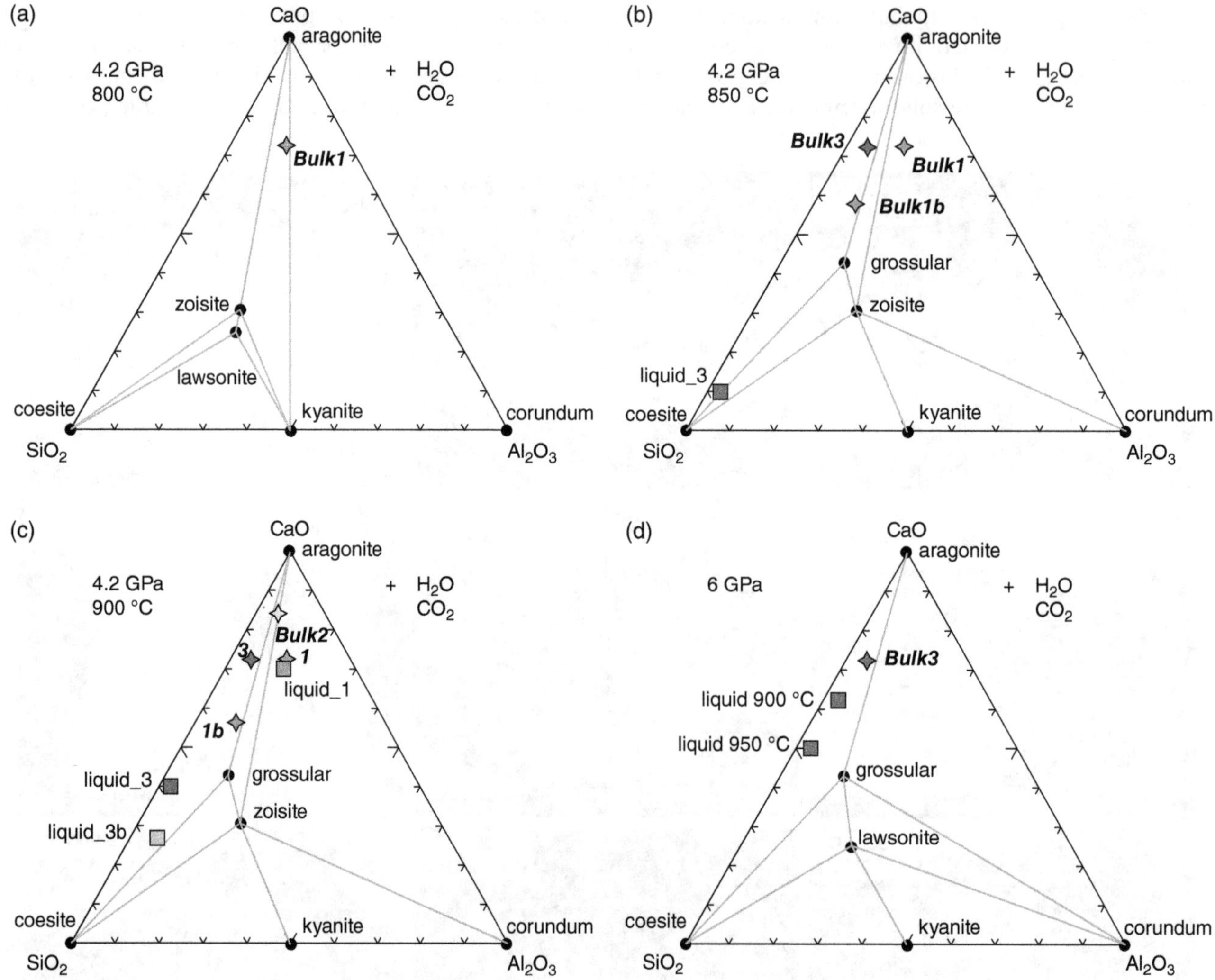

Figure 18.4 Isobaric and isothermal composition diagrams for the system CaO-Al_2O_3-SiO_2-CO_2-H_2O, projected from H_2O and CO_2. Phase assemblages are calculated using thermodynamic data in Holland & Powell (2011) and aragonite properties in Zhao et al. (2019). Diamonds are the bulk compositions investigated here. Note that bulk 3 and 3b differ by H_2O content only. Squares refer to the compositions estimated for the liquid reconstructed from image analysis. See electronic version for color representation of the figures in this book.

recovered, most probably as a result of the high silica proportion in bulk 3. Glass is highly fragmented and vesiculated, as a result of the exsolution of volatile components during quench. Run products on a bulk 3b, with lower H_2O contents (~7 wt. %), include aragonite, grossular, and coesite. Again, the liquid precipitates dendritic carbonates scattered on a glassy, fragmented groundmass characterized by diffuse, small vesicles (Figure 18.3i).

A set of experiments was performed at 6 GPa, at 900 °C and 950 °C, on bulk composition 3. The solid assemblage in this range of P-T is aragonite + grossular + coesite, with inclusions of relic zoisite in poikilitic grossular. At 900 °C, quench products are characterized by an intergrowth of feathery silicates, dendritic carbonate, Al-Si rich spherulites and whiskers (Figure 18.3j). At 950 °C, the segregated liquid precipitates a larger amount of dendritic carbonates, with elongated skeletal and chain-type morphologies, arranged in a rhombohedral intergrowth. Fragmented glass is observed (Figure 18.3 k and l), and poikilitic silica globules include the other components, thus completing the precipitation sequence.

18.4.2. Chemistry of Precipitates and Composition of the Liquid

Since individual solid particles precipitated on quenching are very fine grained, electron probe microanalyses provide a semi-quantitative determination of their chemistry. Therefore, totals reported in Table 18.3 should not be considered entirely compensated by H_2O and CO_2, probably present in all precipitates. The composition of the vapor phase after quench unlikely reflects the H_2O/CO_2

Table 18.3 Composition of quench products (in wt.%).

Run	SPCC3		SPCC8		SPCC9		SPCC10		SPCC11		SPCC13	
P (GPa)	4.2		4.2		4.2		6		6		4.2	
T (°C)	900		900		850		900		950		900	
Bulk	1		3		3		3		3		3b	
	mean	±σ	*mean*	±σ	*mean*	±σ	*mean*	±σ	*mean*	±σ	*mean*	±σ
Droplets												
CaO	0.63	0.13	-	-	0.85*	0.19*	0.31	0.05	3.1	1.79	0.77	0.11
Al_2O_3	2.51	1.68	-	-	2.38*	0.18*	0.98	0.37	3.43	0.66	1.85	0.75
SiO_2	64.06	11.80	-	-	93.32*	3.79*	91.13	3.58	86.97	2.79	88.67	6.58
Total	67.47	13.27	-	-	96.55*	4.11*	92.42	3.61	93.49	1.64	91.29	7.43
X_{Ca}	0.01		-		0.01*		0.00		0.04		0.01	
X_{Si}	0.95		-		0.96*		0.98		0.92		0.97	
X_{Al}	0.04		-		0.03*		0.01		0.04		0.02	
Whiskers												
CaO	0.69	0.47	-	-	-	-	0.32	0.07	-	-	-	-
Al_2O_3	42.96	21.99	-	-	-	-	12.95	4.59	-	-	-	-
SiO_2	38.27	13.88	-	-	-	-	27.56	9.80	-	-	-	-
Total	83.12	9.95	-	-	-	-	40.83	14.62	-	-	-	-
X_{Ca}	0.01		-				0.01		-		-	
X_{Si}	0.45		-		-		0.64		-		-	
X_{Al}	0.54		-		-		0.35		-		-	
Glass												
CaO	-	-	10.85	0.04	-	-	-	-	11.21	4.78	6.42	1.40
Al_2O_3	-	-	2.49	0.14	-	-	-	-	7.24	3.52	4.74	1.38
SiO_2	-	-	59.19	2.81	-	-	-	-	49.01	8.00	61.35	10.02
Total	-	-	72.52	2.96	-	-	-	-	67.46	6.62	72.72	10.11
X_{Ca}	-		0.16		-		-		0.17		0.10	
X_{Si}	-		0.80		-		-		0.70		0.83	
X_{Al}	-		0.04		-		-		0.12		0.08	

Note. X_{Ca} = Ca/(Ca+Si+Al); X_{Si} = Si/(Ca+Si+Al); X_{Al} = Al/(Ca+Si+Al).

ratio in the mobile phase at run conditions. Equation of states for C-O-H fluids by traditional thermodynamic modeling using mixing properties of H_2O and nonpolar species (e.g. Holland & Powell, 1991) predict low CO_2/(H_2O+CO_2) (<0.9) up to 900 °C (see Figure 7 in Poli & Schmidt, 1998). However, complexities in the speciation of complex fluids dissolving both silicates and carbonates (Tumiati et al., 2017; Facq et al., 2016) hinder a reliable estimate of the speciation of volatile species in the investigated liquids at run conditions. Estimates of H_2O-CO_2 abundance in the mobile phase by mass balance would require an estimate of the volume of the mobile phase with respect to solids, but in most experiments vapor/liquid segregates at the hot end of the capsule. Therefore, unless a 3D image is acquired, it is impossible to quantify its volume, and from that to try to reconstruct the relative proportions of voids (volatiles), quench products, and equilibrium solids. We will therefore discuss hereafter the composition of liquids in terms of Ca/Si/Al ratios.

Carbonate particles (spherules, dendrites, chains, whiskery overgrowths) invariably show Ca as the unique cation detected by EPMA (other than carbon, not quantified here), and they are therefore assumed to be $CaCO_3$ in composition. Silica-rich droplets are characterized by Si/(Ca+Si+Al) (X_{Si}) > 0.9, and glass exhibits a wider compositional range, although mean values converge on X_{Si} = 0.7–0.8, X_{Ca} = 0.10–0.17, and X_{Al} = 0.04–0.12. Whiskers grossly fall midway between Si and Al, testifying to significant fractionation of alumina in this phase.

WDS X-ray elemental maps on segregated precipitate intergrowths at the top capsule were acquired at a step size of 0.3 micrometers in order to determine the relative proportions of particles. Image analysis was performed using a customized routine developed as a Mathematica© notebook. The composition of the liquids in terms of Ca, Si, and Al (Table 18.4) was estimated by summing the weighted contribution of carbonate, glass, whiskers, and droplets. Although densities are unknown, we assume that silica droplets have a density in the order of 2/3 that of carbonate, in order to limit carbonate underestimation, which is probably unavoidable, as H_2O is likely to be

Table 18.4 Volume abundance of quench products (as percentage of the total) and reconstructed liquid compositions (in wt.%).

Run	SPCC3	SPCC8	SPCC9	SPCC10	SPCC11	SPCC13
P (GPa)	4.2	4.2	4.2	6	6	4.2
T (°C)	900	900	850	900	950	900
Bulk	1	3	3	3	3	3b
Calcium carbonate	76%	30%	10%	61%	47%	22%
Silica droplets	10%	-	90%	29%	35%	-
Al whiskers	17%	-	-	10%	17%	-
Glass	-	70%	-	-	-	78%
Liquid						
CaO	69.8	38.7	9.7	61.0	48.4	25.4
Al_2O_3	12.4	2.4	2.2	2.7	3.4	5.3
SiO_2	17.8	58.9	88.1	36.3	48.2	69.3

present in both silica-rich droplets as well as in Si-Al whiskers. Even a gross estimate of liquid abundance, and therefore (by mass balance) of total volatiles in the liquid, is further hampered both by the irregular geometry of the meniscus relative to capsule walls, and by difficulty in estimating the actual porosity of the run charge after quench. Nonetheless, assuming that a single volatile-rich phase forms, and that H_2O initially added in the starting material strongly fractionates in the segregated liquid, all liquids should contain more than approximately 30–40 wt.% H_2O.

At 850 °C and 4.2 GPa, it was not possible to clarify from textural evidence whether a solute-rich vapor phase, or volatile-rich liquid already formed. In both cases, as previously stated, we have no indication for two coexisting mobile phases in this run, and the composition of the "liquid" is included in Table 18.4 and plotted in Figure 18.4b. With increasing temperature, the amount of solute content increases, as estimated qualitatively from image analysis, and a larger fraction of calcium carbonate precipitates, leading to a relative enrichment in Ca in Figure 18.4 c and d. It is worth pointing out that it is possible to reconstruct relative abundances only. As a consequence, relative silica enrichment as observed in liquids at 4.2 GPa, 900 °C at bulk 3 or 3b (coesite-bearing) does not imply that absolute amount of dissolved $CaCO_3$ (and then precipitated) is lower than at bulk 1 (corundum-bearing). The experiments performed at 6 GPa suggest an increase in carbonate dissolution and precipitation; however, silicate glass is formed at 6 GPa and 950 °C, but not at 900 °C. This might indicate that isopleths representing silicate dissolution should have a positive dP/dT slope, whereas carbonate dissolution has steeper or negative dP/dT slopes. This is grossly consistent with what is expected from solubility models of $CaCO_3$ and SiO_2 in aqueous fluids at high temperatures and pressures in the coesite field (Caciagli & Manning, 2003; Connolly & Galvez, 2018; Manning, 1994).

18.5. HYDROUS-CARBONATITIC LIQUIDS FORM AT SUB-ARCS

Subducted pelagic carbonates can be modeled as a simple mixture of predominant $CaCO_3$ (nannofosils and foraminifera), minor flint (diatoms, radiolarians), and pelagic clay. The "White Cliffs of Dover" are the most popular example of such a sedimentary sequence. If dry conditions prevail, eutectic melting of calcite, wollastonite, and quartz occurs at temperatures higher than 1300 °C (Huang et al., 1980), therefore indicating that subduction promotes recycling of aragonite into the deep mantle (Figure 18.5). However, most relevant at subduction zones, serpentinized ultramafic lithosphere and altered oceanic crust are known to extensively dehydrate at sub-arc conditions and to release aqueous vapors flushing the slab to mantle interface (Schmidt & Poli, 2014), where calcitic/aragonitic marbles occur.

H_2O strongly depresses the melting temperature of $CaCO_3$ and the wet solidus in CaO-CO_2-H_2O should be in equilibrium with portlandite. Addition of silica (a ubiquitous minor component in limestones, and a major solute in high pressure vapors) introduces numerous mineral phases potentially present: dellaite $Ca_6Si_3O_{11}(OH)_2$, spurrite $Ca_5(SiO_4)_2(CO_3)$, calciochondrodite $Ca_5(SiO_4)_2(OH)_2$, scawtite ($Ca_7Si_6(CO_3)O_{18}$•$2(H_2O)$), tilleyite $Ca_5Si_2O_7(CO_3)_2$ (Boettcher & Wyllie, 1969; Grice, 2005; Liu & Lin, 1995). Such minerals are rare in nature, or overlooked in high-pressure rocks and experimental studies. Boettcher and Wyllie (1969) proposed, on experimental basis, that eutectic melting of hydrous Ca-silicate + $CaCO_3$ + $Ca(OH)_2$ + vapor occurs at very low temperatures and that the second critical end point in the system

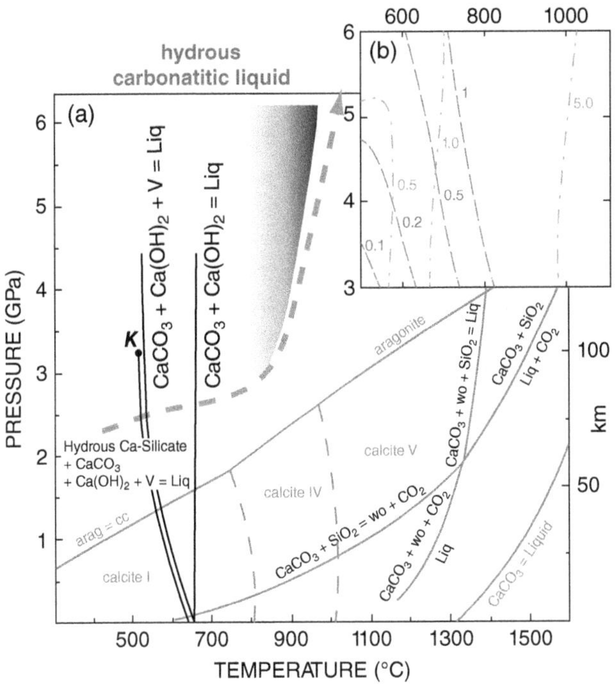

Figure 18.5 (a) Univariant phase relationships, and melting curves, experimentally determined for the systems $CaCO_3$ (gray; Zhao et al., 2019), $CaO-SiO_2-CO_2$ (brown; Huang et al., 1980), and $CaO-SiO_2-H_2O-CO_2$ (black; Wyllie & Boettcher, 1969; Boettcher & Wyllie, 1969). *K* refers to the second critical end point determined in the system CSHC (Boettcher & Wyllie, 1969). The red field represents the condition for generation of hydrous carbonatitic liquids from this study, compared to the P-T path for a slab surface in a warm-type subduction regime (thermal model D80 from Syracuse et al., 2010). (b) Red dashed lines are contours of carbon concentration in weight percent for aqueous fluids saturated in $CaCO_3$ (Kelemen & Manning, 2015); green dashed-dotted lines are contours of total Si molality in aqueous fluid (Connolly & Galvez, 2018). See electronic version for color representation of the figures in this book.

$CaO-SiO_2-H_2O-CO_2$ is located at 515 °C and 3.2 GPa (Figure 18.5), at low $CO_2/(H_2O+CO_2)$. Whether or not this model system is relevant for predicting the behavior of subducted marbles, hydrous carbonatitic liquids are potentially feasible at pressure-temperature conditions routinely attained at the top of subducting slabs (Syracuse et al., 2010).

Clay minerals (ideally $Al_2Si_2O_5[OH]_4$ in the model system investigated here) provide a source of Al_2O_3 to form grossular and kyanite, and of H_2O to form lawsonite, and zoisite, able to promote vapor or liquid on devolatilization. Saturation in silica and in Al-silicates for experiments performed here is observed to prevent the formation of portlandite and of phases recovered in the Al-free system, and most notably causes a shift of the liquidus surface. A diffuse porosity, to be ascribed to a vapor phase, is present at 800 °C, almost 300 °C above the supercritical liquid predicted in the system $CaO-SiO_2-CO_2-H_2O$. Vapor at 4 GPa, 800 °C is predicted to have a $CO_2/(H_2O+CO_2)$ = 0.02 on a thermodynamic basis (Poli & Schmidt, 1998), very similar to what is expected at the second critical end point in the system CSHC. Because evidence of liquid is recovered only from 850 °C, a compositional/thermal barrier necessarily exists between the liquidus surface found by Boettcher and Wyllie (1969) in $CaO-SiO_2-CO_2-H_2O$ and what was found in our study. Such a compositional barrier is possibly represented by aragonite + coesite, as experiments in Boettcher and Wyllie do not show saturation in silica, or by aragonite + coesite + grossular/zoisite, in the case we currently underestimate the Al-content of the liquids. All experiments performed in our study, at varying temperature, pressure, and water content do not show evidence for two coexisting vapor-liquid phases. Although this study was not designed to explicitly investigate the location of the second critical end point in CASHC, we speculate that experimental results suggest attainment of pressures at supercritical conditions.

Temperatures of 850 °C–900 °C at 4.2–6 GPa are predicted to occur in relatively warm subduction zones (Syracuse et al., 2010), e.g. in the Central American and northern Andean subduction zones, where deep-sea drilling sites have shown the presence of large masses of pelagic limestones. Hydrous carbonatitic liquids are therefore expected to be generated at the slab – mantle interface at sub-arc depth. Estimates of P-T paths from orogenic terrains (Penniston-Dorland et al., 2015) might suggest that warmer conditions are the rule rather than the exception, and 800 °C–900 °C at 4 GPa could be a common feature of subduction channels. If the pressure dependence of the dissolution of carbonates in the mobile phase is significantly larger than for silicates, as suggested qualitatively by our experiments (Figure 18.4) and by modeling of silica and $CaCO_3$ dissolution in aqueous fluids (Figure 18.5b), then liquids released by the slab are expected to shift from silicate rich to carbonatitic with increasing pressure. Hydrous carbonatitic liquids promote carbon scavenging from the slab, its transfer to the mantle wedge, and reaction with ultramafic rocks generating carbonated metasomatic wehrlites (Lee & Wyllie, 2000).

ACKNOWLEDGMENTS

This research was supported by the Italian Ministry of Education, University, and Research (MIUR) program PRIN2017 and by the Deep Carbon Observatory (DCO). We are greatly indebted to Andrea Risplendente for careful examination of run charges at the Electron Microprobe.

REFERENCES

Ague, J. J., & Nicolescu, S. (2014). Carbon dioxide released from subduction zones by fluid-mediated reactions. *Nature Geoscience*, *7*, 355–360.

Aiuppa, A., Fischer, T. P., Plank, T., Robidoux, P., & Di Napoli, R. (2017). Along-arc, inter-arc and arc-to-arc variations in volcanic gas CO_2/S_T ratios reveal dual source of carbon in arc volcanism. *Earth-Science Reviews*, *168*, 24–47.

Aubouin, J., von Huene, & the Shipboard Scientific Party (1982). Site 495: Cocos plate—Middle America Trench Outer Slope. *Initial Reports DSDP*, *67*, 79–141.

Boettcher, A. L., & Wyllie, P. J. (1969). The system CaO-SiO_2-CO_2-H_2O-III. Second critical end-point on the melting curve. *Geochimica et Cosmochimica Acta*, *33*, 611–632.

Brey, G. P., Girnis, A.V., Bulatov, V. K., Höfer, H. E., Gerdes, A., & Woodland, A. B. (2015). Reduced sediment melting at 7.5–12 GPa: Phase relations, geochemical signals and diamond nucleation. *Contribution to Mineralagoy and Petrology*, *170*, 1–25.

Caciagli, N. C., & Manning, C. E. (2003). The solubility of calcite in water at 6–16 kbar and 500–800 °C. *Contribution to Mineralogy and Petrology*, *146*, 275–285.

Cann, J. R., Langseth, M. G., Honnorez, J., Von Herzen, R. P., White, S. M., & the Shipboard Scientific Party (1983). 2. Sites 501 and 504: Sediments and ocean crust in an area of high heat flow on the southern flank of the Costa Rica Rift. *Initial Reports DSDP*, *69*, 31–173.

Carter, R. M., McCave, I. N., Richter, C., Carter, L., & the Shipboard Scientific Party (1999). 8. Site 1124: Rekohu drift—from the K/T boundary to the deep western boundary current. *Proceedings of the Ocean Drilling Program, Initial Reports*, *181*, 1–137.

Chou, I-M. (1986). Permeability of precious metals to hydrogen at 2 kb total pressure and elevated temperatures. *American Journal of Science*, *286*, 638–658.

Connolly, J.A.D. (1990). Multivariable phase diagrams: An algorithm based on generalized thermodynamics. *American Journal of Science*, *290*, 666–718.

Connolly, J.A.D., & Galvez, M. E. (2018). Electrolytic fluid speciation by Gibbs energy minimization and implications for subduction zone mass transfer. *Earth and Planetary Science Letters*, *501*, 90–102.

da Mommio, A. (2018). Evoluzione metamorfica delle unità paraderivate nella Finestra dei Tauri Occidentale (Ph.D. thesis, 216 p.). Università degli Studi di Milano.

Dasgupta, R., Hirschmann, M. M., & Withers, A. C. (2004). Deep global cycling of carbon constrained by the solidus of anhydrous, carbonated eclogite under upper mantle conditions. *Earth and Planetary Science Letters*, *227*, 73–85.

Dasgupta, R., & Hirschmann, M. M. (2006). Melting in the Earth's deep upper mantle caused by carbon dioxide. *Nature*, *440*, 659–662.

Donaldson, C. H. (1976). An experimental investigation of olivine morphology. *Contribution to Mineralogy and Petrology*, *57*, 187–213.

Facq, S., Daniel, I., Montagnac, G., Cardon, H., & Sverjensky, D. A. (2016). Carbon speciation in saline solutions in equilibrium with aragonite at high pressure. *Chemical Geology*, *431*, 44–53.

Ferri, F., Poli, S., & Rodríguez-Vargas, A. (2017). Andean volcanoes record carbonatite mantle metasomatism and CO_2 degassing at subduction zones. Paper presented at 27th Goldschmidt Conference, Paris.

Foustoukos, D. I., & Mysen, B. O. (2015). The structure of water-saturated carbonate melts. *American Mineralogist*, *100*, 35–46.

Franzolin, E., Schmidt, M. W., & Poli, S. (2011). Ternary Ca-Fe-Mg carbonates: Subsolidus phase relations at 3.5 GPa and a thermodynamic solid solution model including order/disorder. *Contribution to Mineralogy and Petrology*, *161*, 213–227.

Frezzotti, M. L., Selverstone, J., Sharp, Z. D., & Compagnoni, R. (2011). Carbonate dissolution during subduction revealed by diamond-bearing rocks from the Alps. *Nature Geoscience*, *4*, 703–706.

Garofalo, P. S. (2012). The composition of Alpine marine sediments (Bündnerschiefer Formation, W Alps) and the mobility of their chemical components during orogenic metamorphism. *Lithos*, *128–131*, 55–72.

Genge, M. J., Jones, A. P., & Price, G. D. (1995). An infrared and Raman study of carbonate glasses: Implications for the structure of carbonatite magmas. *Geochimica et Cosmochimica Acta*, *59*, 927–937.

Grassi, D., & Schmidt, M. W. (2011). The melting of carbonated pelites from 70 to 700 km depth. *Journal of Petrology*, *52*(4), 765–789.

Grice, J. D. (2005). The structure of spurrite, tilleyite and scawtite, and relationships to other silicate-carbonate minerals. *The Canadian Mineralogist*, *43*, 1489–1500.

Heirtzler, J.R., Veevers, J. J., Bolli, H. M., Carter, A. N., Cook, P. J., Krasheninnikov, V., et al., & the Shipboard Scientific Party (1974). 3. Site 260. *Initial Reports DSDP*, *27*, 89–127.

Holland, T., & Powell, R. (1991). A Compensated-Redlich-Kwong (CORK) equation for volumes and fugacities of CO_2 and H_2O in the range 1 bar to 50 kbar and 100–1600 °C. *Contribution to Mineralogy and Petrology*, *109*, 265–273.

Holland, T.J.B., & Powell, R. (2011). An improved and extended internally consistent thermodynamic dataset for phases of petrological interest, involving a new equation of state for solids. *Journal of Metamorphic Geology*, *29*, 333–383.

Huang, W. L., Wyllie, P. J., & Nehru, C. E. (1980). Subsolidus and liquidus phase relationships in the system CaO-SiO_2-CO_2 to 30 kbar with geological applications. *American Mineralogist*, *65*, 285–301.

Irving, A. J., & Wyllie, P. J. (1975). Subsolidus and melting relationships for calcite, magnesite and the join $CaCO_3$-$MgCO_3$ to 36 kb. *Geochimica et Cosmochimica Acta*, *39*, 35–53.

Kelemen, P. B., & Manning, C. E. (2015). Reevaluating carbon fluxes in subduction zones, what goes down, mostly comes up. *Proceedings of the National Academy of Sciences U.S.A.*, *112*, E3997–E4006.

Keppler, H. (2003). Water solubility in carbonatite melts. *American Mineralogist*, *88*, 1822–1824.

Kono, Y., Kenney-Benson, C., Hummer, D., Ohfuji, H., Park, C., Shen, G., et al. (2014). Ultralow viscosity of carbonate melts at high pressures. *Nature Communications*, *5*, 1–8.

Lee, W. J., & Wyllie, P. J. (2000). The system CaO-MgO-SiO_2-CO_2 at 1 GPa, metasomatic wehrlites, and primary carbonatite magmas. *Contribution to Mineralogy and Petrology*, *138*, 214–228.

Liu, L-g., & Lin, C-C. (1995). High-pressure phase transformations of carbonates in the system CaO-MgO-SiO_2-CO_2. *Earth and Planetary Science Letters*, *134*, 297–305.

Lofgren, G. (1974). An experimental study of plagioclase crystal morphology: Isothermal crystallization. *American Journal of Science*, *274*, 243–273.

Manning, C. E. (1994). The solubility of quartz in H_2O in the lower crust and upper mantle. *Geochimica et Cosmochimica Acta*, *58*, 4831–4839.

Manning, C. E., Shock, E. L., & Sverjensky, D. A. (2013). The chemistry of carbon in aqueous fluids at crustal and upper-mantle conditions: Experimental and theoretical constraints. *Reviews in Mineralogy & Geochemistry*, *75*, 109–148.

Minarik, W. G., & Watson, E. B. (1995). Interconnectivity of carbonate melt at low melt fraction. *Earth and Planetary Science Letters*, *133*, 423–437.

Paterson, M. S. (1958). The melting of calcite in the presence of water and carbon dioxide. *American Mineralogist*, *43*, 603–606.

Penniston-Dorland, S. C., Kohn, M. J., & Manning, C. E. (2015). The global range of subduction zone thermal structures from exhumed blueschists and eclogites: Rocks are hotter than models. *Earth and Planetary Science Letters*, *428*, 243–254.

Plank, T. (2014). The chemical composition of subducting sediments. In K. Turekian & H. Holland (Eds.), *Treatise on Geochemistry* (2nd ed., vol. 4, pp. 607–629). Elsevier.

Plank, T., & Langmuir, C. H. (1998). The chemical composition of subducting sediment and its consequences for the crust and mantle. *Chemical Geology*, *145*, 325–394.

Poli, S. (2015). Carbon mobilized at shallow depths in subduction zones by carbonatitic liquids. *Nature Geoscience*, *8*, 633–636.

Poli, S., (2016). Melting carbonated epidote eclogites: Carbonatites from subducting slabs. *Progress In Earth and Planetary Science 3*, 27.

Poli, S., & Schmidt, M. W. (1998). The high-pressure stability of zoisite and phase relationships of zoisite-bearing assemblages. *Contribution to Mineralogy and Petrology*, *130*, 162–175.

Schmidt, M. W., & Poli, S. (2014). Devolatilization during subduction. In K. Turekian & H. Holland (Eds.), *Treatise on Geochemistry* (2nd ed., vol. *4*, pp. 669–701). Elsevier.

Skora, S., Blundy, J. D., Brooker, R. A., Green, E.C.R., De Hoog, J.C.M., & Connolly, J.A.D. (2015). Hydrous phase relations and trace element partitioning behaviour in calcareous sediments at subduction-zone conditions. *Journal of Petrology*, *56*(5), 953–980.

Sokol, A. G., Kupriyanov, I. N., & Palyanov, Y. N. (2013). Partitioning of H_2O between olivine and carbonate-silicate melts at 6.3 GPa and 1400 °C: Implications for kimberlite formation. *Earth and Planetary Science Letters*, *383*, 58–67.

Susaki, J., Akaogi, M., Akimoto, S. & Shinomura, O. (1985). Garnet perovskite transformation in $CaGeO_3$: *In situ* X-ray measurements using synchrotron radiation. *Geophysical Research Letters*, *12*, 729–732.

Syracuse, E. M., Van Keken, P. E., & Abers, G. A. (2010). The global range of subduction zone thermal models. *Physics of the Earth and Planetary Interiors*, *183*, 73–90.

Taylor, L. A., Logvinova, A. M., Howarth, G. H., Liu, Y., Peslier, A. H., Rossman, G. R., et al. (2016). Low water contents in diamond mineral inclusions: Proto-genetic origin in a dry cratonic lithosphere. *Earth and Planetary Science Letters*, *433*, 125–132.

Thomsen, T. B., & Schmidt, M. W. (2008). Melting of carbonated pelites at 2.5–5.0 GPa, silicate-carbonatite liquid immiscibility, and potassium-carbon metasomatism of the mantle. *Earth and Planetary Science Letters*, *267*, 17–31.

Tsuno, K., & Dasgupta, R. (2012). The effect of carbonates on near-solidus melting of pelite at 3 GPa: Relative efficiency of H_2O and CO_2 subduction. *Earth and Planetary Science Letters*, *319–320*, 185–196.

Tumiati, S., Tiraboschi, C., Sverjensky, D. A., Pettke, T., Recchia, S., Ulmer, P., et al. (2017). Silicate dissolution boosts the CO_2 concentrations in subduction fluids. *Nature Communications*, *8*, 616.

von der Borch, C.C., and the Shipboard Scientific Party. (1974). Site 212. *Initial Reports DSDP*, *22*, 37–83.

Watson, E. B., Brenan, J. M., & Baker, D. R. (1990). Distribution of fluids in the continental lithospheric mantle. In M. A. Menzies (Ed.), *The Continental Lithospheric Mantle* (pp. 111–125). Oxford: Clarendon.

Wyllie, P. J., & Boettcher, A. L. (1969). Liquidus phase relationships in the system CaO-CO_2-H_2O to 40 kilobars pressure with petrological applications. *American Journal of Science*, *267-A*, 489–508.

Wyllie, P. J., & Tuttle, O. F. (1959). Melting of calcite in presence of water. *American Mineralogist*, *44*, 453–461.

Wyllie, P. J., & Tuttle, O. F. (1960). The system CaO-CO_2-H_2O and the origin of carbonatites. *Journal of Petrology*, *1*, 1–46.

Zhang, J., Li, B., Utsumi, W., & Liebermann, R. C. (1996). *In situ* X-ray observations on the coesite stishovite transition: Reversed phase boundary and kinetics. *Physics and Chemistry of Minerals*, *23*, 1–10.

Zhao, S., Schettino, E., Merlini, M., & Poli, S. (2019). The stability and melting of aragonite: An experimental and thermodynamic model for carbonated eclogites in the mantle. *Lithos*, *324–325*, 105–114.

19

The Viscosity of Carbonate-Silicate Transitional Melts at Earth's Upper Mantle Pressures and Temperatures, Determined by the In Situ Falling-Sphere Technique

Vincenzo Stagno[1,5], Yoshio Kono[2,3], Veronica Stopponi[1], Matteo Masotta[4], Piergiorgio Scarlato[5], and Craig E. Manning[6]

ABSTRACT

The circulation of carbon in Earth's interior occurs through the formation, migration, and ascent of CO_2-bearing magmas throughout the convective mantle. Their chemical composition spans from carbonatitic to kimberlitic as a result of either temperature and pressure variations or local redox conditions at which partial melting of carbonated mantle mineral assemblages occurs. Previous experiments that focused on melting relations of synthetic CO_2-bearing mantle assemblages revealed the stability of carbonate-silicate melts, or transitional melts, that have been generally described to mark the chemical evolution from kimberlitic to carbonatitic melts at mantle conditions. The migration of these melts upward will depend on their rheology as a function of pressure and temperature. In this study, we determined the viscosity of carbonate-silicate liquids (~18 wt% SiO_2 and 22.54 wt% CO_2) using the falling-sphere technique combined with in situ synchrotron X-ray radiography. We performed six successful experiments at pressures between 2.4 and 5.3 GPa and temperature between 1565 °C and 2155 °C. At these conditions, the viscosity of transitional melts is between 0.02 and 0.08 Pa·s; that is, about one order of magnitude higher than what was determined for synthetic carbonatitic melts at similar P-T conditions, likely due to the polymerizing effect of the SiO_2 component in the melt.

19.1. INTRODUCTION

The cycling of carbon in the Earth's interior occurs in part through the ascent of carbonated magmas, in which the presence of variable amounts of CO_2 depends on the degree of partial melting, as controlled by pressure, temperature, and mantle oxidation state (Brey et al., 2008; Brey et al., 2011; Dalton & Presnall, 1998; Dasgupta et al., 2013; Gudfinnsson & Presnall, 2005; Hammouda & Keshav, 2015; Kiseeva et al., 2012, 2013; Litasov & Ohtani, 2009a, 2009b; Litasov & Ohtani, 2010; Stagno & Frost, 2010; Stagno et al., 2013). Experimental studies showed that carbonatitic melts produced by low degrees of partial melting of carbonated mantle rocks evolve to carbonate-silicate melts as the degree of partial melting increases from less than 0.1% to about 1% (Dasgupta et al., 2013). In the present-day convective mantle, carbonate-silicate melts with SiO_2 content between 10 and 25 wt% are likely to form by extensive redox melting of a graphite-bearing peridotite at 75–100 km depth within −1.8 log units < fo_2 (normalized to FMQ buffer) < −2.5 log units (Stagno et al. 2013), while they likely formed at

[1] *Dipartimento di Scienze della Terra, Sapienza Universita' di Roma, Italy*
[2] *Geophysical Laboratory, Carnegie Institution of Washington, Argonne, Illinois, USA*
[3] *Geodynamics Research Center, Ehime University, Ehime, Japan*
[4] *Dipartimento di Scienze della Terra, Università di Pisa, Pisa, Italy*
[5] *Istituto Nazionale di Geofisica e Vulcanologia, Roma, Italy*
[6] *Department of Earth, Planetary and Space Sciences, University of California–Los Angeles, Los Angeles, California, USA*

Carbon in Earth's Interior, Geophysical Monograph 249, First Edition. Edited by Craig E. Manning, Jung-Fu Lin, and Wendy L. Mao.

shallower depths and more reduced fo_2 conditions (between FMQ-2.3 and FMQ-3.5 log units) beneath mid-ocean ridges in the Archaean (Aulbach & Stagno 2016), with important implications for the mobilization of oxidized carbon from the interior of Earth to the atmosphere. There is evidence of the stability of carbonate-silicate melts coexisting with mantle minerals from experimental studies performed within simplified carbonated systems like $CaO-MgO-SiO_2-CO_2$ (Moore & Wood 1998), $CaO-MgO-Al_2O_3-SiO_2-CO_2$ (Dalton & Presnall, 1998; Gudfinnsson & Presnall, 2005; Keshav & Gudfinnsson, 2014), $CaO-MgO-Al_2O_3-SiO_2-Na_2O-CO_2$ (Litasov & Ohtani, 2009b; Litasov & Ohtani, 2010), $CaO-MgO-Al_2O_3-SiO_2-Na_2O-K_2O-Cl-CO_2$ (Litasov & Ohtani, 2009a), $CaO-Al_2O_3-SiO_2(\pm Na_2O)-CO_2$ (Brooker & Kjarsgaard, 2011), and more complex systems representative of peridotite (Brey et al., 2008; Brey et al., 2011; Dasgupta & Hirschmann, 2007; Foley et al., 2009; Ghosh et al., 2014; Rohrbach & Schmidt, 2011; Stagno & Frost, 2010) and eclogite (Kiseeva et al., 2012; Kiseeva et al., 2013) mantle assemblages, respectively. These melts span compositions from ~14 to ~30 wt% SiO_2, CO_2 between 15 and 40 wt%, and Ca# (CaO/(CaO+MgO) mole ratio) of 0.5 on average but with a strong variation (±0.3) due to dependence on the coexisting mineral phases and bulk composition of the starting material, especially the alkali content. Transitional melts fall within the field of carbonate-silicate liquids and are characterized by $SiO_2+Al_2O_3$ varying between ~18 and ~32 wt% (Moussallam et al., 2015). A much narrower chemical variation was proposed by Gudfinnsson and Presnall (2005) with about 15–22 wt% SiO_2, ~0.9–2.6 wt% Al_2O_3, and Ca# of 0.4–0.5. Based on these experimental studies, whose P-T conditions are summarized in Figure 19.1, transitional melts represent a link between near-solidus carbonatitic melts and relatively more SiO_2-rich melilititic and kimberlitic melts (Martin et al., 2013) produced by the melting of CO_2-rich portions of the Earth's mantle at variable depths.

Although P-T conditions required for the stability of transitional melts have been experimentally constrained (Figure 19.1), very little is known about their rheology at depth. The knowledge of the viscosity of carbonate-silicate melts is needed to better constrain their mobility, migration rate (velocity) from the source rock, and the time of melt-rock interaction.

To date, knowledge of viscosity of CO_2-rich melts at high pressure and temperature representative of the mantle is mostly limited to pure carbonate compositions chosen as analogue of natural carbonatitic melts. Dobson

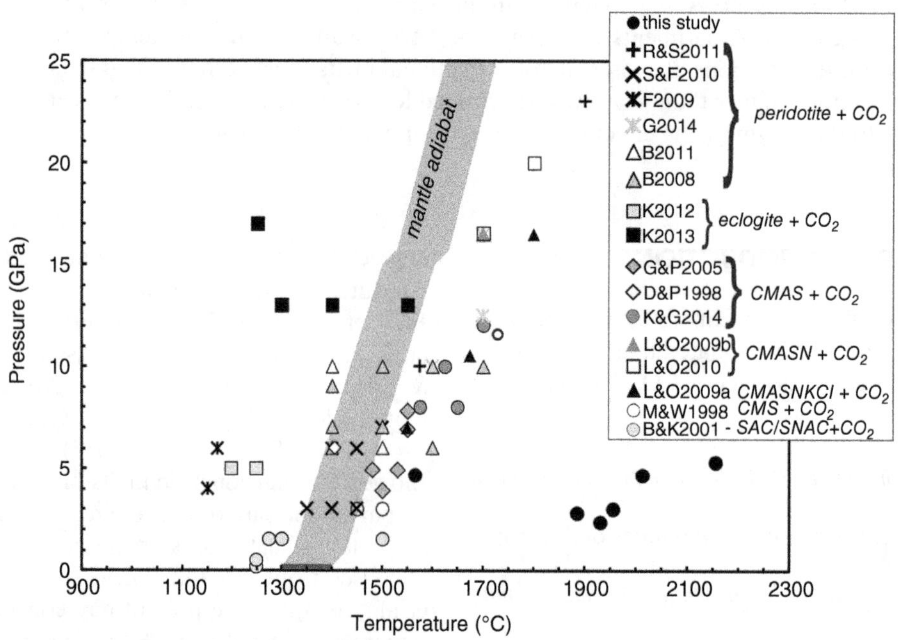

Figure 19.1 Diagram showing temperatures and pressures set for the experiments of this study compared to previous studies where the stability of carbonate-silicate melts with 16–20 wt% SiO_2 was reported, plotted along with the mantle adiabat (Stixrude & Lithgow-Bertelloni, 2007). Black circles are the P-T conditions for the viscosity experiments from this study. Abbreviations for data sources: B2008 = Brey et al., 2008; B2011 = Brey et al., 2011; B&K2001 = Brooker & Kjarsgaard, 2001; D&P2008 = Dalton & Presnall, 1998; F2009 = Foley et al., 2009; G2014 = Ghosh et al., 2014; G&P2005 = Gudfinnsson & Presnall, 2005; K2012 = Kiseeva et al., 2012; K2013 = Kiseeva et al., 2013; K&G2014 = Keshav & Gudfinnsson, 2014; L&O2009a = Litasov & Ohtani, 2009a; L&O2009b = Litasov & Ohtani, 2009b; L&O2010 = Litasov & Ohtani, 2010; M&W1998 = Moore & Wood, 1998; R&S2011 = Rohrbach & Schmidt, 2011; S&F2010 = Stagno & Frost, 2010. See electronic version for color representation of the figures in this book.

et al. (1996) published the first study on viscosity and density measurements of Mg-K and Ca-K carbonate mixtures at conditions of the Earth's mantle obtaining data between 0.06 and 0.036 Pa˙s at 2–5.5 GPa and 800 °C–1500 °C. Kono et al. (2014a) performed an in situ investigation of the viscosity of melts with both calcite ($CaCO_3$) and dolomite $(Mg_{0.40}Fe_{0.09}Ca_{0.51})(CO_3)_2$ compositions up to 6.2 GPa. The study employed an ultrafast synchrotron X-ray camera available at beamline 16 BM-B of Advanced Photon Source (Lemont, Illinois, USA) to record the fall of a platinum sphere at up to 1000 frame˙s^{-1}. Measured viscosities of calcite and dolomite melts were surprisingly low, being in the range of 0.006–0.010 Pa˙s and 0.008–0.010 Pa˙s, respectively. Using the same experimental technique, Stagno et al. (2018) determined even lower viscosities in the range of 0.003–0.007 Pa˙s for Na_2CO_3 melt at mantle pressures and temperatures that raises important questions on the effect of alkali elements on the rheology of carbonatitic magmas. These values are about 2–3 orders of magnitude lower than the measured viscosity of basaltic melts (Sakamaki et al., 2013), suggesting a strong effect of increasing SiO_2 content on the viscosity of magmas. In turn, transitional melts are expected to possess intermediate rheological properties in between basalts and carbonatites. Ghosh and coauthors (Ghosh, Bajgain, et al., 2017; Ghosh & Karki, 2017) investigated the rheological properties of carbonated $MgSiO_3$ melts by first-principle calculations at depths of the core-mantle boundary and temperature up to 5000 K. Extrapolation of their results to upper mantle conditions yields viscosities of ~0.025 Pa˙s. However, this study does not link directly to the chemical composition of carbonate-silicate melts as those obtained in high P-T experiments using the large volume press facilities (Figure 19.1).

We investigated the viscosity of a synthetic carbonate-silicate melt with ~18 wt% SiO_2 and ~22 wt% CO_2 by the falling-sphere technique at high pressure and temperature using the Paris-Edinburgh press combined with in situ synchrotron X-ray radiography at the beamline 16 BM-B of the Advanced Photon Source (Illinois, USA). Our results are compared with those of pure carbonate melts and synthetic basalts with implications for the CO_2 circulation throughout the Earth's interior.

19.2. MATERIALS AND METHODS

Two starting materials were employed in this study, CB2 and CARB2, with composition representative of transitional melts (Gudfinnsson & Presnall, 2005), such as 18.28 wt% SiO_2, 20.43 wt% CaO, 27.50 wt% MgO, 6.72 wt% FeO, 1.95 wt% Al_2O_3, 1.37 wt% Na_2O, 1.21 wt% ClO^-, and 22.54 wt% CO_2. CB2 was prepared by grinding under ethanol oxides and carbonates such as SiO_2, FeO, Al_2O_3, $CaCO_3$, $MgCO_3$, and NaCl to make a nominal composition like the above. CARB2 was prepared by melting CB2 at ~0.3 GPa and 1410(±10) °C for 1 h, using a non-end-load piston cylinder (QUICKPress type) available at the Istituto Nazionale di Geofisica e Vulcanologia (INGV) in Rome (Italy). In the synthesis experiment, the CB2 mixture was loaded into a platinum (Pt) capsule of 5 mm length and 5 mm diameter, welded at both ends. The capsule was loaded into a 19–25 mm crushable MgO-borosilicate glass-NaCl assembly (Masotta et al., 2012). Temperature was constantly monitored during the experiment using a factory calibrated C-type thermocouple, with a maximum error of 5 °C. The run was quenched by shutting down the power. The recovered sample was analyzed by field-emission scanning electron microscopy using a JEOL JSM 6500F, also available at INGV both for textural observations and chemical composition. The sample showed a typical quench texture expected for carbonate-silicate melt compositions, alternating glassy portions with elongated crystals of olivine. Such observation appears in contrast with the well-quenched glass reported by Moussallam et al. (2015) for synthetic transitional melts, and it is likely a consequence of the lower quench rate of the QUICKPress compared to that of the internally heated pressure vessels used by Moussallam et al. However, we decided to use both CARB2 and CB2 oxide mixture to test possible effects of the starting material on viscosity measurements.

Viscosity measurements were conducted using the falling-sphere technique with the Paris-Edinburgh large-volume press at beamline 16-BM-B (Kono et al., 2014b; Kono, 2018). The starting material was placed in a cylindrical graphite capsule with a diameter of 1.2 mm and height of 2 mm. The majority of the experiments were performed using CARB2 starting glass, except in the run at 4.7 GPa/1565 °C, where CB2 oxide/carbonate mixture was employed (Table 19.1). A Pt sphere prepared by arc melting with diameter between ~70 and 190 μm (Table 19.1) was placed in the central portion of the capsule. Some runs were performed using the double-layered probing sphere configuration (Terasaki et al., 2001; runs at 4.7 GPa/1565 °C, 2.8 GPa/1885 °C, 3 GPa/1955 °C), where an additional Pt sphere was placed at the top of the capsule surrounded by $CaCO_3$ powder (calcite, Puratronic®, purity of 99.999%) to delay the fall of the probing sphere at a higher temperature than the sphere placed in the center of the sample, which was used to monitor the onset of melting. Details on the cell assembly used in this study are available in Kono et al. (2014b) and are the same than those used by Stagno et al. (2018). The generated pressure at the sample was constantly monitored using the equation of state of MgO (Kono et al., 2010) by collecting diffraction patterns on the MgO sleeve surrounding the capsule. After compression to target pressure, for each experiment the temperature was set

Table 19.1 Experimental conditions and results.

Run	P[b] (GPa)	T[b] (°C)	ø Pt Sphere (mm)	Terminal Velocity (mm/s)	Recording Time (fps)	Viscosity (Pa·s)	logη	Horizontal shift[a] (mm)
CB2_3_2015	4.7	1565	0.117	1.864(±0.048)	100	0.0529(±0.0007)	−1.277	0.021
CARB2_2_2016	2.8	1885	0.191	4.254(±0.165)	500	0.0510(±0.0016)	−1.292	0.075
CARB2_3_2016	4.7	2012	0.140	3.611(±0.088)	500	0.0368(±0.0014)	−1.434	0.030
CARB2_1_2018	3	1955	0.089	2.687(±0.170)	500	0.0228(±0.0018)	−1.642	0.008
CARB2_2_2018	2.4	1930	0.086	0.749(±0.030)	167	0.0762(±0.0030)	−1.118	0.011
CARB2_5_2018	5.3	2155	0.140	6.628(±0.259)	500	0.0202(±0.0008)	−1.695	0.022

[a] Horizontal shift refers to a lateral shift of the Pt sphere during the fall that might result in decrease of the terminal velocity.
[b] Typical uncertainties on pressure and temperature in the Paris-Edinburgh cell experiments are less than 0.4 GPa and 60 °C (e.g., Kono et al., 2014a; de Grouchy et al., 2017; Hudspeth et al., 2018).

quickly (~100 °C/s) to the value at which the platinum sphere fell down and was estimated using the power vs. temperature calibration curve corrected for the proper hydraulic pressure (Kono et al., 2014b).

Viscosity measurements were conducted using the falling-sphere technique combined with unfocused white X-ray beam for radiographic images captured by a high-speed camera (Photron FASTCAM SA3) with 100–500 frames per second (fps) recording time (Kono et al., 2014b; Kono, 2018). The image resolution of the ultra-fast camera used for the viscosity measurements is 2.5 μm/pixel except for the run at 4.7 GPa/1565 °C (CB2_run3_2015), whose resolution is 5 μm/pixel. The radiographic images acquired from the high-speed camera were used to calculate the probing sphere terminal velocity as a function of elapsed time for each frame by using the Tracker plugin in the ImageJ software (Abramoff et al., 2004); and from the velocity of the falling sphere(s), the melt viscosity was calculated using the Stokes equation,

$$\eta = \frac{gd^2(\rho_s - \rho_l)}{18\upsilon}\frac{F}{E}, \tag{19.1}$$

where ν is the terminal velocity (mm/s) of the probing sphere; ρ_s and ρ_l are the densities (g/cm^3) of the Pt sphere (~19.3 g˙cm^{-3} as calculated from the thermal equation of state; Dorogokupets & Dewaele, 2007) and melt, respectively; d is the diameter of the sphere (mm) determined from the radiographic images using the Prosilica GC1380 high-resolution camera with pixel size of 0.945 μm/pixel (0.850 μm/pixel for runs at 2.4 GPa/1930 °C, 5.3 GPa/2155 °C, 3 GPa/1955 °C); and E and F are correction factors for the wall effect and end effect, respectively. A fixed value of 2.3 g˙cm^{-3} was chosen for the density of carbonate-silicate melts (ρ_l), as determined by numerical simulation by Ghosh et al. (2017). This value is slightly greater than the density of 2 g·cm^{-3} estimated for carbonatitic liquids (Liu & Lange, 2003), while the uncertainty of the density would not produce a significant error in the calculated viscosity because of the large difference in densities between Pt sphere (ρ_s) and the carbonate-silicate melt density (ρ_l).

19.3. RESULTS

The run conditions and each calculated viscosity for synthetic carbonate-silicate melts are shown in Table 19.1. Figure 19.1 summarizes the P-T conditions of our experiments along with literature data where transitional melts were quenched. The short fall time of the Pt sphere results in high terminal velocity ranging between 0.749 and 6.628 mm/s. Figure 19.2 and Figure 19.3 show the fall of the Pt sphere through a series of radiographic images as a function of time, and the terminal velocity used to determine the viscosity through equation (19.1), respectively. The results of the falling distance as function of time and falling velocity for each run are available in the Supplementary Materials. The calculated viscosity of synthetic transitional melts varies from a minimum value of 0.0202(±0.0008) Pa·s at 2155 °C and 5.3 GPa and a maximum value of 0.0762(0.030) Pa˙s at 1930 °C and 2.4 GPa. The viscosity data obtained for melts with ~18 wt% SiO_2 are shown in Figure 19.4, expressed in the logarithm form as function of pressure, along with previous studies on the viscosity of pure carbonate melts and carbonated silicate melts.

We carefully checked not only the falling of the sphere but also the lateral shift of the sphere during its fall, which might cause a decrease in the terminal velocity (Table 19.1). We observed that lateral shift ranges from 0.009 to 0.076 mm during the whole fall. These values are very low compared to the fall distance, where the terminal velocity is achieved (see Supplementary Materials). In contrast, we noted that lateral shift occurs mainly when the starting material begins melting as a consequence of the softening of the powder surrounding the Pt sphere at its initial position. Some experiments performed at $P < 2$ GPa showed the appearance of bubbles due to CO_2

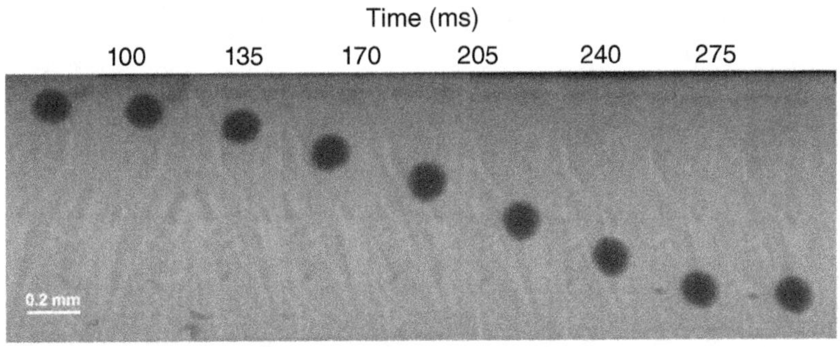

Figure 19.2 X-ray radiography images of the Pt sphere in carbonated silicate melt as function of time for experiment at 4.7 GPa and 2012 °C. See electronic version for color representation of the figures in this book.

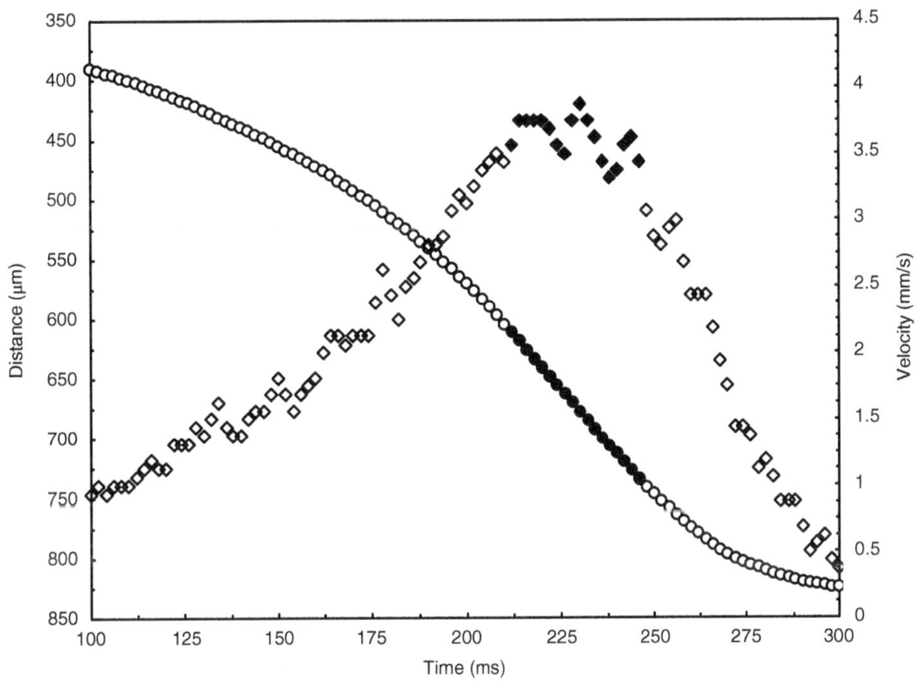

Figure 19.3 Results of the falling distance (circles) as function of time and the falling velocity (diamonds) as function of time for each frame (recording time is 500 fps) for the run at 4.7 GPa and 2012 °C. The terminal velocity reached by the sphere is indicated by filled symbols. See electronic version for color representation of the figures in this book.

exsolution that impeded the fall of the Pt sphere. These experiments are not presented here.

19.4. DISCUSSION

19.4.1. Effect of Pressure and Temperature on the Viscosity Data

Our data show viscosities between 0.0202 and 0.0762 Pa's, which are higher than those determined by Di Genova et al. (2016) and Stagno et al. (2018) for Na_2CO_3 melt, and Kono et al. (2014a) for calcitic and dolomitic melts by about an order of magnitude (Figure 19.4). The viscosity values for transitional melts appear comparable with data by Dobson et al. (1996) for molten $K_2Mg(CO_3)_2$ and $K_2Ca(CO_3)_2$ ranging from 0.006 to 0.036 Pa·s, taking into account the high uncertainty due to the lowest recording time (40 fps) employed by the authors with respect to more recent in-situ measurements. Figure 19.4 includes the viscosity of carbonated silicate melts performed on a $CaCO_3$-$CaSiO_3$ mixture with 22.6 and 31.4 wt% SiO_2, respectively (Kono et al., 2014a). These experiments yielded viscosities of 0.03 and 0.055 Pa's that are in agreement with our data. The viscosity calculated by molecular dynamics simulation by Ghosh and Karki (2017) covers a T range similar to our experimental conditions. Their calculations applied to enstatitic melts with 16 wt% of CO_2 gave a viscosity of 0.025 Pa's at 2200 K

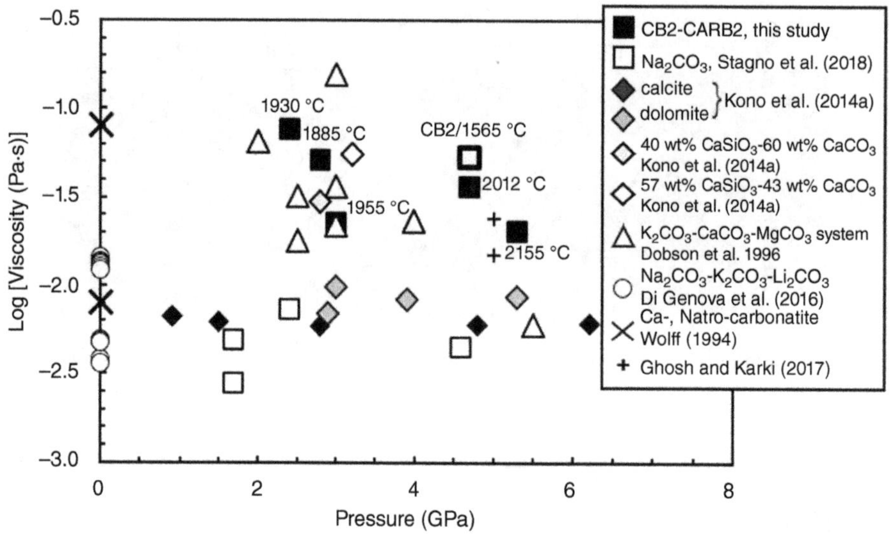

Figure 19.4 Experimentally determined viscosity data plotted as a function of pressure. Data from literature are reported for comparison. The uncertainty in our viscosity measurements is within the symbol size. See electronic version for color representation of the figures in this book.

and 5 GPa. Both experimental (Kono et al., 2014a) and theoretical (Ghosh & Karki, 2017) studies highlight the effect of temperature in lowering the viscosity of carbonated melts. This effect is also observable in Figure 19.4, where our data show a decrease in viscosity at increasing temperature of our experiments rather than a role of pressure. As also highlighted in Figure 19.1, our experiments were performed at much higher temperatures than the mantle adiabat. This is likely due to the higher liquid-glass transition temperature of the starting composition used in this study, linked with the chemical composition of the synthetic starting glassy material, although a possible kinetic effect on the melting process cannot be excluded.

To test the possible effect of the starting material, we performed an experiment (CB2_03_2015) using a mixture of oxides and carbonates. The viscosity determined at 4.7 GPa and only 1565 °C is 0.0529 Pa˙s (Table 19.1), which is higher than the viscosity determined for runs using CARB2. Although melting occurred at lower T, possibly due to the presence of absorbed moisture from the sample, we conclude that the fall of the sphere occurred at (near-)liquidus conditions. The viscosity determined at 4.7 GPa decreases from 0.0529 Pa˙s (run CB2_03_2015 at 1565 °C) to 0.0368 Pa˙s (run CARB2_3_2016 at 2012 °C) within about 400 °C. No obvious pressure effect can be claimed in our determined viscosities, similar to that reported by Stagno et al. (2018) for Na_2CO_3 melt and Kono et al. (2014a) for calcite and dolomite melts. Unfortunately, it remains difficult to collect isobaric data due to the pressure effect on the melting temperature of synthetic starting glasses.

19.4.2. Implications for the Transport of Oxidized Carbon in the Upper Mantle

The presence of small volumes of carbonate-silicate melts in the Earth's upper mantle has been the subject of several experimental studies aimed at understanding the origin of seismic (Hier-Majumder & Tauzin, 2017) and electrical (Gaillard et al., 2008; Yoshino et al., 2012) anomalies detected beneath mid-ocean ridges, as well as the immiscible behavior of CO_2-rich melts during upwelling and as their role in graphite/diamond precipitation or formation by redox melting during the Archaean (Aulbach & Stagno, 2016). Most of these processes imply interaction between migrating melts and host rock, causing CO_2-rich magmas to shift to transitional compositions. We calculated the mobility and migration rate of transitional melts using our determined viscosities. These were calculated using the equation

$$\phi w_0 = \frac{kg\Delta\rho}{\eta}, \tag{19.2}$$

with

$$k = \frac{a^2\phi^n}{C}, \tag{19.3}$$

where k is the permeability, a is the characteristic grain diameter, and ϕ is the melt fraction, while n (equal to 2) and C (equal to 1600) are numerical constants, g is the gravitational acceleration constant, and w_0 is the melt ascent velocity. As representative viscosity value, we considered 0.05 Pa·s, which leads to an average mobility, $\frac{\Delta\rho}{\eta}$, of transitional melts of ~25 g·cm^{-3}·Pa^{-1}·s^{-1}, assuming $\Delta\rho$

(density of the solid mantle-density of carbonate-silicate melt; Ghosh & Karki, 2017) equal to 1.2 g·cm^{-3} in the oceanic lithosphere. This value is between the mobility estimated for basalts of about 0.6 g·cm^{-3}·Pa^{-1}·s^{-1} (Sakamaki et al., 2013) and the high mobility of 148 g·cm^{-3}·Pa^{-1}·s^{-1} for Ca-carbonatitic melts calculated at upper mantle conditions (Kono et al., 2014a), even higher (300 g·cm^{-3}·Pa^{-1}·s^{-1}) in the case of Na_2CO_3 liquid (Stagno et al., 2018). In turn, this implies a dramatic effect of the SiO_2 content of CO_2-rich magmas on their segregation and migration from the silicate mantle at solidus temperatures over time, enhancing, therefore, the possibility of melt-solid interaction and element diffusion. The calculation using equation (19.2) and equation (19.3) results in the melt migration velocity of ~1850 m/yr using *a* of 2 mm and melt fraction of 0.1% in volume, respectively (Keller et al., 2017). The migration velocity thus appears much higher than that proposed for dolomitic melts (80–115 m/yr; Kono et al., 2014a), implying that during the ascent of mantle carbonatitic magmas, as they evolve to more silica-rich compositions, the rheological properties might change dramatically, in turn promoting acceleration of these melts to shallower depth in contrast with a stagnant behavior of pure carbonatitic magmas. In contrast, long residence times are expected for carbonated magmas in the mantle.

19.5. CONCLUSIONS

We determined the viscosity of carbonated-silicate transitional melts with 18 wt% SiO_2 at pressures between 2.4 and 5.3 GPa (~70–160 km of depth) and temperatures between 1565 °C and 2155 °C. Results range between 0.02 and 0.08 Pa˙s with a variation mainly related to the temperature dependence. Our results agree with the sparse data from previous studies on $CaCO_3$-$CaSiO_3$ mixture (Kono et al., 2014a) and confirmed by dynamic simulation models on carbonated-$MgSiO_3$ melts (Ghosh et al., 2017). The higher viscosity compared to carbonatitic melt is therefore shown to correlate with the concentration of SiO_2. The relatively low ascent rate of transitional melts (~1850 m/yr) results in shorter residence time than carbonatitic melts within the upper mantle.

ACKNOWLEDGMENTS

V.S. gratefully acknowledges financial support from the Deep Carbon Observatory and Sapienza University of Rome through "Fondi di Ateneo 2016." This work was performed at HPCAT (Sector 16), Advanced Photon Source (APS), Argonne National Laboratory. HPCAT operation is supported by DOE-NNSA under Award No. DE-NA0001974, with partial instrumentation funding by the National Science Foundation. The Advanced Photon Source is a U.S. Department of Energy (DOE) Office of Science User Facility, operated for the DOE Office of Science by Argonne National Laboratory under Contract No. DE-AC02-06CH11357. The manuscript benefited from the thoughtful review of an anonymous referee and Stefano Poli. The authors also thank Claudia Romano and Brent T. Poe for the stimulating discussions that have supported some ideas in this paper.

SUPPLEMENTAL MATERIAL

CB2 at 4.7 GPa, 1565 °C

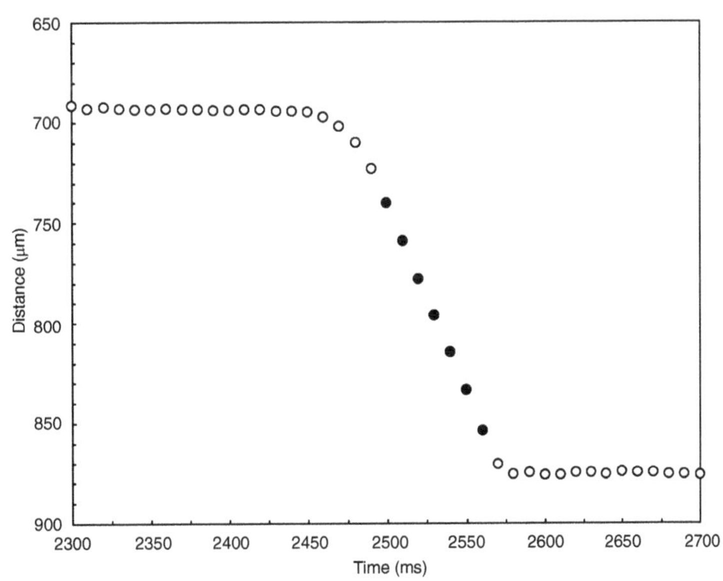

Figure S1 Results of the falling distance of the Pt sphere (117 µm) for the run at 4.7 GPa and 1565 °C as function of time.

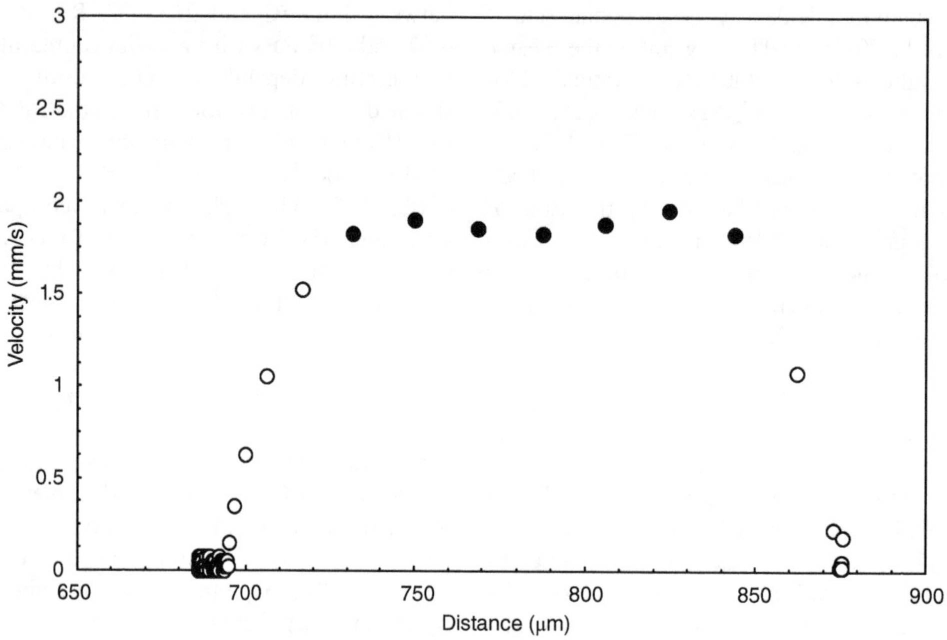

Figure S2 Results of the falling velocity of the Pt sphere for the run at 4.7 GPa and 1565 °C.

CARB2 at 2.8 GPa, 1885 °C

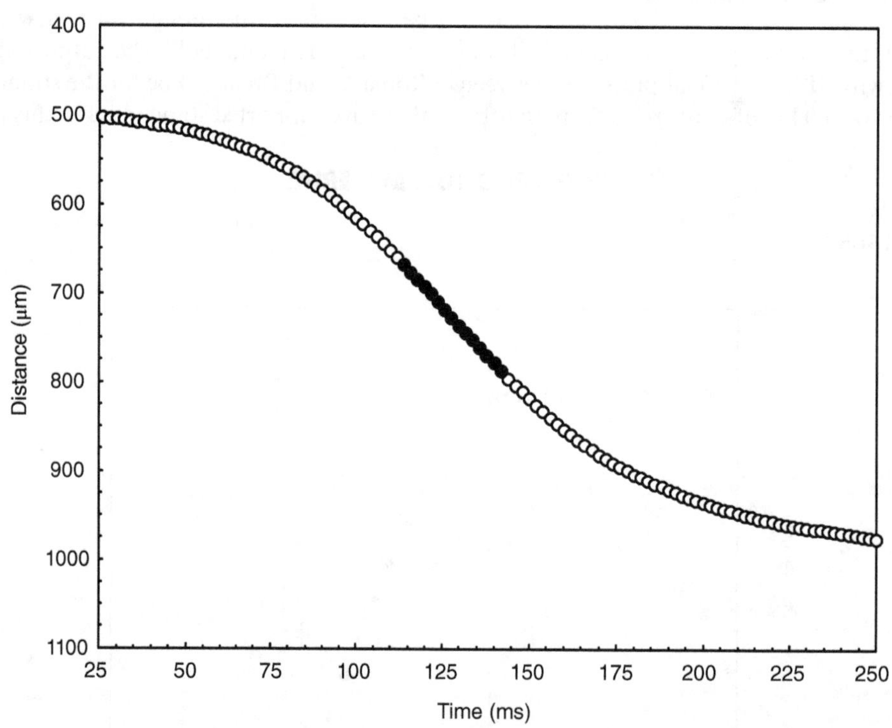

Figure S3 Results of the falling distance of the Pt sphere (191 μm) for the run at 2.8 GPa and 1885 °C as function of time.

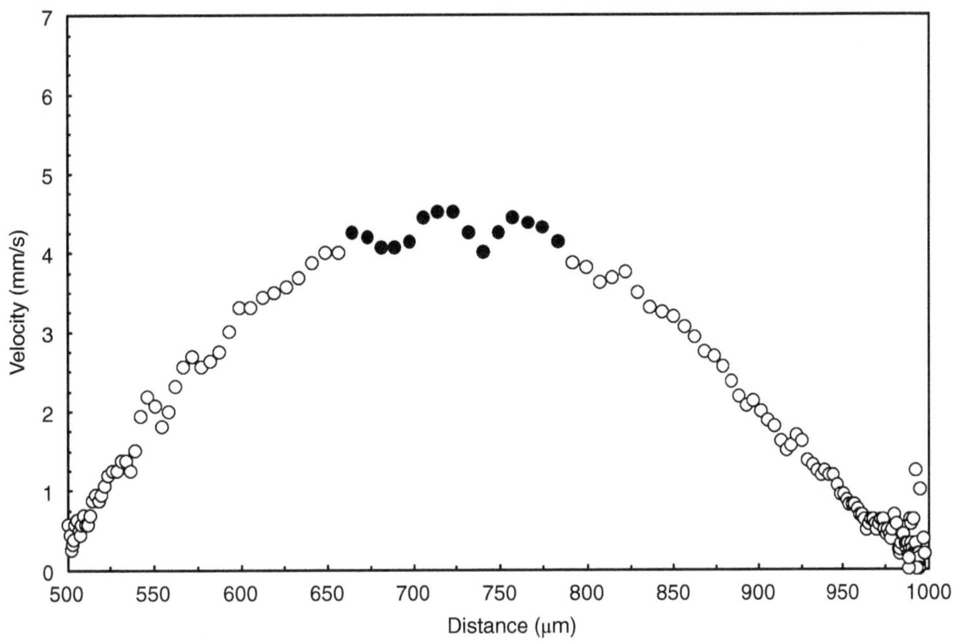

Figure S4 Results of the falling velocity of the Pt sphere for the run at 2.8 GPa and 1885 °C.

CARB2 at 4.7 GPa, 2012 °C

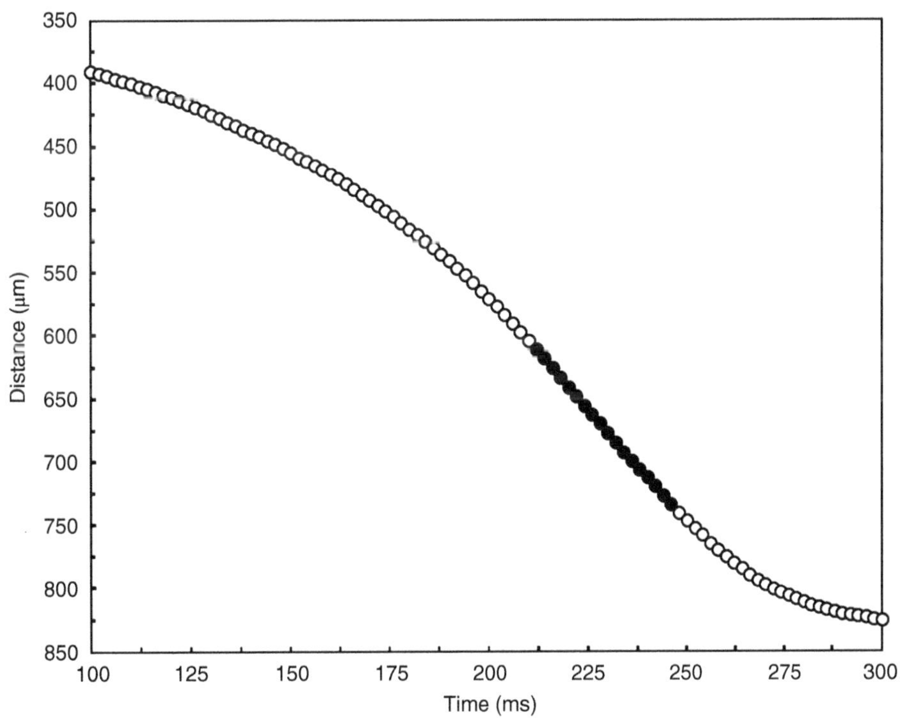

Figure S5 Results of the falling distance of the Pt sphere (140 µm) for the run at 4.7 GPa and 2012 °C as function of time.

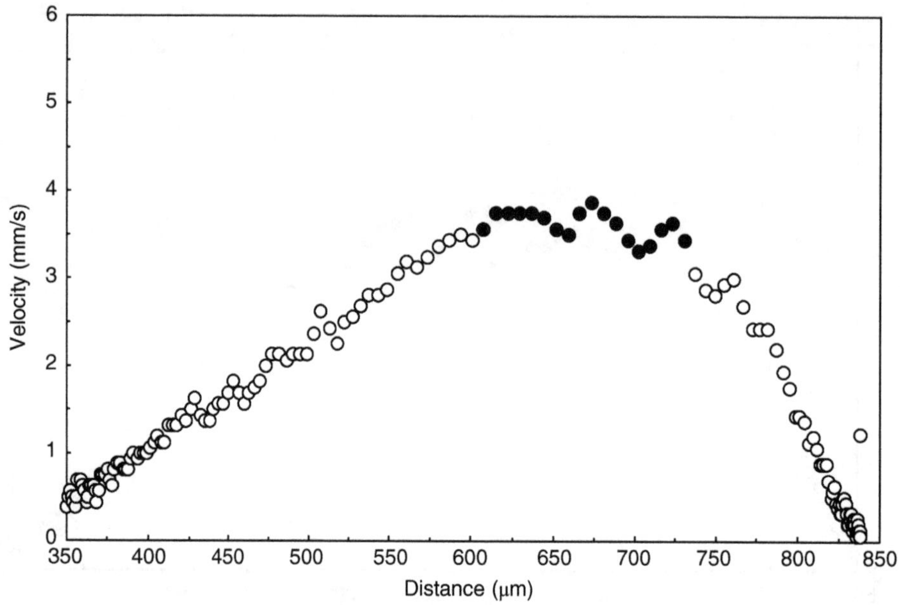

Figure S6 Results of the falling velocity of the Pt sphere for the run at 4.7 GPa and 2012 °C.

CARB2 at 5.3 GPa, 2155 °C

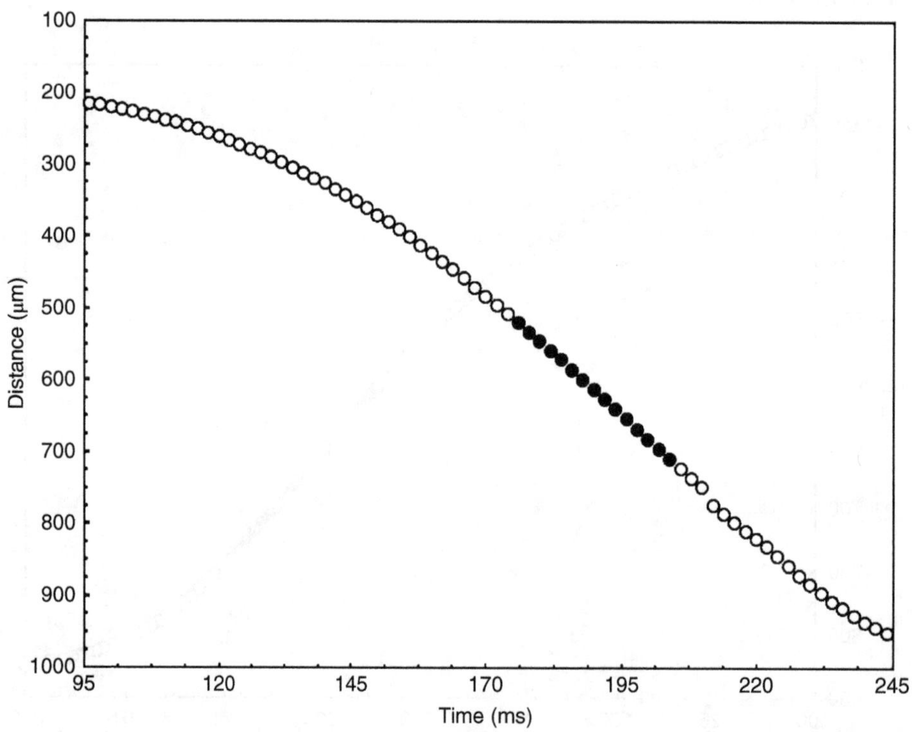

Figure S7 Results of the falling distance of the Pt sphere (140 μm) for the run at 5.3 GPa and 2155 °C as function of time.

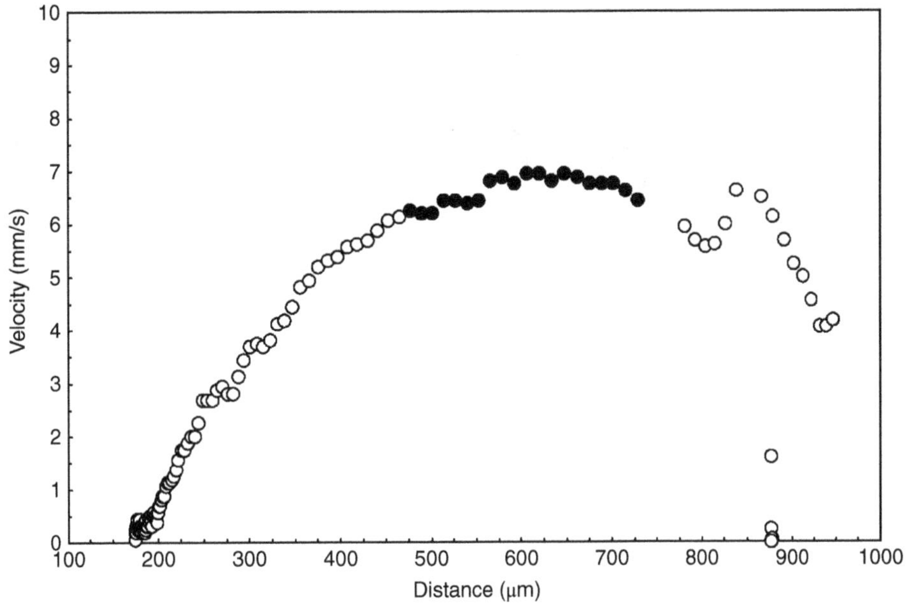

Figure S8 Results of the falling velocity of the Pt sphere for the run at 5.3GPa and 2155 °C.

CARB2 at 2.4 GPa, 1930 °C

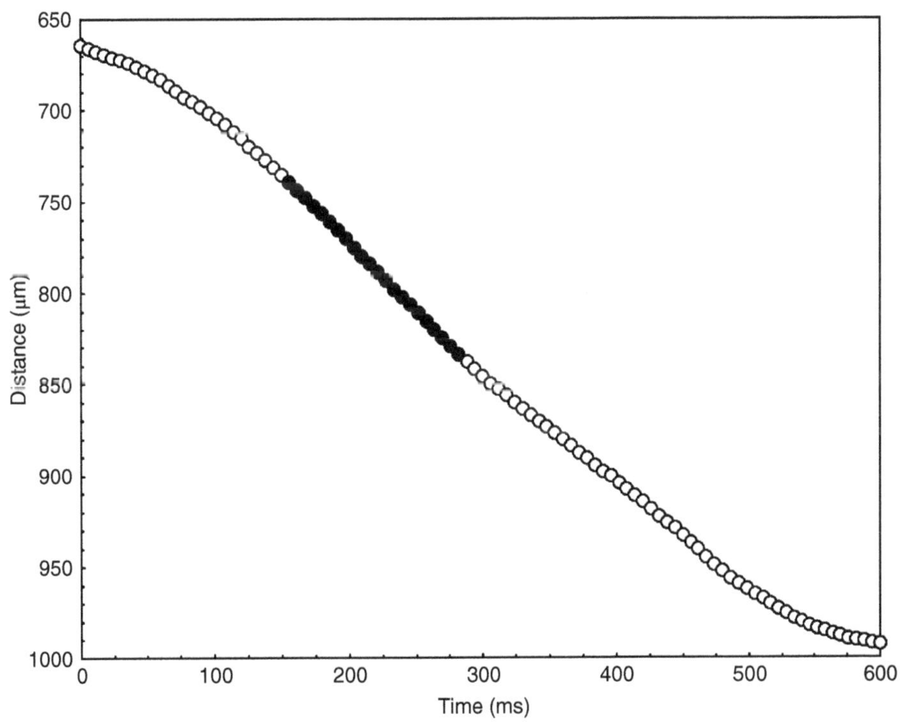

Figure S9 Results of the falling distance of the Pt sphere (86 μm) for the run at 2.4 GPa and 1930 °C as function of time. See electronic version for color representation of the figures in this book.

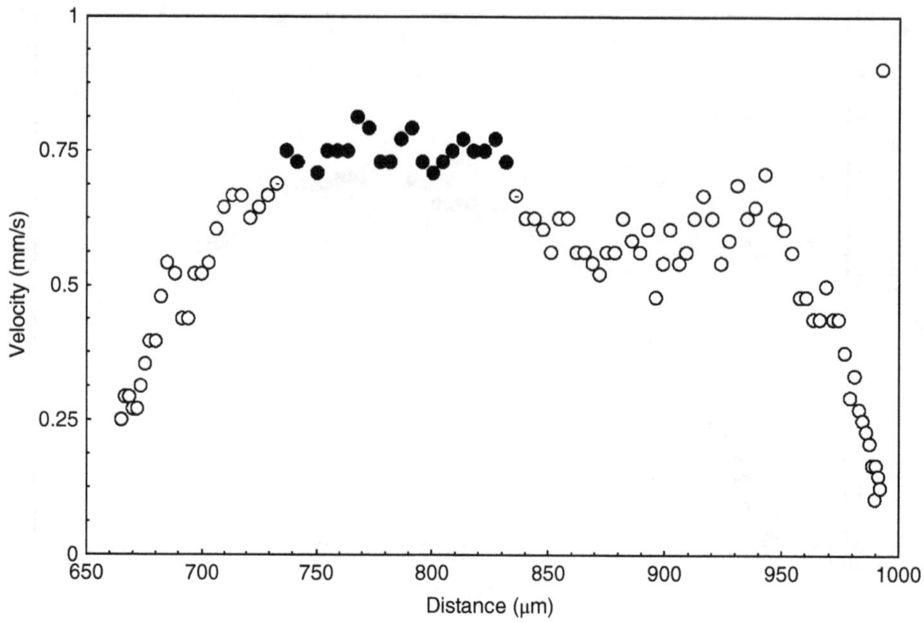

Figure S10 Results of the falling velocity of the Pt sphere for the run at 2.4 GPa and 1930 °C.

CARB2 at 3 GPa, 1955 °C

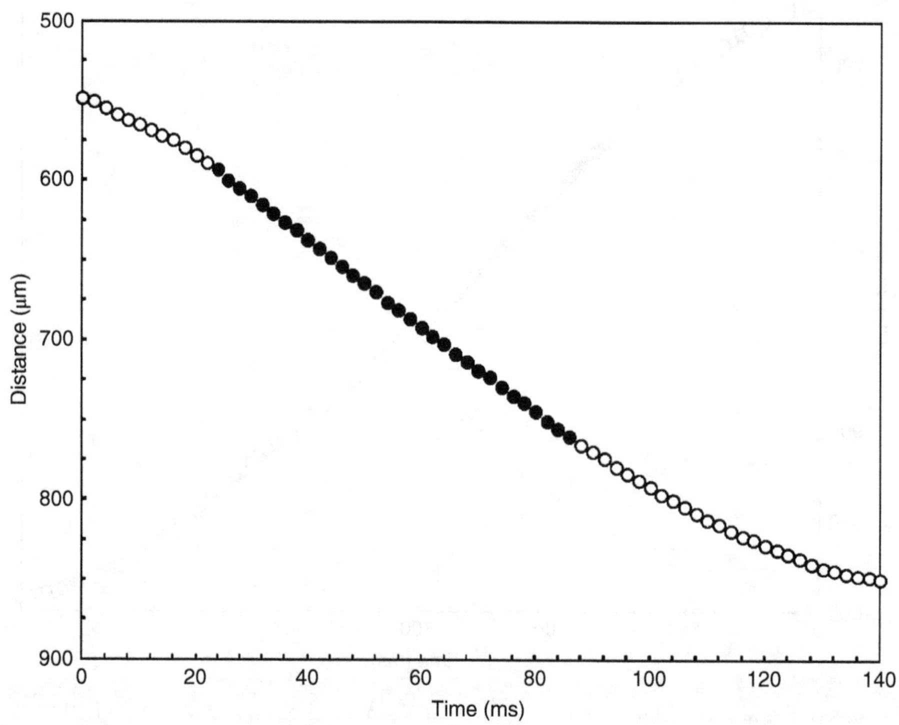

Figure S11 Results of the falling distance of the Pt sphere (89 μm) for the run at 3 GPa and 1955 °C as function of time.

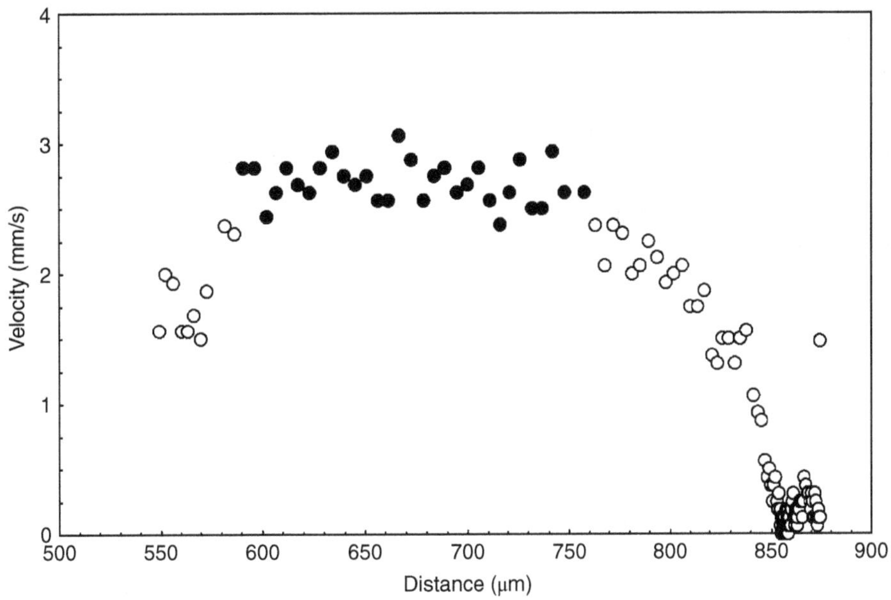

Figure S12 Results of the falling velocity of the Pt sphere for the run at 3 GPa and 1955 °C.

REFERENCES

Abramoff, M. D., Magalhaes, P. J., & Ram, S. J. (2004). Image processing with ImageJ. *Biophotonics International, 11*, 7, 36–42.

Aulbach, S., & Stagno, V. (2016). Evidence for a reducing Archean ambient mantle and its effects on the carbon cycle. *Geology, 44*, 751–754.

Brey, G. P., Bulatov, V. K., Girnis, A. V., & Lahaye, T. (2008). Experimental melting of carbonated peridotite at 6–10 GPa. *J. Petrol. 49*, 4, 797–821.

Brey, G. P., Bulatov, V. K., & Girnis, A. V. (2011). Melting of K-rich carbonated peridotite at 6–10 GPa and the stability of K-phases in the upper mantle. *Chem. Geol., 281*, 333–342.

Brooker, R. A., & Kjarsgaard, B. A. (2011). Silicate–carbonate liquid immiscibility and phase relations in the system SiO_2–Na_2O–Al_2O_3–CaO–CO_2 at 0. 1–2.5 GPa with applications to carbonatite genesis. *J. Petrol., 52*, 7&8, 1281–1305.

Dalton, J. A., & Presnall, D. C. (1998). Carbonatitic melts along the solidus of model lherzolite in the system CaO-MgO-Al_2O_3-SiO_2-CO_2 from 3 to 7 GPa. *Contrib Mineral Petrol, 131*, 123–135.

Dasgupta, R., & Hirschmann, M. M. (2007). Effect of variable carbonate concentration on the solidus of mantle peridotite. *Am. Mineral., 93*, 370–379.

Dasgupta, R., Mallik, A., Tsuno, K., Withers, A. C., Hirth, G., & Hirschmann, M. M. (2013). Carbon-dioxide-rich silicate melt in the Earth's upper mantle. *Nature, 493*, 211–216.

De Grouchy, C. J. L., Sanloup, C., Cochain, B., Drewitt, J. W. E., Kono, Y., & Crépisson, C. (2017). Lutetium incorporation in magmas at depth: Changes in melt local environment and the influence on partitioning behaviour. *Earth and Planetary Science Letters, 464*, 155–165.

Di Genova, D., Cimarelli, C., Hess, K., & Dingwell, D. B. (2016). An advanced rotational rheometer system for extremely fluid liquids up to 1273 K and applications to alkali carbonate melts. *Am. Mineral., 101*, 953–959.

Dobson, D. P., Jones, A. P., Rabe, R., Sekine, T., Kurita, K., Taniguchi, T., et al. (1996). In-situ measurement of viscosity and density of carbonate melts at high pressure. *Earth Planet. Sci. Lett., 143*, 207–215.

Dorogokupets, P. I., & Dewaele, A. (2007). Equations of state of MgO, Au, Pt, NaCl-B1, and NaCl-B2: Internally consistent high-temperature pressure scales. *High Press. Res., 27*, 431–446.

Foley, S. F., Yaxley, G. M., Rosenthal, A., Buhre, S., Kiseeva, E. S., Rapp, R. P., & Jacob, D. E., (2009). The composition of near-solidus melts of peridotite in the presence of CO_2 and H_2O between 40 and 60 kbar. *Lithos, 1125*, 274–283.

Gaillard, F., Malki, M., Iacono-Marziano, G., Pichavant, M., & Scaillet, B. (2008). Carbonatite melts and electrical conductivity in the asthenosphere. *Science, 322*, 1363–1365.

Ghosh, S., Litasov, K., & Ohtani, E. (2014). Phase relations and melting of carbonated peridotite between 10 and 20 GPa: a proxy for alkali- and CO_2-rich silicate melts in the deep mantle. *Contrib Mineral Petrol, 167*, 964.

Ghosh, D. B., Bajgain, S. K., Mookherjee, M., & Karki, B. B. (2017). Carbon-bearing silicate melt at deep mantle conditions. *Sci. Rep., 7*, 848.

Ghosh, D. B., & Karki, B. B. (2017). Transport properties of carbonated silicate melt at high pressure. *Sci. Adv., 3*, e1701840.

Gudfinnsson, G. H., & Presnall, D. C., (2005). Continuous gradations among primary kimberlitic, carbonatitic, melilitic, basaltic, picritic, and komatiitic melts in equilibrium with garnet lherzolite at 3–8 GPa, *J. Petrol., 46*, 1645–1659.

Hammouda, T. & Keshav, S. (2015). Melting in the mantle in the presence of carbon: Review of experiments and discussion on the origin of carbonatites. *Chem. Geol., 418*, 171–188.

Hier-Majumder, S., & Tauzin, B. (2017). Pervasive upper mantle melting beneath the western US. *Earth and Planetary Science Letters, 463*, 25–35.

Hudspeth, J., Sanloup, C., & Kono, Y. (2018). Properties of molten $CaCO_3$ at high pressure. *Geochemical Perspectives Letters*, 17–21.

Keller, T., Katz, R. F., & Hirschmann, M. M. (2017). Volatiles beneath mid-ocean ridges: Deep melting, channelised transport, focusing, and metasomatism. *Earth Planet. Sci. Lett.*, *464*, 55–68.

Keshav, S., & Gudfinnsson, G. H. (2014). Melting phase equilibria of model carbonated peridotite from 8 to 12 GPa in the system $CaO–MgO–Al_2O_3–SiO_2–CO_2$ and kimberlitic liquids in the Earth's upper mantle. *Am. Mineral.*, *99*, 1119–1126.

Kiseeva, E. S., Yaxley, G. M., Hermann, J., Litasov, K. D., Rosenthal, A., & Kamenetsky, V. S. (2012). An experimental study of carbonated eclogite at 3˙5–5˙5 GPa: Implications for silicate and carbonate metasomatism in the cratonic mantle. *J. Petrol.*, *53*, 4, 727–759.

Kiseeva, E. S., Litasov, K. D., Yaxley, G. M., Ohtani, E., & Kamenetsky, V. S. (2013). Melting and phase relations of carbonated eclogite at 9–21GPa and the petrogenesis of alkali-rich melts in the deep mantle. *J. Petrol.*, *50*, 1555–1583.

Kono, Y., Irifune, T., Higo, Y., Inoue, T., & Barnhoorn, A. (2010). P-V-T relation of MgO derived by simultaneous elastic wave velocity and in-situ X-ray measurements: A new pressure scale for the mantle transition region. *Phys. Earth Planet. Inter.*, *183*, 196–211.

Kono, Y., Kenney-Benson, C., Hummer, D., Ohfuji, H., Park, C., Shen, G., et al. (2014a). Ultralow viscosity of carbonate melts at high pressures. *Nat. Commun.*, *5*, 5091.

Kono, Y., Park, C., Kenney-Benson, C., Shen, G., & Wang, Y. (2014b). Toward comprehensive studies of liquids at high pressures and high temperatures: Combined structure, elastic wave velocity, and viscosity measurements in the Paris–Edinburgh cell. *Phys. Earth Planet. Inter.*, *228*, 269–280.

Kono, Y. (2018). Viscosity measurement. In Y. Kono, & C. Sanloup (Eds.), *Magmas under pressure: Advances in high-pressure experiments on structure and properties of melts*. (pp. 261–280). Elsevier.

Litasov, K. D., & Ohtani, E. (2009a). Phase relations in the peridotite-carbonate-chloride system at 7. 0–16. 5 GPa and the role of chlorides in the origin of kimberlite and diamond. *Chemical Geology*, *262*, 29–41.

Litasov, K. D., & Ohtani, E. (2009b). Solidus and phase relations of carbonated peridotite in the system $CaO-Al_2O_3-MgO-SiO_2-Na_2O-CO_2$ to the lower mantle depths: *Physics of the Earth and Planetary Interiors*, *177*, 46–58.

Litasov, K. D., & Ohtani, E. (2010). The solidus of carbonated eclogite in the system $CaO–Al_2O_3–MgO–SiO_2–Na_2O–CO_2$ to 32 GPa and carbonatite liquid in the deep mantle. *Earth Planet. Sci. Lett.*, *295*, 115–126.

Liu, Q., & Lange, R. A. (2003). New density measurements on carbonate liquids and the partial molar volume of the $CaCO_3$ component. *Contrib. Mineral. Petrol.*, *146*, 370–381.

Martin, L.H.J., Schmidt, M. W., Mattsson, H. B., & Guenther, D. (2013). Element partitioning between immiscible carbonatite and silicate melts for dry and H_2O-bearing systems at 1–3 GPa, *Journal of Petrology*, *54*, 11, 2301–2338.

Masotta, M., Freda, C., Paul, T. A., Moore, G. M., Gaeta, M., Scarlato, P., & Troll, V. (2012). Low pressure experiments in piston cylinder apparatus: Calibration of newly designed 25 mm furnace assemblies to P = 150 MPa. *Chem. Geol.*, *312–313*, 74–79.

Moore, K. R., & Wood, B. J. (1998). The transition from carbonate to silicate melts in the $CaO–MgO–SiO_2–CO_2$ system. *J. Petrol.*, *39*, 1943–1951.

Moussallam, Y., Morizet, Y., Massuyeau, M., Laumonier, M., & Gaillard, F. (2015). CO_2 solubility in kimberlite melts. *Chem. Geol.*, *418*, 198–205.

Rohrbach, A., & Schmidt, M. W. (2011). Redox freezing and melting in the Earth's deep mantle resulting from carbon–iron redox coupling. *Nature*, *472*, 209–212.

Sakamaki, T., Suzuki A., Terasaki H., Urakawa S., Katayama Y., Funakoshi K., et al. (2013). Ponded melt at the boundary between the lithosphere and asthenosphere, *Nat. Geosci.*, *6*, 1041–1044.

Stagno, V., & Frost, D. J. (2010). Carbon speciation in the asthenosphere: Experimental measurements of the redox conditions at which carbonate-bearing melts coexist with graphite or diamond in peridotite assemblages. *Earth. Planet. Sci. Lett.*, *30*, 72–84.

Stagno, V., Ojwang, D. O., McCammon, C. A., & Frost, D. J. (2013). The oxidation state of the mantle and the extraction of carbon from Earth's interior. *Nature*, *11679*, 493.

Stagno, V., Stopponi, V., Kono, Y., Manning, C. E., & Irifune, T. (2018). Experimental determination of the viscosity of Na_2CO_3 melt between 1. 7 and 4. 6 GPa at 1200–1700 °C: Implications for the rheology of carbonatite magmas in the Earth's upper mantle. *Chem. Geol.*, *501*, 19–25.

Stixrude, L., & Lithgow-Bertelloni, C. (2007). Influence of phase transformations on lateral heterogeneity and dynamics in Earth's mantle. *Earth Planet. Sci. Lett.*, *263*, 45–55.

Terasaki, H., Kato, T., Urakawa, S., Funakoshi, K. I., Suzuki, A., Okada, T., & Kasai, S. (2001). The effect of temperature, pressure, and sulfur content on viscocity of the Fe-FeS melt. *Earth and Planetary Science Letters*, *190*(1–2), 93–101.

Wolff, J. A. (1994). Physical properties of carbonatite magmas inferred from molten salt data, and application to extraction patterns from carbonatite–silicate magma chambers. *Geol Mag*, *131*, 145–153.

Yoshino, T., McIsaac, E., Laumonier, M., & Katsura, T. (2012). Electrical conductivity of partial molten carbonate peridotite, *Physics of the Earth and Planetary Interiors*, *194–195*, 1–9.

20

Mixed Fluids of Water and Carbon Dioxide

Evan Abramson

ABSTRACT

Equations of state, miscibilities, and speciation in mixed fluids of water and carbon dioxide are discussed. A shift in speciation away from solvated molecules of carbon dioxide increases miscibilities. HKF theory as currently formulated is found incapable of properly handling high pressure data, as partial molal volumes asymptote to a constant value with increasing pressure. Current knowledge of the miscibility surface allows an easy quantification of the fractions of water and carbon dioxide loaded into a diamond anvil cell, and thus should facilitate further research into these mixtures.

20.1. INTRODUCTION

Within Earth's crust and mantle, mixed fluids of water and carbon dioxide are agents for metasomatism, induce melting in the surrounding rock, and may be presumed to be important vehicles for carbon transport both within subducting slabs and into and out of the mantle. Beyond the planetary sciences, high-pressure mixtures of these molecules are involved in many industrial concerns, from oil and gas wells (and possible underground carbon sequestration), to supercritical water oxidation, to explosives technology. As well, both water and carbon dioxide are inexpensive and nontoxic solvents; industrial processes may be envisioned that would profit from these desirable traits coupled with the chemical tunability of mixed-solvent properties through variation of pressure, temperature, and mixing ratios.

Models of the mixed fluids inform our understanding of all the above processes. This chapter explores our knowledge of the thermodynamic properties of solutions of water and carbon dioxide at pressures above, roughly, 0.5 GPa, how such properties have been modeled, and how predictions of the models compare with measurements. While applicability to the deep Earth is obvious, even "high pressure" industrial processes are typically undertaken at pressures less than 0.5 GPa, and so it might be thought that a study restricted to conditions above this limit would be irrelevant to industry. However, desired thermodynamic data often do not exist at pressures intermediate between a few MPa and those considered here, and extrapolations into this range, from lower pressure data, can be erroneous. Interpolations, on the other hand, between the data more easily obtained at a few MPa and and those available above 0.5 GPa may produce useful predictions of properties at the desired intermediate pressures.

Department of Earth and Space Sciences, University of Washington, Seattle, Washington, USA

20.2. AVAILABLE DATA

Mixtures of water and carbon dioxide have been extensively investigated up to pressures of several hundred MPa, with experimental results considerably less available at the higher pressures considered here (see compilations in Duan & Zhang, 2006; Gernert & Span, 2016; Mader, 1991). Data include direct measurements of molar volume, miscibilities, conditions of chemical equilibria with thermodynamically characterized solids, and spectroscopic observations of speciation. Additionally, computer simulations have been used to predict both equations of state (EOS) and speciation (Brodholt & Wood, 1993; Duan & Zhang, 2006; Fu et al., 2017; Pan &

Carbon in Earth's Interior, Geophysical Monograph 249, First Edition. Edited by Craig E. Manning, Jung-Fu Lin, and Wendy L. Mao.

Galli, 2016). Only at quite low pressures, <30 MPa, have specific heats and enthalpies of mixing been reported. Regions in which higher pressure (experimental) data are available are shown in Figure 20.1. With the exception of the rectangle at lowest pressures, each of the depicted regions derives from a single study.

Of particular note are two studies of synthetic inclusions, that of Sterner and Bodnar (1991), up to 0.6 GPa and from 400 °C to 700 °C taken in a quartz host, and of Frost and Wood (1997), between 1 and 2 GPa and from 1100 °C to 1400 °C taken in corundum. Both indicate that excess volumes of mixing are positive and small (<4%), within the explored ranges of pressure and temperature. Volumes deduced from inclusions are uncertain by roughly 1%–2%, only slightly less than the reported excess volumes of mixing, but as pointed out, "The relative incompressibility ... at high pressure (high density) tends to magnify the effect of even subtle non-ideal behavior" (Sterner & Bodnar, 1991).

Chemical potentials have been determined directly from observations of chemical equilibria involving the carboxylation/decarboxylation or hydration/dehydration of various minerals in contact with solutions of defined mole fraction CO_2 (XCO_2). Such experiments have been run at temperatures as low as 310 °C and up to 1200 °C; pressures have typically not exceeded 0.6 GPa, although a few measurements have been made at substantially higher pressures. In particular, Aranovitch and Newton (1999) observed both decarboxylation and dehydration reactions of silicate minerals at pressures up to 1.4 GPa and from 600 °C to 1100 °C, while Eggler et al. (1979) reported on a decarboxylation reaction at 2.6 GPa and 1100 °C.

Speeds of sound in mixed solutions have more recently been measured in a high-pressure diamond anvil cell (DAC) by Brillouin spectroscopy (Qin et al., 2010) and by impulsively stimulated scattering (Abramson, unpublished). A set of such measurements, accurate over a broad range of P and T, would be extremely useful. Unfortunately, at the temperatures of interest, pressure gauges of the usual kind either dissolve, have a poor (e.g., ±0.1 GPa) precision, or both. Thus, the data taken in these preliminary experiments are not likely to be useful.

Observations of miscibility and of speciation taken in the DAC are also included in Figure 20.1, and will be discussed below. What is apparent from this initial diagram, however, is that information relevant to the nature of the CO_2/H_2O fluid above pressures of half a GPa are few, span two separate ranges of pressure and temperature, and that these two ranges comprise quite different types of data. For this reason, the situation has not been greatly advanced since 1991, when Mader wrote, "It is concluded that the data-base available is not yet adequate to derive a reliable equation of state for H_2O-CO_2 mixtures."

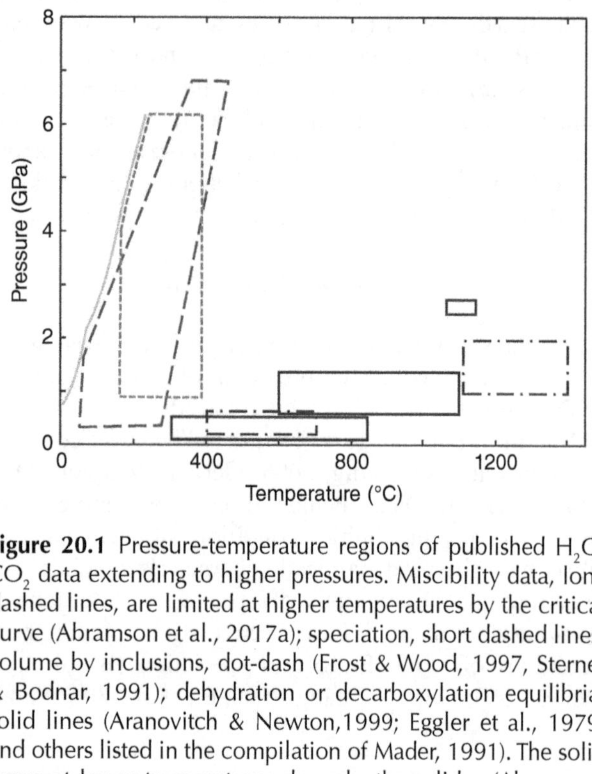

Figure 20.1 Pressure-temperature regions of published H_2O/CO_2 data extending to higher pressures. Miscibility data, long dashed lines, are limited at higher temperatures by the critical curve (Abramson et al., 2017a); speciation, short dashed lines; volume by inclusions, dot-dash (Frost & Wood, 1997, Sterner & Bodnar, 1991); dehydration or decarboxylation equilibria, solid lines (Aranovitch & Newton,1999; Eggler et al., 1979, and others listed in the compilation of Mader, 1991). The solid curve at lower temperatures demarks the solidus (Abramson, 2017; Abramson et al., 2017b). See electronic version for color representation of the figures in this book.

20.3. MISCIBILITIES

A basic question to be asked about the nature of these fluids is, to what degree are water and carbon dioxide miscible at any given pressure and temperature? Until recently, the miscibility surface of this binary had been studied to a maximum pressure of 0.35 GPa (Mather & Franck, 1992; Sterner & Bodnar, 1991; Takenouchi & Kennedy, 1964; Todheide & Franck, 1963). Both Todheide and Franck (1963) and Takenouchi and Kennedy (1964) made direct measurements of the contents of coexisting phases, while Mather and Franck (1992) measured pressures and temperatures of homogenization for (three) samples of known contents, and Sterner and Bodnar (1991) inferred the same from the use of synthetic fluid inclusions. As seen in Figure 20.2, the critical curve descends in temperature from the critical point of pure water, 374 °C (at 22 MPa), to a minimum at roughly 266 °C (at 0.245 GPa) before beginning to rise again with further increase in pressure; the associated critical concentration at the saddle point is around XCO_2 = 42%. The values given in the preceding sentence are those from Todheide and Franck (1963), but though the four works are in loose agreement over the breadth of the measurements, they can still differ considerably, particularly in the region of high XCO_2. For example, in the two-fluid regime at 0.20 GPa and 260 °C, XCO_2 for the carbonic fluid is given as 47%, 57%, 63%, or 66%.

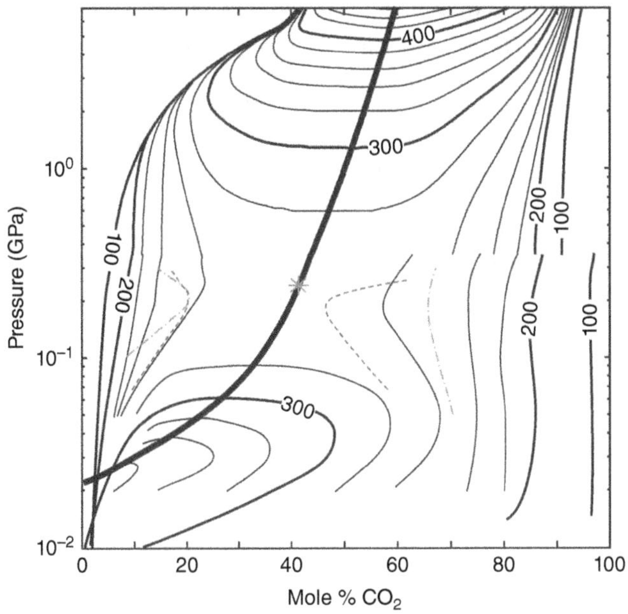

Figure 20.2 Miscibility surface, plotted against mole fraction carbon dioxide and pressure (Abramson et al., 2017a); contours of temperature (°C) are given in 20 °C intervals. Below 0.35 GPa, curves derive from data of Todheide and Franck (1963). The thick line indicates the critical curve, with an asterisk denoting the saddle point. For comparison, the 260° contours from the works of Takenouchi and Kennedy (1964) (dashed lines) and of Sterner and Bodnar (1991)(dot and dash) are also shown (the study of Mather and Frank (1992) comprises only three isopleths, insufficient to define a surface). See electronic version for color representation of the figures in this book

Abramson et al. (2017a) have now measured miscibilities of water–carbon dioxide mixtures, up to a pressure of 7 GPa, through the visual observation of homogenization (mixing) within a DAC. In these experiments, the cell contents were loaded as two separate fluids, the resulting XCO_2 being determined through isotopic doping of the water and subsequent measurement of the amount transferred to the CO_2. These studies are limited to pressures in excess of 0.4 GPa and thus do not overlap those previously cited, although they are in rough agreement with the results of Todheide and Franck (1963). Even with these new results, the critical curve continues to rise to the highest pressure and temperature of the experiments; the conditions under which two immiscible fluids can coexist thus extend yet higher, into unexplored regions.

An interesting feature of the data is the area of closely spaced contours above the saddle-point and centered around XCO_2 = 30%. In this region, an isothermal increase of pressure (e.g. applied to a fluid of XCO_2 = 30%, held at 300 °C, starting at 1 GPa) causes a homogeneous solution first to unmix, then to remix. The initial unmixing was the expected behavior for pressures exceeding that of the saddle-point. The remixing, seen so far only for XCO_2 < 40%, coincides with, and is presumably due to, a pressure-induced shift in speciation.

20.4. SPECIATION

At lower pressures, CO_2 exists as the solvated molecule, CO_2(aq). However, in Raman spectra taken at the higher pressures attained with the DAC (Abramson et al., 2017a), lines associated with CO_2(aq) disappear, to be replaced with a new line, hypothesized to be due to the carbonic acid molecule (Wang et al., 2016), but more likely to represent the appearance of the bicarbonate ion. The widely used Helgeson-Kirkham-Flowers (HKF) formalism (Helgeson et al., 1981) for solutes in aqueous solution, as revised by Shock et al. (1992), can be used to approximate the CO_2(aq)-HCO_3^{-1} equilibrium with results as seen in Figure 20.3. These calculations are necessarily approximate; the required thermodynamic parameters (Facq et al., 2014) for bicarbonate are supported by the bicarbonate-carbonate equilibrium observed at comparable pressure and temperature, but those pertaining to CO_2(aq) only by data at pressures <35 MPa. Additionally, the dielectric constant was taken to be that of pure water, and no attempt was made to calculate activity coefficients for a solution containing 17 mol% CO_2 (which coefficients were assumed to be 1, as for a dilute solution). Even so, the calculated pressures for equimolar solutions, while higher than those measured, are still within a few GPa, and are consistent with the hypothesis that the new species is bicarbonate rather than carbonic acid; data that would allow the corresponding calculation for carbonic acid do not exist. Guidance from theoretical calculations is available only from the work of Pan and Galli (2016), who have used molecular dynamics based on density functional theory to investigate speciation in a dilute (XCO_2 = 1.6%) solution at 11 GPa and 1000 K. Their results show, as expected, no CO_2(aq) to be present, with roughly 80% of the carbon existing as bicarbonate, and the balance either of equal amounts of carbonate and carbonic acid, or mostly carbonic acid, depending on the functional chosen to represent the exchange correlation.

From Figure 20.3, we see that the reaction of CO_2(aq) to form the new species is favored by increasing pressure and decreasing temperature, in accord with the HKF calculations for bicarbonate. The data of Facq et al. (2014) also show, similarly, that higher pressures favor deprotonation of bicarbonate to yield carbonate, while higher temperatures result in the reverse. HKF, however, did not reproduce the observed miscibility equilibrium of fluid carbon dioxide, CO_2(f), with CO_2(aq). Up to roughly 1 GPa experimental data are qualitatively reproduced as pressure rises, with CO_2 first forced into solution and

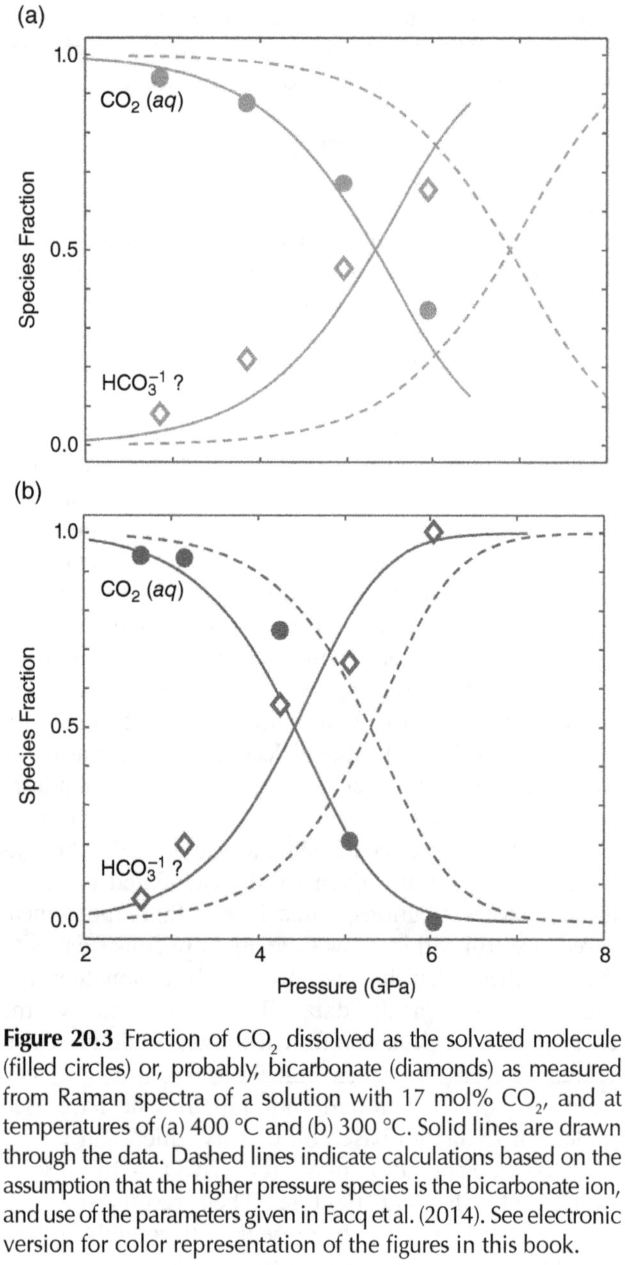

Figure 20.3 Fraction of CO_2 dissolved as the solvated molecule (filled circles) or, probably, bicarbonate (diamonds) as measured from Raman spectra of a solution with 17 mol% CO_2, and at temperatures of (a) 400 °C and (b) 300 °C. Solid lines are drawn through the data. Dashed lines indicate calculations based on the assumption that the higher pressure species is the bicarbonate ion, and use of the parameters given in Facq et al. (2014). See electronic version for color representation of the figures in this book.

then, with further increase in pressure, progressive unmixing. However, as pressure continues to rise, the experimentally observed remixing is not predicted, even while the concentration of CO_2(aq) is predicted (and seen) to diminish in favor of the bicarbonate species.

This divergence of experiment and theory was caused by the use of the current HKF formalism for the solutes in conjunction with experimentally determined fluid volumes for CO_2(f). Within the formalism, partial molar volumes at infinite dilution, V^0, are given by an expression that rapidly asymptotes to a constant value as pressure increases. The formula, developed for lower pressure work, thus results in a V^0 with an unphysical lower limit. Shown in Figure 20.4 are the predicted volumes of both CO_2(aq) and bicarbonate ion, as well as the experimentally determined values for pure CO_2 fluid. The solvated molecule upon compression is calculated to approach a limiting value of ~33 cm³/mol. The volume of the bicarbonate ion actually increases due to a diminution of the electrostrictive effect (as the compressibility of water also diminishes), but it too is predicted to reach a limiting value, of ~32 cm³/mol. In contrast, and as expected, the measured volume of CO_2(f) decreases without apparent limit. The result is that there will always be a pressure high enough that the increases of chemical potential, calculated for both aqueous species as $\int V^0 dP$, exceed that of the pure fluid, resulting in the incorrect prediction that the system will remain unmixed.

In contrast, recent analyses (Brown et al., 2018) of solution data indicate that at pressures near 1 GPa, partial volumes of a solute will approach that of the pure substance (for the same pressure and temperature). Partial volumes of CO_2(aq) and HCO_3^{-1} might then be expected to approximate to those of the pure fluid CO_2 rather than the behavior depicted in Figure 20.4. Although the current formalism can be forced to represent high pressure data (e.g. in Facq et al., 2014), any close fit to observed equilibria is presumably due to compensating errors and should not be expected to remain useful to higher pressures or in conjunction with experimentally derived volumes.

20.5. EQUATION OF STATE

Several equations of state of the binary mixture, taken from the literature (Aranovich & Newton, 1999; Holland & Powell, 2003; Kerrick & Jacobs, 1981; Zhang & Duan, 2010), have been tested against the miscibility surface (Abramson et al., 2017a). Measured and predicted critical curves are shown in Figure 20.5. Three of these EOS were based entirely on the measurements available at relatively low pressures and high temperatures, while Zhang and Duan (2010) supplemented data with molecular dynamic calculations up to 10 GPa, but with the assumption that the carbon dioxide continues to exist as the solvated molecule. Given the large disparity in the pressure-temperature region from which data in support of the various EOS derive, and the region of the miscibility surface where the EOS are being tested (see Figure 20.1), it is surprising that the predictions are even as close to experimental measurements as seen. The EOS of Holland and Powell (2003), for instance, relies on only a single, free fitting parameter for a small excess Gibbs energy and yet parallels (roughly) the observed critical curve.

For all four EOS, the experimentally determined two-phase region lies at higher pressures than predicted. It is

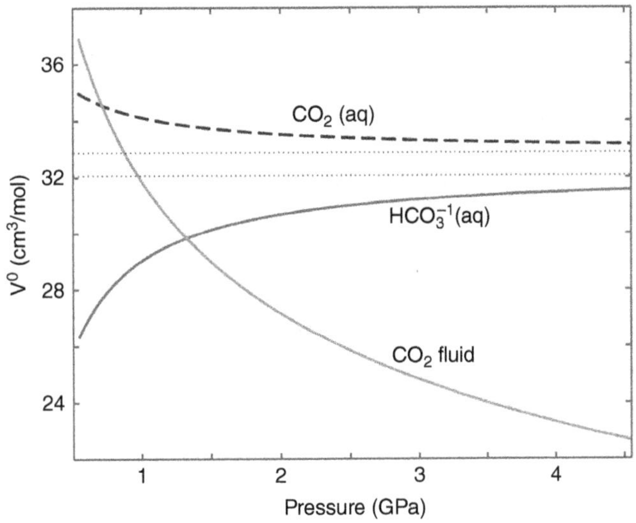

Figure 20.4 Partial molal volumes at infinite dilution, V^0, of solvated CO_2 and bicarbonate ion as calculated, as well as measured volumes of fluid CO_2 (Giordano et al., 2006), are plotted against pressure for a temperature of 300 °C. Dotted lines indicate asymptotes. See electronic version for color representation of the figures in this book.

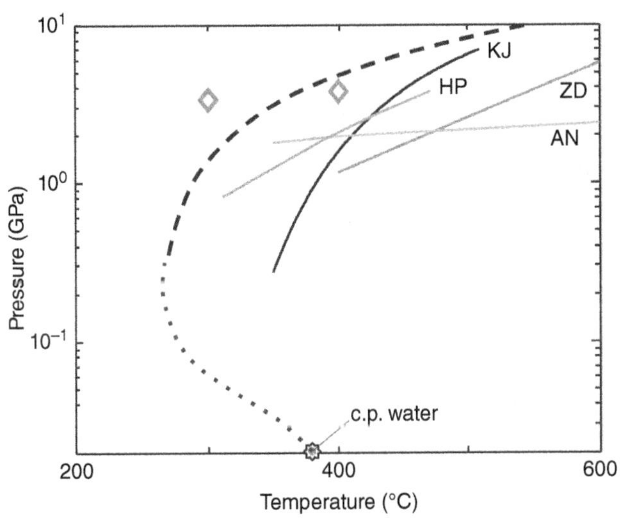

Figure 20.5 Critical pressures are plotted against temperature. Experimental data are from Todheide and Franck (1963) (dots) and Abramson et al. (2017a) (dashes); the critical point of pure water is also indicated. Predictions of four EOS are shown: Kerrick and Jacobs, 1981; Aranovich and Newton, 1999; Holland & Powell, 2003; Zhang & Duan, 2010. The two diamonds indicate points at which rising pressure has resulted in the conversion of roughly 20% of dissolved carbon dioxide from CO_2(aq) to another species (in a solution of XCO_2 = 17%). See electronic version for color representation of the figures in this book.

interesting to speculate as to whether this commonality is due to the observed shift in speciation, which might be expected to delay the onset of unmixing with rising pressure. The two diamonds drawn in the figure indicate where, in the data of Figure 20.3, the fraction of carbon dioxide existing as CO_2(aq) has been reduced by 20%, and might be taken to roughly indicate where the new species begins to significantly affect the EOS. Thus, the nature of the EOS may approximate to an ideal mixing at lower pressures, but less so at higher pressures due to an altered speciation.

20.6. AVENUES FOR PROGRESS

Experiments in the high-pressure DAC require a means of quantifying the fractions of the two independently loaded components. Values of the loaded XCO_2 can be estimated by a visual determination of the relative volumes of the immiscible phases (Qin, 2010); however, the accuracy of this method is dubious. In Abramson et al. (2017a), water was isotopically doped, and subsequent isotopic transfer within the DAC allowed proportions to be determined to ~3% absolute. The process, which is both laborious and time consuming, can only be performed after the end of the main experiment as it requires transferring contents of the DAC into a mass spectrometer. Fortunately, now that the miscibility has been mapped out, the contents of a DAC can be evaluated over much of the possible range of XCO_2 by comparison of the observed fluid-fluid mixing with the miscibility surface; this should facilitate further research on speciation in the binary. Of course, the same comparison might also be used to define CO_2/H_2O ratios in systems with additional components (e.g. SiO_2) as long as they do not dissolve appreciably at the lower temperatures of mixing.

As mentioned earlier, an (accurate) evaluation of speeds of sound would be very useful. Given the ability to determine XCO_2 from observations of mixing and unmixing, it may now be possible to use the divided cell method in the DAC (Abramson & Brown, 2004) with the relatively accurate (but, unfortunately, water soluble) $(Sm)SrB_4O_7$ pressure gauge, and either Brillouin spectroscopy or impulsively stimulated scattering. Traditional acoustic techniques may also be applied in large-volume presses (Song et al., 2011).

ACKNOWLEDGMENTS

This text was prepared with partial funding from National Science Foundation grant EAR-1829147.

REFERENCES

Abramson, E. H. (2017). Three-phase melting curves in the binary system of carbon dioxide and water. *Journal of Physics: Conference Series*, *950*, 042019. https://doi.org/10.1088/1742-6596/950/4/042019

Abramson, E. H., Bollengier, O., & Brown, J. M. (2017a). The water-carbon dioxide miscibility surface to 450 °C and 7 GPa. *American Journal of Science*, *317*, 967–989, https://doi.org/10.2475/09.2017.01

Abramson, E. H., Bollengier, O., & Brown, J. M. (2017b). Water-carbon dioxide solid phase equilibria at pressures above 4 GPa. *Scientific Reports*, *7*, 821. https://doi.org/10.1038/s41598-017-00915-0

Abramson, E. H., & Brown, J. M. (2004). Equation of state of water based on speeds of sound measured in the diamond-anvil cell. *Geochimica et Cosmochimica Acta*, *68*, 1827–1835. https://doi.org/10.1016/j.gca.2003.10.020

Aranovich, L. Y., & Newton, R. C. (1999). Experimental determination of CO_2-H_2O activity-composition relations at 600–1000 °C and 6–14 kbar by reversed decarbonation and dehydration reactions. *American Mineralogist*, *84*, 1319–1332.

Brodholt, J., & Wood, B. (1993). Molecular dynamics simulations of the properties of CO_2-H_2O mixtures at high pressures and temperatures. *American Mineralogist*, *78*, 558–564.

Brown, J. M., Bollengier, O., & Vance, S. (2018). *High-Pressure "Death" of Aqueous Electrolytic Chemistry*. Paper presented at WaterX Exotic Properties of Water under Extreme Conditions, Water X International workshop, La Maddalena, Sardinia (Italy).

Duan, Z., & Zhang, Z. (2006). Equation of state of the H_2O, CO_2, and H_2O–CO_2 systems up to 10 GPa and 2573.15 K: Molecular dynamics simulations with ab initio potential surface. *Geochimica et Cosmochimica Acta*, *70*, 2311–2324.

Eggler, D. H., Kushiro, I., & Holloway, J. R. (1979). Free energies of decarbonation reactions at mantle pressures: I. Stability of the assemblage forsterite-enstatite-mangesite in the system MgO-SiO_2-CO_2-H_2O to 60 kbar. *American Mineralogist*, *64*, 288–293.

Facq, S., Daniel, I., Montagnac, G., Cardon, H., & Sverjensky, D. A. (2014). In situ Raman study and thermodynamic model of aqueous carbonate speciation in equilibrium with aragonite under subduction zone conditions. *Geochimica et Cosmochimica Acta*, *132*, 375–390.

Frost, D. J., & Wood, B. J. (1997). Experimental measurements of the properties of H_20-CO_2 mixtures at high pressures and temperatures. *Geochimica et Cosmochimica Acta*, *61*, 3301–3309.

Fu, J., Zhao, J., Plyasunov, A. B, & Belonoshko, A. B. (2017). Ab initio molecular dynamics study of fluid H_2O-CO_2 mixture in broad pressure-temperature range. *AIP Advances*, *7*, 115217. https://doi.org/10.1063/1.5006131

Gernert, J., & Span, R. (2016). EOS–CG: A Helmholtz energy mixture model for humid gases and CCS gases and CCS mixtures. *The Journal of Chemical Thermodynamics*, *93*, 274–293.

Giordano, V. M., Datchi, F., & Dewaele, A. (2006) Melting curve and fluid equation of state of carbon dioxide at high pressure and high temperature. *The Journal of Chemical Physics*, *125*, 054504.

Helgeson, H. C., Kirkham, D. H., & Flowers, G. C. (1981). Theoretical predictions of the thermodynamic behavior of aqueous electrolytes at high pressures and temperatures: IV. Calculation of activity coefficients, osmotic coefficients, and apparent molal and standard and relative partial molal properties to 600 °C and 5Kb, *American Journal of Science*, *28*, 1249–1516.

Holland, T., & Powell, R. (2003). Activity–composition relations for phases in petrological calculations: An asymmetric multicomponent formulation. *Contributions to Mineralogy and Petrology*, *145*, 492–501.

Kerrick, D. M., & Jacobs, G. K. (1981). A modified Redlich-Kwong equation for H2O, CO2, and H2O–CO2 mixtures at elevated pressures and temperatures. *American Journal of Science*, *281*(6), 735–767. https://doi.org/10.2475/ajs.281.6.735

Mader, U. K. (1991). H_2O-CO_2 mixtures: A review of P-V-T-X data and an assessment from a phase-equilibrium point of view. *Canadian Mineralogist*, *29*, 767–790.

Mather, A. E., & Franck, E. U. (1992). Phase equilibria in the system carbon dioxide-water at elevated pressures. *The Journal of Physical Chemistry*, *96*, 6–8.

Pan, D., & Galli, G. A. (2016). The fate of carbon dioxide in water-rich fluids under extreme conditions. *Science Advances*, *2*, e1601278.

Qin, J., Li, M., Li, J., Chen, R., & Duan, Z. (2010). High temperatures and high pressures Brillouin scattering studies of liquid H2O+CO2 mixtures. *The Journal of Chemical Physics*, *133*, 154513. https://doi.org/10.1063/1.3495972

Shock E. L., Oelkers, E. H., Johnson, J. W., Sverjensky, D. A., & Helgeson, H. C. (1992). Calculation of the thermodynamic properties of aqueous species at high pressures and temperatures: Effective electrostatic radii, dissociation constants and standard partial molal properties to 1000 °C and 5 kbar. *Journal of the Chemical Society, Faraday Transactions*, *88*, 803–826.

Song, W., Liu, Y., Wang, Z., Gong, C., Guo, J., Zhou, W., & Xie, H. (2011). Measurement method for sound velocity of melts in large volume press and its application to liquid sodium up to 2.0 GPa. *Review of Scientific Instruments*, *82*, 086108. https://doi.org/10.1063/1.3625267

Sterner, S. M., & Bodnar, R. J. (1991). Synthetic fluid inclusions. X: Experimental determination of P-V-T-X properties in the CO_2-H_2O system to 6 kb and 700 °C. *American Journal of Science*, *291*, 1–54.

Takenouchi, S., & Kennedy, G. C. (1964). The binary system CO_2-H_2O at high temperatures and pressures. *American Journal of Science*, *262*, 1055–1074.

Todheide, K., & Franck, E. U. (1963). Das Zweiphasengebiet und die Kritische Kurve in System Kohlendioxide-Wasser bis zu Drucken von 3500 Bar. *Zeitschrift fuer Physikalische Chemie (Neue Folge)*, *37*, 387–401.

Wang, H., Zeuschner, J., Eremets, M., Troyan, I., & Willams, J. (2016). Stable solid and aqueous H_2CO_3 from CO_2 and H_2O at high pressure and high temperature. *Scientific Reports*, *6*, 19902.

Zhang, C., & Duan, Z. (2010). GFluid: An Excel spreadsheet for investigating C–O–H fluid composition under high temperatures and pressures. *Computers and Geosciences*, *36*, 569–572.

21

Experimental Determination of Calcite Solubility in H_2O-KCl-NaCl-LiCl Solutions at 700°C and 8 kbar

James Eguchi[1,2], Yuan Li[3], and Craig E. Manning[2]

ABSTRACT

The solubility of calcite in salt-H_2O fluids was determined at 700 °C and 8 kbar using hydrothermal piston-cylinder methods. The investigated salts included NaCl, KCl, LiCl, and CsCl. Results show that, at constant pressure and temperature, calcite solubility increases with increasing concentration of any individual salt and all investigated salt mixtures. Data regression indicates a simple dependence of solubility on the square of salt mole fraction. At a given salt concentration, solubility enhancement increases with decreasing salt cation size. Experiments on fluids with mixtures of multiple salts were used to derive simple relations that can be used to predict calcite solubility in a wide range of salt solutions at the studied conditions. The results provide a basis for extending this approach to different pressures, temperatures, and salt compositions.

21.1. INTRODUCTION

The calcium carbonate minerals calcite and aragonite are the primary reservoirs for oxidized carbon in the continental crust and upper mantle. They are also the dominant carriers of carbon into the Earth's interior (e.g., Dasgupta & Hirschmann, 2010). In subduction settings, models of decarbonation of model lithologies assuming a molecular fluid indicate that relatively little CO_2 is liberated by breakdown of calcite and aragonite at high pressure (e.g., Gorman et al., 2006). Experimental and theoretical studies, on the other hand, reveal strong increases in calcite and aragonite solubility in H_2O with pressure (e.g. Caciagli & Manning, 2003; Ellis, 1963; Facq et al., 2014; Fein & Walther, 1989; Kelemen & Manning, 2015; Newton & Manning, 2002), implying that dissolution as ionic constituents may be important in liberating CO_2 from the subducting slab. This hypothesis is at least locally supported by field evidence for congruent dissolution of calcium carbonate in exhumed subduction complexes (e.g., Ague & Nicolescu, 2014).

Experimental studies have also revealed that salinity plays a critical role in determining calcium carbonate solubility, especially at high P and T. Newton and Manning (2002) and Facq et al. (2016) showed that calcite solubility increases with increasing NaCl concentration, as well as with increasing P and T. These results are important because saline fluids may be common at high P and T (e.g. Andersen & Neumann, 2001; Kawamoto et al., 2013; Kent et al., 2002; Manning & Aranovich 2014; Newton et al. 1998; Philippot & Selverstone, 1991; Scambelluri & Philippot, 2001; Van den Berg & Huizenga, 2001; Manning 2018). These studies also reveal that the

[1] *Department of Earth, Environmental, and Planetary Sciences, Rice University, Houston, Texas, USA*

[2] *Department of Earth, Planetary, and Space Sciences, University of California– Los Angeles, Los Angeles, California, USA*

[3] *State Key Laboratory of Isotope Geochemistry, Guangzhou Institute of Geochemistry, Chinese Academy of Sciences, Guangzhou, China*

Carbon in Earth's Interior, Geophysical Monograph 249, First Edition. Edited by Craig E. Manning, Jung-Fu Lin, and Wendy L. Mao.

salt components in the fluid phase may be a complex mixture of NaCl, KCl, and potentially other solutes. Yet no experimental or theoretical study has previously addressed the relative roles of different salt components on $CaCO_3$ mineral solubility.

The present work was conducted to complement previous studies by Newton and Manning (2002) and Caciagli and Manning (2003) by investigating how different alkali halides and their mixtures affect calcite solubility at high P and T. We investigated calcite solubility in H_2O-salt mixtures at a fixed P and T of 8 kbar and 700 °C. The salt components were combinations of NaCl, KCl, LiCl, and CsCl. The experiments identify and constrain systematic variations in solubility that depend on concentration and salt identity and lead to a model for calcite solubility in H_2O-salt mixtures.

21.2. METHODS

21.2.1. Starting Materials

Experiments were designed to determine the solubility of calcite in aqueous solutions containing NaCl, KCl, LiCl, CsCl, and their mixtures at high P and T. Each experiment contained distilled and deionized H_2O, reagent grade salt (either a single salt or a mixture of salts), and pure natural calcite. Inclusion-free calcite rhombs of natural calcite from Rodeo, Durango, Mexico, weighing 0.5–3.5 mg were smoothed using a diamond file, and then polished using 150 grit sandpaper (Figure 21.1a).

21.2.2. Capsule Assembly

Experiments were performed using a double-capsule technique (Manning & Boettcher, 1994). Inner capsules were cut from 2 mm OD platinum tubes. One end of the capsule was crimped shut and a polished calcite crystal was weighed, then placed inside the inner Pt capsule. The inner Pt capsule was then crimped shut, punctured twice to allow penetration of the solution, and weighed. Outer Pt capsules were made with 3.5 mm OD Pt tubing. After annealing, one end was crimped and welded, then the capsule was weighed. The inner Pt capsule and salts were loaded, weighing after each addition. Prior to loading, all salts were placed in a 300 °C furnace to drive off adsorbed H_2O. H_2O was finally added; any H_2O in excess of the target value was allowed to evaporate until the desired mass was reached. Salts and H_2O were added in this fashion instead of as a premixed solution in order to achieve precisely measured salt concentrations that were often higher than salt saturation at ambient conditions. After loading, the open end of the capsule was crimped and welded. Weld integrity was checked by reweighing the capsule after holding at 110 °C furnace for 15 minutes; any capsules that lost mass were discarded.

The masses of the starting calcite crystal and inner Pt capsule were determined on a Mettler Toledo ultramicrobalance with $1\sigma = 0.2$ μg. The masses of the outer Pt capsule and all materials were measured on a Mettler M3 microbalance with $1\sigma = 2$ μg. When propagating 1σ uncertainty through each weighing step, resulting 1σ uncertainty in $X_{\text{total salt}}$ (see equation [21.2]) $< 1.7 \times 10^{-3}$ and $m_{CaCO3} < 1.9 \times 10^{-3}$ molal.

21.2.3. Experimental Apparatus and Setup

All experiments were conducted in a 2.54 cm diameter end-loaded piston cylinder apparatus, using NaCl-graphite furnaces. Capsules were loaded with long axes horizontal to minimize the temperature gradients. All experiments were conducted at 700 °C and 8 kbar. Temperature was monitored and controlled using Type S $Pt\text{-}Pt_{90}Rh_{10}$ thermocouples, with an uncertainty of ±3 °C. Experiments were first pressurized to ~6 kbar, then heated. Thermal expansion initially brought the sample to the target pressure of 8 kbar, and pressure was held to within ±150 bars of the desired pressure throughout the duration of the experiment by bleeding pressure when necessary. At final run conditions, the NaCl cells are frictionless and require no pressure correction (Manning & Boettcher 1994). All experiments were held at 700 °C, 8 kbar, for at least 4 hours, then quenched by cutting power. Temperatures dropped rapidly to <100 °C in <1 min.

21.2.4. Solubility Determination

Quenched capsules were retrieved from the furnace assembly, soaked in distilled water for 5–10 minutes to dissolve any adhering NaCl, and dried at 110 °C. After drying, the capsule was weighed to check for fluid loss during the run. A second test to assess whether H_2O was lost during the experiment was to puncture the capsule and place it stepwise in 110 °C and 300 °C furnaces to drive off all water, then the dried capsule was reweighed. The mass difference between the full and dried capsule corresponds to the "water out" in Tables 21.1 and 21.2 and provides second a check on fluid retention during an experiment. During heating to drive off H_2O, salts were sometimes lost from the punctured outer capsule due to vigorous boiling of H_2O; therefore, this measurement could not be made for all experiments. The outer Pt capsule was then cut open with a razor blade. The outer capsule contained precipitated salts and quenched skeletal calcite crystals. The inner Pt capsule was then retrieved and inspected for any accidental damage caused by the

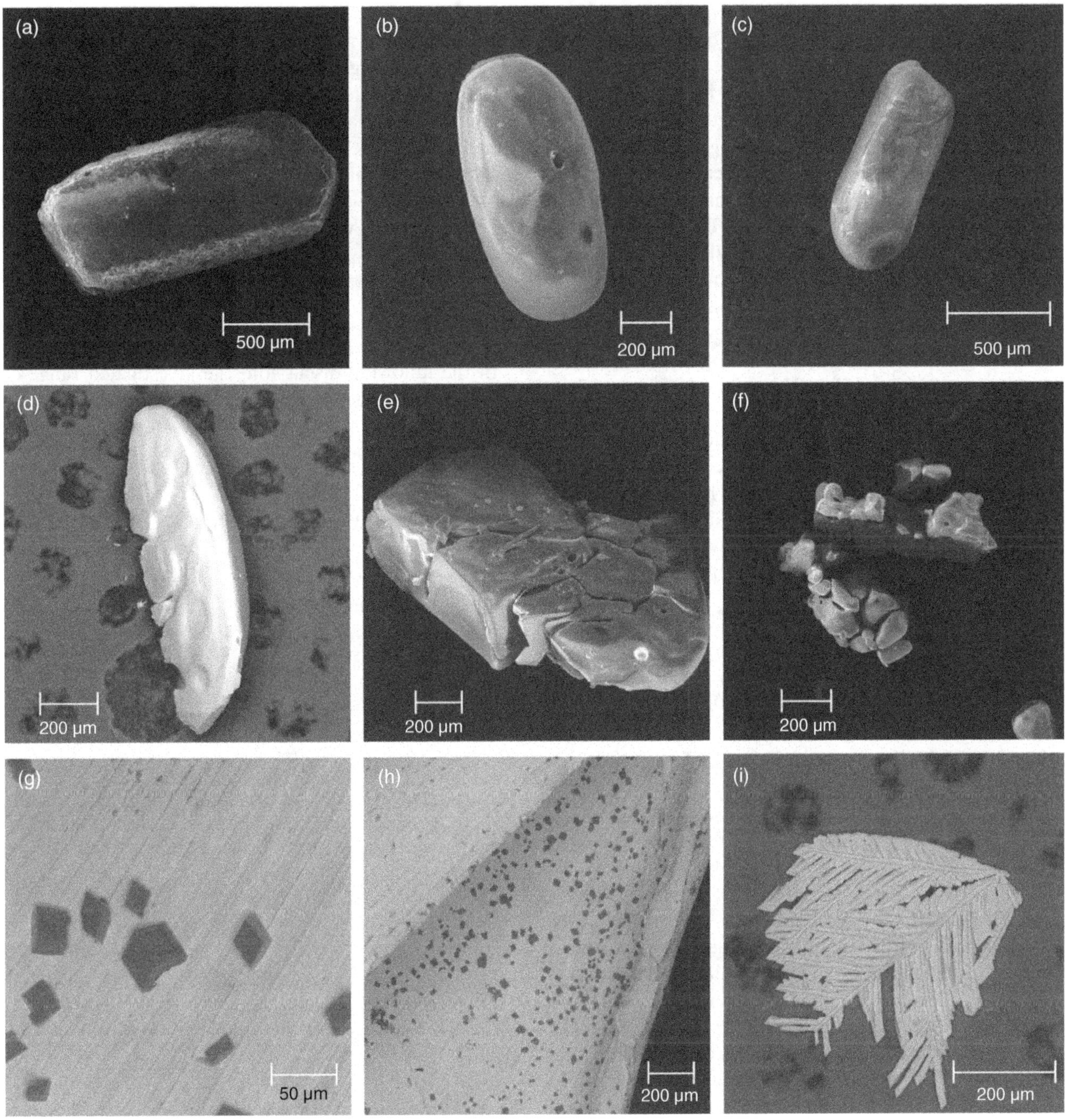

Figure 21.1 SEM images of starting crystal and run products. (b)–(f) are ordered from lowest to highest calcite solubility to illustrate the progression of textures with increasing dissolved $CaCO_3$. Values in parentheses refer to experiment number and SEM mode. (a) Starting crystal (SE). (b) Dissolution-rounded calcite crystal (L-19; SE). (c) Dissolution-rounded calcite crystal (MT-4; SE). (d) Dissolution-rounded calcite crystal showing incipient pits and grooves (NL-1; BSE). (e) Dissolution-rounded calcite crystals displaying deep pits and grooves (L-10; SE). (f) Aggregate of dissolution-rounded calcite crystals (L-11; SE). (g) Small, euhedral calcite quench crystal on inner surface of inner Pt capsule (NL-3; BSE). (h) Same as (g) but lower magnification, showing distribution of calcite quench crystals (BSE). (i) Skeletal calcite quench crystals observed only in runs with very high solubility, interpreted to be due to strong oversaturation upon quenching (run L-14; BSE). See electronic version for color representation of the figures in this book.

razor blade when opening the outer Pt capsule. Experiments were discarded if inner Pt capsule had been damaged and/or appeared to have lost any Pt.

The inner capsule was placed in a beaker of distilled water which was then stirred and heated for ~5 minutes. The capsule was then dried and weighed. This process was repeated until the inner capsule came to a constant mass, which signaled that all salts had been removed, leaving behind only insoluble calcite. This procedure has a negligible effect on the final result due to calcite's extremely low solubility in H_2O at ambient conditions. Calcite dissolves congruently in H_2O and H_2O-salt solutions (Caciagli & Manning, 2003; Newton & Manning, 2002), so the weight change of a crystal yields a direct measurement of solubility. However, increasing calcite solubility raises the possibility that the calcite crystal will break into small pieces, which made direct weighing challenging because it became difficult to retrieve all the calcite fragments quantitatively. This can be circumvented by determining the weight change of the inner capsule itself, provided that all dissolved calcite precipitated in the outer capsule upon quenching. Calcite solubility is calculated with

$$m_{CaCO_3} = \frac{1000\left(w_{ic}^{i} - w_{ic}^{f}\right)}{100.078\, w_{H_2O}}, \tag{21.1}$$

where m_{CaCO_3} is the concentration of $CaCO_3$ in solution in units of molality (moles of $CaCO_3$ per kg H_2O), w_{ic}^{i} and w_{ic}^{f} are the initial and final masses of the inner Pt capsules, w_{H_2O} is the mass of water (all w_i in mg), and 100.078 is the molecular weight of $CaCO_3$ (g/mol).

If dissolved calcite precipitated in the inner capsule during quenching, the solubility determined by equation (21.1) gives a minimum value. The accuracy of solubility determined in this way was therefore checked in selected experiments by directly weighing calcite crystals where they could be recovered from the inner Pt capsule. The inner capsule contained undissolved calcite and small amounts of quenched skeletal calcite that could be distinguished based on texture under a binocular microscope. If the undissolved calcite crystal was unbroken or broken into just a few pieces, the fragments were separated from the quench crystals and weighed to give a maximum calcite solubility. Maximum calcite solubility can be determined with equation (21.1) by replacing w_{ic}^{i} and w_{ic}^{f} with the initial and final masses of the calcite crystal. This solubility measurement is treated as a maximum, because it is possible that some undissolved fragments of the calcite crystal went unnoticed and unweighed. Minimum and maximum calcite solubility measurements are in close agreement when both could be measured, and maximum values are always greater than minimum values (Tables 21.1 and 21.2). The difference in measurements of minimum and maximum solubility was always much larger than the propagated 1σ weighing error in any individual solubility measurement. However, minimum and maximum solubility could not be measured in all cases. We therefore used the average range between minimum and maximum of 0.028 molal ($n = 18$) as a conservative estimate of uncertainty when plotting experiments for which only minimum solubility was determined.

21.2.5. Salt Concentration Notation

In the following discussion, the investigated range of salts and mixtures requires several alternate ways of portraying salt concentrations. The total salt mole fraction in a solution is defined as

$$X_{\text{total salt}} = \frac{\sum_i n_i}{\sum_i n_i + n_{H_2O}}, \tag{21.2}$$

where n_i is the number of moles of alkali-halide i added to the aqueous solution. The mole fraction of salt i in a solution is

$$X_i = \frac{n_i}{\sum_i n_i + n_{H_2O}}. \tag{21.3}$$

For mixed-salt solutions, the molar abundance of salt i (N_i) relative to all salts (excluding dissolved $CaCO_3$) is

$$N_i = \frac{n_i}{\sum_i n_i}. \tag{21.4}$$

21.3. RESULTS

Results for all experiments are given in Tables 21.1 and 21.2.

21.3.1. Textures

Run products from the inner capsule were collected for textural analysis on a Tescan Vega 3 scanning electron microscope at UCLA. Figure 21.1 shows SEM images of a starting calcite crystal and several run products displaying different textures. In Figure 21.1, parts b–f are ordered from lowest solubility to highest solubility. At lowest solubilities, partly dissolved calcite crystals preserve original shapes but become more rounded. At highest solubilities, run product crystals are highly irregular in shape and display wormy pits and grooves. These textures are observed regardless of salt identity, suggesting that tex-

Table 21.1 Calcite solubility measurements at 700 °C and 8 kbar in H_2O-single-salt solutions.

Run No.	t (hrs)	Salt In (mg)	H_2O In (mg)	H_2O Out (mg)	$X_{total\ salt}$ In	Cap. In (mg)	Cap. Out (mg)	CC In (mg)	CC Out (mg)	Minimum $CaCO_3$ Solubility (molal)	Maximum $CaCO_3$ Solubility (molal)
Pure H_2O											
W-1	23	0.000	16.996	17.043	0.000	49.529	49.502	0.479	-	0.016	-
KCl											
K-1	22	2.751	22.723	-	0.028	56.615	56.522	0.468	-	0.041	-
Y-5	27	6.580	29.810	-	0.051(1)	61.706	61.452	3.277	-	0.085(1)	-
Y-1	24	13.874	30.213	-	0.100	64.243	63.961	2.445	2.018	0.093(1)	0.141(1)
Y-2	72	30.573	30.486	-	0.195	60.946	60.474	3.142	2.527	0.155(1)	0.202(1)
MT-4	17	15.557	15.178	-	0.199	52.500	52.271	0.615	-	0.151(1)	-
Y-3	23	53.238	29.985	-	0.300	70.881	70.174	3.578	2.854	0.236	0.241
Y-4	25	41.432	14.596	-	0.407	62.145	61.628	3.945	-	0.354(1)	-
J-4	25	47.575	16.352	-	0.413	55.567	54.971	1.271	-	0.364(1)	-
J-8	22	54.224	16.175	-	0.448	49.054	48.371	1.077	0.380	0.422(1)	0.431(1)
Y-6	22	61.794	15.558	-	0.490	62.313	61.661	2.327	-	0.419(1)	-
NaCl											
J-6	24	11.201	30.453	-	0.102	53.870	53.501	1.003	0.600	0.121(1)	0.132(1)
MT-1	23	16.220	20.029	20.031	0.200	49.171	48.662	1.269	-	0.253(1)	-
J-7	25	32.768	31.394	-	0.243	56.207	55.147	1.906	0.800	0.337	0.352
J-10	22	53.457	30.282	-	0.352	52.475	60.529	2.690	0.673	0.642	0.665
J-12	21	32.428	14.606	-	0.406	57.934	56.826	2.490	-	0.758	-
J-16	24	42.328	14.748	-	0.469	53.505	62.346	2.069	-	0.785	-
J-18	24	40.268	9.343	-	0.571	47.659	46.942	1.551	-	0.767(1)	-
LiCl											
L-27	24	1.144	21.296	21.836	0.022(2)	50.562	50.374	0.646	-	0.088(2)	-
L-9	4	1.444	25.520	26.009	0.023(1)	48.969	48.756	1.186	0.975	0.084(1)	0.083(1)
L-19	8	1.918	33.985	34.511	0.023(1)	52.696	52.426	0.997	0.695	0.079(1)	0.086(1)
L-26	30	1.296	23.679	24.268	0.023(1)	54.670	54.455	0.638	-	0.091(1)	-
L-18	24	1.658	20.896	-	0.033(1)	50.114	49.863	0.773	-	0.120(1)	-
L-7	5	3.371	25.162	25.551	0.054	60.837	60.461	3.666	-	0.149(1)	-
L-15	24	3.718	16.154	-	0.089	48.517	48.009	1.206	-	0.314(1)	-
L-8	5	10.023	24.393	24.819	0.149	63.979	62.589	3.140	1.602	0.569	0.630
L-16	23	3.905	9.070	9.299	0.155	61.366	60.694	1.743	-	0.740(1)	-
L-14	23	9.831	18.545	-	0.184	52.850	50.961	2.107	-	1.018	-
L-17	23	4.487	8.425	-	0.185	53.897	53.018	1.625	-	1.043(1)	-
L-10	23	6.244	11.298	12.020	0.190	59.114	57.950	1.822	-	1.029	-
L-11	5	5.585	8.170	8.435	0.225	62.703	61.628	2.900	-	1.314(1)	-
L-12	17	6.390	7.360	-	0.270	65.337	63.735	2.365	-	2.174(1)	-
L-13	19	9.796	10.263	-	0.289	60.497	57.553	1.763	-	2.865	-
CsCl											
CS-1	23	15.341	16.944	17.551	0.088	53.028	52.897	0.060	-	0.079	-

Note. "Cap." is the final Pt capsule containing solution and calcite crystal (CC). "In" and "out" refer to measurements made before and after experiment, respectively. "H_2O in" measurements were used in all calculations of calcite solubility. The 1σ weighing uncertainty was propagated through each weighing step. Parenthetical numbers reflect 1σ uncertainty in last digit. Any 1σ < 0.0005 was excluded from table.

Table 21.2 Calcite solubility measurements at 700 °C and 8 kbar in mixed-salt solutions.

Run No.	t (hrs)	KCl In (mg)	NaCl In (mg)	LiCl In (mg)	H_2O In (mg)	H_2O Out (mg)	X_{KCl} In	X_{NaCl} In	X_{LiCl} In	$X_{total\ salt}$ In	Cap. In (mg)	Cap. Out (mg)	CC In (mg)	CC Out (mg)	Minimum $CaCO_3$ solubility (molal)	Maximum $CaCO_3$ Solubility (molal)
NaCl-KCl																
J-3	25	2.187	1.576	0.000	27.393	-	0.019	0.017	0.000	0.036(1)	50.995	50.848	1.312	-	0.054(2)	-
Y-7	20	6.935	5.798	0.000	29.026	-	0.052	0.055	0.000	0.107	62.555	62.229	1.795	-	0.112(1)	-
Y-8	23	15.840	13.082	0.000	34.988	-	0.089	0.094	0.000	0.183	62.355	61.721	2.712	2.040	0.181(1)	0.192
MT-3	31	13.533	4.160	0.000	20.256	20.340	0.132	0.052	0.000	0.184	47.981	47.646	0.937	-	0.165(1)	-
MT-2	25	4.044	7.959	0.000	15.219	15.322	0.052	0.132	0.000	0.184(1)	48.694	48.378	0.714	-	0.207(1)	-
J-1	24	13.292	10.678	0.000	16.539	-	0.139	0.143	0.000	0.282	61.508	61.018	1.628	-	0.296(1)	-
J-2	24	20.679	16.343	0.000	16.283	-	0.190	0.191	0.000	0.381	59.804	59.122	2.354	1.504	0.418(1)	0.522
J-5	23	28.267	22.331	0.000	16.318	-	0.227	0.229	0.000	0.457	58.197	57.308	2.459	1.543	0.544	0.561
J-9	22	47.463	37.080	0.000	14.280	-	0.308	0.307	0.000	0.616	58.562	56.963	3.038	-	1.119	-
J-11	23	48.678	37.801	0.000	11.224	-	0.340	0.336	0.000	0.676	59.059	57.560	2.901	1.313	1.334	1.414
J-14	24	37.513	29.691	0.000	5.745	-	0.378	0.382	0.000	0.760	72.378	71.479	2.663	-	1.563(1)	-
J-17	24	41.974	32.778	0.000	3.301	-	0.431	0.429	0.000	0.860	66.214	65.668	0.000	-	1.653(2)	-
KCl-LiCl																
KL-4	23	5.858	0.000	3.250	21.096	20.715	0.059	0.000	0.058	0.117(1)	51.085	50.529	1.111	-	0.263(1)	-
KL-1	23	5.402	0.000	3.389	15.403	-	0.072	0.000	0.079	0.151(1)	59.937	59.397	1.458	-	0.350(1)	-
MT-8	23	10.267	0.000	2.048	14.959	-	0.136	0.000	0.048	0.183(1)	51.887	51.418	0.925	-	0.313(1)	-
MT-9	24	4.346	0.000	6.247	16.491	16.730	0.052	0.000	0.131	0.183(1)	52.775	51.596	1.717	-	0.714	-
KL-3	19	12.569	0.000	6.765	26.291	26.583	0.094	0.000	0.089	0.184	60.342	59.043	1.884	0.537	0.494	0.512
NaCl-LiCl																
NL-3	20	0.000	2.917	2.017	25.917	26.477	0.000	0.032	0.031	0.063(1)	59.489	59.108	0.854	-	0.147(1)	-
NL-2	23	0.000	4.702	2.861	20.023	-	0.000	0.064	0.054	0.117(1)	59.219	58.678	0.775	-	0.270(1)	-
NL-1	30	0.000	5.735	4.265	20.650	20.877	0.000	0.073	0.075	0.148(1)	65.955	65.161	1.846	-	0.384	-
MT-5	29	0.000	8.018	2.410	15.465	15.777	0.000	0.130	0.054	0.184(1)	51.737	51.104	0.968	-	0.409(1)	-
MT-6	15	0.000	3.392	5.906	15.689	-	0.000	0.054	0.130	0.185(1)	53.360	52.264	1.570	-	0.698	-
NL-4	22	0.000	5.638	4.409	14.173	14.431	0.000	0.098	0.105	0.203(1)	57.891	56.906	1.735	-	0.694	-
NaCl-LiCl-KCl																
KLN-7	21	3.452	3.273	1.742	20.057	20.481	0.037	0.045	0.033	0.114(1)	58.236	57.856	1.589	-	0.189(1)	-
KLN-8	22	6.120	5.165	1.296	20.890	21.209	0.060	0.065	0.022	0.148(1)	53.088	52.644	1.111	-	0.212(1)	-
KLN-9	21	5.150	1.666	3.163	15.626	16.012	0.066	0.027	0.072	0.166(1)	62.990	62.356	2.229	-	0.405(1)	-
KLN-3	19	9.214	3.313	3.643	21.408	21.448	0.085	0.039	0.059	0.183(1)	50.523	49.737	1.279	0.417	0.367	0.403
KLN-5	22	1.978	2.491	5.401	15.744	-	0.025	0.040	0.119	0.184(1)	58.991	57.978	2.046	-	0.643	-
KLN-2	25	2.027	5.646	2.633	14.873	14.720	0.027	0.096	0.061	0.184(1)	61.029	60.408	0.898	0.262	0.417(1)	0.427(1)
KLN-1	8	4.847	3.671	2.628	15.105	15.076	0.063	0.061	0.060	0.185(1)	55.800	55.254	1.275	-	0.361(1)	-
KLN-4	23	11.304	1.455	3.795	21.111	21.544	0.105	0.017	0.062	0.185(1)	51.901	51.154	1.491	0.723	0.354(1)	0.363

Note. "Cap." is the final Pt capsule containing solution and calcite crystal (CC). "In" and "out" refer to measurements made before and after experiment, respectively. "H_2O in" measurements were used in all calculations of calcite solubility. The 1σ weighing uncertainty was propagated through each weighing step. Parenthetical numbers reflect 1σ uncertainty in last digit. Any 1σ < 0.0005 was excluded from table.

ture is controlled by magnitude of solubility rather than solution composition. In low solubility experiments, a single, large grain was typically recovered (Figure 21.1 b–d), whereas high solubility experiments typically yielded aggregates of small crystals (Figure 21.1 e, f). As solubility increased, dissolution pits and grooves likely cut channels through the entire width of the crystal, resulting in the aggregates of smaller crystals seen in Figure 21.1 e and f.

The small (~20 μm) euhedral rhombs shown in Figure 21.1g are interpreted to be quench material due to their uniform size, high nucleation density, and even distribution on all surfaces of the Pt capsules (Figure 21.1h), including the original calcite crystal (Caciagli & Manning, 2003). Due to their small size, the mass of these crystals has a negligible effect on solubility determinations, especially in light of the generally high solubilities at investigated conditions; they were therefore noted where present, but we did not attempt to account for them in solubility calculations. Large, skeletal crystals (Figure 21.1i) are observed in highest solubility experiments. These are interpreted to be quench crystals, consistent with their rapid-growth textures, which can be expected from the extreme oversaturation upon quenching from very high solubility conditions.

21.3.2. Solubility in Single-Salt Solutions

The first set of experiments was conducted using aqueous solutions containing a single salt. Multiple runs conducted at constant $X_{LiCl} = 0.023$ showed no systematic variation of calcite solubility with time longer than 4 hr (Table 21.1). Caciagli and Manning (2003) showed that in pure H_2O, equilibrium was reached by 12 hours; therefore, all other experiments were run for a minimum of 12 hours. Figure 21.2a shows variation of calcite solubility with $X_{total\ salt}$ for aqueous solutions containing KCl, NaCl, and LiCl. In all cases, calcite solubility displayed

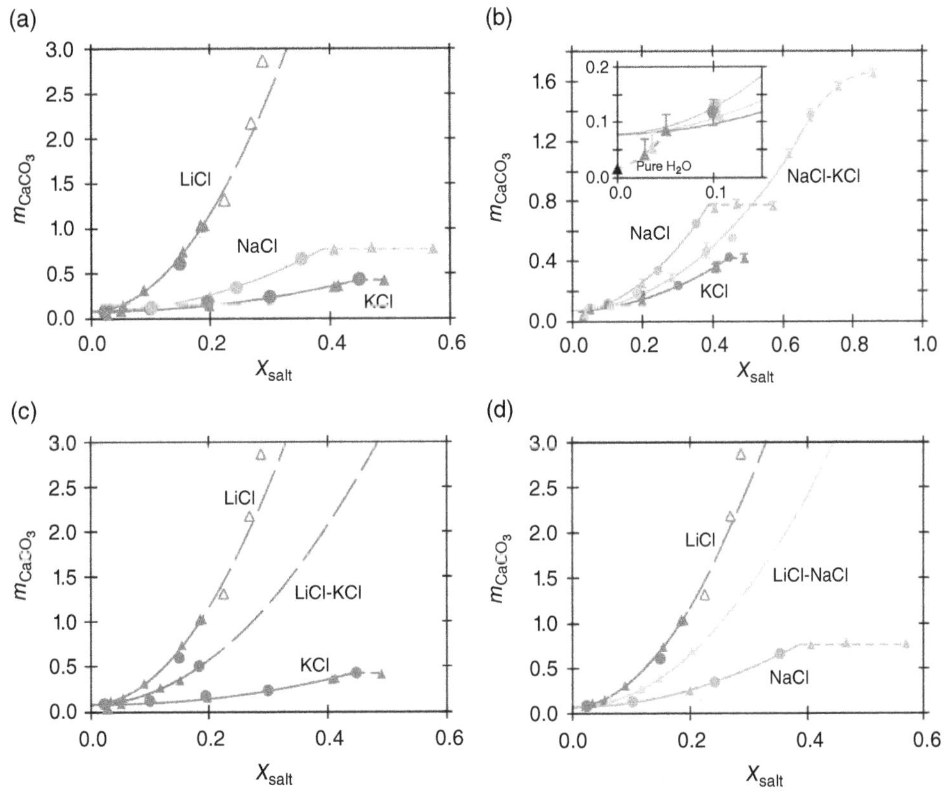

Figure 21.2 $CaCO_3$ solubility (molality) vs. $X_{total\ salt}$ in single-salt solutions (a) and 2-salt solutions (b)–(d). Circles represent average of measured solubility. Triangles represent experiments for which only minimum solubilities were recorded. Error bars are shown where larger than symbol; those for average solubilities reflect the range between minimum and maximum values, whereas those for minimum solubilities reflect the average difference between minimum and maximum values for all experiments in which a maximum value could be measured (see text). Only positive error bars are included on triangles as the triangles themselves represent the minimum value. Experiments represented with unfilled triangles were not considered in data regression due to significant scatter (see text). Solid lines are based on regressions given in text (equations [21.5]–[21.7]). Short-dashed lines signify salt saturation. Long-dashed signify extrapolation of curves. Inset in (b) is shown to demonstrate deviation from trend at low salt concentrations. See electronic version for color representation of the figures in this book.

quadratic growth with $X_{total\ salt}$. Experiments at high LiCl concentration yielded high calcite solubility and display significantly more scatter, likely due to increased amounts of quench calcite precipitating in the inner capsule at high solubility.

In NaCl and KCl solutions, calcite solubility increased until, at high $X_{total\ salt}$, no further change in calcite solubility was observed, consistent with alkali-halide-saturation. Where $X_{total\ salt}$ in the fluid is greater than the saturating concentration, no further solubility increase occurs because the solution composition must remain constant, and all additional salt simply increases the volume of halite or sylvite grown during the experiment (e.g. Aranovich & Newton, 1996). Saturation values of NaCl and KCl inferred from our experiments are $X_{NaCl} \approx 0.40$ and $X_{KCl} \approx 0.45$, which are in good agreement with previous studies: $X_{NaCl} \approx 0.40$ $X_{KCl} \approx 0.50$ (Aranovich & Newton, 1996, 1997). Figure 21.2a shows that LiCl saturation was not attained.

A single experiment was conducted in a CsCl solution. Calcite solubility in CsCl is lower than that in LiCl, NaCl, or KCl at the same $X_{total\ salt}$ (Table 21.1).

21.3.3. Solubility in Mixed-Salt Solutions

A second set of experiments examined how mixing salts affects calcite solubility (Table 21.2). Mixtures included 1:1 molar ratios of two salts: NaCl-KCl, NaCl-LiCl, and KCl-LiCl, in which calcite solubility increased quadratically with increasing $X_{total\ salt}$ (Figure 21.2 b–d). At any given $X_{total\ salt}$, calcite solubility in the mixed-salt solution lies at a value between that in the end-member single-salt solutions of which the solution is composed (Figure 21.2 b–d).

In the NaCl-KCl solution (Figure 21.2b), calcite solubility continued to increase until $X_{total\ salt} > 0.8$. Figure 21.2a shows that in single-salt solutions, NaCl and KCl saturate at about $X_{NaCl} = 0.40$ and $X_{KCl} = 0.45$, respectively. Therefore, in a solution with a 1:1 molar ratio of NaCl-KCl, we might expect to see NaCl saturate at $X_{total\ salt} \approx 0.80$ ($X_{NaCl} = 0.40$; $X_{KCl} = 0.40$) and KCl to saturate at $X_{total\ salt} \approx 0.90$ ($X_{NaCl} = 0.45$; $X_{KCl} = 0.45$). Figure 21.2b shows that at around $X_{total\ salt} = 0.70$, the curve changes concavity, which we interpret as signaling NaCl saturation. As more NaCl and KCl are added beyond $X_{total\ salt} = 0.70$ in a 1:1 molar ratio, KCl concentration in the solution increases while any additional NaCl simply precipitates as halite, resulting in the behavior seen in Figure 21.2b. The flattening of the curve around $X_{total\ salt} = 0.90$ is likely due to KCl saturation.

Another set of experiments was conducted in two-salt solutions holding $X_{total\ salt}$ constant ($X_{total\ salt} = 0.18$) while varying the proportion of the two salts (Figure 21.3). Figure 21.3a shows variation of calcite solubility with NaCl/(NaCl + KCl). As NaCl/(NaCl + KCl) varies from 0 to 1, calcite solubility changes from calcite solubility in a pure KCl solution to that in a pure NaCl solution. Figure 21.3 shows that calcite solubility as a function of NaCl/(NaCl+KCl) is better fit by a second order polynomial rather than a linear function. Experiments along the LiCl/(LiCl+NaCl) and LiCl/(LiCl+KCl) exhibit similar behavior.

Experiments were also conducted with solutions comprising three salts (KCl, NaCl, LiCl). Figure 21.3a shows experiments in which the mole fraction of LiCl relative to total salts (N_{LiCl}) is kept constant at 0.33, while varying NaCl/(NaCl + KCl) at $X_{total\ salt} = 0.18$. Similar experiments were also conducted along LiCl-NaCl and LiCl-KCl joins (Figure 21.3 b, c). Mixing experiments in three-salt solutions behaved similarly to those in two-salt solutions.

21.4. DISCUSSION

21.4.1. Solubility at $X_{total\ salt} < 0.05$

The inset in Figure 21.2b shows calcite solubility at low $X_{total\ salt}$. Three experiments at $X_{total\ salt} < 0.05$ yielded solubility lower than that predicted by the fitted trends, and instead appear to lie on a separate trend projecting towards calcite solubility in pure H_2O. This suggests that there are likely two different regions where the dominant species differ. Fein and Walther (1989) studied calcite solubility and speciation in NaCl-H_2O solutions at 400 °C–600 °C and 2 kbar and inferred that solutes were fully dissociated at low salt concentrations, but at least partly associated (e.g. $CaCl^+$) at higher salt concentrations. More detailed studies of the solubility patterns at low $X_{total\ salt}$ in KCl and other salt solutions are required to more quantitatively evaluate this possible transition in speciation. The following discussion applies only to solubilities at $X_{total\ salt} > 0.05$.

21.4.2. Dependence of Calcite Solubility on $X_{total\ salt}$

The rise in calcite solubility with total salt mole fraction (Figure 21.2) suggests a quadratic dependence of calcite solubility on $X_{total\ salt}$ (Newton & Manning, 2002). Least-squares regressions of the salt-undersaturated solubility results give

$$m_{CaCO_3,KCl} = 1.715 X_{KCl}^2 + 0.079 \quad (21.5)$$

$$m_{CaCO_3,NaCl} = 4.581 X_{NaCl}^2 + 0.079 \quad (21.6)$$

$$m_{CaCO_3,LiCl} = 27.043 X_{LiCl}^2 + 0.079. \quad (21.7)$$

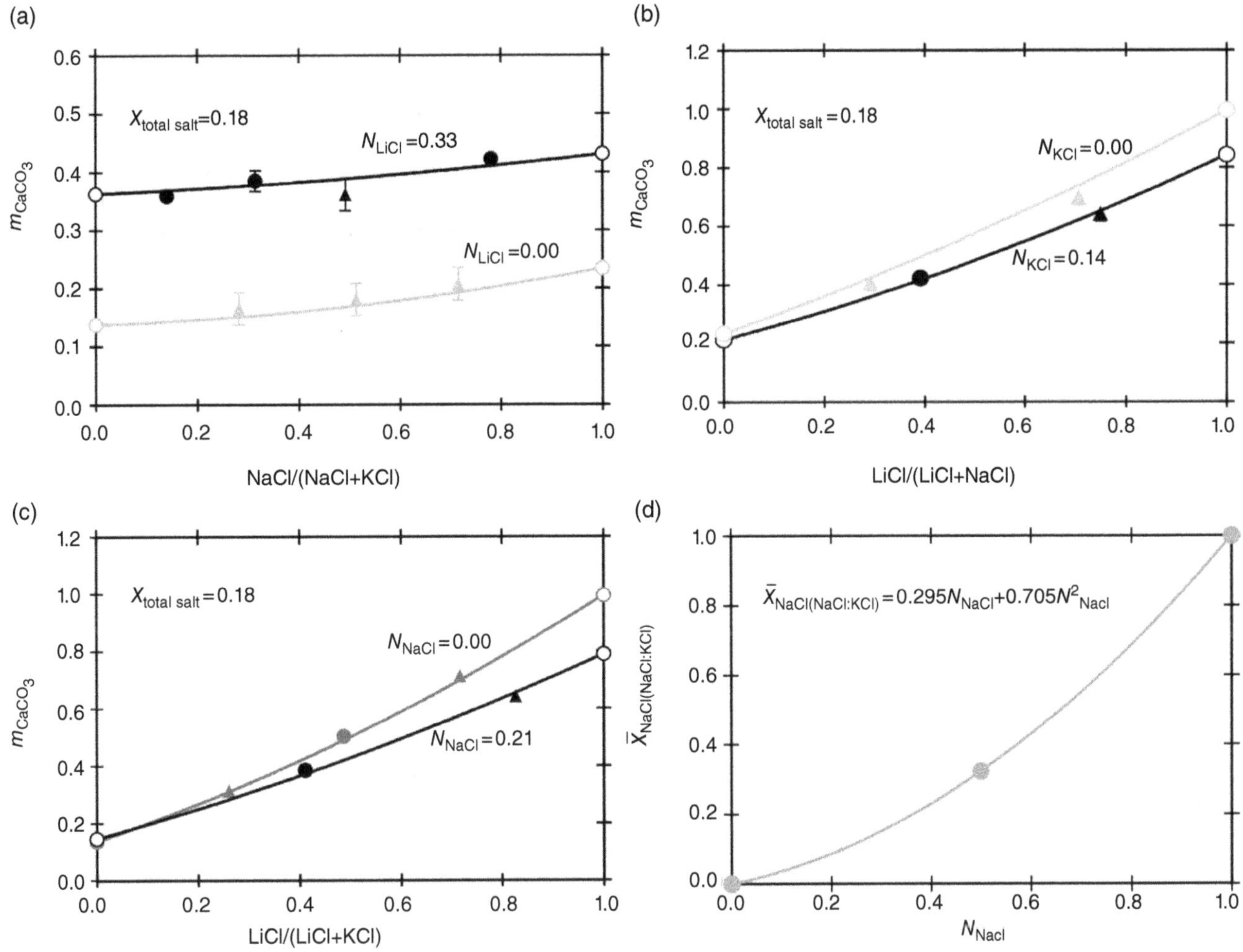

Figure 21.3 Calcite solubility (molality) variations as a function of varying salt ratios. (a) Solubility variations with molar NaCl/(NaCl + KCL). (b) Solubility variations with molar LiCl/(LiCl + NaCL). (c) Solubility variations with molar LiCl/(LiCl + KCL). Curves are calculated using model discussed in text. Symbols and error bars as in Figure 21.2, except for unfilled circles, which represent calcite solubility in one and two-salt solutions calculated using equations (21.5)–(21.7) and (11). (d) Variation in $\bar{x}_{NaCl(NaCl:KCl)}$ with N_{NaCl} in NaCl-KCl solutions. Fitted second order polynomials of the type shown were used to derive $\bar{x}_{i(A:B)}$ for all two-salt solutions (see text). See electronic version for color representation of the figures in this book.

The coefficients of determination (r^2) for equations (21.5)–(21.7) are 0.985, 0.999, and 0.995, respectively, with standard errors of estimates of 0.015 molal, 0.006 molal, and 0.029 molal, respectively. Intercepts were fixed at 0.079, which is the average intercept derived from regressing each set of results separately, with a standard deviation of 0.008. Intercepts were fixed so that regressions could be employed with mixing models discussed below. The high coefficients of determination demonstrate that fixing the intercepts at 0.079 has negligible effects on fitting of curves. An intercept of 0.079 would suggest that at 700 °C and 8 kbar in pure H_2O m_{CaCO3} = 0.079. However, the change in behavior at $X_{total\ salt} \approx 0.05$ (Figure 21.2b) suggests that m_{CaCO3} = 0.079 is an overestimation of calcite solubility in pure H_2O, consistent with the salt-free experiment (W-1, Table 21.1). Three experiments with the highest X_{LiCl} were excluded from the regression due to the higher amount of scatter, as noted above; however, extrapolation shows good agreement between the predicted and measured solubilities (Figure 21.2). It should be noted that the empirical equations presented here are only applicable from $0.05 < X_{total\ salt}$ < alkali-halide saturation.

21.4.3. Dissolution Mechanism

Experimental results show that calcite solubility increases with increasing alkali-halide concentration until alkali-halide saturation (Figure 21.2), regardless of the salt or salt mixture. This behavior suggests that H_2O is not principally responsible for solubility and speciation reactions because as H_2O activity decreases, calcite solubility increases (Manning et al., 2013). The fact that H_2O activity has minimal effects on calcite solubility suggests the calcite-solution equilibrium can likely be described by a reaction that does not involve H_2O (Newton & Manning, 2002):

$$CaCO_3 + 2\,NaCl_{(aq)} = CaCl_{2(aq)} + Na_2CO_{3(aq)}. \quad (21.8)$$

This reaction predicts that C speciation in hydrous fluids at this P-T is dominated by Na^+-CO_3^{2-} complexes, which is supported by the ab initio calculations of Pan and Galli (2016). If the dissolution reaction of the form of equation (21.8) is applicable, calcite solubility should be proportional to the square of $X_{total\ salt}$, which is demonstrated by equations (21.5)–(21.7). Behavior of calcite solubility in KCl and LiCl solutions is similar to that in NaCl solutions, which suggests equations analogous to equation (21.8) can be written for KCl and LiCl.

Figure 21.4 shows that at a given $X_{total\ salt}$, calcite solubility increases with decreasing cation radius of the participating salt. Aranovich and Newton (1997) investigated the activity of KCl in H_2O solutions at high pressures and temperatures and found them to be lower than NaCl activities when pressure, temperature, and salt mole fraction were equal. Decreasing salt activity with decreasing cation radius may explain the trend seen in Figure 21.4; however, activity-composition relations are not known for the other salts. Nevertheless, the increase in calcite solubility with decreasing cation size suggests the stability of the aqueous complexes in order from least to greatest would be $Cs_2CO_{3(aq)} < K_2CO_{3(aq)} < Na_2CO_{3(aq)} < Li_2CO_{3(aq)}$. This trend in stability can be predicted using Pearson's hard-soft acid-base principles, which state that hard acids form stronger bonds with hard bases and soft acids form stronger bonds with soft bases. The carbonate ion is a hard base, so it should form stronger bonds with harder acids. Parr and Pearson (1983) defined chemical hardness as

$$h = \frac{I - A}{2}, \quad (21.9)$$

where h is hardness, I is ionization energy, and A is electron affinity. Figure 21.4b shows calcite solubility increases with the hardness of the salt cation. Thus, the increased calcite solubility in solutions containing smaller cations is likely due in part to the harder cations forming more stable complexes with the hard carbonate ion. This suggests calcite solubility in aqueous alkali chloride solutions may be predictable using the relationship between calcite solubility and chemical hardness of the alkali.

21.4.4. Predicting Calcite Solubility in Multisalt Solutions

The experimental data can be used to derive general relations for calcite solubility in two- and three-salt solutions. Below we develop a simple empirical scheme for the solubility of calcite at 700 °C, 8 kbar, $X_{total\ salt} > 0.05$, and a range of salt ratios.

21.4.4.1. Two-salt Solutions

If the effects of alkali-halide mixing on calcite solubility affect calcite solubility in an ideal fashion, then calcite solubility can be calculated as a simple linear combination of calcite solubilities in the single-salt solutions:

$$m_{A:B} = N_A m_A + N_B m_B, \quad (21.10)$$

where $m_{A:B}$ is calcite solubility in the mixed-salt solution, N_i is the molar abundance of salt i relative to other salts in solution, and m_i is the calcite solubility in the single i-H_2O salt solution. If $A{:}B$ is 1:1, then $N_A = N_B = 0.5$ in

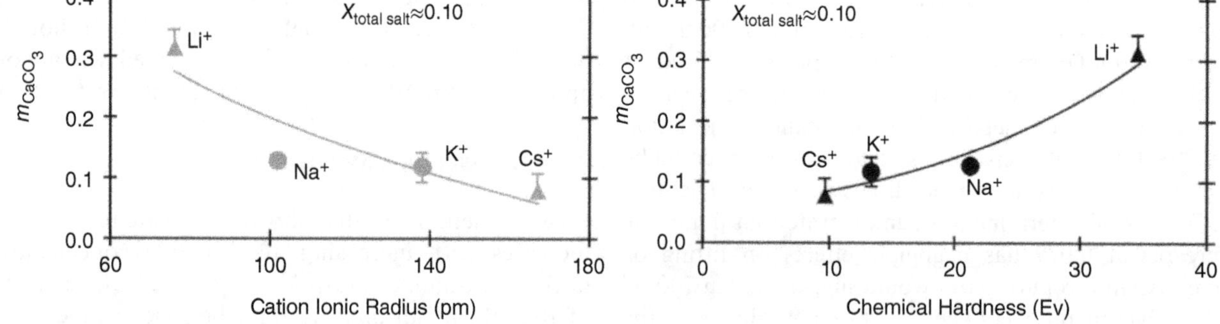

Figure 21.4 (a) Calcite solubility (molality) vs. ionic radius of salt cation. Experiments show trend of increasing calcite solubility with decreasing cation size. (b) Calcite solubility (molality) vs. chemical hardness of salt cation. Data show trend of increasing calcite solubility with increasing chemical hardness of salt cation. Symbols and error bars as in Figure 21.2. See electronic version for color representation of the figures in this book.

equation (21.10). However, Figure 21.2b shows that although the molar NaCl:KCl ratio is 1:1, calcite solubility at a given $X_{\text{total salt}}$ is not one half the difference between that in the respective single-salt solutions, so equation (21.10) cannot be used. Instead, an "effective mole fraction" $\bar{x}_i$ can be defined for any two-salt mixture such that

$$m_{A:B}=\bar{x}_{A(A:B)}m_A+\bar{x}_{B(A:B)}m_B, \quad (21.11)$$

where $\bar{x}_{i(i:j)}$ is the effective mole fraction of salt i in an i-j salt solution. In an A-B solution, $\bar{x}_{A(A:B)}$ varies from 0 at $N_A = 0$ ($N_B = 1$) to 1 at $N_A = 1$ ($N_B = 0$), and

$$\bar{x}_{B(A:B)}=1-\bar{x}_{A(A:B)}. \quad (21.12)$$

Combining the constraints from end-member solutions with the assumption that $\bar{x}_{A(A:B)}$ is described by a simple second-order polynomial of the form

$$\bar{x}_{A(A:B)}=a_1N_A+a_2N_A^2 \quad (21.13)$$

yields a function describing the variation of $\bar{x}_{NaCl(NaCl:KCl)}$ with N_{NaCl} in all NaCl-KCl solutions. Regressions were made by fitting a polynomial to $\bar{x}_{NaCl(NaCl:KCl)}$ in a pure NaCl solution, pure KCl solution, and 1:1 NaCl:KCl solution at a constant $X_{total\ salt}$ (Figure 21.3d). For NaCl-KCl solutions, this equation is

$$\bar{x}_{NaCl(NaCl:KCl)}=0.295N_{NaCl}+0.705N_{NaCl}^2. \quad (21.14)$$

At any N_{NaCl} in NaCl-KCl solutions, $\bar{x}_{KCl(NaCl:KCl)}$ can be derived from equation (21.12).

Using this approach, functions describing $\bar{x}_{A(AB)}$ in the other two-salt solutions were derived:

$$\bar{x}_{LiCl(LiCl:NaCl)}=0.793N_{LiCl}+0.207N_{LiCl}^2 \quad (21.15)$$

$$\bar{x}_{LiCl(LiCl:KCl)}=0.695N_{LiCl}+0.305N_{LiCl}^2. \quad (21.16)$$

For each two-salt system, values of $\bar{x}_{B(A:B)}$ can be calculated using equation (21.12).

21.4.4.2. Three-salt Solutions

The success of the predictive scheme developed for two-salt solutions suggests the same approach can be generalized to three-salt solutions. If correct, we can write

$$m_{CaCO_3}=\sum_i\bar{x}_i m_i, \quad (21.17)$$

where m_i is calcite solubility in a solution containing only salt i and $\bar{x}_i$ is the effective mole fraction. A simple approach to deriving values of $\bar{x}_{A(ABC)}$ is to use $\bar{x}_{A(AB)}$ in a C-free solution and then correct for the effect of addition of salt C. That is:

$$\bar{x}_{A(ABC)}=\bar{x}_{A(AB)} \quad (21.18)$$

$$\bar{x}_{C(ABC)}=N_{A(AB)}\bar{x}_{C(AC)}+N_{B(AB)}\bar{x}_{C(BC)} \quad (21.19)$$

$$\bar{x}_{B(ABC)}=1-\bar{x}_{A(ABC)}-\bar{x}_{C(ABC)}. \quad (21.20)$$

Calcite solubility was calculated in three-salt solutions using equations (21.5)–(21.7) and (21.17)–(21.20), with $A \equiv$ NaCl, $B \equiv$ KCl, and $C \equiv$ LiCl. Figure 21.5 shows measured vs. predicted calcite solubility in one-, two-, and three-salt solutions. The tight clustering around the 1:1 line shows that this method of predicting calcite solubility accurately reproduces the experimental data. The success of the predictive scheme developed above implies that calcite solubility in solutions comprising H_2O-NaCl-KCl-LiCl can be predicted over a wide range of total salt mole fractions. Figure 21.6 shows ternary diagrams with contours of constant calcite solubility calculated with the scheme described above. Figure 21.6 shows that calcite solubility greatly increases with increasing $X_{\text{total salt}}$, increasing N_{LiCl}, and also that the range of calcite solubility greatly increases with increasing salt mole fraction.

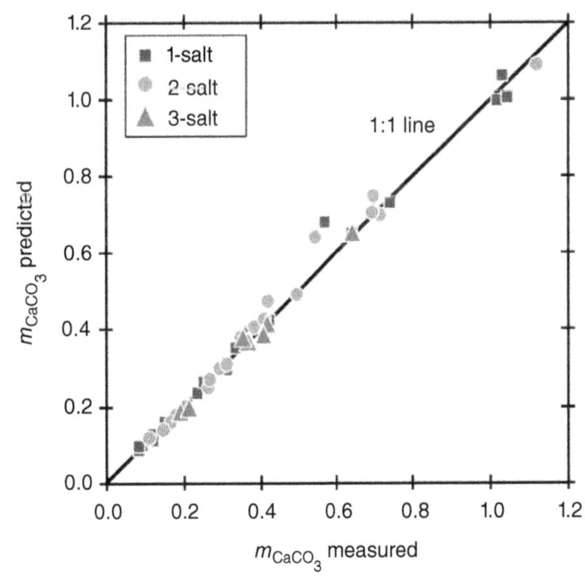

Figure 21.5 Predicted calcite solubility vs. measured calcite solubility in salt solutions. Data cluster tightly around 1:1 line, demonstrating accuracy presented in salt-solution calcite solubility model. See electronic version for color representation of the figures in this book.

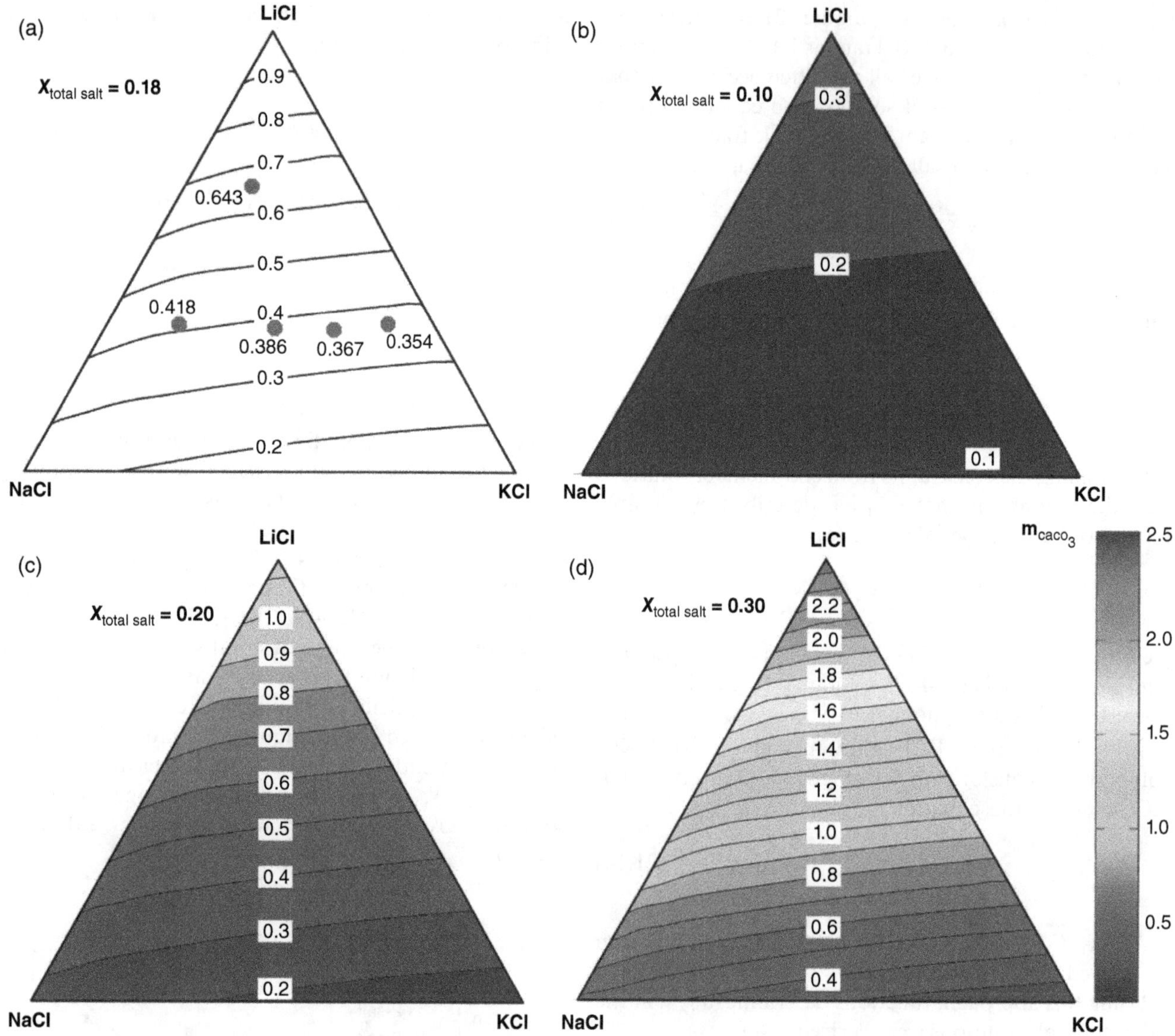

Figure 21.6 Ternary diagram with contours of constant calcite solubility (molality) at $X_{total\ salt}$ = 0.18, calculated with method described in text. (a) Points are experiments labeled with measured calcite solubility. (b)–(d) Ternary diagrams with contours of constant calcite solubility (molality), calculated with method described in text. See electronic version for color representation of the figures in this book.

21.5. CONCLUSIONS

The results of this study present several conclusions:

1. Calcite solubility follows a consistent, predictable behavior with increasing salt concentration in aqueous KCl, NaCl, and LiCl solutions.

2. Calcite solubility in mixed salt solutions is not accurately predicted by assuming ideal mixing of salts in solution; however, the present study provides a method to make corrections for mixing and accurately predict calcite solubility in mixed-salt solutions.

3. Calcite solubility in saline solutions is enhanced relative to pure H_2O by complexation of the Ca^{2+} and CO_3^{2-} ions with the dissolved salt ions. The observed trend of calcite solubility increasing with decreasing cation size, i.e. CsCl < KCl < NaCl < LiCl, is likely due to the CO_3^{2-} ion forming more stable complexes with harder ions as predicted by Pearson's hard-soft acid-base principles.

These findings allow for prediction of calcite solubility over a wide range of fluid compositions relevant to natural systems. However, calcite dissolution in dilute salt solutions ($X_{total\ salt}$ < 0.05) deviates from the trends extrapolated from $X_{total\ salt}$ > 0.05; therefore, more studies should be conducted on low salt concentration fluids. Furthermore, the present experiments were performed at a single P-T condition; therefore, experiments over a large range of temperature and pressure are required to better constrain carbon transport in geologic fluids. In addition, geologic fluids are not composed solely of alkali halides; $CaCl_2$, $MgCl_2$, and $FeCl_2$ are commonly found salts, and the effects of salts such as these should also be studied. The results of the present study suggest that mixing of salts in aqueous fluids is nonideal, and while the present study provides an empirical method to correct for nonideality and accurately predict calcite solubility in mixed salt solutions, it does not provide a physical explanation of nonideality upon mixing. Therefore, nonideal mixing of salts in solution may also deserve more attention.

ACKNOWLEDGMENTS

The authors thank Robert Newton, Michael Huh, and Adam Makhluf for their help in the experimental laboratory, as well as Margo Odlum for her help on the SEM. Research supported by the Deep Carbon Observatory and National Science Foundation grants EAR 1347987 and 1732256. Y.L. appreciates the support from the Elite Network Bavaria (ENB) program of Germany during his stay in UCLA.

REFERENCES

Ague, J. J., & Nicolescu S. (2014). Carbon dioxide released from subduction zones by fluid-mediated reactions. *Nature Geoscience*, *7*, 355–360.

Andersen, T., & Neumann, E. R. (2001). Fluid inclusions in mantle xenoliths. *Lithos*, *55*(1–4), 301–320.

Aranovich, L. Y., & Newton, R. C. (1996). H_2O activity in concentrated NaCl solutions at high pressures and temperatures measured by the brucite-periclase equilibrium. *Contributions to Mineralogy and Petrology*, *125*, 200–212.

Aranovich, L. Y., & Newton, R. C. (1997). H_2O activity in concentrated KCl and KCl-NaCl solutions at high temperatures and pressures measured by the brucite-periclase equilibrium. *Contributions to Mineralogy and Petrology*, *127*, 261–271.

Caciagli, N. C., & Manning, C. E. (2003). The solubility of calcite in water at 6–16 kbar and 500–800 °C. *Contributions to Mineralogy and Petrology*, *146*, 275–285.

Dasgupta, R., & Hirschmann, M. M. (2010). The deep carbon cycle and melting in Earth's interior. *Earth and Planetary Science Letters*, *298*, 1–13.

Ellis, A. J. (1963). The solubility of calcite in sodium chloride solutions at high temperatures. *American Journal of Science*, *261*(3), 259–276.

Facq, S., Daniel, I., Montagnac, G., Cardon, H., & Sverjensky, D. A. (2014). In situ Raman study and thermodynamic model of aqueous carbonate speciation in equilibrium with aragonite under subduction zone conditions. *Geochimica et Cosmochimica Acta*, *132*, 375–390.

Facq, S., Daniel, I., Montagnac, G., Cardon, H., & Sverjensky, D. A. (2016). Carbon speciation in saline solutions in equilibrium with aragonite at high pressure. *Chemical Geology*, *431*, 44–53.

Fein, J. B., & Walther, J. V. (1989). Calcite solubility and speciation in supercritical NaCl-HCl aqueous fluids. *Contributions to Mineralogy and Petrology*, *103*, 317–324.

Gorman, P. J., Kerrick, D. M., & Connolly, J.A.D. (2006). Modeling open system metamorphic decarbonation of subducting slabs. *Geochemistry, Geophysics, Geosystems*, *7*, Q04007.

Kawamoto, T., Yoshikawa, M., Kumagai, Y., Mirabueno, M.H.T., Okuno, M., & Kobayashi, T. (2013). Mantle wedge infiltrated with saline fluids from dehydration and decarbonation of subducting slab. *Proceedings of the National Academy of Sciences*, *110*, 9663–9668.

Kelemen, P. B., & Manning, C. E. (2015). Reevaluating carbon fluxes in subduction zones, what goes down, mostly comes up. *Proceedings of the National Academy of Sciences*, *112*, E3997–E4006.

Kent, A.J.R., Peate, D. W., Newman, S., Stolper, E. M., & Pearce, J. A. (2002). Chlorine in submarine glasses from the Lau Basin: Seawater contamination and constraints on the composition of slab-derived fluids. *Earth and Planetary Science Letters*, *202*(2), 361–377.

Manning, C. E. (2018). Fluids of the Lower Crust: Deep is Different. *Annual Reviews of Earth and Planetary Sciences*, *46*, 67–97.

Manning, C. E., & Aranovich, L. Y. (2014). Brines at high pressure and temperature: Thermodynamic, petrologic, and geochemical effects. *Precambrian Research*, *253*, 6–16.

Manning, C. E., & Boettcher, S. L. (1994). Rapid-quench hydrothermal experiments at mantle pressures and temperatures. *American Mineralogist*, *79*, 1153–1158.

Manning, C. E., Shock, E. L., & Sverjensky, D. A. (2013). The chemistry of carbon in aqueous fluids at crustal and upper-mantle conditions: Experimental and theoretical constraints. *Reviews in Mineralogy and Geochemistry*, *75*, 109–114.

Newton, R. C., Aranovich, L. Y., Hansen, E. C., & Vandenheuvel, B. A. (1998). Hypersaline fluids in Precambrian deep-crustal metamorphism. *Precambrian Research*, *91*, 41–63.

Newton, R. C., & Manning, C. E. (2002). Experimental determination of calcite solubility in H_2O-NaCl solutions at deep crust/upper mantle pressures and temperatures: Implications for metasomatic processes. *American Mineralogist*, *87*, 1401–1409.

Pan, D., & Galli, G. (2016). The fate of carbon dioxide in water-rich fluids at extreme conditions. *Science Advances*, *2*, e1601278.

Parr, R. G., & Pearson, R. G. (1983). Absolute hardness: companion parameter to absolute electronegativity. *Journal of the American Chemical Society*, *105*, 7512–7516.

Phillipot, P., & Selverstone, J. (1991). Trace-element rich brines in eclogitic veins: Implications for fluid composition and transport during subduction. *Contributions to Mineralogy and Petrology*, *106*, 417–430.

Scambelluri, M., & Phillipot, P. (2001). Deep fluids in subduction zones. *Lithos*, *55*(1–4), 213–227.

Van den Berg, R., & Huizenga, J. (2001). Fluids in granulites of the Southern Marginal Zone of the Limpopo belt, South Africa. *Contributions to Mineralogy and Petrology*, *141*, 529–545.

22

The Changing Character of Carbon in Fluids with Pressure: Organic Geochemistry of Earth's Upper Mantle Fluids

Dimitri Sverjensky[1], Isabelle Daniel[2], and Alberto Vitale Brovarone[3,4]

ABSTRACT

Decades of research have now firmly established that aqueous fluids in Earth's crust from groundwater to deep basinal brines to shallow mid-ocean ridge hydrothermal systems can contain metastable equilibria involving C-species in which methane does not participate. It now appears, however, that this situation does not extend to the upper mantle. Instead, field evidence from theoretical models, experimental diamond-anvil cell and fluid inclusion studies, and high-pressure metamorphic rocks all indicate a wide variety of aqueous C-species at high pressures. At pressures above about 2.0–3.0 GPa, methane and all other aqueous C-species with different oxidation states between -4.0 and +4.0 can coexist in variable proportions. Furthermore, the aqueous fluids can coexist with immiscible hydrocarbon fluids. As a consequence, C-species with a mixture of oxidation states in aqueous fluids and hydrocarbon fluids at high pressures in the upper mantle can influence the oxidation state of their upper mantle environment.

22.1. INTRODUCTION

A wide variety of aqueous carbon species occurs in fluids located at the surface of Earth to at least upper mantle depths. Interestingly, however, the modeling treatment of aqueous carbon species tends to depend on the geochemical environment in which they are found. For example, in seawater, aqueous inorganic carbon speciation has been treated using the classical inorganic C-species CO_2, HCO_3^-, and CO_3^{2-}, and Ca and Mg complexes of the ions. Such models have been very thoroughly studied under surficial conditions, where their relative abundances generally reflect thermodynamic equilibria and are crucial to an understanding of seawater chemistry, life in the oceans, and therefore climate change. The same species in various proportions are important in rainwater, soil water, and shallow groundwaters, together with a wide range of organic C-species that are completely out of thermodynamic equilibrium with the inorganic species. However, in deep groundwaters, such as those in sedimentary basins where sampled fluids reach temperatures on the order of 150 °C, some aqueous organic carbon species (e.g. CH_3COO^-) are in chemical communication with the inorganic ones, forming metastable thermodynamic equilibria, as will be discussed below (Shock, 1988, 1990). At the even higher temperatures of hot-spring and mid-ocean ridge environments (up to about 400 °C), a wide array of metastable equilibria may exist (Manning et al., 2013; McCollom, 2013).

In contrast, in crustal metamorphic environments at temperatures much greater than 400 °C, a very different treatment of aqueous carbon speciation has prevailed.

[1] *Department of Earth and Planetary Sciences, Johns Hopkins University, Baltimore, Maryland, USA*

[2] *Université Lyon, Université Lyon 1, Ens de Lyon, CNRS, UMR 5276, Laboratoire de Géologie de Lyon, Villeurbanne, France*

[3] *Dipartimento di Scienze della Terra, Università degli studi di Torino, Torino, Italy*

[4] *Sorbonne Université, Museum National d'Histoire Naturelle, UMR CNRS 7590, IRD, Institut de Minéralogie, des Physique de Matériaux et de Cosmochimie, Paris, France*

Carbon in Earth's Interior, Geophysical Monograph 249, First Edition. Edited by Craig E. Manning, Jung-Fu Lin, and Wendy L. Mao.

Crustal metamorphic fluids have long been generally treated as simple molecular mixtures, e.g. CO_2, CO, CH_4, H_2, and H_2O. For simplicity, models of the fluids are often referred to as COH fluids (Zhang & Duan, 2009). Three assumptions are inherent in COH fluid models. First, the fluids are assumed to contain no other solutes, e.g. no Na, K, Mg, Fe, Ca, Al, or Si, even though they are generally regarded as being in equilibrium with their host rocks and every mineral in the rocks has a definite solubility under these conditions. Second, only molecular species such as those listed above are in the model fluids. There are no ions. As a consequence, there can be no concept of pH, which is such an important parameter in the near-surface environment. Third, the fugacity of oxygen is the only chemical link between the fluids and rocks, except for the molecules listed above, typically CO_2, CH_4, and H_2O. The same overall approach has been widely used for fluids in subduction zones, even though decades of experimental studies have shown that at temperatures of about 700 °C to 1,200 °C and pressures approaching 2.0 GPa, the solubilities of minerals in aqueous fluids become substantial (Manning, 2004; Manning, 2018).

In upper mantle environments such as the subcratonic lithospheric mantle, at temperatures of about 800 °C–1,400 °C and pressures of about 4.0 to 6.0 GPa, the COH fluid model has again been the default model for aqueous C-species in the deep Earth. However, fluid inclusion and experimental studies have demonstrated the importance of much higher solute concentrations in the fluids than even in crustal metamorphic environments. Models of the C-speciation vary from high-pressure extensions of the COH fluid model to discussions of a class of fluids termed high-density fluids, or HDFs (Navon, 1999), which are often equated with melts. Such models are very constrained in their ability to explain the causes of diamond formation and how the fluids may also be involved in mantle metasomatic processes that influence the geochemical and mineralogical evolution of the subcratonic lithospheric mantle. The reason is that the COH fluid models have relied almost entirely on invoking redox changes as the cause of precipitation of diamond. In fact, either oxidation of CH_4 or reduction of CO_2 to form diamond are the only chemical processes that can be invoked using a COH fluid model. The HDF model relies more on inferring potential processes by using trace element and isotopic tracers as proxies for fluid and/or melt interaction with upper mantle rocks.

Recent developments in theoretical and experimental aqueous geochemistry enable a new treatment of aqueous C-speciation in deep crustal and upper mantle fluids that is analogous to the classical models of surficial and shallow crustal fluids. A continuous modeling capability from surficial to upper mantle conditions is now possible with the Deep Earth Water (DEW) model (Sverjensky, Harrison, et al., 2014; Huang & Sverjensky, 2019). Predictive speciation models of subcratonic lithospheric mantle fluids involving aqueous ions and metal complexes have been developed (Huang et al., 2017; Mikhail et al., 2017; Mikhail & Sverjensky, 2014; Sverjensky, Harrison, et al., 2014; Sverjensky, Stagno, et al., 2014) based on Raman spectroscopic data for C-bearing ions (Facq et al., 2016; Facq et al., 2014), ab initio theoretical results (Pan & Galli, 2016; Pan et al., 2013), and measured mineral solubilities (Manning, 1994; Tropper & Manning, 2007). The fundamental model does not change from shallow to deep conditions in the Earth. In contrast to the COH and HDF models, the DEW model enables examination of changes in equilibrium aqueous speciation from surficial to deep Earth conditions, and treats a much greater variety of species, including ions and metal complexes. As a result, it is possible to examine patterns of change in aqueous fluid speciation from shallow to deep conditions in a way that was not previously possible.

As an example of recent advances in the modeling of deep Earth fluids, it has been predicted that aqueous inorganic and organic C-species can be in complete thermodynamic equilibrium in deep Earth fluids, in marked contrast to geochemical systems at lower temperatures (Sverjensky, Stagno, et al., 2014). Furthermore, recent ab initio calculations of speciation in aqueous fluids have demonstrated that under upper mantle temperatures and pressures, new species become important. At 1,000 K and 10.0 GPa, ab initio and thermodynamic calculations show that CO_2^0 is not an important species. Instead, $H_2CO_3^0$ is the major oxidized neutral carbon species (Pan & Galli, 2016). This is a major difference to the traditional COH fluid model for metamorphic and deep Earth fluids. Experimental studies have also indicated changes in the character of aqueous C-species at elevated pressures. Aqueous methane forms rapidly and easily at high pressures (Huang et al., 2017; Scott et al., 2004) and immiscible aliphatic hydrocarbon fluids coexisting with oxidized species of carbon have been documented (Li, 2017; Huang et al., 2017). Together, these advances in theory and experiments represent a first step towards understanding the potential role of aqueous C-species during fluid-rock interactions in the deep Earth.

Recent field discoveries have also indicated marked differences in the behavior of C-species between conditions in the mid-ocean ridge hydrothermal systems and those in subduction zones at upper mantle conditions. The amounts of aqueous methane in mid-ocean ridge hydrothermal systems are small. They may be of biologic origin, or inherited from preexisting fluid inclusions in minerals in the igneous oceanic crust (McDermott et al., 2015), or related to catalysis by chromitite minerals (Etiope et al., 2018). In contrast, abundant immiscible abiotic methane has now been documented in alpine ultramafic massifs previously equilibrated at upper mantle pressures (Vitale Brovarone et al., 2017). The effects of hydrocarbon immiscibility on the mobility and reactivity of deep carbon remain poorly constrained. Hydrocarbon immiscibility phenomena likely affect the wettability of porous media at depth, and thereby the mobility of deep carbon. This feature can potentially lead to unexpected reactivity of mineral surfaces and localized-fluid-rock equilibria.

In the present study, we wish to draw attention to evidence for the changing character of aqueous C-species with temperature and pressure. Based on the discussion above, two major regimes with very different behavior of carbon species are now clear. Aqueous C-species appear to behave completely differently in the shallow crustal environments compared with the upper mantle. In order to emphasize the transition between the two environments, we summarize below the key evidence for these two major regimes.

22.2. CARBON SPECIATION IN SHALLOW CRUSTAL FLUIDS

Long-term metastability of aqueous C-species on geologic timescales was first proposed decades ago based on theoretical interpretation of the measured abundances of short-chain aliphatic acids in oil-field brines (Shock, 1988). As an example, it can be seen in Figure 22.1a that Gulf Coast brines systematically cluster about a line with a theoretical slope of 1.5 consistent with the equilibrium

$$3CH_3COOH^0 + 2H_2 = 2CH_3CH_2COOH^0 + 2H_2O \quad (22.1)$$

and a rearrangement of the corresponding Law of Mass Action represented by

$$\log a_{CH_3CH_2COOH^0} = +1.5\log a_{CH_3COOH^0} + (0.5\log K - 2\log f_{H2}) \quad (22.2)$$

The correlation strongly suggests that there is an equilibrium between the organic acid anions.

The intercepts of the lines defined by equation (22.2), together with theoretically calculated values of log$K$, enable calculation of values of f_{H2} for the oil field brines. The values of logf_{H2} are shown by the symbols in Figure 22.1b. The contours on the plot represent theoretically calculated values of logf_{CO2} calculated from the hypothetical equilibrium

$$2CO_2^0 + 4H_2 = CH_3COOH^0 + 2H_2O \quad (22.3)$$

Values of logf_{CH4} were calculated from another hypothetical equilibrium:

$$CH_3COOH^0 + 4H_2 = 2CH_4 + 2H_2O. \quad (22.4)$$

The hatched area in Figure 22.1b is limited by an upper value of f_{CO2} equal to less than the total pressure in the system. It can be seen that the symbols in Figure 22.1b clearly overlap with the logf_{CO2} values, i.e. the calculated f_{H2} values of the brines are consistent with f_{CO2} values less than the total pressure. However, corresponding values of the calculated f_{H2} values of the brines are not consistent

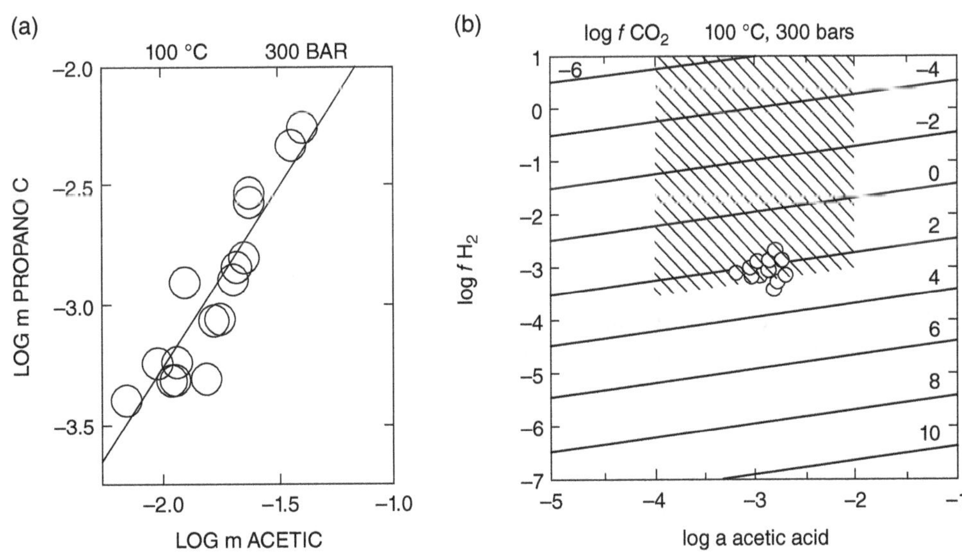

Figure 22.1 Metastable equilibria in oil-field brines (Shock, 1988, 1990). (a) Logarithms of the molalities of aliphatic acids in oil-field brines relative to a line with a theoretical slope of 1.5. (b) Values of the logarithm of the hydrogen gas fugacity for oil-field brines such as those defined by (a) and their acetic acid concentrations. The contours represent the hypothetical fugacity of CO_2 in metastable equilibrium with the acetic acid. See electronic version for color representation of the figures in this book.

with f_{CH4} values less than the total pressure. Instead, they suggest hypothetical f_{CH4} values on the order of 10,000 bars, which is completely unreasonable. Overall, Figures 22.1 a and b strongly suggest that a metastable equilibrium between aliphatic acids and CO_2 is possible, but that CH_4 is out of equilibrium with the other carbon-bearing species.

Subsequent experimental studies have confirmed the validity of the concept of metastable equilibrium between aqueous C-species when methane is kinetically hindered from reacting under shallow crustal conditions. The first example involved the demonstration of equilibrium at 325 °C and 350 bars between ethene and ethane (Seewald, 1994), which can be written

$$2C_2H_4^0 + 2H_2O + 2H_2 = 2CH_3CH_3^0 + 2H_2O, \quad (22.5)$$

with f_{H2} buffered by the mineral assemblage pyrite-pyrrhotite-magnetite according to the equilibrium

$$3FeS_2 + Fe_3O_4 + 4H_2 = 6FeS + 4H_2O. \quad (22.6)$$

Over a period of hundreds of hours, a steady state of the aqueous species was reached that is well within uncertainties of the predicted thermodynamic metastable equilibrium. Numerous subsequent studies have confirmed reversible hydrothermal metastable equilibria among many types of aqueous organic species (McCollom & Seewald, 2003a, 2003b; Seewald et al., 2006), including aromatic aqueous C-species (Shipp et al., 2013; Shock et al., 2013; Yang et al., 2012).

Confirmation of the existence of metastable equilibria in the field has been more elusive (Etiope & Sherwood Lollar, 2013). The outstanding example remains that of oil-field brines discussed above. Numerous studies have suggested that small amounts of methane in mid-ocean ridge systems have been formed by abiotic synthesis facilitated by mineral catalysis. But this is difficult to demonstrate beyond reasonable doubt. Other studies have suggested that methane in gas seeps and springs associated with ultramafic bodies has an abiotic origin catalyzed by Cr or Ru in rocks (Etiope et al., 2018). The strongest evidence of a metastable equilibrium is the recent study of serpentinite in the mid-ocean ridge Atlantis Massif. The study demonstrated that the amino acid tryptophan formed from a natural abiotic synthesis catalyzed by iron-rich clay minerals (Ménez et al., 2018). This remarkable discovery has wide-ranging implications for the metastable organic geochemistry of mid-ocean ridge hydrothermal systems, as the formation of amino acids and other biomolecules has long been predicted theoretically. Interestingly, tryptophan had previously been predicted to be one of the two most likely amino acids to occur by abiotic synthesis in a wide range of mid-ocean ridge hydrothermal systems (Shock & Canovas, 2010). Figures 22.2 a and b show predictions of the chemical affinity for tryptophan or phenylalanine to form during mixing of many different mid-ocean ridge hydrothermal fluids with cold seawater under conditions of metastable equilibrium. The very large positive chemical affinities indicate a strong thermodynamic driving force favoring the formation of the organic molecule in each case in the absence of methane. This prediction agrees well with the Atlantis Massif discovery of abiotic tryptophan.

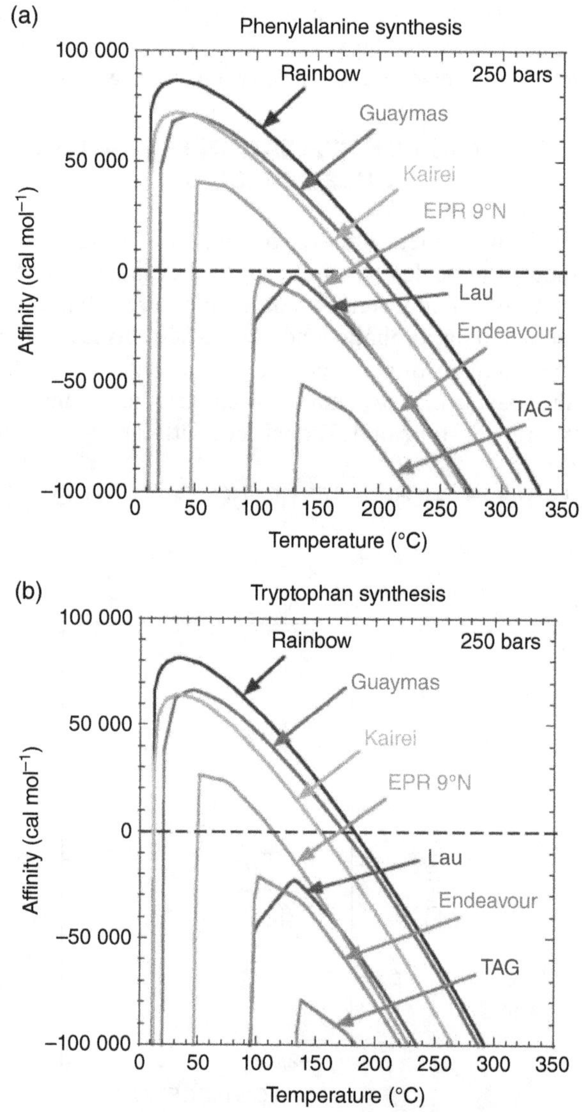

Figure 22.2 Predicted thermodynamic driving force (affinity) for amino acid synthesis in mid-ocean ridge fluids mixing with ambient seawater (Shock & Canovas, 2010). (a) Phenylalanine. (b) Tryptophan. See electronic version for color representation of the figures in this book.

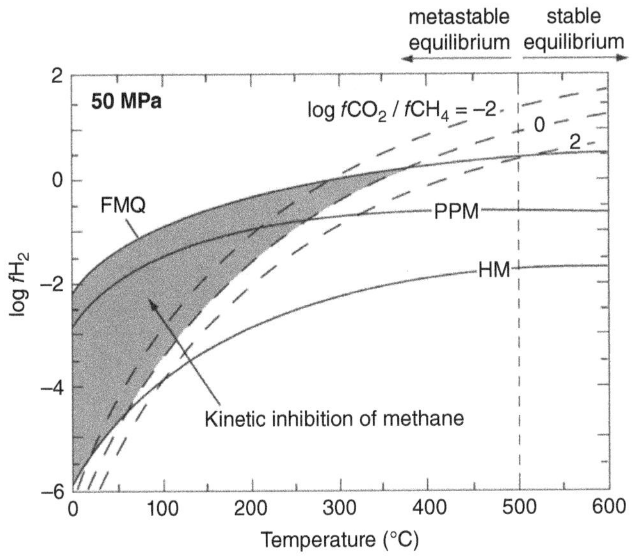

Figure 22.3 Prediction of the conditions of metastable equilibria under shallow crustal conditions where methane formation is kinetically inhibited (shaded area) (Manning et al., 2013) and metastable equilibria of other C-bearing species is favored. Solid curves represent mineral hydrogen buffers and dashed curves represent contours of the fugacity ratio of CO_2 and CH_4 gases. See electronic version for color representation of the figures in this book.

Overall, the kinetic reluctance of aqueous methane to participate in reactions with other aqueous C-species opens up the possibility of a rich set of metastable equilibria between C-bearing species in crustal hydrothermal systems (Figure 22.3). The solid curves in the figure refer to the mineral reactions that might buffer the fugacity of H_2. The potential range of conditions under which metastable equilibria are most likely to be detected is summarized by the shaded area in Figure 22.3 (Manning et al., 2013, after Shock, 1992), i.e. at temperatures less than about 400 °C. It is at these temperatures that the kinetics of formation of CH_4 are expected to be so slow that natural fluids in the shallow crust will contain a wide variety of aqueous organic carbon species in metastable equilibrium with CO_2, HCO_3^-, and CO_3^{2-}, as well as carbonate minerals. The likely upper temperature limit for metastable equilibria is suggested to be about 500 °C. Above this temperature, it is suggested that the kinetics of reversible reduction of CO_2 to CH_4 are sufficiently fast that a full thermodynamic equilibrium in aqueous solution will exist. Such fluids should contain CO_2, CH_4, HCO_3^-, CO_3^{2-}, and related metal complexes of the ions. They should not contain other aqueous organic species in any significant concentrations (Manning et al., 2013). It should be emphasized that Figure 22.3 refers to a pressure of 500 bars. However, as will be discussed below, at much higher pressures, it is possible that a completely different situation exists.

22.3. CARBON SPECIATION IN UPPER MANTLE FLUIDS

22.3.1. Theoretical Studies

22.3.1.1. The End of Metastablity: The Aqueous Organic Geochemistry of the Upper Mantle

The COH fluid model has been widely used to address the expected species of aqueous carbon under upper mantle conditions. As an example, calculated variations in the mole fractions of CO_2 and CH_4 in fluids in equilibrium with graphite or diamond as a function of depth in the Earth (Zhang & Duan, 2009) only include these species and H_2O, CO, H_2, and C_2H_6. There are no ions. It is assumed that the f_{O2} is imposed exclusively by the intramineral equilibrium in rocks sampled as xenoliths from the subcratonic lithospheric mantle. Between depths of about 100 and 200 km, H_2O-rich fluids are predicted to transition from having CO_2 as the predominant C-species to CH_4 as the main C-species (Zhang & Duan, 2009). In other words, the transition between about 100 and 200 km is from fluids with high CO_2/ CH_4 ratios to fluids with high CH_4/ CO_2 ratios. This conclusion about C-speciation in upper mantle fluids is widely used in discussions of diamond formation (Shirey et al., 2013). The simplicity of the COH model for C-speciation at upper mantle conditions stands in contrast to the complexity of metastable C-speciation possible below about 500 °C under shallow crustal conditions indicated by Figure 22.3 and the crustal oil-field type brines discussed earlier. The contrast has traditionally been taken to indicate that at shallow crustal conditions metastability prevails, with many different aqueous C-species with different oxidation states, but that at upper mantle conditions a simple equilibrium C-speciation exists, with the two main oxidation states of C, i.e. CH_4 and CO_2, and not much in between. This situation persisted for decades until it became possible to use a different model for upper mantle fluids.

At crustal pressures and at temperatures greater than about 500 °C, under which CH_4 can be assumed to freely react, the expected thermodynamic equilibrium speciation of carbon should be simple, as illustrated in Figure 22.4a at 600 °C and 5.0 kb. It can be seen that equilibrium between CH_4, CO_2, HCO_3^-, and CO_3^{2-} is depicted. No aqueous ionic C-species with intermediate oxidation states have predominance fields on the diagram. In contrast, it can be seen in Figure 22.4b that consideration of acetic acid and acetate at 600 °C and the much higher pressure of 50 kb results in the prediction of a substantial predominance field for acetate and a very small one for acetic acid

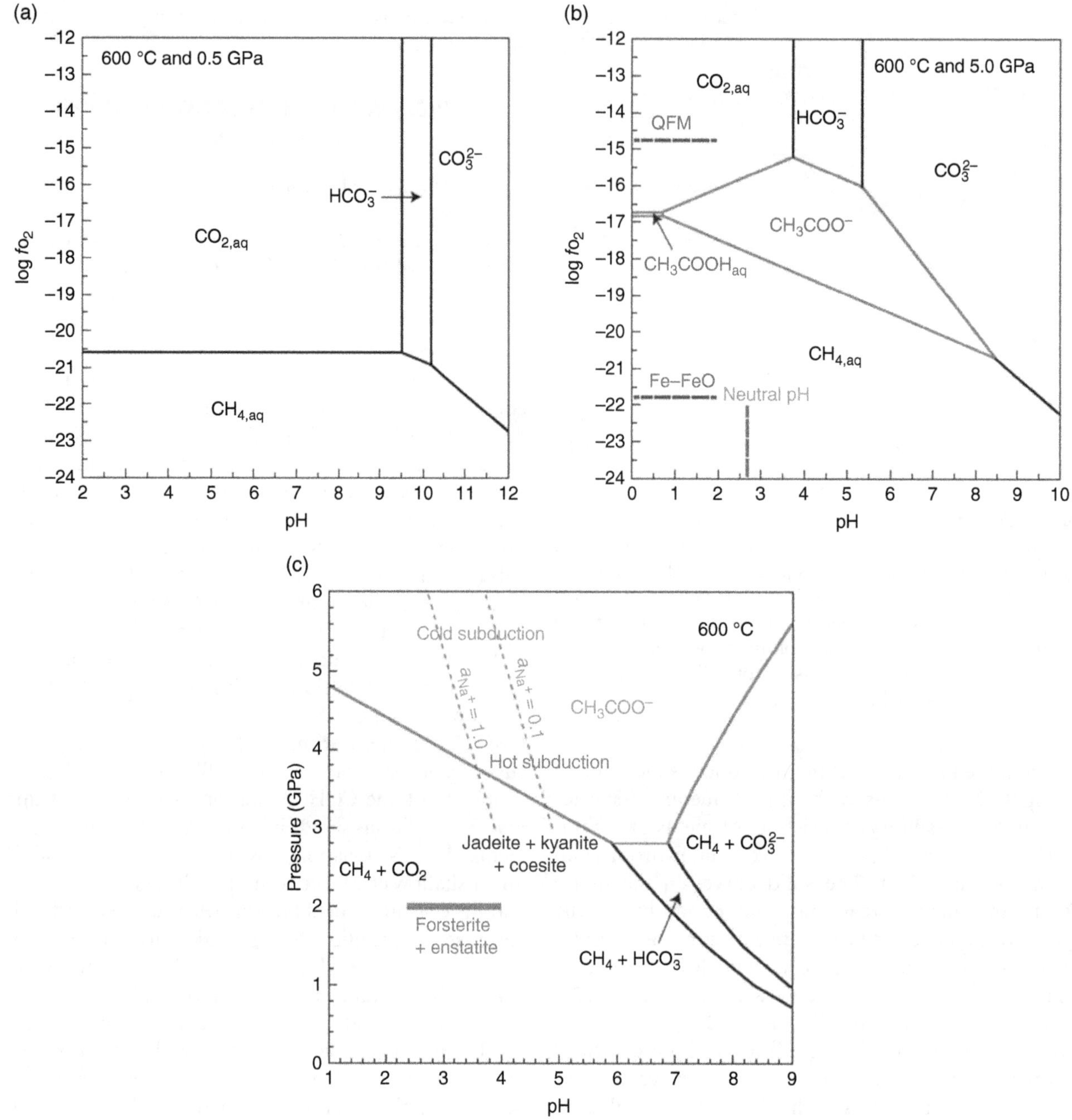

Figure 22.4 Predicted aqueous C-speciation at 600 °C (Sverjensky, Stagno, et al., 2014). (a) Low pressure. (b) High pressure. (c) Predicted aqueous C-speciation as a function of pressure and pH. Expected pH ranges are shown for forsterite+enstatite (peridotite) and jadeite+kyanite+coesite (metasedimentary eclogite). See electronic version for color representation of the figures in this book.

(Sverjensky, Stagno, et al., 2014). Calculations such as these, using the DEW model, give a different picture of C-speciation in upper mantle fluids to that offered by the COH fluid model. More generally, it is predicted that in upper mantle fluids there is the possibility of a thermodynamically stable, rich variety of aqueous C-species with a wide range of oxidation states of carbon between those of carbon in CH_4 and CO_2. Furthermore, all the C-species are thermodynamically linked to the major element compositions of the silicate rocks. The richness of the aqueous organic C-speciation in upper mantle fluids is analogous to that in the shallow crust. The major difference, however, is that at upper mantle pressures there is complete thermodynamic equilibrium between all the aqueous C-species, including CH_4, whereas in shallow crustal fluids metastable equilibria exists.

It is clear from Figures 22.4 a and b that fluids at 50.0 kb are predicted to behave differently than fluids at 5.0 kb. At what pressure does this transition take place? In the specific case of acetate, model calculations indicate the appearance of significant quantities of acetate relative to C(IV) species at about 29.0 kb at 600 °C, as shown in Figure 22.4c. At 800 °C, the transition pressure is predicted to be about 39.0 kb, and by 1000 °C there is no significant amount of acetate relative to oxidized C-species (Sverjensky, Stagno, et al.; 2014). It should be emphasized that the calculations reported in Figures 22.4 a–c considered only acetate and acetic acid as organic C-species. Other organic C-species, including metal-organic complexes, are likely to be more important (Huang & Sverjensky, 2019). Overall, these results indicate that thermodynamic equilibrium between all aqueous C-species, both inorganic and organic, is a strong function of the pressure of the fluid.

22.3.1.2. $H_2CO_3^0$: The Forgotten Molecule in Upper Mantle Fluids

All the models discussed above have assumed that the predominant electrically neutral oxidized C-species in water is CO_2^0, as at ambient conditions. However, a theoretical ab initio molecular dynamics study of the $Na_2CO_3 - H_2O$ system at conditions up to 727 °C and 10.0 GPa (Pan & Galli, 2016) revealed two surprising and very important results. First, a variety of $NaHCO_3$ and $Na - CO_3$ complexes dominate the aqueous C-speciation, and second, $H_2CO_3^0$ is a much more abundant molecule than CO_2^0. The latter is the complete opposite to the situation under ambient conditions in water.

The discovery by ab initio molecular dynamics of the importance of $H_2CO_3^0$ relative to the traditionally used CO_2^0 in high pressure aqueous fluids made by Pan and Galli (2016) raises an important question. Under which conditions does the transformation takes place? Experimental determinations of the thermodynamic stability of $H_2CO_3^0$ at near ambient conditions have long been reported (Soli & Byrne, 2002; Wissbrun et al., 1954). And in situ spectroscopic indications of the existence of $H_2CO_3^0$ at low and high temperatures have been reported (Abramson et al., 2017; Dubessy et al., 1999; Lam et al., 2014). The trace amounts of $H_2CO_3^0$ relative to CO_2^0 in aqueous solutions at ambient conditions have made it convenient for geochemists to ignore the hydrate $H_2CO_3^0$. Under crustal hydrothermal and metamorphic conditions extending into the upper mantle, numerous studies of $CO_2 - H_2O$ fluids exist using mixtures of the generic molecules CO_2 and H_2O (e.g. Zhang & Duan, 2009). Quantifying the temperature and pressure conditions of the transformation of CO_2^0 to $H_2CO_3^0$ is important because it will potentially affect the modeling of all deep Earth fluids, as well as our understanding of carbon transport in the deep carbon cycle.

22.3.2. Experimental Studies

Recent experimental studies have started to test some of the predictions discussed above about aqueous C-species under upper mantle conditions. Specifically, the prediction that organic acid anions might be stable at pressures above about 30.0 kb in high-density fluids led to a diamond anvil cell study of the stability of Na-acetate solutions (Huang et al., 2017) at 300 °C and 30.0 kb. A very high concentration of Na-acetate (1.0 molal) was used to focus on acetate transformations. In situ analysis of the experimental products was made with Raman spectroscopy. Acetate was partly transformed into bicarbonate, and the bicarbonate to acetate ratio reached a steady state that persisted for the whole experiment, up to about 60 hours. However, the most dramatic result of the experiment was not predicted. Within a couple of hours of reaching the peak temperature and pressure in the experiments, it was discovered that immiscible isobutane fluid formed along with CH_4 and HCO_3^- and eventually appeared to reach a steady state, coexisting with aqueous acetate and Na-carbonate crystals (Figure 22.5 a, b). The immiscibility could not be predicted, owing to a lack of adequate thermodynamic characterization at high pressures of even simple hydrocarbon fluids such as isobutane.

The results shown in Figures 22.5 a and b are also of interest because at 30 kb, CH_4 formed easily and rapidly in the experiments, even at a temperature as low as 300 °C. This behavior contrasts sharply with Figure 22.3, according to which, at crustal pressures, CH_4 formation will be kinetically hindered at 300 °C. Consequently, it appears that high pressure facilitates the formation of CH_4 even at only 300 °C. In order to gain a more fundamental understanding of the kinetics of methane formation and the conditions under which immiscibility might occur, it will be essential to do more experiments and to develop theoretical models that will enable extrapolations to different temperatures and pressures and chemically more complex systems relevant to nature.

Immiscible behavior of hydrocarbon fluids was also discovered in another recent experimental study using synthetic fluid inclusions trapped in quartz (Li, 2017). An initial fluid consisting of 3 wt % NaCl and 18.5 wt% HCOOH in 78.5 wt % H_2O was buffered over a range of different oxidation states from Fe-FeO to Re-ReO_2 at 600 °C and 700 °C and 15.0 to 25.0 kb. The synthetic fluid inclusions showed evidence of trapping two immiscible fluids, one aqueous and one methane rich, denoted Type 1 and Type 2 in Figures 22.5 c and d. Other organic compounds were detected in both types of fluid inclusions, and H_2 was also detected in methane-rich fluid inclusions from the most reducing conditions (i.e. with the Fe-FeO buffer). A remaining uncertainty for the latter inclusions

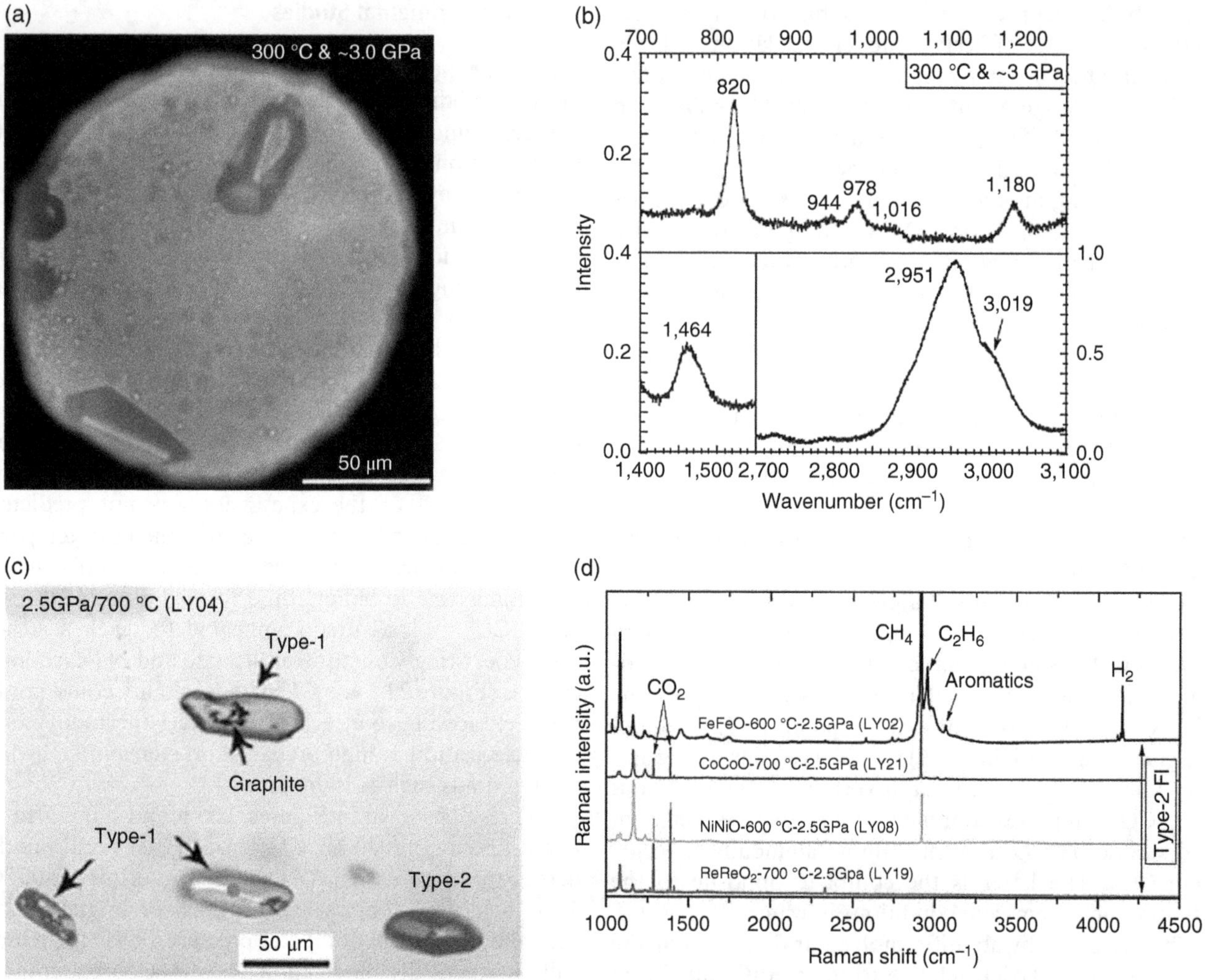

Figure 22.5 Experimental evidence for immiscible hydrocarbon fluids. (a) Immiscible isobutane and aqueous fluid plus Na-carbonate crystals. (b) Raman spectrum of immiscible isobutane fluid (Huang et al., 2017). (c) Synthetic fluid inclusions in quartz showing two immiscible fluids. (d) Raman spectra (Li, 2017). See electronic version for color representation of the figures in this book.

is the nature of the methane-rich fluid at the elevated temperatures of trapping in quartz. It was clearly a hydrocarbon fluid, but it may have been a different hydrocarbon fluid before cooling to room temperature. It should be noted, however, that the experimental results shown in Figures 22.5 a–d refer to two different starting materials, Na-acetate and formic acid, as well as to very different temperatures and experimental durations that could have allowed the immiscible fluids to behave differently in the fluid inclusions. Nevertheless, both resulted in immiscibility between aqueous fluids and hydrocarbon fluids.

22.3.3. Field Studies

Immiscible hydrocarbon-rich fluids have now been documented in high-pressure metamorphic rocks. The Lanzo massif in the Italian Western Alps is an example of an exhumed segment of high-pressure subducted oceanic lithosphere. The massif includes carbonated ultramafic rocks bearing hydrocarbon-rich fluid inclusions (Vitale Brovarone et al., 2017). Microstructures and stable isotope data constrained the genesis of the hydrocarbons to high-pressure events of fluid-rock interactions leading to a reduction of subducted carbonate minerals. In these rocks, two clearly different yet microstructurally coexisting fluid populations can be observed (Figure 22.6a). One population is present as negative crystal-shaped to rounded inclusions and consists of a single phase CH_4 + H_2 fluid at ambient conditions, as identified by Raman spectroscopy (Figure 22.6b). The second population consists of two-phase (liquid-gas) inclusions and contains CH_4 + H_2O + H_2 at ambient conditions (Figure 22.6b). Figure 22.6a shows a trail of

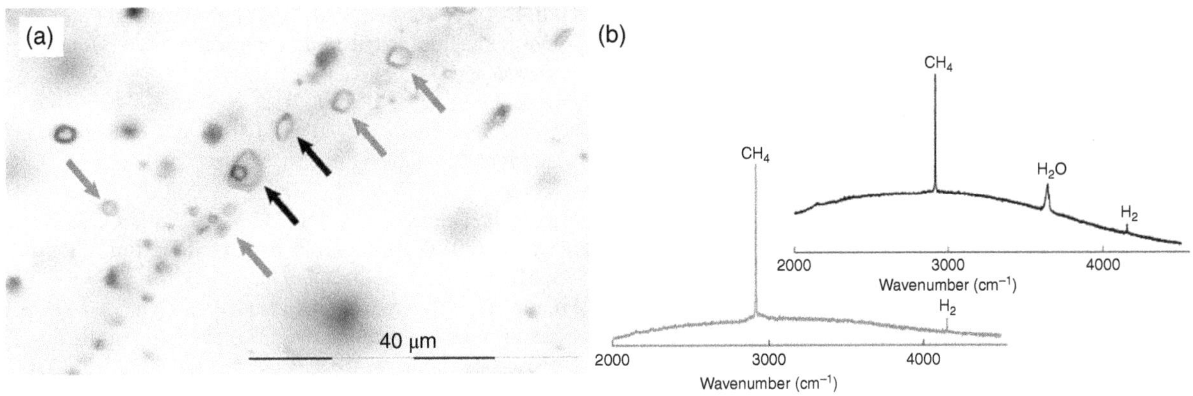

Figure 22.6 Natural immiscible hydrocarbon and aqueous fluids in high-pressure rocks from the Lanzo massif, Italian Western Alps. (a) Fluid inclusions of two coexisting types. (b) Raman spectra of the two coexisting populations (Vitale Brovarone et al., 2017). See electronic version for color representation of the figures in this book.

pseudosecondary fluid inclusions containing both populations. The microstructure shows that the two fluids were both present during the trail formation, a strong indication of fluid immiscibility. Mineralogical features suggest that the formation and entrapment of these fluids happened at conditions of ≤400 °C and 1.0 GPa (Vitale Brovarone et al., 2017).

Documentation of natural high-pressure immiscible hydrocarbons is still scarce. The identification of natural immiscible hydrocarbons in the rock record, the quantification of the parameters controlling the immiscibility, and the definition of specific immiscible phases are hampered by two main features. The first one is the postentrapment mechanical reworking of the fluid inclusions, which can lead to partial or even selective loss of the initial fluid composition (Bakker & Jansen, 1994; Hall & Sterner, 1993). As an example, Bakker and Jansen (1994) demonstrated that preferential H_2O leakage can occur in H_2O - CO_2 fluid inclusions along dislocations and planar defects in the host mineral. Slight compositional variations related to leakage may render the identification of coexisting immiscible fluids particularly challenging. Comparison between multiple samples might therefore be necessary, yet not always possible. The second one is the respeciation of the fluid inclusions during cooling and decompression from their formation conditions. Methane appears to be the most common hydrocarbon species present in fluid inclusions in high-pressure rocks and analyzed at ambient conditions (Arai et al., 2012; Herms et al., 2012; Shi et al., 2005; Song et al., 2009). However, it is possible that the CH_4 present in these inclusions, or part of it, results from the respeciation of more complex hydrocarbon fluids. In the Lanzo case study, the effect of at least local respeciation is suggested by the presence of daughter graphite in some individual fluid inclusions (Vitale Brovarone et al., 2017), which clearly indicates carbon redistribution within the inclusion later than their formation (Cesare, 1995). Nevertheless, the resemblance of the natural fluid inclusions with the experimental results from Li (2017) under reducing conditions is clear: immiscible C-O-H and C-H fluids coexisted at subduction zone conditions. From this perspective, modern thermodynamic tools represent a powerful means to estimate the original speciation of natural hydrocarbon-rich fluids at high-pressure conditions.

22.4. CONCLUDING REMARKS

First, we have emphasized the long-standing evidence of metastable equilibria among aqueous C-species with different oxidation states under shallow crustal conditions. A great variety of aqueous C-species can coexist in this way when methane is prevented from forming, or reacting, by sluggish kinetics. Outstanding examples of this behavior include the short-chain aliphatic acids in brines in sedimentary basins and the amino acid tryptophan in a mid-ocean ridge environment, as well as a growing number of experimental studies of hydrocarbons, alcohols, and alkenes under hydrothermal conditions.

Second, we describe emerging evidence for a very different behavior for aqueous C-species at higher temperatures and pressures. It has long been assumed that temperatures greater than about 500 °C facilitate the ease of formation of methane in aqueous fluids, so that complete thermodynamic equilibrium among C-species can exist. For example, at 600 °C and 5.0 kb, the expected speciation is very simple: CH_4, CO_2, $H_2CO_3^0$, HCO_3^-, and CO_3^{2-}. In contrast, at 600 °C and at pressures greater than about 30.0 kb, additional organic aqueous C-species may become thermodynamically stable, e.g. aliphatic acid anions in particular. Consequently, at upper mantle temperatures and pressures, there could be a rich C-speciation present in fluids with a great variety of oxidation states of C between −IV and +IV.

Third, we summarize theoretical, experimental, and field evidence that immiscible hydrocarbon fluids may coexist with aqueous fluids at upper mantle temperatures and pressures. Hydrocarbon fluids in subduction zones could represent a new mode of carbon transport in the deep carbon cycle and influence the oxidation state of their environments.

ACKNOWLEDGMENTS

The authors wish to thank their collaborators in the Deep Carbon Observatory for numerous ideas and suggestions, particularly Mark Ghiorso, Craig Manning, Sami Mikhail, Vincenzo Stagno, and Everett Shock. Financial support is acknowledged from the Deep Carbon Observatory (Daniel, Sverjensky), and NSF Awards EAR-1624325 and ACI-1550346 (Sverjensky), and ANR T-ERC grant, CNRS INSU-SYSTER and MIUR Rita Levi Montalcini (Vitale Brovarone).

REFERENCES

Abramson, E. H., Bollengier, O., & Brown, J. M. (2017). Water-carbon dioxide solid phase equilibria at pressures above 4 GPa. *Scientific Reports*, *7*, 821.

Arai, S., Ishimaru, S., & Mizukami, T. (2012). Methane and propane micro-inclusions in olivine in titanoclinohumite-bearing dunites from the Sanbagawa high-P metamorphic belt, Japan: Hydrocarbon activity in a subduction zone and Ti mobility. *Earth Planet. Sci. Lett.*, *353–354*, 1–11.

Bakker, R. J., and Jansen, J.B.H. (1994). A mechanism for preferential H2O leakage from fluid inclusions in quartz, based on TEM observations. *Contributions to Mineralogy and Petrology*, *116*, 7–20.

Cesare, B. (1995). Graphite precipitation in C-O-H fluid inclusions: Closed system compositional density changes, and thermobarometric implications. *Contributions to Mineralogy and Petrology*, *122*, 25–33.

Dubessy, J., Moissette, A., Bäkker, R. J., Frantz, J. D., & Zhang, Y.-G. (1999). High-temperature Raman spectroscopic study of H2O-CO2-CH4 mixtures in synthetic fluid inclusions: First insights on molecular interactions and analytical implications. *European Journal of Mineralogy*, 23–32.

Etiope, G., Ifandi, E., Nazzari, M., Procesi, M., Tsikouras, B., Ventura, G., et al. (2018). Widespread abiotic methane in chromitites. *Scientific Reports*, *8*, 8728.

Etiope, G., & Sherwood Lollar, B. (2013). Abiotic methane on Earth. *Reviews of Geophysics*, *51*, 276–299.

Facq, S., Daniel, I., Montagnac, G., Cardon, H., & Sverjensky, D. A. (2016). Carbon speciation in saline solutions in equilibrium with aragonite at high pressure. *Chemical Geology*, *431*, 44–53.

Facq, S., Daniel, I., & Sverjensky, D. A. (2014). *In situ* Raman study and thermodynamic model of aqueous carbonate speciation in equilibrium with aragonite under subduction zone conditions. *Geochim. Cosmochim. Acta*, *132*, 375–390.

Hall, D. L., & Sterner, S. M. (1993). Preferential water loss from synthetic fluid inclusions. *Contributions to Mineralogy and Petrology*, *114*, 489–500.

Herms, P., John, T., Bakker, R. J., & Shenk, V. (2012). Evidence for channelized external fluid flow and element transfer in subducting slabs (Raspas Complex, Ecuador). *Chemical Geology*, *310*, 79–96.

Huang, F., Daniel, I., Cardon, H., Montagnac, G., & Sverjensky, D. A. (2017). Immiscible hydrocarbon fluids in the deep carbon cycle. *Nature Communications*, *8*.

Huang, F., & Sverjensky, D. A. (2019). Extended Deep Earth water model for predicting major element mantle metasomatism. *Geochimica et Cosmochimica Acta*, *254*, 192–230.

Lam, R. K., England, A. H., Sheardy, A. T., Shih, O., Smith, J. W., Rizzuto, A. M., et al. (2014). The hydration structure of aqueous carbonic acid from X-ray absorption spectroscopy. *Chem. Phys. Lett.*, *614*, 282–286.

Li, Y. (2017). Immiscible CHO fluids formed at subduction zone conditions. *Geochem. Pespect. Lett*, *3*, 12–21.

Manning, C. E. (1994). The solubility of quartz in H_2O in the lower crust and upper-mantle. *Geochim. Cosmochim. Acta*, *58*, 4831–4839.

Manning, C. E. (2004). The chemistry of subduction-zone fluids. *Earth Planet. Sci. Lett. 223*, 1–16.

Manning, C. E. (2018). Fluids of the lower crust: Deep is different. *Annu. Rev. Earth Planet. Sci.*, *46*, 67–97.

Manning, C. E., Shock, E. L., & Sverjensky, D. A. (2013). The chemistry of carbon in aqueous fluids at crustal and upper-mantle conditions: Experimental and theoretical constraints. *Reviews in Mineralogy and Geochemistry*, *75*, 108–148.

McCollom, T. M. (2013). Laboratory simulations of abiotic hydrocarbon formation in Earth's deep subsurface. *Rev Mineral Geochem*, *75*, 467–494.

McCollom, T. M., & Seewald, J. S. (2003a). Experimental constraints on the hydrothermal reactivity of organic acids and acid anions: I. Formic acid and formate. *Geochim. Cosmochim. Acta*, *67*, 3625–3644.

McCollom, T. M., & Seewald, J. S. (2003b). Experimental study of the hydrothermal reactivity of organic acids and acid anions: II. Acetic acid, acetate, and valeric acid. *Geochim. Cosmochim. Acta*, *67*, 3645–3664.

McDermott, J. M., Seewald, J. S., German, C. R., & Sylva, S. P. (2015). Pathways for abiotic organic synthesis at submarine hydrothermal fields. *Proceedings of the National Academy of Sciences*, *112*, 7668–7672.

Ménez, B., Pisapia, C., Andreani, M., Jamme, F., Vanbellingen, Q. P., Brunelle, A., et al. (2018). Abiotic synthesis of amino acids in the recesses of the oceanic lithosphere. *Nature*, *564*, 59–63.

Mikhail, S., Barry, P. H., & Sverjensky, D. A. (2017). The relationship between mantle pH and the deep nitrogen cycle. *Geochim. Cosmochim. Acta*, *209*, 149–160.

Mikhail, S., & Sverjensky, D. A. (2014). Nitrogen speciation in upper mantle fluids and the origin of Earth's nitrogen-rich atmosphere. *Nat. Geosci.*, *7*, 816–819.

Navon, O. (1999). Diamond formation in the Earth's mantle, *Proceedings of the 7th International Kimberlite Conference*. Cape Town: Red Roof Design, pp. 584–604.

Pan, D., & Galli, G. A. (2016). The fate of carbon dioxide in water-rich fluids in the Earth's mantle. *Science Advances*, *2*, e1601278–e1601278.

Pan, D., Spanu, L., Harrison, B., Sverjensky, D. A., & Galli, G. (2013). The dielectric constant of water under extreme conditions and transport of carbonates in the deep Earth. *Proceedings of the National Academy of Sciences*, *110*, 6646–6650.

Shock, E. L. (1992). Chemical environments of submarine hydrothermal systems. *Orig Life Evol Biosph*, *22*, 67–107.

Scott, H. P., Hemley, R. J., Mao, H., Hershbach, D. R., Fried, L. E., Howard, W. M., & Bastea, S. (2004). Generation of methane in the Earth's mantle: In situ high pressure–temperature measurements of carbonate reduction. *Proc. Natl. Acad. Sci*, *101*, 14023–14026.

Seewald, J. S. (1994). Evidence for metastable equilibrium between hydrocarbons under hydrothermal conditions. *Nature*, *370*, 285–287.

Seewald, J. S., Zolotov, M., & McCollom, T. M. (2006). Experimental investigation of single carbon compounds under hydrothermal conditions. *Geochim. Cosmochim. Acta*, *70*, 446–460.

Shi, G. U., Tropper, P., Cui, W., Tan, J., & Wang, C. (2005). Methane (CH4)-bearing fluid inclusions in the Myanmar jadeitite. *Geochemical Journal*, *39*, 503–516.

Shipp, J., Gould, I. R., Herckes, P., Shock, E. L., Williams, L. B., & Hartnett, H. E. (2013). Organic functional group transformations in water at elevated temperature and pressure: Reversibility, reactivity, and mechanisms. *Geochim. Cosmochim. Acta*, *104*, 194–209.

Shirey, S. B., Cartigny, P., Frost, D. J., Keshav, S., Nestola, F., Nimis, P., et al. (2013). Diamonds and the geology of mantle carbon. *Rev Mineral Geochem*, *75*, 355–421.

Shock, E. L. (1988). Organic acid metastability in sedimentary basins. *Geology*, *16*, 886–890.

Shock, E. L. (1990). Geochemical constraints on the origin of organic compounds in hydrothermal systems. *Origins of Life and the Evolution of the Biosphere*, *20*, 331–367.

Shock, E. L., & Canovas, P. (2010). The potential for abiotic organic synthesis and biosynthesis at seafloor hydrothermal systems. *Geofluids*, *10*, 161–192.

Shock, E. L., Canovas, P., Yang, Z., Boyer, G., Johnson, K., Robinson, K., et al. (2013). Thermodynamics of organic transformations in hydrothermal fluids. *Reviews in Mineralogy and Geochemistry*, *76*, 311–350.

Soli, A. L., & Byrne, R. H. (2002). CO2 system hydration and dehydration kinetics and the equilibrium CO2/H2CO3 ratio in aqueous NaCl solution. *Mar. Chem.*, *78*, 65–73.

Song, S., Su, L., Niu, Y., Lai, Y., & Zhang, L. (2009). CH 4 inclusions in orogenic harzburgite: Evidence for reduced slab fluids and implication for redox melting in mantle wedge. *Geochim. Cosmochim. Acta*, *73*, 1737–1754.

Sverjensky, D. A., Harrison, B., & Azzolini, D. (2014). Water in the deep Earth: The dielectric constant and the solubilities of quartz and corundum to 60 kb and 1,200 °C. *Geochim. et Cosmochim. Acta*, *129*, 125–145.

Sverjensky, D. A., Stagno, V., & Huang, F. (2014). Important role for organic carbon in subduction-zone fluids in the deep carbon cycle. *Nat. Geosci.*, *7*, 909–913.

Tropper, P., & Manning, C. E. (2007). The solubility of corundum in H_2O at high pressure and temperature and its implications for Al mobility in the deep crust and upper mantle. *Chemical Geology*, *240*, 54–60.

Vitale Brovarone, A., Martinez, I., Elmaleh, A., Compagnoni, R., Chaduteau, C., Ferraris, C., & Esteve, I. (2017). Massive production of abiotic methane during subduction evidenced in metamorphosed ophicarbonates from the Italian Alps. *Nature Communications*, *8*, 14134.

Wissbrun, K. F., French, D. M., & Patterson Jr, A. (1954). The true ionization constant of carbonic acid in aqueous solution from 5 to 45. *The Journal of Physical Chemistry*, *58*, 693–695.

Yang, Z., Gould, I. R., Williams, L. B., Hartnett, H. E., & Shock, E. L. (2012). The central role of ketones in reversible and irreversible hydrothermal organic functional group transformations. *Geochim. Cosmochim. Acta*, *98*, 48–65.

Zhang, C., & Duan, Z. (2009). A model for C-O-H fluid in the Earth's mantle. *Geochim. Cosmochim. Acta*, *73*, 2089–2102.

23

Free Energies of Reaction for Aqueous Glycine Condensation Chemistry at Extreme Temperatures

Matthew Kroonblawd[1] and Nir Goldman[1,2]

ABSTRACT

We have performed high throughput quantum molecular dynamics simulations to determine the free energy surface for aqueous glycine condensation reactions at conditions similar to oceanic hydrothermal vents (1 g cm^{-3} and temperatures ranging from 400 K to 1000 K). Our simulations identify significant changes in the free energy surface topology and subsequent chemical reactivity with increasing temperature. We predict that temperatures at 400 K and below favor dipeptide formation, whereas higher temperatures facilitate the reverse hydrolysis reaction, with solvated glycine molecules showing greater stability. This change in favorability is correlated with a shift in the location and characteristics of specific reaction bottlenecks or barriers. Simultaneously, we observe that relative free energy barriers (total energy plus entropic contributions) for both condensation and hydrolysis reactions generally decrease with increasing temperature. Our results indicate that relatively modest temperatures near 400 K may best facilitate formation of oligoglycine molecules in oceanic systems related to the synthesis of life-building compounds.

23.1. INTRODUCTION

Polypeptide synthesis in prebiotic environments is thought to be a pathway for the formation of more complicated molecules needed for the origins of life (Barbier, Visscher, & Schwartz, 1993; Brack, 2007; Danger, Plasson, & Pascal, 2012). Condensation reactions involving the simplest protein-forming amino acid glycine (Gly) serve as an important model for understanding the formation of more complicated polypeptides and proteins (Borsook, 1953; Faisal et al., 2005; Imai, Honda, Hatori, Brack, et al., 1999; Imai, Honda, Hatori, & Matsuno, 1999; Kitadai, 2014; Lemke, Rosenbauer, & Bird, 2009; Sakata, Kitadai, & Yokoyama, 2010; Schreiner et al., 2009; Shock, 1992; Sugahara, & Mimura, 2014; Van Dornshuld, Vergenz, & Tschumper, 2014). Glycine has been recovered from cometary comas and meteorites and could have been delivered to ancient Earth through an impact event (Altwegg et al., 2016; Blank et al., 2001; Chyba & Sagan, 1992; Goldman et al., 2010; Martins et al., 2013; Sugahara & Mimura, 2014, 2015). Oligoglycine chains can be created during condensation reactions

$$Gly_n + Gly \rightarrow Gly_{n+1} + H_2O, \qquad (23.1)$$

through formation of C–N peptide bonds and the elimination of water. Peptide bonds in oligoglycine are quite stable once formed, with a 100+ year half-life for the

[1] *Physical and Life Sciences Directorate, Lawrence Livermore National Laboratory, Livermore, California, USA*

[2] *Department of Chemical Engineering, University of California–Davis, California, USA*

Carbon in Earth's Interior, Geophysical Monograph 249, First Edition. Edited by Craig E. Manning, Jung-Fu Lin, and Wendy L. Mao.

reverse (hydrolysis) reaction at room temperature and pressure without a catalyst (Wolfenden, 2011). One possibility for the formation of polypeptides or precursor molecules on early Earth is the presence of extreme temperatures and/or pressures (Blank et al., 2001; Goldman et al., 2010; Goldman & Tamblyn, 2013; Martins et al., 2013; Sugahara & Mimura, 2014, 2015). Extreme conditions are a known means to access otherwise unlikely and slow chemical synthetic routes and can prompt the formation of new and exotic products and phases (Goldman et al., 2010; Goldman & Tamblyn, 2013; Imai, Honda, Hatori, Brack, et al., 1999; Imai, Honda, Hatori, & Matsuno, 1999; Koziol & Goldman, 2015; Lemke et al., 2009; Martins et al., 2013; Schreiner et al., 2009; Sugahara & Mimura, 2014, 2015). This could have conceivably involved cycling between high and low temperatures in submarine hydrothermal vents, which can exhibit temperature variations of up to ≈400 K, often reaching $T > 600$ K (Imai, Honda, Hatori, Brack, et al., 1999; Imai, Honda, Hatori, & Matsuno, 1999; Kitadai, 2014; Lemke et al., 2009; Shock, 1992; Schreiner et al., 2009; Tivey et al., 1995; Yanagawa et al., 1990). Individual prebiotic synthesis scenarios thus can span a wide range of relevant states, potentially resulting in large changes to chemical kinetics and equilibria for even simple reactions such as peptide oligomerization. However, determining the specific temperatures and pressures or thermodynamic paths that might favor oligoglycine synthesis through laboratory experiments alone can be slow and expensive.

Quantum-based molecular dynamics (QMD) coupled with enhanced sampling methods such as metadynamics (Laio & Parrinello, 2002) or umbrella sampling (Torrie & Valleau, 1974)) can provide an independent method for predicting prebiotic chemistry in these systems (Kroonblawd et al., 2018; Pietrucci & Saitta, 2015; Schreiner et al., 2009). Ab initio methods such as Kohn-Sham Density Functional Theory (Kohn & Sham, 1965) (DFT) can yield accurate information but are too computationally intensive to simultaneously probe a wide range of thermodynamic thermodynamic conditions. Force-matched semiempirical models (Goldman et al., 2018; Goldman et al., 2015; Kroonblawd et al., 2018) based on the density functional tight binding (DFTB) method (Elstner et al., 1998; Koskinen, & Mäkinen, 2009; Porezag et al., 1995) offer an efficient alternative that is orders of magnitude less computationally intensive while retaining a high degree of accuracy. The computational efficiency of DFTB allows for running many independent simulations concurrently to generate trajectories that can approach chemical equilibrium timescales (Cawkwell et al., 2015; Kroonblawd, & Goldman, 2018; Kroonblawd et al., 2018).

Very recently, we developed a force-matched DFTB model for glycine that was tailored to reproduce DFT predictions for aqueous glycine condensation reactions at $\rho = 1$ g cm^{-3} and $T = 300$ K (Kroonblawd, et al., 2018). In that case, the efficiency of DFTB was found to be critical for performing the multinanosecond long free energy (umbrella sampling) simulations needed to converge predictions for the free energy surface (energy plus entropy) of reaction. Here, we extend these efforts to sample temperatures relevant to potential hydrothermal vent synthesis (i.e. 400 K $\leq T \leq$ 1000 K). Accurate free energy calculations are essential to determine the temperature effects on condensed phase chemical reactivity, which can be significant at the elevated conditions discussed throughout this work. We then use a chemical analysis of the underlying trajectories to obtain clear connections between features on the reaction free energy surface and particular reaction intermediates.

23.2. METHODS

In this section, we detail the number of parameters and variables that were set in our calculations. In addition to brief description of the DFTB method, we include definitions of a reaction-path-specific set of variables that were necessary in order to efficiently monitor chemistry in such a high-dimensionality system, and the schedule of different biasing potentials that were used to determine the underlying free energy surfaces. We also describe simplified analytical tools that were employed to analyze the different types of carbon-nitrogen-oxygen (CNO)-containing backbones produced over the course of our simulations.

23.2.1. Simulation Details

Molecular dynamics (MD) simulations of aqueous glycine condensation were performed using the self-consistent charge DFTB method. DFTB (Elstner et al., 1998; Koskinen & Mäkinen, 2009; Porezag et al., 1995) is a semi-empirical quantum simulation approach that yields a high degree of computational efficiency while potentially retaining the accuracy of the computationally intensive Kohn-Sham DFT. The formalism for DFTB with self-consistent charges has been discussed in detail elsewhere (Aradi et al., 2007; Gaus et al., 2011; Goldman, 2015; Goldman et al., 2018; Goldman et al., 2015; Koskinen & Mäkinen, 2009). Briefly, the method assumes neutral, spherically symmetric charge densities on the atoms and expands the DFT Hamiltonian to second-order in charge fluctuations. The DFTB total energy is expressed as

$$E_{\text{DFTB}} = E_{\text{BS}} + E_{\text{Coul}} + E_{\text{Rep}}. \tag{23.2}$$

Here, E_{BS} is the electronic band structure energy, which is evaluated within a tight-binding framework (Slater & Koster, 1954), E_{Coul} captures charge transfer between atoms, and E_{Rep} is an empirical term that accounts for ionic repulsion and Kohn-Sham double counting terms. Parameters for E_{BS} and E_{Coul} were taken from the mio-1-1 parameterization (available at http://www.dftb.org), a typical off-the-shelf parameter set for organic systems. Most of the generic E_{Rep} interactions were replaced with force-matched ones (Kroonblawd, et al., 2018) that were specifically tuned to accurately predict the glycine condensation free energy surface at 300 K and 1 g cm^{-3}. The specific parameterization used here is the equal-weight variant of the force-matched DFTB model that is described in detail in Kroonblawd, et al. (2018).

All simulations were performed using a cubic, three-dimensionally periodic simulation cell at constant density 1 g cm^{-3}. The simulation cell had a side length of 12.374 Å and contained two glycine molecules solvated in 55 H_2O molecules, which is equivalent to one diglycine molecule in 56 H_2O molecules. Simulation parameters follow our previous results for glycine condensation at T = 300 K. Deuterium masses were substituted for hydrogen masses and the electronic structure was evaluated without spin polarization at the Γ-point only. Umbrella sampling simulations were driven using the PLUMED 1.3 plugin (Bonomi et al., 2009) using collective coordinates that are described in detail in section 23.2.2.

DFTB-MD trajectories were integrated using Extended Lagrangian Born-Oppenheimer dynamics (Niklasson, 2008; Niklasson et al., 2009; Niklasson et al., 2006; Zheng et al., 2011) driven by the LAMMPS simulation software suite (Plimpton, 1995) with forces and stresses evaluated by the DFTB+ code (Aradi et al., 2007). The time step was set to 0.20 fs and isochoric-isothermal (i.e., *NVT*) sampling was performed with a Nosé-Hoover-style thermostat (Hoover, 1985; Nosé, 1984). The electronic structure was evaluated using four self-consistent charge cycles per time step and Fermi-Dirac thermal smearing (Mermin, 1965) with the electronic temperature set equal to the instantaneous ionic kinetic temperature.

The DFT-MD simulations discussed here were performed using Car-Parrinello dynamics (Car & Parrinello, 1985) driven by the Quantum Espresso ab initio software package (Giannozzi et al., 2009). Similar to previous work (Kroonblawd et al., 2018; Schreiner et al., 2009), the time step and electron mass were set to 0.145 fs (6.0 au) and 700 au, respectively, and both the ionic and electronic degrees of freedom were coupled to a Nosé-Hoover thermostat (Hoover, 1985; Nosé, 1984). We used the Perdew-Burke-Ernzerhof (Perdew et al., 1996) generalized-gradient approximation functional with ultrasoft pseudopotentials (Vanderbilt, 1990) and a 25 Ry planewave cutoff.

23.2.2. Free Energy Calculations

The initial glycine condensation reaction 2Gly → Gly_2 + H_2O was sampled at selected temperatures using the same path collective coordinates that we previously applied to this reaction at T = 300 K in Kroonblawd et al. (2018). Path coordinates are particularly useful for describing complicated reactions such as glycine condensation because they significantly reduce the dimensionality of the sampled collective coordinate space. Two path coordinates are defined (Branduardi et al., 2007) in terms of the known reactant and product species as

$$s(t) = \frac{\sum_k k \exp\left(-\lambda D\left[\mathbf{R}(t),\mathbf{R}_k\right]\right)}{\sum_k \exp\left(-\lambda D\left[\mathbf{R}(t),\mathbf{R}_k\right]\right)}, \tag{23.3}$$

and

$$z(t) = -\frac{1}{\lambda}\ln\left\{\sum_k \exp(-\lambda D[\mathbf{R}(t),\mathbf{R}_k])\right\}, \tag{23.4}$$

where $D[\mathbf{R}(t), \mathbf{R}_k]$ is some appropriate *distance metric* to measure the displacement of an instantaneous atomic configuration $\mathbf{R}(t)$ from a sequence of reference configurations $\mathbf{R}_k$, and λ is a constant to be specified later. Coordinates $s(t)$ and $z(t)$ are respectively interpreted as measuring the progress along and deviation from the path defined by the sequence of $\mathbf{R}_k$. The distance metric used here,

$$D\left[\mathbf{R}(t),\mathbf{R}_k\right] = \sum_{i\beta}\left[C_{i\beta}(t) - C_{i\beta}^k\right]^2, \tag{23.5}$$

is a function of the local coordination environment for particular atoms (Pietrucci & Saitta, 2015) that is quantified as

$$C_{i\beta}(t) = \sum_{j\in\beta}\frac{1-\left[\frac{r_{ij}(t)}{d_{\alpha\beta}}\right]^6}{1-\left[\frac{r_{ij}(t)}{d_{\alpha\beta}}\right]^{12}}. \tag{23.6}$$

Here, $C_{i\beta}(t)$ measures the number of atoms of type β coordinated to atom i of type α at time t. The sum is taken over all atoms j of type β in the system, r_{ij} is the separation distance between atoms i and j, and $d_{\alpha\beta}$ are constants (typically <2 Å) set for each pair type.

Parameter choices for the path variables exactly follow our earlier work on aqueous glycine condensation. Equation (23.5) was evaluated for the four C and two N atoms that form the diglycine backbone, tracking the coordination of those atoms to all other C, N, O, and H atoms in the system. Two reference configurations define the reaction path, with endpoints at $k = 1$ for the two neutral aqueous glycine reactant molecules and $k = 2$ for the single aqueous neutral diglycine product molecule. The $C^k_{i\beta}$ corresponding to these endpoints were set using time averages obtained at 300 K from short unbiased MD simulations of neutral glycine and diglycine in explicit H_2O solvent performed using DFTB with the parameter set mio-1-1. The parameters $d_{\alpha\beta}$ were set to 1.8 Å for X–X pairs and 1.5 Å for X–H pairs, where atom type X is either C, N, or O. The parameter λ governs the surface topology and is usually chosen so that $\lambda D[\mathbf{R}_k, \mathbf{R}_{k+1}] \approx 2.3$, which for the present case is satisfied when $\lambda = 0.70$. Neutral glycine and diglycine configurations are approximately mapped to the (s, z) positions (1.1, −0.1) and (1.9, −0.1), and zwitterionic forms are in regions with $z > 0.2$.

Free energy surfaces were computed using umbrella sampling (Torrie, & Valleau, 1974), in which harmonic bias potentials

$$V_{\text{Bias}}(s,z) = \frac{1}{2}K_s(s-s_0)^2 + \frac{1}{2}K_z(z-z_0)^2, \quad (23.7)$$

were applied to the collective coordinates s and z. The free energy

$$F(s,z) = -k_B T \ln\langle \rho(s,z)\rangle, \quad (23.8)$$

was obtained within an arbitrary constant from an ensemble of independent biased simulations, each with different s_0 and z_0. We used the weighted histogram analysis method (WHAM) (Kumar et al., 1992; Roux, 1995) to compute the average unbiased distribution function $\langle\rho(s, z)\rangle$ from our ensemble of simulations. Histogram bin widths were set to 0.01 for both the s and z dimensions. The self-consistent WHAM equations were evaluated following the discussion in Roux (1995), with the maximum allowed per-cycle change in the free energy constant for each biased distribution set to 10^{-5} kcal mol^{-1}.

Based on our previous experience applying path coordinates to glycine condensation reactions, we were able to devise an efficient schedule of bias force constants and centroids to adequately sample s and z space. Sampling the s dimension is made somewhat complicated by a particularly large gradient along s in the free energy surface in $1.1 < s < 1.3$. We divided s-space into three separate zones covered by 23 simulations for a given z_0, with each zone having a particular grid spacing for s_0 and force constant K_s. The first zone had $s_0 = 1.050$ and 1.100 with $K_s = 1000$ kcal mol^{-1}. The second zone covered the high-gradient region and had s_0 set in 0.025 increments in $1.125 \le s_0 \le 1.275$ with $K_s = 4000$ kcal mol^{-1}. The third zone had s_0 set in 0.050 increments in $1.300 \le s_0 \le 1.950$ with $K_s = 1000$ kcal mol^{-1}. Gradients along z were more uniform, so we set z_0 in 0.100 increments in $-0.200 \le z_0 \le +0.600$ with $K_z = 1000$ kcal mol^{-1}. Independent initial thermal configurations were prepared at $s_0 = 1.950$, $z_0 = 0.200$ for each temperature. From these starting points, we first expanded along s by seeding new umbrella sampling simulations with configurations taken from simulations with adjacent s_0 to yield a "slice" of the surface. This "slice" was then expanded along z, similarly taking the initial configuration for each new simulation from a simulation with adjacent s_0 and z_0 to yield the full surface.

23.2.3. Chemical Analysis

The diversity of chemical configurations sampled in our umbrella simulations was analyzed through a multipart approach. In the first part of the analysis, the trajectory for each independent simulation was passed through an in-house molecule recognition code to identify underlying stable and transient chemical structures sampled during the simulation. We removed all hydrogen atoms from our trajectories prior to the analysis to limit the search to only unique CNO backbone structures in order to simplify our analysis. Consequently, neutral and zwitterionic forms of glycine and diglycine are treated as equivalent. Molecules sampled in the umbrella simulations were identified using both a bond distance and lifetime criterion. Two atoms were considered connected in each instant of a given trajectory if they were separated by no more than 1.9 Å. The instantaneous connections were used to construct a set of all unique molecules, and those molecules were considered stable if their interatomic connections persisted for at least 40 fs. Molecules persisting for less than 40 fs were classified as transients. We note that our overall conclusions are not affected by these chosen parameter values. Time averages for the concentration of each unique CNO backbone species i were obtained for each umbrella sampling simulation j, with the normalizing volume taken to correspond to the simulation cell. We denote these concentrations as $c_{i,j}$, where each $c_{i,j}$ is a scalar with units mol L^{-1}.

The second part of the analysis combined the information from the independent trajectories to obtain a map of the concentration for each chemical species $c_i(s, z)$ as a function of the path variables s and z. Our approach is shown schematically for two simulations in Figure 23.1, although in practice the analysis was performed on the 207 independent simulations in each umbrella sampling ensemble. First, we mapped the concentrations $c_{i,j}$ to the path variable space (s, z) through the following heuristic. Each umbrella sampling simulation j has associated with it a biased distribution function $\rho_j(s, z)$ corresponding to the region of (s, z) space

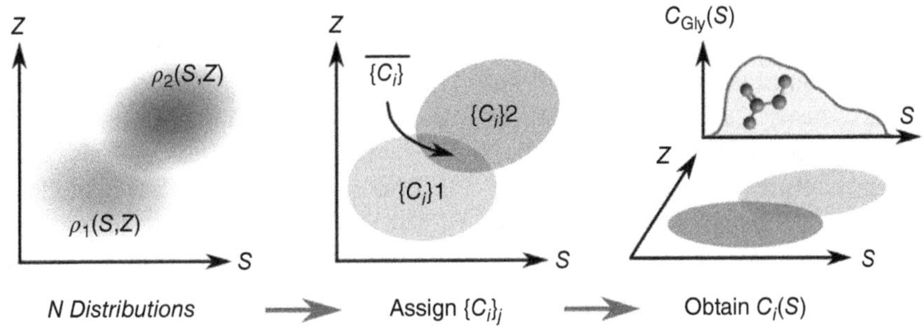

Figure 23.1 Schematic for mapping concentration information generated in two umbrella sampling trajectories to collective coordinate space. See electronic version for color representation of the figures in this book.

that it sampled. We assigned the scalar $c_{i,j}$ uniformly to the exact region of space spanned by $\rho_j(s, z)$ for each simulation j to obtain $c_{i,j}(s, z)$. (That is, $c_{i,j}[s, z] = c_{i,j}$ where $\rho_j[s, z] > 0$ and $c_{i,j}[s, z] = 0$ where $\rho_j[s, z] = 0$.) Overall maps for $c_i(s, z)$ were obtained by combining all of the $c_{i,j}(s, z)$ generated in the ensemble of umbrella sampling simulations. By design, there are regions of (s, z) space that were sampled by multiple simulations, so we took $c_i(s, z)$ to be the arithmetic average of all $c_{i,j}(s, z)$ from simulations sampling such regions. The concentrations $c_i(s, z)$ thus obtained did not vary significantly with respect to the z-coordinate, so we averaged $c_i(s, z)$ over z to obtain the concentration $c_i(s)$ as a function of s only. It should be noted that $\rho_j(s, z)$ and $c_i(s, z)$ are both computed on the same histogram grid employed for computing the the free energy surface $F(s, z)$.

23.3. RESULTS AND DISCUSSION

We begin this section with validation of our DFTB model through comparison to DFT-computed results. In particular, we observe close correspondence between the two methods for the energetics of the reactants (aqueous glcyine) and products (diglycine). We also consider the potential difficulties in converging condensed phase free energy calculations in part due to the large manifold of hydrogen-bonding rearrangements that can occur. We then discuss the temperature dependence of our results and focus on the transition from an exergonic (spontaneous release of free energy) to an endergonic (absorption of free energy) process that occurs between 400 and 500 K. Finally, we determine the different chemical pathways and energetics that are present over the range of temperatures studied here.

23.3.1. Validation and Convergence of Our Calculations

In order to provide a computationally feasible validation of our DFTB model, we have computed the average potential energy difference (e.g. excluding entropy effects) between the reactants and products with comparison to results from DFT (Figure 23.2). Here, values are compared between our DFTB model and DFT results for a system of two glycine molecules (E_{2Gly}) solvated in 55 H_2O molecules vs. a single diglycine molecule (E_{Gly_2}) solvated in 56 H_2O molecules. These results were readily determined from single, unbiased NVT simulations run for 5 ps for each system at each temperature. Uncertainty in the average potential energy for each state was taken to be the standard deviation of the energy fluctuations, and those uncertainties were propagated when computing the average difference $E_{2Gly} - E_{Gly_2}$. Both DFT and DFTB predict that the reactants and products are nominally isoenergetic over the entire temperature interval considered. Although DFTB consistently predicts the energy difference to be more negative than does DFT, our results agree within uncertainty. Similarities between the DFT and DFTB predictions lend confidence in the transferability of our DFTB model, which was originally developed for glycine condensation reactions at 300 K.

Free energy surfaces $F(s, z)$ for glycine condensation were computed as functions of the path coordinates s and z using DFTB and umbrella sampling simulations following the approach described in section 23.2.2. Calculations were performed over temperatures from 400 K ≤ T ≤ 1000

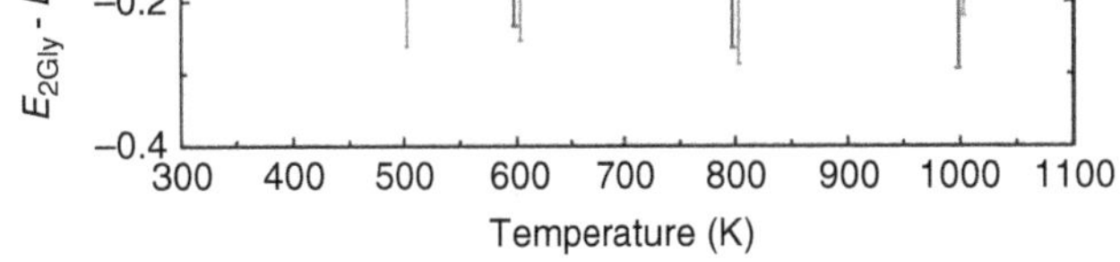

Figure 23.2 Average potential energy difference between reactant and product configurations computed as a function of temperature obtained using DFT and DFTB. Error bars correspond to propagated uncertainties. See electronic version for color representation of the figures in this book.

K in 100 K increments. A total of 207 independent 20 ps simulations were performed to generate the surface at each temperature (≈4.1 ns combined trajectory per surface), which corresponds to nearly 29 ns of combined trajectory overall. $F(s, z)$ was computed in the domain $1.0 \le s \le 2.0$ and $-0.2 \le z \le 0.6$ and captures both neutral and zwitterionic forms for the reactants and products. Convergence of our free energy calculations was assessed through a blocking analysis (Figure 23.3). Our blocked predictions for the free energy of reaction ΔF_{rxn} exhibit drift towards more positive values on the order of a few kcal mol^{-1} at 400, 800, 900, and 1000 K, though the predicted barriers are perhaps more steady. Despite the very long combined QMD simulation time, our analysis reveals that our predictions are quasi-steady over the last ten time blocks.

23.3.2. Temperature Dependence

Temperature dependence for ΔF_{rxn} and the reaction barrier were assessed by computing the free energy surfaces using the last 10 ps of each biased umbrella sampling trajectory. Figure 23.4 shows the temperature dependence for ΔF_{rxn} in panel (a) and for the forward and backward reaction barriers in units of kcal mol^{-1} and $k_B T$ in panels (b) and (c), respectively. Values at $T = 300$ K were previously obtained with this model in Kroonblawd et al. (2018). Uncertainties in our calculations were computed as the standard deviation of the mean obtained from the last 10 ps in the blocking analysis described above and were propagated (where appropriate) when isolating

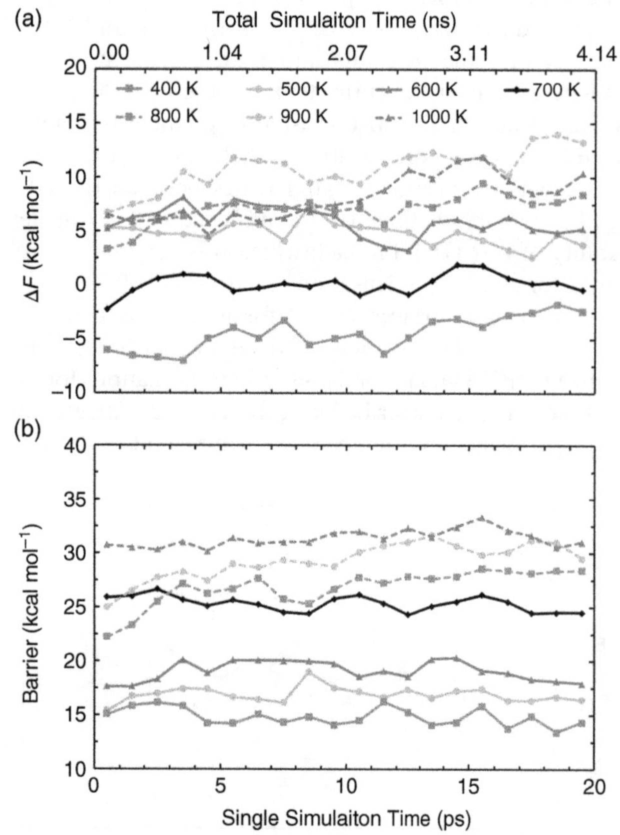

Figure 23.3 Blocking analysis for the predicted free energy difference ΔF_{rxn} and barrier height of the glycine condensation reaction $2Gly \rightarrow Gly_2 + H_2O$. Each data point in a given panel corresponds to a single $F(s, z)$ surface generated with the weighted histogram analysis method using biased histograms computed in nonoverlapping 1 ps windows. See electronic version for color representation of the figures in this book.

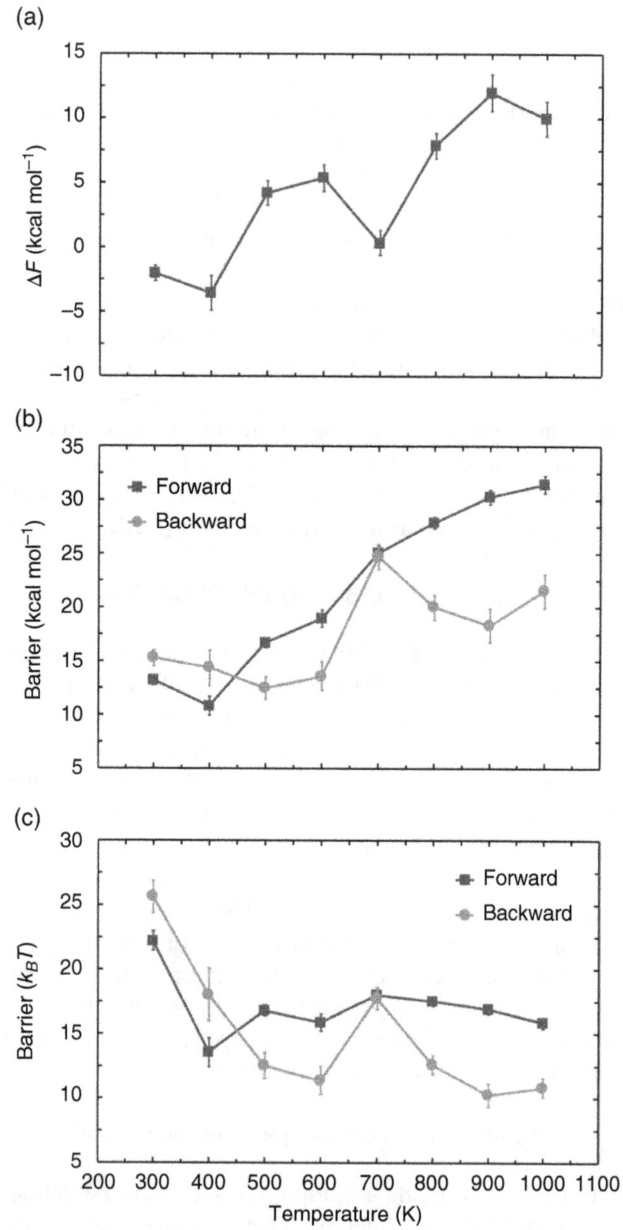

Figure 23.4 Temperature dependence of (a) the free energy difference ΔF_{rxn} and the reaction barrier in units of (b) kcal mol^{-1} and (c) $k_B T$. See electronic version for color representation of the figures in this book.

forward/backward reaction barriers. We note that due to the quasi-steady nature of the computed free energy surfaces, the reported uncertainties may not fully capture fluctuations anticipated in the limit of infinite simulation time.

Several qualitative trends are apparent in the isochoric temperature dependence of the free energy surface. The first trend is that ΔF_{rxn} generally increases with increasing temperature. A transition from a nominally exergonic ($\Delta F_{rxn} < 0$) process at low temperatures to an endergonic ($\Delta F_{rxn} > 0$) one at high temperatures occurs between 400 K and 500 K. Possible explanations for the abrupt decrease in ΔF_{rxn} at 700 K are discussed in section 23.3.4 in the context of the underlying chemical pathways. In terms of absolute values, the reaction barrier for both the forward and backward reaction generally increase with increasing temperature. However, in terms of $k_B T$ (the inherent kinetic energy of a system at a given temperature), the barrier decreases substantially going from 300 K to 400 K, after which it more or less forms a plateau. Commensurate with the transition from an exergonic to endergonic process, the barrier for the back reaction is lower than the barrier to the forward reaction at elevated temperatures.

Glycine condensation reactions have been studied under a diverse set of reaction conditions, with variations in temperature, pressure/density, pH, presence of contaminants or catalysts, and under dynamically changing states (Borsook, 1953; Faisal et al., 2005; Imai, Honda, Hatori, Brack, et al., 1999a; Imai, Honda, Hatori, & Matsuno, 1999b; Kitadai, 2014; Kroonblawd et al., 2018; Lemke et al., 2009; Radzicka & Wolfenden, 1996; Sakata et al., 2010; Schreiner et al., 2009; Shock, 1992; Sugahara & Mimura, 2014; Van Dornshuld et al., 2014). In drawing comparisons, it is important to consider that the present study focuses on the isochoric temperature dependence of glycine condensation at neutral pH in pure water. The average pressure predicted by DFTB at the highest temperature considered here is ≈15 kbar (due to thermal pressurization), although the small system size leads to significant fluctuations of a similar magnitude. With few exceptions, most experimental and theoretical treatments that specifically discuss the free energy difference between glycine and diglycine have found ΔF_{rxn} to be nominally zero within ≈5 kcal mol^{-1} across a wide range of conditions (Borsook, 1953; Danger et al. 2012; Kitadai, 2014; Lemke et al., 2009; Shock, 1992; Wolfenden, 2011). Thus, our results for ΔF_{rxn} are generally consistent in magnitude with the established literature.

The typical maximum bulk temperature and pressure at hydrothermal vents are $T_{max} \approx 660$ K and $P_{max} \approx 0.2$ kbar, which are both well below the maximums considered here. Several studies found ΔF_{rxn} to slightly decrease with increasing temperature at sub-kbar pressures (Kitadai, 2014; Lemke et al., 2009; Shock, 1992). However, equation of state models predict that ΔF_{rxn} increases with increasing pressure under isothermal conditions from the saturation pressure up to at least 5 kbar (Shock, 1992). A subset of hydrothermal experiments performed at 0.2 kbar and 513 K ≤ T ≤ 573 K showed spontaneous and nearly complete hydrolysis of diglycine into glycine and other products (Faisal et al., 2005), consistent with the results presented here. Similar condensation reaction barrier heights to those computed here have been found in other quantum molecular dynamics calculations (Schreiner et al., 2009), computed over conditions of 300 K, 1.00 g cm^{-3} and 500 K, 0.85 g cm^{-3}. However, the absolute values reported here for the reverse or back reaction (hydrolysis) are significantly lower due to the very large exergonic free energy change ($\Delta F_{rxn} = -20$ kcal mol^{-1}) predicted in that study. In terms of $k_B T$ units, both studies predict that an increase in temperature will lower the effective barrier.

23.3.3. Exergonic-to-Endergonic Transition

We now focus on detailed results of the transition from a spontaneous process to one that requires heat absorption, between 400 and 500 K (Figure 23.5). Once again, free energy surfaces were computed using the last 10 ps of the umbrella trajectories that correspond to a nominal steady state. Maps of regions accessible at an energy equivalent to the minimum barrier height were generated by "filling" the free energy surface to the corresponding energy level. No significant differences in chemical concentrations were observed with respect to our z path coordinate, so we averaged results to obtain the concentration $c_i(s)$ as a function of s only. Our analysis yielded five unique CNO backbone species during the course of reactivity, including glycine **1**, diglycine **2**, a linear intermediate **3** that we previously identified at 300 K, and two different cyclic intermediates **4** and **5**. The concentration of **1** is roughly twice that of **2** because there are two glycine reactant molecules and a single diglycine product molecule. At 400 K, the general sequence for the reaction is **1** → **3** → **4** → **2**. A similar sequence is seen at 500 K, but with small quantities of **5** detected early in the transition to diglycine. The elimination of water occurs during the transition from **3** to **4**.

The two starkest changes in the free energy surface with temperature are the location of the absolute minimum and the overall reaction barrier (or bottleneck) in relation to the underlying chemistry. At 400 K, the free energy minimum corresponds to zwitterionic diglycine and the barrier is near the onset of the transition from glycine to intermediate **3**. In contrast, at 500 K, the free energy minimum of the reaction is neutral glycine and the barrier corresponds to the elimination of water from **3** and the subsequent transition to a mixture of **2** and **4**. Another key difference is that the

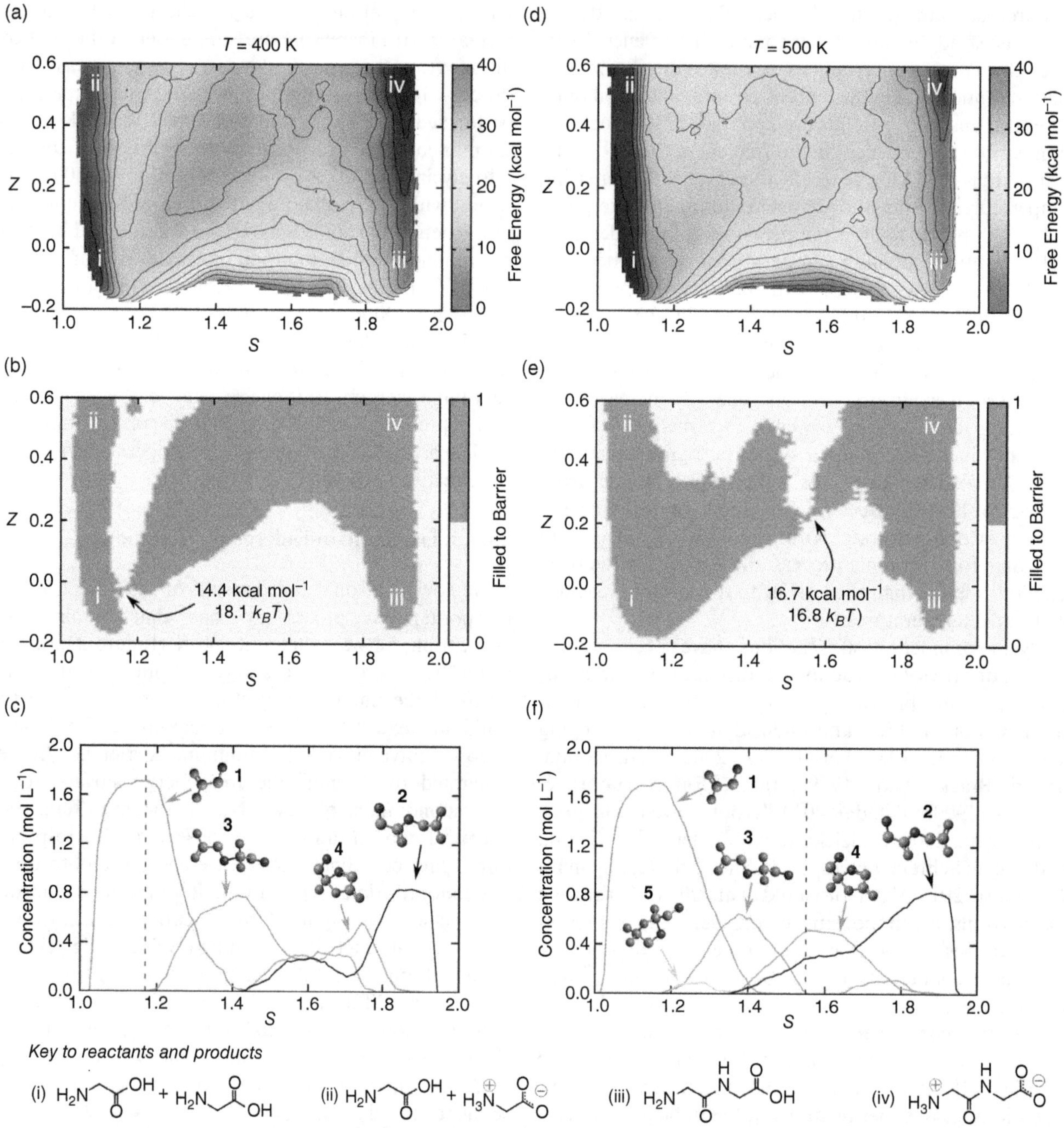

Figure 23.5 Comparison of free energy surface characteristics at 400 K and 500 K. The first column corresponds to 400 K and shows (a) the free energy surface $F(s, z)$, (b) approximate minimum free energy barrier heights, and (c) the concentrations $c_i(s)$ of unique CNO backbone species. The second column shows those same properties at 500 K in panels (d) through (f). Contour lines on $F(s, z)$ are drawn in 2.5 kcal mol^{-1} increments, and locations on $F(s, z)$ that correspond to particular neutral and zwitterionic reactant and product species are indicated by roman numerals. The s-coordinate for the minimum barrier location (or bottleneck) is demarcated in the concentration plots with vertical dashed lines. Atom colors are grey, blue, and red for C, N, and O. See electronic version for color representation of the figures in this book.

400 K surface has an apparent free energy minimum corresponding roughly to intermediate **3** centered at approximately $s = 1.4$, $z = 0.4$. A second local maximum at $s \approx 1.6$ is close in value to the bottleneck at $s = 1.17$ and has similar underlying chemistry to the 500 K bottleneck. One possibility is that solute/solvent interactions are strong enough to stabilize **3** at 400 K but not at 500 K. Previous theoretical studies have shown that even modest increases

in temperature can significantly disrupt the hydrogen bonding network in liquid water (Sahle et al., 2013), thus affecting solubility and the stability of charged species.

23.3.4. Characteristics at Higher Temperatures

We assessed the the changes in the free energy landscape for glycine condensation in 100 K increments up to a maximum temperature of 1000 K. As mentioned, the maximum temperature at hydrothermal vents is typically near 700 K. The higher temperatures considered here approach conditions that are relevant to more extreme prebiotic scenarios, such as during cometary impacts. Free energy surfaces were computed using the last 10 ps of our umbrella sampling trajectories and are shown for each temperature in Figure 23.6. Although not shown, the product and reactant species (i–iv) identified in Figure 23.5 map to the same analogous regions of $F(s, z)$ as at lower temperatures.

A distinct transition in the relative free energies for the diglycine end product itself occurs as the temperature in increases from 500 K to 700 K (see Figure 23.5d for the 500 K surface). At 500 K, the zwitterionic form (iv) is ≈5 kcal mol^{-1} lower in free energy than the neutral form (iii). When the temperature is increased to 600 K, these two forms have practically the same free energy and there is no appreciable barrier to interchange between them. Finally, at 700 K the neutral form becomes more favorable, being >5 kcal mol^{-1} lower in free energy than the zwitterion. Concurrently, there is a subtle increase in the favorability of neutral glycine (i) over its zwitterionic form (ii) with increasing temperature. One possible cause for the apparent minimum in the global $\Delta F_{rxn}(T)$ previously noted at 700 K in Figure 23.4a could be due to these changes and the overall energetic favorability of neutral molecules. The apparent minimum in $\Delta F_{rxn}(T)$ is relatively shallow (≈5 kcal mol^{-1}), which is on the same order of magnitude as the relative differences in free energy between the neutral and zwitterionic forms. Above 700 K, the surface topology does not change substantially with increasing temperature, consistent with the approximately monotonic behavior of $\Delta F_{rxn}(T)$ in 700 K ≤ T ≤ 1000 K.

A chemical concentration analysis was performed on the biased sampling trajectories for each temperature in 600 K ≤ T ≤ 1000 K and is shown in Figure 23.7. The qualitative chemical transformation process (**1** → **3** → **4** → **2**) and location of the bottleneck are similar to the results obtained at 500 K in Figure 23.5f. Perhaps the most significant outlier is the response at 800 K, which exhibits a significantly reduced concentration of linear intermediate **3** relative to the other temperatures. Given that **3** and cyclic intermediate **4** appear similar in terms of the biased collective coordinates, the apparent scatter in their relative concentrations as a function of temperature is likely a consequence of limited sampling trajectory. Similarly, cyclic intermediate **5** is present at both 500 K and for 700 K ≤ T ≤ 1000 K, but it is absent from our simulations at 600 K. When present, the concentration of **5** is

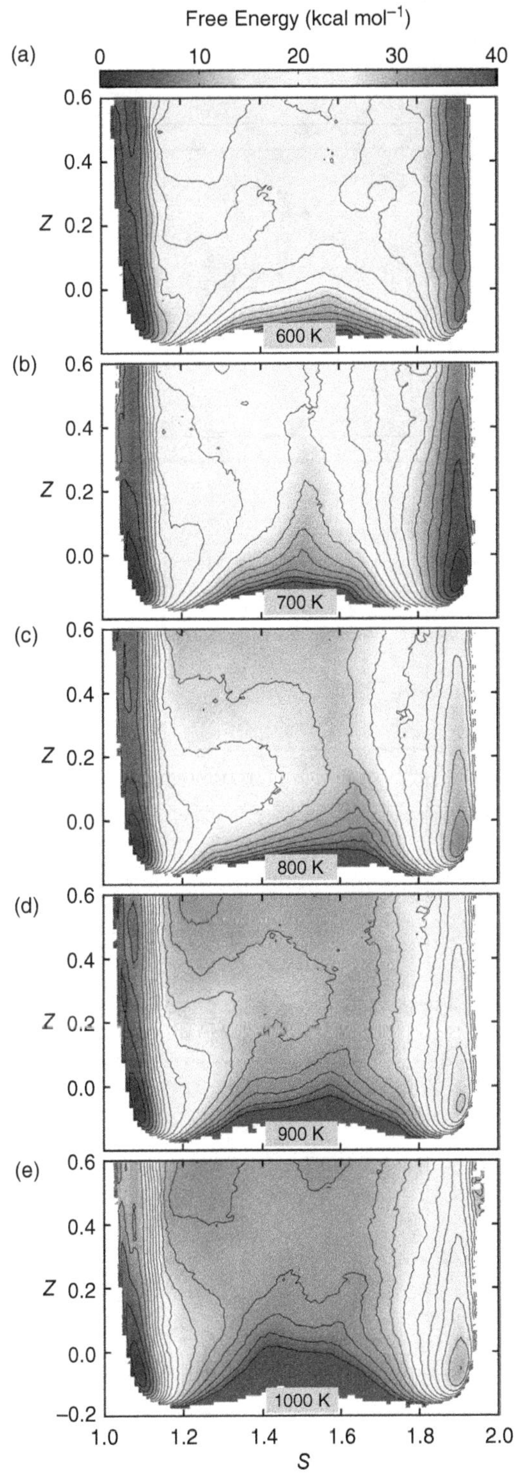

Figure 23.6 Free energy surfaces for glycine condensation at temperatures in 600 K ≤ T ≤ 1000 K. See electronic version for color representation of the figures in this book.

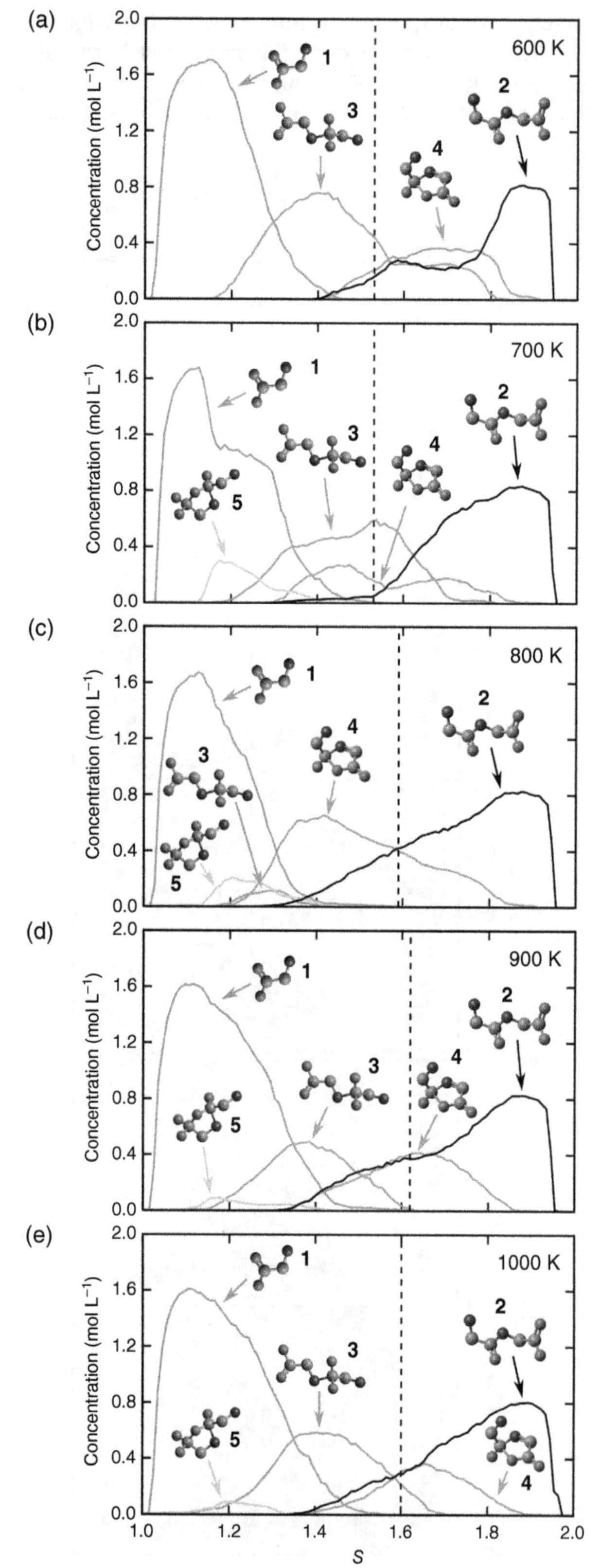

Figure 23.7 Concentration analysis of CNO backbone species at temperatures in 600 K ≤ T ≤ 1000 K. See electronic version for color representation of the figures in this book.

generally quite small compared to the other intermediates. However, the combined trajectory length generated here is substantially longer than would presently be obtainable with a purely ab initio approach such as DFT.

23.4. CONCLUSIONS

We have performed quantum-based molecular dynamics simulations to predict free energy surfaces (i.e. the manifold of temperature effects on chemical reactivity) for aqueous glycine condensation reactions at 1.0 g cm^{-3} and at temperatures ranging from 400 K up to 1000 K. Diglycine (glycylglycine) is predicted to be lower in free energy than glycine at ≈400 K and below, but higher temperatures are predicted to promote hydrolysis to form glycine as the thermodynamically preferred product. Our calculations show that the changes in the relative free energies between glycine and diglycine are likely entropic (temperature driven) in origin. In terms of a factor of $k_B T$ (e.g. normalizing energies relative to the system temperature), the free energy barriers for the forward (condensation) and reverse (hydrolysis) reaction are predicted to decrease substantially with increasing temperature, promoting overall chemical reactivity. Chemical analysis of the underlying trajectories reveals that the nature of our determined reaction barrier changes between 400 K and 500 K, coinciding with an exergonic to endergonic transition for the total reaction. At 400 K, it is likely that hydrogen-bonding networks can help stabilize the formation of different charged intermediates, thus facilitating diglycine synthesis. In contrast, these solvent-mediated effects are less prevalent at higher temperatures, promoting hydrolysis of the diglycine peptide. Our results indicate that elevated temperatures near 400 K might best promote facile formation of peptide bonds, while more extreme temperatures (and pressures) may result in the potential destruction of these types of life-building molecules. Our detailed chemical free energy analysis can be useful for developing coarse-grain models such as equation of state and kinetic simulation methods for larger scale computations. Our overall goal with these efforts is to provide a "bottom-up" approach for understanding organic chemistry under extreme conditions, where there is a strong need for knowledge of the kinetics and thermodynamics of chemical bond formation and breaking over extended periods of time

ACKNOWLEDGMENTS

The authors wish to acknowledge the support and collaboration of the Deep Carbon Observatory. This work was supported by NASA Astrobiology: Exobiology and Evolutionary Biology Program Element NNH14ZDA001N.

This work was performed under the auspices of the U.S. Department of Energy by Lawrence Livermore National Laboratory under Contract DE-AC52-07NA27344.

REFERENCES

Altwegg, K., Balsiger, H., Bar-Nun, A., Berthelier, J.-J., Bieler, A., Bochsler, P., et al. (2016). Prebiotic chemicals—amino acid and phosphorus—in the coma of comet 67p/Churyumov-Gerasimenko. *Sci. Adv.*, *2*(5). Retrieved from http://advances.sciencemag.org/content/2/5/e1600285. doi: 10.1126/sciadv.1600285

Aradi, B., Hourahine, B., & Frauenheim, T. (2007). Dftb+, a sparse matrix-based implementation of the dftb method. *J. Phys. Chem. A*, *111*(26), 5678–5684. Retrieved from http://dx.doi.org/10.1021/jp070186p (PMID: 17567110). doi: 10.1021/jp070186p

Barbier, B., Visscher, J., & Schwartz, A. W. (1993). Polypeptide-assisted oligomerization of analogs in dilute aqueous solution. *J. Mol. Evol.*, *37*(6), 554–558. Retrieved from https://doi.org/10.1007/BF00182740. doi: 10.1007/ BF00182740

Blank, J. G., Miller, G. H., Ahrens, M. J., & Winans, R. E. (2001). Experimental shock chemistry of aqueous amino acid solutions and the cometary delivery of prebiotic compounds. *Orig. Life Evol. Biosph.*, *31*(1), 15–51. Retrieved from https://doi.org/10.1023/A:1006758803255 doi: 10.1023/A:1006758803255

Bonomi, M., Branduardi, D., Bussi, G., Camilloni, C., Provasi, D., Raiteri, P., et al. (2009). PLUMED: A portable plugin for free-energy calculations with molecular dynamics. *Comput. Phys. Commun.*, *180*(10), 1961–1972. Retrieved from http://www.sciencedirect.com/science/article/pii/S001046550900157X (PLUMED is available at http://www.plumed.org/). doi: https://doi.org/10.1016/j.cpc.2009.05.011

Borsook, H. (1953). Peptide bond formation. *Adv. Protein Chem.*, *8*, 127–174. Retrieved from http://www.sciencedirect.com/science/article/pii/S0065323308600923. doi: http://dx.doi.org/10.1016/S0065-3233(08)60092-3

Brack, A. (2007). From interstellar amino acids to prebiotic catalytic peptides: A review. *Chem. Biodivers.*, *4*(4), 665–679. Retrieved from http://dx.doi.org/10.1002/cbdv.200790057. doi: 10.1002/cbdv.200790057

Branduardi, D., Gervasio, F. L., & Parrinello, M. (2007). From a to b in free energy space. *J. Chem. Phys.*, *126*(5), 054103. Retrieved from http://dx.doi.org/10.1063/1.2432340, doi: 10.1063/1.2432340

Car, R., & Parrinello, M. (1985). Unified approach for molecular dynamics and density-functional theory. *Phys. Rev. Lett.*, *55*, 2471–2474. Retrieved from https://link.aps.org/doi/10.1103/PhysRevLett.55.2471 doi: 10.1103/ PhysRevLett.55.2471

Cawkwell, M. J., Niklasson, A.M.N., & Dattelbaum, D. M. (2015). Extended lagrangian Born-Oppenheimer molecular dynamics simulations of the shock-induced chemistry of phenylacetylene. *The Journal of Chemical Physics*, *142*(6), 064512. Retrieved from https://doi.org/10.1063/1.4907909. doi: 10.1063/1.4907909

Chyba, C., & Sagan, C. (1992). Endogenous production, exogenous delivery and impact-shock synthesis of organic molecules: An inventory for the origins of life. *Nature*, *355*, 125–132. Retrieved from http://dx.doi.org/10.1038/355125a0. doi: 10.1038/355125a0

Danger, G., Plasson, R., & Pascal, R. (2012). Pathways for the formation and evolution of peptides in prebiotic environments. Chem. Soc. Rev., *41*, 5416–5429. Retrieved from http://dx.doi.org/10.1039/C2CS35064E. doi: 10.1039/C2CS35064E

Elstner, M., Porezag, D., Jungnickel, G., Elsner, J., Haugk, M., Frauenheim, T., et al. (1998). Self-consistent-charge density-functional tight-binding method for simulations of complex materials properties. *Phys. Rev. B: Condens. Matter Mater. Phys.*, *58*, 7260–7268. Retrieved from https://link.aps.org/doi/10.1103/PhysRevB.58.7260. doi: 10.1103/PhysRevB.58.7260

Faisal, M., Sato, N., Quitain, A. T., Daimon, H., & Fujie, K. (2005). Hydrolysis and cyclodehydration of dipeptide under hydrothermal conditions. *Ind. Eng. Chem. Res.*, *44*(15), 5472–5477. Retrieved from http://dx.doi.org/10.1021/ie0500568. doi: 10.1021/ie0500568

Gaus, M., Cui, Q., & Elstner, M. (2011). DFTB3: Extension of the self-consistent-charge density-functional tight-binding method (SCC-DFTB). *J. Chem. Theory Comput.*, *7*, 931.

Giannozzi, P., Baroni, S., Bonini, N., Calandra, M., Car, R., Cavazzoni, C., et al. (2009). Quantum espresso: A modular and open-source software project for quantum simulations of materials. *J. Phys. Condens. Matter*, *21*(39), 395502. Retrieved from http://stacks.iop.org/0953-8984/21/i=39/a=395502 (Quantum Espresso is available at http://www.quantum-espresso.org/).

Goldman, N. (2015). Multi-center semi-empirical quantum models for carbon under extreme thermodynamic conditions. *Chem. Phys. Lett.*, *622*, 128–136.

Goldman, N., Aradi, B., Lindsey, R. K., & Fried, L. E. (2018). Development of a multicenter density functional tight binding model for plutonium surface hydriding. *Journal of Chemical Theory and Computation*, *14*(5), 2652–2660. Retrieved from https://doi.org/10.1021/acs.jctc.8b00165. doi: 10.1021/acs.jctc.8b00165

Goldman, N., Fried, L. E., & Koziol, L. (2015). Using force-matched potentials to improve the accuracy of density functional tight binding for reactive conditions. *J. Chem. Theory Comput.*, *11*(10), 4530–4535. Retrieved from http://dx.doi.org/10.1021/acs.jctc.5b00742. doi: 10.1021/acs.jctc.5b00742

Goldman, N., Reed, E. J., Fried, L. E., Kuo, I.-F. W., & Maiti, A. (2010). Synthesis of glycine-containing complexes in impacts of comets on early earth. *Nat. Chem.*, *2*(11), 949–954. Retrieved from http://dx.doi.org/10.1038/nchem.827. doi: 10.1038/nchem.827

Goldman, N., & Tamblyn, I. (2013). Prebiotic chemistry within a simple impacting icy mixture. *J. Phys. Chem. A*, *117*(24), 5124–5131. Retrieved from http://dx.doi.org/10.1021/jp402976n. doi: 10.1021/jp402976n

Hoover, W. G. (1985). Canonical dynamics: Equilibrium phase-space distributions. *Phys. Rev. A: At., Mol., Opt. Phys.*, *31*, 1695–1697. Retrieved from https://link.aps.org/doi/10.1103/PhysRevA.31.1695. doi: 10.1103/ PhysRevA.31.1695

Imai, E.-I., Honda, H., Hatori, K., Brack, A., & Matsuno, K. (1999). Elongation of oligopeptides in a simulated submarine hydrothermal system. *Science*, *283*(5403), 831–833. Retrieved from http://science.sciencemag.org/content/283/5403/831. doi: 10.1126/science.283.5403.831

Imai, E.-I., Honda, H., Hatori, K., & Matsuno, K. (1999). Autocatalytic synthesis of oligoglycine in a simulated submarine hydrothermal system. *Orig. Life Evol. Biosph.*, *29*, 249–259.

Kitadai, N. (2014). Thermodynamic prediction of glycine polymerization as a function of temperature and pH consistent with experimentally obtained results. *J. Mol. Evol.*, *78*(3), 171–187. Retrieved from https://doi.org/10.1007/s00239-014-9616-1. doi: 10.1007/s00239-014-9616-1

Kohn, W., & Sham, L. J. (1965). Self-consistent equations including exchange and correlation effects. *Phys. Rev.*, *140*, A1133–A1138. Retrieved from https://link.aps.org/doi/10.1103/PhysRev.140.A1133. doi: 10.1103/PhysRev.140.A1133

Koskinen, P., & Mäkinen, V. (2009). Density-functional tight-binding for beginners. *Comput. Mater. Sci.*, *47*(1), 237–253. Retrieved from http://www.sciencedirect.com/science/article/pii/S0927025609003036. doi: http://doi.org/10.1016/j.commatsci.2009.07.013

Koziol, L., & Goldman, N. (2015). Prebiotic hydrocarbon synthesis in impacting reduced astrophysical icy mixtures. *The Astrophysical Journal*, *803* (2), 91. Retrieved from http://stacks.iop.org/0004-637X/803/i=2/a=91.

Kroonblawd, M. P., & Goldman, N. (2018, May). Mechanochemical formation of heterogeneous diamond structures during rapid uniaxial compression in graphite. *Phys. Rev. B*, *97*, 184106. Retrieved from https://link.aps.org/doi/10.1103/PhysRevB.97.184106. doi: 10.1103/PhysRevB.97.184106

Kroonblawd, M. P., Pietrucci, F., Saitta, A. M., & Goldman, N. (2018). Generating converged accurate free energy surfaces for chemical reactions with a force-matched semiempirical model. *Journal of Chemical Theory and Computation*, *14*(4), 2207–2218. Retrieved from https://doi.org/10.1021/acs.jctc.7b01266. doi: 10.1021/acs.jctc.7b01266

Kumar, S., Rosenberg, J. M., Bouzida, D., Swendsen, R. H., & Kollman, P. A. (1992). The weighted histogram analysis method for free-energy calculations on biomolecules: I. the method. *J. Comput. Chem.*, *13*(8), 1011–1021. Retrieved from http://dx.doi.org/10.1002/jcc.540130812. doi: 10.1002/jcc.540130812

Laio, A., & Parrinello, M. (2002). Escaping free-energy minima. *Proc. Natl. Acad. Sci. U.S.A.*, *99*(20), 12562–12566. doi: 10.1073/pnas.202427399

Lemke, K. H., Rosenbauer, R. J., & Bird, D. K. (2009). Peptide synthesis in early earth hydrothermal systems. *Astrobiology*, *9*(2), 141.

Martins, Z., Price, M. C., Goldman, N., Sephton, M. A., & Burchell, M. J. (2013). Shock synthesis of amino acids from impacting cometary and icy planet surface analogues. *Nat. Geosci.*, *6*(12), 1045–1049. Retrieved from http://dx.doi.org/10.1038/ngeo1930 doi: 10.1038/ngeo1930

Mermin, N. D. (1965). Thermal properties of the inhomogeneous electron gas. *Phys. Rev.*, *137*, A1441–A1443. Retrieved from https://link.aps.org/doi/10.1103/PhysRev.137.A1441. doi: 10.1103/PhysRev.137.A1441

Niklasson, A.M.N. (2008). Extended Born-Oppenheimer molecular dynamics. *Phys. Rev. Lett.*, *100*, 123004. Retrieved from https://link.aps.org/doi/10.1103/PhysRevLett.100.123004. doi: 10.1103/PhysRevLett.100.123004

Niklasson, A.M.N., Steneteg, P., Odell, A., Bock, N., Challacombe, M., Tymczak, C. J., et al. (2009). Extended lagrangian born–oppenheimer molecular dynamics with dissipation. *J. Chem. Phys.*, *130*(21), 214109. Retrieved from http://dx.doi.org/10.1063/1.3148075. doi: 10.1063/1.3148075

Niklasson, A.M.N., Tymczak, C. J., & Challacombe, M. (2006, Sep). Time-reversible Born-Oppenheimer molecular dynamics. *Phys. Rev. Lett.*, *97*, 123001. Retrieved from https://link.aps.org/doi/10.1103/PhysRevLett.97.123001. doi: 10.1103/PhysRevLett.97.123001

Nosé, S. (1984). A unified formulation of the constant temperature molecular dynamics methods. *J. Chem. Phys.*, *81*(1), 511–519. Retrieved from http://dx.doi.org/10.1063/1.447334. doi: 10.1063/1.447334

Perdew, J. P., Burke, K., & Ernzerhof, M. (1996, Oct). Generalized gradient approximation made simple. *Phys. Rev. Lett.*, *77*, 3865–3868. Retrieved from https://link.aps.org/doi/10.1103/PhysRevLett.77.3865. doi: 10.1103/PhysRevLett.77.3865

Pietrucci, F., & Saitta, A. M. (2015). Formamide reaction network in gas phase and solution via a unified theoretical approach: Toward a reconciliation of different prebiotic scenarios. *Proc. Natl. Acad. Sci. U.S.A.*, *112*(49), 15030–15035. Retrieved from http://www.pnas.org/content/112/49/15030.abstract doi: 10.1073/pnas.1512486112

Plimpton, S. (1995). Fast parallel algorithms for short-range molecular dynamics. *J. Comput. Phys.*, *117*(1), 1–19. Retrieved from http://www.sciencedirect.com/science/article/pii/S002199918571039X (LAMMPS is available at http://lammps.sandia.gov). doi: http://dx.doi.org/10.1006/jcph.1995.1039

Porezag, D., Frauenheim, T., Köhler, T., Seifert, G., & Kaschner, R. (1995). Construction of tight-binding-like potentials on the basis of density-functional theory: Application to carbon. *Phys. Rev. B: Condens. Matter Mater. Phys.*, *51*, 12947–12957. Retrieved from https://link.aps.org/doi/10.1103/PhysRevB.51.12947. doi: 10.1103/PhysRevB.51.12947

Radzicka, A., & Wolfenden, R. (1996). Rates of uncatalyzed peptide bond hydrolysis in neutral solution and the transition state affinities of proteases. *Journal of the American Chemical Society*, *118*(26), 6105–6109. Retrieved from https://doi.org/10.1021/ja954077c. doi: 10.1021/ja954077c

Roux, B. (1995). The calculation of the potential of mean force using computer simulations. *Comput. Phys. Commun.*, *91*(1), 275–282. Retrieved from http://www.sciencedirect.com/science/article/pii/001046559500053I. doi: http://dx.doi.org/10.1016/0010-4655(95)00053-I

Sahle, C. J., Sternemann, C., Schmidt, C., Lehtola, S., Jahn, S., Simonelli, L., et al. (2013). Microscopic structure of water at elevated pressures and temperatures. *Proceedings of the National Academy of Sciences*, *110*(16), 6301–6306.

Sakata, K., Kitadai, N., & Yokoyama, T. (2010). Effects of pH and temperature on dimerization rate of glycine: Evaluation of favorable environmental conditions for chemical evolution of life. *Geochim. Cosmochim. Acta*, *74*(23), 6841–6851. Retrieved from http://www.sciencedirect.com/science/article/pii/S0016703710004813. doi: http://dx.doi.org/10.1016/j.gca.2010.08.032

Schreiner, E., Nair, N. N., & Marx, D. (2009). Peptide synthesis in aqueous environments: The role of extreme conditions on peptide bond formation and peptide hydrolysis. *J. Am. Chem. Soc.*, *131*(38), 13668–13675. Retrieved from http://dx.doi.org/10.1021/ja9032742. doi: 10.1021/ja9032742

Shock, E. L. (1992). Stability of peptides in high-temperature aqueous solutions. *Geochim. Cosmochim. Acta*, *56*(9), 3481–3491. Retrieved from http://www.sciencedirect.com/science/article/pii/001670379290392V. doi: http://dx.doi.org/10.1016/0016-7037(92)90392-V

Slater, J. C., & Koster, G. F. (1954). Simplified LCAO method for the periodic potential problem. *Phys. Rev.*, *94*, 1498–1524. Retrieved from https://link.aps.org/doi/10.1103/PhysRev.94.1498. doi: 10.1103/PhysRev.94.1498

Sugahara, H., & Mimura, K. (2014). Glycine oligomerization up to triglycine by shock experiments simulating comet impacts. *Geochem. J.*, *48*(1), 51–62. doi: 10.2343/geochemj.2.0285

Sugahara, H., & Mimura, K. (2015). Peptide synthesis triggered by comet impacts: A possible method for peptide delivery to the early Earth and icy satellites. *Icarus*, *257*, 103–112. Retrieved from http://www.sciencedirect.com/science/article/pii/S0019103515001724. doi: http://dx.doi.org/10.1016/j.icarus.2015.04.029

Tivey, M. K., Humphris, S. E., Thompson, G., Hannington, M. D., & Rona, P. A. (1995). Deducing patterns of fluid flow and mixing within the tag active hydrothermal mound using mineralogical and geochemical data. *Journal of Geophysical Research: Solid Earth*, *100*(B7), 12527–12555. Retrieved from https://agupubs.onlinelibrary.wiley.com/doi/abs/10.1029/95JB00610. doi: 10.1029/95JB00610

Torrie, G. M., & Valleau, J. P. (1974). Monte Carlo free energy estimates using non-Boltzmann sampling: Application to the sub-critical Lennard-Jones fluid. *Chem. Phys. Lett.*, *28*(4), 578–581. Retrieved from http://www.sciencedirect.com/science/article/pii/0009261474801090. doi: http://dx.doi.org/10.1016/0009-2614(74)80109-0

Vanderbilt, D. (1990). Soft self-consistent pseudopotentials in a generalized eigenvalue formalism. *Phys. Rev. B: Condens. Matter Mater. Phys.*, *41*, 7892–7895. Retrieved from https://link.aps.org/doi/10.1103/PhysRevB.41.7892. doi: 10.1103/PhysRevB.41.7892

Van Dornshuld, E., Vergenz, R. A., & Tschumper, G. S. (2014). Peptide bond formation via glycine condensation in the gas phase. *J. Phys. Chem. B*, *118*(29), 8583–8590. Retrieved from http://dx.doi.org/10.1021/jp504924c. doi: 10.1021/jp504924c

Wolfenden, R. (2011). Benchmark reaction rates, the stability of biological molecules in water, and the evolution of catalytic power in enzymes. *Annu. Rev. Biochem.*, *80*(1), 645–667. Retrieved from https://doi.org/10.1146/annurev-biochem-060409-093051. doi: 10.1146/annurev-biochem-060409-093051

Yanagawa, H., Kojima, K., Ito, M., & Handa, N. (1990, Sep 01). Synthesis of polypeptides by microwave heating: I. Formation of polypeptides during repeated hydration-dehydration cycles and their characterization. *J. Mol. Evol.*, *31*(3), 180–186. Retrieved from https://doi.org/10.1007/BF02109494. doi: 10.1007/BF02109494

Zheng, G., Niklasson, A. M. N., & Karplus, M. (2011). Lagrangian formulation with dissipation of Born-Oppenheimer molecular dynamics using the density-functional tight-binding method. *J. Chem. Phys.*, *135* (4), 044122. Retrieved from http://dx.doi.org/10.1063/1.3605303 doi: 10.1063/1.3605303

24

Predicted Speciation of Carbon in Subduction Zone Fluids

Meghan Guild[1] **and Everett L. Shock**[1,2,3,4]

ABSTRACT

High pressure-temperature aqueous fluids are essential to melt generation, element cycling, and fluid-melt-rock reactions occurring in subduction zones. Recent advances in theoretical thermodynamic modeling help facilitate calculations at a range of pressure conditions relevant to subduction zones. Here we explore stable and metastable equilibrium speciation of C_1 and C_2 aqueous carbon species along a theoretical slab surface pressure-temperature path. These calculations reveal a thermodynamic drive to stabilize small organic compounds at elevated pressures and temperatures, with pH buffered by the diopside-antigorite-forsterite mineral assemblage. At stable equilibrium, oxidized forms of aqueous carbon dominate the speciation at and above oxidation conditions set by the fayalite-magnetite-quartz (FMQ) assemblage. Under conditions more reduced than FMQ, a larger variety of aqueous carbon species are stabilized. If metastability were to persist along the path targeted by this study, it is predicted that a plethora of C_1 and C_2 aqueous species would be stabilized, especially under reduced conditions. These results point the way for theoretical geochemical modeling in the pressure-temperature-composition space of subduction zone fluids and provide new constraints on forms of deep carbon.

24.1. INTRODUCTION

Subduction zones transfer material from the shallow crust of the Earth to the upper mantle, including volatiles, such as H_2O, CO_2, and organic compounds, that are carried within hydrous and other volatile-rich phases of the downgoing sediment, slab, and mantle. Along the path of subduction, volatiles move through the subducted material, are transferred into the overlying mantle wedge. Volatiles injected into the mantle wedge can drive fluid-melt-rock reactions, including dissolution, precipitation, oxidation-reduction, melting, and crystallization processes. Those not transferred through subduction into the mantle are released at the surface in fluids venting at trenches, and in gases and melts erupted by arc volcanoes.

The release of volatiles from the subducting slab into the mantle wedge is linked to the pressure-temperature-dependent breakdown reactions of hydrous minerals (e.g. Fumagalli & Poli, 2005; Grove et al., 2006; Schmidt & Poli, 1998; Till et al., 2012; Ulmer & Trommsdorff, 1995). Models constraining the thermal structure of subduction zones (e.g. Abers et al., 2006; Peacock 2003; Syracuse et al., 2010) and slab surface temperature (e.g. Penniston-Dorland et al., 2015; Syracuse et al., 2010) help us predict where these breakdown reactions occur (Cerpa et al., 2017; Hacker, 2008; van Keken et al., 2002); however, the compositions of the released fluids remain elusive. Subduction fluids have been characterized by direct sampling (e.g. Fryer et al., 1990; Kimura et al., 1997; Manning, 2004) from fluid inclusions in high-pressure metamorphic rocks (e.g. Vitale Brovarone et al., 2017; Frezzotti et al., 2011; Hermann et al., 2006; Scambelluri & Philippot, 2001) and are constrained by experimental (e.g. Caciagli & Manning 2003; Facq et al., 2014, 2016; Li, 2017) and theoretical studies (e.g. Manning et al., 2013; Facq et al.,

[1] *School of Earth and Space Exploration, Arizona State University, Tempe, Arizona, USA*

[2] *Group Exploring Organic Processes in Geochemistry (GEOPIG), Arizona State University, Tempe, Arizona, USA*

[3] *School of Molecular Sciences, Arizona State University, Tempe, Arizona, USA*

[4] *Center for Fundamental and Applied Microbiomics, Arizona State University, Tempe, Arizona, USA*

Carbon in Earth's Interior, Geophysical Monograph 249, First Edition. Edited by Craig E. Manning, Jung-Fu Lin, and Wendy L. Mao.

2014, 2016; Huang et al., 2017). Altogether, these studies indicate that shallow subduction zone fluids are relatively dilute with salinities similar to seawater, but that calcite and aragonite become highly soluble at elevated conditions, and methane or larger C-compounds may be immiscible at high pressures. Theoretical studies suggest a thermodynamic drive to stabilize small organic compounds at high pressure-temperature (P-T) conditions (Sverjensky, Stagno, et al., 2014). These recent advances in the geochemical thermodynamics of high-pressure aqueous solutions (Sverjensky, Harrison, et al., 2014) facilitate predictive studies of subduction zone fluid compositions at elevated pressures and temperatures, and inspired the research reported here. The P-T ranges now accessible by aqueous thermodynamic modeling are shown in Figure 24.1, where they are compared to slab surface temperature trajectories from model D80 in Syracuse et al. (2010). Now that these P-T conditions are accessible, we can further explore the role of pH and oxidation state on fluid composition.

In the subduction setting, fluids released from the slab are called upon to explain the oxidized nature of the mantle wedge (e.g. Brounce et al., 2015; Kelley & Cottrell, 2009; Parkinson & Arculus, 1999; Wood et al., 1990), although the dominant oxidizing agent is uncertain (i.e. carbon, iron, sulfur). Fe and S may be involved in the oxidation of the mantle wedge, but aqueous carbon species are sufficiently abundant to influence the oxidation of the mantle wedge according to Sverjensky, Stagno, et al. (2014). When looking at the range of oxygen fugacities in the sub-arc mantle relative to the fayalite-magnetite-quartz (FMQ) mineral buffer, discussed in detail below, sub-arc mantle xenoliths record a relatively narrow range of oxygen fugacities from ΔFMQ 0 to +1.5 (Figure 24.2; Bénard et al., 2018; Brandon et al., 1996; Parkinson & Arculus, 1999). Arc lavas are generally more oxidized, with positive values relative to FMQ that are thought to be caused by oxidizing slab fluids (e.g. Brounce et al., 2015; Kelley & Cottrell, 2009). Some suggest oxidation is via crystallization during ascent and crustal storage (e.g. Lee et al., 2005; 2010), but these models are not supported by experimental studies (Waters & Lange, 2016). Relatively more reduced, the MORB mantle oxygen fugacities are constrained from abyssal peridotites, with a wide range of oxygen fugacities from ΔFMQ –2.5 to +1.4 (Figure 24.2; Birner et al., 2018; Bryndzia & Wood, 1999).

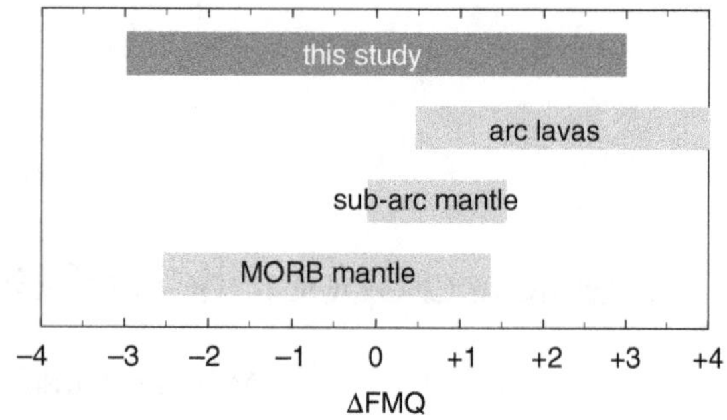

Figure 24.2 Oxygen fugacity of relevant Earth reservoirs relative to the fayalite-magnetite-quartz (FMQ) mineral buffer. This study targets calculations from FMQ–3 to FMQ+3. Arc lava values constrained by Rowe et al. (2009), Jugo et al. (2010), and Evans et al. (2012). Sub-arc mantle values from arc xenoliths constrained by Brandon et al. (1996), Parkinson and Arculus (1999), and Bénard et al. (2018). MORB mantle values from abyssal peridotites constrained by Bryndzia and Wood (1990) and Birner et al. (2018). See electronic version for color representation of the figures in this book.

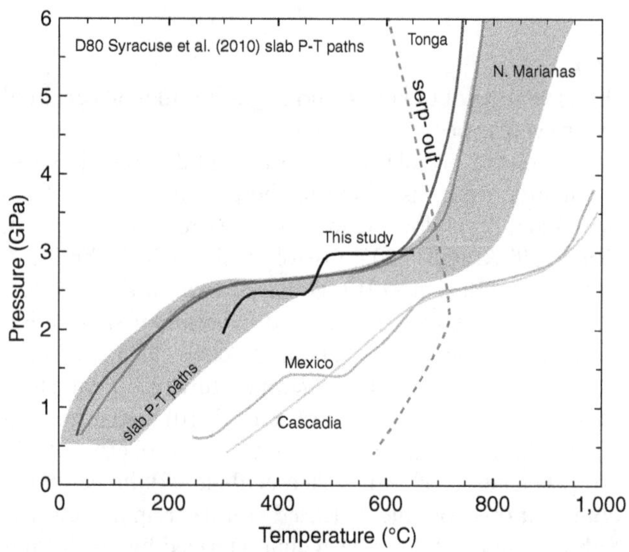

Figure 24.1 P-T trajectories of slab surface temperatures in subduction zones. The range of pressures and temperature displayed is accessible by the Deep Earth Water model (DEW). The trajectory adopted for this study is shown in black. Examples of cold subduction zones include Tonga and N. Marianas. Cascadia and Mexico are examples of hot subduction zones. The gray field shows a summary of slab surface P-T trajectories from Syracuse et al. (2010), model D80. "Serpentine-out" boundary represents the P-T stability of antigorite serpentine for a hydrated mantle composition (Ulmer & Trommsdorff, 1995). See electronic version for color representation of the figures in this book.

Gas fluxes at arc volcanoes are rarely characterized from direct measurement of CO_2 concentrations; instead, SO_2 and CO_2/SO_2 measurements help constrain these emissions (Kelemen & Manning, 2015). Volcanic gas flux measurements of ^3He together with $^3He/CO_2$ measurements of gases, fluids, and volcanic glasses, also help constrain the concentrations of CO_2 emitted from arc volcanoes (Kelemen & Manning, 2015). Nevertheless, the gases emitted from arc volcanoes are oxidized (e.g. CO_2, CO, SO_2, etc.), consistent with the oxidized nature of the eruptive products at arcs. Diffuse degassing in the forearc of subduction zones is also a source of CO_2 at convergent margins. At depth, the bulk composition, redox state, pressure, and temperature of the system combine to control the fluid/volatile chemistry, all of which can be visualized through thermodynamic modeling. For example, the reduction of CO_2 to CH_4 can be represented by

$$CO_{2(aq)} + 2H_2O = CH_{4(aq)} + 2O_{2(g)}, \quad (24.1)$$

where the reduction of C is tied to the oxidation of H_2O. This is an appealing representation for petrologic applications because it can be linked with values of the fugacity of O_2 (fO_2) obtained from mineral equilibria. The disproportionation of H_2O represented by

$$2H_{2(aq)} + O_{2(g)} = 2H_2O \quad (24.2)$$

can be combined with reaction (1), leading to

$$CO_{2(aq)} + 4H_{2(aq)} = CH_{4(aq)} + 2H_2O, \quad (24.3)$$

which can be linked to aqueous reactions in which activities of H_2 (aH_2) can be used to represent the oxidation state of the system. If equilibrium, with respect to reaction (2), is attained, then reactions (1) and (3) are interchangeable. It follows that equilibria at low values of fO_2 (and therefore high values of aH_2) will be dominated by CH_4, and those at high values of fO_2 (and therefore low values of aH_2) will be dominated by CO_2.

These familiar consequences can be represented by a relative predominance diagram such as that shown in Figure 24.3. This diagram shows the equilibrium

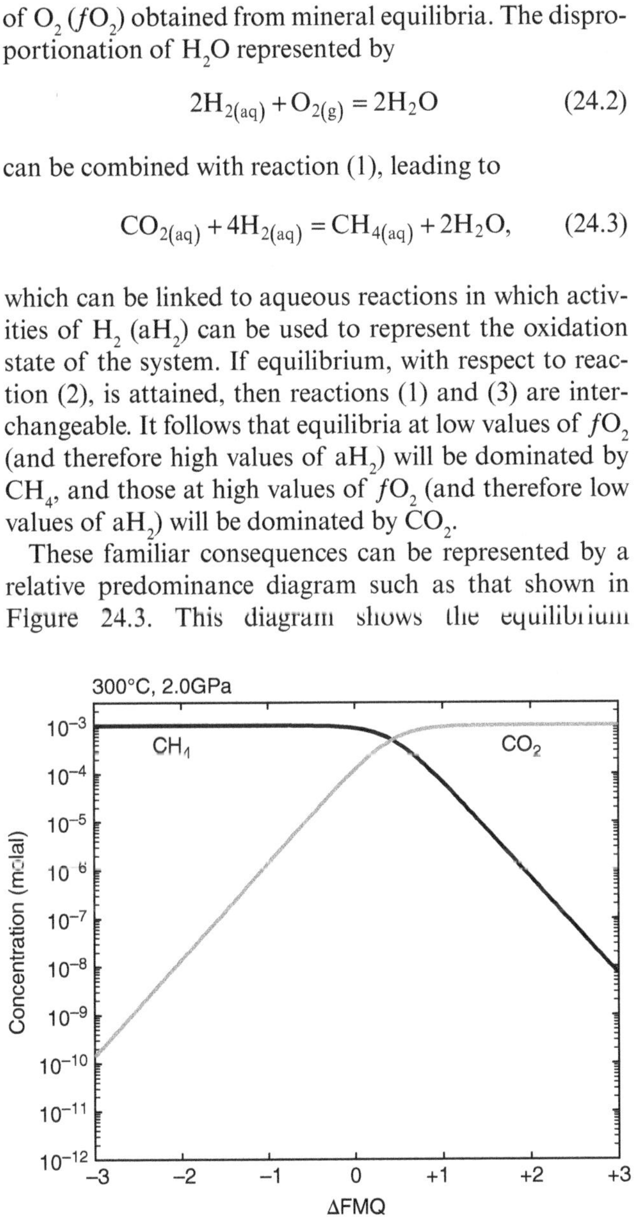

Figure 24.3 Speciation of aqueous methane (CH_4) and aqueous carbon dioxide (CO_2) at 300 °C and 2.0 GPa over a range of oxygen fugacities from FMQ–3 to FMQ+3. Concentration of carbon in the system fixed at 0.001 molal. See electronic version for color representation of the figures in this book.

concentrations of aqueous CH_4 and CO_2 as functions of the oxidation state of a system at 300 °C and 2 GPa for a system containing 10^{-3} m total dissolved carbon and only these two aqueous species. The oxidation state is represented by differences (Δ) from that set by the FMQ assemblage according to

$$3Fe_2SiO_4 + O_{2(g)} = 2Fe_3O_4 + 3SiO_2 \quad (24.4)$$

in units of log fO_2. Note that at a slightly positive value of ΔFMQ, i.e. a value of fO_2 somewhat greater than the value set by FMQ at this temperature and pressure, the distribution of dissolved carbon shifts from aqueous CH_4 to aqueous CO_2.

By considering only reaction (3), it would be possible to conclude that these two aqueous species represent the speciation of carbon in a fluid found in a shallow subduction system. The assemblage of minerals hosting that fluid in the subduction channel would impart some chemical characteristics on the fluid. At equilibrium, the mineral assemblage (i.e. rock) would lead to the formation of aqueous complexes with the major rock-forming elements in the fluid. In fact, the rock would be able to buffer the pH and other solute activities. The potential for the fluid to have a pH value immediately raises the possibility that aqueous CO_2 is accompanied by other fully oxidized forms of carbon such as aqueous bicarbonate (HCO_3^-) and aqueous carbonate (CO_3^{2-}). The likelihood that other solutes derived from the rock may also be present leads to the possibility that inorganic metal-bicarbonate and metal-carbonate complexes could also be present. Once we allow these considerations, we can rapidly conclude that the actual speciation of carbon in a subduction zone fluid may be more complicated than represented by Figure 24.3. If we also consider the possibility that aqueous carbon species other than the redox endmembers CO_2 and CH_4 might be present, including compounds such as CO, or formic acid (HCOOH) or methanol (CH_3OH), etc., we may wonder what sorts of true complexity may be glossed over by an idealized representation such as that in Figure 24.3.

Here we present a series of stable equilibrium speciation calculations for aqueous carbon along a model subduction fluid trajectory, including results for aqueous C_1 and C_2 organic species. In addition, we explore the response of inorganic and organic carbon speciation at metastable equilibrium conditions, where methane formation from other forms of aqueous carbon is inhibited, in an effort to depict the potential complexity of aqueous organic carbon speciation near the slab-mantle interface. It should be kept in mind that seeking answers to questions through thermodynamic modeling yields two types of responses: no and maybe. *No* is a definitive response, and *maybe* means that the potential for transformations may be realized if mechanisms exist. The need for reaction mechanisms is acutely felt when confronted by unknown or unfamiliar transformations of organic compounds in aqueous solutions at high pressures and temperatures and is a research area ripe for experimental and theoretical developments (Shock et al., 2019).

24.2. SPECIATION CALCULATIONS FOR HIGH P-T AQUEOUS FLUIDS

Prior to the development of the Deep Earth Water (DEW) model (Sverjensky, Harrison, et al., 2014; see also Facq et al., 2014), speciation and mass-transfer calculations involving aqueous fluids were limited to 1000 °C and 0.5 GPa (Shock et al., 1992, 1997; Sverjensky et al., 1997). The DEW model extended the pressure range of such calculations to 6.0 GPa, which has allowed new predictions for the lower crust, upper mantle, and subduction zones. In the present study, we expanded on the variety of aqueous organic solutes and metal-ligand complexes included in the DEW model and used the EQ3 speciation code (Wolery & Jarek, 2003) with a completely customized database of equilibrium constants generated with the DEW model. A complete list of aqueous species included in these calculations is shown in Table 24.1, together with the abbreviations used in subsequent figures and the average oxidation state of carbon in each compound (Z_C). DEW parameters for newly added species are listed in Table 24.2. Note that standard state thermodynamic properties at the reference conditions of 298.15 K and 0.1 MPa for these aqueous species are taken from sources that maintain internal consistency with minerals and aqueous species used in the DEW model. As in the case of other applications of the DEW model, activity coefficients are set equal to 1, and activities are equated with molalities (Sverjensky, Stagno, et al., 2014). Results may differ once a suitable activity coefficient model is developed.

Initial fluid compositions used in the speciation calculations are selected to be simple yet representative of general features of subduction zone fluids that would allow for experimental testing of the predictions. Manning (2004) showed that sampled subduction zone fluids are similar in composition to seawater; thus, a seawater-like salinity was chosen for these calculations. Na^+ and Cl^- are both input with concentrations of 0.5 m, similar to the composition of seawater, and the total dissolved inorganic carbon (DIC) was set to 0.001 m, which is a conservative estimate of the DIC in seawater (Manning, 2004; McCollom, 2008; Shock & Canovas, 2010).

All speciation calculations were conducted at equilibrium with diopside, forsterite, and antigorite to represent a hydrated mantle assemblage relevant for mineralogies observed above the slab-mantle interface, within the subduction channel. As a consequence, the activities of Ca^{2+}, Mg^{2+}, and $SiO_{2(aq)}$ in the modeled subduction zone fluids were fixed by this assemblage in a manner analogous to fluid-rock reactions that control the pH and activities of major rock-forming elements within subduction zone fluids. Predicted total concentrations of dissolved calcium, magnesium, and silica are shown in Figure 24.4. Note that these total concentrations are calculated by summing the molal abundances of all aqueous species containing each of these constituents. Also shown in Figure 24.4 are calculated pH values in equilibrium with the diopside-forsterite-antigorite assemblage, which

Table 24.1 C_1 and C_2 compounds included in speciation calculations arranged according to Z_C, the average oxidation state of carbon in each compound.

Name	Symbol	Z_C
Methane	CH_4	−4
Acetaldehyde*	CH_3COH	−3
Ethane	C_2H_6	−3
Ethanol	CH_3CH_2OH	−2
Ethylene	C_2H_4	−2
Methanol	CH_3OH	−2
Calcium formate*	$CaCH_2O^+$	−1
Ethyne*	C_2H_2	−1
Magnesium formate*	$MgCH_2O^+$	−1
Sodium formate*	$NaCH_2O$	−1
Acetate	CH_3COO^-	0
Acetic acid	CH_3COOH	0
Calcium acetate*	$CaCH_3COO^+$	0
Formaldehyde*	CH_2O	0
Magnesium acetate*	$MgCH_3COO^+$	0
Sodium acetate	$NaCH_3COO$	0
Calcium glycolate*	$C_2H_3CaO_3^+$	+1
Glycolate	$HOCH_2CO_2H$	+1
Glycolic acid	$C_2H_4O_3$	+1
Magnesium glycolate*	$C_2H_3MgO_3^+$	+1
Sodium glycolate*	$Na(CH_3OCO_2)$	+1
Calcium diformate*	$Ca(CHO_2)_2$	+2
Carbon monoxide	CO	+2
Formate	$HCOO^-$	+2
Formic acid	$HCOOH$	+2
Magnesium diformate*	$Mg(CHO_2)_2$	+2
Sodium diformate*	$Na(CHO_2)_2^-$	+2
Oxalate*	$C_2O_4^{2-}$	+3
Oxalic acid*	$HOOCCOOH$	+3
Bicarbonate	HCO_3^-	+4
Calcium bicarbonate	$Ca(HCO_3)^+$	+4
Calcium carbonate	$CaCO_3$	+4
Carbon dioxide	CO_2	+4
Carbonate	CO_3^{2-}	+4
Magnesium bicarbonate	$Mg(HCO_3)^+$	+4
Magnesium carbonate	$MgCO_3$	+4
Sodium bicarbonate	$NaHCO_3$	+4
Sodium carbonate	$NaCO_3^-$	+4

* Species has been added to the Deep Earth Water (DEW) model as part of the current study.

Table 24.2 Nonsolvation parameters for species added to the DEW model in the current study. Estimated in accordance with Sverjensky et al. (2014).

		Z_c	$a_1 \times 10$	$a_2 \times 10^{-2}$	a_3	$a_4 \times 10^{-4}$	c_1	$c_2 \times 10^{-4}$
Name	Symbol		cal mol^{-1} bar^{-1}	cal mol^{-1}	cal K mol^{-1} bar^{-1}	cal K mol^{-1}	cal mol^{-1} K^{-1}	cal K mol^{-1}
Acetaldehyde[c]	CH_3COH	−3	9.8306	5.0765	0.5815	−2.9889	23.2243	4.0745
Calcium formate[d]	$CaCH_2O^+$	−1	4.6033	0.0997	4.7545	−2.7831	27.4191	3.2190
Ethyne[a]	C_2H_2	−1	10.8770	6.0729	−0.2540	−3.0301	33.0425	7.7615
Magnesium formate[d]	$MgCH_2O^+$	−1	4.0399	−0.4367	5.2043	−2.7609	34.0760	4.2375
Sodium formate[d]	$NaCH_2O$	−1	8.0897	3.4191	1.9712	−2.9203	19.2524	1.6098
Calcium acetate[b]	$CaCH_3COO^+$	0	7.3852	2.7483	2.5337	−2.8926	58.2181	13.8929
Formaldehyde[c]	CH_2O	0	6.5616	1.9642	3.1912	−2.8602	11.8522	0.1635
Magnesium acetate[b]	$MgCH_3COO^+$	0	6.8123	2.2028	2.9911	−2.8701	64.6340	14.8910
Calcium glycolate[d]	$C_2H_3CaO_3^+$	+1	7.2320	2.6025	2.6559	−2.8866	57.8082	13.9947
Magnesium glycolate[d]	$C_2H_3MgO_3^+$	+1	6.6165	2.0164	3.1473	−2.8624	63.4637	15.0132
Sodium glycolate[d]	$Na(CH_3OCO_2)$	+1	10.7698	5.9708	−0.1684	−3.0258	50.2551	12.3855
Calcium diformate[d]	$Ca(CHO_2)_2$	+2	11.5661	6.7289	−0.8040	−3.0572	42.6363	9.7374
Magnesium diformate[d]	$Mg(CHO_2)_2$	+2	10.7310	5.9338	−0.1374	−3.0243	48.3211	11.7133
Sodium diformate[d]	$Na(CHO_2)_2^-$	+2	15.6445	10.6120	−4.0599	−3.2177	32.2630	6.1319
Oxalate[e]	$C_2O_4^{2-}$	+3	8.8670	4.1591	1.3508	−2.9509	−8.7110	−17.9658
Oxalic acid[e]	HOOCCOOH	+3	10.9784	6.1694	−0.3348	−3.0340	16.7460	1.5894

[a] Shock and Helgeson (1990).
[b] Shock and Koretsy (1993).
[c] Schulte and Shock (1993).
[d] Shock and Koretsky (1995).
[e] Shock (1995).

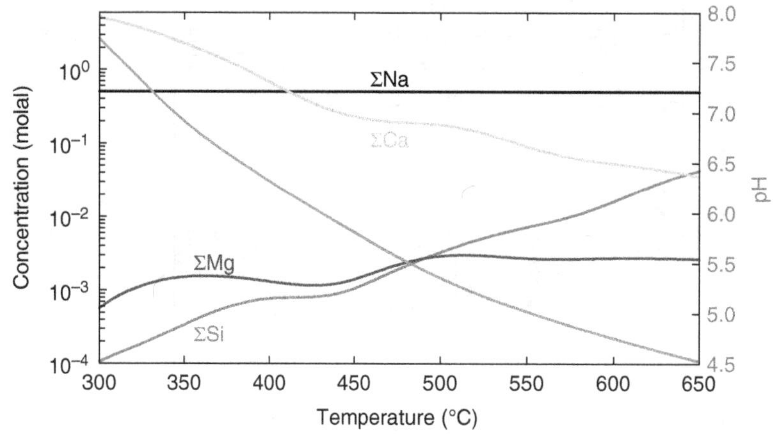

Figure 24.4 ΣCa, ΣMg, ΣSi, ΣNa, and pH vs. temperature for the calculations presented in Figures 24.5 through 24.9. The summed concentrations of Ca, Mg, Si, and Na are on the left *y*-axis on a log scale from 10^0 to 10^{-4} molal. pH is plotted on the right *y*-axis in red. See electronic version for color representation of the figures in this book.

Table 24.3 Input parameters for speciation calculations. Log fO_2 and neutral pH values were calculated using the DEW model. Neutral pH is shown here for comparison to pH controlled by antigorite-diopside-forsterite.

Temperature (°C)	Pressure (GPa)	Log fO_2	Neutral pH
300	2.0	−32.5	3.77
350	2.5	−28.2	3.50
400	2.5	−25.3	3.42
450	2.5	−22.7	3.37
500	3.0	−20.0	3.17
550	3.0	−18.2	3.14
600	3.0	−16.5	3.11
650	3.0	−15.1	3.10

are mildly alkaline at these pressures and temperatures (7.7 to 4.5) and similar to those observed in mafic lithologies modeled by Galvez et al. (2016). Neutral pH along this P-T path is shown in Table 24.3, along with log fO_2 fixed in these calculations, at the FMQ mineral buffer. An abbreviated range of concentrations is shown in Figures 24.5 through 24.9 (10^{-3} to 10^{-12} molal).

24.3. EQUILIBRIUM SPECIATION OF AQUEOUS ONE-CARBON (C_1) COMPOUNDS

The results presented in Figure 24.5 summarize calculations for a variety of aqueous C_1 species, including inorganic carbon oxides and metal complexes, as well as organic compounds and methane ($CH_{4\,(aq)}$). The three panels in Figure 24.5 contain plots for speciation calculations along the subduction fluid trajectory of Figure 24.1 at oxygen fugacities relative to FMQ. The left panel shows the calculated equilibrium speciation of carbon at three log fO_2 units below FMQ (FMQ-3), and the center and right panels show results for FMQ (FMQ0) and three log fO_2 units above FMQ (FMQ+3), respectively. As a frame of reference, predicted concentrations for aqueous CO_2 and aqueous CH_4 are shown as slightly bolder curves than the other aqueous species. Any C_1 aqueous species listed in Table 24.1 that fail to attain predicted concentrations of at least 10^{-12} m do not appear in these diagrams.

The center panel in Figure 24.5 at FMQ provides a framework for interpreting the results in all three panels. Predicted equilibrium concentrations of various C_1 aqueous species are shown as functions of temperature at the pressures indicated in Figure 24.1. At FMQ, the equilibrium speciation of aqueous carbon is dominated by oxidized species, especially carbonate (CO_3^{2-}), bicarbonate (HCO_3^-), and their calcium complexes at the pH values and total Ca abundances of the modeled fluid. The aqueous $CaCO_3$ complex is predicted to dominate the speciation of dissolved carbon below ~450 °C at these conditions. The Mg complexes of bicarbonate ($Mg(HCO_3)^+_{(aq)}$) and carbonate ($MgCO_{3(aq)}$) are predicted to be at much lower abundances than their Ca counterparts owing to the difference in total concentration of Mg relative to Ca (see Figure 24.4). In contrast, the Na complexes are predicted to be less abundant than Ca complexes owing to their lower stability, despite the greater abundance of total Na in the model fluid (see Figure 24.4). Note that the combined effects of pH values set by equilibrium between the diopside-forsterite-antigorite mineral assemblage and the fluid, together with the shifting pK values for the carbonic acid system, lead to the prediction that relative abundances of bicarbonate and carbonate aqueous species tend to decrease with increasing temperature, with the exception of $Ca(HCO_3)^+$, and that aqueous CO_2 becomes more abundant than bicarbonate or carbonate ions above about 550 °C. In

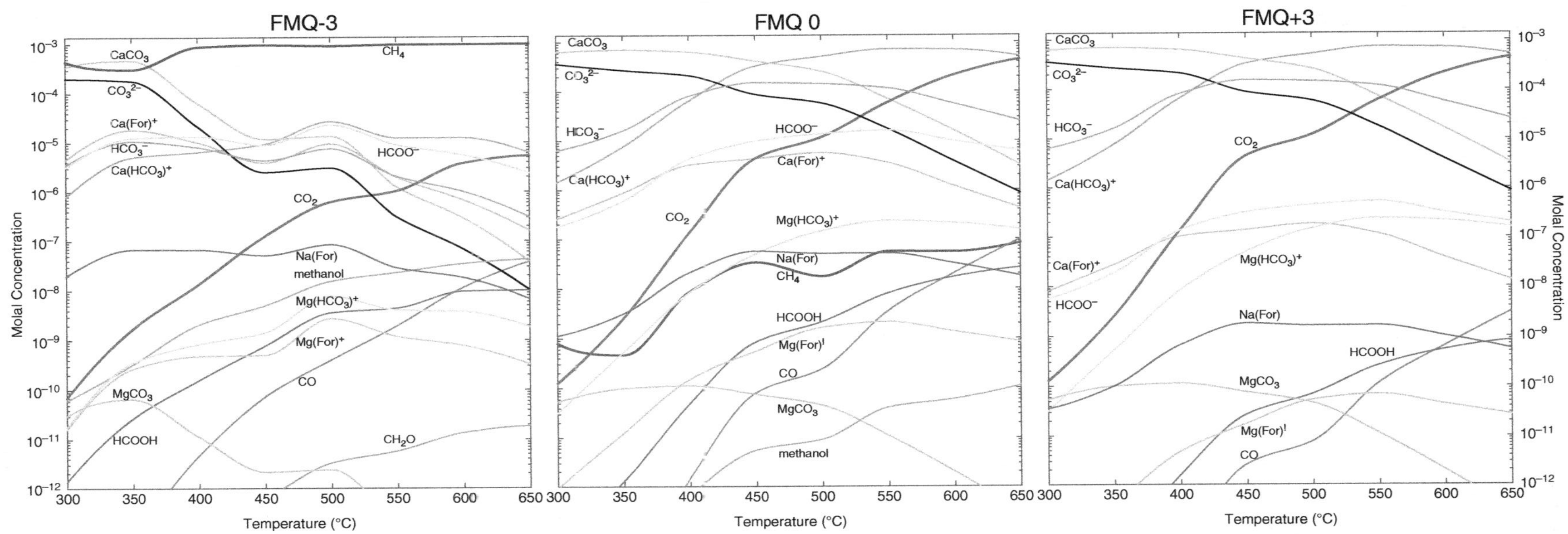

Figure 24.5 Equilibrium speciation calculation of C_1 aqueous spec es for the P-T path outlined in Figure 24.1. pH values are fixed by diopside, forsterite, and antigorite activities. Calculations performed at FMQ–3, FMQ0, and FMQ+3. Colors and labels denote species. "For" = formate = $HCOO^-$; "Ac" = acetate = CH_3COO^-; "Glyc" = glycolate = $CH_3OCO_2^-$. See electronic version for color representation of the figures in this book.

effect, the decrease in pH with increasing temperature moves from more alkaline conditions where CO_3^{-2} > HCO_3^- > $CO_{2(aq)}$ at temperatures below ~420 °C to more acidic conditions where $CO_{2(aq)}$ > HCO_3^- > CO_3^{-2} above ~560 °C. As temperature decreases, the predicted concentration of $CO_{2(aq)}$ plunges over six orders of magnitude, reaching nearly 10^{-10} m at 300 °C.

Turning to redox relations, the curves showing calculated concentrations of $CH_{4(aq)}$ and $CO_{2(aq)}$ cross at about 325 °C and FMQ0, as could be anticipated from the simplified system depicted in Figure 24.3. However, in contrast with the streamlined version in Figure 24.3, neither $CO_{2(aq)}$ nor $CH_{4(aq)}$ are expected to be major contributors to the speciation of carbon at these conditions. Where the curves cross, they both represent concentrations $<10^{-9}$ m. The sinuous curve showing the predicted concentration of $CH_{4(aq)}$ increases by ~two orders of magnitude as temperature increases. This shift reveals that, even though the predicted speciation of aqueous carbon is progressively dominated by $CO_{2(aq)}$, the oxidation state set by FMQ shifts to conditions that are slightly more reduced relative to speciation in the C-O-H system as temperature increases at the varying pressures along the subduction fluid trajectory. This subtle shift is also the reason why calculated concentrations of aqueous CO, methanol, and formic acid increase with increasing temperature. Note that the predicted concentrations of aqueous formate and the metal-formate complexes show maxima with increasing temperature, which reflects the shift in the calculated pH relative to the predicted speciation of formic acid.

There is considerable similarity between the more oxidized conditions at FMQ+3 shown in the right panel of Figure 24.3 and the moderate redox conditions (FMQ0) of the center panel. As an example, the speciation of the most oxidized forms of aqueous carbon (i.e. $CaCO_3$, CO_3^{2-}, HCO_3^- $Ca(HCO_3)^-$, CO_2, $MgCO_3$) are virtually unchanged at FMQ+3 compared with FMQ0. In contrast, $CH_{4(aq)}$ is predicted to attain extremely low concentrations (~10^{-16}–10^{-14} m) at FMQ+3, which fall below the lowest concentration shown. The relative positions of formate to formic acid and of formate to its complexes remain the same, but the curves have shifted by 1.5 orders of magnitude to lower concentrations at FMQ+3. Likewise, $CO_{(aq)}$ is predicted to attain lower concentrations than in the FMQ0 case and reaches concentrations above 10^{-12} m only above 440 °C. The methanol curve has shifted to concentrations below the plotted range (10^{-19} to 10^{-15} m).

At strongly reduced conditions (FMQ-3) shown in the left panel of Figure 24.3, $CH_{4(aq)}$ is the dominant aqueous carbon species at all temperatures above ~365 °C, and thus no crossover between $CH_{4(aq)}$ to $CO_{2(aq)}$ is observed over the range of temperatures considered. However, the interplay between the shifts in pH, association constants for carbonic acid species and complexes, and the relative effects of the redox state on the C-O-H system means that carbonate and especially the aqueous $CaCO_3$ complex are predicted to decrease in abundance as temperature increases. With increasing temperature, the predicted concentrations of these forms of oxidized aqueous carbon decrease as $CH_{4(aq)}$ dominates the carbon speciation. It can be seen that the pH-driven temperature transitions in carbonic acid speciation described for FMQ0 and FMQ+3 persist at FMQ-3, though at lower relative concentrations, especially at high temperatures. Reduced conditions at FMQ-3 enhance the concentrations of formic acid, formate, and its complexes at lower temperatures compared with FMQ0, but the overwhelming predominance of $CH_{4(aq)}$ with increasing temperature drives several of the formate species to lower concentrations than at corresponding temperatures at FMQ0. The distribution of aqueous carbon among minor species is predicted to be particularly complicated between ~400 °C and ~500 °C, where six different aqueous species attain concentrations around 10^{-5} m, including forms of carbonate, bicarbonate, and formate. The interplay of subtle shifts in fO_2 and pH, together with the redox- and pH-dependence of reactions between C_1 species, leads to undulations in the predicted concentration curves, including several local maxima around 500 °C. Note that the predicted concentrations of methanol are considerably enhanced at FMQ-3 compared with FMQ0, and that formaldehyde concentrations actually attain values within the plotted concentration range.

In summary, results in Figure 24.5 demonstrate that the speciation of C_1 compounds can be considerably more complex than a binary choice between $CO_{2(aq)}$ and $CH_{4(aq)}$. Several oxidized forms of aqueous carbon (i.e. $CaCO_3$, CO_3^{2-}, HCO_3^-, $Ca(HCO_3)^-$) are more abundant than $CO_{2(aq)}$ along this subduction fluid trajectory. In particular, the speciation results for both FMQ0 and FMQ+3, where oxidized forms of aqueous carbon dominate, $CO_{2(aq)}$ never reaches predicted concentrations as high as aqueous $CaCO_3$ at low temperatures or $Ca(HCO_3)^+$ at high temperatures. At FMQ-3, where $CH_{4(aq)}$ dominates the predicted speciation at most temperatures, formate and $Ca(For)^+$ concentrations rival those of carbonic acid species, indicating a thermodynamic drive for transfer of carbon between inorganic and organic C_1 forms. Experimental confirmations of such transformations are provided by McCollom and Seewald (2003a, 2003b) and Seewald et al. (2006).

24.4. EQUILIBRIUM SPECIATION INCLUDING AQUEOUS TWO-CARBON (C_2) COMPOUNDS

The making and breaking of carbon-carbon bonds influences the complexity of carbon speciation in subduction fluids. While mechanisms for these reactions are underconstrained at present, a thermodynamic analysis can reveal conditions where C_2 compounds may exist together with the C_1 compounds considered above. Speciation results including both C_1 and C_2 aqueous species are shown in Figure 24.6 for the subduction fluid trajectory adopted in this study. The C_1 speciation depicted in Figure 24.5 is largely unchanged in the calculations of both C_1 and C_2 species in Figure 24.6.

At FMQ0, as shown in the center panel of Figure 24.6, $Ca(Ac)^+$, acetate, and $Ca(For)_2$ appear at concentrations below 10^{-10} m. The concentrations of these C_2 species maximize between 400 °C and 550 °C, with broad plateauing curves. In contrast, at the more oxidized conditions (FMQ+3) shown in the right panel, C_2 compounds are not observed above concentrations of 10^{-12} m; thus, the speciation of aqueous carbon is predicted to be the same as the C_1 case shown in the right panel of Figure 24.5.

In the reduced case (FMQ-3), nine new forms of aqueous carbon with concentrations greater than 10^{-12} m join the speciation results: acetate, $Ca(Ac)^+$, $Mg(Ac)^+$, Na(Ac), acetic acid, ethane, ethanol, $Ca(For)_2$, and $Ca(Glyc)^+$. This is a consequence of the average oxidation state of carbon (Z_C) being lower for C_2 compounds than corresponding C_1 compounds. As an example, Z_C for acetic acid is 0 and Z_C for formic acid is +2. Curves showing predicted concentrations for all of these compounds exhibit a similar positive excursion at 500 °C, observed for the C_1-only results in Figure 24.5. The highest concentration curves are for acetate and $Ca(Ac)^+$. These curves are nearly parallel with $Ca(Ac)^+$, predicted to be around an order of magnitude higher in concentration. The Na(Ac) curve is similar in shape to that of acetate and $Ca(Ac)^+$ but falls off with a steeper slope to lower concentrations at temperatures below 400 °C. Predicted concentrations of ethane and methanol switch in relative abundance across the diagram, with ethane > methanol at lower temperatures and methanol > ethane at higher temperatures. Methanol is predicted to reach higher concentrations than any of the C_2 compounds above ~550 °C. The ethanol concentration curve increases more than two orders of magnitude between 300 °C and 500 °C, after which it plateaus, varying in concentration between 2 and 3×10^{-10} m.

Compared with the results for C_1 shown in Figure 24.5, speciation results that include C_2 compounds in Figure 24.6 indicate the potential for three C_2 species to attain equilibrium concentrations > 10^{-12} m at FMQ0, and at FMQ-3, six additional C_2 species can reach such concentrations. In the FMQ+3 case, no C_2 compounds attain equilibrium concentrations above 10^{-12} m. It is worth emphasizing that including the C_2 species has negligible effects on the predicted concentrations of the C_1 species. Instead, the calculated results attain a greater richness owing to the inclusion of many more aqueous species. We anticipate that including additional species with higher carbon numbers will cause small changes in C_1 or C_2 speciation. In general, the control of pH and fO_2 by water-rock reactions dictates how the predicted concentration curves are distributed. The main influence on relative concentrations is the Z_C values of the various aqueous organic species, which largely determines how they will respond to changes in fO_2. The abundance of C_2 (and higher carbon number) compounds will also correlate positively with the total dissolved carbon in the system. Superimposed on these patterns are variations driven by pH changes that affect relative abundances of the organic acids and their complexes.

24.5. SENSITIVITY OF AQUEOUS CARBON CHEMISTRY TO OXYGEN FUGACITY

The oxidation states selected for Figure 24.6 frame the dependence of aqueous carbon speciation on fO_2, and the sensitivity can be explored more thoroughly in the isothermal results in Figure 24.7 that show how predicted concentrations depend on fO_2. By examining the 300 °C panel, for example, it can be seen that the most oxidized forms of aqueous carbon are nearly insensitive to changes in fO_2, with curves that gently slope to lower concentrations as the oxidation state is progressively more reduced than FMQ. A ranking of oxidized forms of carbon shows that aqueous CO_2 is among the least abundant forms, owing to the pH of the fluid and the relative stabilities of aqueous carbonate and bicarbonate complexes. Note that the predicted concentration of aqueous CH_4 increases dramatically, indicating that its speciation is highly sensitive to small changes in fO_2. The various formate-containing species also increase with decreasing fO_2, although less dramatically than aqueous CH_4. Note that C_2 compounds attain concentrations > 10^{-12} m only at the most reduced conditions considered. The trajectories of the acetate-bearing compounds are nearly parallel with that of aqueous CH_4. Ethane is the steepest curve in the plot. At 400 °C, shifts in pH and cation concentrations lead to shifts in the relative abundances of the most oxidized forms of aqueous carbon, as well as subtle shifts among the formate-containing species. Aqueous CO_2 is more abundant than in the calculations for 300 °C, but it is still a minor contributor to the speciation of oxidized carbon in these fluids. The concave downward curvature of the oxidized aqueous carbon species toward more reducing conditions (negative ΔFMQ values) is directly related to the dramatic increase in predicted concentrations of aqueous methane. Likewise, there is more curvature in the formate curves at 400 °C than at 300 °C. Note that the C_2 compounds appearing in the lower left portion of the plot attain higher concentrations at 400 °C than at 300 °C, all else being equal. All of these

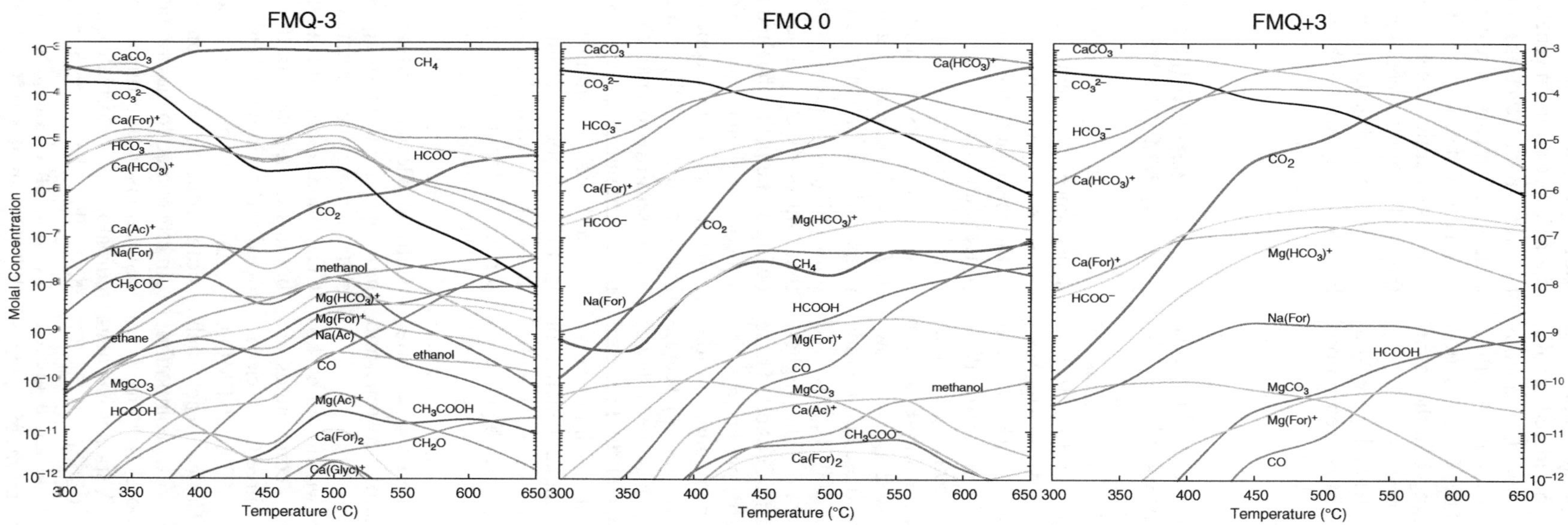

Figure 24.6 Equilibrium speciation calculations including C_1 and C_2 aqueous species for the P-T path outlined in Figure 24.1. pH values are fixed by diopside, forsterite, and antigorite activities. Calculations performed at FMQ–3, FMQ0, and FMQ+3. "For" = formate = $HCOO^-$; "Ac" = acetate = CH_3COO^-; "Glyc" = glycolate = $CH_3OCO_2^-$. See electronic version for color representation of the figures in this book.

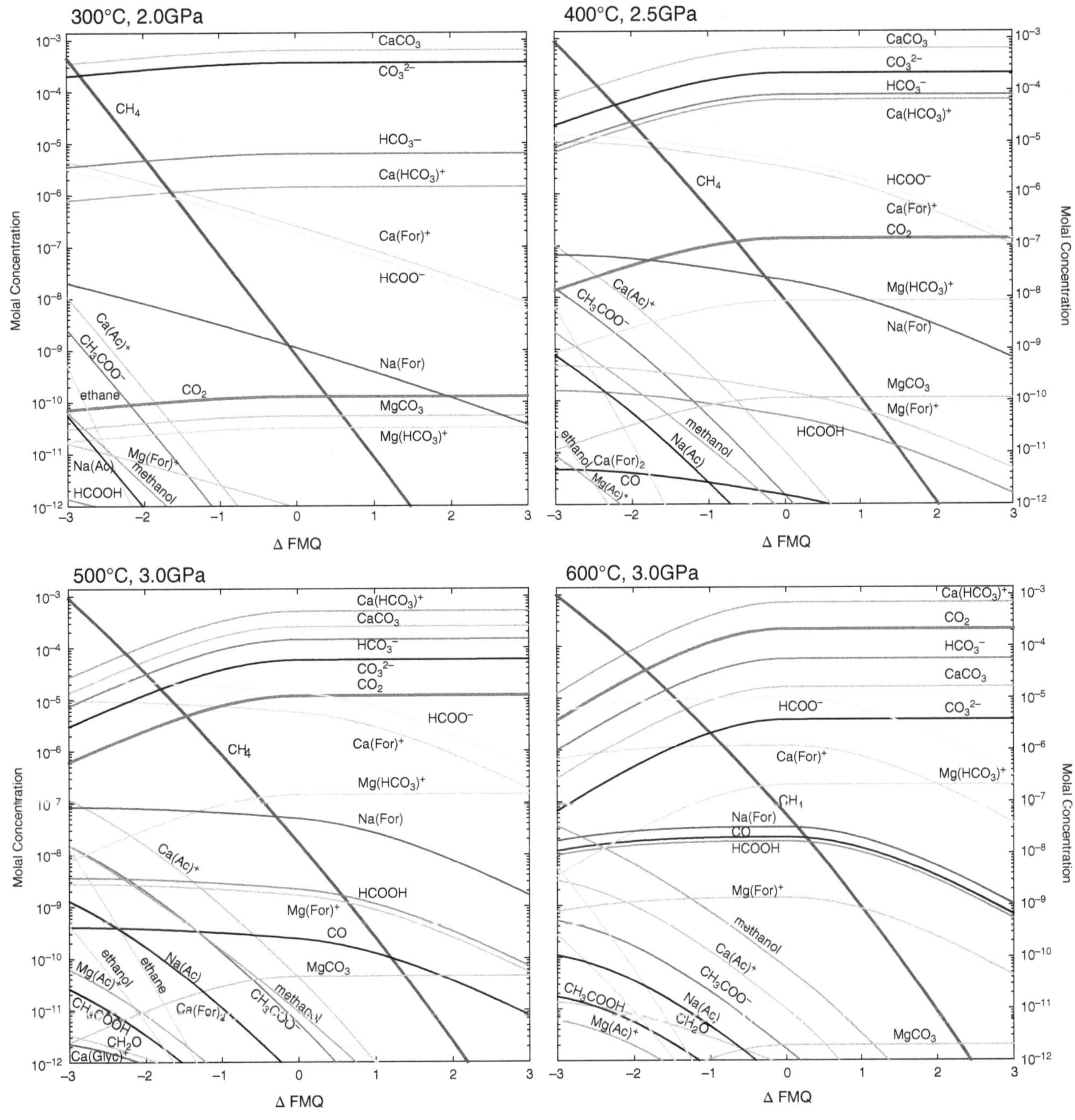

Figure 24.7 Equilibrium concentrations of aqueous C_1 and C_2 species as functions of ΔFMQ at isothermal and isobaric steps along the subduction path shown in Figure 24.1. Results fill gaps among the plots shown in Figure 24.6. Colors and abbreviations are consistent between figures. See electronic version for color representation of the figures in this book.

observations are consequences of the interplay between oxidation states set by mineral equilibria, Z_C of the various aqueous carbon species, and the shifts in equilibrium constants with temperature and pressure.

The contrast between results at 300 °C and 400 °C in Figure 24.7 sets precedents for observable differences in the plots for 500 °C and 600 °C. Aqueous CO_2 climbs up in a ranking of the relative abundances of oxidized carbon species with increasing temperature, again because of shifts in pH and cation abundances. Increasing curvature is evident for the fully oxidized carbon species and formate-bearing compounds throughout. Even the curves showing predicted abundances of aqueous CH_4 show curvature as temperature increases, which results from the interplay of stabilities of aqueous CH_4 and the other reduced forms of carbon, including the C_2 species. Note, however, that the C_2 species are comparatively more abundant in the results for 400 °C and 500 °C than they are at 600 °C, which is a consequence of the shift to rock-dominated conditions that are more oxidized overall at the highest temperature relative to the stability of carbon-bearing compounds. To summarize, it is *not* that C_2 and other reduced-carbon species are inherently less stable than C_1 compounds at higher temperatures, but the fact that rock-influenced oxidation states shift to relatively more oxidized conditions, allowing fully oxidized forms of carbon to reach higher abundances than their reduced counterparts as temperature increases.

24.6. DISEQUILIBRIUM AND ITS EFFECTS ON THE SPECIATION OF AQUEOUS CARBON

The results summarized in Figures 24.5 through 24.7 apply to systems that have reached stable thermodynamic equilibrium. Therefore, the predicted concentrations of organic compounds are all *minimum* values, and there is no thermodynamic drive for them to be lower than shown. However, in some settings, invoking stable equilibrium may not be valid. Disequilibrium between CO_2 and CH_4 is observed in crustal fluids and is more pronounced at temperatures below 500 °C (e.g. Charlou et al., 1998, 2000; Janecky & Seyfried, 1986; Manning et al., 2013; McCollom & Seewald 2001; McDermott et al., 2015; Shock, 1988, 1989, 1990, 1992, 1994; Shock & Schulte, 1998; Shock et al., 2019; Wang et al., 2015, 2018). In order for CO_2 and CH_4 to reach equilibrium (i.e. react reversibly), a kinetic barrier, involving breaking bonds and transferring eight electrons, must be overcome (Manning et al., 2013). Stable equilibrium between CO_2 and CH_4 may only be attained at high pressures and temperature (Manning et al., 2013), but constraints on kinetic rates at these conditions are lacking.

The conditions where metastable equilibrium states can form in the aqueous C-O-H system correspond to low-grade metamorphic conditions and shallow, cold subduction zone paths (e.g. Tonga, N. Marianas; Figure 24.1). Therefore, it is possible that reaction pathways from dissolved inorganic carbon compounds to small organic solutes are available during subduction, even though complete reduction to methane is inhibited. If metastability persists, even transiently, along cooler P-T paths of subduction zones, then fluids in such systems could contain aqueous organic solutes produced abiotically during subduction. We simulated the presence of kinetic barriers to methane production by removing $CH_{4(aq)}$ from our thermodynamic database in a set of calculations. These results illustrate the potential for greater diversity in speciation of carbon in subduction zone fluids under conditions that permit disequilibria.

Suppressing methane to simulate metastability leaves the speciation of aqueous carbon for the FMQ+3 shown in Figure 24.6 unaffected, as shown in Figure 24.8. The only visible difference at FMQ0 is the lack of a curve showing $CH_{4(aq)}$ concentrations in Figure 24.8. Because the amount of carbon represented by $CH_{4(aq)}$ at stable equilibrium is quite small, the other curves are not dramatically affected by its absence from the model. In contrast, suppression of aqueous CH_4 to simulate metastable equilibrium at the most reduced conditions (FMQ-3) drives all concentrations of organic and inorganic aqueous carbon species higher, and several species appear in the plot in Figure 24.8 that are not present in the corresponding plot in Figure 24.6. Furthermore, the slopes in the FMQ-3 plot in Figure 24.8 tend to be steeper, especially at temperatures above 400 °C, where $CH_{4(aq)}$ dominates the stable equilibrium speciation shown in Figure 24.6. The changes in both the position and slopes of the curves at FMQ-3 for metastable equilibrium compared with stable equilibrium are explained by the redistribution of the dissolved carbon from $CH_{4(aq)}$ to other species.

Comparison of the FMQ-3 plots in Figures 24.6 and 24.8 also shows that the diversity of carbon species is greater in the metastable equilibrium case. In the absence of $CH_{4\ (aq)}$, the speciation of the fully oxidized forms of aqueous carbon (i.e. $CaCO_3$, CO_3^{2-}, HCO_3^-, $Ca(HCO_3)^-$) in the FMQ-3 plot have the same speciation pattern observed at FMQ0 in either the stable (Figure 24.6) or metastable (Figure 24.8) examples. Comparison of the plots for FMQ-3 and FMQ0 in Figure 24.8 reveals that the curves from formate-bearing aqueous species plot at higher concentrations at FMQ-3. Note that in this plot the $Ca(For)^+$ and formate curves are superimposed upon the bicarbonate curve at low temperatures (below 400 °C), and that the $Ca(For)^+$ and HCO_3^- curves cross one another three times as temperature increases. Formate is predicted to be more abundant than $Ca(HCO_3)^-$ at temperatures <~450 °C, and nearly as abundant as $Ca(HCO_3)^-$ between 450 °C and 500 °C. $Ca(Ac)^+$ concentrations, which are calculated to exceed those of aqueous CO_2 up to ~525 °C, maximize in the range of 10^{-5} m, representing a two-orders-of-magnitude positive shift in concentration relative to the equilibrium case. All acetate-bearing species show increasing concentrations with increasing temperature at FMQ-3, and all but acetic acid (CH_3COOH) maximize above 550 °C, owing to the shift in pH relative to the speciation of acetic acid. Methanol and ethanol are predicted to increase in abundance with increasing temperature at FMQ-3, with the calculated abundance of ethanol approaching that of methanol at higher temperatures. Calculated abundances of aqueous ethane rival those of aqueous CO_2 at lower temperatures and continue to increase at the highest temperatures

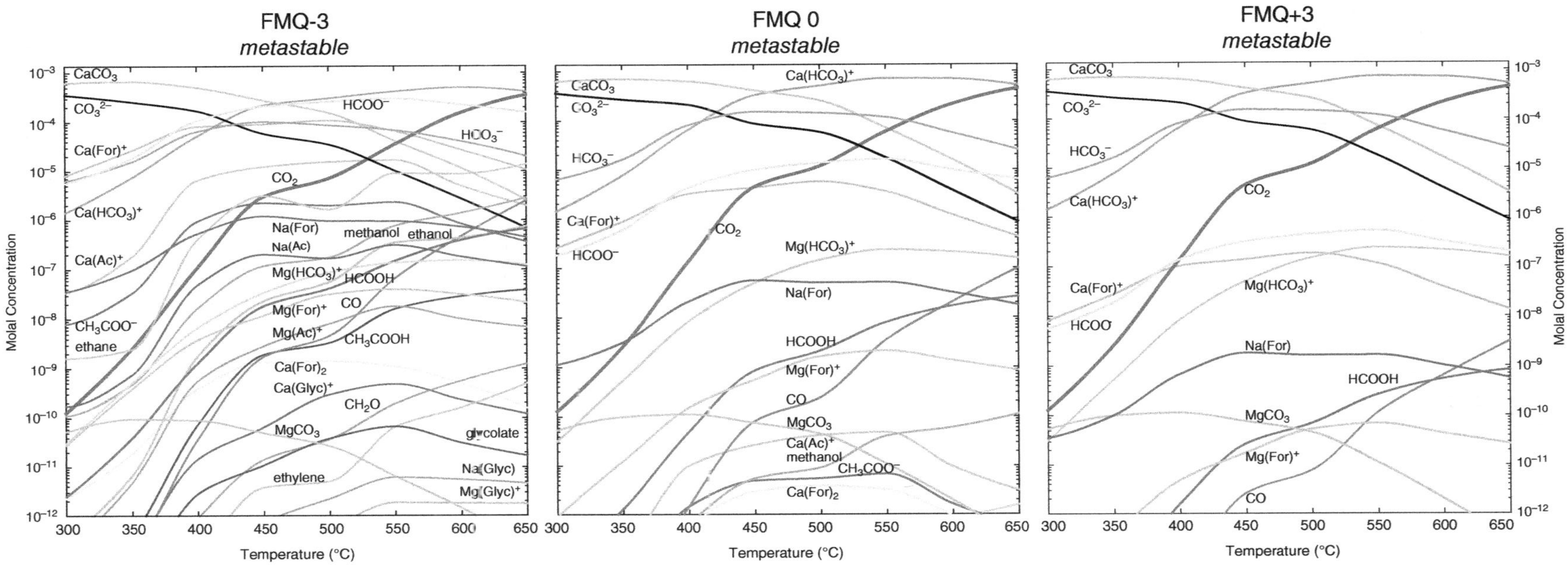

Figure 24.8 Metastable speciation calculations including C_1 and C_2 species for the P-T path outlined in Figure 24.1. Methane ($CH_{4(aq)}$) is suppressed in these calculations to simulate metastability. pH of calculations fixed by antigorite-forsterite-diopside activities. Calculations performed at FMQ-3, FMQ0, and FMQ+3. Colors and abbreviations are consistent between figures. See electronic version for color representation of the figures in this book.

shown, despite the overall shift to relatively more oxidizing conditions favoring CO_2. These same conditions allow species with Z_C values of +1 (i.e. glycolate, Na(Glyc), $Ca(Glyc)^+$, and $Mg(Glyc)^+$) to appear at concentrations greater than 10^{-12} m. Calculated $Ca(Glyc)^+$ and glycolate concentrations maximize at 550 °C, with similarly shaped curves, while Na(Glyc) and $Mg(Glyc)^+$ plateau at 550 °C and maintain concentrations of 4×10^{-11} m and 2×10^{-11} m, respectively, as temperature increases.

The effects of temperature and fO_2 on metastable distributions of aqueous carbon species can be explored further with the plots in Figure 24.9, which shows metastable companions to the stable equilibrium plots in Figure 24.7. Comparison of the plots in Figures 24.7 and 24.9 reveals that the relative positions among the fully oxidized carbon species are the same at the selected temperatures, as are the relative positions among the formate-bearing species and among the acetate-bearing

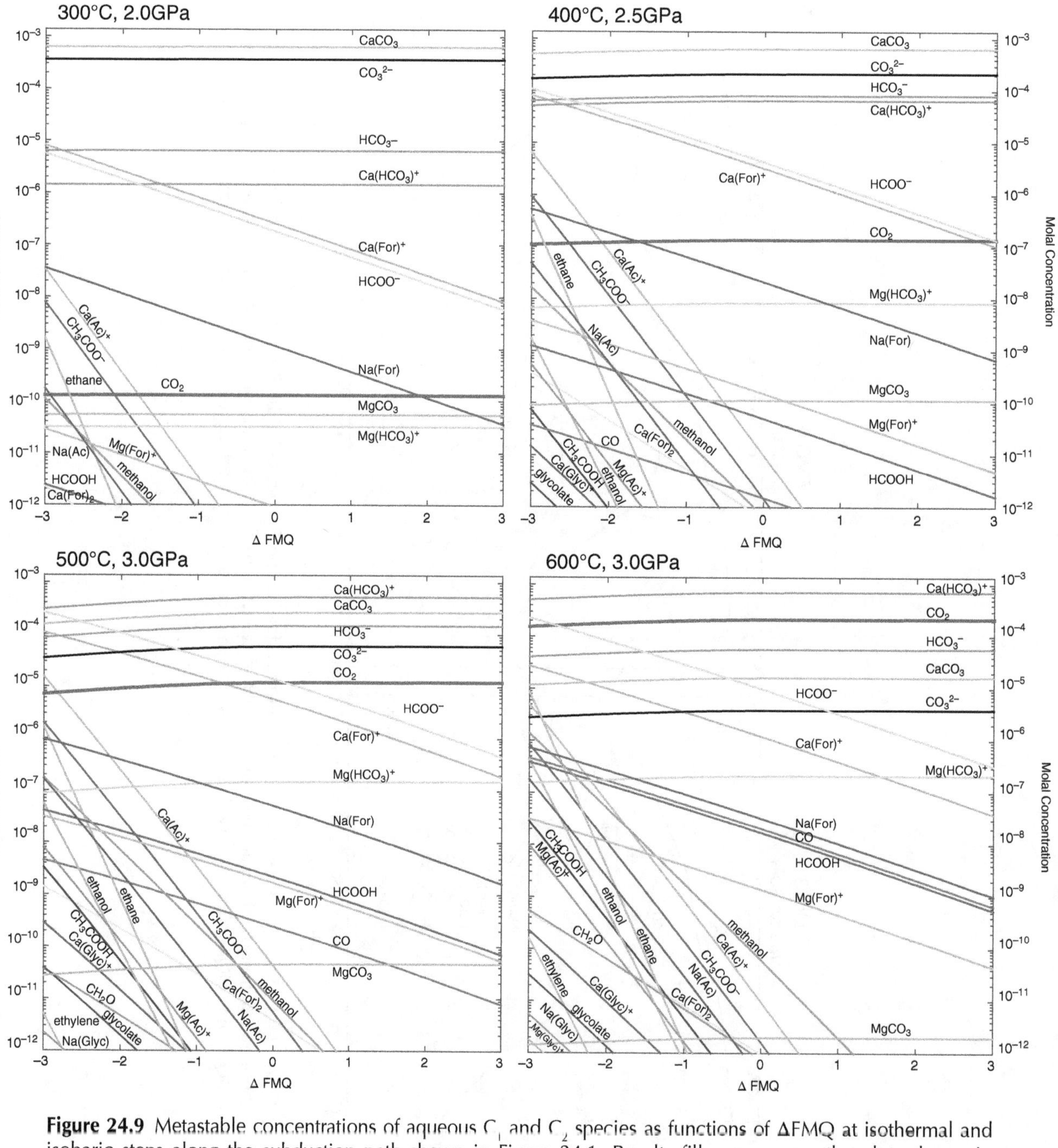

Figure 24.9 Metastable concentrations of aqueous C_1 and C_2 species as functions of ΔFMQ at isothermal and isobaric steps along the subduction path shown in Figure 24.1. Results fill gaps among the plots shown in Figure 24.8. Colors and abbreviations are consistent between figures. See electronic version for color representation of the figures in this book.

species. Unaffected by the suppression of methane, the relative positions of the curves are maintained despite an increase in total predicted abundances. At 300 °C, the distribution of fully oxidized carbon species is predicted to be virtually independent of fO_2 in the metastable equilibrium case. The suppression of aqueous methane formation eliminates the curvature shown in the 300 °C plot in Figure 24.7. The corresponding curves in the plots for higher temperatures in Figure 24.9 exhibit small amounts of curvature with decreasing fO_2, corresponding to the increasing abundances of various formate-bearing species. Note that the elimination of aqueous methane from the model also means that the diagonal curves showing calculated concentrations of the various formate-bearing species show negligible curvature in Figure 24.9. In all cases, the relative abundances of C_2 and reduced C_1 compounds are calculated to be greater in the metastable state than in the stable equilibrium state shown in Figure 24.7. In addition, there is a greater variety of aqueous carbon species in the metastable equilibrium results than in the stable equilibrium results, which is expected based on the greater variety of compounds shown at FMQ-2 in Figure 24.8 compared to the corresponding plot in Figure 24.6. In summary, disequilibrium conditions inhibiting the path to aqueous CH_4 at reduced conditions (FMQ-3) enhances the aqueous organic carbon speciation, including a strong thermodynamic drive to stabilize ~20 aqueous organic carbon species.

24.7. CONCLUDING REMARKS

In conceptual models of subduction, aqueous carbon-bearing fluids are given critical jobs like triggering partial melting (Dasgupta, 2013), oxidizing the mantle wedge (Evans, 2012; Frost & McCammon, 2008), mantle metasomatism, and diamond formation (Frost & McCammon 2008; Sverjensky, Stagno, et al., 2014), and all of these are the motivations for the predictions described above. Some efforts have been made to model C-O-H systems by using a mixture of neutral gases, including CO_2, CH_4, and H_2O, without consideration of aqueous ions or organic compounds (Connolly, 2005; Zhang & Duan, 2009). In reality, these supercritical aqueous fluids contain ions and neutral species whose speciation is dependent upon the coexisting mineral assemblage (i.e. the rock). Researchers are beginning to explore these systems through theoretical and experimental studies (e.g. Dvir et al., 2011, 2013; Facq et al., 2014, 2016; Huang et al., 2017; Kessel et al., 2005, 2015; Sverjensky, Stagno, et al., 2014); however, much of the accessible pressure-temperature-compositional space remains unexplored.

The mineral assemblage coexisting with the fluid controls, to some extent, the bulk composition of the fluid and buffers the pH and fO_2 of the fluid, especially in the subduction setting where the fluid-to-rock ratio is low (<<1). From a petrologic perspective, the FMQ mineral buffer is often assumed to be a constant frame of reference; however, from the perspective of the fluid, the meaning of FMQ changes as a function of temperature and pressure. An example of this shifting frame of reference for the fluid can be seen in Figure 24.7. At FMQ0 in the 300 °C panel, the concentration of aqueous CH_4 is higher than aqueous CO_2 as it is in the simple speciation shown in Figure 24.3. This has to be the case at equilibrium regardless of how many additional aqueous species are considered in Figure 24.7. Nevertheless, it is useful to see in Figure 24.7 that many oxidized forms of carbon are more abundant than aqueous CH_4 at FMQ0, which is not the impression left by Figure 24.3. Examining the other panels in Figure 24.7 reveals that at FMQ0 the concentration of aqueous CO_2 is higher than that of aqueous CH_4, reflecting the trend of carbon oxidation in the fluid as temperature increases. This change in meaning of FMQ from the fluid perspective is also reflected in numerous other changes in carbon speciation. Similarly, when the pH is buffered by a mineral assemblage, it imposes changes in fluid speciation because the mineral-buffered pH alters the speciation of carbonic, formic, acetic, and other acids, which also affects their complexes. As demonstrated in the predictions outlined above, metal-ligand complexes can be dominant forms of carbon in subduction fluids, raising the possibility that such complexes may be intimately involved in metasomatism. Small shifts in pH and fO_2 can have dramatic consequences in the speciation of aqueous carbon, as demonstrated by the results summarized above.

The calculations presented here capture only a small range within the accessible pressure-temperature-composition regime. It is our hope that the framework and methodology presented here will empower others to investigate rock-fluid interactions along pressure-temperature paths, with fluid compositions, mineral assemblages, and fO_2 values that reflect their interests in calculations or experiments. Future directions may include

- In-depth interrogations of changes in oxidation state that can be recorded in mineral assemblages sampled from subduction zones. This study reveals that small changes in fO_2 may result in dramatic changes in carbon speciation in aqueous fluids. Each time pressure-temperature-fO_2 results are obtained from petrologic studies, they can be used to evaluate the composition of the fluid that was simultaneously present. This can be particularly useful as primary fluids are rarely pristine, if preserved at all.
- A renewed search for organic compounds in fluid inclusions and at grain boundaries of rock samples from subduction zones. Calculations of the type presented here provide guidance to find the samples with the greatest potential for organic constituents and can provide prioritized lists of species to target. It should be kept in mind that the present study included only C_1 and C_2

organic compounds, and none containing nitrogen or sulfur. Vastly expanding predictions for the DEW model could further transform our appreciation for the complexity of subduction zone fluid compositions.

• Consideration of the influence of metastable states on the potential for abiotic organic synthesis in subduction zones, which could represent nutrients for the deepest parts of the biosphere that may reside in shallow portions of subduction zones.

• New models might include interplay among Fe, S, and C species in fluids, leading to quantitative tests of which redox-sensitive aqueous species are the most effective oxidizing agents in diverse subduction zone settings.

ACKNOWLEDGMENTS

We thank James Leong, Tucker Ely, and Dimitri Sverjensky for their help with coding, databases, and interpretation of the results generated in this study. Thank you to Christy Till, EPIC, and GEOPIG for helpful discussions during every iteration of this work. This work was supported in part with funds from the Sloan Foundation through grants in support of the Extreme Physics and Chemistry community of the Deep Carbon Observatory.

REFERENCES

Abers, G. A., van Keken, P. E., Kneller, E. A., Ferris, A., & Stachnik, J. C. (2006). The thermal structure of subduction zones constrained by seismic imaging: Implications for slab dehydration and wedge flow. *Earth. Planet. Sci. Lett.*, *241*, 387–397. https://doi.org/10.1016/j.epsl.2005.11.055

Bénard, A., Klimm, K., Woodland, A. B., Arculus, R. J., Wilke, M., Botcharnikov, R. E., et al. (2018). Oxidising agents in sub-arc mantle melts link slab devolatilisation and arc magmas. *Nat. Commun.*, *9*(3500). https://doi.org/10.1038/s41467-018-05804-2

Birner, S. K., Cottrell, E., Warren, J. M., Kelley, K. A., & Davis, F. A. (2018). Peridotites and basalts reveal broad congruence between two independent records of mantle fO_2 and despite local redox heterogeneity. *Earth. Planet. Sci. Lett.*, *494*, 172–189. https://doi.org/10.1016/j.epsl.2018.04.035

Brandon, A. D., & Draper, D. S. (1996). Constraints on the origin of the oxidation state of mantle overlying subduction zones: An example from Simcoe, Washington, USA. *Geochim. Cosmochim. Acta*, *60*, 1739–1749. https://doi.org/10.1016/0016-7037(96)00056-7

Brounce, M., Kelley, K. A., Cottrell, E., & Reagan, M. K. (2015). Temporal evolution of mantle wedge oxygen fugacity during subduction initiation. *Geology*, *43*(9), 775–778. https://doi.org/10.1130/G36742.1

Vitale Brovarone, A., Martinez, I., Elmaleh, A., Compagnoni, R., Chaduteau, C., Ferraris, C., & Esteve, I. (2017). Massive production of abiotic methane during subduction evidenced in metamorphosed ophicarbonates from the Italian Alps. *Nat. Commun.*, *8*, 14134. https://doi.org/10.1038/ncomms14134

Bryndzia, L. T., & Wood, B. J. (1990). Oxygen thermobarometry of abyssal spinel peridotites: The redox state and C-O-H volatile composition of the Earth's sub-oceanic upper mantle. *Am. J. Sci.*, *290*, 1093–1116. doi:10.2475/ajs.290.10.1093

Caciagli, N., & Manning, C. E. (2003). The solubility of calcite in water at 6–16 kbar and 500–800 °C. *Contrib. Mineral. Petrol.*, *146*, 275–285. https://doi.org/10.1007/s00410-003-0501-y

Cerpa, N. G., Wada, I., & Wilson, C. R. (2017). Fluid migration in the mantle wedge: Influence of mineral grain size and mantle compaction. *J. Geophys. Res. Solid Earth*, *122*(8), 6247–6268. https://doi.org/10.1002/2017JB014046

Charlou, J. L., Donval, J. P., Douville, E., Jean-Baptiste, P., Radford-Knoery, J., Fouquet, Y., et al. (2000). Compared geochemical signatures and the evolution of Menez Gwen (37°50′N) and Lucky Strike (37°17′N) hydrothermal fluids, south of the Azores Triple Junction on the Mid-Atlantic Ridge. *Chem Geol.*, *171*(1–2), 49–75. https://doi.org/10.1016/S0009-2541(00)00244-8

Charlou, J. L., Fouquet, Y., Bougault, H., Donval, J. P., Etoubleau, J., Jean-Baptiste, P., et al. (1998). Intense CH4 plumes generated by serpentinization of ultramafic rocks at the intersection of the 15°20′N fracture zone and the Mid-Atlantic Ridge. *Geochim. Cosmochim. Acta*, *62*(13), 2323–2333. https://doi.org/10.1016/S0016-7037(98)00138-0

Connolly, J.A.D. (2005). Computation of phase equilibria by linear programming: A tool for geodynamic modeling and its application to subduction zone decarbonation. *Earth. Planet. Sci. Lett.*, *236*(1–2), 524–541. https://doi.org/10.1016/j.epsl.2005.04.033

Dasgupta, R. (2013). Ingassing, storage, and outgassing of terrestrial carbon through geologic time. *Rev. Mineral Geochem.*, *75*(1), 183–229. https://doi.org/10.2138/rmg.2013.75.7

Dvir, O., Pettke, T., Fumagalli, P., & Kessel, R. (2011). Fluids in the peridotite-water system up to 6 GPa and 800 °C: New experimental constrains on dehydration reactions. *Contributions to Mineralogy and Petrology*, *161*(6), 829–844. https://doi.org/10.1007/s00410-010-0567-2

Dvir, O., Angert, A., & Kessel, R. (2013). Determining the composition of C-H-O liquids following high-pressure and high-temperature diamond-trap experiments. *Contributions to Mineralogy and Petrology*, *165*(3), 593–599. https://doi.org/10.1007/s00410-012-0825-6

Evans, K. A. (2012). The redox budget of subduction zones. *Earth-Sci. Rev.*, *113*(1–2), 11–32. https://doi.org/10.1016/j.earscirev.2012.03.003

Facq, S., Daniel, I., Montagnac, G., Carbon, H., & Sverjensky, D. A. (2014). In situ Raman study and thermodynamic model of aqueous carbonate speciation in equilibrium with aragonite under subduction zone conditions. *Geochim. Cosmochim. Acta*, *132*, 375–390. https://doi.org/10.1016/j.gca.2014.01.030

Facq, S., Daniel, I., Montagnac, G., Cardon, H., & Sverjensky, D. A. (2016). Carbon speciation in saline solutions in equilibrium with aragonite at high pressure. *Chem. Geol.*, *431*, 44–53. https://doi.org/10.1016/j.chemgeo.2016.03.021

Frezzotti, M. L., Selverstone, J., Sharp, Z. D., & Compagnoni, R. (2011). Carbonate dissolution during subduction revealed by diamond-bearing rocks from the Alps. *Nat. Geosci.*, *4*, 703–706. https://doi.org/10.1038/ngeo1246

Frost, D. J., & McCammon, C. A. (2008). The redox state of Earth's mantle. *Annu. Rev. Earth Planet Sci.*, *36*, 389–420. https://doi.org/10.1146/annurev.earth.36.031207.124322

Fryer, P., Pearce, J. A., Stokking, L. B., et al. (1990). *Proc. ODP, Init. Repts.*, *125*, 1092. College Station, TX (Ocean Drilling Program). https://doi.org/10.2973/odp.proc.ir.125.1990

Fumagalli, P., & Poli, S. (2005). Experimentally determined phase relations in hydrous peridotites to 6.5 GPa and their consequences on the dynamics of subduction zones. *J. Petrol.*, *46*(3), 555–578. https://doi.org/10.1093/petrology/egh088

Galvez, M. E., Connolly, J.A.D., & Manning, C. E. (2016). Implication for metal and volatile cycles from the pH of subduction zone fluids. *Nature*, *539*, 420–424. https://doi.org/10.1038/nature20103

Grove, T. L., N. Chatterjee, S. W. Parman, & Medard, E. (2006). The influence of H_2O on mantle wedge melting, *Earth Planet. Sci. Lett.*, *249*(1–2), 74–89. https://doi.org/10.1016/j.epsl.2006.06.043

Hacker, B. R. (2008). H_2O subduction beyond arcs, *Geochem. Geophys. Geosyst.*, *9*, Q03001. https://doi.org/10.1029/2007GC001707

Hermann, J., Spandler, C., Hack, A., & Korsakov, A. V. (2006). Aqueous fluids and hydrous melts in high-pressure and ultra-high pressure rocks: Implications for element transfer in subduction zones. *Lithos*, *92*(3–4), 399–417. https://doi.org/10.1016/j.lithos.2006.03.055

Huang, F., Daniel, I., Cardon, H., Montagnac, G., & Sverjensky D. A. (2017). Immiscible hydrocarbon fluids in the deep carbon cycle. *Nat. Commun.*, *8*, 15798. https://doi.org/10.1038/ncomms15798

Janecky, D. R., & Seyfried, W.E. (1986). Hydrothermal serpentinization of peridotite within the oceanic-crust: Experimental investigations of mineralogy and major element chemistry. *Geochim. Cosmochim. Acta*, *50*(7), 1375–1378. https://doi.org/10.1016/0016-7037(86)90311-X

Jugo, P. J., Wilke, M., & Botcharnikov, R. E. (2010). Sulfur K-edge XANES analysis of natural and synthetic basaltic glasses: Implications for S speciation and S content as function of oxygen fugacity. *Geochim. Cosmochim. Acta*, *74*(20), 5926–5938. https://doi.org/10.1016/j.gca.2010.07.022

Kelemen, P. B., & Manning, C. E. (2015). Reevaluating carbon fluxes in subduction zones, what goes down, mostly comes up. *Proc. Natl Acad. Sci.*, *112*(30), E3997–E4006. https://doi.org/10.1073/pnas.1507889112

Kelley, K. A., & Cottrell, E. (2009). Water and the oxidation state of subduction zone magmas. *Science*, *325*(5940), 605–607. https://doi.org/10.1126/science.1174156

Kessel, R., Pettke, T., & Fumagalli, P. (2015). Melting of metasomatized peridotite at 4–6 GPa and up to 1200 °C: An experimental approach. *Contributions to Mineralogy and Petrology*, *169*(4), 37. https://doi.org/10.1007/s00410-015-1132-9

Kessel, R., Schmidt, M. W., Ulmer, P., & Pettke, T. (2005). Trace element signature of subduction-zone fluids, melts and supercritical liquids at 120–180 km depth. *Nature*, *437*(7059), 724–727. https://doi.org/10.1038/nature03971

Kimura, G., Silver, E. A., Blum, P., et al. (1997). *Proc. ODP, Init. Repts.*, *170*, 458. College Station, TX (Ocean Drilling Program). https://doi.org/10.2973/odp.proc.ir.170.1997

Lee, C.T.A., Leeman, W. P., Canil, D., & Li, Z.X.A. (2005). Similar V/Sc systematics in MORB and arc basalts: Implications for the oxygen fugacities of their mantle source regions. *J. Petrol.*, *46*(11), 2313–2336. https://doi.org/10.1093/petrology/egi056

Lee, C.T.A., Luffi, P., Le Roux, V., Dasgupta, R., Albarède, F., & Leeman, W.P. (2010). The redox state of arc mantle using Zn/Fe systematics. *Nature*, *468*, 681–685. https://doi.org/10.1038/nature09617

Li, Y. (2017). Immiscible C-H-O fluids formed at subduction zone conditions. *Geochem. Perspective Letters*, *3*(1), 12–21. https://doi.org/10.7185/geochemlet.1702

Manning, C. E. (2004). The chemistry of subduction-zone fluids. *Earth Planet. Sci. Lett.*, *223*(1–2), 1–16. https://doi.org/10.1016/j.epsl.2004.04.030

Manning, C. E., Shock, E. L., & Sverjensky, D. A. (2013). The chemistry of carbon in aqueous fluids at crustal and upper mantle conditions: Experimental and theoretical constraints. In R. M. Hazen, J. A. Baross, A. P. Jones (Eds.), *Carbon in Earth, Reviews in Mineralogy & Geochemistry* (Vol. 75, pp. 109–148). Mineralogical Society of America.

McCollom, T. (2008). Observational, experimental, and theoretical constraints on carbon cycling in mid–ocean ridge hydrothermal systems. In R. P. Lowell, J. S. Seewald, A. Metaxas, M. R. Perfit (Eds.), *Magma to Microbe: Modeling Hydrothermal Processes at Ocean Spreading Centers, Geophysical Monograph Series*, (Vol. 178, pp. 193–213). Washington, DC: American Geophysical Union.

McCollom, T. M., & Seewald, J. S. (2001). A reassessment of the potential for reduction of dissolved CO_2 to hydrocarbons during serpentinization of olivine. *Geochim. Cosmochim. Acta*, *65*(21), 3769–3778. https://doi.org/10.1016/S0016-7037(01)00655-X

McCollom, T. M., & Seewald, J. S. (2003a). Experimental constraints on the hydrothermal reactivity of organic acids and acid anions: I. Formic acid and formate. *Geochim. Cosmochim. Acta.*, *67*(19),3625–3644.https://doi.org/10.1016/S0016-7037(03)00136-4

McCollom, T. M., & Seewald, J. S. (2003b). Experimental study of the hydrothermal reactivity of organic acids and acid anions. II. Acetic acid, acetate, and valeric acid. *Geochim. Cosmochim. Acta*, *67*(19), 3645–3664. https://doi.org/10.1016/S0016-7037(03)00135-2

McDermott, J. M., Seewald, J. S., German, C. R., & Sylva, S. P. (2015). Pathways for abiotic organic synthesis at submarine hydrothermal fields. *Proc. Natl Acad. Sci.*, *112*(25), 7668–7672. https://doi.org/10.1073/pnas.1506295112

Parkinson, I. J., & Arculus, R. J. (1999). The redox state of subduction zones: Insights from arc-peridotites. *Chemical Geology*, *160*(4),409–423.https://doi.org/10.1016/S0009-2541(99)00110-2

Peacock, S. M. (2003). Thermal structure and metamorphic evolution of subducting slabs. In J. M. Eiler (Ed.), *Inside the Subduction Factory, Geophysical Monograph Series*, (Vol. *138*, pp. 7–22). Washington, DC: American Geophysical Union.

Penniston-Dorland, S. C., Kohn, M. J., & Manning, C. E. (2015). The global range of subduction zone thermal structures from exhumed blueschists and eclogites: Rocks are hotter than models. *Earth Planet. Sci. Lett.*, *428*, 243–254. https://doi.org/10.1016/j.epsl.2015.07.031

Rowe, M. C., Kent, A.J.R., & Nielsen, R. L. (2009). Subduction influence on oxygen fugacity and trace and volatile elements in basalts across the Cascade volcanic arc. *J. Petrol.*, *50*(1), 61–91. https://doi.org/10.1093/petrology/egn072

Scambelluri, M., & Philippot, P. (2001). Deep fluids in subduction zones. *Lithos*, *55*(1–4), 213–227. https://doi.org/10.1016/S0024-4937(00)00046-3

Schmidt, M. W., & Poli, S. (1998). Experimentally based water budgets for dehydrating slabs and consequences for arc magma generation. *Earth Planet. Sci. Lett.*, *163*(1–4), 361–379. https://doi.org/10.1016/S0012-821X(98)00142-3

Schulte, M. D., & Shock, E. L. (1993). Aldehydes in hydrothermal solution: Standard partial molal thermodynamic properties and relative stabilities at high temperatures and pressures. *Geochim. Cosmochim. Acta*, *57*(16), 3835–3846. https://doi.org/10.1016/0016-7037(93)90337-V

Seewald, J. S., Zolotov, M. Y., & McCollom, T. (2006). Experimental investigation of single carbon compounds under hydrothermal conditions. *Geochim. Cosmochim. Acta*, *70*(2), 446–460. https://doi.org/10.1016/j.gca.2005.09.002

Shock, E. L. (1988). Organic acid metastability in sedimentary basins. *Geology*, *16*(10), 886–890. https://doi.org/10.1130/0091-7613(1988)016<0886:OAMISB>2.3.CO;2

Shock, E. L. (1989). Corrections to "Organic acid metastability in sedimentary basins." *Geology*, *17*(6), 572–573. https://doi.org/10.1130/0091-7613(1989)017<0572:CTOAMI>2.3.CO;2

Shock, E. L. (1990). Geochemical constraints on the origin of organic compounds in hydrothermal systems. *Origins of Life and Evolution of the Biosphere*, *20*, 331–367.

Shock, E. L. (1992). Chemical environments in submarine hydrothermal systems. In N. Holm (Ed.), *Marine Hydrothermal Systems and the Origin of Life*, *Origins of Life and Evolution of the Biosphere (special issue)*, *22*, 67–107.

Shock, E. L. (1994). Application of thermodynamic calculations to geochemical processes involving organic acids. In M. Lewan & E. Pittman (Eds.), *The Role of Organic Acids in Geological Processes*, (pp. 270–318). Berlin, Heidelberg: Springer.

Shock, E. L. (1995). Organic acids in hydrothermal solutions: Standard molal thermodynamic properties of carboxylic acids and estimates of dissociation constants at high temperatures and pressures. *Am. J. Sci.*, *295*, 496–580. https://doi.org/10.2475/ajs.295.5.496

Shock, E., Bockisch, C., Estrada, C., Fecteau, K., Gould, I., Hartnett, H., et al. (2019). Earth as organic chemist. In B. Orcutt, I. Daniel, & R. Dasgupta (Eds.), *Whole Earth Carbon*. Cambridge University Press.

Shock, E., & Canovas, P. (2010). The potential for abiotic organic synthesis and biosynthesis at seafloor hydrothermal systems. *Geofluids*, *10*, 161–192. https://doi.org/10.1111/j.1468-8123.2010.00277.x

Shock, E. L., & Helgeson, H. C. (1990). Calculation of the thermodynamic and transport properties of aqueous species at high pressure and temperature: Standard partial molal properties of organic species. *Geochim. Cosmochim. Acta*, *54*(4), 915–945. https://doi.org/10.1016/0016-7037(90)90429-O

Shock, E. L., & Koretsky, C. M. (1993). Metal-organic complexes in geochemical processes: Calculation of standard partial molal thermodynamic properties of aqueous acetate complexes at high pressures and temperatures. *Geochim. Cosmochim. Acta*, *57*(20), 4899–4922. https://doi.org/10.1016/0016-7037(93)90128-J

Shock, E. L., & Koretsky, C. M. (1995). Metal-organic complexes in geochemical processes: Estimation of standard partial molal thermodynamic properties of aqueous complexes between metal cations and monovalent organic acid ligands at high pressures and temperatures. *Geochim. Cosmochim. Acta*, *59*(8), 1497–1532. https://doi.org/10.1016/0016-7037(95)00058-8

Shock, E. L., Oelkers, E. H., Johnson, J. W., Sverjensky, D. A., & Helgeson, H. C. (1992). Calculation of the thermodynamic and transport properties of aqueous species at high pressures and temperatures: Effective electrostatic radii to 1000 °C and 5 kb. *J. Chem. Soc., Faraday Trans.*, *88*, 803–826. https://doi.org/10.1039/FT9928800803

Shock, E. L., Sassani, D. C., Willis, M., & Sverjensky, D. A. (1997). Inorganic species in geologic fluids: Correlations among standard molal thermodynamic properties of aqueous ions and hydroxide complexes. *Geochim. Cosmochim. Acta*, *61*(5), 907–950. https://doi.org/10.1016/S0016-7037(96)00339-0

Shock, E. L., & Schulte, M. D. (1998). Organic synthesis during fluid mixing in hydrothermal systems. *J. Geophys. Res.*, *103*(12), 28513–28527. https://doi.org/10.1029/98JE02142

Sverjensky, D. A., Harrison, B., & Azzolini, D. (2014). Water in the deep Earth: The dielectric constant and the solubilities of quartz and corundum. *Geochim. Cosmochim. Acta*, *129*, 125–145. https://doi.org/10.1016/j.gca.2013.12.019

Sverjensky D. A., Shock, E. L., & Helgeson, H. C. (1997). Prediction of the thermodynamic properties of aqueous metal complexes to 1000 °C and 5 kb. *Geochim. Cosmochim. Acta*, *61*(7), 1359–1412. https://doi.org/10.1016/S0016-7037(97)00009-4

Sverjensky, D. A., Stagno, V., & Huang, F. (2014). Important role of organic carbon in subduction-zone fluids in the deep carbon cycle. *Nat. Geosci.*, *7*, 9–13.

Syracuse, E. M., van Keken, P. S., & Abers, G. A. (2010). The global range of subduction zone thermal models. *Phys. Earth Planet. In.*, *183*(1–2), 73–90. https://doi.org/10.1016/j.pepi.2010.02.004

Till, C. B., Grove, T., & Withers, A. C. (2012). The beginnings of hydrous mantle wedge melting. *Contrib. Min. Petrol.*, *163*(4), 669–688. https://doi.org/10.1007/s00410-011-0692-6

Ulmer, P., & Trommsdorff, V. (1995). Serpentine stability to mantle depths and subduction-related magmatism. *Science*, *268*(5212), 858–861. https://doi.org/10.1126/science.268.5212.858

van Keken, P. E., Kiefer, B., & Peacock, S. M. (2002). High-resolution models of subduction zones: Implications for mineral dehydration reactions and the transport of water into the deep mantle. *Geochem. Geophys. Geosys.*, *3*(10), 1056. https://doi.org/10.1029/2001GC000256

Wang, D. T., Gruen, D. S., Lollar, B. S., Hinrichs, K-U., Stewart, L. C., Holden, J. F., et al. (2015). Nonequilibrium clumped isotope signals in microbial methane. *Science*, *348*(6233), pp. 428–431. https://doi.org/10.1126/science.aaa4326

Wang, D. T., Reeves, E. P, McDermott, J. M., Seewald, J. S., Ono, S. (2018). Clumped isotopologue constraints on the origin of methane at seafloor hot springs. *Geochim. Cosmochim. Acta*, *223*, 141–158. https://doi.org/10.1016/j.gca.2017.11.030

Waters, L. E., & Lange, R. A. (2016). No effect of H_2O degassing on the oxidation state of magmatic liquids. *Earth Planet. Sci. Lett.*, *447*, 48–59. https://doi.org/10.1016/j.epsl.2016.04.030

Wolery, T. J., & Jarek, R. L. (2003). Software User's Manual: EQ3/6, Version 8.0. Software Document No. 10813-UM-8.0-00. Albuquerque, NM: Sandia National Laboratories.

Wood, B. J., Bryndzina, L. T., & Johnson, K. E. (1990). Mantle oxidation state and its relationship to tectonic environment and fluid speciation. *Science*, *248*(4953), 337–345. https://doi.org/10.1126/science.248.4953.337

Zhang, C., & Duan, Z. (2009). A model for C-H-O fluid in the Earth's mantle. *Geochim. Cosmochim. Acta.*, *73*(7), 2089–2102. https://doi.org/10.1016/j.gca.2009.01.021

25

Energetics of the Citric Acid Cycle in the Deep Biosphere

Peter A. Canovas, III[1,2] and Everett L. Shock[1,2,3,4]

ABSTRACT

Constraints on the internal composition of microbial cells are used together with standard state thermodynamic data to evaluate energy demands associated with the citric acid cycle and its individual steps to explore geobiochemical processes in the deep subsurface biosphere. Two pressure-temperature ranges are considered: up to 200 °C and to ~2 kb with the revised Helgeson-Kirkham-Flowers equations of state, and up to 200 °C from 10 to 60 kilobars with the Deep Earth Water model. The former pressure-temperature ranges encompass conditions for known life in natural systems, and the latter push the upper pressures for life into those of subduction zones using guidance from laboratory experiments. The neutral solutes $H_2(aq)$ and $CO_2(aq)$ can diffuse freely across cell membranes and impose external conditions into the composition of microbial cells, and as a result there is a range of chemical affinities for citric acid cycle reactions that prevail throughout the deep subsurface biosphere. These ranges raise the possibility that the citric acid cycle releases energy when run in either the forward or reverse directions depending on the affinities involved. The results of this theoretical study support the notion that life may extend far deeper into subduction zones than is generally appreciated.

25.1. PHYSICAL CONDITIONS OF THE DEEP BIOSPHERE

It is estimated that 10%–33% of the total number of living cells on Earth reside within the subsurface deep biosphere on the continents and in the seafloor (Colman et al., 2017; Hoshino & Inagaki, 2019; Inagaki et al., 2015; Ino et al., 2016; Kallmeyer et al., 2012; Magnabosco et al., 2018; Trembath-Reichert et al., 2017; Whitman et al., 1998). These estimates are limited by sampling, which means we are still just beginning to explore the extent, biodiversity, and potential for novel metabolic pathways in the deep biosphere. Our goal is to assist that exploration by providing a geochemical framework for integrating the deep biosphere into geologic processes to facilitate predictions. One way to approach this goal is to determine energetic demands of biochemical pathways and cycles that are met in the deep biosphere by geochemical energy sources. In this chapter, we combine our standard state thermodynamic data for aqueous species involved in the citric acid cycle (Canovas & Shock, 2016) with estimates of their intercellular abundances to evaluate thermodynamic affinities for reactions within the citric acid cycle at elevated pressures and temperatures. We also evaluate new parameters for the same aqueous species consistent with the Deep Earth Water (DEW) model (Sverjensky, Harrison, et al., 2014), which allows us to extrapolate the energetics of the citric acid cycle into conditions of shallow subduction zones. At present, there are no data to indicate that shallow portions of subduction zones are populated with microbes, but existing field and laboratory data indicate that temperatures and pressures in these systems are not necessarily inhibitory for life. We propose that shallow reaches of subduction zones represent the deepest habitats of the deep biosphere.

[1] School of Earth and Space Exploration, Arizona State University, Tempe, Arizona, USA

[2] Group Exploring Organic Processes in Geochemistry (GEOPIG), Arizona State University, Tempe, Arizona, USA

[3] School of Molecular Sciences, Arizona State University, Tempe, Arizona, USA

[4] Center for Fundamental and Applied Microbiomics, Arizona State University, Tempe, Arizona, USA

Carbon in Earth's Interior, Geophysical Monograph 249, First Edition. Edited by Craig E. Manning, Jung-Fu Lin, and Wendy L. Mao.

Abundant evidence suggests that extant microbial communities inhabit the subsurface to depths of at least 5 km with well-documented microbial communities flourishing at 4.2 km beneath the Earth's surface (Figure 25.1). Microbes are known to thrive at temperatures in excess of 100 °C at elevated pressures, and at least one strain lives at 121 °C (Takai et al., 2008). These results are summarized in Figure 25.1a, together with conversions of depth to pressure (Figure 25.1b), high-pressure evidence from lab experiments (Figure 25.1c), and the pressure-temperature framework of the present study (Figure 25.1d). Although not all reported depths can be converted to pressures owing to incomplete information in primary sources, many pressures can be estimated for the deep biosphere, and pressures used in laboratory microbial growth experiments extend well beyond those of currently explored subsurface habitats.

Depths and temperatures reported for subsurface samples of microbial life are summarized in Figure 25.1a, which also shows trajectories of crustal geothermal gradients appropriate for continental (a.k.a. terrestrial, 20 °C km^{-1}), oceanic (a.k.a. marine, 35 °C km^{-1}), and hydrothermal (100 °C km^{-1}) settings. Note that some temperatures where subsurface life thrives are <0 °C. While depth may be a physical barrier for microbes to overcome, it is not necessarily a physiochemical barrier. Therefore, pressures were estimated from depths for those cases where sufficient data were reported, as plotted in Figure 25.1b.

Studies of marine sediment and rock samples often include a seawater depth, which we converted to pressure, assuming that a 10.3 m column of seawater would exert 1 atm or 1.013 bar pressure. For terrestrial (continental) samples, pressure was calculated as overburden using either lithostatic or hydrostatic pressure, whichever was appropriate for the system. Overburden (P_{ob}, in Pa) was calculated as

$$P_{ob} = \rho * g * d, \tag{25.1}$$

where ρ represents density in kg m^{-3}, g indicates the gravitational constant 9.806 m s^{-2}, and d stands for the depth in meters. Equation (25.1) provides a close estimate of the pressure beyond that which the atmosphere is exerting on a sample at depth. If the density of a particular rock type was unavailable for a sampling site from literature sources, an estimated average density was used (i.e. 2.65 g cm^{-3} for granite, 2.55 g cm^{-3} for sandstone, and 0.92 g cm^{-3} for ice, etc.). Additional estimates were needed for in situ down-borehole pressures for some groundwater sample locations if the available information was limited to the depth and the artesian head. As an example, one sampling site came from a depth of 1270 m from an artesian well with a head of 466 ft or ~142 m, indicating the ability of the well pressure to sustain a column of water 142 m above the surface, and making the total pressure at the sampling locale that of a column of water under hydrostatic pressure with an apparent depth of 1412 m plus one bar of atmospheric pressure at that location or a total of ~140 bars.

Laboratory experiments confirming microbial growth summarized from the literature shown in Figure 25.1c were conducted at temperatures and pressures throughout and beyond those reached in known subsurface habitats. Presently, laboratory experiments show that microbes are capable of persisting to somewhat higher temperatures than where they are found in nature, as indicated by the highest temperatures in Figure 25.1c. Researchers have also shown that microbes can survive and thrive at pressures of tens of kilobars, which are considerably higher than those known from natural systems. These pressures are reached at several tens of kilometers and suggest that the deep biosphere may extend much deeper than previously thought.

Field and laboratory evidence depicted in Figures 25.1 a–c allows the pressure-temperature (P-T) reference frame for the deep biosphere shown in Figure 25.1d, in which all known habitable P-T space is encompassed in the gray zone. In the study summarized here, we extended the temperature range to 200 °C and pressures to 60 kilobars (6 GPa) to be inclusive. As shown in Figure 25.1d, it is not presently possible to conduct theoretical calculations over all of this P-T space, and the results shown below are truncated by the vapor-liquid saturation (boiling) curve for H_2O at low pressures and high temperatures, and by the liquid-ice phase boundary at high pressures and low temperatures.

We tackled thermodynamic calculations for the citric acid cycle in two stages. First, we assessed chemical affinities for the overall cycle and its component reactions at temperatures and pressures covered by the revised Helgeson-Kirkham-Flowers (HKF) equation of state (Shock et al., 1992), which permits calculations to 5 kb (0.5 GPa). Second, higher pressure calculations were accomplished by estimating an additional set of revised-HKF parameters for the DEW model (Sverjensky, Harrison, et al., 2014), and then completing the extrapolation of chemical affinities to higher pressures. These developments are described below, after the following discussion of the thermodynamic framework employed.

25.2. THERMODYNAMIC FRAMEWORK AND COMPUTATIONAL METHODS

A first step in assessing the thermodynamic viability of a process in the citric acid cycle is to calculate the standard Gibbs energy, ΔG_r^o, of the reactions involved, which relate to the equilibrium constants, K_r, via

$$\Delta G_r^o \equiv -RT \ln K_r, \tag{25.2}$$

where R represents the gas constant and T stands for temperature in Kelvin. Both standard Gibbs energies and

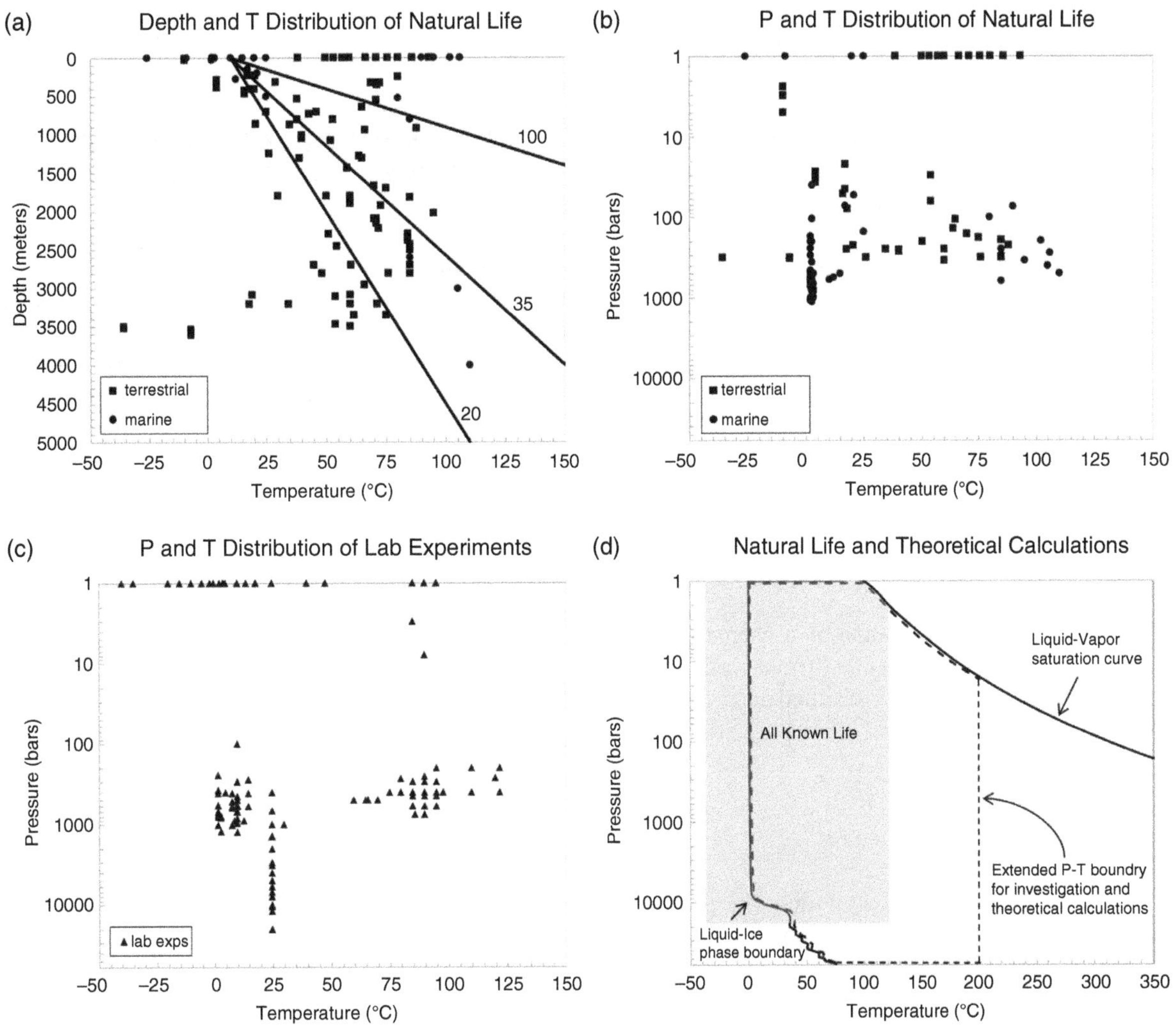

Figure 25.1 The known extent of the Earth's biosphere in depth, temperature, and pressure using evidence from natural systems and laboratory experiments, together with the ranges of thermodynamic calculations conducted in this study. (A) Subsurface habitats and geothermal gradients (in °C km^{-1}) with symbols showing the depths (excluding that of seawater) and temperatures of subsurface samples. Note that geotherms are populated to 4 km depth. (B) Result of converting depths from (A) into pressure, as possible. Many reported depths had no readily available pressure data correlated with them. Our inability to estimate pressures means that several samples shown in (A) were left out of (B). (C) Results of microbial growth laboratory experiments at elevated temperatures and pressures. Note that ranges of temperature and pressure in (C) exceed those in (B). (D) The temperature and pressure ranges of calculations conducted in this study (dashed lines), which encompass most of those established by studies of natural systems and laboratory experiments (gray zone). References for data and pressure estimates include Amend and Shock (2001), Apps (2010), Arcuri and Ehrlich (1977), Ask et al. (2001), Bakermans et al. (2006), Bale et al. (1997), Bartlett (2002), Bernhardt et al. (1988), Bonch-Osmolovskaya et al. (2003), Böning et al. (2004), Brazelton and Baross (2009), Breezee et al. (2004), Byers et al. (1998), Canganella et al. (1997), Carvalho (2013), Chivian et al (2008), Collins et al. (2010), D'Elia et al. (2008), Daumas et al. (1986, 1988), DeLong et al. (1997), Delwiche et al. (1996), Deming and Baross (1986), Deming et al. (1988), Ehrlich et al. (1972), Ekendahl and Pedersen (1994), Ellis and Ege (1975), Erauso et al. (1993), Fardeau et al. (2000), Fell (1967), Grabowski et al. (2005), Greene et al. (1997), Grossman and Shulman (1995), Haldeman et al. (1993), Havig et al. (2011), Inagaki et al. (2003), Johnson et al. (1992), Kaksonen, Plumb, et al. (2006), Kaksonen, Spring, et al. (2006), Kato et al. (1995, 1996, 1997, 1998), Kieft et al. (1999, 2005), Kimura et al. (2005), Klouche et al. (2007), Kotelnikova and Pedersen (1998), Kotelnikova et al. (1998), Kotlar et al. (2011), L'Haridon et al. (1995), H. Li et al. (2006, 2007), X. Li et al. (2010), Liesack et al. (1991), Liu et al. (1997), Loiacono et al. (2012), Love et al. (1993), Marteinsson et al. (1997, 1999, 2013), Meersman et al. (2013), Meyer-Dombard et al. (2011), Miller et al. (1988), Miyoshi et al. (2005), Mochimaru et al. (2007), Mori et al. (2002), Morita and ZoBell (1955), Motamedi and Pedersen (1998), Nakai et al. (2011), Nazina et al. (2007), Nilsen and Torsvik (1996), Nilsen, Torsvik, et al. (1996), Nilsen, Beeder et al. (1996), Nogi et al. (1998), Nunoura et al. (2005), Olson et al. (1981), Onstott et al. (1997, 1998), Panikov and Sizova (2007), Parkes et al. (1994), Pedersen (1997), Pedersen and Ekendahl (1990), Pope et al. (1975), Price and Sowers (2004), Rivkina et al. (2000), Rosnes et al. (1991), Russell et al. (1994), Sahl et al. (2008), Salamatin et al. (1998), Salisbury et al. (2002), Schippers and Neretin (2006), Scholander et al. (1965), Schwarz and Colwell (1975a, 1975b), Schwarz et al. (1975), Sharma et al. (2002), Shi et al. (1997), Shock (2009), Stetter et al. (1993), Steurer and Underwood (2003), Stevens et al. (1993), Stevens and McKinley (1995), Szewzyk et al. (1994, 1997), Takai, Moser, Onstott, et al. (2001), Takai, Moser, DeFlaun, et al. (2001), Takai et al. (2002, 2005, 2008), Tanaka et al. (2001), Tardy-Jacquenod et al. (1998), Tobal (1993), Trimarco et al. (2006), Turley (2000), Vanlint et al. (2011), Wanger et al. (2012), Winnock and Pontalier (1970), Wouters et al. (2013), Wynter et al. (1996), Yanagibayashi et al. (1999), Yayanos et al. (1981, 1982), Yayanos and Dietz (1982), Yayanos (1986), Yoshioka et al. (2009), Zhang et al. (2005, 2006), Zimov et al. (2006), Zink et al. (2003), ZoBell (1952, 1958), and ZoBell and Morita (1957). See electronic version for color representation of the figures in this book.

equilibrium constants are functions of temperature, pressure, and choice of standard state.* The overall Gibbs energy of a reaction, ΔG_r, can be calculated from

$$\Delta G_r = \Delta G_r^o + RT\,ln Q_r, \tag{25.3}$$

where Q_r stands for the activity product given by

$$Q_r = \prod_i a_i^{v_{i,r}}, \tag{25.4}$$

where a_i stands for the activity of the *i*th chemical species in the reaction and $v_{i,r}$ stands for the stoichiometric reaction coefficient of the *i*th chemical species in the *r*th reaction, which is positive for products and negative for reactants.

The overall Gibbs energy change of a chemical system (ΔG) depends on the progress of the various reactions possible in that system. Focusing on individual reactions is facilitated by evaluating their chemical affinities, A_r, defined as

$$A_r \equiv -\left(\frac{\partial \Delta G}{\partial \xi_r}\right)_{P,T,\xi_k}, \tag{25.5}$$

where ξ_r represents the progress of the *r*th reaction of interest and ξ_k stands for the progress of all other reactions in the system (Helgeson, 1979). According to equation (25.5), as a reaction proceeds and lowers the Gibbs energy of the system, its chemical affinity will be positive and reach zero at equilibrium. Negative affinity values mean that the reverse reaction is favored to proceed. The above relations can be combined to yield

$$A_r = RT\,ln\left(\frac{K_r}{Q_r}\right). \tag{25.6}$$

In practice, we evaluate equilibrium constants for reactions in the citric acid cycle with equations, data, and parameters from Canovas and Shock (2016), together with a new set of parameters developed in this study for the DEW model. Activity products are calculated from compositional constraints from natural or laboratory systems. In the present study, we used this approach to evaulate thermodynamic drives (a.k.a. chemical affinities) for individual steps in the citric acid cycle, as well as the overall cycle over ranges of temperature and pressure that prevail in subsurface environments.

*The aqueous solution standard state adopted in this study is a hypothetical one molal solution referenced to infinite dilution at any temperature and pressure. The standard state for gases is the pure gas at any temperature and 1 bar, that for liquid H_2O is the pure liquid at any temperature and pressure, and that for minerals is the pure crystalline solid at any temperature and pressure.

25.3. EVALUATING CHEMICAL AFFINITIES FOR THE CITRIC ACID CYCLE WITHIN MICROBIAL CELLS

Evaluating chemical affinities requires equilibrium constants and activity products, as demonstrated in the preceding discussion. The revised-HKF equations of state together with data and parameters provided by Canovas and Shock (2016) make it possible to estimate standard state thermodynamic data, including equilibrium constants, for species involved in the citric acid cycle as illustrated in Figure 25.2. The stepwise reactions indicated in Figure 25.2 are listed in Box 25.1, together with definitions of abbreviations. Standard Gibbs energies of these reactions are illustrated at elevated temperatures and up to 5 kb by Canovas and Shock (2016). It should be kept in mind that microbial metabolisms sometimes involve reactions as shown in Figure 25.2, and sometimes the reverse of these reactions. The traditional forward direction of the citric acid cycle reflects human metabolism. Also, we have adopted the convention common to biochemistry in which metabolic reactions are written as if involving only the fully dissociated ionic forms of acids and other compounds, which rarely reflects the actual speciation of these compounds at natural conditions whether inside or outside cells (Canovas & Shock, 2016).

In addition to equilibrium constants, activity products are also needed to evaluate chemical affinities, and to do so requires constraints on the abundances of all of the chemical species shown in Figure 25.2 and linked through the reactions listed in Box 25.1. Data for most of these compounds are scarce from natural environments, but many of their concentrations are constrained within microbial cells. However, these constraints are chiefly from *Escherichia coli* and not from any of the thermophiles or barophiles that thrive in the deep biosphere. Nevertheless, assuming that data from microbes provide better constraints than setting values to arbitrary ranges, we have adopted the data listed in Table 25.1 for the purpose of estimating a set of chemical affinities for the citric acid cycle within cells throughout the deep biosphere. Note that we list these values as apparent activities, as a reminder that we have yet to conduct a comprehensive speciation of these and other solutes within microbial cells. When that becomes possible, and when a greater variety of solute concentrations are obtained on a wider variety of organisms through advances in metabolomics, improvements in the calculated affinities of metabolic processes can be anticipated.

A few notes about the data in Table 25.1 are warranted. As indicated, most of these values come from studies of *Escherichia coli*, which is one of the most thoroughly studied bacteria and from which much has been learned about microbial metabolism and biochemical pathways. Nevertheless, we expect that additional investigations

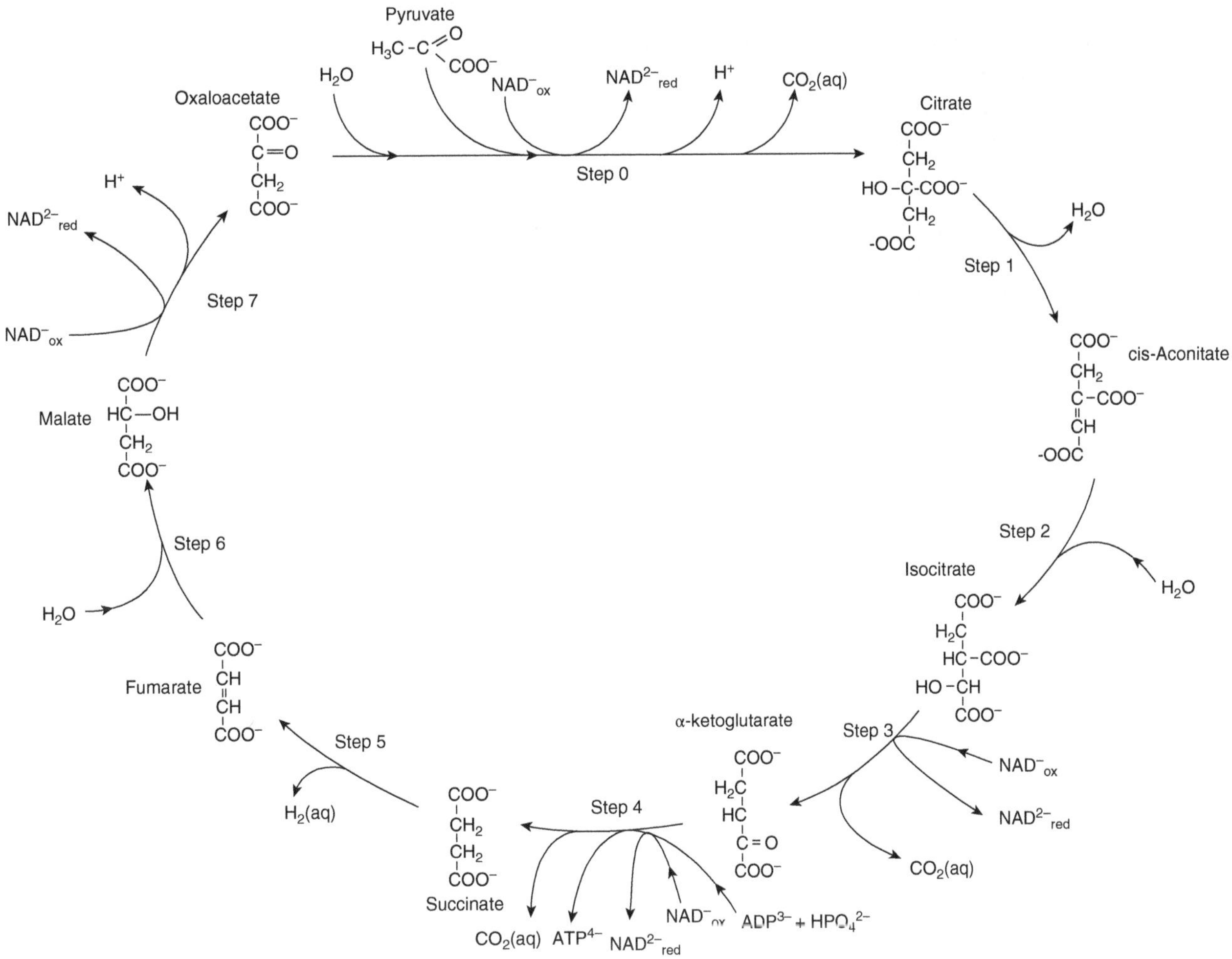

Figure 25.2 (after Canovas & Shock, 2016) Thermodynamic depiction of the citric acid cycle used to assess affinities associated with individual steps, as well as the affinities for the overall metabolic pathway. Some of the steps represent condensed or alternative versions of reactions given in Box 25.1. See electronic version for color representation of the figures in this book.

would yield variations in the values selected. In the case of HPO_4^{2-}, more than one concentration is reported in the literature, so we allowed its activity to vary between the minimum and maximum values reported in Table 25.1. In contrast, we were unable to find measurements on two solutes that appear repeatedly in the citric acid cycle, $H_2(aq)$ and $CO_2(aq)$. One reason is that neutral solutes diffuse more easily across cellular membranes than do ions, which are typically pumped through special ion channels. We have taken advantage of these properties of neutral solutes to assert that their activities within cells are set by their activities in the external environment. In the case of $H_2(aq)$, we chose a lower value set by the concentration in bottom seawater (McCollom, 2007; Shock & Canovas, 2010), and a higher value set by the concentration in hydrothermal fluid venting at the Rainbow hydrothermal field on the mid-Atlantic ridge (Charlou et al., 2002; Shock & Canovas, 2010). The lower value for $CO_2(aq)$ is again taken from bottom seawater (McCollom, 2007; Shock & Canovas, 2010), while the higher value was measured during active fermentation (Merlin et al., 2003) and therefore has somewhat more to do directly with microbial metabolism.

As an example of how affinities are calculated, consider the reaction for step 3 in the citric acid cycle, given by

$$\text{isocitrate}^{3-} + \text{NAD}^{-}_{\text{ox}} \Rightarrow \alpha\text{-ketoglutarate}^{2-} + \text{NAD}^{2-}_{\text{red}} + \text{CO}_2(\text{aq}). \tag{25.7}$$

The version of equation (25.4) for the activity product of reaction (7), Q_7, corresponds to

$$Q_7 = \frac{\left(a\alpha\text{-ketoglutarate}^{2-}\right)\left(a\ NAD^{2-}_{red}\right)\left(aCO_2(aq)\right)}{\left(a\,isocitrate^{3-}\right)\left(a\,NAD^{-}_{ox}\right)}, \tag{25.8}$$

which can be evaluated with data in Table 25.1 to yield $Q_7 = 10^{-4.45}$ using the minimum activity for $CO_2(aq)$.

Box 25.1 Individual reactions representing forward steps in the citric acid cycle, balanced by mass and charge, as illustrated in Figure 25.2 using the biochemical convention of fully dissociated anions.

1. citrate^{3-} → *cis*-aconitate^{3-} + H_2O
2. *cis*-aconitate^{3-} + H_2O → isocitrate^{3-}
3. isocitrate^{3-} + NAD^-_{ox} → α-ketoglutarate^{2-} + NAD^{2-}_{red} + $CO_2(aq)$
4. α-ketoglutarate^{2-} + NAD^-_{ox} + ADP^{3-} + HPO_4^{2-} → succinate^{2-} + NAD^{2-}_{red} + $CO_2(aq)$ + ATP^{4-}
   4A. α-ketoglutarate^{2-} + CoA + NAD^-_{ox} → succinyl-CoA$^-$ + NAD^{2-}_{red} + $CO_2(aq)$
   4B. succinyl-CoA$^-$ + ADP^{3-} + HPO_4^{2-} → CoA + ATP^{4-} + succinate^{2-}
5. succinate^{2-} → fumarate^{2-} + $H_2(aq)$
   5A. succinate^{2-} + FAD → fumarate^{2-} + $FADH_2$
6. fumarate^{2-} + H_2O → malate^{2-}
7. malate^{2-} + NAD^-_{ox} → oxaloacetate^{2-} + NAD^{2-}_{red} + H^+
0. pyruvate$^-$ + NAD^-_{ox} + oxaloacetate^{2-} + H_2O → citrate^{3-} + NAD^{2-}_{red} + $CO_2(aq)$ + H^+
   0A. pyruvate$^-$ + NAD^-_{ox} + CoA → acetyl-CoA + NAD^{2-}_{red} + $CO_2(aq)$
   0B. acetyl-CoA + oxaloacetate^{2-} + H_2O → citrate^{3-} + CoA + H^+

Overall forward citric acid cycle resulting in pyruvate oxidation, NAD and FAD reduction, and ATP generation:
pyruvate$^-$ + 4NAD^-_{ox} + ADP^{3-} + HPO_4^{2-} + 2H_2O + FAD → 3$CO_2(aq)$ + 4NAD^{2-}_{red} + 2H^+ + ATP^{4-} + $FADH_2$
Substituting reaction (5) for 5A as a proxy for the reduction of FAD, as in Figure 25.4:
pyruvate$^-$ + 4NAD^-_{ox} + ADP^{3-} + HPO_4^{2-} + 2H_2O → 3$CO_2(aq)$ + 4NAD^{2-}_{red} + 2H^+ + ATP^{4-} + $H_2(aq)$

acetyl-CoA = acetyl-coenzyme A
ATP = adenosine triphosphate
ADP = adenosine diphosphate
CoA = coenzyme A
FAD = oxidized flavin adenine dinucleotide
$FADH_2$ = reduced flavin adenine dinucleotide
NAD^-_{ox} = oxidized nicotinamide adenine dinucleotide
NAD^{2-}_{red} = reduced nicotinamide adenine dinucleotide

Combining this value for Q_7 in equation (25.6) with values of the equilibrium constant for reaction (7), K_7, calculated with data and parameters from Canovas and Shock (2016) yields the affinities for reaction (7) shown in Figure 25.3 for step 3 of the citric acid cycle.

The contoured plots of affinity in Figure 25.3 show minimum values as calculated using values of $H_2(aq)$, $CO_2(aq)$, and HPO_4^{2-} from Table 25.1, which affect the affinities calculated for steps 0, 3, 4, 5, and the overall cycle. Maximum affinities for these steps using alternate data from Table 25.1 are shown in Figure 25.4. Note that the pressure-temperature diagrams in these figures have inverted pressure axes to allow interpretations of changes with depth downward on the plots. The plots cover temperatures from 0 °C to 200 °C, and pressures up to 2 kb. The contours are labeled in kcal mol^{-1} for the reactions as written in Box 25.1. In the case of the overall reactions, these are the calculated affinities for one complete turn of the forward cycle. One striking feature of the majority of the plots shown in Figures 25.3 and 25.4 is that the affinities of most of the reactions in the citric acid cycle depend little on changes in pressure over the pressure range that encompasses all of the known natural occurrences of microbial life, as indicated in Figure 25.1a. Perhaps a useful assumption for most steps of the citric acid cycle in the known subsurface biosphere is that affinities shift meaningfully with temperature but very little with pressure.

Examination of Figure 25.3 reveals that minimum affinities for step 0, in which pyruvate from outside the cycle and oxaloacetate from inside the cycle are combined with H_2O to form citrate and $CO_2(aq)$ coupled with reduction of NAD (nicotinamide adenine dinucleotide), are positive and decrease with increasing temperature at all pressures considered. Positive values of affinity mean that energy would be released as the reaction proceeds as written. Small positive values of affinity are associated with the dehydration of citrate to form *cis*-aconitate in step 1, and they become slightly more positive as temperature increases at all pressures. In contrast, the hydration of *cis*-aconitate to form isocitrate in step 2 is accompanied by small negative affinities that become increasingly negative with increasing temperatures. The results obtained here indicate that this step requires the input of energy throughout the biosphere. Minimum values of affinity for step 3 shown in Figure 25.3, and given by reaction (7) above, are positive and approximately twice the magnitude of the minimum affinities for step 1. Apparently, the oxidative decarboxylation of isocitrate to yield α-ketoglutarate and $CO_2(aq)$, which is coupled to the reduction of NAD, can be an energy-yielding process

Table 25.1 Chemical species in and associated with the citric acid cycle and their apparent activities within microbial cells used in the present study (see text).

Chemical Species	Apparent Activity
citrate^{-3}	$2 * 10^{-3}$[a]
cis-aconitate^{-3}	$1.6 * 10^{-5}$[a]
isocitrate^{-3}	$2 * 10^{-3}$[b]
α-ketoglutarate^{-2}	$4.4 * 10^{-4}$[a]
succinate^{-2}	$5.7 * 10^{-4}$[a]
fumarate^{-2}	$1.2 * 10^{-4}$[a]
malate^{-2}	$1.7 * 10^{-3}$[a]
oxaloacetate^{-2}	$3 * 10^{-5}$[c]
pyruvate^{-}	$9 * 10^{-4}$[d]
NAD^{-}_{ox}	$2.6 * 10^{-3}$[a]
NAD^{-2}_{red}	$8.3 * 10^{-3}$[a]
ADP^{-3}	$5.6 * 10^{-4}$[a]
ATP^{-4}	$9.6 * 10^{-3}$[a]
FAD	$1.7 * 10^{-4}$[a]
$FADH_2$	$3.8 * 10^{-5}$[e]
$H_2(aq)$	$1.6 * 10^{-2}$ $(4 * 10^{-10})$[f]
$CO_2(aq)$	$2 * 10^{-2}$ $(4.93 * 10^{-5})$[g]
HPO_4^{-2}	$5 * 10^{-3}$ $(2 * 10^{-2})$[h]
H^+	$1 * 10^{-7}$[i]
H_2O	1[j]

[a] Bennett et al. (2009)
[b] Set equal to that of citrate because no in situ measurement is available
[c] Peng et al. (2004)
[d] Sundararaj et al. (2004) and Phillips et al. (2008)
[e] Ishii et al. (2007)
[f] There are no measurements, but easily diffuses across membranes, and intracellular concentration and apparent activity are assumed to mimic that of the extracellular environment (see text). Value without parentheses corresponds to the dissolved hydrogen concentration from vent fluid at the Rainbow hydrothermal field from Charlou et al. (2002), while the value in parentheses corresponds to that of bottom seawater from McCollom (2007).
[g] Value without parentheses was measured by Merlin et al. (2003) during active fermentation and is used as the value that provides the lowest calculated apparent affinity, while that in parentheses corresponds to the concentration of $CO_2(aq)$ of bottom seawater and therefore aligns with statement by Lu et al. (2009) that the internal $CO_2(aq)$ concentration of cells is essentially that of the external environment.
[h] The value outside of the parentheses corresponds to the concentration from Sundararaj et al. (2004) and Phillips et al. (2008) and provides the lowest calculated apparent affinity, while that in parentheses is the value assumed by Bennett et al. (2009).
[i] Set to neutrality at 25 °C and 1 bar to simplify the calculations, though it should be noted that neutrality depends on temperature and pressure.
[j] Set to unity to simplify the calculations, though it should be noted that the actual activity of water could deviate from this value depending on salinity.

throughout the known biosphere. Minimum affinity contours for step 4 are shaped differently than contours for all other steps in the citric acid cycle, revealing a broad shallow trough of small positive values in pressure and temperature associated with the conversion of α-ketoglutarate to succinate and $CO_2(aq)$ coupled to production of adenosine triphosphate (ATP) from adenosine diphosphate (ADP) and reduction of NAD. The positive values of affinity for steps 3 and 4 are consistent with the notion that the $CO_2(aq)$-producing steps of the forward citric acid cycle are associated with energy release coupled to biosynthesis. In contrast, step 5, in which succinate is oxidized to form the double bond in fumarate, is energetically costly, as revealed by the negative values of affinity at all temperatures and pressures shown in Figure 25.3. However, these results show that the energetic cost of this step decreases with increasing temperature. The hydration of fumarate to malate in step 6 requires energy as indicated by the negative affinity values at all temperatures and pressures, but the magnitudes are much lower than those associated with step 5. Note that the energy cost of step 6 is predicted to increase slightly with increasing temperature at all pressures. The only step in the citric acid cycle that exhibits a change in the sign of its affinity is step 7, in which malate oxidation to oxaloacetate is coupled to NAD reduction. Step 7 requires the input of energy at low temperatures, is accompanied by a release of energy at high temperatures, and is predicted to be the least pressure-dependent of all of the affinities shown in Figure 25.3.

Finally, the plot in the lower right-hand corner of Figure 25.3 shows minimum affinities for the overall forward citric acid cycle in which pyruvate is oxidized to $CO_2(aq)$ and coupled to the reduction of NAD and production of ATP. The minimum affinity values shown for the other steps combine to yield contours that are closely spaced at high temperature and approach an increasingly flat plateau as temperatures decrease, suggesting that energy from the overall citric acid cycle at low temperatures is nearly independent of pressure. They also yield the surprising result that the affinity for the overall forward cycle can be negative at low temperatures and positive at high temperatures while having only a slight pressure dependence. This implies that the citric acid cycle run in reverse would release energy at temperatures below 100 °C at all pressures when using minimum affinity values. Portions of the reverse citric acid cycle are often recognized as aspects of microbial metabolism, and results shown in Figure 25.3 are consistent with the *release of energy* during CO_2 fixation into larger organic compounds via the citric acid cycle under these conditions. This combination of CO_2 fixation into organic forms with the release of energy is an example of how biochemical cycles may have emerged from geochemical precursors (Shock & Boyd, 2015).

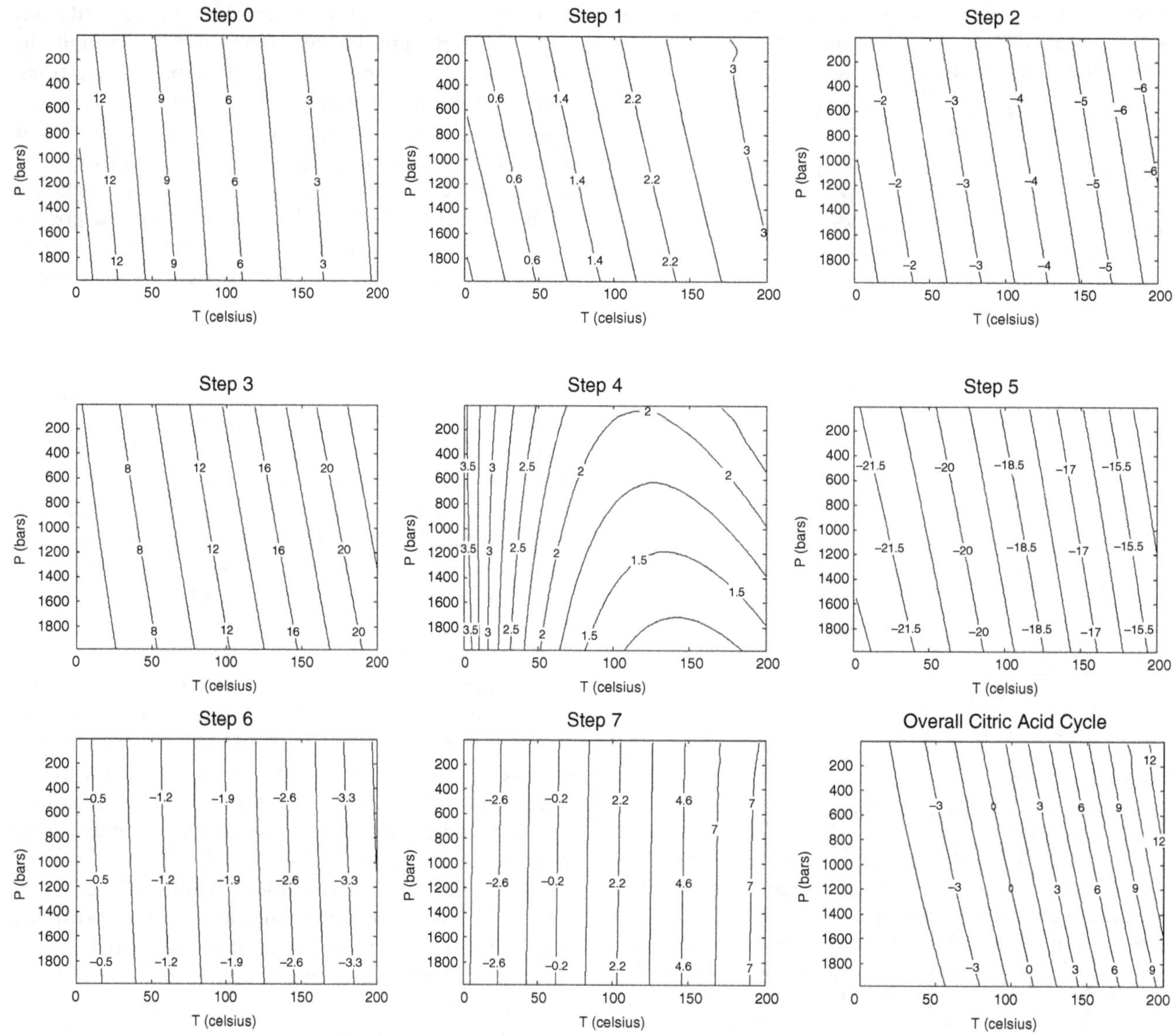

Figure 25.3 Contours of minimum affinity values (in kcal mol^{-1}) as functions of temperature and pressure for steps in the citric acid cycle as given in Box 25.1 and calculated with equation (25.6) using the compositional data listed in Table 25.1 and equilibrium constants calculated with equations, data, and parameters from Canovas and Shock (2016). See electronic version for color representation of the figures in this book.

The notion that the overall forward citric acid cycle would cost energy to run is at odds with its role in human metabolism, but it must be kept in mind that the minimum affinity values from *Escherichia coli* and geochemical constraints used to generate the plot in Figure 25.3 are unlikely to correspond to conditions within human cells. In fact, the notion of the overall forward cycle costing energy at environmental conditions may be challenged by the maximum affinity values shown in Figure 25.4. As mentioned above, steps 0, 3, 4, and 5 all involve either CO_2(aq) or H_2(aq) that can diffuse freely across microbial membranes and impose external environmental conditions inside cells. The plots for these steps in Figure 25.4 reflect the maximum affinities enabled by the ranges of values shown in Table 25.1. The consequences of external control of CO_2(aq) and H_2(aq) are that affinities for steps 0, 3, and 5 can be about 10 kcal mol^{-1} more positive at the maximum values shown in Figure 25.4. In the case of step 5, which is energetically costly regardless of compositional constraints, maximum affinities are considerably less costly than the minimum values shown in Figure 25.3. Note also that the differently shaped contours for step 4 exhibit considerably more positive, energy-releasing values in Figure 25.4 than in Figure 25.3, and that the trough has shifted to lower temperatures. The combined effects of all of these differences cause the maximum affinities for the overall forward cycle to be large and positive at all temperatures and pressures, as shown in Figure 25.4. If these conditions for H_2(aq) and CO_2(aq) are attained in subsurface environments, then the reverse citric acid cycle would

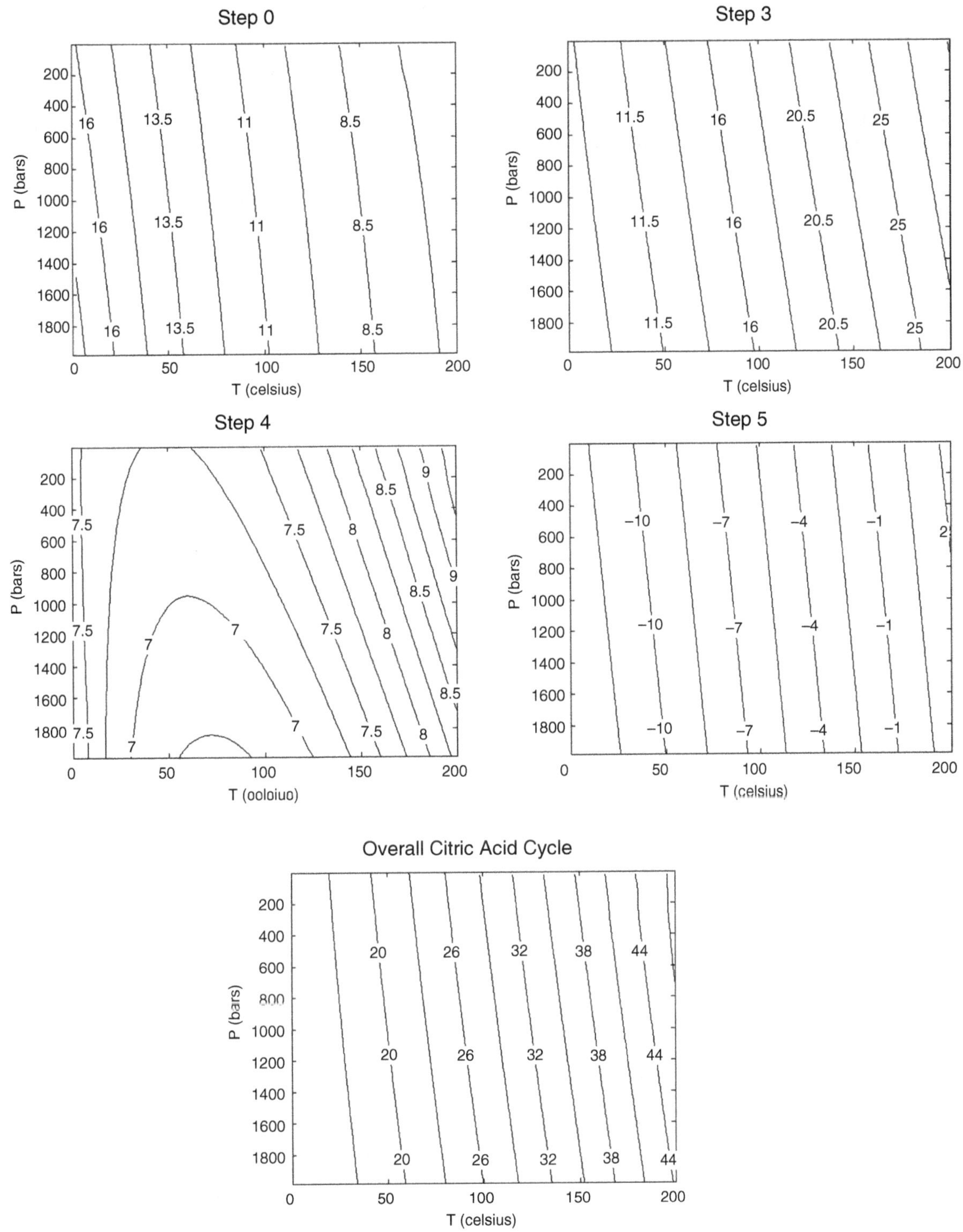

Figure 25.4 Contours of maximum affinity values (in kcal mol^{-1}) as functions of temperature and pressure calculated as in Figure 25.3 for steps in the citric acid cycle affected by variations in $H_2(aq)$, $CO_2(aq)$, and HPO_4^{2-} from Table 25.1. See electronic version for color representation of the figures in this book.

not be a source of energy as it evidently is when minimum affinity values are considered. The consequences of the differences between results in Figures 25.3 and 25.4 underscore the need for development of clever methods to determine solute abundances within microbial cells in laboratory experiments and in natural ecosystems.

25.4. THE CITRIC ACID CYCLE IN A SUBDUCTION ZONE DEEP BIOSPHERE

Does the deep biosphere extend only to the depths we have explored? We doubt it. Laboratory experiments summarized in Figure 25.1c indicate that microbes survive pressures of tens of kilobars. Coupled with the success of life at temperatures >100 °C, these data suggest a deep biosphere that is far more extensive than revealed by observations of natural systems. Specifically, high pressures and low temperatures found in some subduction zones overlap the conditions of extreme laboratory growth experiments. Based on evidence from these experiments, depths up to 50 km at many continental margins where subduction zones inject cold, altered oceanic crust into the mantle could provide stable and habitable zones where microorganisms could thrive. Habitable depths might be even deeper at fast subduction zones that are also characterized by old, cold slabs with high slab dip angles, such as those occurring at Tonga.

Geophysical models of the relations between depth, pressure, and temperature in the Tonga subduction zone (Syracuse et al., 2010) are shown in Figure 25.5. The warmer slab surface intersects the 121 °C isotherm at about 50 km depth, corresponding to a pressure of about 15 kb in all models. The cooler Moho, which is deeper in the insulated interior of the slab, would intersect the 121 °C isotherm at a little more than 100 km depth and a pressure of about 35 kb. These observations lead to the conclusion that the highest temperatures and pressures of laboratory microbial experiments are attained within the upper 100 km, or so, of the Tonga subduction zone. In this study, we conducted calculations up to 200 °C to include additional conditions where life may thrive in excess of the currently established upper temperature. Note that the slab surface intersects the 200 °C isotherm at about 20 kb in Figure 25.5. The models for conditions at the Moho diverge as temperatures increase and intersect the 200 °C isotherm at pressures from ~44 kb to ~52 kb. These conditions are within the pressure-temperature limits of the DEW geochemical model (Sverjensky, Harrison, et al., 2014), which allows us to calculate standard state thermodynamic data throughout this deepest of potential subsurface biospheres.

The DEW model accounts for experimentally determined properties of water and aqueous solutions to 60 kb (6.0 GPa) and has enabled new perspectives on the progress of water-rock-organic reactions in subduction zone fluids and the upper mantle (Facq et al., 2014; Guild & Shock, 2019; Huang et al., 2017; Sverjensky, Stagno, et al., 2014; Tao et al., 2018). The development of the DEW model is accompanied by changes in how some of the revised-HKF equation of state parameters are estimated. Specifically, extrapolations to high pressures required revision of the way that standard partial molal volumes of aqueous species are calculated with the revised-HKF parameters. Correlation methods provided by Sverjensky, Harrison, et al. (2014) allow estimates of these parameters in the absence of experimental data, which is the case for everything involved in the citric acid cycle other than H^+ and $CO_2(aq)$. These methods were used together with existing standard state data from Canovas and Shock (2016) to obtain the revised-HKF equation of state parameters in Table 25.2 that can be

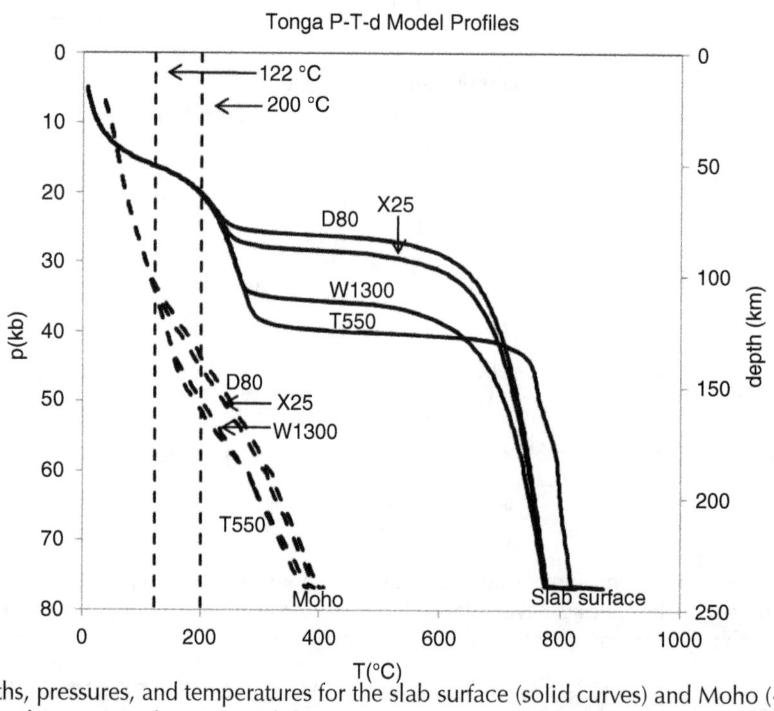

Figure 25.5 Depths, pressures, and temperatures for the slab surface (solid curves) and Moho (~7 km below the slab surface, dashed curves) in the Tonga subduction zone using geothermal gradient modeling data from the D80, T550, W1300, and X25 models of Syracuse et al. (2010). Dashed vertical lines indicate the highest known temperature for life (121 °C), as well as the highest temperature used in citric acid cycle affinity calculations (200 °C). Note that all models of the subduction zone produce depths and pressures that are essentially the same at 121 °C, and that model differences at slightly higher temperatures affect the pressure/depth relations at 200 °C. See electronic version for color representation of the figures in this book.

Table 25.2 Summary of standard partial molal thermodynamic data at 25 °C and 1 bar for aqueous species in the citric acid cycle (Canovas & Shock, 2016) and related biomolecules that take part in reactions in the citric acid cycle (LaRowe & Helgeson, 2006a, 2006b) that are not included in the Deep Earth Water (DEW) database, along with equation of state parameters required to calculate the corresponding properties at high temperatures and pressures between 10 and 60 kb. Unless otherwise indicated, revised-HKF equation of state parameters were estimated using the software from DEW model version 11.0.1 (Sverjensky, Harrison, et al., 2014).

Species	$\Delta\bar{G}_f^{o}$ [a]	$\Delta\bar{H}_f^{o}$ [a]	$\bar{S}^{o}$ [b]	$\bar{C}_p^{o}$ [b]	$\bar{V}^{o}$ [c]	$a_1 \times 10^{1}$ [d]	$a_2 \times 10^{-2}$ [d]	a_3 [e]	$a_4 \times 10^{-4}$ [f]	c_1 [b]	$c_2 \times 10^{-4}$ [f]	$\omega_e \times 10^{-5}$ [a]
Citric acid	−297180.	−364527.	78.89	73.47	112.98	23.4335	18.0278	−10.2781	−3.5243	48.6718	11.9312	−0.06[g]
H_2-citrate$^-$	−292912.	−363530.	67.92	47.04	102.48	22.1425	16.7987	−9.2475	−3.4735	47.5542	6.5474	1.50[g]
H-citrate^{2-}	−286417.	−362947.	48.09	7.71	91.08	20.3986	15.1383	−7.8553	−3.4048	33.5330	−1.4641	2.48[g]
Citrate^{3-}	−277690.	−363750.	16.13	−43.16	74.08	17.2265	12.1182	−5.3229	−3.2800	6.2075	−11.8263	2.75[g]
Cis-aconitic acid	−236200.	−291400.	63.9	64.0	108.7	22.5524	17.1890	−9.5747	−3.4896	42.1636	10.0022	−0.1640[h]
H_2-*cis*-aconitate$^-$	−233600.	−290480.	58.2	37.6	98.2	20.9505	15.6638	−8.2958	−3.4265	35.0910	4.6245	0.7477[h]
H-*cis*-aconitate^{2-}	−227700.	−289900.	40.4	−1.7	86.8	19.6254	14.4022	−7.2380	−3.3744	29.1329	−3.3809	2.6010[h]
Cis-aconitate^{3-}	−218970.	−290700.	8.4	−52.6	69.8	17.3231	12.2102	−5.4000	−3.2838	18.4954	−13.7492	4.6843[h]
Isocitric acid	−295880.	−363960.	76.4	74.5	114.3	23.6793	18.2619	−10.4743	−3.5339	49.0727	12.1411	−0.0820[h]
H_2-isocitrate$^-$	−291390.	−362960.	64.7	43.1	103.8	21.9908	16.6543	−9.1264	−3.4675	40.3381	6.7634	0.6493[h]
H-isocitrate^{2-}	−284960.	−362380.	45.1	3.8	92.4	20.6788	15.4052	−8.0790	−3.4158	34.6306	−1.2420	2.5298[h]
Isocitrate^{3-}	−276230.	−363180.	13.1	−42.1	75.4	18.3766	13.2132	−6.2410	−3.3252	23.9941	−11.6104	4.6132[h]
α-ketoglutaric acid	−201800.	−245700.	76.1	39.9	95.7	20.0660	14.8217	−7.5897	−3.3917	28.7766	5.0930	−0.0840[h]
H-α-ketoglutarate$^-$	−198800.	−242300.	77.1	2.9	88.9	19.0070	13.8134	−6.7443	−3.3500	12.1179	−2.4439	0.4615[h]
α-ketoglutarate^{2-}	−191800.	−240400.	60.1	−49.9	82.7	18.6860	13.5078	−6.4881	−3.3374	−1.8644	−13.1992	2.3026[h]
Succinic acid	−177800.	−218000.	62.3	53.3	82.44	17.4474	12.3285	−5.4993	−3.2887	35.7970	7.8226	−0.1744[h]
H-succinate$^-$	−172060.	−217350.	45.2	9.3	69.99	15.5663	10.5375	−3.9975	−3.2146	20.3194	−1.1402	0.9446[h]
Succinate^{2-}	−164380.	−217350.	19.5	−50.5	56.32	13.8575	8.9105	−2.6333	−3.1474	3.4443	−13.3215	2.9170[h]
Fumaric acid	−154820.	−186260.	60.62	47.00	77.9	16.5601	11.4838	−4.7909	−3.2537	31.9979	6.5393	−0.1860[h]
H-fumarate$^-$	−150600.	−186150.	46.83	10.0	65.4	14.6630	9.6775	−3.2764	−3.1791	20.5021	−0.9976	0.9199[h]
Fumarate^{2-}	−144320.	−186830.	23.48	−42.8	51.7	12.9315	8.0290	−1.8941	−3.1109	7.4061	−11.7530	2.8572[h]
Malic acid	−213530.	−259310.	68.21	56.43	82.22	17.4231	12.3054	−5.4799	−3.2877	37.9851	8.4602	−0.1360[h]
H-malate$^-$	−208810.	−258610.	54.75	19.4	75.40	16.5476	11.4718	−4.7809	−3.2532	24.9065	0.9172	0.8000[h]
Malate^{2-}	−201860.	−258890.	30.48	−33.4	69.19	16.2774	11.2146	−4.5662	−3.2426	11.9385	−9.8382	2.7512[h]
Oxaloacetic acid	−200000.	−235100.	72.9	29.1	79.1	16.8320	11.7426	−5.0080	−3.2644	22.2537	2.8931	−0.1050[h]
H-oxaloacetate$^-$	−196600.	−231300.	74.0	−7.9	72.3	15.8056	10.7654	−4.1886	−3.2240	6.2205	−4.6438	0.5084[h]
Oxaloacetate^{2-}	−190600.	−230300.	57.2	−60.7	66.1	15.4832	10.4584	−3.9312	−3.2114	−7.7894	−15.3992	2.3465[h]

(Continued)

Table 25.2 (Continued)

Species	$\Delta\bar{G}_f^{o}$ [a]	$\Delta\bar{H}_f^{o}$ [a]	$\bar{S}^{o}$ [b]	$\bar{C}_p^{o}$ [b]	$\bar{V}^{o}$ [c]	$a_1 \times 10^{1}$ [d]	$a_2 \times 10^{-2}$ [d]	a_3 [e]	$a_4 \times 10^{-4}$ [f]	c_1 [b]	$c_2 \times 10^{-4}$ [f]	$\omega_e \times 10^{-5}$ [a]
Pyruvic acid	−117000.	−140300.	62.1	36.3	64.6	13.9819	9.0291	−2.7327	−3.1523	25.8192	4.3597	−0.1760[h]
$Pyruvate^-$	−113600.	−137400.	60.4	−3.7	51.5	11.8649	7.0134	−1.0426	−3.0689	10.5798	−3.7883	0.7144[h]
NAD^{2-}_{red}	−524441.	−783126.	137.2	177.2	335.8	70.7652	63.0924	−48.0643	−5.3872	187.4039	33.0608	8.4[i]
NAD^{-}_{ox}	−529811.	−776066.	147.6	169.7	335.9	70.7846	63.1109	−48.0798	−5.3880	183.0056	31.5321	8.4[i]
ADP^{3-}	−452649.	−624106.	52.17	41.5	191.7	40.6730	34.4416	−24.0408	4.2028	67.3629	5.4271	4.0[i]
ATP^{4-}	−657038.	−859264.	47.96	21.3	196.8	42.1334	35.8320	−25.2067	−4.2603	64.7373	1.3124	5.0[i]

[a] cal mol^{-1},
[b] cal mol^{-1} K^{-1},
[c] cm^3 mol^{-1},
[d] cal mol^{-1} bar^{-1},
[e] cal K mol^{-1} bar^{-1},
[f] cal K mol^{-1},
[g] LaRowe and Helgeson (2006a),
[h] Canovas and Shock (2016),
[i] LaRowe and Helgeson (2006b).

used with the DEW model, which already includes H^+, $H_2(aq)$, $CO_2(aq)$, and HPO_4^{2-}. It should be noted that these parameters can be used reliably at pressures between 10 kb and 60 kb, but not below 10 kb.

Equilibrium constants for the reactions in Box 25.1 were calculated with the DEW model for temperatures up to 200 °C and pressures from 10 to 60 kb. At the lower temperatures and higher pressures considered, a polymorph of ice is stable rather than liquid water, so equilibrium constants for reactions involving aqueous solutes were not calculated at these conditions. Values of activity products were calculated as described above using data from Table 25.1 and combined with the DEW-based equilibrium constants to evaluate affinities throughout low-temperature conditions of subduction zone fluids in the Tonga model shown in Figure 25.5. The results are shown in Figure 25.6 for minimum affinity values and in Figure 25.7 for maximum affinity values, using the same constraints as in the construction of Figures 25.3 and 25.4.

The plots in Figure 25.6 show minimum affinity contours on P-T plots that extend from 10 to 60 kb and from 0 °C to 200 °C. These plots also show depths on the right-hand ordinate. The contours are truncated where they hit the edge of the ice stability field. Also shown in each plot are traces of the slab surface (solid curve) and the Moho (dashed curve) from the Tonga subduction models adopted from Figure 25.5. A striking feature of the plots in Figure 25.6 is that the affinity contours have considerably lower slopes than the corresponding low-pressure plots shown in Figure 25.3. In fact, rather than being able to conclude that the minimum affinities for the citric acid cycle and most of its individual steps change with temperature but are nearly independent of pressure, which is the case of the low-pressure predictions, at the higher pressures of Figure 25.6, affinities depend strongly on both pressure and temperature. In several cases, the trajectory of the slab surface in pressure and temperature is roughly parallel with the affinity contours of individual steps, suggesting that slab surface conditions are energetically similar for these reactions regardless of depth. In contrast, the trajectory of the Moho cuts across the affinity contours. The implication is that the energetic consequences for the citric acid cycle at the Moho are considerably more variable than at the slab surface.

Minimum affinities for step 0 in Figure 25.6 are positive and increase with increasing pressure, indicating that this reaction yielding citrate and $CO_2(aq)$ from pyruvate and oxaloacetate is predicted to release energy throughout shallow conditions in steeply dipping subduction zones. Affinities for step 0 at the conditions of the slab surface vary from 17 kcal mol^{-1} at the initiation of subduction to 12 kcal mol^{-1} at 200 °C and depth >60 km, which is equivalent to the affinity at 25 °C and ~1 kb for step 0 in Figure 25.3. In contrast to the subparallel trajectory of the slab surface, the trajectory for the Moho crosses several affinity contours and is consistent with greater energy releases with increasing depth, maximizing at about 29 kcal mol^{-1} at 200 °C and nearly 160 km. Although the slopes of the affinity contours for step 1 are similar to those for step 0, the magnitudes of the affinities are opposite in sign and become more negative with increasing depth and pressure. This means that the dehydration of citrate to form *cis*-aconitate is not favored at these conditions and would require the input of energy. The energy requirements at slab surface conditions are considerably milder than along the trace of the Moho. These results contrast with affinities for step 1 at low pressures shown in Figure 25.3, which are small and positive at all pressures and temperatures considered.

The slab-surface trajectory is also subparallel to the minimum affinity contours for steps 2, 3, and 5 as shown in Figure 25.6. Those for step 2 hover around zero, those for step 3 are slightly negative, and those for step 5 are strongly negative, which means that the hydration of *cis*-aconitate to isocitrate is predicted to effectively break even, the oxidation of isocitrate to α-ketoglutarate and $CO_2(aq)$ requires a small input of energy, and the oxidation of succinate to fumarate is cost intensive. Comparison with analogous plots in Figure 25.3 shows that differences in the magnitudes of affinities between the low-pressure and high-pressure calculations are least for step 2 and considerably more dramatic for steps 3 and 5. The effects of pressure are particularly noteworthy for step 3, as strongly negative affinities can be encountered at subduction zones of Figure 25.6 that are not seen at the lower pressures in Figure 25.3. It should be kept in mind that negative affinities for forward reactions also mean that the reverse reactions would release energy. It is particularly notable that the reduction of fumarate to succinate, the reverse version of step 5, would release considerable energy at slab-surface conditions. The Moho trajectory for step 2 crosses the affinity contours much like it does for steps 0 and 1, and like step 0 becomes more energetically favorable with depth. In contrast, the Moho trajectories for steps 3 and 5, which also cross many affinity contours, extend into conditions where these reactions are increasingly costly with increasing depth. Again, this means that the reverse reactions would be energy releasing, and both are capable of releasing impressive amounts of energy along the Moho trajectory.

The minimum affinity contours shown in Figure 25.6 for steps 4, 6, and 7 are steeper than the trajectories for the other steps, and as a consequence, the changes along slab-surface and Moho trajectories appear to be more dramatic. However, the spacings between contours for steps 4 and 6 are considerably smaller than in the plots for steps 0, 1, 3, and 5, and those for step 7 are comparable to those for step 2. In the case of step 4, in which the

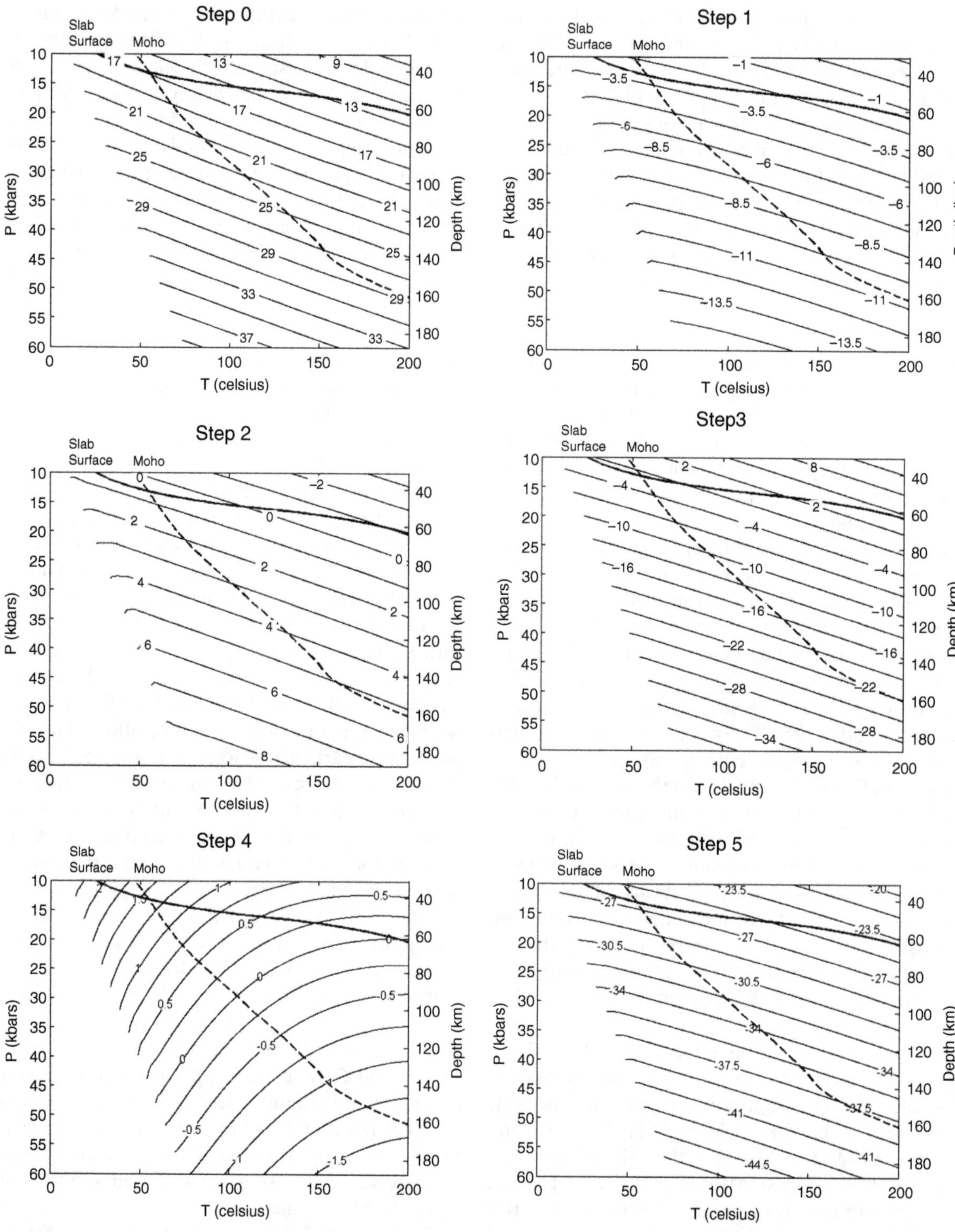

Figure 25.6 Contours of minimum affinity values (in kcal mol^{-1}) as functions of temperature and pressure for steps in the citric acid cycle as given in Box 25.1 and calculated with equation (25.6) using the compositional data listed in Table 25.1 and equilibrium constants calculated with the DEW model using data and parameters from Table 25.2. Slab-surface and Moho trajectories for the Tonga subduction zone shown in Figure 25.5 are indicated. See electronic version for color representation of the figures in this book.

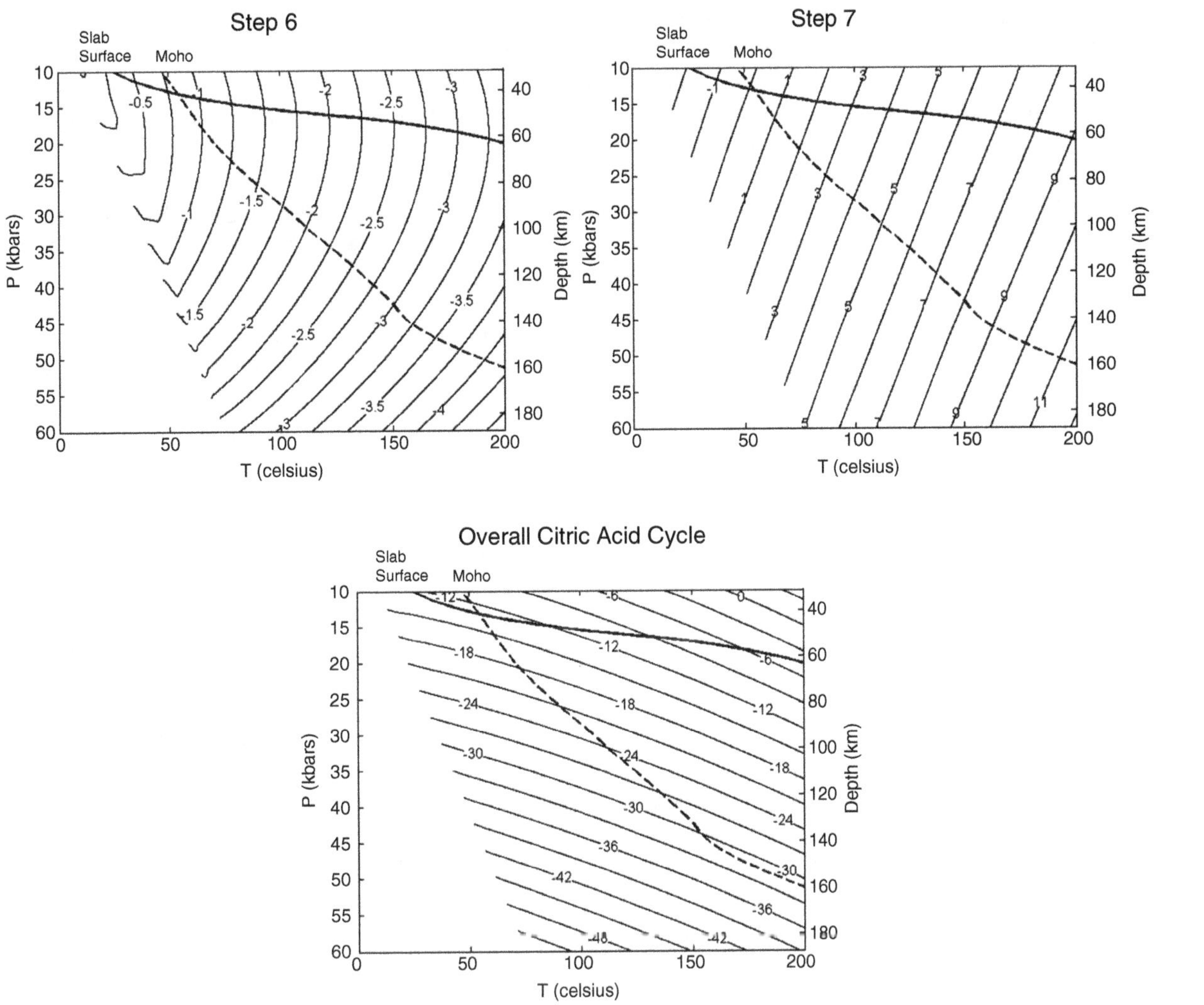

Figure 25.6 (Continued)

oxidation of α-ketoglutarate to succinate and $CO_2(aq)$ is coupled to the production of ATP from ADP, minimum affinities change along the slab surface from slightly positive to zero with increasing depth. The Moho trajectory for step 4 crosses into negative affinity values, consistent with energy release if the reaction were to proceed in the reverse direction of how it is written in Box 25.1, leading to production of α-ketoglutarate. Small positive affinity values accompany step 4 at the lower pressure plots in Figure 25.3 as well. As in the case of step 4, slab-surface and Moho trajectories for step 6 in Figure 25.6 also move to somewhat more negative values with depth, which means that hydration of fumarate to malate becomes increasingly costly, and the reverse reaction more energetically favorable over the course of shallow subduction. Note that the magnitudes of the affinities for step 6 at the high-pressure conditions shown in Figure 25.6 are similar to the those at much lower pressures shown in Figure 25.3. In contrast to the high-pressure behavior shown by step 6, the Moho and slab-surface trajectories for step 7 move to more positive affinities with increasing depth, indicating that the oxidation of malate to oxaloacetate yields increasing energy with depth. The steep contours for step 7 indicate that changes in temperature tend to have more profound effects on the minimum affinities than do changes in pressure, which is a conclusion reached for most steps in the citric acid cycle at low pressures illustrated in Figure 25.3.

Minimum affinity contours for the overall citric acid cycle shown in Figure 25.6 almost all show negative values, indicating that the forward citric acid cycle costs energy and that the reverse citric acid cycle would be capable of releasing energy. The slab-surface trajectory moves to less negative values with increasing depth, but the Moho trajectory moves steadily to more negative values with increasing depth before becoming essentially parallel to the affinity contours at depths greater than

140 km. Note that deep Moho trajectories for steps 0, 1, 2, 3, and 5 also approach constant affinity values at these depths, which is why the overall cycle affinities show this behavior. All subduction zone affinities for the overall cycle in Figure 25.6 are more negative than any of the affinity contours for the overall cycle in Figure 25.3. In terms of minimum affinities, this means that increasing depth increasingly favors the reverse citric acid cycle.

As in the low-pressure examples discussed above, variations in activities of $H_2(aq)$, $CO_2(aq)$, and HPO_4^{2-} from Table 25.1 will affect the affinities of steps 0, 3, 4, and 5, and therefore the affinities of the overall cycle as well. Maximum affinities attainable owing to these activity variations are shown for these steps in Figure 25.7 for high-pressure conditions. Note that affinities for step 0 in Figure 25.7 are about 6 kcal mol^{-1} more positive than in Figure 25.6. Similarly, those for step 3 are about 4 kcal mol^{-1} more positive, and those for step 5 are about 15 kcal mol^{-1} more positive. The curved contours for step 4 fall in somewhat different positions in Figure 25.7 than they do in Figure 25.6, and affinities are generally between 3 and 6 kcal mol^{-1} more positive in Figure 25.7. The consequences of these more positive affinities in the maximum case are dramatic for the affinities of the overall cycle shown in Figure 25.7. Neither the slab-surface nor the Moho trajectories are predicted to cross into negative affinity territory, although the Moho trajectory reaches a broad minimum of about 2 kcal mol^{-1} below about 120 km. In contrast, the slab-surface trajectory is predicted to cross into more and more positive affinity values with depth. Unlike the minimum affinity values shown in Figure 25.6, which are consistent with the reverse citric acid cycle releasing energy throughout shallow subduction conditions, maximum affinity values lead to the prediction that the forward citric acid cycle would release energy. These dramatic differences underscore the influence of changes in the composition of the external geochemical environment on how basic biochemical processes can operate, which can be anticipated from the principles of geobiochemistry (Shock & Boyd, 2015).

The two pressure-temperature ranges described above do not overlap, owing to constraints of the revised-HKF and DEW models, and the vertical axes vary enormously in range. Nevertheless, it is possible to envision how contours connect from the lower pressure plots in Figure 25.3 and 25.4 to the higher pressure plots in Figures 25.6 and 25.7. The nearly vertical contours in Figures 25.3 and 25.4 must bend at higher pressures to become the much flatter contours in Figure 25.6 and 25.7, reflecting the increasing influence of changes in reaction volumes as pressure increases. It is useful to take the scales of the figures into account. As an example, the $A_r = -2$ kcal mol^{-1} contour toward the left of the step 2 plot in Figure 25.3 has a slope of ~0.125 °C bar^{-1}, and the same $A_r = -2$ kcal mol^{-1} contour in the upper right corner of the Step 2 plot of Figure 25.6 has a slope of 0.007 °C bar^{-1}. In the eight-kilobar gap between the two figures, this contour apparently moves from <50 °C at 2 kb to ~ 150 °C at 10 kb, with a change in slope of ~95%.

25.5. GEOCHEMICAL INFLUENCE ON THE POTENTIAL ENERGY YIELD OF THE CITRIC ACID CYCLE

In an effort to emphasize how geochemical compositional differences affect the energy yield of the citric acid cycle and even which direction yields energy, we have assembled the four overall citric acid cycle plots from Figures 25.3, 25.4, 26.6, and 25.7 in Figure 25.8 and used color scales to illustrate the differences in affinities. It should be kept in mind that it is the variations in $H_2(aq)$, $CO_2(aq)$, and HPO_4^{2-}, all of which depend on differences in geochemical processes, that drive the differences in affinity ranges in these figures.

The top two plots in Figure 25.8 show affinities for the overall cycle for the minimum case on the left and the maximum case on the right, and both use the same color coding indicated by the upper right bar. Using the same scale allows the observation that the plot for minimum affinity conditions is considerably flatter, varying from about 10 to about −5 kcal mol^{-1}, than the maximum affinity version, in which affinity values range from about 10 to about 45 kcal mol^{-1}. Note that the equilibrium, $A_r = 0$, contour appears on the minimum affinity plot and not on the maximum affinity plot. It follows that all affinities in the maximum case are positive and consistent with energy release through the citric acid cycle by its operation in the forward direction. In the minimum affinity case, most of the left side of the plot is in negative affinity space, and much of that negative affinity space encompasses the known subsurface biosphere as illustrated in Figure 25.1d. Therefore, this analysis is consistent with the citric acid cycle dominantly operating to release energy in the reverse direction at minimum affinity ranges controlled by geochemical processes in the subsurface.

A similar story is revealed by comparing the minimum and maximum affinity plots for subduction zone pressures and temperatures shown in the lower two plots in Figure 25.8, which use the color coding indicated by the lower right bar. As shown in the plot in the lower right, maximum affinity values are consistent with positive affinities for the overall citric acid cycle throughout conditions attained along the slab-surface and Moho trajectories, although the latter approaches the $A_r = 0$ contour at the greatest depths. Geochemical compositions leading to these maximum affinity values would be consistent with energy release through the forward citric acid cycle in subduction zone habitats. In contrast, the lower left

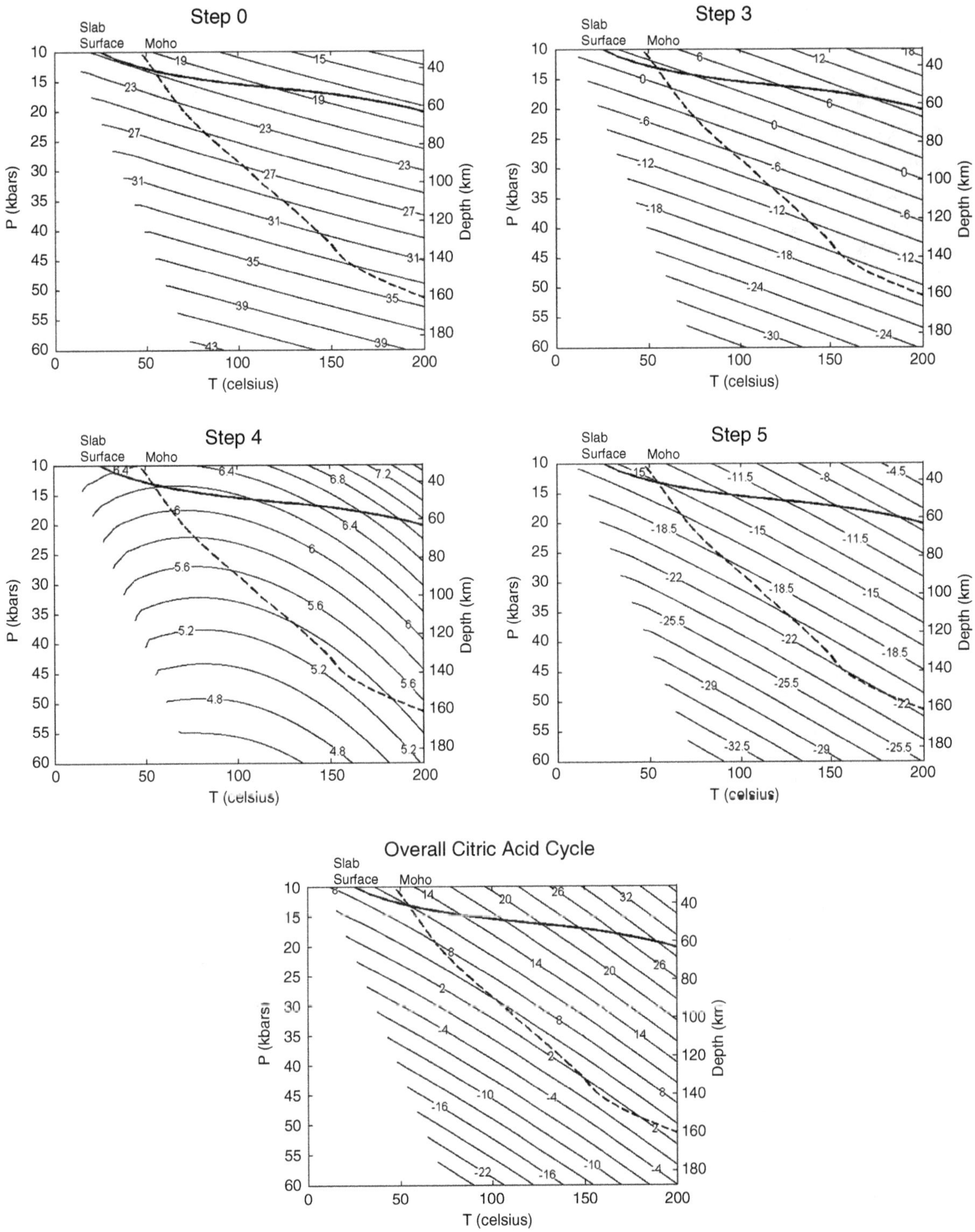

Figure 25.7 Contours of maximum affinity values (in kcal mol^{-1}) as functions of temperature and pressure calculated as in Figure 25.6 for steps in the citric acid cycle affected by variations in H_2(aq), CO_2(aq) and HPO_4^{2-} from Table 25.1. See electronic version for color representation of the figures in this book.

plot indicates that minimum affinity values are consistent with negative affinities over most of the pressures and temperatures considered, which means that the reverse citric acid cycle would be associated with energy release in subduction habitats. In general, conditions along the slab-surface trajectory fall at about −10 kcal mol^{-1}, while those along the Moho trajectory reach −30 kcal mol^{-1} at the greatest depths.

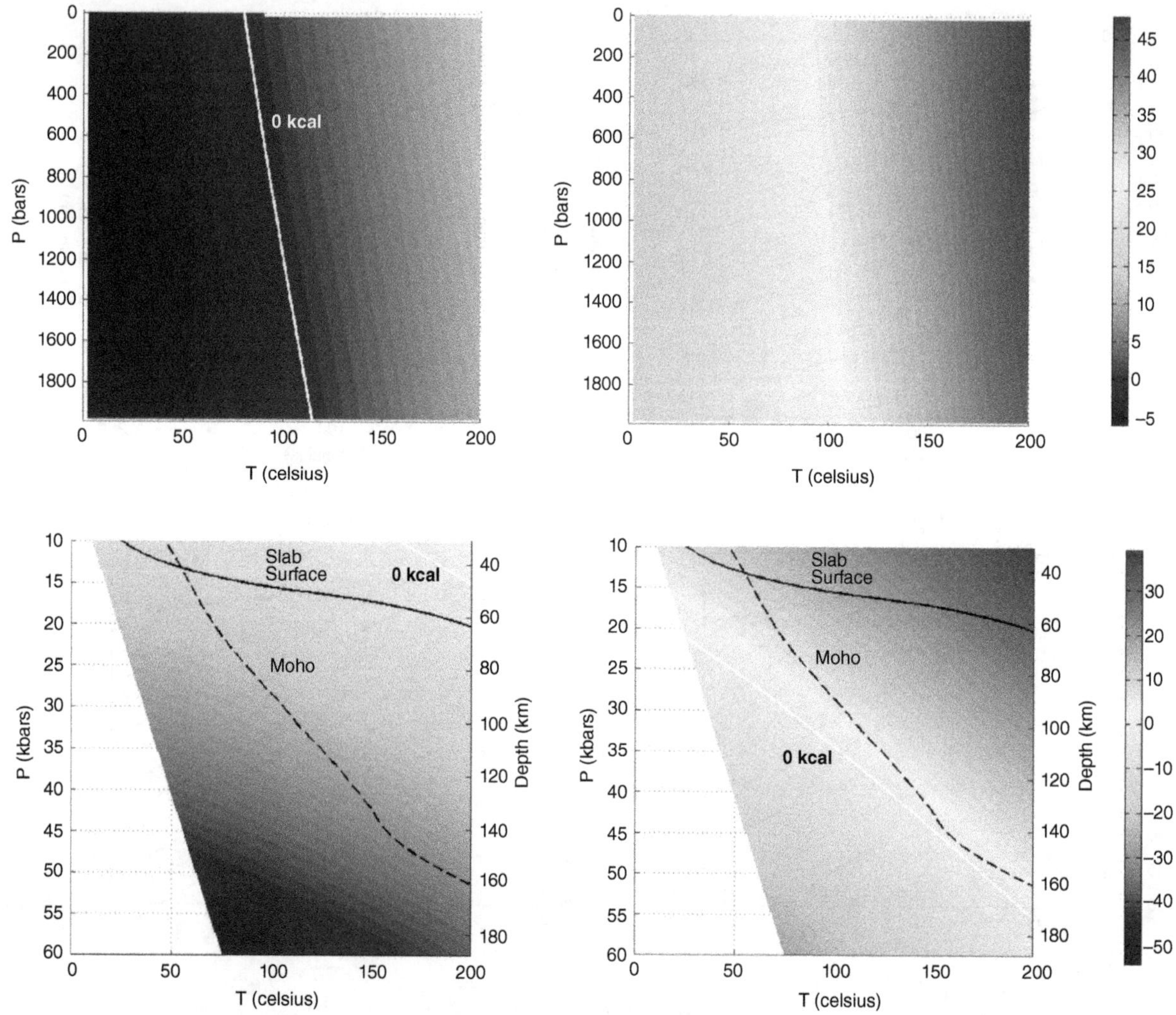

Figure 25.8 Color maps of affinity (in kcal mol^{-1}) across temperature and pressure for the overall forward citric acid cycle as written in Box 25.1 at both low-pressure (top) and high-pressure conditions. In each pressure version, the minimum affinity values are shown on the left and maximum affinity values on the right. High-pressure plots contain slab-surface and Moho trajectories for the Tonga subduction zone from Figure 25.5. Equilibrium contours, where $A = 0$, are shown by the white lines. At conditions where $A > 0$, energy would be released by running the citric acid cycle in the forward direction, and where $A < 0$, energy would be released by running the cycle in the reverse direction. Note the shift in location for the $A = 0$ contour between the minimum and maximum affinity plots at higher pressures. See electronic version for color representation of the figures in this book.

25.6. CONCLUDING REMARKS

Uncertainties in the internal compositions of microbial cells inhibit full assessments of affinities for the citric acid cycle and its individual steps throughout the deep biosphere. We have tried to limit ambiguity by selecting a single set of compositions for most of the compounds in the cycle. At the same time, we allowed activities of $H_2(aq)$ and $CO_2(aq)$ to vary in response to plausible ranges of geochemical compositions. These uncharged solutes are likely to diffuse more easily across cell membranes than ions, which means their activities can be imposed on microbial cells from external geochemical processes. The inescapable influence of external processes on these constituents of the citric acid cycle raises the possibility that geochemical processes can determine whether energy is released as the citric acid cycle operates in either the forward or reverse directions in the deep biosphere.

We find from this analysis that our proposal that conditions in shallow ranges of subduction zones may be conducive to life is tenable. The deep biosphere may indeed extend much further into the subsurface than we presently have access. There is even the possibility that these results could refresh thinking about early life on Earth. Genetic evidence from all known life sends a strong message that use of chemical energy sources preceded the use of sunlight to power life on Earth (Shock & Boyd, 2015). The inescapable conclusion from this existing evidence is that there is no compelling reason to think that life emerged at the surface or that early habitats should be equated to those at the surface. Instead, life itself seems to indicate that it had geologic origins that were plausibly internal to the planet. The energetic analysis performed in this study shows that one of the central biochemical cycles that is deeply engrained in biological energy transfer may be intimately tied to geochemical compositions that can vary enormously with depth.

We anticipate that calculations of this kind can be improved upon. Here are some ways:

• The main source of constraints for the abundances of compounds involved in the citric acid cycle is studies of *Escherichia coli*, which is hardly a denizen of the deep biosphere. Perhaps the ongoing metabolomic revolution will provide economical ways to inventory internal cellular compositions of a wider variety of microbes, including deep residents. The results could be used in a far more comprehensive analysis of how compositions affect affinities and therefore the directions that the citric acid cycle can be run for an energetic profit in subsurface habitats, and the magnitudes of those energy releases. Additional advances in theoretical estimations could open other biochemical pathways to investigation.

• Subsurface exploration for life can expand through scientific drilling, especially if those drilling efforts focus on extreme conditions in the accessible subsurface that may be most similar to those that persist in the far less accessible reaches of subduction zones that may indeed be habitable. Intensive sampling and analysis of subsurface fluids for inorganic and organic solutes, together with live capture of subsurface residents, will improve the compositional framework adopted here.

• The standard state properties for the citric acid cycle used here come from our critique of the available data (Canovas & Shock, 2016), and we refer to the review there for extensive and obvious gaps in current knowledge. While we are optimistic that our predictions are applicable to high pressure and temperature extremes of the biosphere, we would also not be surprised to find that crucial refinements are possible. Specifically, measurements of partial molar properties of solutes involved in the citric acid cycle at high temperatures and pressures would revolutionize the accuracy of predictions of the type attempted here.

• We adopted pressure-temperature-depth relations for the Tonga subduction zone because it is steep, which allows subducted materials to reach high pressures while still in the relatively low temperature range compatible with known life. Other pressure-temperature-depth models are likely to lead to other conclusions about how chemical energy can be accessed through the citric acid cycle in other specific locations.

ACKNOWLEDGMENTS

We thank Tucker Ely, Grayson Boyer, Alta Howells, Kirt Robinson, Kris Fecteau, Chris Glein, Kristin Johnson, Meghan Guild, Dimitri Sverjensky, and Eric Boyd for helpful conversations during the course of this research. This work was supported by grants from the Sloan Foundation to the Extreme Chemistry and Physics community of the Deep Carbon Observatory. Particular thanks are owed to Craig Manning and Wendy Mao for their patience while we prepared the results of this study for publication.

REFERENCES

Amend, J. P., & Shock, E. L. (2001). Energetics of overall metabolic reactions of thermophilic and hyperthermophilic Archaea and Bacteria. *FEMS Microbiology Reviews*, *25*(2), 175–243.

Apps, J. (2010). Geohydrological studies for nuclear waste isolation at the Hanford Reservation: Vol. I. Executive summary; Vol. II. Final report. *Lawrence Berkeley National Laboratory*.

Arcuri, E. J., & Ehrlich, H. L. (1977). Influence of hydrostatic pressure on the effects of the heavy metal cations of manganese, copper, cobalt, and nickel on the growth of three deep-sea bacterial isolates. *Applied and Environmental Microbiology*, *33*(2), 282–288.

Ask, D., Stephansson, O., & Cornet, F.H. (2001). Integrated stress analysis of hydraulic stress data in the Äspö region, Sweden. Analysis of hydraulic fracturing stress data and hydraulic tests on pre-existing fractures (HTPF) in boreholes KAS02, KAS03, and KLX02. *SKB International Progress Report*, IPR-01-26, Stockholm.

Bakermans, C., Ayala-del-Río, H. L., Ponder, M. A., Vishnivetskaya, T., Gilichinsky, D., Thomashow, M. F., & Tiedje, J. M. (2006). *Psychrobacter cryohalolentis* sp. nov. and *Psychrobacter arcticus* sp. nov., isolated from Siberian permafrost. *International Journal of Systematic and Evolutionary Microbiology*, *56*(6), 1285–1291.

Bale, S. J., Goodman, K., Rochelle, P. A., Marchesi, J. R., Fry, J. C., Weightman, A. J., & Parkes, R. J. (1997). *Desulfovibrio profundus* sp. nov., a novel barophilic sulfate-reducing bacterium from deep sediment layers in the Japan Sea. *International Journal of Systematic Bacteriology*, *47*(2), 515–521.

Bartlett, D. H. (2002). Pressure effects on in vivo microbial processes. *Biochimica et Biophysica Acta (BBA)-Protein Structure and Molecular Enzymology*, *1595*(1), 367–381.

Bennett, B. D., Kimball, E. H., Gao, M., Osterhout, R., Van Dien, S. J., & Rabinowitz, J. D. (2009). Absolute metabolite concentrations and implied enzyme active site occupancy in *Escherichia coli*. *Nature Chemical Biology*, *5*(8), 593–599.

Bernhardt, G., Jaenicke, R., Lüdemann, H. D., König, H., & Stetter, K. O. (1988). High pressure enhances the growth rate of the thermophilic archaebacterium *Methanococcus thermolithotrophicus* without extending its temperature range. *Applied and Environmental Microbiology*, *54*(5), 1258–1261.

Bonch-Osmolovskaya, E. A., Miroshnichenko, M. L., Lebedinsky, A. V., Chernyh, N. A., Nazina, T. N., Ivoilov, V.

S., et al. (2003). Radioisotopic, culture-based, and oligonucleotide microchip analyses of thermophilic microbial communities in a continental high-temperature petroleum reservoir. *Applied and Environmental Microbiology*, *69*(10), 6143–6151.

Böning, P., Brumsack, H. J., Böttcher, M. E., Schnetger, B., Kriete, C., Kallmeyer, J., & Borchers, S. L. (2004). Geochemistry of Peruvian near-surface sediments. *Geochimica et Cosmochimica Acta*, *68*(21), 4429–4451.

Brazelton, W. J., & Baross, J. A. (2009). Abundant transposases encoded by the metagenome of a hydrothermal chimney biofilm. *The ISME journal*, *3*(12), 1420–1424.

Breezee, J., Cady, N., & Staley, J. T. (2004). Subfreezing growth of the sea ice bacterium "*Psychromonas ingrahamii*". *Microbial Ecology*, *47*(3), 300–304.

Byers, H. K., Stackebrandt, E., Hayward, C., & Blackall, L. L. (1998). Molecular investigation of a microbial mat associated with the Great Artesian Basin. *FEMS Microbiology Ecology*, *25*(4), 391–403.

Canganella, F., Gonzalez, J. M., Yanagibayashi, M., Kato, C., & Horikoshi, K. (1997). Pressure and temperature effects on growth and viability of the hyperthermophilic archaeon *Thermococcus peptonophilus*. *Archives of Microbiology*, *168*(1), 1–7.

Canovas, P. C. III, & Shock, E. L. (2016). Geobiochemistry of metabolism: Standard state thermodynamic properties of the citric acid cycle. *Geochimica et Cosmochimica Acta*, *195*, 293–322.

Carvalho, V.M.L. (2013). Metagenomic Analysis of Mariana Trench Sediment Samples. Master's thesis, University of Copenhagen.

Charlou, J. L., Donval, J. P., Fouquet, Y., Jean-Baptiste, P., & Holm, N. (2002). Geochemistry of high H_2 and CH_4 vent fluids issuing from ultramafic rocks at the Rainbow hydrothermal field (36° 14′ N, MAR). *Chemical Geology*, *191*(4), 345–359.

Chivian, D., Brodie, E. L., Alm, E. J., Culley, D. E., Dehal, P. S., DeSantis, T. Z., et al. (2008). Environmental genomics reveals a single-species ecosystem deep within Earth. *Science*, *322*(5899), 275–278.

Collins, R. E., Rocap, G., & Deming, J. W. (2010). Persistence of bacterial and archaeal communities in sea ice through an Arctic winter. *Environmental Microbiology*, *12*(7), 1828–1841.

Colman, D. R., Poudel, S., Stamps, B.W., Boyd, E. S., & Spear J. R. (2017). The deep, hot biosphere: Twenty-five years of retrospection. *Proceedings of the National Academy of Sciences*, *114*(27), 6895–6903.

Daumas, S., Cord-Ruwisch, R., & Garcia, J. L. (1988). *Desulfotomaculum geothermicum* sp. nov., a thermophilic, fatty acid-degrading, sulfate-reducing bacterium isolated with H_2 from geothermal ground water. *Antonie van Leeuwenhoek*, *54*(2), 165–178.

Daumas, S., Lombart, R., & Bianchi, A. (1986). A bacteriological study of geothermal spring waters dating from the Dogger and Trias period in the Paris basin. *Geomicrobiology Journal*, *4*(4), 423–433.

D'Elia, T., Veerapaneni, R., & Rogers, S. O. (2008). Isolation of microbes from Lake Vostok accretion ice. *Applied and Environmental Microbiology*, *74*(15), 4962–4965.

DeLong, E. F., Franks, D. G., & Yayanos, A. A. (1997). Evolutionary relationships of cultivated psychrophilic and barophilic deep-sea bacteria. *Applied and Environmental Microbiology*, *63*(5), 2105–2108.

Delwiche, M., Colwell, F. S., Tseng, H.-Y., Gao, G., & Onstott, T. C. (1996). Pressure and temperature adaptation of a bacterium recovered from 2.8 kilometers beneath the surface of the earth. Abstr. 96th Annual Meeting American Society for Microbiology, New Orleans.

Deming, J. W., & Baross, J. A. (1986). Solid medium for culturing black smoker bacteria at temperatures to 120 °C. *Applied and Environmental Microbiology*, *51*(2), 238–243.

Deming, J. W., Somers, L. K., Straube, W. L., Swartz, D. G., & Macdonell, M. T. (1988). Isolation of an obligately barophilic bacterium and description of a new genus, *Colwellia* gen. nov. *Systematic and Applied Microbiology*, *10*(2), 152–160.

Ehrlich, H. L., Ghiorse, W. C., & Johnson G. L. (1972). Distribution of microbes in manganese modules from the Atlantic and Pacific oceans. *Developments in Industrial Microbiology*, *13*, 57–65.

Ekendahl, S., & Pedersen, K. (1994). Carbon transformations by attached bacterial populations in granitic groundwater from deep crystalline bed-rock of the Stripa research mine. *Microbiology*, *140*(7), 1565–1573.

Ellis, W. L., & Ege, J. R. (1975). Determination of *in situ* stress in U12g tunnel, Rainier Mesa, Nevada Test Site, Nevada. *U.S. Geological Survey Report RSGS* 474–219, 18 p.

Erauso, G., Reysenbach, A. L., Godfroy, A., Meunier, J. R., Crump, B., Partensky, F., et al. (1993). *Pyrococcus abyssi* sp. nov., a new hyperthermophilic archaeon isolated from a deep-sea hydrothermal vent. *Archives of Microbiology*, *160*(5), 338–349.

Facq, S., Daniel, I., Montagnac, G., Carbon, H., & Sverjensky, D. A. (2014). *In situ* Raman study and thermodynamic model of aqueous carbonate speciation in equilibrium with aragonite under subduction zone conditions. *Geochimica et Cosmochimica Acta*, *132*, 375–390.

Fardeau, M. L., Magot, M., Patel, B. K., Thomas, P., Garcia, J. L., & Ollivier, B. (2000). *Thermoanaerobacter subterraneus* sp. nov., a novel thermophile isolated from oilfield water. *International Journal of Systematic and Evolutionary Microbiology*, *50*(6), 2141–2149.

Fell, J. W. (1967) Distribution of yeasts in the Indian Ocean. *Bull. Mar. Sci.*, *17*, 454–470.

Grabowski, A., Tindall, B. J., Bardin, V., Blanchet, D., & Jeanthon, C. (2005). *Petrimonas sulfuriphila* gen. nov., sp. nov., a mesophilic fermentative bacterium isolated from a biodegraded oil reservoir. *International Journal of Systematic and Evolutionary Microbiology*, *55*(3), 1113–1121.

Greene, A. C., Patel, B. K., & Sheehy, A. J. (1997). *Deferribacter thermophilus* gen. nov., sp. nov., a novel thermophilic manganese- and iron-reducing bacterium isolated from a petroleum reservoir. *International Journal of Systematic Bacteriology*, *47*(2), 505–509.

Grossman, D., & Shulman, S. (1995). The biosphere below. *Earth (Waukesha, Wis.)*, *4*(3), 34–40.

Guild, M., & Shock, E. L. (2019). Carbon speciation in subduction zone fluids. In W. Mao & C. Manning (Eds.), *Deep Carbon* (this volume).

Haldeman, D. L., Amy, P. S., Ringelberg, D., & White, D. C. (1993). Characterization of the microbiology within a 21 m^3 section of rock from the deep subsurface. *Microbial Ecology*, *26*(2), 145–159.

Havig, J. R., Raymond, J., Meyer-Dombard, D. R., Zolotova, N., & Shock E. L. (2011). Merging isotopes and community genomics in a siliceous sinter-depositing hot spring, *Journal of Geophysical Research, 116*, G01005, doi:10.1029/2010JG001415.

Helgeson, H. C. (1979) Mass transfer among minerals and hydrothermal solutions. In H. L. Barnes (Ed.), *Geochemistry of Hydrothermal Ore Deposits* (pp. 568–610). New York: John Wiley.

Hoshino, T., & Inagaki, F. (2019). Abundance and distribution of Archaea in the subseafloor sedimentary biosphere. *The ISME Journal*, *13*, 227–231.

Huang, F., Daniel, I., Cardon, H., Montagnac, G., & Sverjensky D. A. (2017). Immiscible hydrocarbon fluids in the deep carbon cycle. *Nature Communications*, *8*, 15798. https://doi.org/10.1038/ncomms15798

Inagaki, F., Hinrichs, K.-U., Kubo, Y. Bowles, M. W., Heuer, V. B., Hong, W.-L., et al. (2015). Exploring deep microbial life in coal-bearing sediment down to ~2.5 km below the ocean floor. *Science*, *349*(6246), 420–424.

Inagaki, F., Takai, K., Hirayama, H., Yamato, Y., Nealson, K. H., & Horikoshi, K. (2003). Distribution and phylogenetic diversity of the subsurface microbial community in a Japanese epithermal gold mine. *Extremophiles*, *7*(4), 307–317.

Ino, K., Konno, U., Kouduka, M., Hirota, A., Togo, Y. S., Fukuda, A., et al. (2016). Deep microbial life in high-quality granitic groundwater from geochemically and geographically distinct underground boreholes. *Environmental Microbiology Reports*, *8*, 285–294. doi: 10.1111/1758-2229.12379

Ishii, N., Nakahigashi, K., Baba, T., Robert, M., Soga, T., Kanai, A., et al. (2007). Multiple high-throughput analyses monitor the response of *E. coli* to perturbations. *Science*, *316*(5824), 593–597.

Johnson, J. W., Oelkers, E. H., & Helgeson, H. C. (1992). SUPCRT92: A software package for calculating the standard molal thermodynamic properties of minerals, gases, aqueous species, and reactions from 1 to 5000 bar and 0 to 1000 C. *Computers & Geosciences*, *18*(7), 899–947.

Kaksonen, A. H., Plumb, J. J., Robertson, W. J., Spring, S., Schumann, P., Franzmann, P. D., & Puhakka, J. A. (2006). Novel thermophilic sulfate-reducing bacteria from a geothermally active underground mine in Japan. *Applied and Environmental Microbiology*, *72*(5), 3759–3762.

Kaksonen, A. H., Spring, S., Schumann, P., Kroppenstedt, R. M., & Puhakka, J. A. (2006). *Desulfotomaculum thermosubterraneum* sp. nov., a thermophilic sulfate-reducer isolated from an underground mine located in a geothermally active area. *International Journal of Systematic and Evolutionary Microbiology*, *56*(11), 2603–2608.

Kallmeyer, J., Pockalny, R., Adhikari, R. R., Smith, D. C., & D'Hondt, S. (2012). Global distribution of microbial abundance and biomass in sub-seafloor sediment. *Proceedings of the National Academy of Sciences*, *109*, 16213–16216.

Kato, C., Li, L., Nogi, Y., Nakamura, Y., Tamaoka, J., & Horikoshi, K. (1998). Extremely barophilic bacteria isolated from the Mariana Trench, Challenger Deep, at a depth of 11,000 meters. *Applied and Environmental Microbiology*, *64*(4), 1510–1513.

Kato, C., Li, L., Tamaoka, J., & Horikoshi, K. (1997). Molecular analyses of the sediment of the 11000-m deep Mariana Trench. *Extremophiles*, *1*(3), 117–123.

Kato, C., Masui, N., & Horikoshi, K. (1996). Properties of obligately barophilic bacteria isolated from a sample of deep-sea sediment from the Izu-Bonin Trench, *Journal of Marine Biotechnology*, *4*, 96–99.

Kato, C., Sato, T., & Horikoshi, K. (1995). Isolation and properties of barophilic and barotolerant bacteria from deep-sea mud samples. *Biodiversity & Conservation*, *4*(1), 1–9.

Kieft, T. L., Fredrickson, J. K., Onstott, T. C., Gorby, Y. A., Kostandarithes, H. M., Bailey, T. J., et al. (1999). Dissimilatory reduction of Fe (III) and other electron acceptors by a Thermus isolate. *Applied and Environmental Microbiology*, *65*(3), 1214–1221.

Kieft, T. L., McCuddy, S. M., Onstott, T. C., Davidson, M., Lin, L. H., Mislowack, B., et al. (2005). Geochemically generated, energy-rich substrates and indigenous microorganisms in deep, ancient groundwater. *Geomicrobiology Journal*, *22*(6), 325–335.

Kimura, H., Sugihara, M., Yamamoto, H., Patel, B. K., Kato, K., & Hanada, S. (2005). Microbial community in a geothermal aquifer associated with the subsurface of the Great Artesian Basin, Australia. *Extremophiles*, *9*(5), 407–414.

Klouche, N., Fardeau, M. L., Lascourrèges, J. F., Cayol, J. L., Hacene, H., Thomas, P., & Magot, M. (2007). *Geosporobacter subterraneus* gen. nov., sp. nov., a spore-forming bacterium isolated from a deep subsurface aquifer. *International Journal of Systematic and Evolutionary Microbiology*, *57*(8), 1757–1761.

Kotelnikova, S., Macario, A. J., & Pedersen, K. (1998). *Methanobacterium subterraneum* sp. nov., a new alkaliphilic, eurythermic and halotolerant methanogen isolated from deep granitic groundwater. *International Journal of Systematic Bacteriology*, *48*(2), 357–367.

Kotelnikova, S., & Pedersen, K. (1998). Microbial oxygen consumption in Äspö tunnel environment, SKB-PR-HRL-98-1, Swedish Nuclear Fuel Waste Management Company, Stockholm, Sweden.

Kotlar, H. K., Lewin, A., Johansen, J., Throne-Holst, M., Haverkamp, T., Markussen, S., et al. (2011). High coverage sequencing of DNA from microorganisms living in an oil reservoir 2.5 kilometres subsurface. *Environmental Microbiology Reports*, *3*(6), 674–681.

LaRowe, D. E., & Helgeson, H. C. (2006a). Biomolecules in hydrothermal systems: Calculation of the standard molal thermodynamic properties of nucleic-acid bases, nucleosides, and nucleotides at elevated temperatures and pressures. *Geochimica et Cosmochimica Acta*, *70*, 4680–4724.

LaRowe, D. E. & Helgeson, H. C. (2006b). The energetics of metabolism in hydrothermal systems: Calculation of the standard molal thermodynamic properties of magnesium complexed adenosine nucleotides and NAD and NADP at elevated temperatures and pressures. *Thermochimica Acta*, *448*, 82–106.

L'Haridon, S., Reysenbach, A. L., Glenat, P., Prieur, D., & Jeanthon, C. (1995). Hot subterranean biosphere in a continental oil reservoir. *Nature*, *377*(6546), 223–224.

Li, H., Yang, S. Z., Mu, B. Z., Rong, Z. F., & Zhang, J. (2006). Molecular analysis of the bacterial community in a continental high-temperature and water-flooded petroleum reservoir. *FEMS Microbiology Letters*, *257*(1), 92–98.

Li, H., Yang, S. Z., Mu, B. Z., Rong, Z. F., & Zhang, J. (2007). Molecular phylogenetic diversity of the microbial community associated with a high-temperature petroleum reservoir at an offshore oilfield. *FEMS Microbiology Ecology*, *60*(1), 74–84.

Li, X. L., Bastiaens, W., Van Marcke, P., Verstricht, J., Chen, G. J., Weetjens, E., & Sillen, X. (2010). Design and development of large-scale in-situ PRACLAY heater test and horizontal high-level radioactive waste disposal gallery seal test in Belgian HADES. *Journal of Rock Mechanics and Geotechnical Engineering*, *2*(2), 103–110.

Liesack, W., Weyland, H., & Stackebrandt, E. (1991). Potential risks of gene amplification by PCR as determined by 16S rDNA analysis of a mixed-culture of strict barophilic bacteria. *Microbial Ecology*, *21*(1), 191–198.

Liu, Y., Karnauchow, T. M., Jarrell, K. F., Balkwill, D. L., Drake, G. R., Ringelberg, D., Clarno, R. & Boone, D. R. (1997). Description of two new thermophilic *Desulfotomaculum* spp., *Desulfotomaculum putei* sp. nov., from a deep terrestrial subsurface, and *Desulfotomaculum luciae* sp. nov., from a hot spring. *International Journal of Systematic and Evolutionary Microbiology*, *47*(3), 615–621.

Loiacono, S. T., Meyer-Dombard, D.A.R., Havig, J. R., Poret-Peterson, A. T., Hartnett, H. E., & Shock, E. L. (2012). Evidence for high-temperature in situ nifH transcription in an alkaline hot spring of Lower Geyser Basin, Yellowstone National Park. *Environmental Microbiology*, *14*(5), 1272–1283.

Love, C. A., Patel, B.K.C., Nichols, P. D., & Stackebrandt, E. (1993). *Desulfotomaculum australicum*, sp. nov., a thermophilic sulfate-reducing bacterium isolated from the Great Artesian Basin of Australia. *Systematic and Applied Microbiology*, *16*(2), 244–251.

Lu, S., Eiteman, M. A., & Altman, E. (2009). Effect of CO_2 on succinate production in dual-phase *Escherichia coli* fermentations. *Journal of Biotechnology*, *143*(3), 213–223.

Magnabosco, C., Lin, L.-H., Dong, H., Bomberg, M., Ghiorse, W., Stan-Lotter, H., et al. (2018). The biomass and biodiversity of the continental subsurface. *Nature Geoscience*, *11*, 707–717.

Marteinsson, V. T., Birrien, J. L., Reysenbach, A. L., Vernet, M., Marie, D., Gambacorta, A., et al. (1999). *Thermococcus barophilus* sp. nov., a new barophilic and hyperthermophilic archaeon isolated under high hydrostatic pressure from a deep-sea hydrothermal vent. *International Journal of Systematic and Evolutionary Microbiology*, *49*(2), 351–359.

Marteinsson, V. T., Moulin, P., Birrien, J., Gambacorta, A., Vernet, M., & Prieur, D. (1997). Physiological responses to stress conditions and barophilic behavior of the hyperthermophilic vent archaeon *Pyrococcus abyssi*. *Applied and Environmental Microbiology*, *63*(4), 1230–1236.

Marteinsson, V. T., Runarsson, A., Stefánsson, A., Thorsteinsson, T., Johannesson, T., Magnusson, S. H., et al. (2013). Microbial communities in the subglacial waters of the Vatnajökull ice cap, Iceland. *The ISME Journal*, *7*(2), 427–437.

McCollom, T. (2007) Geochemical constraints on sources of metabolic energy for chemolithoautotrophy in ultramafic-hosted deep-sea hydrothermal systems. *Astrobiology*, *7*, 933–950.

Meersman, F., Daniel, I., Bartlett, D. H., Winter, R., Hazel, R., & McMillan, P. F. (2013). High-pressure biochemistry and biophysics. *Reviews in Mineralogy and Geochemistry*, *75*, 607–648.

Merlin, C., Masters, M., McAteer, S., & Coulson, A. (2003). Why is carbonic anhydrase essential to *Escherichia coli*? *Journal of Bacteriology*, *185*(21), 6415–6424.

Meyer-Dombard, D.A.R., Swingley, W., Raymond, J., Havig, J., Shock, E. L., & Summons, R. E. (2011). Hydrothermal ecotones and streamer biofilm communities in the Lower Geyser Basin, Yellowstone National Park. *Environmental Microbiology*, *13*(8), 2216–2231.

Miller, J. F., Shah, N. N., Nelson, C. M., Ludlow, J. M., & Clark, D. S. (1988). Pressure and temperature effects on growth and methane production of the extreme thermophile *Methanococcus jannaschii*. *Applied and Environmental Microbiology*, *54*(12), 3039–3042.

Miyoshi, T., Iwatsuki, T., & Naganuma, T. (2005). Phylogenetic characterization of 16S rRNA gene clones from deep-groundwater microorganisms that pass through 0.2-micrometer-pore-size filters. *Applied and Environmental Microbiology*, *71*(2), 1084–1088.

Mochimaru, H., Yoshioka, H., Tamaki, H., Nakamura, K., Kaneko, N., Sakata, S., Imachi, H., Sekiguchi, Y., Uchiyama, H. & Kamagata, Y. (2007). Microbial diversity and methanogenic potential in a high temperature natural gas field in Japan. *Extremophiles*, *11*(3), 453–461.

Mori, K., Hanada, S., Maruyama, A., & Marumo, K. (2002). *Thermanaeromonas toyohensis* gen. nov., sp. nov., a novel thermophilic anaerobe isolated from a subterranean vein in the Toyoha Mines. *International Journal of Systematic and Evolutionary Microbiology*, *52*(5), 1675–1680.

Morita, R. Y., & ZoBell, C. E. (1955). Occurrence of bacteria in pelagic sediments collected during the Mid-Pacific Expedition. *Deep Sea Research (1953)*, *3*(1), 66–73.

Motamedi, M., & Pedersen, K. (1998). Note on *Desulfovibrio aespoeensis* sp. nov., a mesophilic sulfate-reducing bacterium from deep groundwater at Äspö hard rock laboratory, Sweden. *International Journal of Systematic Bacteriology*, *48*(1), 311–315.

Nakai R., Abe T., Takeyama H., & Naganuma T. (2011). Metagenomic analysis of 0.2-mm-passable microorganisms in deep-sea hydrothermal fluid. *Marine Biotechnology*, *13*, 900–908.

Nazina, T. N., Grigor'Yan, A. A., Shestakova, N. M., Babich, T. L., Ivoilov, V. S., Feng, Q., et al. (2007). Microbiological investigations of high-temperature horizons of the Kongdian petroleum reservoir in connection with field trial of a biotechnology for enhancement of oil recovery. *Microbiology*, *76*(3), 287–296.

Nilsen, R. K., & Torsvik, T. (1996). *Methanococcus thermolithotrophicus* isolated from North Sea oil field reservoir water. *Applied and Environmental Microbiology*, *62*(2), 728–731.

Nilsen, R. K., Beeder, J., Thorstenson, T., & Torsvik, T. (1996). Distribution of thermophilic marine sulfate reducers in

North Sea oil field waters and oil reservoirs. *Applied and Environmental Microbiology*, *62*(5), 1793–1798.

Nilsen, R. K., Torsvik, T., & Lien, T. (1996). *Desulfotomaculum thermocisternum* sp. nov., a sulfate reducer isolated from a hot North Sea oil reservoir. *International Journal of Systematic Bacteriology*, *46*(2), 397–402.

Nogi, Y., Masui, N., & Kato, C. (1998). *Photobacterium profundum* sp. nov., a new, moderately barophilic bacterial species isolated from a deep-sea sediment. *Extremophiles*, *2*(1), 1–8.

Nunoura, T., Hirayama, H., Takami, H., Oida, H., Nishi, S., Shimamura, S., et al. (2005). Genetic and functional properties of uncultivated thermophilic crenarchaeotes from a subsurface gold mine as revealed by analysis of genome fragments. *Environmental Microbiology*, *7*(12), 1967–1984.

Olson, G. J., Dockins, W. S., McFeters, G. A., & Iverson, W. P. (1981). Sulfate-reducing and methanogenic bacteria from deep aquifers in Montana. *Geomicrobiology Journal*, *2*(4), 327–340.

Onstott, T. C., Phelps, T. J., Colwell, F. S., Ringelberg, D., White, D. C., Boone, D. R., et al. (1998). Observations pertaining to the origin and ecology of microorganisms recovered from the deep subsurface of Taylorsville Basin, Virginia. *Geomicrobiology Journal*, *15*(4), 353–385.

Onstott, T. C., Tobin, K., Dong, H., DeFlaun, M. F., Fredrickson, J. K., Bailey, T., et al. (1997). Deep gold mines of South Africa: Windows into the subsurface biosphere. *Optical Science, Engineering and Instrumentation '97* (344–357). International Society for Optics and Photonics.

Panikov, N. S., & Sizova, M. V. (2007). Growth kinetics of microorganisms isolated from Alaskan soil and permafrost in solid media frozen down to −35 C. *FEMS Microbiology Ecology*, *59*(2), 500–512.

Parkes, R. J., Cragg, B. A., Bale, S. J., Getliff, J. M., Goodman, K., Rochelle, P. A., et al. (1994). Deep bacterial biosphere in Pacific Ocean sediments. *Nature*, *371*(6496), 410–413.

Pedersen, K. (1997). Microbial life in deep granitic rock. *FEMS Microbiology Reviews*, *20*(3-4), 399–414.

Pedersen, K., & Ekendahl, S. (1990). Distribution and activity of bacteria in deep granitic groundwaters of southeastern Sweden. *Microbial Ecology*, *20*(1), 37–52.

Peng, L., Arauzo-Bravo, M. J., & Shimizu, K. (2004). Metabolic flux analysis for a ppc mutant *Escherichia coli* based on ^{13}C-labelling experiments together with enzyme activity assays and intracellular metabolite measurements. *FEMS Microbiology Letters*, *235*(1), 17–23.

Phillips, R., Kondev, J., Theriot, J., & Garcia, H. (2008). *Physical Biology of the Cell*. New York: Garland Science.

Pope, D. H., Smith, W. P., Swartz, R. W., & Landau, J. V. (1975). Role of bacterial ribosomes in barotolerance. *Journal of Bacteriology*, *121*, 664–669.

Price, P. B., & Sowers, T. (2004). Temperature dependence of metabolic rates for microbial growth, maintenance, and survival. *Proceedings of the National Academy of Sciences of the United States of America*, *101*(13), 4631–4636.

Rivkina, E. M., Friedmann, E. I., McKay, C. P., & Gilichinsky, D. A. (2000). Metabolic activity of permafrost bacteria below the freezing point. *Applied and Environmental Microbiology*, *66*(8), 3230–3233.

Rosnes, J. T., Torsvik, T., & Lien, T. (1991). Spore-forming thermophilic sulfate-reducing bacteria isolated from North Sea oil field waters. *Applied and Environmental Microbiology*, *57*(8), 2302–2307.

Russell, C. E., Jacobson, R., Haldeman, D. L., & Amy, P. S. (1994). Heterogeneity of deep subsurface microorganisms and correlations to hydrogeological and geochemical parameters. *Geomicrobiology Journal*, *12*(1), 37–51.

Sahl, J. W., Schmidt, R., Swanner, E. D., Mandernack, K. W., Templeton, A. S., Kieft, T. L., et al. (2008). Subsurface microbial diversity in deep-granitic-fracture water in Colorado. *Applied and Environmental Microbiology*, *74*(1), 143–152.

Salamatin, A. N., Lipenkov, V. Y., Barkov, N. I., Jouzel, J., Petit, J. R., & Raynaud, D. (1998). Ice core age dating and paleothermometer calibration based on isotope and temperature profiles from deep boreholes at Vostok Station (East Antarctica). *Journal of Geophysical Research: Atmospheres (1984–2012)*, *103*(D8), 8963–8977.

Salisbury, M. H., Shinohara, M., Richter, C., Araki, E., Barr, S. R., D'Antonio, M., et al. (2002). Leg 195 summary. In M. H. Salisbury, et al. (Eds.), *Proceedings of the Ocean Drilling Program, Initial Reports* (Vol. 195, pp. 1–63). College Station, Texas: Ocean Drilling Program.

Schippers, A., & Neretin, L. N. (2006). Quantification of microbial communities in near-surface and deeply buried marine sediments on the Peru continental margin using real-time PCR. *Environmental Microbiology*, *8*(7), 1251–1260.

Scholander, P. F., Bradstreet, E. D., Hemmingsen, E. A., & Hammel, H. T. (1965). Sap pressure in vascular plants: Negative hydrostatic pressure can be measured in plants. *Science*, *148*(3668), 339–346.

Schwarz, J. R., & Colwell, R. R. (1975a). Macromolecular biosynthesis in Pseudomonas bathycetes at deep-sea pressure and temperature. In Proceedings, 75th annual meeting of the American Society for Microbiology, American Society for Microbiology, Washington DC, p. 162.

Schwarz, J. R., & Colwell, R. R. (1975b). Heterotrophic activity of deep-sea sediment bacteria. *Applied Microbiology*, *30*(4), 639–649.

Schwarz, J. R., Walker, J. D., & Colwell, R. R. (1975). Deep-sea bacteria: Growth and utilization of n-hexadecane at in situ temperature and pressure. *Canadian Journal of Microbiology*, *21*(5), 682–687.

Sharma, A., Scott, J. H., Cody, G. D., Fogel, M. L., Hazen, R. M., Hemley, R. J., & Huntress, W. T. (2002). Microbial activity at gigapascal pressures. *Science*, *295*(5559), 1514–1516.

Shi, T., Reeves, R. H., Gilichinsky, D. A., & Friedmann, E. I. (1997). Characterization of viable bacteria from Siberian permafrost by 16S rDNA sequencing. *Microbial Ecology*, *33*(3), 169–179.

Shock, E. L. (2009). Minerals as energy sources for microorganisms. *Economic Geology*, *104*(8), 1235–1248

Shock, E. L., & Boyd, E. S. (2015). Principles of geobiochemistry. *Elements*, *11*, 395–401.

Shock, E. L., & Canovas, P. C. (2010). The potential for abiotic organic synthesis and biosynthesis at seafloor hydrothermal systems. *Geofluids*, *10*, 161–192.

Shock, E. L., Oelkers, E. H., Johnson, J. W., Sverjensky, D. A., & Helgeson, H. C. (1992). Calculation of the thermodynamic properties of aqueous species at high pressures and temperatures: Effective electrostatic radii, dissociation constants, and standard partial molal properties to 1000 °C and 5 kb. *Journal of the Chemical Society, Faraday Transactions*, *88*, 803–826.

Stetter, K. O., Huber, R., Blöchl, E., Kurr, M., Eden, R. D., Fielder, M., et al. (1993). Hyperthermophilic archaea are thriving in deep North Sea and Alaskan oil reservoirs. *Nature*, *365*(6448), 743–745.

Steurer, J. F., & Underwood, M. B. (2003). Data report: The relation between physical properties and grain-size variations in hemipelagic sediments from Nankai Trough. In *Proceedings of the Ocean Drilling Program, Scientific Results*, *190*(196), 1–25.

Stevens, T. O., & McKinley, J. P. (1995). Lithoautotrophic microbial ecosystems in deep basalt aquifers. *Science*, *270*(5235), 450–455.

Stevens, T. O., McKinley, J. P., & Fredrickson, J. K. (1993). Bacteria associated with deep, alkaline, anaerobic groundwaters in southeast Washington. *Microbial Ecology*, *25*(1), 35–50.

Sundararaj, S., Guo, A., Habibi-Nazhad, B., Rouani, M., Stothard, P., Ellison, M., & Wishart, D. S. (2004). The Cybercell Database (CCDB): A comprehensive, self-updating, relational database to coordinate and facilitate *in silico* modeling of *Escherichia coli*. *Nucleic Acids Research*, *32*(Database issue), D293–D295.

Sverjensky, D. A., Harrison, B., & Azzolini, D. (2014). Water in the deep Earth: The dielectric constant and the solubilities of quartz and corundum. *Geochimica et Cosmochimica Acta*, *129*, 125–145.

Sverjensky, D. A., Stagno, V., & Huang, F. (2014). Important role of organic carbon in subduction-zone fluids in the deep carbon cycle. *Nature Geoscience*, *7*, 9–13.

Syracuse, E. M., van Keken, P. E., & Abers, G. A. (2010). The global range of subduction zone thermal models. *Physics of the Earth and Planetary Interiors*, *183*(1), 73–90.

Szewzyk, U., Szewzyk, R., & Stenström, T. A. (1994). Thermophilic, anaerobic bacteria isolated from a deep borehole in granite in Sweden. *Proceedings of the National Academy of Sciences*, *91*(5), 1810–1813.

Szewzyk, U., Szewzyk, R., & Stenstroem, T. A. (1997). Thermophilic fermentative bacteria from a deep borehole in granitic rock in Sweden. In *Optical Science, Engineering and Instrumentation '97* (pp. 330–334). International Society for Optics and Photonics.

Takai, K., Hirayama, H., Sakihama, Y., Inagaki, F., Yamato, Y., & Horikoshi, K. (2002). Isolation and metabolic characteristics of previously uncultured members of the order Aquificales in a subsurface gold mine. *Applied and Environmental Microbiology*, *68*(6), 3046–3054.

Takai, K., Moser, D. P., DeFlaun, M., Onstott, T. C., & Fredrickson, J. K. (2001). Archaeal diversity in waters from deep South African gold mines. *Applied and Environmental Microbiology*, *67*(12), 5750–5760.

Takai, K., Moser, D. P., Onstott, T. C., Spoelstra, N., Pfiffner, S. M., Dohnalkova, A., & Fredrickson, J. K. (2001). *Alkaliphilus transvaalensis* gen. nov., sp. nov., an extremely alkaliphilic bacterium isolated from a deep South African gold mine. *International Journal of Systematic and Evolutionary Microbiology*, *51*(4), 1245–1256.

Takai, K., Moyer, C. L., Miyazaki, M., Nogi, Y., Hirayama, H., Nealson, K. H., & Horikoshi, K. (2005). *Marinobacter alkaliphilus* sp. nov., a novel alkaliphilic bacterium isolated from subseafloor alkaline serpentine mud from Ocean Drilling Program Site 1200 at South Chamorro Seamount, Mariana Forearc. *Extremophiles*, *9*(1), 17–27.

Takai, K., Nakamura, K., Toki, T., Tsunogai, U., Miyazaki, M., Miyazaki, J., et al. (2008). Cell proliferation at 122 °C and isotopically heavy CH_4 production by a hyperthermophilic methanogen under high-pressure cultivation. *Proceedings of the National Academy of Sciences*, *105*(31), 10949–10954.

Tanaka, T., Burgess, J., & Wright, P. (2001). High-pressure adaptation by salt stress in a moderately halophilic bacterium obtained from open seawater. *Applied Microbiology and Biotechnology*, *57*(1–2), 200–204.

Tao, R., Zhang, L., Tian, M., Zhu, J., Liu, X., Liu, J., et al. (2018). Formation of abiotic hydrocarbon from reduction of carbonate in subduction zones: Constraints from petrological observation and experimental simulation. *Geochimica et Cosmochimica Acta*, *239*, 390–408.

Tardy-Jacquenod, C., Magot, M., Patel, B.K.C., Matheron, R., & Caumette, P. (1998). *Desulfotomaculum halophilum* sp. nov., a halophilic sulfate-reducing bacterium isolated from oil production facilities. *International Journal of Systematic Bacteriology*, *48*(2), 333–338.

Tobal, G. M. (1993). Purification and characterization of a malate dehydrogenase from the marine bacterium *Shewanella* SC2. Master's Thesis, Scripps Institute of Oceanography, University of California–San Diego, La Jolla, CA.

Trembath-Reichert, E., Morono, Y., Ijira, A., Hoshino, T., Dawson, K. S., Inagaki, F. & Orphan, V. J. (2017). Methyl-compound use and slow growth characterize microbial life in 2-km-deep subseafloor coal and shale beds. *Proceedings of the National Academy of Sciences*, *114*, E9206–E9215.

Trimarco, E., Balkwill, D., Davidson, M., & Onstott, T. C. (2006). *In situ* enrichment of a diverse community of bacteria from a 4–5 km deep fault zone in South Africa. *Geomicrobiology Journal*, *23*(6), 463–473.

Turley, C. (2000). Bacteria in the cold deep-sea benthic boundary layer and sediment–water interface of the NE Atlantic. *FEMS Microbiology Ecology*, *33*(2), 89–99.

Vanlint, D., Mitchell, R., Bailey, E., Meersman, F., McMillan, P. F., Michiels, C. W., & Aertsen, A. (2011). Rapid acquisition of gigapascal-high-pressure resistance by *Escherichia coli*. *Mbio*, *2*(1), e00130–10.

Wanger, G., Moser, D., Hay, M., Myneni, S., Onstott, T. C., & Southam, G. (2012). Mobile hydrocarbon microspheres from > 2-billion-year-old carbon-bearing seams in the South African deep subsurface. *Geobiology*, *10*(6), 496–505.

Whitman, W. B., Coleman, D. C., & Wiebe, W. J. (1998). Prokaryotes: The unseen majority. *Proceedings of the National Academy of Sciences*, *95*, 6578–6583.

Winnock, E., & Pontalier, Y. (1970). Lacq gas field, France. In *AAPG Memoir 14: Geology of Giant Petroleum Fields* (pp. 370–387). Tulsa, OK: American Association of Petroleum Geologists.

Wouters, K., Moors, H., Boven, P., & Leys, N. (2013). Evidence and characteristics of a diverse and metabolically active microbial community in deep subsurface clay borehole water. *FEMS Microbiology Ecology*, *86*(3), 458–473.

Wynter, C., Patel, B.K.C., Bain, P., Jersey, J. D., Hamilton, S., & Inkerman, P. A. (1996). A novel thermostable dextranase from a *Thermoanaerobacter* species cultured from the geothermal waters of the Great Artesian Basin of Australia. *FEMS Microbiology Letters*, *140*(2-3), 271–276.

Yanagibayashi, M., Nogi, Y., Li, L., & Kato, C. (1999). Changes in the microbial community in Japan Trench sediment from a depth of 6292 m during cultivation without decompression. *FEMS Microbiology Letters*, *170*(1), 271–279.

Yayanos, A. A. (1986). Evolutional and ecological implications of the properties of deep-sea barophilic bacteria. *Proceedings of the National Academy of Sciences*, *83*(24), 9542–9546.

Yayanos, A. A., & Dietz, A. S. (1982). Thermal inactivation of a deep-sea barophilic bacterium, isolate CNPT-3. *Applied and Environmental Microbiology*, *43*(6), 1481–1489.

Yayanos, A. A., Dietz, A. S., & Van Boxtel, R. (1981). Obligately barophilic bacterium from the Mariana Trench. *Proceedings of the National Academy of Sciences*, *78*(8), 5212–5215.

Yayanos, A. A., Dietz, A. S., & Van Boxtel, R. (1982). Dependence of reproduction rate on pressure as a hallmark of deep-sea bacteria. *Applied and Environmental Microbiology*, *44*(6), 1356–1361.

Yoshioka, H., Sakata, S., Cragg, B. A., Parkes, R. J., & Fujii, T. (2009). Microbial methane production rates in gas hydrate-bearing sediments from the eastern Nankai Trough, off central Japan. *Geochemical Journal*, *43*(5), 315–321.

Zhang, G., Dong, H., Jiang, H., Xu, Z., & Eberl, D. D. (2006). Unique microbial community in drilling fluids from Chinese continental scientific drilling. *Geomicrobiology Journal*, *23*(6), 499–514.

Zhang, G., Dong, H., Xu, Z., Zhao, D., & Zhang, C. (2005). Microbial diversity in ultra-high-pressure rocks and fluids from the Chinese Continental Scientific Drilling Project in China. *Applied and Environmental Microbiology*, *71*(6), 3213–3227.

Zimov, S. A., Schuur, E. A., & Chapin III, F. S. (2006). Permafrost and the global carbon budget. *Science*, *312*(5780), 1612–1613.

Zink, K. G., Wilkes, H., Disko, U., Elvert, M., & Horsfield, B. (2003). Intact phospholipids—microbial "life markers" in marine deep subsurface sediments. *Organic Geochemistry*, *34*(6), 755–769.

ZoBell, C. E. (1952). Bacterial life at the bottom of the Philippine Trench. *Science*, *115*(2993), 507–508.

ZoBell, C. E. (1958). Ecology of sulfate reducing bacteria. *Producers Monthly*, *22*(7), 12–29.

ZoBell, C. E., & Morita, R. Y. (1957). Barophilic bacteria in some deep sea sediments. *Journal of Bacteriology*, *73*(4), 563.

26

Deep Hydrocarbon Cycle: An Experimental Simulation

Vladimir Kutcherov[1,2], Kirill Ivanov[3], Elena Mukhina[4], and Aleksandr Serovaiskii[2]

ABSTRACT

The concept of a deep hydrocarbon cycle is proposed based on results of experimental modeling of the transformation of hydrocarbons under extreme thermobaric conditions. Hydrocarbons immersed in the subducting slab generally maintain stability to a depth of 50 km. With deeper immersion, the integrity of the traps is disrupted and the hydrocarbon fluid contacts the surrounding ferrous minerals, forming a mixture of iron hydride and iron carbide. This iron carbide transported into the asthenosphere by convective flows can react with hydrogen or water and form an aqueous hydrocarbon fluid that can migrate through deep faults into the Earth's crust and form multilayer oil and gas deposits. Other carbon donors in addition to iron carbide from the subducting slab exist in the asthenosphere. These donors can serve as a source of deep hydrocarbons that participate in the deep hydrocarbon cycle, as well as an additional feed for the general upward flow of the water-hydrocarbon fluid. Geological data on the presence of hydrocarbons in ultrabasites squeezed from a slab indicate that complex hydrocarbon systems may exist in a slab at considerable depths. This confirms our experimental results, indicating the stability of hydrocarbons to a depth of 50 km.

26.1. INTRODUCTION

Descriptions of the global carbon cycle are generally limited to processes occurring in the oceans, the atmosphere, and in the surface sedimentary layer of the Earth's crust. Knowledge of the deep carbon cycle is fragmentary, despite the fact that the deep layers of the Earth may contain up to 90% of all planetary carbon (Javoy, 1997). It is assumed that the upward flux of carbon is formed mainly from CO_2 and CH_4 during volcanic eruptions and the downward flow is realized in subduction zones (Manning, 2014). Subduction processes play a key role in the evolution of the continental crust and upper mantle (Dasgupta et al., 2013; Stern, 2002).

Despite the significant progress in understanding the processes of subduction, the potential role of hydrocarbons in subduction zones is not well understood and has long been debated (Manning, 2004). Numerous major petroleum deposits have been located close to subduction zones (Hessler & Sharman, 2018; Mann et al., 2003). Hydrocarbons located in these zones sink with the subducting slab while being exposed to extremely high temperatures and pressures. Taking into consideration the ~44,500 km length of global subduction zones, Kelemen and Manning (2015) suggest that the total amount of immersed hydrocarbons is likely significant.

[1] *KTH Royal Institute of Technology, Stockholm, Sweden*

[2] *Gubkin University, Moscow, Russia*

[3] *The Zavaritsky Institute of Geology and Geochemistry, Ekaterinburg, Russia*

[4] *Skolkovo Institute of Science and Technology, Moscow, Russia*

Carbon in Earth's Interior, Geophysical Monograph 249, First Edition. Edited by Craig E. Manning, Jung-Fu Lin, and Wendy L. Mao.

Furthermore, hydrocarbons in the ocean crust sinking with the subducting slab could have been generated inside the slab at a comparably shallow depth. The experimental results presented in Mukhina et al. (2017) demonstrated that the formation of complex hydrocarbon mixtures can occur at a depth of 70–80 km in the subducting slabs. Petrological observation and experimental simulation also confirmed that carbonates and water sinking with the subducting slab are transformed into graphite and light saturated hydrocarbons (Tao et al., 2018). Estimations of the carbon content of subduction-zone fluids show that organic carbon species may also be present in the deep Earth (Sverjensky et al., 2014).

The abiogenic formation of complex hydrocarbon systems in the upper and lower mantle and their upward migration to the Earth's crust was investigated in a series of studies (Belonoshko et al., 2015; Kenney et al., 2002; Kolesnikov et al., 2009; Kutcherov et al., 2010; Sokol et al., 2017; Sonin et al., 2014). Although Kolesnikov et al. (2017) conducted a thorough review of theoretical and experimental studies on the possible pathways for the formation of complex hydrocarbons at pressures and temperatures of the upper mantle, the behavior of hydrocarbons in subducting slabs is still not well understood.

In the subduction zone, at a depth of 50–80 km, the temperature at the boundary of the plates coincides with the continental geotherm and reaches 600°–700 °C (Karato, 2013). If pressure increases to 0.5–1 GPa, the melting temperature for the majority of silicates in the presence of water drops to 600°–700 °C (Bezmen et al., 2005; Shimada, 1969). Water-saturated carbonates, the predominant carbon-bearing phases commonly present in the subducting slab, behave similarly (Cooper et al., 1975). Concurrently, alumosilicates and water-saturated carbonates also melt at depths of 50–80 km. Thus, we suggest that hydrocarbon traps retain their integrity at depths of 50–80 km. With further immersion, the integrity of the traps is disrupted and the hydrocarbon fluid begins to contact the surrounding iron-bearing minerals.

26.2. MATERIALS AND METHODS

Two different types of high-pressure equipment, described below, were employed: diamond anvil cells and the Toroid-type large reactive volume unit.

26.2.1. Diamond Anvil Cells

Diamond anvil cells with a culet diameter of 250 μm were used with steel or rhenium gaskets (250 μm thickness) with the 125 μm hole drilled in the center. The hole was filled with the liquid hydrocarbon system that also served as a pressure medium in the experiments. Cell temperature and pressure were measured by the fluorescence of the ruby (Cr-doped Al_2O_3) and Sm:YAG (Sm-doped $Y_3Al_5O_{12}$) chips loaded into the chamber together with the sample (Mao et al., 1986; Trots et al., 2013). Two different heating methods were applied in the experiments: resistive and laser heating.

The resistive heating method was used for the experiments with temperatures up to 450 °C, carried out by a Pt heater installed inside the body of the cell. The temperature was measured by the ruby fluorescence. A Pt/Pt-Rh (10%) thermocouple was mounted on the side surface of the diamond for additional temperature control during heating. Prior to heating, the sample was loaded into the chamber and the cell was closed and pressurized using the screws on the body of the cell. The temperature was increased at a rate of 50°–60 °C per hour, according to the thermocouple readings. The temperature and pressure inside the cell were monitored two to three times per hour by measuring the fluorescence of ruby and Sm:YAG. The pressure was kept constant during the heating. After the required temperature was reached, the cell was kept at the required thermobaric parameters for several hours. Then the heating was stopped and the sample was quenched. The sample analysis was carried out by means of the Raman spectrometer LabRam coupled with the He-Ne laser (514.5 nm wavelength and the power 0.001–0.6 W).

In the laser heating experiments, the systems, consisting of a mixture of saturated hydrocarbons C_{15}-C_{40} (99.9%, Merch KGaA, EMD Millipore Chemical 1.07160.1000) and ^{57}Fe-enriched wüstite $Fe_{0.94}O$ (in the first series) and pyroxene-like glass $(Mg_{0.91}Fe_{0.09})(Si_{0.91}Al_{0.09})O_3$ (in the second series), were heated by the portable laser heating setup, with two optical fibre-based opposite-mounted 50 and 100 W (900°–1500 °C) power lasers (Kupenko et al., 2012). The temperature of the heated surface of the sample was measured by means of multiwavelength spectroradiometry of the sample radiation (Dubrovinsky & Saxena, 1999). The temperature radiation spectra were collected in the wavelength range of visible light and near infrared light (600–900 nm); afterwards, the spectra were fitted by means of the Planck function. The heating experiment continued for 5 minutes with constant movement of the focal point of the laser for equal heating of the surface of the sample. The analysis was carried out by means of conventional Mössbauer spectroscopy (with the source ^{57}Co) and Raman spectroscopy at experimental pressure and ambient temperature before and after heating. The isomer shift and the velocity scale were calibrated relative to α-Fe. The transmission Mössbauer spectra were fitted to Lorentzian line-shapes by MossA software (Prescher et al., 2012).

26.2.2. The Toroid-Type Large Reactive Volume Unit

The Toroid-type large reactive volume unit enabled the investigation of sample transformation at pressures up to 8 GPa and temperatures up to 1800 K with the time exposure of the required conditions from one second up to several days. An experimental assemblage included a couple of hard-alloy anvils, also known as anvils-with-hole, one toroid-type chamber, one cylindrical sample container, and two graphitic resistive heaters. The hydraulic pressure applied to pistons of the device was transferred to the anvils and, finally, to the sample container. The Toroid-type chamber served as a pressure medium, transmitting the pressure equally to the surface of the sample container. One of the main advantages of the Toroid-type chamber is the possibility of using large-volume containers, up to 0.3 cm^3, that enable ex situ chromatography analysis to be carried out.

The mixture of water and iron carbide (Fe_3C, American Elements, CAS# 12011-67-5) was loaded into a sealed steel container and then placed inside the Toroid-type chamber. Water was loaded in excess amount to reach the full transformation of iron carbide. Two resistive heaters (made of the mixture graphite-Al_2O_3) were placed above and below the container. Then the sample assemblage was placed between two hard-alloy anvils. Pressurizing the carefully calibrated internal and external diameters of the cylindrical container produced a sealed zone inside the container. Water served as a pressure medium for the container, preventing it from being crushed.

When experimental pressure was reached, the heating was turned on. All the variables (the hydraulic pressure, electrical power, duration of the exposure, and further quenching) were adjusted using the press control software. The temperature and pressure in the container were measured using calibration curves, made for particular containers, chambers, and samples by phase transitions of reference compounds (Bi, PbSe, PbTe for pressure calibration, Pb, Sn, Ti, Cu for temperature calibration). After the sample was exposed for a certain time, the heating was switched off and the sample was quenched. Once the temperature had dropped to ambient, the container was depressurized and recovered from the chamber. The pressure-sealed steel container maintained chemical product integrity. The products were analyzed by means of gas chromatography.

A "Chromatek 5000" gas chromatograph was used for analysis thanks to a specific gas-extracting device. The sealed gas-extracting cell was used to remove the gaseous product from the container by going down a sharp steel stock, making a hole in the container and recovering the products. Helium as carrier gas was supplied from the system of the chromatograph, taking the gaseous product to the chromatograph. The chromatograph was equipped with a capillary column and two flame ionisation detectors. This analytical equipment enabled light hydrocarbons and inorganic gases to be detected and measured. The details of the experimental method are described in Mukhina et al. (2017). Following chromatography analysis, the solid products were recovered from the steel container and analyzed by powder X-ray diffraction. A Seifert MZ III powder X-ray diffractometer, employing CuK$_\alpha$ radiation (1.5405981 Å) in the Bragg-Brentano geometry, was used.

26.3. EXPERIMENTAL RESULTS

The results of three series of experimental investigation are presented below. Each experiment in each series was repeated three times to confirm results.

26.3.1. Behavior of the Hydrocarbon Systems at Depths Down to 50 km

In the first series of the investigation, the model hydrocarbon system, similar to natural gas condensate, was used (density ρ_{20}^{4} 794.7 kg/m^3; Table 26.1). The experiments were carried out in diamond anvil cells with resistive heating at temperatures of 320°–450 °C and pressures of 0.7–1.4 GPa with various exposure times (Table 26.2). Raman spectroscopy was employed for product analysis. The experimental results demonstrated the thermal stability of the hydrocarbon system at the thermobaric conditions applied. The example of the Raman spectra of the model hydrocarbon system (before and after heating at 450 °C and 1.4 GPa) is presented in Figure 26.1.

The composition of the hydrocarbon system remained constant in all three experiments of the current series of the investigation. All peaks of the spectra, corresponding to hydrocarbons of the model system, kept their relative intensity, shape, and Raman wavelength shift. No new peaks appeared in the spectra after heating, signifying the absence of new components in the sample. Increasing exposure time from 3 to 12 hours did not influence the composition of the hydrocarbon system (Experiment 3, Table 26.2).

The results of the first series of the experimental investigation demonstrated that the hydrocarbon system,

Table 26.1 Characteristics of the model hydrocarbon system.

Compound	Content, %
Aromatics	55
Paraffins	40
Naphthenes	5
Asphaltenes	0
Sulphur	0

Table 26.2 Conditions of the experiments with model hydrocarbon system.

Exp.	Pressure, (±0.2) GPa	Temperature, (±20) °C	Corresponding depth, km	Exposure time, hours
1	0.7	320	20–30	3
2	1.2	420	30–40	3
3	1.4	450	40–50	12

similar to natural gas condensate, retained its qualitative and quantitative composition in the thermobaric range 320°–450 °C and 0.7–1.4 GPa, corresponding to a depth down to 50 km.

26.3.2. Hydrocarbon System Transformation at a Subduction Depth Down to 290 km

In the second series of the experimental investigation, the behaviour of the system, consisting of a mixture of saturated hydrocarbons C_{15}-C_{40} and iron-bearing minerals, was investigated. As iron-bearing minerals, ^{57}Fe-enriched pyroxene-like glass $(Mg_{0.91}Fe_{0.09})(Si_{0.91}Al_{0.09})O_3$ and wüstite $Fe_{0.94}O$ composition, synthesized by Kantor et al. (2004), were used. The powder of ^{57}Fe-enriched pyroxene-like glass was synthesised from a pelletized mixture of Fe_2O_3, $Mg(OH)_2$, SiO_2 and $Al(OH)_3$ using a 1 atmosphere box furnace. Two experiments were carried out in diamond anvil cells with the laser heating in this series (Table 26.3).

The Mössbauer spectra of the samples, collected before and after heating, show a mixture of iron hydride FeH (Narygina et al., 2011) and iron carbide Fe_7C_3 (Prescher et al., 2015) (with trace amount of α-Fe and an unidentified iron compound) detected in both experiments as a result of the chemical interaction between the hydrocarbon system and the iron-bearing minerals (Figure 26.2). Raman spectroscopy analysis demonstrated the formation of graphite (Tuinstra & Koenig, 1970), water in the form of ice VII (Walrafen et al., 1982), and methane (Kolesnikov et al., 2009; Sterin et al., 2013) (Figure 26.3). Pure hydrogen was not detected in the sample after the heating.

According to the experimental results obtained, the chemical reaction may be described as follows:

$$7.5\,\mathrm{GPa{:}FeO(wüstite)} + C_nH_{2n+2} \rightarrow FeH + Fe_7C_3 + C + H_2O + CH_4 \qquad (26.1)$$

$$8.8\,\mathrm{GPa{:}FeSiO_3\,(in\ the\ form\ of\ pyroxene)} + C_nH_{2n+2} \rightarrow FeH + Fe_7C_3 + C + H_2O + CH_4 \qquad (26.2)$$

Summarizing our observations, we conclude that heating of hydrocarbons with iron oxides or silicates to temperatures above 1000 °C and pressures above 7 GPa results in

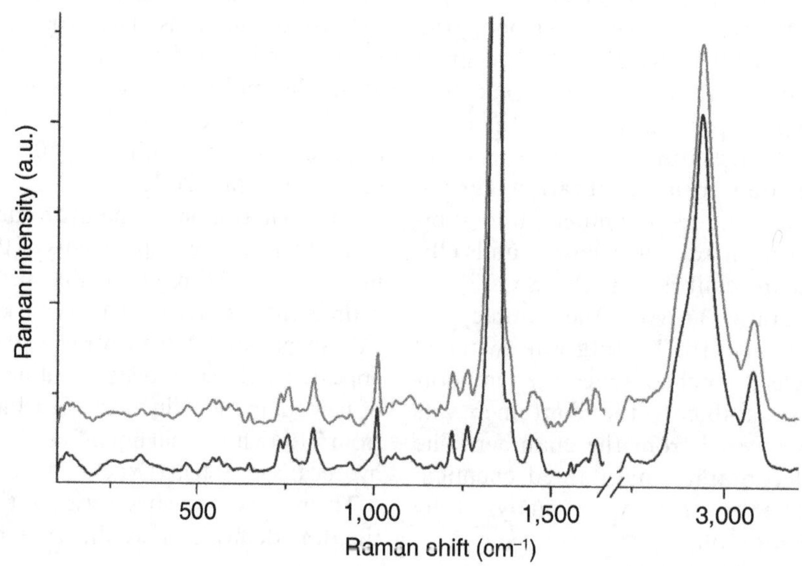

Figure 26.1 The Raman spectra of the model hydrocarbon system at 1.4 GPa before heating (blue curve) and after 12 hours of heating at 450 °C (green curve). See electronic version for color representation of the figures in this book.

Table 26.3 Experimental details.

Exp #	System	Final P, (±0.2)GPa	T, (±100) °C	Mössbauer Spectroscopy Results	Raman Results
1	Paraffin oil + wüstite $Fe_{0.94}O$	7.5	1100	$FeH + Fe_7C_3$	$C + H_2O + CH_4$
2	Paraffin oil + pyroxene-like glass $(Mg_{0.91}Fe_{0.09})(Si_{0.91}Al_{0.09})O_3$	8.8	1300	$FeH + Fe_7C_3$ + remaining pyroxene glass	$C + H_2O + CH_4$

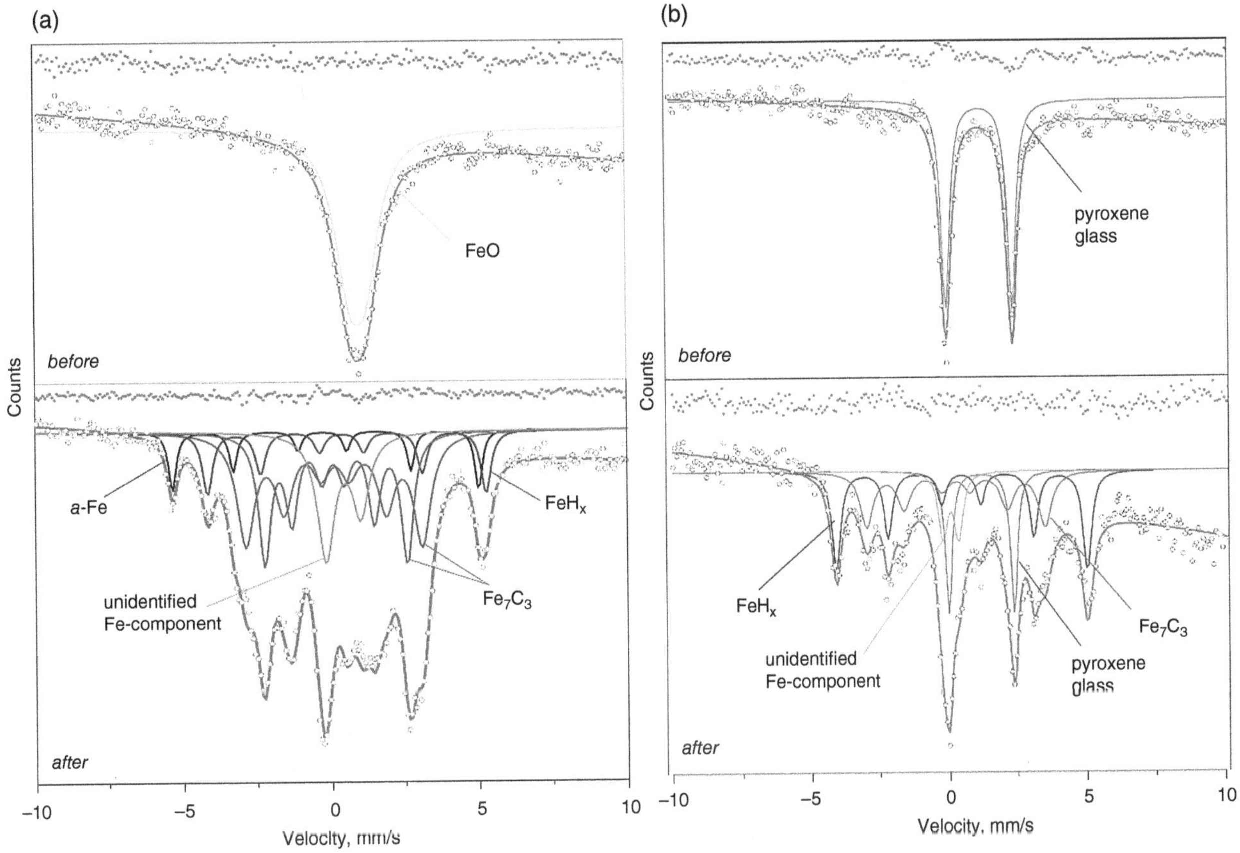

Figure 26.2 The Mössbauer spectra of the samples: (a) At 7.5 GPa before heating: doublet of $Fe_{0.94}O$ with a narrow quadrupole splitting; after laser heating: produced Fe_xH (blue sextet), Fe_7C_3 (two purple doublets), α-Fe (black sextet), and an undefined iron component. (b) At 8.8 GPa before heating: doublet of pyroxene glass; after heating: produced Fe_xH (blue sextet), Fe_7C_3 (purple doublet), remained pyroxene glass and an undefined iron component. See electronic version for color representation of the figures in this book.

formation of a mixture of iron hydride and iron carbide in the subducting slab. It is notable that formation of this mixture was observed independently of whether iron-bearing silicate or oxides were used as starting materials.

The ability of iron carbide to act as a carbon donor in the deep abiogenic synthesis of hydrocarbon systems was investigated on the third series of experiments.

26.3.3. Chemical Interaction of Iron Carbide and Water at the Thermobaric Conditions of the Asthenosphere

In the third series of the current experimental investigation, the chemical reaction between iron carbide (Fe_3C) and water was studied in the thermobaric range, corresponding to depths of 100–150 km. Two experiments were carried out: the first one at 3.5 GPa and 750 °C, and the second one at 4.5 GPa and 850 °C. The experiments were carried out in the Toroid-type large reactive volume unit. The gas chromatograph "Chromatech-5000" was employed for the reaction products analysis with the further analysis of solid products by powder X-ray diffraction data. The X-ray diffraction analysis demonstrated full transformation of iron carbide (the distinguishable peaks of Fe_3C were absent on the spectra after the heating; Kumari et al., 2016; Williams et al., 2016) with the formation of wustite FeO (Khurshid et al., 2013) (Figure 26.4). The results of the experiments demonstrated the chemical interaction

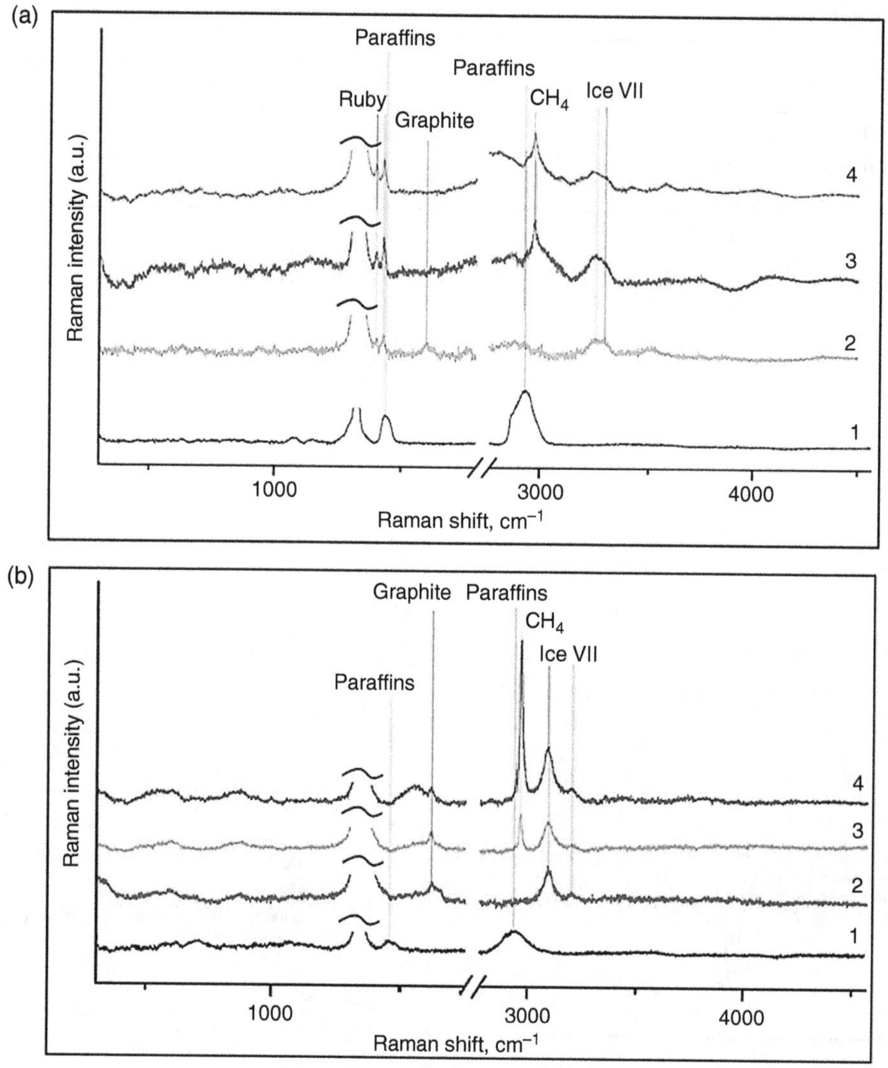

Figure 26.3 The Raman spectra of the sample before and after laser heating at 7.5 GPa (a) and 8.8 GPa (b). Number on the spectra: 1 = before heating; 2–4 = after heating, from the less heated zone on the border of DAC (2) to the more heated zone in the center of DAC (3, 4). See electronic version for color representation of the figures in this book.

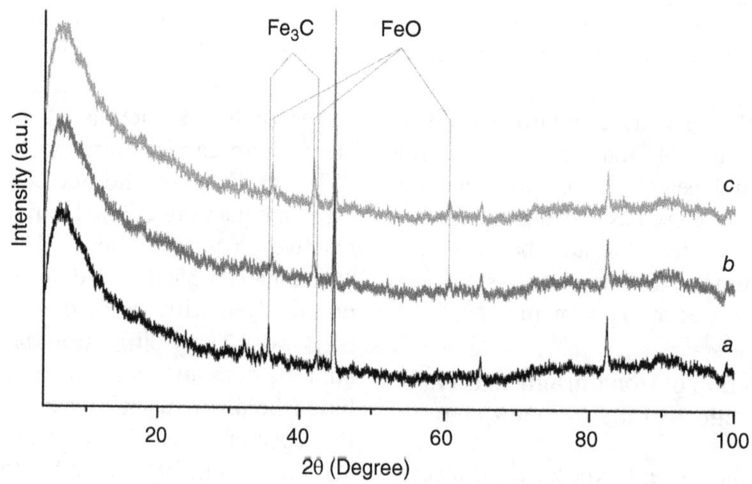

Figure 26.4 XRD pattern of pure iron carbide Fe_3C (a), solid products of the chemical reaction between Fe_3C and H_2O at 750 °C and 3.5 GPa (b), solid products of the chemical reaction between Fe_3C and H_2O at 850 °C and 4.5 GPa (c). See electronic version for color representation of the figures in this book.

between iron carbide and water under the above-mentioned thermobaric conditions with the formation of complex hydrocarbon mixtures, consisting of light paraffins and naphtenes (Figure 26.5). As can be seen in Figure 26.5, the formation of heavier hydrocarbons is prevalent at lower thermobaric conditions, corresponding to upper levels of the mantle. According to the experimental results, the chemical reaction proceeded according to the following scheme:

$$Fe_3C + H_2O \rightarrow FeO + C_nH_{2n+2} + C_mH_{2m} \quad (26.3)$$

The thermobaric conditions for all the experiments discussed above are shown in Figure 26.6, which also presents pressure-temperature profiles of the coldest and hottest subducting slabs and Earth's geotherm.

26.4. GEOLOGICAL OBSERVATIONS

Our experimental results demonstrate that hydrocarbons in the slab maintain their stability to a depth of 50 km. In this study, we compared the experimental results with geological observations from a well-studied ancient island-arc system, the Urals. Signs of subduction are clearly manifested in the Urals. Signs of subduction are clearly manifested in the Urals. The Main Uralian Fault (MUF) is the paleozone of subduction (K. Ivanov, 2001). It extends in the sub-meridional direction for more than 2000 km and divides the Urals fold belt into two sectors, the western (paleocontinental) and the eastern (paleoisland-arc). Here, the oceanic plate of the Early Paleozoic Ural Paleoocean sinks under the Irendyk island arc (Zonenshain et al., 1991). Structural and paleomagnetic data indicate that subduction

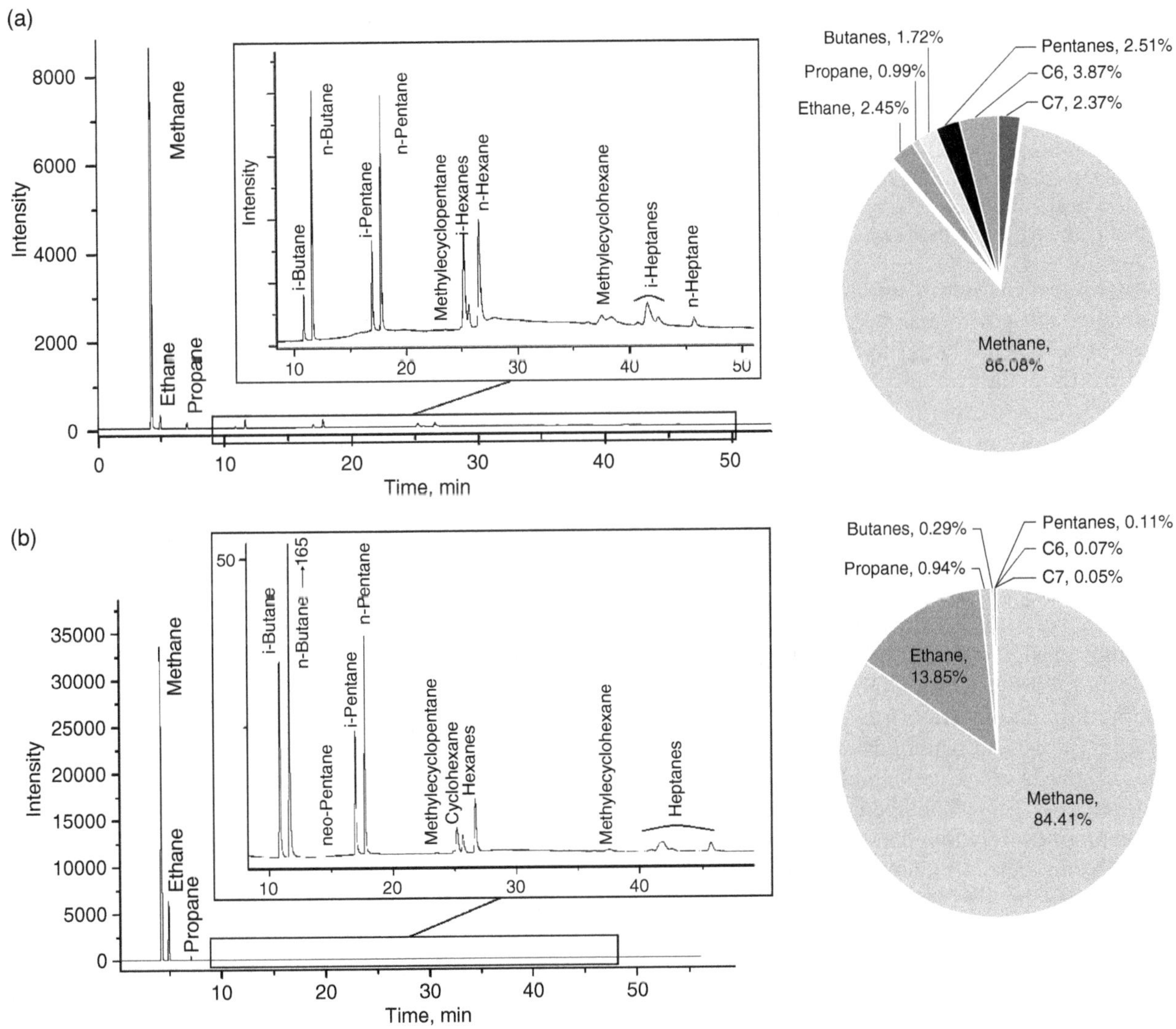

Figure 26.5 The chromatogram (left) and the fraction composition (right) of the hydrocarbon system, formed from the chemical interaction between iron carbide and water at 750 °C and 3.5 GPa (a), and 850 °C and 4.5 GPa (b). See electronic version for color representation of the figures in this book.

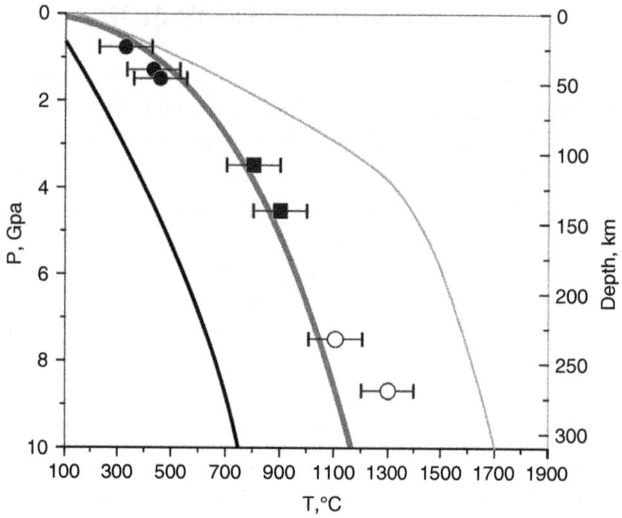

Figure 26.6 Experimental thermobaric conditions. Black circles = experiments with model hydrocarbon system. Open circles = pyroxene-like glass $(Mg_{0.91}Fe_{0.09})(Si_{0.91}Al_{0.09})O_3$ + paraffin oil. Squares = iron oxide (II) $Fe_{0.94}O$ + paraffin oil. Pressure-temperatures profiles of the coldest (blue solid line) and the hottest (red solid line) subducting slabs are taken from Karato (2013). Earth's geotherm (black solid line) is modified after Pollack & Chapman (1977), and the black dashed line is the melting curve of γ-FeH_x (Sakamaki et al., 2009). See electronic version for color representation of the figures in this book.

(and subsequent collision) in the Urals in the Late Paleozoic was oblique and not frontal, accompanied by significant movement of the Ural blocks to the north.

Hydrocarbon data in ultrabasic rocks is especially pertinent alongside experimental results on the behavior of hydrocarbons in the subducted oceanic crust. In the Urals, there are two main types of ultrabasites (S. Ivanov et al., 1975). The first type is alpinotype (ophiolitic), the largest massifs of which are located in the zone of the MUF, extending more than 2,000 km. The second one is platiniferous: area-based dunite-clinopyroxenite-gabbroic massifs located in the suprasubduction zone (K. Ivanov et al., 2007) on the east of the MUF in the Middle and Northern Urals. Different forms of carbon, including bituminous (from traces to 100 g/t), gaseous CO_2, CH_4 forms, and diamonds were found in the ultrabasites of the Urals (Chashchukhin & Votyakov, 2009; Koshkina et al., 1974; Shteinberg & Lagutina, 1984).

The diamonds were found in both types of Ural ultrabasites. Numerous fine diamonds from the ophiolitic massif Rai-Iz of the Polar Urals were studied and described (Yang et al., 2015). There is also a description of the diamond in the dunites of the Kamenushinsky massif of the Platinum Belt of the Urals (Kaminsky, 2007). We believe that the presence of diamonds in ultrabasites indicates their transformation at high pressures at considerable depths in the subduction zone. The intermittent high pressure–low temperature belt of eclogite-glaucophane metamorphic rocks (sometimes containing small diamonds as well) stretches for almost 2000 km in the western part of the MUF. A study of eclogite and glaucophane rocks from the above-mentioned belt show that their formation occurred under pressure of 15 kbar, which corresponds to a depth of 50 km (Lennykh et al., 1995).

The bitumen content found in ultrabasites (serpentinized dunites of different ophiolitic allochthons) is from 1 to 63 g/t. Polycyclic aromatic hydrocarbons, varying from 5 to 103 g/t, were found in these rocks. $C_{14}H_{10}$, $C_{20}H_{12}$, $C_{22}H_{12}$, and $C_{24}H_{12}$ were also identified (Chashchukhin & Votyakov, 2009). The content of heavy alkanes was determined in the ultramafites of the largest alpinotypic massif: Voikar-Synya (Polar Urals): 1.3 g/ton (δ13C = −23.4 ‰) in harzburgite, 1.6 g/ton (δ13C = −26.5 ‰) in wehrlite, 1 g/t in websterite, and 2.3 g/t (δ13C = −26.7 ‰) in pyroxenite. In composition, alkanes in pyroxenite vary from $C_{18}H_{38}$ to $C_{33}N_{68}$ (with a maximum at $C_{22}N_{46}$). They also include pristane ($S_{19}N_{40}$) and phytane ($C_{20}H_{42}$) (Sugisaki & Mimura, 1994). Thus, geological data on the presence of hydrocarbons in ultrabasites squeezed from a slab indicates that complex hydrocarbon systems may exist in a slab at considerable depths. This confirms our experimental results, indicating the stability of hydrocarbons to a depth of 50 km.

Another type of geological observation deals with a result of a study of the primary fluid inclusion in diamonds and garnets, the mantle origin of which is beyond doubt. The composition of the primary fluid inclusion was studied by mass spectrometry in seven native Arkansas diamonds. The result of the investigation has confirmed the presence of different kinds of hydrocarbon in all samples (Melton & Giardini, 1974). Raman and infrared spectroscopy were used to study the composition of the primary fluid in garnet from the Udachnaya kimberlite pipe, (Yakutia, Russia). Primary fluids extracted from garnets comprise saturated hydrocarbons from CH_4 to C_6H_{14} (Tomilenko et al., 2009).

The obtained data on fluid inclusions in natural diamonds and garnets, together with the results of our experiments, have provided conclusive evidence that a mixture of hydrocarbons of similar composition to the main components of natural petroleum can form in the Earth's mantle.

26.5. CONCLUSION

Our experimental results are compatible with the presence of a deep Earth hydrocarbon cycle (Figure 26.7), which can be described as follows.

1. Hydrocarbons accumulated in the traps in the Earth's crust, together with sedimentary rocks immersed in the subducting slab, maintain their stability to a depth of 50 km. The presence of hydrocarbons in ultrabasites

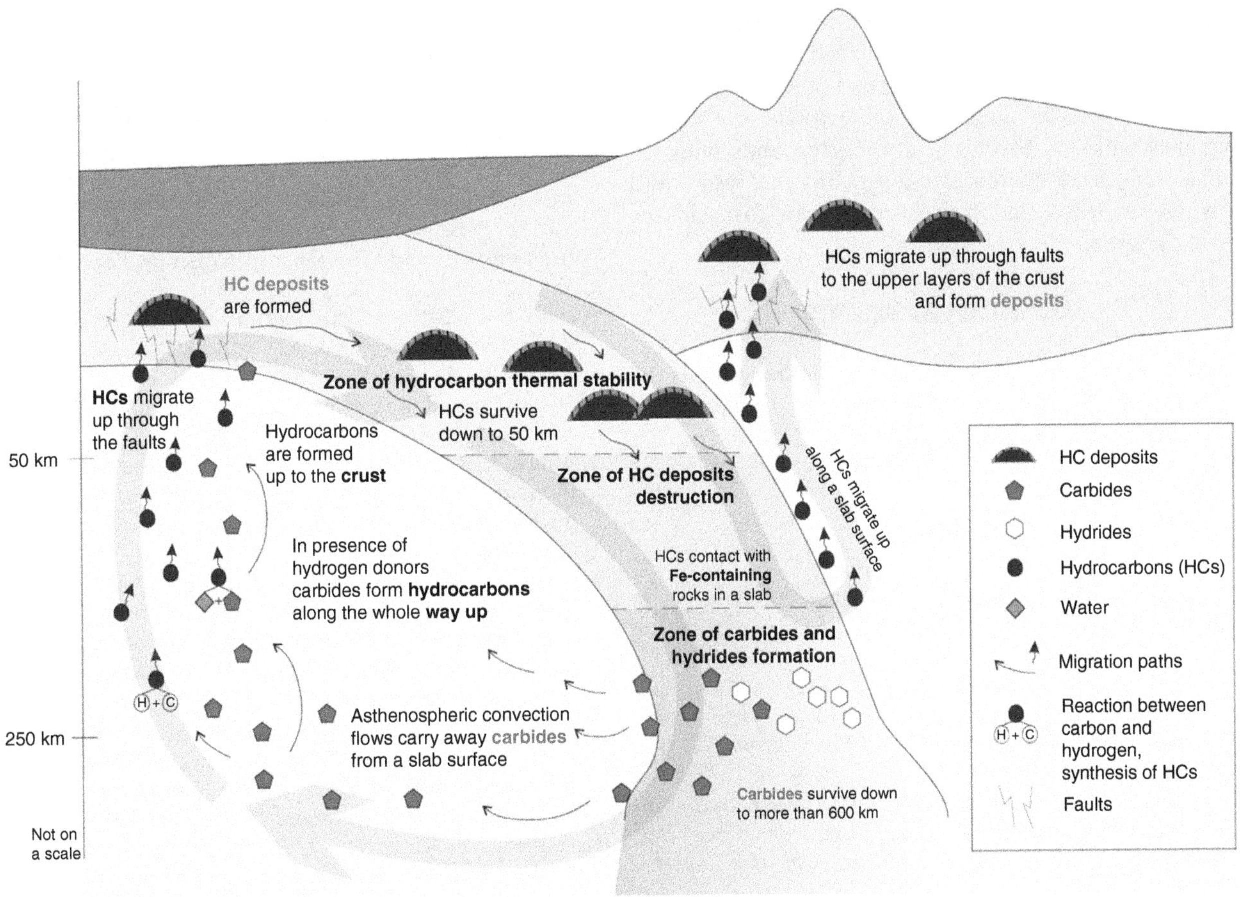

Figure 26.7 Deep hydrocarbon cycle. Hydrocarbons trapped in subducting slabs remain stable at depths of 50 km. At depths of 210–290 km, these hydrocarbons react with surrounding ferrous minerals to form a mixture of iron hydride and iron carbide. This iron carbide transported into the asthenosphere by convective flows reacts with hydrogen or water and form an aqueous hydrocarbon fluid. Water-hydrocarbon mantle fluids can migrate upward through deep faults in the crust, forming multilayer accumulations of oil and gas. Other carbon donors existing in the asthenosphere can also serve as a source of deep hydrocarbons, additional feed for the general upward flow of the water-hydrocarbon fluid. See electronic version for color representation of the figures in this book.

from the Main Uralian Fault, the paleozone of subduction, confirms existence of hydrocarbons in a slab at considerable depths. This observation corresponds with our experimental results and confirms the stability of hydrocarbons to a depth of 50 km.

2. At depths of 50–80 km, the integrity of the traps is disrupted and hydrocarbon fluid enters into contact with surrounding iron-bearing minerals. Lightly saturated hydrocarbons are generated inside the slab from descending carbonates and water.

3. Further immersion transforms the hydrocarbon fluid. At a depth of 80 km, the system consists of hydrocarbon phase, graphite, and water. Methane and other light hydrocarbons can migrate up along the slab–continental plate border.

4. At depths of 210–290 km, light hydrocarbons react with iron-bearing minerals presented in the slab and form a mixture of iron hydride and iron carbide.

5. Iron carbide transported in the asthenosphere by convective flows can react with hydrogen or water present in the asthenosphere and form an aqueous hydrocarbon fluid that can migrate through deep faults into the Earth's upper crust and form multilayer oil and gas deposits in rocks of any lithological composition, genesis, and age.

6. There are other carbon donors in the asthenosphere, in addition to iron carbide coming from the subducting slab. These donors can also serve as a source of deep hydrocarbons. Theoretical calculations (Huang et al., 2017; Karpov et al., 1998; Kenney et al., 2002; Spanu et al., 2011) and experimental results (Kutcherov et al., 2002; Kutcherov et al., 2010; Sokol et al., 2017) show that abiogenic synthesis of complex hydrocarbon systems is possible at temperatures of 900°–1700 °C in the 3–7 GPa pressure range. Similar conditions exist in the layer of the Earth's asthenosphere at depths of 100–250 km. The amount of these hydrocarbons could be significant (Kutcherov et al., 2010).

These hydrocarbons also participate in the deep hydrocarbon cycle, being an additional feed for the general upward flow of the water-hydrocarbon fluid. The presence of hydrocarbons in diamonds and garnets, the mantle origin of which is beyond doubt, corresponds with the above-mentioned theoretical and experimental results and provides evidence that hydrocarbons can form in the Earth's mantle.

ACKNOWLEDGMENTS

V.K., E.M., and A.S. gratefully acknowledge the support of Flotten AB (Sweden) and the Sloan Foundation (USA). K.I. is grateful to RFBR (project 18-05-70016) for its financial support.

REFERENCES

Belonoshko, A. B., Lukinov, T., Rosengren, A., Bryk, T., & Litasov, K. D. (2015). Synthesis of heavy hydrocarbons at the core-mantle boundary. *Scientific Reports*, *5*, 18382.

Bezmen, N. I., Zharikov, V. A., Zevelsky, V. O., & Kalinichev, A. G. (2005). Melting of alkali aluminosilicate systems under hydrogen-water fluid pressure, P_{tot} = 2 kbar. *Petrology*, *13*(5), 407.

Chashchukhin, I. S., & Votyakov, S. L. (2009). Behavior of iron-group elements, oxybarometry, and genesis of unique chromite deposits in the Kempirsai massif. *Geol. Ore Deposits*, *51*(2), 123–138.

Cooper, B. S., Coleman, S. H., Barnard, P. C., & Butterworth, J. S. (1975). Palaeotemperatures in the northern North Sea basin. *Petroleum and the Continental Shelf of North West Europe*, *1*, 487–492.

Dasgupta, R., Mallik, A., Tsuno, K. Withers, A. C., Hirth, G., & Hirschmann, M. M. (2013). Carbon-dioxide-rich silicate melt in the Earth's upper mantle. *Nature*, *493*(7431), 211–222.

Dubrovinsky, L. S., & Saxena, S. K. (1999). Emissivity measurements on some metals and oxides using multiwavelength spectral radiometry. *High Temperatures–High Pressures*, *31*(4), 393–399.

Hessler, A. M., & Sharman, G. R. (2018). Subduction zones and their hydrocarbon systems. *Geosphere*, *14*(5), 2044–2067.

Huang, F., Daniel, I., Cardon, H., Montagnac, G., & Sverjensky, D. A. (2017). Immiscible hydrocarbon fluids in the deep carbon cycle. *Nature Communications*, *8*, 15798.

Ivanov, K. S. (2001). Estimation of paleovelocities of subduction and collision during the formation of the Urals. *Dokl. Earth Sci.*, *377*, 164–167.

Ivanov, K. S., Volchenko, Y., & Koroteev, V. (2007). Nature of the Urals platinum belt and its chromite-platinum metal deposits. *Dokl. Earth Sci.*, *417A*(9), 1304–1307.

Ivanov, S. N., Perfiliev, A. S., Efimov, A. A., Smirnov, G. A., Necheukhin, V. M., & Fershtater, G. B. (1975). Fundamental features in the structure and evolution of the Urals. *Am. J. Sci.*, *275*, 107–130.

Javoy, M. (1997). The major volatile elements of the Earth: Their origin, behavior, and fate. *Geophysical Research Letters*, *24*(2), 177–180.

Kaminsky, F. V. (2007). Non-kimberlitic diamondiferous igneous rocks: 25 years on. *Geological Society of India*, *69*(3), 557.

Kantor, A. P., Jacobsen, S. D., Kantor, I. Y., Dubrovinsky, L. S., McCammon, C. A., Reichmann, H. J., & Goncharenko, I. N. (2004). Pressure-induced magnetization in FeO: Evidence from elasticity and Mössbauer spectroscopy. *Phys. Rev. Lett.*, *93*, 215502.

Karato, S.-I. (2013). *Physics and chemistry of the deep Earth*. Ames, IA: John Wiley & Sons.

Karpov, I. K., Zubkov, V. S., Stepanov, A. N., & Bychinsky, V. A. (1998), Chekaliuk's thermodynamic model of the C-H system: A remake. *Dokl. Earth Sci.*, *358*(1), 30–33.

Kelemen, P. B., & Manning, C. E. (2015). Reevaluating carbon fluxes in subduction zones: What goes down, mostly comes up. *Proceedings of the National Academy of Sciences*, *112*(30), 3997–4006.

Kenney, J. F., Kutcherov, V. A., Bendeliani, N. A., & Alekseev, V. A. (2002). The evolution of multicomponent systems at high pressures: VI. The thermodynamic stability of the hydrogen-carbon system: The genesis of hydrocarbons and the origin of petroleum. *PNAS*, *99*, 10976–10981.

Khurshid, H., Li, W., Chandra, S., Phan, M.-H., Hadjipanayis, G. C., Mukherjee, P., & Srikanth, H. (2013). Mechanism and controlled growth of shape and size variant core/shell FeO/Fe_3O_4 nanoparticles. *Nanoscale*, *5*(17), 7942–7952.

Kolesnikov, A., Kutcherov, V. G., & Goncharov, A. F. (2009). Methane-derived hydrocarbons produced under upper-mantle conditions. *Nature Geoscience*, *2*(8), 566–570.

Kolesnikov, A. Y., Saul, J. M., & Kutcherov, V. G. (2017). Chemistry of hydrocarbons under extreme thermobaric conditions. *ChemistrySelect*, *2*(4), 1336–1352.

Koshkina, T. M., Lagutina, M. V., & Shapiro, V. A. (1974). Cohenite diagnosis in south Urals ultrabasites. *J. Geophys.*, *40*(4), 565–569.

Kumari, R., Krishnia, L., Kumar, V., Singh, S., Singh, H. K., Kotnala, R. K., et al. (2016). Fe_3C-filled carbon nanotubes: Permanent cylindrical nanomagnets possessing exotic magnetic properties. *Nanoscale*, *8*(7), 4299–4310.

Kupenko, I., Dubrovinsky, L., Dmitriev, V., & Dubrovinskaia, N. (2012). In situ Raman spectroscopic study of the pressure induced structural changes in ammonia borane. *J. Chem. Phys.*, *137*(7).

Kutcherov, V. G., Bendeliani, N. A., Alekseev, V. A. & Kenney, J. F. (2002). Synthesis of hydrocarbons from minerals at pressures up to 5 GPa. *Dokl. Phys. Chem.*, *387*, 328–330.

Kutcherov, V. G., Kolesnikov, A. I., Dyugheva, T. I., Kulikova, L. F., Nikolaev, N. N., Sazanova, O. A., & Braghkin, V. V. (2010). Synthesis of complex hydrocarbon systems at temperatures and pressures corresponding to the Earth's upper mantle conditions. *Dokl. Phys. Chem.*, *433*, 132–135.

Lennykh, V. I., Valizer, P. M., Beane, R., Leech, M., & Ernst, W. G. (1995). Petrotectonic evolution of the Maksyutov complex, Southern Urals, Russia: Implications for ultrahigh-pressure metamorphism. *International Geology Review*. *37*(7), 584–600.

Mann, P., Gahagan, L., & Gordon, M. B. (2003). Tectonic setting of the world's giant oil and gas fields. In M. T. Halbouty (Ed.), *AAPG memoir* (Vol. *78*, pp. 15–105). AAPG.

Manning, C. E. (2004). The chemistry of subduction-zone fluids. *Earth and Planetary Science Letters*, *223*(1–2), 1–16.

Manning, C. E. (2014). Geochemistry: A piece of the deep carbon puzzle. *Nature Geosci*, *7*(5), 333–334.

Mao, H. K., Xu, J., & Bell, P. M. (1986). Calibration of the ruby pressure gauge to 800 kbar under quasi-hydrostatic conditions. *J. Geophys. Res. B*, *91*, 4673–4676.

Melton, C. E., & Giardini, A. A. (1975). Experimental results and theoretical interpretation of gaseous inclusions found in Arkansas natural diamonds. *Geochim. Cosmochim. Acta*, *60*(56), 413–417.

Mukhina, E., Kolesnikov, A., & Kutcherov, V. (2017). The lower pT limit of deep hydrocarbon synthesis by $CaCO_3$ aqueous reduction. *Scientific Reports*, *7*(1), 5749.

Narygina, O., Dubrovinsky, L. S., McCammon, C. A., Kurnosov, A., Kantor, I. Y., Prakapenka, V. B. & Dubrovinskaia, N. A. (2011). X-ray diffraction and Mössbauer spectroscopy study of fcc iron hydride FeH at high pressures and implications for the composition of the Earth's core. *Earth and Planetary Science Letters*, *307*(3–4), 409–414.

Prescher, C., Dubrovinsky, L., Bykova, E., Kupenki, I., Glazyrin, K., Kantor, A., et al. (2015). High Poisson's ratio of Earth's inner core explained by carbon alloying. *Nature Geosci*, *8*(3), 220–223.

Prescher, C., McCammon, C., & Dubrovinsky, L. (2012). MossA: A program for analyzing energy-domain Mossbauer spectra from conventional and synchrotron sources. *J. Appl. Crystallogr.*, *45*, 329–331.

Sakamaki, K., Takahashi, E., Nakajima, Y., Nishihara, Y., Funakoshi, K., Suzuki, T., & Fukai, Y. (2009). Melting phase relation of FeH_x up to 20 GPa: Implication for the temperature of the Earth's core. *Phys. Earth Planet. Inter.*, *174*(1–4), 192–201.

Shimada, M. (1969). Melting of albite at high pressures in the presence of water. *Earth and Planetary Science Letters*, *6*(6), 447–450.

Shteinberg, D. S., & Lagutina, M. V. (1984). *Carbon in ultrabasites and basites*. Moscow: Nauka.

Sokol, A. G., Tomilenko, A. A., Bul'bak, T. A., & Sobolev, N. V. (2017). Synthesis of hydrocarbons by CO_2 fluid conversion with hydrogen: Experimental modeling at 7.8 GPa and 1350 °C. *Dokl. Earth Sci.*, *477*, 1483–1487.

Sonin, V. M., Bul'bak, T. A., Zhimulev, E. I., Tomilenko, A. A., Chepurov, A. I., & Pokhilenko, N. P. (2014). Synthesis of heavy hydrocarbons under P-T conditions of the Earth's upper mantle. *Dokl. Earth Sci.*, *454*(1), 32–36.

Spanu, L., Donadio, D., Hohl, Schwegler, E., & Galli, G. (2011). Stability of hydrocarbons at deep Earth pressures and temperatures. *Proceedings of the National Academy of Sciences*, *108*(17). 6843–6846.

Sterin, K. E., Aleksanyan, V., & Zhizhin, G. N. (2013). *Raman spectra of hydrocarbons*. Elsevier.

Stern, R. J. (2002). Subduction zones. *Reviews of Geophysics*, *40*(4), 1–38.

Sugisaki, R., & Mimura, K. (1994). Mantle hydrocarbons: Abiotic or biotic? *Geochim. Cosmochim. Acta*, *58*(11), 2527–2542.

Sverjensky, D. A., Stagno, V., & Huang, F. (2014). Important role for organic carbon in subduction-zone fluids in the deep carbon cycle. *Nature Geoscience*, *7*(12), 909–913.

Tao, R., Zhang, L., Tian, M., Zhu, J., Liu, X., Liu, J., et al. (2018). Formation of abiotic hydrocarbon from reduction of carbonate in subduction zones: Constraints from petrological observation and experimental simulation. *Geochim. Cosmochim. Acta*, *239*, 390–408.

Tomilenko, A.A. et al. (2009). Primary hydrocarbon inclusions in garnet of diamondiferous eclogite from the Udachanaya kimberlite pipe, Yakutia. *Doklady Earth Sciences*, *426*(4), 695–698.

Trots, D. M., Kurnosov, A., Ballaran, T. B., Tkachev, S., Zhuravlev, K., Prakapenka, V., et al. (2013). The Sm:YAG primary fluorescence pressure scale. *Journal of Geophysical Research: Solid Earth*, *118*(11), 5805–5813.

Tuinstra, F., & Koenig, J. L. (1970). Raman spectrum of graphite. *The Journal of Chemical Physics*, *53*(3), 1126–1130.

Walrafen, G. E., Abebe, M., Mauer, F. A., Block, S., Piermarini, G. J., & Munro, R. (1982). Raman and X-ray investigations of ice VII to 36.0 GPa. *The Journal of Chemical Physics*, *77*(4), 2166–2174.

Williams, B., Clifford, D., El-Gendy, A. A., & Carpenter, E. E. (2016). Solvothermal synthesis of Fe_7C_3 and Fe_3C nanostructures with phase and morphology control. *Journal of Applied Physics*, *120*(3), 033904.

Yang, J., Robinson, P. T., & Dilek, Y. (2015). Diamond-bearing ophiolites and their geological occurrence. *Episodes*, *38*(4), 344–364.

Zonenshain, L. P., Kuzmin, M. I., & Bocharova, N. Y. (1991). Hot-field tectonics. *Tectonophysics*, *199*(2), 165–192.

27

Diamondoids Under Pressure

Sulgiye Park[1], Yu Lin[2], and Wendy L. Mao[1,3]

ABSTRACT

Due to their outstanding properties, diverse geometry, and ability to be manipulated and functionalized, diamondoids have gained interest as highly attractive targets as molecular building blocks for applications in biomedicine, materials science, and nanotechnology, in addition to the field of petroleum engineering (Clay et al., 2009; Mcintosh et al., 2004; Sasagawa & Shen, 2008; Willey et al., 2006; Zhang et al., 2016). Diamondoid molecules are also a useful system for basic science studies. They display atomic-level uniformity and a systematic series of sizes and geometries, making them ideal materials for exploring how different parameters can be used to tune properties. In this chapter, the pressure-induced structural modifications in a range of diamondoids are reported. In section 27.2, the critical role of molecular geometry in determining the phase transition pressures and bulk moduli of diamondoids is discussed, followed by the sensitivity of diamondoids to hydrostaticity and deviatoric stress under compression. Section 27.3 examines recent work investigating functionalized diamondoids at high pressure and concludes with a perspective for future work and exciting potential directions for studying diamondoids at extreme conditions.

27.1. INTRODUCTION

Carbon-based nanomaterials, such as graphene, fullerenes, and carbon nanotubes, have attracted broad interest in both science and industry for their unique and outstanding properties, which make them attractive for a wide range of nanotechnology applications (Stauss & Terashima, 2017; Yang et al., 2016). Here, another exceptional member in the carbon nanomaterial family, diamondoid molecules and their functionalized derivatives, are discussed in terms of how their structures and properties evolve under extreme conditions.

Diamondoid molecules are hydrogen-terminated carbon nanomaterials in the form of $C_{4n+6}H_{4n+12}$ (Mansoori, 2008; Marchand, 2003). With size variations of 0.5–2 nm in diameter, diamondoids represent the ultimate limit in reducing the size of diamond. The most basic unit of diamondoids, adamantane ($C_{10}H_{16}$), is composed of ten-carbon atoms which form a tetra-cyclic cage that can be superimposed on a cubic diamond lattice, with larger diamondoids consisting of varying numbers of these cage units (Figure 27.1a). The lower diamondoids, ada-, dia- and tria-mantane (when $n \leq 3$), each have one isomer. Higher diamondoids (when $n > 3$) have multiple structural isomers depending on the relative directions of face-fused single cage adamantane units (Clay et al., 2009; Mcintosh et al., 2004; Sasagawa & Shen, 2008; Willey et al., 2006; Zhang et al., 2016). Entirely constituted by sp^3-bonded carbons arranged in a three-dimensional (3D) network with covalent interatomic interactions, diamondoid molecules are highly stable and dense. Their resemblance to bulk diamond means that diamondoids similarly have high thermal stability and rigidity.

In nature, diamondoid molecules can be found in deep fossil fuel reservoirs, such as coal, petroleum, gas condensate, and natural gas, formed by hydrocarbon rearrangement reactions under acidic conditions (Fort & Schleyer

[1]Department of Geological Sciences, Stanford University, Stanford, California, USA
[2]Stanford Institute for Materials and Energy Sciences, SLAC National Accelerator Laboratory, Menlo Park, California, USA
[3]Photon Science, SLAC National Accelerator Laboratory, Menlo Park, California, USA

Carbon in Earth's Interior, Geophysical Monograph 249, First Edition. Edited by Craig E. Manning, Jung-Fu Lin, and Wendy L. Mao.

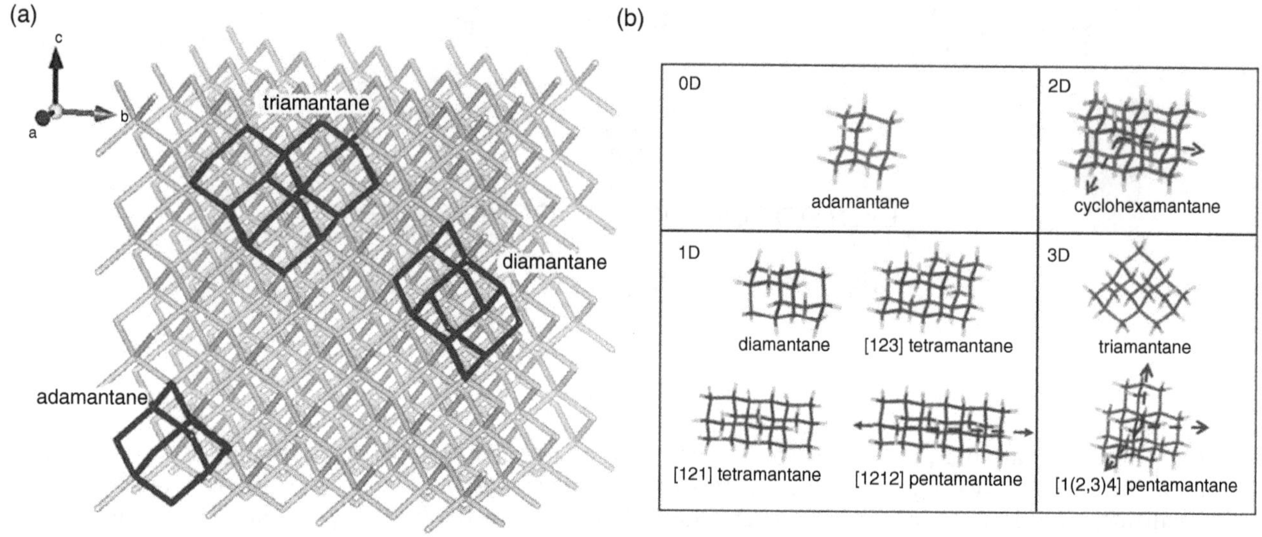

Figure 27.1 (a) Molecular structure of lower diamondoids superimposed on a cubic diamond lattice. (b) Molecular structures of the eight diamondoids studied; relative directions of face-fused adamantane cages in polymantane homologues are indicated by dashed arrows. (Adapted from Yang et al., 2016.) See electronic version for color representation of the figures in this book.

1964; Katz et al., 2008). Due to their similarity to the bulk diamond structure, diamondoids are extremely resistant to weathering, which enables them to be isolated from petroleum by hydrocracking oil to gas (Dahl et al., 2010; Ma, 2016; Ma et al., 2017; Stout & Douglas, 2004). Bulk concentration of the petroleum diamondoids is adamantane, which is generally accompanied by small amounts of diamantane and triamantane (Katz et al., 2008; Mckervey, 1980; Wingert, 1992). Higher diamondoids also exist in minute quantities in a gas condensate produced from very deep (~6.8 km below the Earth's surface) petroleum reservoirs (Dahl et al., 2003; R. Lin & Wilk, 1995). Diamondoids represent an important geochemical tool for characterizing reservoir fluids. Applied solely and/or in conjunction with well-known biomarkers, diamondoids are used for correlation, biodegradation, and maturity assessments involved in petroleum exploration, quality evaluation, and chemical fingerprinting of light petroleum and gasoline (Ekweozor & Telnaes, 1990; Mackenzie et al., 1980; Requejo et al., 1992; Seifert & Moldowan, 1986; Van Graas, 1990). For example, by taking advantage of the rigidity and concentration effect of diamondoids, one can reliability constrain the extent of thermal cracking and the depth at which it occurs in a given reservoir (Dahl et al., 1999).

Due to their outstanding properties, diverse geometry, and ability to be manipulated and functionalized, diamondoids have gained interest as highly attractive targets as molecular building blocks for applications in biomedicine, materials science, and nanotechnology, in addition to the field of petroleum engineering (Clay et al., 2009; Mcintosh et al., 2004; Sasagawa & Shen, 2008; Willey et al., 2006; Zhang et al., 2016). Furthermore, diamondoid molecules are a useful system for basic science studies. They display atomic-level uniformity and a systematic series of sizes and geometries, making them ideal materials for exploring how structural and chemical parameters affect properties.

In this chapter, the pressure-induced structural modifications in a range of diamondoids are reported. In section 27.2, the critical role of molecular geometry in determining the phase transition pressures and bulk moduli of diamondoids is discussed, followed by the sensitivity of diamondoids to hydrostaticity and deviatoric stress under compression. Section 27.3 examines recent work investigating functionalized diamondoids at high pressure, followed by a conclusion on perspective for future work and exciting potential directions for studying diamondoids at extreme conditions.

27.2. ROLE OF MOLECULAR GEOMETRY ON DIAMONDOIDS AT HIGH PRESSURE

The high-pressure behavior of carbon allotropes has long been a subject of intense investigation, fueled by the dramatic changes in physical and chemical properties at extreme conditions. Diamondoids exhibit interesting structural stability coupled with configurational changes in their molecular packing at high pressures. Here we report pressure-induced modifications in eight selected diamondoid crystals with varying molecular geometries: adamantane, diamantane, triamantane, [121]tetramantane, [123]tetramantane, [1212]pentamantane, [1234] pentamantane, and cyclohexamantane (Figure 27.1) (Yang et al., 2016). A summary of the structural information and physical parameters for the eight diamondoids studied is given in Table 27.1.

To understand the high-pressure behavior of diamondoids, diamond-anvil cell was utilized to reach pressures up to ~40 GPa. Structural evolution as a function of pressure was then probed using both Raman spectroscopy and angle-dispersive synchrotron X-ray powder diffraction (XRD). The XRD results were used to produce pressure-volume curves, which were fit to an isothermal third-order Birch-Murnaghan equation of state (EOS) (Birch, 1947). Pressure was increased incrementally under both hydrostatic and nonhydrostatic conditions, and measurements were collected both during compression and decompression cycles.

At ambient conditions, diamondoid molecules exhibit multiple characteristic vibrational modes, appearing at three different wavenumber ranges: C-C-C bending and C-C stretching at lower Raman shifts ~200–900 cm^{-1}, C-H wagging/C-H_2 twisting/C-H_2 scissoring at midrange Raman shifts ~900–1600 cm^{-1}, and C-H stretching at higher Raman shifts ~2800–3200 cm^{-1} (Filik, 2010; Filik et al., 2006a, 2006b). Upon compression, the only vibrational region undergoing a continuous, monotonic change is the C-C stretching mode, while other vibrational regions display complex peak splitting and merging (Figure 27.2a) (Yang et al., 2016). The C-C stretching region, which blue shifts as a function of pressure, originates from a breathing vibration across a single adamantane cage (Filik et al., 2006a). The pressure-induced change in peak position is attributed to the parabolic relationship from the changes in the C-C bond force constant under compression. In some diamondoid molecules (adamantane, diamantane, and [121]tetramantane), the evolution of this breathing mode at high pressure undergoes an obvious discontinuity, due to the alteration in intermolecular packing, which in turn, modifies the intramolecular vibrations (Figure 27.2b). Despite the discontinuity, however, the breathing mode persists even up to the highest pressure reached, indicating that the diamond-like cages can remain intact even up to high pressures. Upon pressure release, the changes observed in all Raman spectra are completely reversible. The observation provides evidence against the reconstructive mechanisms widely described for other carbon-containing molecular solids, wherein decomposition and amorphization are readily discussed in terms of the breaking and formation of new bonds at high pressures. The comparable pressure range used in our studies to those reported previously for carbon-based solids suggests that only modifications of molecular packing are observed for diamondoids at high pressure due to the particularly rigid diamond-like molecular cages of diamondoids (Yang et al., 2016).

In agreement with the Raman spectroscopy, XRD patterns show a completely reversible trend upon pressure release in all the diamondoids studied. Additionally, strong hysteresis effects are observed upon decompression. As observed in Figure 27.2c, [123]tetramantane with an ambient structure of P1 exhibits a new peak (marked with an asterisk) at a pressure ~11.8 GPa that persists up to ~15 GPa. Upon release, the metastable high-pressure phase remains until pressure is completely reversed, indicating a strong hysteresis effect under hydrostatic condition.

The relative pressure-induced volume changes for the low-pressure phases were fit to a third-order Birch-Murnaghan EOS, and the results are plotted in Figure 27.3. The small ambient bulk moduli (K_0) of all diamondoid crystals are attributed to their weak intermolecular van der Waals interactions, wherein the compression at low pressures occurs through simple volume reduction of the intermolecular spacing. At higher pressures, the volume reduction becomes increasingly difficult due to the dominant Coulomb repulsion and stiff intramolecular covalent bonds, resulting in rapid increase in bulk moduli and large pressure derivatives (K_0') (Yang et al., 2016).

The bulk moduli of the diamondoids illustrate a systematic correlation with their molecular geometry. Zero-dimensional adamantane with all dimensions measured within the nanoscale has the lowest bulk modulus, while 3D [1(2,3)4]pentamantane and 2D cyclohexamantane have the highest bulk moduli (Figure 27.3b). It is interesting to note that 2D cyclohexamantane, which is confined to one dimension at nanoscale, has similar bulk modulus as 3D [1(2,3)4] pentamantane, which resembles bulk nanomaterials that are unconfined to the nanoscale in any dimension. One-dimensional diamondoid molecules that are confined to two (x,y) dimensions fall somewhere in between 0D adamantane and 2D–3D diamondoids. For comparisons, the EOS of C_{60} fullerene falls within a similar region in the pressure-volume space as diamondoid crystals, which can be attributed to its comparable molecular size, geometry, and intermolecular interactions (Yang et al., 2016).

For the pressure range studied, the compressibility of the diamondoid crystals is influenced by the hydrogen interactions between adjacent molecules, which help govern the molecular geometry and packing orientations. As summarized in Table 27.1, a mutual feature across the diamondoid crystals is the fast onset of pressure-induced structural transitions. Except for triamantane, which persists its structure up to ~18 GPa, all other diamondoids undergo at least one structural phase transition. Interestingly, the phase transition pressure in diamondoids is found to be positively correlated to the hydrogen separation distance (H–H distance), or the distance between the nearest hydrogen atoms in adjacent molecules (Yang et al., 2016). The smaller H–H distances, such as those of adamantane (2.37 Å) and [1(2,3)4]pentamantane (2.30 Å), suggest that the adjacent molecules are close and relatively

Table 27.1 Structural information for the eight investigated diamondoid crystals.

Diamondoid	Molecular Formula	Molecular Weight	Molecular Volume (Å³)	Unit Cell Volume (Å³)	Density [g/cm³]	Ambient Phase	Dimensionality	H—H Distance (Å)	K_0 (GPa)	K_0' (GPa)
Adamantane	$C_{10}H_{16}$	136.125	N/A	844	1.08	*Fm3m*	0D	2.37	3.2(0.2)	19(1.0)
Diamantane	$C_{14}H_{20}$	188.314	N/A	1033	1.21	*Pa3*	1D	2.65	5.0(0.6)	19.5(2.8)
Triamantane	$C_{18}H_{24}$	240.390	N/A	5124	1.24	*Fddd*	3D	2.81	13.9(1.6)	9.6(1.5)
[121]tetramantane	$C_{22}H_{28}$	292.466	284	768	1.27	$P2_1/n$	1D	2.53	5.5(0.6)	18.8(2.3)
[123]tetramantane	$C_{22}H_{28}$	292.466	—	369	1.31	*P1*	1D	2.52	5.0(1.3)	20.0(1.1)
[1(2,3)4]pentamantane	$C_{26}H_{32}$	344.542	329	1815	1.26	*Pnma*	3D	2.30	16.0(1.2)	11.8(1.0)
[1212]pentamantane	$C_{26}H_{32}$	344.542	—	1763	1.30	$P2_12_12_1$	1D	2.48	8.1(0.8)	8.5(1.5)
Cyclohexamantane	$C_{26}H_{30}$	342.526	319	1239	1.38	*-R3*	2D	2.63	13.5(1.5)	11.9(1.8)

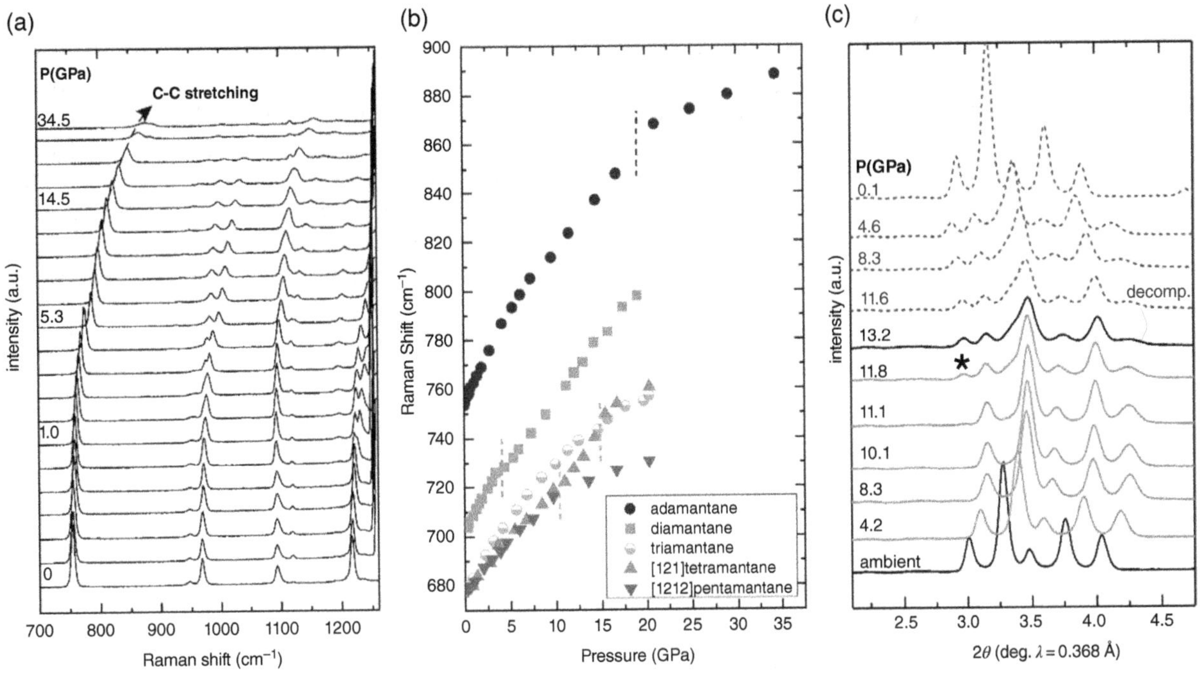

Figure 27.2 (a) Raman spectra of adamantane as a function of increasing pressure. The C-C stretching mode, or the carbon cage breathing vibration, exhibits a smooth shift to higher Raman shift with increasing pressure. (b) Evolution of carbon cage breathing mode for representative diamondoid molecules as a function of increasing pressure. Dashed vertical lines indicate distinctive discontinuities in the development trend of the breathing mode. (c) Representative XRD patterns of [123]tetramantane as a function of compression and decompression (decomp). Asterisk denotes a new peak that indicates a phase transformation. (Parts a and b reproduced from Yang et al., 2016.) See electronic version for color representation of the figures in this book.

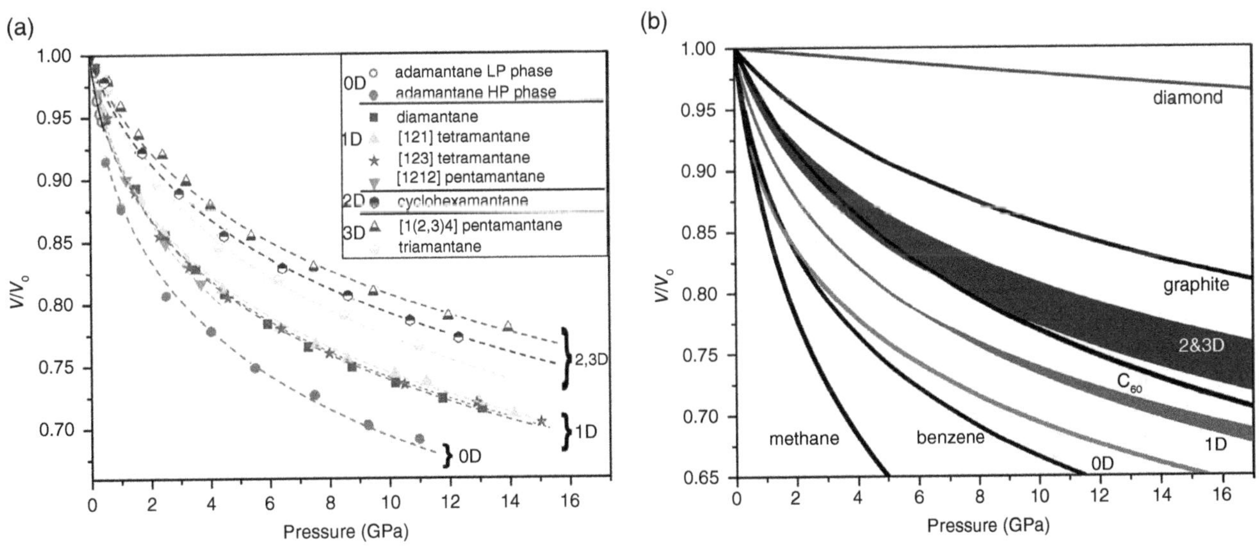

Figure 27.3 (a) Normalized unit cell volume changes as a function of increasing pressure in various diamondoid crystals. Dashed lines are pressure-volume curves fitted to the third-order Birch-Murnaghan equations of state (EOSs). (b) The EOS distributions of diamondoids categorized by the different dimensionalities. For comparisons, EOSs of other carbon allotropes and hydrocarbons are plotted. (Adapted from Yang et al., 2016.) See electronic version for color representation of the figures in this book.

compact, leading to quick buildup of repulsive forces. Hence, those with smaller H–H distances require less compression to trigger phase change, resulting in low phase-transition pressures. Contrarily, large H–H distances imply that a greater compression is needed to reach a critical transition point and prompt a phase transformation, as evidenced by high phase-transition pressure of triamantane, which has ~20% higher H–H distance (2.81 Å) compared with adamantane and [1(2,3)4]pentamantane (Yang et al., 2016).

In addition to the sensitivity to H–H distance, phase transitions in diamondoids exhibit a strong dependency on hydrostaticity. By subjecting diamantane ($C_{14}H_{20}$) to different pressure media at high pressure, it was found that its crystal structure is extremely sensitive to deviatoric stress (Yang et al., 2014a). Diamantane transforms almost immediately (~0.15 GPa) when external pressure is applied under nonhydrostatic conditions (Yang et al., 2014a). This transition pressure, when silicone oil is applied, is elevated to 7 GPa, which coincides with the hydrostatic limit of silicone oil (Klotz et al., 2009). Further experiments using helium as a pressure medium confirmed that the discrepancy in transition pressure is associated with the pressure medium quality, which underlines the critical role of deviatoric stress in triggering phase transitions in diamondoids (Yang et al., 2014a, 2014b). Indeed, deviatoric stress plays a critical role in inducing phase transformations, and in some cases, greatly reduces phase-transition pressure (Kailer et al., 1998; Y. Lin et al., 2011). Under hydrostatic conditions, diamondoids preserve their initial high-symmetry phases until the distortion of the unit-cell geometry becomes too high and eventually results in the observed phase changes.

Diamondoids are very robust in response to multiple rounds of compression and decompression. Even after 20 repeated pressure cycles up to ~5 GPa, diamondoids were able to retain their initial structures (Figure 27.4). This exquisite recoverability, combined with small ambient bulk moduli and high structural stability, makes diamondoids promising candidates for cushioning devices. In fact, diamondoids' recoverability far exceeds that of more conventional materials, such as foamy polystyrene and recently proposed carbon nanotube cushion materials (Liu et al., 2008). The energy absorption density (at ~2 GPa) of lower diamondoids is on the order of 10^2 kJ/kg, which is at least one order of magnitude greater than those of conventional cushion materials (Liu et al., 2008). In addition to possible applications in cushioning devices, the unique properties of diamondoids offer several other applications. For example, the geometry-dependent physical property of diamondoids suggest a possibility in designing nanoscale structures with tunable mechanical strength. The robust diamondoid networks also make them ideal materials to synthesize acentric metal-organic

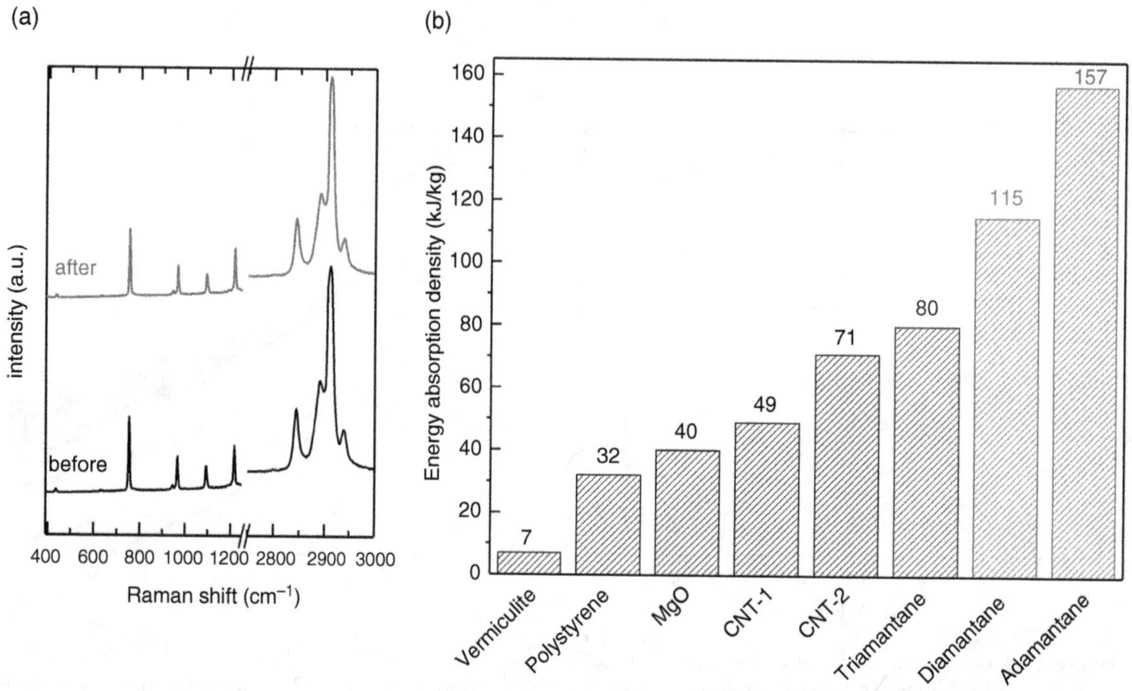

Figure 27.4 (a) Raman spectra of adamantane before and after 20 pressure cycles up to ~5 GPa. (b) Energy absorption density of conventional cushioning materials compared with three lower diamondoids. CNT-1 and CNT-2 refer to the microagglomerate and superagglomerate carbon nanotubes, respectively (Liu et al. 2008). (Reproduced from Yang et al. 2016.) See electronic version for color representation of the figures in this book.

diamondoid frameworks or supramolecular systems (Evans & Lin, 2001; Evans et al., 1999; Zaworotko, 1994).

27.3. CURRENT AND FUTURE PROSPECTS AND DIRECTIONS

Motivated by the unique properties of diamondoids, including their systematic sizes and geometries, a large body of work has recently been devoted to selectively separate, manipulate, and functionalize diamondoids (Gunawan et al., 2014). A wide range of functionalized diamondoid molecules have been reported with great potential for applications in molecular electronics, mechanics, biology, and chemistry (e.g. Archibald et al., 1991; Malik et al., 1991; Mcintosh et al., 2004; Patel et al., 2008; Shinisha & Sunoj 2005; Tewari et al., 2005; Wang et al., 2011). These studies take advantage of the fact that diamondoids are excellent building blocks to construct complex ordered nanoelements with molecular precision for their superior stability and rigidity (Garcia et al., 2009; Gunawan et al., 2014; Merkle, 2000). However, the diamondoid units on their own can be brittle due to full hybridization with hydrogen and weak intermolecular interactions (Garcia et al., 2009, Yan et al., 2018). Functionalization of such diamondoid molecules aims to solve several of the limitations by generating chemically active sites and manipulating the intermolecular/intramolecular interactions for fine tuning of the nanostructures. Such functionalization encourages self-assembly of the building blocks, eventually driving the system to optimized configurations (Garcia et al., 2009).

Recently, novel mechanochemistry through isotropic compression by engineering molecular structures that translate macroscopic isotropic stress into molecular-level anisotropic strain was reported (Yan et al., 2018). The "steric blockage" scenario, where rigid ligands in steric contact hampers relative motion and hinders reactivity, is demonstrated using copper(1) adamantane-1-thiolate (Cu-S-Ada), which is proposed for mechanochemistry of specific mechanosynthesis (Yan et al., 2018).

Cu-S-Ada crystals are composed of the mechanophore, a Cu-S nanowire, with a three-atom cross-section that is surrounded by a ligand shell consisting of adamantyl groups (Figure 27.5) (Yan et al., 2017; Yan et al., 2018). The sterically hindered ligand shell of Cu-S-Ada, caused by the rigid ligands, impedes relative movement of the adamantyl groups and deformation of Cu-S mechanophore. Density functional theory calculations show that the structure of adamantyl groups in Cu-S-Ada crystals remains rigid up to 20 GPa with less than 0.01 Å change in the average C-C bond length (Yan et al., 2018). Upon pressure release, the initial structure is recovered, indicating that the deformation is elastic. Such results are vastly dissimilar to the comparable building block system of copper(1) *m*-carborane-9-thiolate (Cu-S-M9) crystals (where m = meta positions of the carbon atoms in the carborane), wherein the cage-like nature and substantially large molecular distance between the M9 groups enable relatively free motions without mutual steric hindrance under compression (Yan et al., 2018). The free motions of Cu-S-M9 leads to a reduction of copper (Cu[1] → Cu[0]) after compression up to ~8 GPa, driven by the anisotropic deformation of Cu_4S_4 core (Yan et al., 2018). This reduction process, however, is inhibited when adamantane replaces the main building block, and the sterically hindered ligands limit the density differences near mechanophore, prohibiting any chemical reaction to occur (Yan et al., 2018).

The possible mechanochemistry driven by the relative motion of the rigid ligands of diamondoid cages functionalized with transition metals and chalcogens opens further exciting avenues for highly specific mechanosynthesis. The behavior of specific properties tuned from the functionalization of diamondoids should be further constrained by investigating the physical and chemical evolution under extreme conditions.

The exceptional properties of diamondoids and functionalized diamondoids point towards many potential applications. As discussed above, the stability of diamondoids in response to repeated pressure cycling makes them attractive candidates for cushioning devices.

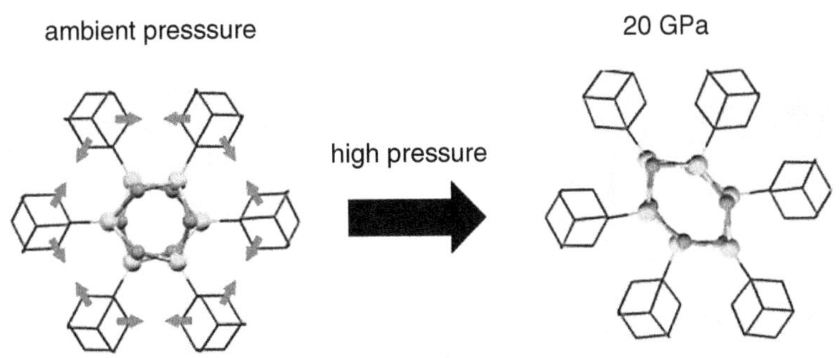

Figure 27.5 Structure of Cu-S-Ada crystals at ambient conditions. Red arrows denote the direction of motion for adamantyl groups under compression. (Reproduced from Yan et al., 2018.) See electronic version for color representation of the figures in this book.

Furthermore, rigid ligands from molecular engineering and functionalization diamondoids make them attractive for mechanochemistry and mechanosynthesis.

The evolution of structures and properties in diamondoid molecules upon compression warrants further investigations involving additional stimuli, such as high temperature. The simultaneous exposure of diamondoids to static high pressure and temperature could potentially cause spontaneous dehydrogenation, which in turn could drive the materials to form novel phases or unique structures not easily accessible using alternative pathways or precursor systems. A potential outcome of this is diamond. Given the resemblance of the diamondoids to the bulk diamond, diamondoid molecules may serve as optimal seed materials for forming diamond at high pressure and temperature. Additionally, time-resolved studies can be conducted to visualize the melting and reconstruction of shock-compressed diamondoids. Such studies can be performed by coupling the extremely bright X-ray pulses from X-ray free-electron lasers like the Linac Coherent Light Source (LCLS) at SLAC National Accelerator Laboratory with a high power optical laser, which can generate a shockwave in the material that results in high-pressure/high-temperature conditions. By probing material modifications in a nanosecond time regime, the dynamic pathways driving transitions in diamondoids can be evaluated, including the intermediate state that may not be accessible through static compression. These results will not only provide insight into the transformation of hydrocarbons into diamond or other carbon allotropes, but also help in determining physical and chemical parameters for synthesizing desirable carbon phases.

ACKNOWLEDGMENTS

The static compression experimental work was supported by the Department of Energy under Contract DE-AC02-76SF00515. The computational work and ongoing dynamic compression experiments are supported by the National Science Foundation under award ECCS-1542152, as well as the Deep Carbon Observatory.

REFERENCES

Archibald, T. G., Malik A. A., Baum, K. and Unroe, M. R (1991). Thermally stable acetylenic adamantane polymers. *Macromolecules*, *24*(19), 5261–5265.

Birch, F. (1947). Finite elastic strain of cubic crystals. *Physical Review*, *71*(11), 809–824.

Clay, W. A., Liu, Z., Liu, Z., Yang, W., Fabbri, J. D., et al. (2009). Origin of the monochromatic photoemission peak in diamondoid monolayers. *Nano Letters*, *9*(1), 57–61. doi:10.1021/nl802310k

Dahl, J. E., Liu, S. G., & Carlson, R.M.K. (2003). Isolation and structure of higher diamondoids, nanometer-sized diamond molecules. *Science*, 299(5603), 96–99. doi:10.1126/science.1078239

Dahl, J.E.P., Moldowan, J. M., Peters, K E., Claypool, G. E., Rooney, M. A., Michael, G. E., et al. (1999). Diamondoid hydrocarbons as indicators of natural oil cracking. *Nature*, *399*, 54–57. doi:10.1038/19953

Dahl, J.E.P., Moldowan, J. M., Wei, Z., Lipton, P. A., et al. (2010). Synthesis of higher diamondoids and implications for their formation in petroleum. *Angew. Chem. Int. Ed.*, *49*, 9881–9885. doi:10.1002/anie.201004276

Ekweozor, C. M., & Telnaes, N. (1990). Oleanane parameter: Verification by quantitative study of the biomarker occurrence in sediments of the Niger Delta. *Organic Geochemistry*, *16*(1-3), 401–413. doi:10.1016/0146-6380(90)90057-7

Evans, O. R, & Lin, W. (2001). Crystal engineering of nonlinear optical materials based on interpenetrated diamondoid coordination networks. *Chem Mater*, *13*, 2705–2712. doi:10.1021/cm010301n

Evans, O. R., Xiong, R.-G., Wang, Z.-Y., Wong, G. K., & Lin, W. (1999). Crystal engineering of acentric diamondoid metal-organic coordination networks. *Angew. Chem. Int. Ed.*, *38*(4), 536–538.

Filik, J., Harvey, J. N., Allan, N. L., May, P. W., Dahl, J.E.P., Liu, S., & Carlson, R. M. K. (2006). Raman spectroscopy of diamondoids. *Spectrochimica Acta Part A: Molecular and Biomolecular Spectroscopy*, *64*(3), 681–692. doi:10.1016/j.saa.2005.07.070

Filik, J., Harvey, J. N., Allan, N. L., May, P. W., Dahl, J.E.P., Liu, S., & Carlson, R. M. K. (2006). Raman spectroscopy of nanocrystalline diamond: An ab initio approach. *Physical Review B*, *74*(035423). doi:10.1103/PhysRevB.74.035423

Filik, J. (2010). Diamondoid hydrocarbons. In N. Ali, A. Öchsner, & W. Ahmed (Eds.), *Carbon-based nanomaterials* (chap. 1). Zurich; Enfield, NH: Trans Tech Publications. http://www.academia.edu/495358/Diamondoid_Hydrocarbons

Fort, R. C. Jr., & Schleyer, P.V.R. (1964). Adamantane: Consequences of the diamondoid structure. *Chem. Rev.*, *64*, 277–300.

Garcia, J. C., Justo, J. F., Machado, W.V.M., & Assali, L.V.C. (2009). Functionalized adamantane: Building blocks for nanostructure self-assembly. *Physical Review B*, *80*(125421). doi:10.1103/PhysRevB.80.125421

Gunawan, M. A., Hierso, J. C., Poinsot, D., Fokin, A. A., Fokina, N. A., Tkachenko, B. A., and Schreiner, P. R. (2014). Diamondoids: Functionalization and subsequent applications of perfectly defined molecular cage hydrocarbons. *New J. Chem*, *38*, 28–41. doi:10.1039/c3nj00535f.

Kailer, A., Gogotsi, Y G., & Nickel. K. G. (1998). Phase transformations of silicon caused by contact loading. *Journal of Applied Physics*, *81*(7), 3057–3063. doi:10.1063/1.364340

Katz, B. J., Mancini, E. A., & Kitchka, A. A. (2008). A review and technical summary of the AAPG Hedberg Research Conference on "Origin of petroleum—Biogenic and/or abiogenic and its significance in hydrocarbon exploration and production. *AAPG Bulletin*, *92*(5), 549–556. doi:10.1306/01210808006

Klotz, S., Chervin, J-C., Munsch, P., & Marchand, G. L. (2009). Hydrostatic limits of 11 pressure transmitting media. *Journal of Physics D: Applied Physics*, *42*(7), 075413. doi:10.1088/0022-3727/42/7/075413

Lin, R., & Wilk. Z. A. (1995). Natural occurrence of tetramantane (C22H28), pentamantane (C26H32) and hexamantane (C30H36) in a deep petroleum reservoir. *Fuel*, *74*(10), 1512–1521.

Lin, Y., Yang, Y., Ma, H. W., Cui, Y., & Mao, W. L. (2011). Compressional behavior of bulk and nanorod LiMn 2 O 4 under nonhydrostatic stress. *J. Phys. Chem. C*, *115*, 9844–9849. doi:10.1021/jp112289h

Liu, Y., Qian, W., Zhang, Q., Cao, A., Li, Z., Zhou, W., et al. (2008). Hierarchical agglomerates of carbon nanotubes as high-pressure cushions. *Nano Letters*, *8*(5), 1323–1327. doi:10.1021/nl0733785

Ma, A. (2016). Advancement in application of diamondoids on organic geochemistry. *Journal of Natural Gas Geoscience*, *1*(4), 257–265. doi:10.1016/J.JNGGS.2016.09.001

Ma, A., Jin, Z., Zhu, C., & Bai, Z. (2017). Cracking and thermal maturity of Ordovician oils from Tahe Oilfield, Tarim Basin, NW China. *Journal of Natural Gas Geoscience*, *2*(4), 239–252. doi:10.1016/J.JNGGS.2017.12.001

Mackenzie, A. S., Patience, R. L., Maxwell, J. R., Vandenbroucke, M., & Durand, B. (1980). Molecular parameters of maturation in the Toarcian Shales, Paris Basin, France: I. Changes in the configurations of acyclic isoprenoid alkanes, steranes and triterpanes. *Geochimica et Cosmochimica Acta*, *44*(11), 1709–1721. doi:10.1016/0016-7037(80)90222-7

Malik, A. A., Archibald, R. G., Baum, K., & Unroe, M. R. (1991). New high-temperature polymers based on diamantane. *Macromolecules*, *24*, 5266–5268.

Mansoori, G. A. (2008). Diamondoid molecules. *Advances in Chemical Physics*, *136*, 207–58. doi:10.1002/9780470175422.ch4

Marchand, A. P. (2003). Diamondoid hydrocarbon: Delving into nature's bounty. *Science*, *299*(5603), 52–53. doi:10.1126/science.1078239

Mcintosh, G. C, Yoon, M., Berber, S., & Tománek, D. (2004). Diamond fragments as building blocks of functional nanostructures. *Physical Review*, *70*(045401). doi:10.1103/PhysRevB.70.045401

Mckervey, M. A. (1980). Synthetic approaches to large diamondoid hydrocarbons. *Tetrahedron*, *36*(8), 971–992. doi:10.1016/0040-4020(80)80050-0

Merkle, R. C. (2000). Molecular building blocks and development strategies for molecular nanotechnology. *Nanotechnology*, *11*, 89–99.

Patel, K., Angelos, S., Dichtel, W. R., Coskun, A., Yang, Y.-W., Zink, J. I., & Stoddart, J. F. (2008). Enzyme-responsive snap-top covered silica nanocontainers. *Jour. Amer. Chem. Soc.*, *130*, 2382–83. doi:10.1021/ja0772086

Requejo, A. G., Gray, N. R., Freund, H., Thomann, H., Melchior, M. R., Gebhard, L. A., et al. (1992). Maturation of petroleum source rocks: 1. Changes in kerogen structure and composition associated with hydrocarbon generation. *Energy & Fuels*, *6*(2), 203–214. doi:10.1021/ef00032a015

Sasagawa, R., Shen, Z-X. (2008). A route to tunable direct band-gap diamond devices: Electronic structures of nanodiamond crystals. *Journal of Applied Physics*, *104*(7), 073704. doi:10.1063/1.2986637

Seifert, W. K., & Moldowan, J. M. (1986). Use of biological markers in petroleum exploration. *Meth. in Geochem. and Geophy.*, *24*, 261–290.

Shinisha, C. B., & Sunoj, R. B. (2005). On the origins of kinetic resolution of cyclohexane-1,2-diols through stereoselective acylation by chiral tetrapeptides. *Tetrahedron: Asymmetry*, *44*(2), 3242–3245. doi:10.1021/ol9011822

Stauss, S., & Terashima, K. (2017). *Diamondoids: Synthesis, Properties and applications*, 1st ed. CRC Press.

Stout, S. A., & Douglas. G. S. (2004). Diamondoid hydrocarbons: Application in the chemical fingerprinting of natural gas condensate and gasoline. *Environmental Forensics*, *5*(4), 225–235. doi:10.1080/15275920490886734

Tewari, A., Hein, M., Zapf, A., & Beller, M. (2005). Efficient palladium catalysts for the amination of aryl chlorides: A comparative study on the use of phosphium salts as precursors to bulky, electron-rich phosphines. *Tetrahedron*, *61*(41), 9705–9709. doi:10.1016/J.TET.2005.06.067

Van Gass, G. W. (1990). Biomarker maturity parameters for high maturities: Calibration of the working range up to the oil/condensate threshold. *Org. Geochem.*, *16*(4-6), 1025–1032.

Wang, Y., Tkachenko, B. A., Schreiner, P. R., & Marx. A. (2011). Diamondoid-modified DNA. *Org. Biomol. Chem*, *9*(7482). doi:10.1039/c1ob05929g

Willey, T. M., Bostedt, C., Buuren, T. V., Dahl, J.E.P., Liu, S. G., Carlson, R.M.K., et al. (2006). Observation of quantum confinement in the occupied states of diamond clusters. *Physical Review B*, *74*(205432). doi:10.1103/PhysRevB.74.205432

Wingert, W. S. (1992). G.c.-m.s. analysis of diamondoid hydrocarbons in smackover petroleums. *Fuel*, *71*(1), 37–43. doi:10.1016/0016-2361(92)90190-Y

Yan, H., Hohman, J. N., Li, F. H., Jia, C. J., Solis-Ibarra, D., Wu, B., et al. (2017). Hybrid metal-organic chalcogenide nanowires with electrically conductive inorganic core through diamondoid-directed assembly. *Nature Materials*, *16*(3), 349–355. doi:10.1038/nmat4823

Yan, H., Yang, F., Pan, D., Lin, Y., Hohman, J. N., Solis-Ibarra, D., Li, H. et al. (2018). Sterically controlled mechanochemistry under hydrostatic pressure. *Nature*, *554*, 505–510. doi:10.1038/nature25765

Yang, F., Lin, Y., Baldini, M., Dahl, J.E.P., Carlson, R.M.K., & Mao, W. L. (2016). Effects of molecular geometry on the properties of compressed diamondoid crystals. *J. Phys. Chem. Lett*, *7*, 4641–4647. doi:10.1021/acs.jpclett.6b02161

Yang, F., Lin, Y., Dahl, J.E.P., Carlson, R.M.K., & Mao, W. L. (2014a). Deviatoric stress-induced phase transitions in diamantane. *The Journal of Chemical Physics*, *141*(15), 154305. doi:10.1063/1.4897252

Yang, F., Lin, Y., Dahl, J.E.P., Carlson, R.M.K., & Mao, W. L. (2014b). High pressure Raman and X-ray diffraction study of [121] tetramantane. *The Journal of Physical Chemistry C*, *118*, 7683–7689. doi:10.1021/jp500431k

Zaworotko, M. J. (1994). Crystal engineering of diamondoid networks. *Chemical Society Reviews*, *23*(4), 283. doi:10.1039/cs9942300283

Zhang, J. L., Ishiwata, H., Babinec, T. M., Radulaski, M., Mü K., Lagoudakis, K. G., et al. (2016). Hybrid Group IV nanophotonic structures incorporating diamond silicon-vacancy color centers. *Nano Lett*, *16*, 212–217. doi:10.1021/acs.nanolett.5b03515

INDEX

Carbon in Earth's Interior, Geophysical Monograph 249, First Edition. Edited by Craig E. Manning, Jung-Fu Lin, and Wendy L. Mao.

www.ingramcontent.com/pod-product-compliance
Lightning Source LLC
LaVergne TN
LVHW062143200926
848484LV00007B/1849